# Microscopy of Semiconducting Materials, 1985

# Microscopy of Semiconducting Materials, 1985

Proceedings of the Royal Microscopical Society Conference held in St Catherine's College, Oxford, 25–27 March 1985

Edited by A G Cullis and D B Holt

M $\overset{\text{S}}{\phantom{x}}$ M
IV

Institute of Physics
Conference Series Number 76
Adam Hilger Ltd, Bristol and Boston

CODEN IPHSAC 76 1–530 (1985)

*British Library Cataloguing in Publication Data*

Royal Microscopical Society, *Conference (1985: Oxford)*
  Microscopy of semiconducting materials 1985:
  proceedings of the Royal Microscopical Society
  conference held in St Catherine's College, Oxford,
  25–27 March 1985.—(Conference series/Institute
  of Physics; ISSN 0305-2346; no. 76)
  1. Semiconductors—Research     2. Electron
  microscopy
  I. Title     II. Cullis, A. G.     III. Holt, D. B.
  IV. Series
  537.6'22              QC611.4

    ISBN 0-85498-167-5
    ISSN 0305-2346

Conference Co-Chairmen
        A G Cullis and D B Holt

RMS Local Organiser
        Lt Col P G Fleming

Honorary Editors
        A G Cullis and D B Holt

Published on behalf of The Institute of Physics by Adam Hilger Ltd
Techno House, Redcliffe Way, Bristol BS1 6NX, England
PO Box 230, Accord, MA 02018, USA

Printed in Great Britain by J W Arrowsmith Ltd, Bristol

# Preface

This volume contains invited and contributed papers presented at the conference on 'Microscopy of Semiconducting Materials' which took place at St Catherine's College, Oxford on 25–27 March 1985. The conference was the fourth in the series which highlights progress in microscopical studies of semiconductors and was organised under the auspices of the Materials Section of the Royal Microscopical Society with co-sponsorship by the Electron Microscopy and Analysis Group of The Institute of Physics and the Materials Research Society. The emphasis of the meeting was strongly international and more than 200 scientists from the UK and fourteen other countries participated.

The conference addressed advances which have taken place in all areas of semiconductor microscopy ranging from basic materials research to the assessment of electronic device structures. The 81 papers provided in this volume have been grouped into sections according to their topic coverage. Special attention is focused on developments in the application of high resolution transmission electron microscopy, the study of dislocations and the investigation of transient processing phenomena. The basic materials work reported here extends from the characterisation of bulk, as-grown crystals to the evaluation of homoepitaxial and heteroepitaxial layers and superlattices. While a variety of standard electron microscopy techniques are employed in these studies, there is selective discussion of the use of beam-induced conductivity and cathodoluminescence measurements in the scanning microscope, in addition to the application of a number of microanalytical methods. Ion backscattering and channelling analyses together with x-ray diffraction and topography studies are also featured. The later sections give many examples of the assessment of device processed silicon and include descriptions of electron beam testing of finished electronic devices.

Manuscripts of conference papers were submitted for publication in camera-ready format. Each manuscript was reviewed by at least one referee and modified accordingly. The editors are most grateful to the following referees for their speedy and meticulous work:

P D Augustus, H Baumgart, D Bensahel, G R Booker, A Bourret, G T Brown, L M Brown, M R Brozel, C B Carter, C Claeys, S M Davidson, J M Gibson, P Haasen, J Heydenreich, J L Hutchison, D C Joy, D M Maher, J W Mayer, H Oppolzer, S J Pennycook, F A Ponce, R C Pond, D J Stirland, B K Tanner and C A Warwick

Finally, it is a pleasure to thank colleagues in our laboratories and officers of the RMS for assistance provided with the many tasks involved in the organisation of this conference. Special thanks are due to Mrs D M Handley for expert secretarial work which was indispensable for the success of this event.

**June 1985**                                                                                           **A G Cullis**
                                                                                                        **D B Holt**

# Contents

†Invited

†Invited

†Invited

**Section 7: Superlattices**

**Section 8: Scanning EBIC and CL**

†Invited

†Invited

**Section 12: Device testing by scanning microscopy**

†Invited

*Inst. Phys. Conf. Ser. No. 76: Section 1*
*Paper presented at Microsc. Semicond. Mater. Conf., Oxford, 25–27 March 1985*

# Structure of microdefects in semiconducting materials

F A Ponce

Xerox Palo Alto Research Center, Palo Alto, CA 94304, U.S.A.

Abstract    Microdefects are common in semiconducting materials of the highest quality.  Some are present in as-grown materials, others appear during subsequent processing, especially when subjected to thermal treatments.  Their identification has not been possible until very recently because their size is relatively too small to be observed by conventional electron microscopy.  A review of recent applications of HREM to the study of microdefects in silicon and GaAs is presented in this paper.

## 1. Introduction

With the advent of high resolution transmission electron microscopy (HRTEM) it is now possible to investigate the structure of semiconducting materials with resolutions close to the atomic level.  Whereas conventional electron microscopy provides with point resolutions of the order of a couple of nanometers, modern microscopes have the power to resolve point separations below 1.8 A.  In recent years, this capability has allowed the understanding of the microstructure of semiconducting materials with improved levels of accuracy.

Lattice defects in semiconducting materials can be categorized as point defects, extended defects and microdefects.  The structure of point defects such as vacancies and interstitials, with dimensions of less than .5 nm, cannot generally be observed by transmission electron microscopy (TEM).  Extended defects such as dislocations and stacking faults have dimensions of the order of 100 nm and are easily observed by conventional electron microscopy and by other techniques such as chemical etching.  Somewhere in between is another category of lattice defects named microdefects.

Microdefects are common in semiconducting materials.  They are associated with the growth process and depend strongly on the thermal history of the material.  Because as-grown crystals are usually not in thermodynamic equilibrium immediately after their growth, microdefects often develop as the system tries to minimize its free energy.  Since their sizes are of the order of tens of nanometers or less, they are typically not observed by chemical etching or X-ray topographic techniques, and their identification by conventional electron microscopic techniques is cumbersome.  It is in the identification of the nature of microdefects that HRTEM has had a strong impact.  A review of recent applications of this technique to the study of microdefects in silicon and gallium arsenide is presented in this paper.

2.    Structure of Thermally Induced Microdefects in Czochralski Silicon

The Czochralski (CZ) technique is widely used to grow silicon crystals for various integrated circuit technologies.  Immediately following the growth process, these materials exhibit very few defects, if any, when examined using X-ray topography or chemical etching techniques.  Microdefects are typically generated during subsequent materials processing involving thermal anneals.  Due to their characteristic spiral (swirl-like) distribution in planes perpendicular to the growth direction, these defects are known as swirl defects in analogy with similar patterns observed in float zone (FZ) materials (de Kock 1977).  The origin of these defects, however, need not be the same in CZ as in FZ materials.  The structure associated with the precipitation of oxygen in silicon has been studied using conventional transmission electron microscopy (TEM) (Tan and Tice 1976, Maher et al 1976, Tempelhoff et al 1979).  Recently, HRTEM has been used to directly image the lattice structure of these microdefects.

The microstructure associated with oxygen precipitation depends strongly on the thermal history of the material.  For an isothermal process, oxygen precipitation occurs in three stages:  nucleation, intermediate and quasi-equilibrium.  Various models exist for the nucleation stage suggesting homogeneous and heterogeneous nucleation, the second one being associated with the presence of carbon.  In the intermediate stage, the precipitate becomes three dimensional with the development of oxide precipitates and extrinsic stacking faults (Ponce, Yamashita and Hahn 1983, Ponce and Hahn 1984).  The presence of extrinsic faults in the vecinity of precipitates is necessary because the oxygen precipitation process requires the genearation of silicon interstitials.  Each $SiO_2$ molecule formed in the silicon lattice occupies approximately the equivalent volume of two silicon atoms.  Therefore, each $SiO_2$ molecule formed in the lattice requires either the presence of a silicon vacancy or the generation of an interstitial.  The silicon interstitials incorporate into the lattice by the formation of an extra plane.  The final stages occur after very prolonged heat treatments involving up to hundreds of hours, during which the system searches for a

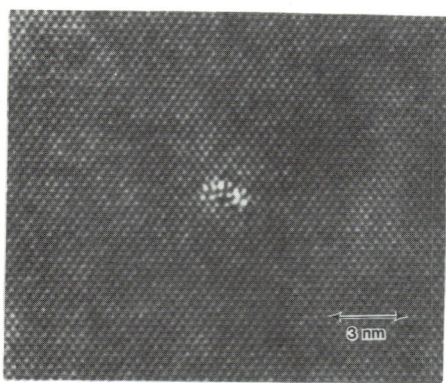 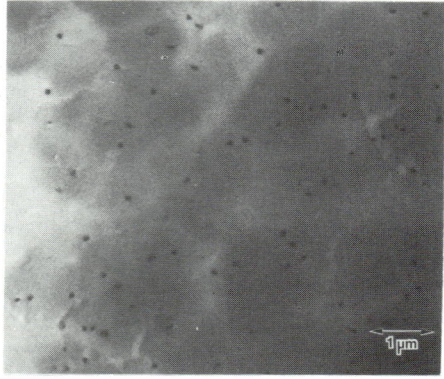

Fig.  1.    Ribbon-like  precipitate after heat treatment at 450°C for 210 hours.

Fig.  2.    Bright  field  micrograph showing  microdefect  distribution following  annealing at 750°C.

quasi-thermodynamic equilibrium. In the final stages of precipitation, the precipitates acquire well defined morphologies which, depending on the temperature, fall into three categories: a) the low temperature regime (450-650°C), b) intermediate temperatures (650-950°C) and c) high temperatures (above 950°C).

In the low temperature regime, rod-like defects were first observed by Tempelhoff et al 1977. Recently, Bourret et al (1984) identified the structure of these precipitates as that of coesite, the crystalline high-pressure form of $SiO_2$, and thereby identified their chemical composition. Fig. 1 shows the cross section view of a rod-like precipitate, which typically extend in length up to the order of a micron, and has a cross-section of a few nanometers.

In the intermediate temperature range, precipitation occurs in the shape of particles rather than rods as shown in Fig. 2. After several hours at 750°C, the resulting microdefects consist of thin platelets of amorphous structure alligned along (100) silicon planes and small extrinsic loops in the vecinity of the platelet as observed in Fig. 3 (Ponce, Hahn et al 1983). Further heat treatments at 750°C will cause the formation of well defined, larger, extrinsic stacking faults and (100) amorphous platelets as shown in Fig. 4. These platelets typically have thicknesses of 1 to 4 nm, and diameters of 30 to 50 nm.

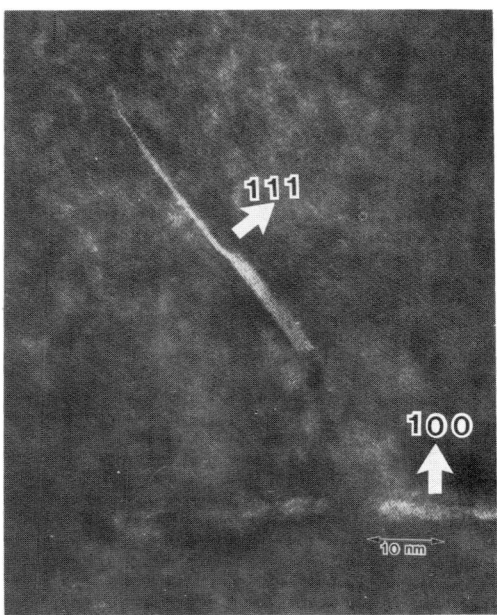

Fig. 3. Lattice image of microdefect after anneal at 750°C. {111} and {100} defects are observed.

In the high temperature regime (T > 950°C), the rate of oxygen precipitation is very low, and typically few, if any, precipitates are observed. The precipitation of oxygen, however, can be maximized by the introduction of a nucleation step at an intermediate temperature (~700°C), followed by a high temperature anneal. The microstructure after a high temperature anneal consists of small polyhedral precipitates (Ponce, Hahn et al 1983; Ponce, Yamashita and Hahn 1983) consisting of amorphous silica and bound by {111} and {100} silicon planes, as shown in Fig. 5. These polyhedral particles have diameters of about 10nm, their density is determined by the nucleation step, and they originate in the {100} platelets produced during the intermediate temperature step. We have also demonstrated that the native oxide when buried in chemically deposited polycrystalline silicon, goes through a transition from a planar morphology to polyhedral particles identical to those found in the bulk due to oxygen precipitation. It can therefore be inferred that the observed morphologies are inherent properties of the Si-SiO$_2$ system. Tiller et al. have shown that the needle-platelet-sphere morphology is expected solely from thermodynamical considerations of precipitation under various strain field constraints. In the high temperature case, the polyhedral shape is a consequence of the anisotropy of the Si/SiO$_2$ interface energy.

Even though there is agreement on the microdefect morphology after prolonged heat treatments, there is still much to be learned about the initial and intermediate stages of oxygen precipitation and, due of its technological importance, this is a challenge for the years to come.

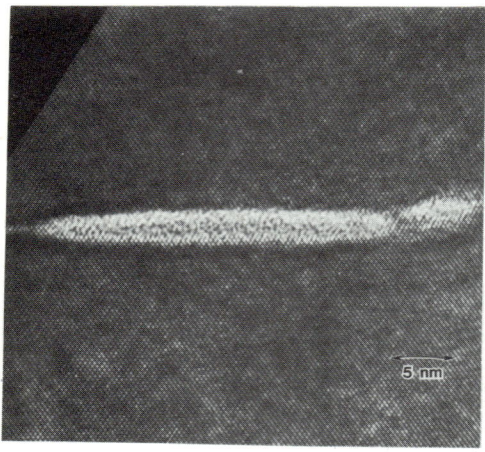

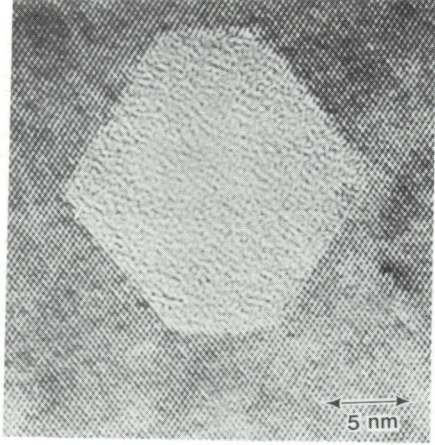

Fig. 4. Amorphous silica platelet parallel to {100} silicon planes. After anneal at 850°C for 60 hrs.

Fig. 5. Polyhedral silica precipitate following heat treatment at 1200°C for 64 hours after a 750°C nucleation step for 100 hrs.

## 3.  Microdefects in Semi-Insulating LEC GaAs

In recent years there has been increasing interest in semi-insulating (SI)
liquid encapsulated Czochralski (LEC) GaAs materials. Their high electron
mobility makes them suitable for high speed integrated circuit applications.
Typical IC technologies, however, require high quality, homogeneous and
uniform materials. Tajima (1982) demonstrated that there is a close relation
between etch pit densities and electrical and optical parameters such as
resistivity, Hall mobility and photoluminescence efficiency.

The structure of high purity semi-insulating materials has been studied by
HRTEM (Ponce, Wang et al 1984). Fig. 6 shows a transmission X-ray topograph of
a <110> slice cut normal to the crystal growth axis, close to the seed end of
the crystal.  This crystal had been grown using a low pressure (2 atm) LEC
method with *in situ* synthesis.  The X-ray intensity follows the familiar W-
shape reported in the literature (e.g.: Jacob G, 1982). This X-ray topograph
suggests a cellular structure, where cell dimensions range from less than 0.1
up to a few mm, the largest occuring at $r \sim \frac{1}{2}R$.  A transmission electron
micrograph of the region close to a cell boundary is shown in Fig. 7.
Typically the cell boundaries appear to be flat and do not necessarily exhibit
the presence of dislocations.   Dislocations have been observed in the
vecinity of these cell but are not believed to be constituents of the cell
boundary itself.   Cell  boundaries  do  not  always  follow  a  particular
crystallographic plane but usually appear aligned along the growth direction.
Fig. 8 shows the lattice structure of a cell boundary in a very thin region of
the crystal. The cell boundary is composed of small particles whose structure
appears to be amorphous, the typical dimensions of the particles being 1 to 10
nm.  In Fig. 8 the microdefects are aligned along the 100 growth direction.
When the boundary deviates from such direction, larger precipitates are
occasionally observed, as shown in Fig. 9.

Microdefects are also observed in the cell interior.  We have observed
precipitates with dimensions of less than 10 nm as shown in Fig. 10,

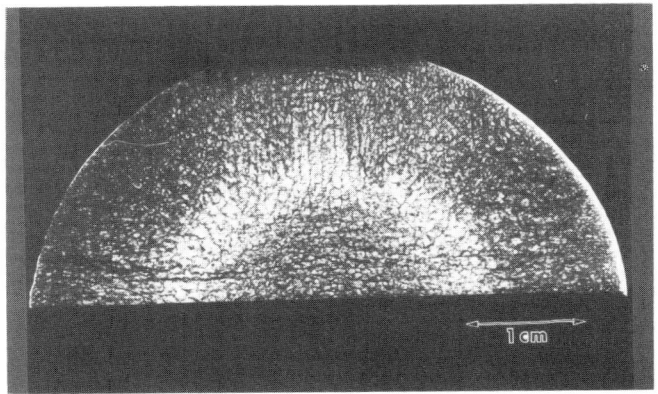

Fig. 6.  Transmission X-ray topograph
of <100> section perpendicular to the
growth direction.

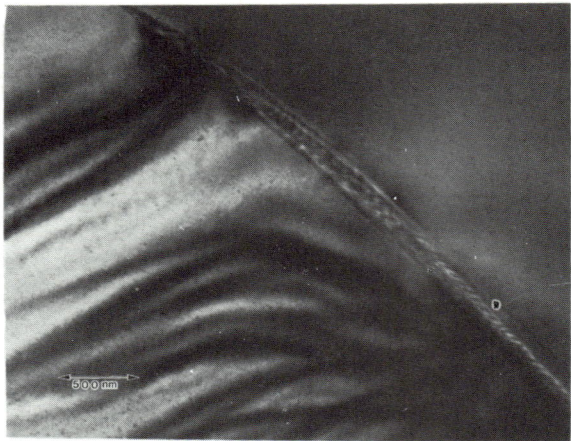

Fig. 7. Bright-field micrograph of a
cell boundary.

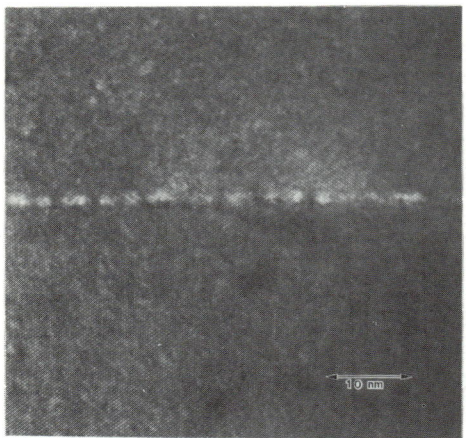

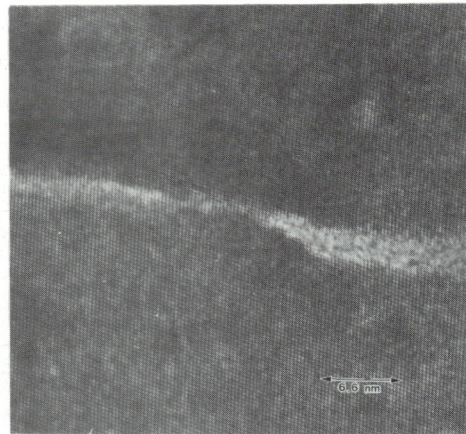

Fig. 8. High resolution lattice
image of region containing the cell
boundary. Small coherent particles
(~1-2 nm in diameter) are observed.

Fig. 9. Amorphous precipitate along
a cell boundary.

especially in the center ($r < \frac{1}{4}R$) and in the periphery of the crystal ($r > \frac{3}{4}R$).
These precipitates have a structure that appears amorphous in HRTEM, and we
have been unable to obtain evidence of a crystalline phase. These
precipitates appear to be coherent in the sense that there are no dislocations
associated with them and that the strain field is weak. Burger's circuits
around them are complete ($b=0$), and therefore they cannot be associated with
decorated dislocations. Further chemical analysis is not possible because of
the small size of the precipitate and because of the instability under

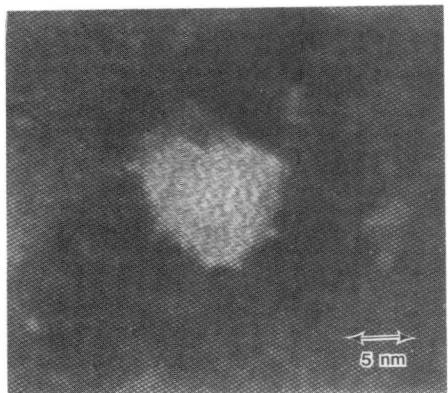

Fig. 10. Amorphous precipitate in the interiour of a cell close to the edge of the crystal.

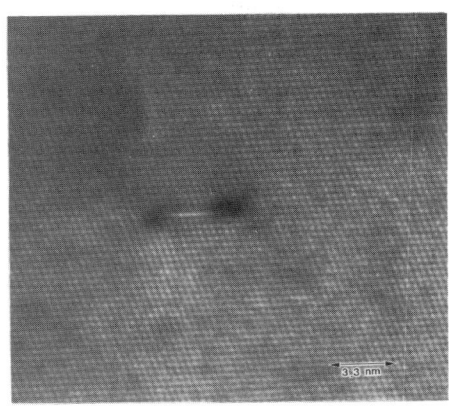

Fig. 11. Small insterstitial loop. The loop is actually contained in the sample and its real structure is not clealy visible.

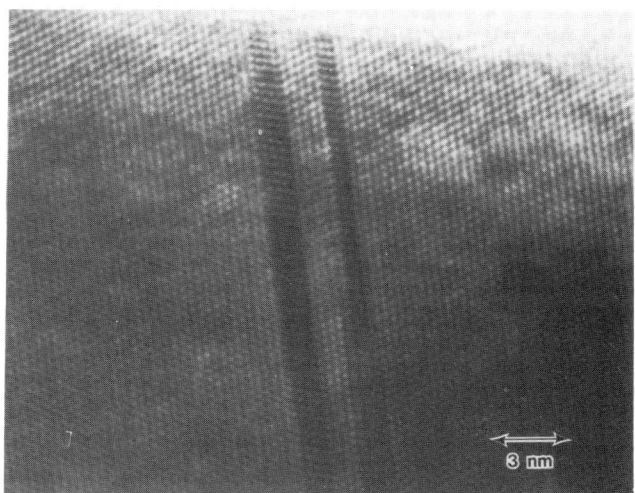

Fig. 12. Multiple stacking faults (microtwin) associated with crystal-lographic slip.

electron beam irradiation. They may be Ga particles incorporated during growth and be associated with deviations from stoichiometry, intrinsic of the crystal growth process.

Another type of microdefect present in the interior of the cells are small loops of both the interstitial and vacancy types. They lie along (111) planes and have diameters ranging from 2 to 5 nm. They are typically smaller than the sample thickness, thus it is difficult to obtain a clear image from them. Fig. 11 shows an interstitial loop about 3 nm in diameter. Because of their size, it is not possible to do direct chemical analysis of these microdefects. Figure 12 shows another common defect present in this type of materials. It corresponds to a specimen taken from the region close to the ingot surface and consists of multiple staking faults caused by crystallographic slip.

The main point to be raised in the discussion of the microstructure of SI LEC GaAs is that the defect structure is more complex than expected. It was commonly believed that the cellular boundaries were but a polygonal array of dislocations introduced during growth and after, either directly from the seed of by the incorporation of encapsulant which would later crack the crystal. Our investigations have shown that the cell boundaries have very little to do with dislocation networks, as can be determined from the absence of net displacements (Burger's vector) around most of the microdefects found at or near the cell boundaries. Thus the dislocation generation and multiplication model does not seem to be consistent with our observations.

A second mechanism which could explain the observed cellular structure is constitutional supercooling (see e.g. Laudise 1970). Deviations from stoichiometry in the melt could lead to a formation of a thin layer which is rich in one of the contituents. Small fluctuations could lead to the breakdown of the growth interface and to the formation of a cellular structure. Thus the cell boundaries and the observed amorphous phases could be explained. Constitutional supercooling would be enhanced by a lowering of the thermal conductivity in the enriched layer in the liquid. There is, however, little data at the present time that could allow an accurate model from basic principles.

Before dismissing dislocations altogether from being responsible for the microstructure observed in as-grown GaAs, it is necessary to first understand the structure of clean dislocations. Recently, Ponce et al. (1985) have investigated the structure of plastically deformed GaAs crystals. The HREM lattice images in Fig. 12 show the existence of perfect edge dislocations and that 60° dislocations exist in both the dissociated and undissociated form. Much work remains to be done in order to understand the thermal stability and electrical activity of dislocations.

It must be noted that not all GaAs crystal ought to be similar to the material which was subject of the study in this section. For instance, crystals grown in high pressure pullers are known to have some marked structural differences. Even within the same materials, properties vary significantly from the seed-end to the tail-end, and from the center to the edge of the ingot. There is much to learn in this regard, and the most sophisticated electron microscopic techniques will constantly be needed in order to bring new light into these complex problems.

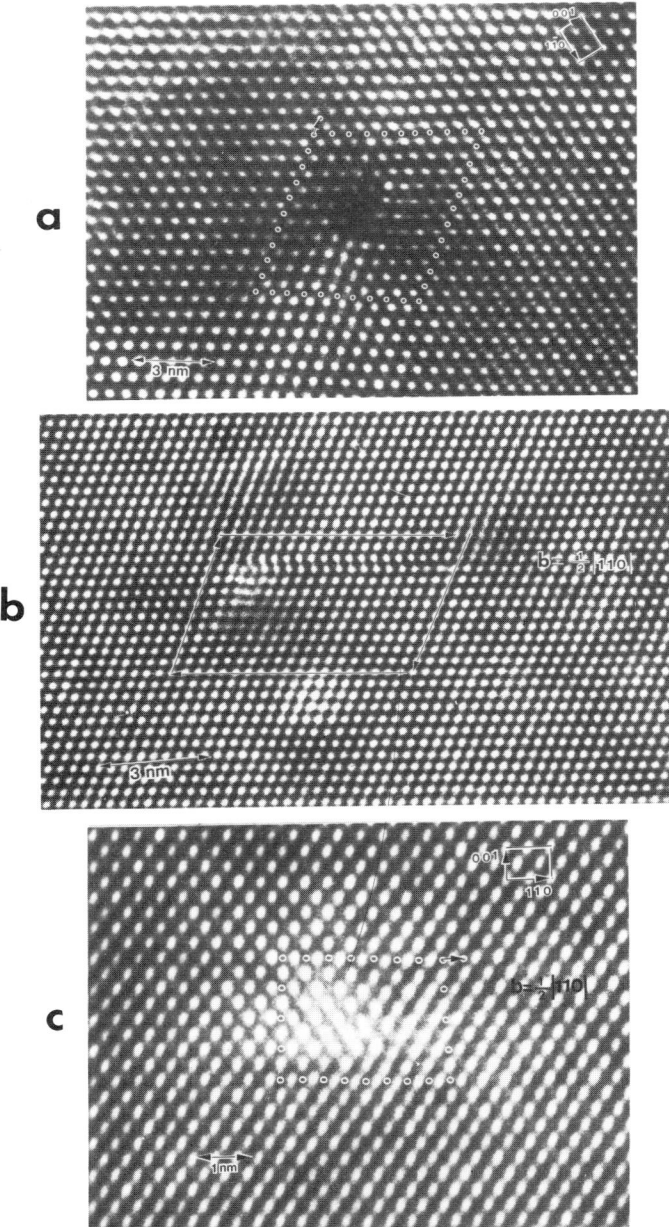

Fig. 13. Dislocations in plastically deformed GaAs. All have Burger's vector b=½a$_o$[110]. 60° dislocations are observed which are a) undissociated and b) dissociated into a 30° Shockley partial and a 90° partial dislocation. c) Perfect edge dislocation. (Ponce, Anderson, Haasen and Brion 1985).

References

Bourret A, Thibault-Desseaux J and Seidman D N 1984 J. Appl. Phys. 55 825
Bourret A 1985 Thirteenth International Conference on Defects in Semi-
    conductors eds L C Kimerling and J M Parsey Jr (Warrendale, Metallurgical
    Society of AIME) pp 129-146
de Kock A J R 1977 Crystal Growth and Materials eds E Kaldis and H J Scheel
    (Amsterdam: North Holland) pp 662-670
Jacob G 1982 Semi-Insulating III-V Materials, Evian eds S Makram-Ebeid and
    B Tuck (Nantwich: Shiva) pp 2-18
Laudise R A 1970 The Growth of Single Crystals (Prentice-Hall) pp 104-9
Maher D M, Studinger A and Patel J R 1976 J. Appl. Phys. 47 3813
Ponce F A, Anderson G B, Haasen P and Brion H to be published
Ponce F A and Hahn S 1984 Mat. Res. Soc. Symp. Proc. Vol 31:153
Ponce F A, Hahn S, Yamashita T, Scott M and Carruthers J R 1983 Microscopy
    of Semiconducting Materials 1983 eds A G Cullis, S M Davidson and
    G R Booker (Bristol: Inst. Phys.) pp 65-70
Ponce F A, Yamashita T and Hahn S 1983 Appl. Phys. Lett. 43 1051
Ponce F A, Wang F-C and Hiskes R 1984 Semi-Insulating III-V Materials,
    Kahnee-ta eds D C Look and J S Blakemore (Nantwich: Shiva) pp 68-75
Tan T Y and Tice W K 1976 Philos. Mag. 34 615
Tempelhoff K and Spiegelber F 1977 Semiconductor Silicon eds H Huff and
    E Sirtl (Pennington: Electrochemical Soc.) pp 585-595
Tempelhoff K, Spiegelber F, Gleichmann R and Wruck D 1979 Phys. Status
    Solidi A26 213
Tiller W A, Hahn S and Ponce F A, unpublished.

*Inst. Phys. Conf. Ser. No. 76: Section 1*
*Paper presented at Microsc. Semicond. Mater. Conf., Oxford, 25–27 March 1985*

# Enhanced oxygen diffusion and precipitation in silicon

W  Bergholz, J L Hutchison and P Pirouz

Department of Metallurgy and Science of Materials,
Oxford University, Parks Road, Oxford OX1 3PH.

Abstract   Prolonged annealing of CZ-silicon at 485°C results in the formation of ribbon-like oxygen precipitates on $\{311\}^-$ planes interpreted as the coesite phase.  From this an enhancement of the oxygen diffusion coefficient by more than three orders of magnitude is inferred.  Excess selfinterstitials are accommodated in extrinsic dislocation loops on $\{111\}^-$planes ("loopites") and probably also in 'blob'-like defects visible in high resolution micrographs with little detectable lattice strain.  Evidence is presented that coesite ribbons can nucleate on $\{311\}^-$ steps on loopites and vice versa.

## 1. Introduction

In recent years the extrinsic gettering process has become increasingly important in silicon-device technology.  As this technique essentially amounts to the controlled precipitation of oxygen, a thorough understanding of the diffusion and precipitation of oxygen is particularly important;  recent evidence from stress induced dichroism measurements (Stavola et al. 1983) suggests that the single diffusion jump rate of interstitial oxygen atoms can be enhanced by about one order of magnitude at 485°C by preannealing at 900°C for 2h.  It is not clear whether this enhancement is just a local phenomenon or results in an accelerated long range diffusion of oxygen in silicon.

In this paper we report the observation of oxygen precipitation by TEM at 485°C providing evidence for an enhanced long range migration of oxygen atoms.  The importance of excess selfinterstitials produced by the oxygen precipitation for the mechanism of accelerated diffusion and the nucleation of extended defects is discussed.

## 2. Experimental

The starting material was in the form of 3 inch Czochralski grown $\langle 011 \rangle$ boron-doped ($C_{bor} = 1.3 \times 10^{15} \text{cm}^{-3}$) silicon wafers from Wacker Chemitronic, with an initial oxygen concentration of $8 \times 10^{17} \text{cm}^{-3}$ and a carbon concentration below the detection limit of $2 \times 10^{16} \text{cm}^{-3}$.  To suppress contamination by transition elements, the specimens were annealed in a double-walled furnace in a flow of nitrogen boiled off a liquid nitrogen reservoir.  As an additional precaution, the annealing was carried out without a thermocouple within the furnace walls.  Subsequent calibration showed the annealing temperature to be 485°C±5°C rather than the nominal temperature of 450°C quoted in an earlier paper (Bergholz et al. 1984). $\{011\}$ foils were $Ar^+$-ion beam thinned and examined in a Jeol 200CX, a Jeol 100B and a Philips EM300 microscope. $\langle 011 \rangle$ -lattice images shown have been taken from regions of the specimen less than 200 Å thick, close to Scherzer defocus and are from 'as received' specimens except for fig.3a and fig.6. The oxygen concentrations were determined by infrared spectroscopy using the calibration of Graff et al (1973).

3. Results

Annealing at 485°C for 921h leads to a loss of about 40% of the dissolved oxygen (Fig.1) both for as-received and for the 925°C, 2h pre-annealed material. Note that a significant effect of the pre-anneal is only detectable for annealing times less than 100h.

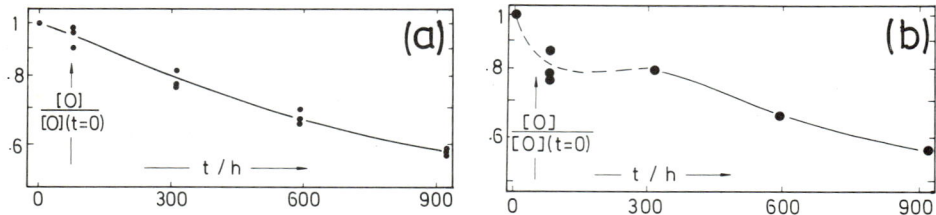

Fig.1 Normalised oxygen concentration $[0]$ as a function of annealing time t. (a) as received    (b) 925°C, 2h pre-anneal.

Electron microscopy reveals the presence of three types of defects: (i) ribbon-like defects, elongated along $\langle 110 \rangle$ directions (Fig.2), having a  x$\langle 100 \rangle$ Burgers vector.

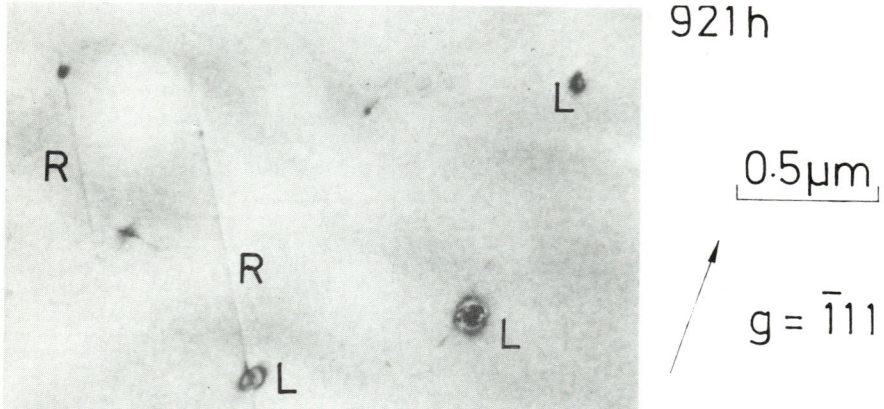

921h

0·5μm

$g = \bar{1}11$

Fig.2 Bright-field two-beam micrograph showing ribbon-like precipitates R and loopites L. Annealing time 921h.

High resolution lattice images (Fig.3) reveal a ribbon-like cross section of up to 1x6nm on $\{311\}$ habit planes. Although no clear crystal structure within the defect is resolved (this may be due to the large surface to volume ratio or severe buckling of the foil near the defect) as compared to those produced after annealing at 635°C (Bergholz et al. 1985) or 650°C (Bourret et al. 1984; Bender 1984), it appears that the defects are precipitates with a remarkably high aspect ratio of $10^3$-$10^4$.

(ii) Loop-like defects ("Loopites") with diameters up to 150nm (Fig.2) which have {111} as their habit planes (Fig.4).

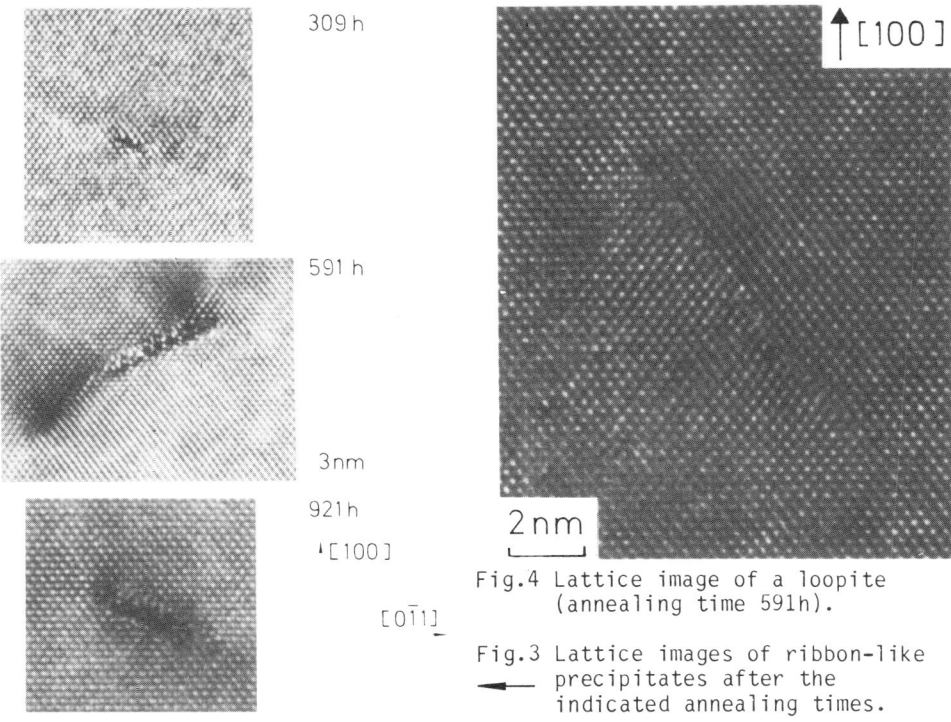

309 h

591 h

3nm

921h

↓[100]

[0̄1̄1]

↑[100]

2nm

Fig.4 Lattice image of a loopite (annealing time 591h).

Fig.3 Lattice images of ribbon-like precipitates after the indicated annealing times.

For small defects an extra plane is clearly visible; larger defects tend to have steps in the {111} habit plane which preclude any clearly visible image of an extra plane in a lattice image.

(iii) "Blob"-like defects with diameters between 2 and 15 nm appear as dark regions in lattice images with little detectable lattice strain. An outstanding feature of these defects is their instability under an electron beam (Fig.5).

## 4. Discussion

### 4.1 Interpretation of the Defects

The ribbon-like precipitates display most of the salient features of the ribbon-like precipitates observed at 635°C and 650°C, the major difference being a {113} habit plane rather than the {100}. However, coesite precipitates on {100} habit planes tend to have sections along {311}, and conversely ribbon-precipitates on {311} habit planes formed at 485°C sometimes show indications of continuing growth along {100}. This suggests that we are dealing with the same defect, namely coesite; {311} being the preferred habit plane during the initial stages of growth, with the {100} becoming the more favourable habit plane in order to minimise the strain energy for larger precipitates. Indeed, an estimate of the oxygen content of the precipitates (assuming an oxygen concentration of

$5.8 \times 10^{22} \text{cm}^{-3}$ for coesite) from the average ribbon cross section of $6.7 \pm 1.3 \times 10^{-14} \text{cm}^2$ and the line density of $1.0 \pm 0.3 \times 10^{8} \text{cm}^{-2}$ for t = 921h yields a value closely equal to the loss of oxygen from solution, as detected by IR-spectroscopy. In conclusion, there is strong evidence that the ribbon-like precipitates are coesite in an embryonic stage.

Defects of a similar appearance but narrower in cross section, have been observed after irradiation of CZ silicon and germanium with electrons or ions at elevated temperatures (Tan et al 1979, Pasemann et al. 1983, Desseaux-Thibault et al.1983) and have been either interpreted as $\{311\}$ stacking faults or "Ge-coesite" in the germanium case. Our results also indicate, as suggested by the latter authors, that charged particle irradiation of CZ-silicon can lead to the formation of coesite. This would eliminate the need for rather complicated defects made up from selfinterstitials to account for the large defect thickness.

With an extra $\{111\}$ plane of atoms, loopites with a density of about $3 \times 10^{12} \text{cm}^{-3}$ and an average diameter of $530 \text{Å}$ can accommodate about $7 \times 10^{16} \text{cm}^{-3}$, or more than 60% of the excess interstitials produced by the precipitation of $3.4 \times 10^{17} \text{cm}^{-3}$ oxygen atoms as coesite. It is noteworthy that the loopites have substituted the extrinsic $\{113\}$ dislocation dipoles as sinks for excess selfinterstitials during oxygen precipitation around 650°C.

The rapid disintegration of "blob" defects under the electron beam indicates that they are made up of highly mobile point defects, an obvious candidate being selfinterstitials. Another link of "blob" defects to selfinterstitials is the observation of blobs with clear

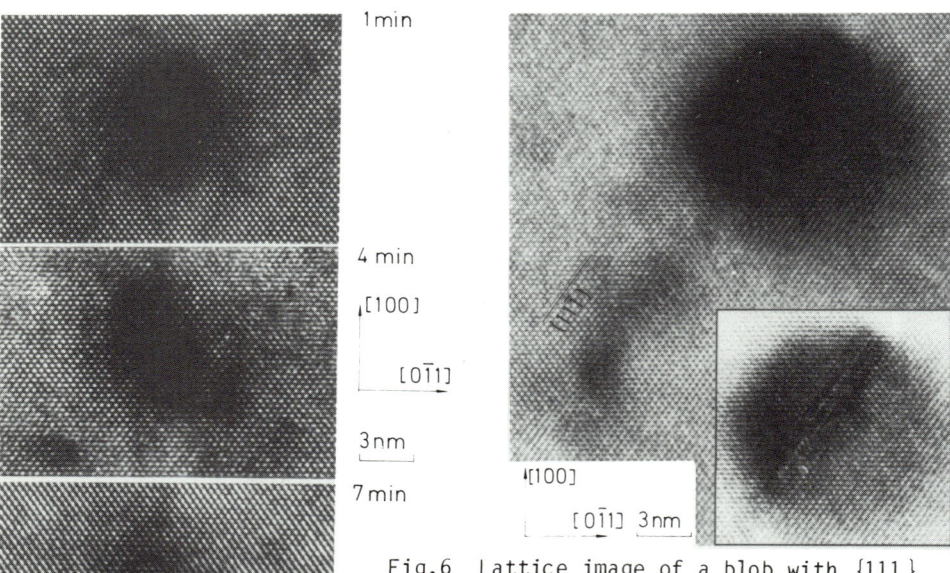

Fig.6 Lattice image of a blob with $\{111\}$ features (annealing time 311h). Insert: printed with a shorter exposure time.
Fig.5 Lattice image of a blob after irradiation times indicated (annealing time 591h).

features along {111} planes (Fig.6), suggesting that a transformation of a blob into a loopite is taking place. Note however that Bourret et al. (1983) have observed similar "blob"-defects and interpreted them as amorphous $SiO_2$ precipitates. Since for short annealing times we find that more oxygen disappears from solution than reappears in detectable ribbons, we cannot rule out that the blobs also contain oxygen.

## 4.2 Enhanced Oxygen Diffusion

The large aspect ratio of the ribbon-like precipitates means that the oxygen migration to these precipitates can be approximately treated as the diffusion to an array of cylindrical sinks of a density of $10^8 cm^{-2}$. As described in detail elsewhere (Bergholz et al. 1985), by applying Ham's theory (Ham 1956) the effective oxygen diffusion coefficient is estimated as $D_{eff}$ (485°C) = $1.5 \times 10^{-14} cm^2 s^{-1}$, whereas the diffusion coefficient obtained from D(T) = $0.17 cm^2 s^{-1}$ exp(-2.54eV/kT) (Stavola et al 1983) yields D(485°C) = $2 \times 10^{-18} cm^2 s^{-1}$, i.e. there is an enhancement of nearly 4 orders of magnitude for the long range migration of oxygen.

The diffusion of oxygen via oxygen-selfinterstitial pairs has been proposed by Ourmazd et al. (1984) as an enhancement mechanism. This proposal fits several qualitative experimental observations:
(i) According to the analysis by Ourmazd et al. (1984) only $5 \times 10^{14} cm^{-3}$ excess selfinterstitials are needed to account for an enhancement factor of 10 observed after a pre-anneal for 2 h at 900°C. We can reasonably expect a much higher number of excess selfinterstitials to build up during precipitation at 485°C, as 40% of the oxygen precipitates compared to about one percent at 900°C. Hence a much larger enhancement is expected, as observed. Once the oxygen precipitation at 485°C starts, the effect of the preanneal would be negligible, in accordance with experiment.
(ii) During the initial stages of precipitation the effective diffusion coefficient should show a sharp increase with time since the increase in the concentration of selfinterstitials by precipitation amounts to a positive feedback situation. For a time-independent diffusion coefficient a plot of the cross section should show a decreasing slope ("convex" shape), whereas the experimental curve (Fig.7), despite large error bars, suggests an increasing slope over the observed time range, indicating an increase in the effective diffusion coefficient.

Although two alternatives, namely, the diffusion by oxygen-vacancy pairs (Oates et al. 1984) or by oxygen molecules (Gosele et al. 1983) cannot account in an obvious manner for (i) and (ii), more evidence is needed before definite conclusions about the mechanism can be reached.

## 4.3 Nucleation of Ribbon Precipitates

It has been suggested by Bourret et al (1984) that thermal donors nucleate coesite precipitates. We have found evidence that an alternative mechanism is (also) operative: As remarked earlier, loopites tend to have steps in their {111} habit planes, which can be multiple steps along {311} planes (Fig.8a).These steps can be decorated by ribbon precipitates on {311} planes (Fig.8b). This and the frequent association of one or more ribbon-like precipitates with individual loopites indicates that at least some of the coesite precipitates have nucleated on {311} steps in the {111} habit planes of the loopites.

However the reverse process seems to occur as well since ribbon precipitates with two or more loopites along the $\langle 110 \rangle$ axis have also been observed. This finding emphasizes the special role of $\{311\}$ planes for the nucleation of extended defects, from interstitial point defects, the full implications of which have yet to be explored.

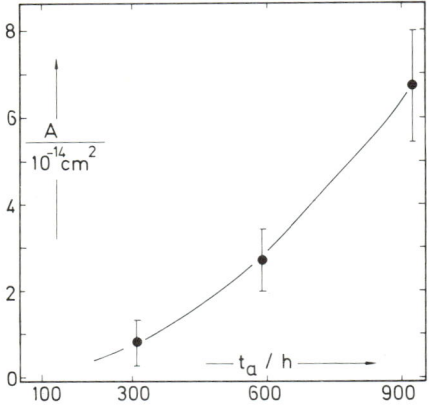

Fig.7 Ribbon cross section A as a function of annealing time t.

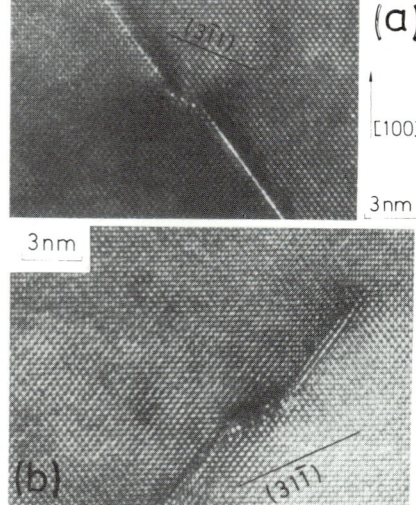

Fig.8 Undecorated ((a), t=311h) and decorated((b), t=591h) $\{311\}$ steps of a loopite.

Acknowledgements: This work was financially supported by the Royal Society - D.F.G., SERC and BP Venture Research Unit. The authors would like to thank Professor Sir Peter Hirsch for provision of laboratory facilities and Drs. G.R.Booker and I.G.Salisbury for discussions.

References
Bender H 1984 phys. stat. sol. (a) 86 245
Bergholz W, Pirouz P and Hutchison J L 1984 J. Electron. Mat. 14a 717
    (Note that in this paper nominal temperatures are quoted (450°C, 600°C,
    900°C), which correspond to 485°C, 635°C and 925°C (calibrated))
Bergholz W, Hutchison J L and Pirouz P 1985 J. Appl. Phys. in press
Bourret A, Thibault-Desseaux J and Seidmann B N 1984 J. Appl. Phys. 55
    825
Desseaux-Thibault J, Bourret A and Penisson J M 1983 Inst. Phys. Conf.
    Ser. 67 ,Section 2 71
Gosele V and Tan T Y 1981 "Defects in Semiconductors", eds S. Mahajan and
    S W Corbett, (North Holland, New York) p753
Graff K, Grallath E, Ades S, Goldbach G and Tolg G 1973 Solid-St.
    Electron 16 887.
Ham F S 1956 J. Appl. Phys. 30 915
Pasemann M, Hoehl P, Aseev A L and Pchelyakov O P 1983 phys. stat. sol.
    (a) 80 135
Oates A S, Newman R C and Tucker J H 1984 J. Electron Mat. 14a, 709
Ourmazd A, Schroter W and Bourret A 1984 J. Appl. Phys. 55 825
Stavola M, Patel J R, Kimmerling L C and Freeland P E 1983 Appl. Phys.
    Lett. 42 73
Tan T Y, Foell H, Mader S and Krakow W 1981 "Defects in Semiconductors"
    Vol 2, J. Narayan and T Y Tan eds, Proc. Mat. Res. Symp. 2, p179.

*Inst. Phys. Conf. Ser. No. 76: Section 1*
*Paper presented at Microsc. Semicond. Mater. Conf., Oxford, 25–27 March 1985*

# HREM investigation of the precipitation of antimony during thermal annealing of ion implanted silicon

H Bender

IMEC, c/o Universiteit Antwerpen, RUCA, Groenenborgerlaan 171, B-2020
Antwerpen, Belgium

   Abstract   The results are discussed of a study by means of high resolution electron microscopy and optical diffraction of the precipitation
   of antimony in ion implanted silicon wafers. The wafers are investigated as-implanted and after subsequent thermal annealing steps. The in-
   depth distribution and defect characterization are discussed.

## 1. Introduction

   Ion implantation of different dopant elements is widely used for the
formation of p-n junctions in integrated circuits. In the case of a high
dose, the near surface layer of the wafers turns amorphous and solid phase
epitaxial regrowth occurs during subsequent thermal annealing. Depending
on the experimental conditions, lattice defects and precipitates can be
introduced in the silicon matrix.

The implantation and subsequent precipitation of antimony has been studied
by means of electron microscopy techniques by Narayan and Holland (1982)
and Pennycook et al (1983, 1984a,b). They identified the precipitates as
partially coherent antimony with an icositetrahedral shape.

In this paper the results are discussed of an investigation by means of
high resolution electron microscopy and optical diffraction of antimony
implanted silicon wafers with [001] and [011] orientation.

## 2. Experimental procedure

Czochralski grown [001] or [011] oriented silicon wafers are implanted with
200 keV $Sb^+$ ions to a dose of $4.4 \times 10^{15}$ $cm^{-2}$ without substrate cooling. The
wafers are subsequently heat treated in a nitrogen ambient at 550°C for
20 min, at 900°C for 40 min or in two-steps : 550°C 20 min + 900°C 40 min.
The as-implanted and annealed wafers are studied in plan-view as well as
in cross-section. The HREM investigations are performed with a JEM 200 CX
electron microscope equipped with a ± 10° goniometer stage.

## 3. Results

Cross-section micrographs taken after the different treatment steps are
shown in fig. 1 for the [001] oriented wafers. Due to the ion implantation
the surface layer has become amorphous (fig. 1a). The amorphous-crystalline interface is very rough (fig. 2) and is situated at a depth of 160 nm
and 175 nm for respectively the [001] and [011] oriented wafers. Below this

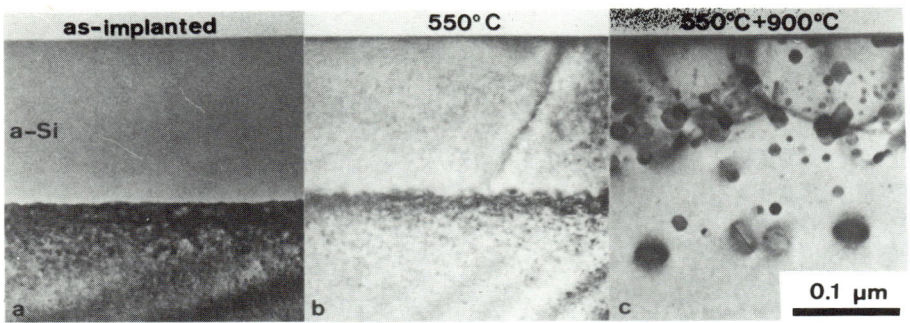

Fig. 1   Cross-section micrographs taken after the different treatments for
[001] oriented wafers.

interface defects are present which are only well defined for the [001]
oriented wafers. They are similar to the rod-like defects as previously
reported in the case of electron irradiation or ion implantation (Desseaux-
Thibault et al  1983, Bartsch et al  1984).

After a heat treatment of 20 min at 550°C the amorphous region has recrystallized but the defect region which was present under the amorphous-
crystalline interface remains unaltered (fig. 1b, 3). Due to this disturbed
region, the epitaxy of the regrown layer is imperfect and therefore contains dislocations.

The defect configuration after the treatment at 900°C and after the two-
step treatment 550°C + 900°C (fig. 1c) is similar.  In the top layer a
large density of precipitates is present, which is maximal between 40 and
120 nm under the wafer surface.  In the [001] wafers still a small density
of precipitates occurs below this depth up to the original amorphous-
crystalline interface (fig. 1c). The region with precipitates contains also
perfect dislocations, of which the density is larger for the [001] than for
the [011] wafers.  Furthermore stacking fault tetrahedra are observed in this
layer.  They have edges of 6 - 10 nm.  The characterization of this type of
defect  will be fully discussed elsewhere (Coene et al 1985).  In the layer
which originally contained implantation defects, dislocations are formed.
In the [001] oriented wafers only small loops occur (fig. 1c), while in
the [011] wafers a dislocation network is present at this depth.

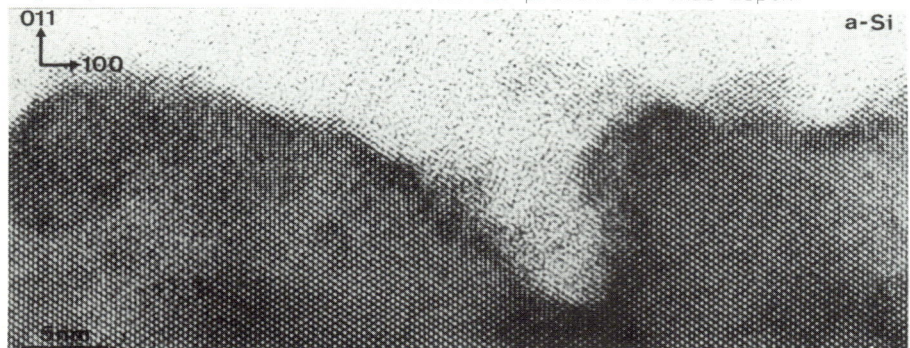

Fig. 2   The amorphous-crystalline interface after the ion implantation.

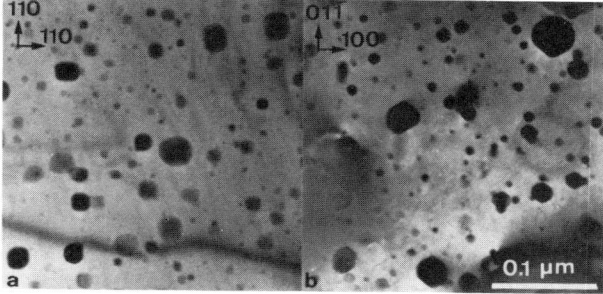

Fig. 3 Implantation de-
fect present under the
recrystallized layer after
an anneal at 550°C.

Fig. 4 Plan-view micrographs showing the [001]
and [011] projection of the precipitates and
their size variation.

The size of the precipitates varies between 5 and 40 nm (fig. 4). The projections along <001>, <011> and <111> are respectively nearly squares,elong-
ated and regular hexagons.  These projections are consistent with a trun-

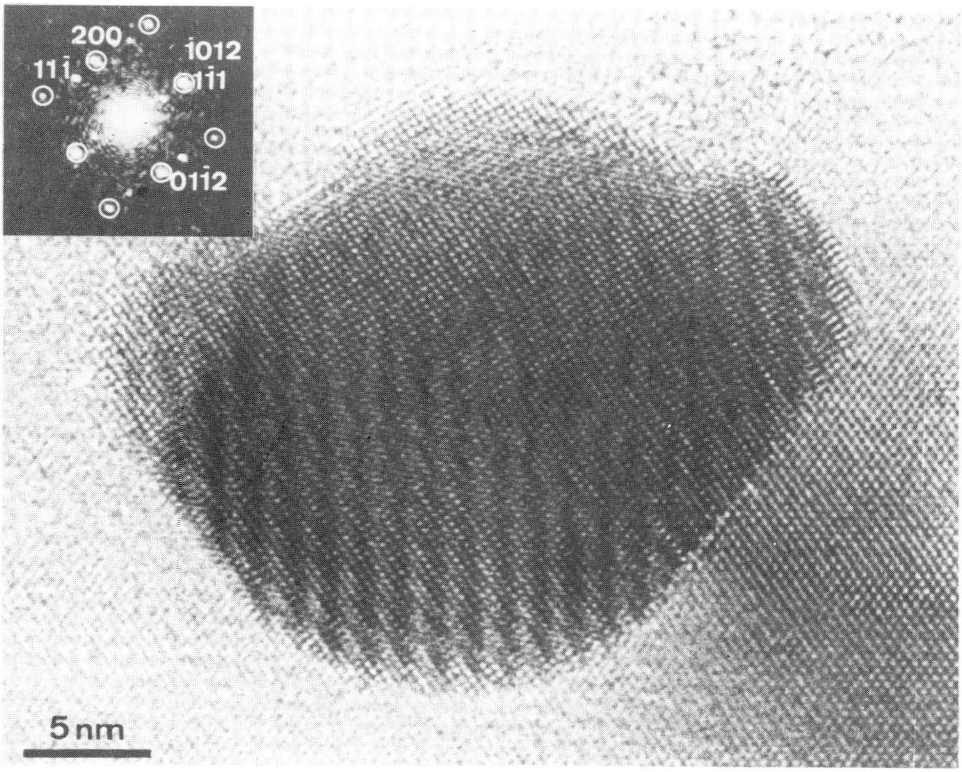

Fig. 5 HREM image and optical diffraction pattern demonstrating the orientation relationship between the Sb-precipitates and the Si-matrix. Bravais
and Miller indices refer respectively to the precipitate and to the matrix.

cated octahedron as also found for oxide precipitates (Bender 1984) and
are not consistent with an icositetrahedron as proposed by Pennycook et al
(1983) because for that body the <011> projection is not an elongated hexa-
gon. The truncated octahedron corresponds with the smallest surface energy
possible with low index planes. The shape of the precipitates often deviates from the ideal regular truncated octahedron. This is most pronounced
for those which are pinned on dislocations.

The orientation relationship as has been deduced by Pennycook et al (1983)
by means of microdiffraction, is directly verified by simultaneous structure imaging of the silicon matrix and the precipitate. Fig. 5 shows a
precipitate near the specimen edge for the silicon oriented some degrees off
the[011] orientation. Therefore the imaging of the silicon structure is
not perfect. The precipitate shows Moiré fringes in the thicker part of
the specimen and a regular nearly square-like pattern in the thinner parts
with an angle of 87° between the lattice planes. This corresponds to the
[2$\bar{2}$01] projection of the trigonal R $\bar{3}$ m Sb-structure. The coherency of
the (1$\bar{1}$1)$_{Si}$ and ($\bar{1}$012)$_{Sb}$ planes can directly be seen on the image of fig.5.
It can also be deduced from the optical diffractogram shown in the inset.
One therefore has :

$$\{\bar{1}012\}_{Sb} \; // \; \{1\bar{1}1\}_{Si} \quad \text{coherent}$$

$$\text{and} \quad <2\bar{2}01>_{Sb} \; // \; <011>_{Si}$$

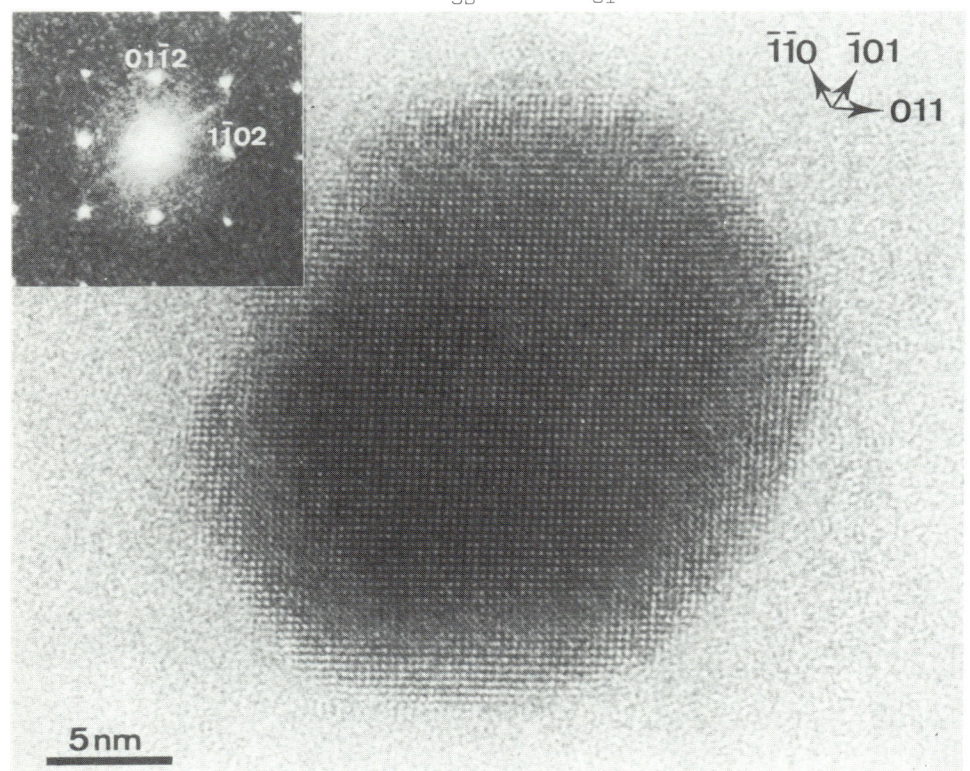

Fig. 6   HREM image of a [$\bar{2}$021]$_{Sb}$ oriented precipitate observed for the specimen tilted ∿ 3° off the [1$\bar{1}$1]$_{Si}$ orientation around the [231]$_{Si}$ axis.

which implies that the Sb-precipitates are partially coherent, embedded in the silicon matrix. From this orientation relationship, the different possibilities to obtain resolvable antimony orientations with respect to the low index silicon ones can be derived. The observed Sb orientations are summarized in table 1. It is seen that the $<2\bar{2}01>_{Sb}$ orientation can be reached by suitable tilting away from the $<001>_{Si}$ and $<111>_{Si}$ orientations. From the latter case, an example is shown in fig. 6. As the silicon lattice cannot be resolved, it appears with phase contrast and no Moiré

Table 1 : Observed Sb orientations

| Sb orientation | Si orientation |
|---|---|
| $<2\bar{2}01>$ | $\sim <0\bar{1}1>$ |
| | $\sim 3°$ off $<111>$ tilted around $<2\bar{3}1>$ |
| | $\sim 6°$ off $<001>$ tilted around $<110>$ |
| | $\sim <211>$ |
| $<0\bar{1}11>$ | $\sim 8°$ off $<001>$ tilted around $<1\bar{1}0>$ |
| | $\sim 15°$ off $<011>$ tilted around $<1\bar{1}1>$ |
| $<0001>$ | $\sim 10°$ off $<001>$ tilted around $<210>$ |
| $<2\bar{4}23>$ | $\sim 10°$ off $<001>$ tilted around $<110>$ |

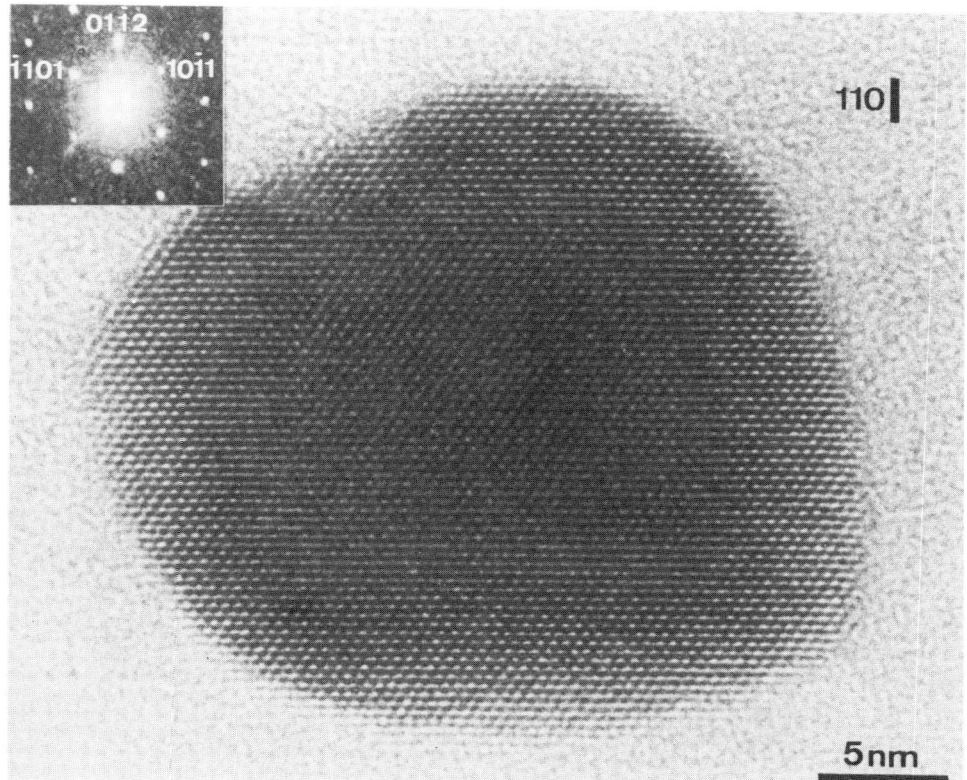

Fig. 7 A $[0\bar{1}11]_{Sb}$ oriented precipitate observed for the specimen tilted $\sim 8°$ around $[110]_{Si}$ away from the $[001]_{Si}$ orientation.

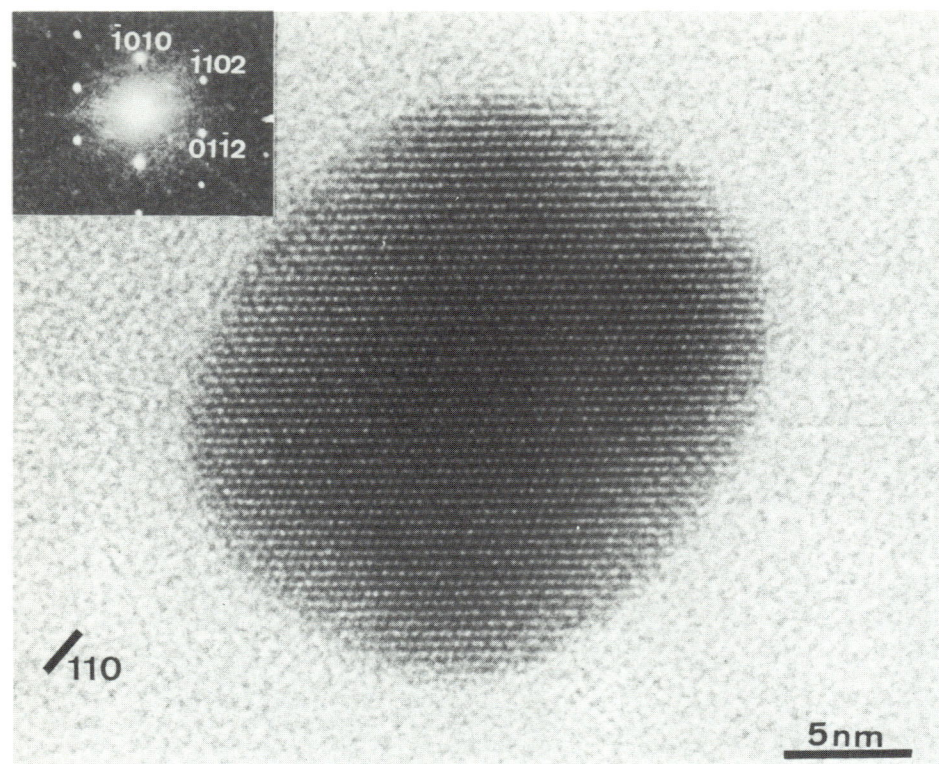

Fig. 8   A HREM image of a $[2\bar{4}23]_{Sb}$ oriented precipitate obtained for the matrix tilted $\sim 10°$ around $[110]_{Si}$ off the $[001]_{Si}$ orientation.

fringes now disturb the image.  Also along the $<211>_{Si}$ orientation, $<2\bar{2}01>_{Sb}$ oriented precipitates are observed.  Other antimony poles which have been found are : $[0\bar{1}11]_{Sb}$ (fig. 7), $[2\bar{4}23]_{Sb}$ (fig. 8) and $[0001]_{Sb}$ .  All these orientations can consistently be explained by assuming that the precipitates consist of antimony and they also confirm the deduced orientation relationship.

References

Bartsch H, Hoehl D and Kästner G 1984  phys. stat.sol. (a) 83  543
Bender H 1984  phys.stat.sol. (a) 86 245
Coene W, Bender H and Amelinckx S 1985  to be published in Phil. Mag. A
Desseaux-Thibault J, Bourret A and Penisson J M 1983  Inst. Phys. Conf. Ser. 67  71
Narayan J and Holland O W 1982 phys. stat. sol. (a) 73 225
Pennycook S J, Narayan J and Holland O W 1983 J. Appl. Phys. 54 6875
Pennycook S J, Narayan J and Holland O W 1984a J. Appl. Phys. 55  837
Pennycook S J, Narayan J and Holland O W 1984b Appl. Phys. Lett. 44 547

*Inst. Phys. Conf. Ser. No. 76: Section 1*
*Paper presented at Microsc. Semicond. Mater. Conf., Oxford, 25–27 March 1985*

# HREM determination of the structure of the {211} Σ = 3 twin in germanium

A. Bourret, L. Billard and M. Petit[+]

DRF/Service de Physique and DMG[+], Centre d'Etudes Nucléaires de Grenoble 85 X – 38041 Grenoble Cédex, France

Abstract. The structure of the same {211} Σ = 3 twin in germanium has been observed by HREM along two different common axes <011> and <231> in order to determine the 3-dimensional relaxation pattern. Relaxations have a cm symmetry with b = 2 <111> and c = <011> giving direct evidence for reconstruction along <011> axis. The rigid body translation is also measured. Among the three reconstructed models already proposed, the model due to Papon and Petit is the closest to the experimental observations. Energy calculation confirms that this model is the most stable. However detailed comparison between observed and computer simulated images shows that an atom pair at one of the reconstruction sites is not correctly positioned in this model.

## 1. Introduction

The structure of {211} twins in semiconductors has been for several years subject to both experimental and theoretical studies. This twin is often observed in ribbons or ingots of polycristalline silicon. Recent observations by EBIC have shown that this twin is electrically active in contrast to the coherent twin (Sharko et al. 1982). Since the origin of this activity might be associated with an intrinsic structural effect, there is considerable experimental interest in the determination of the structure of this twin. Vlachavas and Pond (1981), have demonstrated that a large rigid body translation (RBT) along <111> occurs in Silicon. Fontaine and Smith (1982) and later Labidi and Rocher (1985) found a non neglibible component of the RBT along <211>. Bacmann (1982) measured similar RBT in germanium showing that the structure should be identical in both materials. Two models involving tetracoordinated atoms were proposed : the first is symmetrical with respect to the interface (RBT = 0 along <111>) and is closely related to the Möller model (1981) except that dangling bonds are reconstructed along <011> as proposed by Pond et al (1983) ; the second model involves a reconstructed translation twin as originally introduced by Fontaine and Smith (1982). However energy calculation using a modified Keating potential (Pond et al 1983) has demonstrated that the symmetrical model has the lowest energy contrary to the observations. Moreover a recent electron diffraction experiment (Papon and Petit 1985) on a {211} twin in germanium has given a cm planar symmetry group and a unit cell twice as large as the coincident site lattice (csl) in two directions. A new model (hereafter referred to as the P and P model) was then proposed by Papon and Petit (1985) : it contains a very asymmetric structure with two reconstructed bonds along the <011> common axis per csl period. The reconstruction occurs alternately up and down along <011> giving a large unit cell (fig. 1).

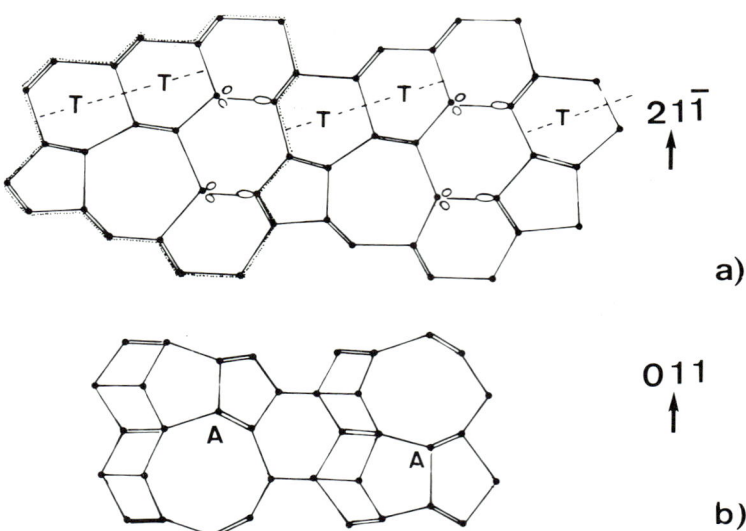

Fig. 1 Atomic model proposed by Papon and Petit (1985) as projected along
a) <011> axis b) <211> axis.

This paper describes HREM observations which confirm this recent electron
diffraction data and which give a direct measure of the RBT. Atomic positions
after relaxations are theoretically calculated and introduced as initial
positions for image simulation. Simulated and experimental images are then
compared in order to test the modelling.


2. HREM observations

Germanium bicrystals prepared by the CZ method are cut in {011} and {231}
slices in order to be observed along two different common axes. After
mechanical thinning specimens are ion milled. High resolution lattice
imaging is performed at 200 keV with the following experimental parameters :
$C_s = 1.05$ mm, beam divergence = 0.015 mrd, objective aperture = 0.5 $\mathring{A}^{-1}$. Two
defocusing distances ($\Delta z$) are systematically recorded corresponding to black
and white atoms respectively ; the intermediate position with a minimum of
contrast and double frequencies is employed as a test of appropriate
orientation and astigmatism correction.

Along the <011> axis very reproducible patterns are obtained for $\Delta z \simeq - 700 \mathring{A}$
and $- 1100 \mathring{A}$ (fig. 2). The apparent periodicity along the interface is <111>.

The RBT components in the {011} plane are measured direclty from HREM images
within an accuracy of 10 %. Table I gives a comparison between the previous
determination using the α-fringe method and the present determination using
HREM. It should be noted that the <211> component is found to vary along the
interface and to depend on the number of defects contained in the {211} twin.
The periodic and planar portions of this twin are often bounded by various
defects : steps, secondary dislocation, local decomposition of the twin in
two GBs... In the vicinity of these defects and particularly at {111} twin
steps, the <211> RBT tends to increase.

Table I. Measured component of the RBT on a {211} twin in Silicon and Germanium expressed in fraction of the corresponding vector.

| References | Material | <211> | <111> | <011> |
|---|---|---|---|---|
| Fontaine and Smith (1982) | Si | +1/18 | 1/9 | O |
| Vlachavas and Pond (1981) | Si | +1/100 | 1/8 | O |
| Labidi and Rocher (1985) | Si | +1/18 | 1/8 | O |
| Bacmann (1982) | Ge | > O | ? | O |
| This work | Ge | +1/35 to 1/17 | 1/12 to 1/10 | O |

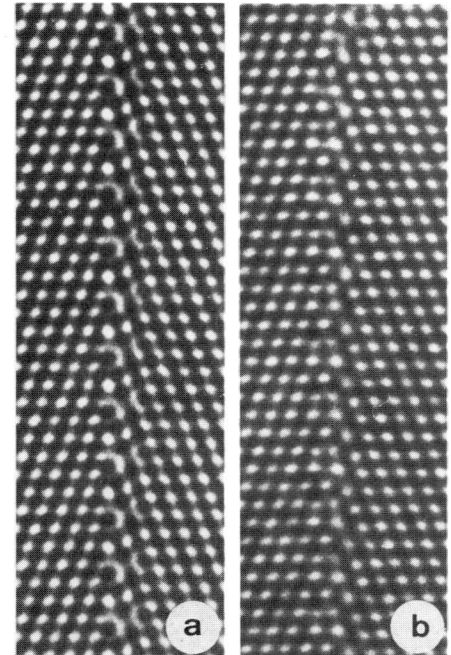
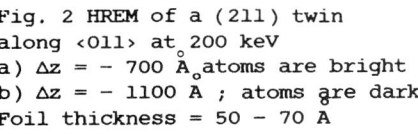
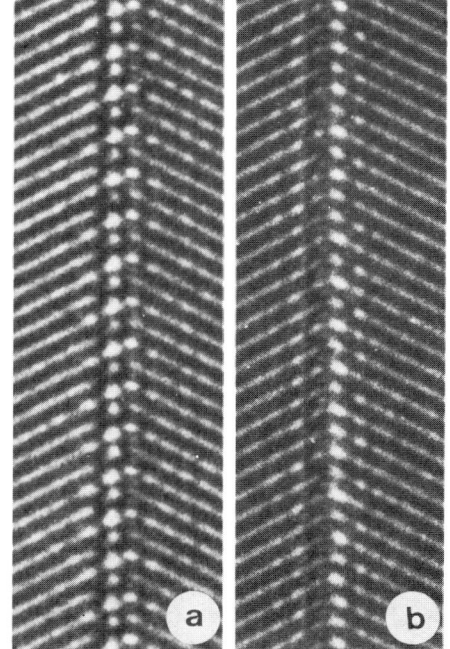

Fig. 2 HREM of a (211) twin along <011> at 200 keV
a) Δz = – 700 Å atoms are bright
b) Δz = – 1100 Å ; atoms are dark
Foil thickness = 50 – 70 Å

Fig. 3 HREM of a (211) twin along <231> axis at 200 keV
a) Δz = – 550 Å ; atoms are dark
b) Δz = –1300 Å; atoms are bright
Foil thickness = 100 – 120 Å

When observed along the <231> common axis (fig. 3) different patterns are obtained depending on the defocusing distance. Close to the Scherzer defocus a clear double periodicity is visible with a 7.39 Å repeat distance in good accordance with the space group and the unit cell determined by electron diffraction. Therefore the relaxations are periodic with a centered unit cell b = 2 <111> c = <011>. The RBT component along <011> should in principle be measurable on a <231> projection. However due to the fact that only one set of {111} planes is visible in each crystal, this determination is not accurate enough.

The combination of the two HREM observations enables us to conclude unambiguously that : i) reconstruction along the <011> axis occurs. It doubles the periodicity along this direction. This reconstruction is similar to the reconstructed models of the 30° partial dislocation (Marklund 1979) and is directly shown for the first time. ii) along the interface the successive reconstructed rows are shifted by 1/2<011> forming the centered pattern as described in the P and P model. iii) The RBT component along <111> is in good agreement with previous measurements. However along <211> the experimental values are more scattered and depend on the defect density : this may explain the different results obtained by previous authors.

## 3. Computer modelling of the {211} twin structure

Valence potential functions like the Keating potential are used for calculating the equilibrium position of the atoms in different models. The program developed by Lançon (1984) is an extension of a package used for studying metallic amorphous structures. Two parameters are employed to describe the force constant for bond length variation and bond stretching (see for example Bourret et al. 1983) and all atoms are supposed to be tetracoordinated. Three reconstructed models have been calculated in the following manner : 174 atoms with 4 nearest neighbours in accordance with the topology of the choosen model are free to move inside a parallelepiped with two periodic conditions at the {111} faces, and two perfect crystallites free to move in any direction at the external {211} faces. The box size is (25,7 x 19,6 x 8) Å³. The energy minimization giving the final atomic coordinates and the RBT is performed by the method of conjugate gradients. Computer results are given in Table II for three models. It can be concluded that the only

Table II. Calculated total energy and RBT for
three different reconstructed models

|  | Total energy (J m$^{-2}$) | <211> | <111> | <011> |
|---|---|---|---|---|
| Symmetrical model Pond et al (1983) | 0,30 | 1/99<211> | 0 | 0 |
| Translation twin Fontaine et al (1982) | 0,60 | 1/58<211> | 1/6 <111> | 0 |
| P and P model (1985) | 0,29 | 1/50<211> | 1/11<111> | 0 |

model compatible with the experience is the P and P model : it has the lowest total energy, the correct order of magnitude for the RBT and moreover the correct periodicity. Therefore this model is used for image simulation.

## 4. Image simulation

Both <011> and <231> projections of the previously calculated atomic positions are introduced in a multislice programm to simulate images at different thicknesses and defocusing distances. The sampling forms an image with 128 x 128 points on a square area with a dimension $3a\sqrt{3}$ for <011> projection and $8a\sqrt{3}/\sqrt{7}$ for <231> projection. These calculated images are compared with experimental images (fig. 4 and 5).

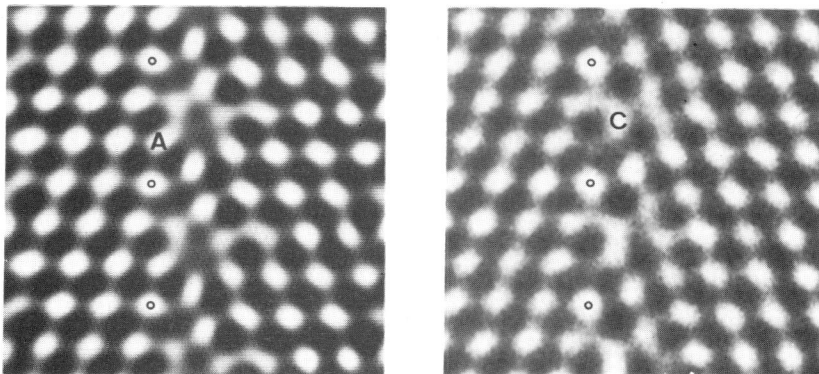

Fig. 4 Comparison between calculated and experimental images for <011> projections using the P and P model. $\Delta z = -760$ Å ; thickness = 52.4 Å ; 200 keV. Note the elongated C spot not reproduced in this simulation and the absence of the A spot in the experimental image.

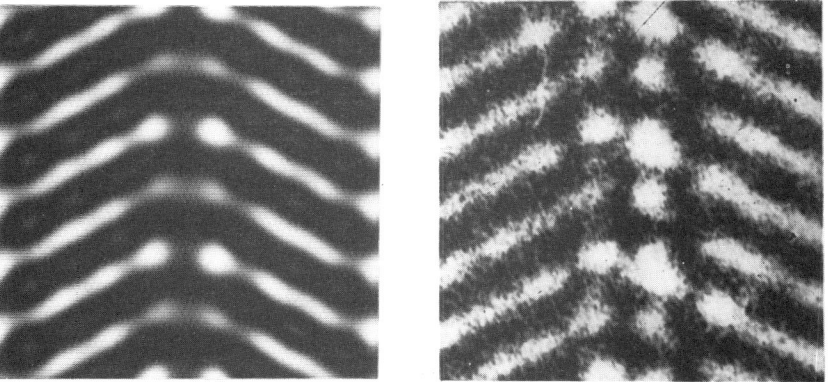

Fig. 5 Comparison between calculated and experimental images for <231> projection using the P and P model. $\Delta z = -512$ Å ; thickness = 105 Å ; 200 keV. Note the double periodicity along the GB due to reconstruction.

It is apparent from fig. 4 and 5 (and is confirmed at other defocusing distances) that there is a qualitative agreement between the model and the true {211} twin structure. In particular the two 6-atoms rings typical of the {111} twin on one side of the interface are well imaged. However on the other side the reconstruction involving an atom pair at A is badly simulated. The elongated dot C observed experimentally does not correspond to atomic positions in the model. This difference is systematically observed and could show that one        reconstruction is not correctly described by the P and P model. It is important to note that these atoms have the maximum energy after relaxation due to highly distorted bonds around them. Therefore, to conclude, it is suggested that further stabilization could occur either through electronic rearrangement destroying the tetracoordination, or by preferential impurity segregation at this site.

## Acknowledgements

The authors wish to thank Dr. J.J. Bacmann for his constant interest during these experiments and for supplying the bicrystals. Technical support from C. Bouvier and C. Closse is gratefully acknowledged.

## References

BACMANN J.J. 1982, J. de Physique C6-93
BOURRET A., DESSEAUX J., LANCON F. 1983 J. de Physique, C4-15
SHARKO R., GERVAIS A., TEXIER-HERVO C. 1982 J. de Physique, C1-129
FONTAINE C., SMITH D.A. 1982 Appl. Phys. Lett. 40 (2) 153
MARKLUND S. 1979 Phys. Stat. Sol. (b) 83
LABIDI M., ROCHER A. 1985 J. de Physique Appl. in press
LANCON F. 1984, Grenoble, Thesis
MOLLER H.J; 1981 Phil. Mag. A 43 1053
PAPON A.M., PETIT M.  1985 Scripta Met. in press
POND R.C., BACON D.J., BASTAWEESY A.M. 1983 Inst. Phys. Conf. n° 67 253
VLACHAVAS D.S., POND R.C. 1981 Inst. Phys. Conf. n° 60 159

*Inst. Phys. Conf. Ser. No. 76: Section 1*
*Paper presented at Microsc. Semicond. Mater. Conf., Oxford, 25–27 March 1985*

# HREM studies of II−VI heteroepitaxial layers

A G Cullis, N G Chew, J L Hutchison*, S J C Irvine and J Giess

Royal Signals and Radar Establishment, St Andrews Road,
Malvern, Worcs. WR14 3PS
*Department of Metallurgy and Science of Materials, University of Oxford,
Parks Road, Oxford OX1 3PH

Abstract   The structure of CdTe and $Cd_xHg_{1-x}Te$ layers grown on InSb and GaAs substrates are studied by high resolution, cross-sectional transmission electron microscopy.   The CdTe/InSb misfit is very small and there are few defects in epitaxial layers grown under optimum conditions. The interface can, nevertheless, be delineated in high resolution images by careful adjustment of the electron-optical conditions.   The CdTe/GaAs and $Cd_xHg_{1-x}Te$/GaAs systems exhibit large misfit with interfacial dislocations spaced by ⪞3nm.

## 1. Introduction

Heteroepitaxial semiconductor materials are assuming increasing importance for electronic device fabrication.   Many different layer/substrate combinations are under investigation and the present article will focus on three specific II-VI/III-V combinations, namely CdTe on InSb and GaAs and $Cd_xHg_{1-x}Te$ on GaAs.   In each case, the crystallographic perfection of the epitaxial layers is a crucial consideration.   For example, defects may nucleate at or near the heterointerface during initial growth and then propagate outwards into subsequently grown material. Thus, it is important to fully characterise layer defects especially in interfacial regions. This is achieved in the present work by lattice imaging of cross-sectional specimens using high resolution transmission electron microscopy.

The three heteroepitaxial systems under study exhibit different interfacial structures.   This is due to the widely different lattice mismatches:   only 0.05% for CdTe/InSb but nearly 14% for CdTe/GaAs and $Cd_xHg_{1-x}Te$/GaAs.   A large difference in the spacing of interfacial dislocations is observed although, in each system, high quality single crystal epitaxial layers can be produced under optimum growth conditions.

## 2. Experimental Methods

Epitaxial layer growth was carried out by molecular beam epitaxy (MBE) in ultra-high vacuum for the CdTe layers on InSb (Williams et al 1985) whereas metal-organic chemical vapour deposition (MOCVD) was used to prepare the CdTe and $Cd_xHg_{1-x}Te$ layers on GaAs (Tunnicliffe et al 1984). The crystallographic orientation of all substrates was (001).   The structure of the heteroepitaxial samples was studied in the transmission electron microscope (TEM).   Specimens were thinned in cross-section to electron transparency by sequential mechanical polishing and low voltage (3-6kV) ion milling.

Reactive I$^+$ ion bombardment was employed in order to minimise spurious damage formation in the II-VI semiconductor layers (Chew and Cullis 1984, Cullis et al 1985).  Diffraction contrast studies of the specimen foils were carried out in a JEM 120C TEM operated at 120kV.  A portion of the high resolution lattice imaging work was performed using a JEM 200CX, 200kV TEM equipped with an LaB$_6$ cathode and an objective lens with C$_s$ of 1.2mm, giving an optimum resolution at Scherzer defocus of 0.24nm.  Additional high resolution observations employed a JEM 4000EX, 400kV TEM, again equipped with an LaB$_6$ cathode and also having an objective lens of C$_s$ ~1mm yielding a Scherzer defocus resolution of better than 0.18nm.

## 3. Results and Discussion

### 3.1 CdTe Layers on (001) InSb

The structure of CdTe layers grown by MBE depends critically both upon the InSb substrate surface preparation and upon the substrate temperature.  The surface must first be cleaned of impurities, usually by ion bombardment and heat treatment.  However, free metallic In should not be allowed to accumulate during this processing since it can disrupt subsequent growth (Chew et al 1984b, Wood et al 1984).  The effect of residual surface impurities upon CdTe growth is illustrated in Fig. 1a where it is seen that they give rise to layer defects, including dislocations and microtwins.  For an optimally cleaned substrate surface, if CdTe growth is initiated at a temperature <150°C polycrystalline material interspersed with Te precipitates is produced (Chew et al 1984a).  For high temperature growth above ~200°C layer defects can also form with some reaction between the CdTe and InSb, but a deposition temperature of ~180°C allows the formation of high quality, single crystal CdTe with few extended crystallographic defects.  A layer of this type is shown in Fig. 1b, where the only feature visible is a line of dark contrast at the interface with the substrate.

A high resolution, [110] pole projection cross-sectional image of the interfacial region of the sample in Fig. 1b is presented in Fig. 2a.  The grid of contrast points is produced by crossed 111 and 002 lattice fringes with spacings of 0.37nm and 0.32nm, respectively (the diffracted beams contributing to the image were 000, 4x111-type and 2x002-type). The crystal lattice extending across the interface is essentially perfect and no defects are evident.  Indeed, since the CdTe/InSb misfit is only ~0.05% and there will be residual layer strain, any resulting interfacial dislocations will

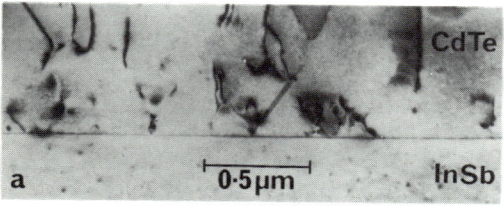

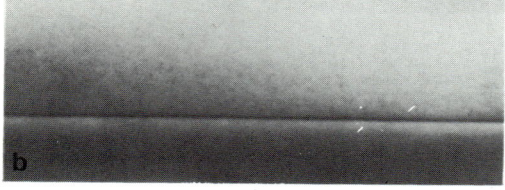

Fig. 1   Cross-sectional, strong-beam, transmission electron images of CdTe layers on InSb: a) growth on contaminated substrate; b) growth under optimum conditions.

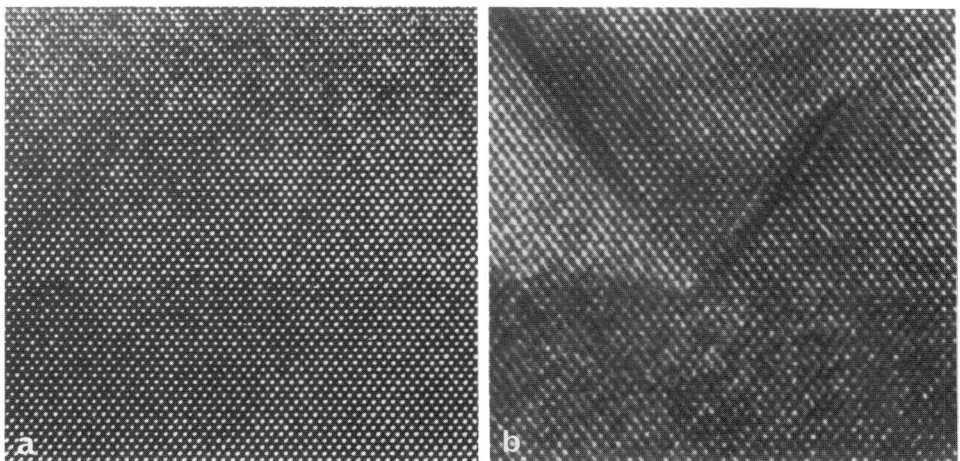

Fig. 2    Cross-sectional, high resolution, transmission electron
lattice fringe images of CdTe/InSb interfaces:    a) defect-free
material;   b) defect-containing CdTe layer.

be spaced ≯1µm apart so that they are unlikely to be found in this type of
small area image.   However, a dark line of contrast is faintly visible
along the interface and may be produced by lattice relaxation due to the
local change in atomic bonding or by strain introduced by residual, dispersed impurity.   In addition, a difference in overall image contrast is
clearly seen to characterise the two different materials across the interfacial plane – InSb is dark , CdTe is light. This may seem surprising due
to the similarity in the mean atomic weights of the two compounds. Nevertheless, detailed contrast calculations (Hutchison et al 1985) demonstrate
that this contrast difference does occur over a restricted range of imaging
conditions due, in part, to a variation in reflection extinction distances.

Figure 2b shows a high resolution, lattice fringe image of the interfacial
region of a layer sample produced under non-optimum conditions.  A pair of
inclined defects in the CdTe originating from the interface plane is
visible.  The position of defect initiation is extremely localised and, for
example, no large scale oxide or impurity film is present.  Thus, defects
can be formed during layer growth at very small (atomic dimension) interface disturbances such as, for example, small aggregates of impurities or
antisite defects.

3.2 CdTe and $Cd_xHg_{1-x}Te$ Layers on (001) GaAs

When CdTe layers are grown by MOCVD under optimum reactant flow rate
conditions and at a GaAs substrate temperature of 410°C, good single crystal
material is formed.  Usually, however, the CdTe contains dislocations as
shown in Fig. 3.   The dislocation density varies with depth and, while it
is high within ∿300nm of the GaAs interface, the density rapidly falls as
the layer surface is approached. The total defect density within ∿30nm of
the heterointerface is extremely high, so that overlapping contrast precludes conventional imaging.  Accordingly, Fig. 4 shows a high resolution,
lattice fringe image of this region and it is clear that many inclined
microtwin lamellae originate during growth at the interface. However, these

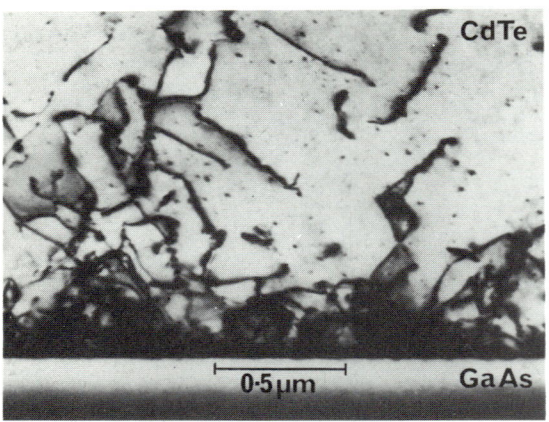

Fig. 3 Cross-sectional, strong-beam, transmission electron image of CdTe layer on GaAs.

defects usually penetrate only a short distance into the overlying CdTe layer. The CdTe/GaAs interface itself is imaged at higher magnification in Fig. 5 and it is found that an array of misfit dislocations is present. The (projected) Burgers vectors of these dislocations are all of type a/2 <110>, which is consistent with them being undissociated 60°-type dislocations. Their mean separation, measured as ∿2.8nm, is in good agreement with that predicted for the 14% misfit.

The growth of $Cd_xHg_{1-x}Te$ (x=0.2) layers upon (001) GaAs by MOCVD can also give good single crystal material. Again, the predominant layer defects are dislocations which fall in number density with distance from the GaAs. In addition, however, as shown in Fig. 6 there is a densely packed band of defects extending ∿250nm into the ternary alloy beyond the interface. The

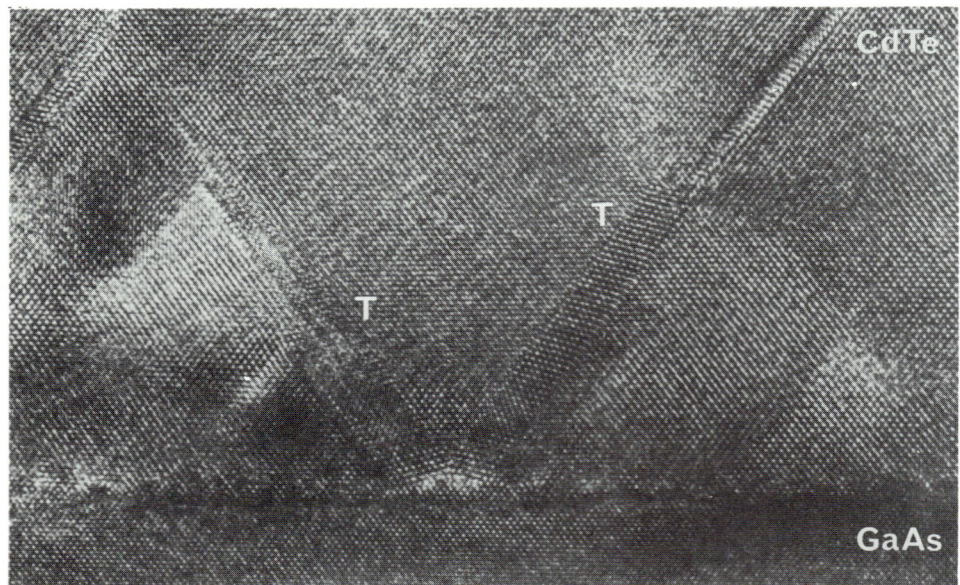

Fig. 4     Cross-sectional, high resolution, transmission electron lattice fringe image of defects in CdTe layer on GaAs.

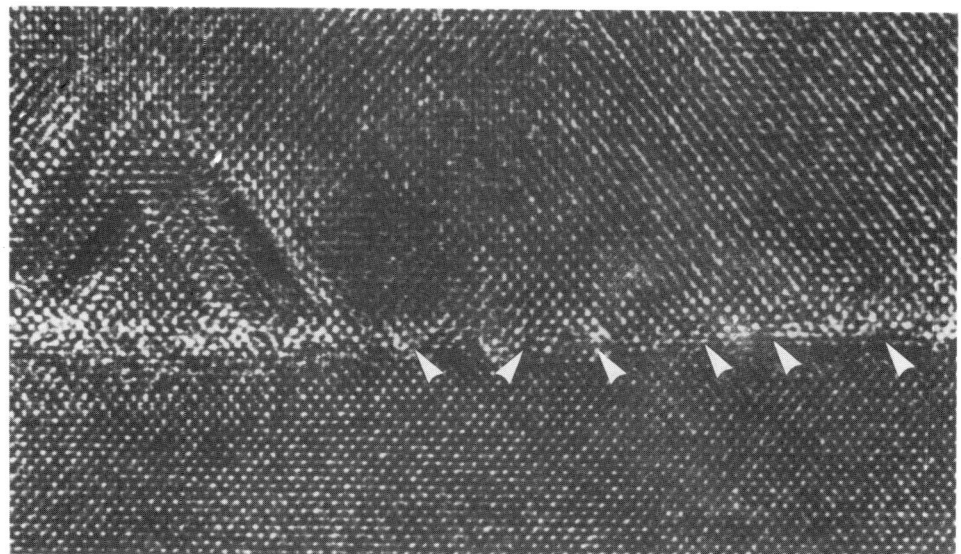

Fig. 5   Cross-sectional, high resolution, transmission electron lattice fringe image of CdTe/GaAs interface showing misfit dislocations (arrowed).

band is bounded at its upper surface by a raft of mainly 60°-type dislocations and contains numerous threading dislocations together with some misoriented material and cavities. The interface between the layer and the GaAs substrate once again is characterised by an array of closely spaced misfit dislocations - see Fig. 7.

The reason for formation of the defect band adjacent to the $Cd_xHg_{1-x}Te$/GaAs interface (Fig. 6) is not yet clear. However, if required, it can be eliminated by growth of an initial buffer layer of CdTe before incorporation of the Hg. The resulting layer structure is then similar to that shown in Fig. 3: work on this approach is continuing.

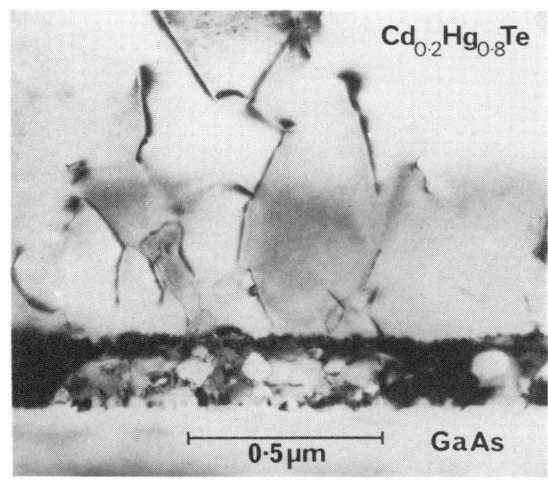

$Cd_{0.2}Hg_{0.8}Te$

0·5 μm    GaAs

Fig. 6 Cross-sectional, strong-beam, transmission electron image of $Cd_{0.2}Hg_{0.8}Te$ layer on GaAs.

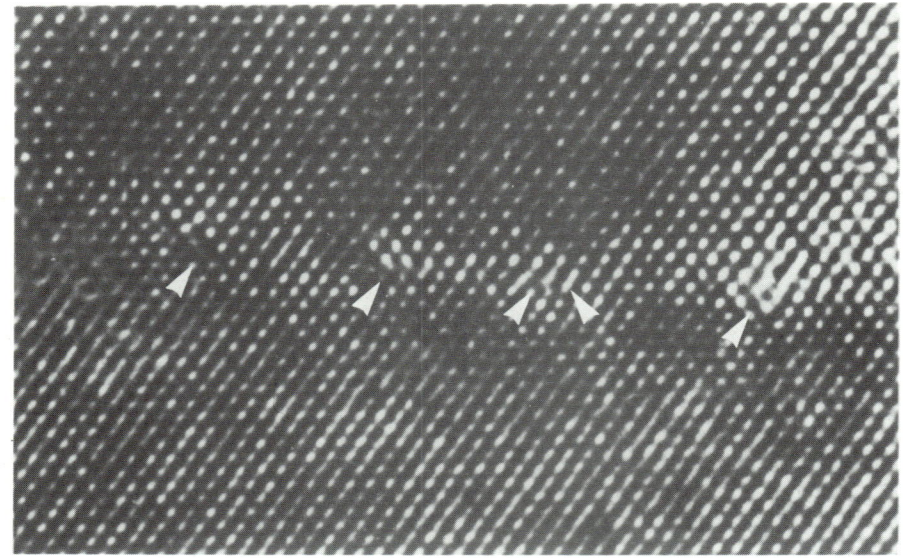

Fig. 7    Cross-sectional, high resolution, transmission electron lattice fringe image of $Cd_{0.2}Hg_{0.8}Te$/GaAs interface showing misfit dislocations (arrowed).

4. Conclusions

Defect-free CdTe can be grown on InSb by MBE at ~180°C with correct surface preparation. High resolution microscopy has shown that very small (sub-monolayer) aggregates of, for example, surface impurities can lead to the formation of layer defects. The large (~14%) misfit between CdTe and $Cd_xHg_{1-x}Te$ layers and GaAs leads to the formation of closely (~2.8nm) spaced misfit dislocations. High resolution studies are essential to observe the individual dislocations in these networks. In order to minimise defect formation in the region of the interface, alloy layers of $Cd_xHg_{1-x}Te$ grown on GaAs by MOCVD are best prepared using buffer layers of CdTe.

Acknowledgement

The authors would like to thank G M Williams for preparing the MBE CdTe layers on InSb. JLH thanks SERC for support. © HMSO, London, 1985.

References

Chew N G and Cullis A G 1984 Appl. Phys. Lett. 44 142
Chew N G, Cullis A G and Williams G M 1984 Appl. Phys. Lett. 45 1090
Chew N G, Williams G M and Cullis A G 1984 Electron Microscopy and Analysis
     1983 ed P Doig (Bristol: Institute of Physics) pp 437-440
Cullis A G, Chew N G and Hutchison J L 1985 Ultramicroscopy
Hutchison J L, Waddington G, Chew N G and Cullis A G in preparation
Tunnicliffe J. Irvine S J C, Dosser O D and Mullin J B 1984 J. Crystal
     Growth 68 245
Williams G M, Whitehouse C R, Chew N G, Blackmore G W and Cullis A G 1985
     J. Vac. Sci. Technol. B3 704
Wood S, Greggi J Jr, Farrow R F C, Takei W J, Shirland F A and Noreika A J
     1984 J. Appl. Phys. 55 4225

*Inst. Phys. Conf. Ser. No. 76: Section 1*
*Paper presented at Microsc. Semicond. Mater. Conf., Oxford, 25–27 March 1985*

35

# Investigation of the early growth of epitaxial silicon-on-sapphire using high resolution transmission electron microscopy

K C Paus[1], J C Barry[1], G R Booker[1], T B Peters[2] and M G Pitt[2]

[1]Department of Metallurgy & Science of Materials,
University of Oxford, Parks Road, Oxford OX1 3PH
[2]GEC Research Laboratories, Hirst Research Centre,
East Lane, Wembley, Middlesex HA9 7PP

Abstract    Epitaxial silicon was grown on (01$\bar{1}$2) sapphire ($\alpha$-Al$_2$O$_3$) substrate slices at 850, 940 and 1050°C using SiH$_4$ chemical vapour deposition. The growth was terminated when separate Si centres, often <20nm across, were still present on the Al$_2$O$_3$ substrate surface. TEM lattice-image examinations using the cross-section method showed that most of the centres were single- or multiply-twinned, and that the amount of twinning decreased as the growth temperature increased.

## 1. Introduction

CMOS devices fabricated in epitaxial silicon layers grown on single-crystal sapphire ($\alpha$-Al$_2$O$_3$) substrate slices have the advantage, compared with analogous devices fabricated in bulk Si slices, of high speed, low power consumption and radiation hardness. For such fabrication, (01$\bar{1}$2) Al$_2$O$_3$ slices are generally used because these give (100) Si layers, the complete epitaxial relationship being (100)Si$||$(01$\bar{1}$2) Al$_2$O$_3$ and [001]Si$||$[2$\bar{1}$$\bar{1}$0] Al$_2$O$_3$. There is a large lattice mismatch between the Si layer and Al$_2$O$_3$ substrate, and this gives rise to high concentrations of crystallographic defects in the Si layer. TEM examinations (Abrahams et al 1981, Hutchison et al 1981) of cross-section specimens using the lattice-imaging method showed the presence of micro-twin lamallae and stacking faults on inclined {111} planes running from the interface into the layer, together with small areas, up to ~10nm across, in the interface region that were multiply-twinned. The present paper describes a continuation of this earlier work. In particular, analogous TEM examinations have been made on specimens in which the Si deposition was stopped before the initial growth centres had joined up to give a continuous layer. This was done in order to see whether the twinning had already occurred at this early stage of growth.

## 2. Experimental

Polished Al$_2$O$_3$ substrate slices orientated within 0.5° of (01$\bar{1}$2), obtained from either Kyocera Co., Japan, or Union Carbide Inc., USA, were given a standard chemical cleaning treatment. They were placed in a SiH$_4$ chemical vapour deposition reactor and heated at 1200°C in hydrogen. The temperature was decreased to either 850, 940 or 1050°C and then Si was deposited. Several specimens were prepared corresponding to various combinations of SiH$_4$ flow-rate (0.06 to 0.46 $\ell$/min) and deposition times (1.2 to 6.6s).

SEM examinations were performed on bulk specimens using a JEOL 100C with a
scanning attachment.   TEM examinations were performed on cross-section
specimens corresponding to (011) Si and prepared by $Ar^+$ ion thinning.   The
specimens were examined in a JEOL 200CX and tilted so that the electron
beam was parallel to $[011]$ Si.   Lattice images were obtained using 7 Si
beams and 9 $Al_2O_3$ beams as described previously (Hutchison et al 1981).

## 3. 940°C Specimens

SEM examinations of a series of such specimens deposited at 940°C showed
that the initial Si growth took the form of separate circular centres ~20
to 25nm across (corresponding to ~15% area coverage).   It continued as
circular and elongated centres up to ~100nm across and ~250nm long (~70%
coverage).   It eventually gave an almost continuous layer comprising elongated  structures  separated  by  channels  up  to  ~15nm  across  (~90%
coverage).

The TEM examinations showed the following.   The $Al_2O_3$ substrate surface
was planar corresponding to (01$\overline{1}$2) $Al_2O_3$, except for single atomic steps
that occurred at intervals along the surface.   For the initial stage of
growth, individual Si centres could be seen in cross-section with $Al_2O_3$
areas in between which were free from Si.   For each centre, the Si/$Al_2O_3$
interface almost always showed the presence of one, and sometimes two,
atomic steps (e.g. S in Fig.3).   This suggests that the initial nucleation
events for the individual centres took place at surface steps on the $Al_2O_3$
substrate slice.   A small number of centres consisted entirely of single-
crystal material corresponding to the standard epitaxial relationship (M
in Fig.1), i.e.   these centres were untwinned.   The majority of centres
comprised a mixture of untwinned and singly-twinned material, the individual regions being separated by planar {111} twin boundaries.   For
example, Fig.2 shows a centre with the left side untwinned (M) and the
right side singly-twinned $(T_1)$.   Fig.3 shows a centre with untwinned
material ($M_1$, $M_2$ and $M_3$), singly-twinned material $(T_1)$ and stacking faults
(SF).   A small number of centres comprised multiply-twinned material.   For
example, Fig.4 shows a centre with the left and right sides untwinned ($M_1$
and $M_2$), the upper part of the middle portion singly-twinned $(T_1)$, and the
lower part of the middle portion doubly-twinned $(T_2)$.   These regions are
separated by planar {111} twin boundaries $M_1/T_1$, $T_1/M_2$ and $T_1/T_2$, and an
incoherent boundary $T_2/M_2$.   For those centres that were twinned, the
places where the twin boundaries met the Si/$Al_2O_3$ interface corresponded
in some cases to an atomic step at the interface, but not in most cases.

For the initial stage of growth, the cross-section specimens showed that
the upper surface of the individual Si centres was generally curved,
although crystallographic facets were beginning to develop.   As growth
proceeded (Fig.5), the facets became more pronounced, {111} being most
common, but {311} and {100} also occurring.   When the channels between the
Si centres began to fill in, additional twinning often occurred.   For
example, Fig.6 shows complex multiple-twinning arising at the edge of a
growth centre ($M_3$).   A preliminary analysis indicates that in addition to
planar {111} twin boundaries, there are three non-coherent twin boundaries
($M_2/T_{2a}$, $T_{2b}/M_3$ and $T_5/M_3$).   Fig.7 shows what are considered to be two
centres ($M_1$ and $M_2$), containing micro-twin lamellae that subsequently
joined together.   This joining gave two doubly-twinned regions ($T_{2a}$ and
$T_{2b}$) with two incoherent boundaries $M_1/T_{2a}$ and $T_{2b}/M_2$ and one planar {111}
twin boundary $T_{2a}/T_{2b}$.

## 4. 850/940/1050°C Specimens

SEM examinations of specimens deposited at 850, 940 and 1050°C (0.06 $\ell$/min, 3.0s) showed mainly circular centres with number densities ~1000, 500 and 135 /($\mu$m)$^2$ respectively, and mean widths ~22, 35 and 60nm respectively. TEM examinations showed for the 850°C specimen, compared with the 940°C specimen, that the condition of the $Al_2O_3$ substrate surface was similar, and the amount of twinning in the centres was greater (Fig.8). For the 1050°C specimen, compared with the 940°C specimen, the $Al_2O_3$ substrate surface was heavily eroded with irregular steps present often as high as ~4nm (Figs.9 and 10). The amount of twinning in the centres was less by ~4X. The facetting of the surfaces of the centres was more pronounced. Relatively long lengths of highly irregular $Al_2O_3$ substrate surface were often grown over by Si centres without generating any twin defects (e.g. XY in Fig.10).

## Acknowledgments

We should like to thank Dr J L Hutchison for helpful discussions and to acknowledge support by the Scientific and Engineering Research Council and the Alvey Directorate.

## References

Abrahams M S, Hutchison J L and Booker G R 1981 Phys. Stat. Sol. (a) 63 K3
Hutchison J L, Booker G R and Abrahams M S 1981 Inst. Phys. Conf. Ser. No.67, p139

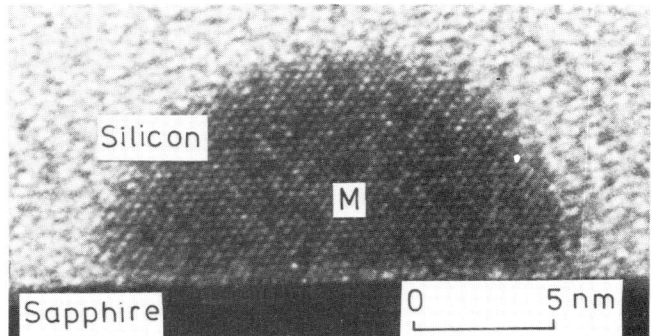

Fig.1 Defect-free early growth centre (940°C).

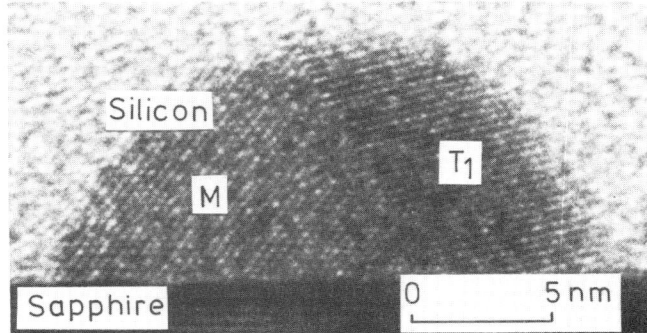

Fig.2 Singly twinned early growth centre (940°C).

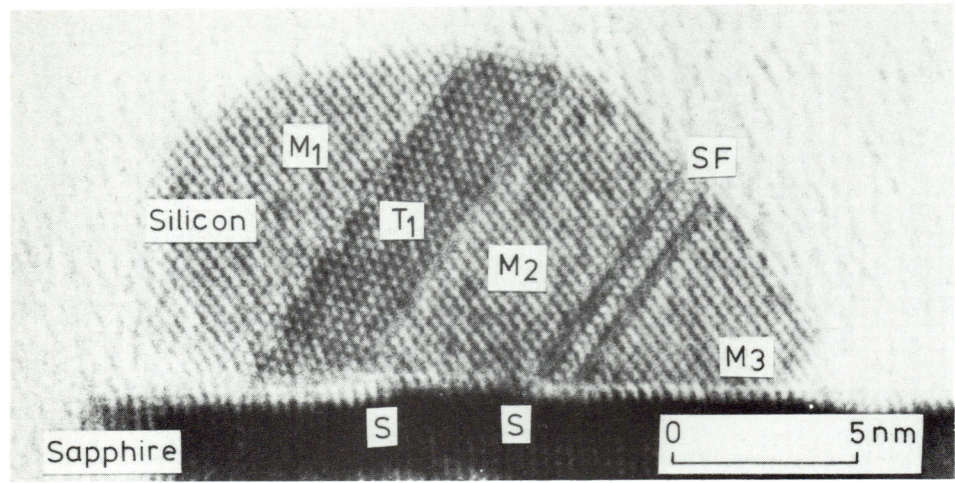

Fig.3 Growth centre with a microtwin ($T_1$) two stacking faults (SF) and substrate surface steps (S) (940°C).

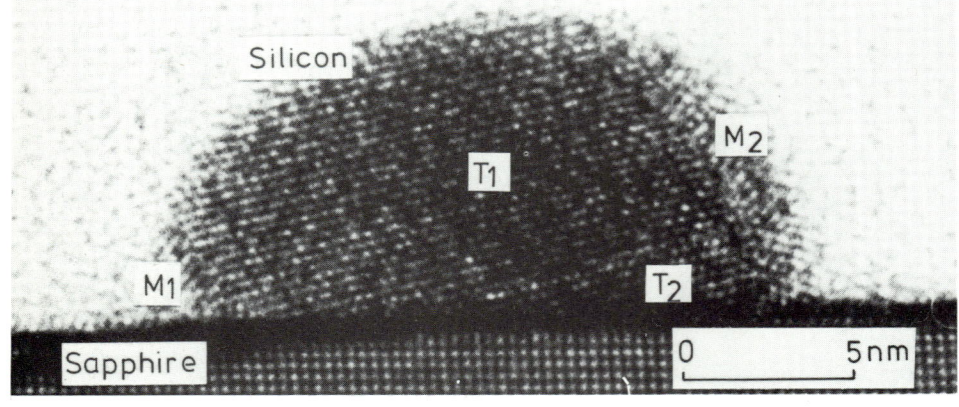

Fig.4 Multiply twinned growth centre (940°C).

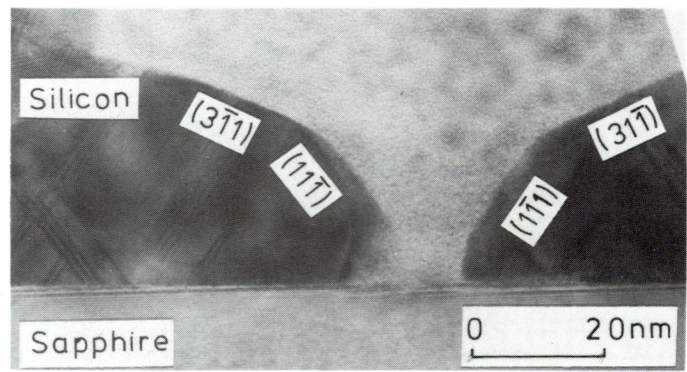

Fig.5 Later stages of growth showing {111} and {311} facets (940°C).

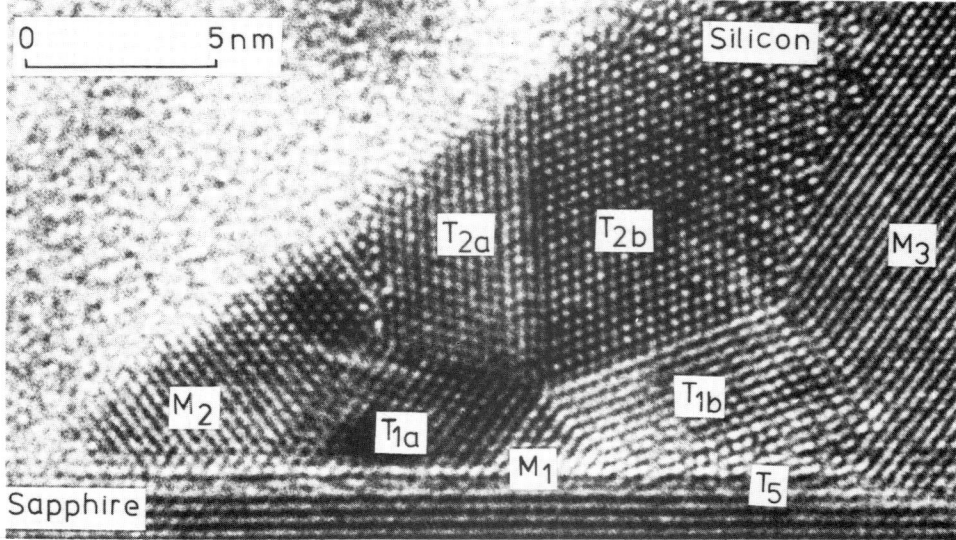

Fig.6 Complex multiple twinning at the edge of a growth centre (940°C).

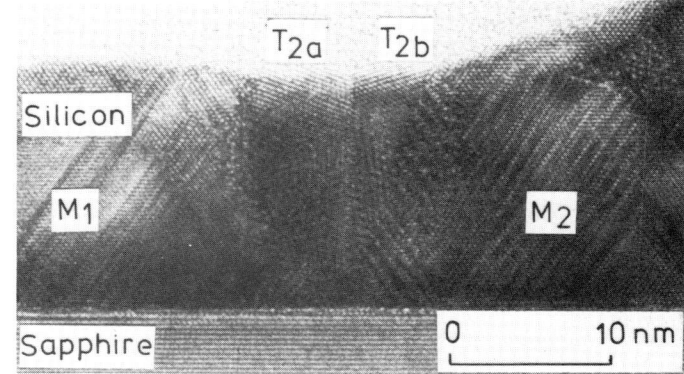

Fig.7 Coalescence of
two growth centres
(940°C).

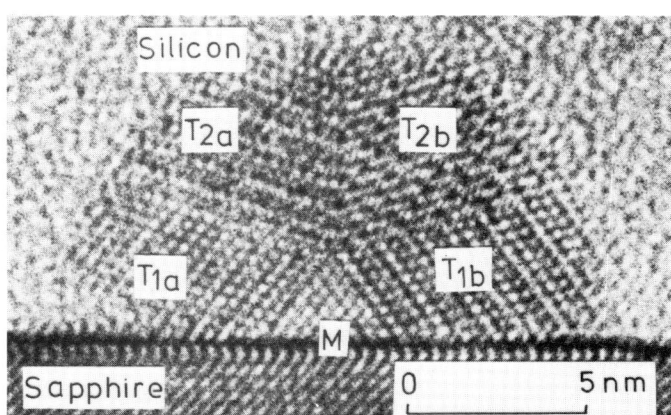

Fig.8 Low temperature
(850°C) growth centre,
multiply twinned.

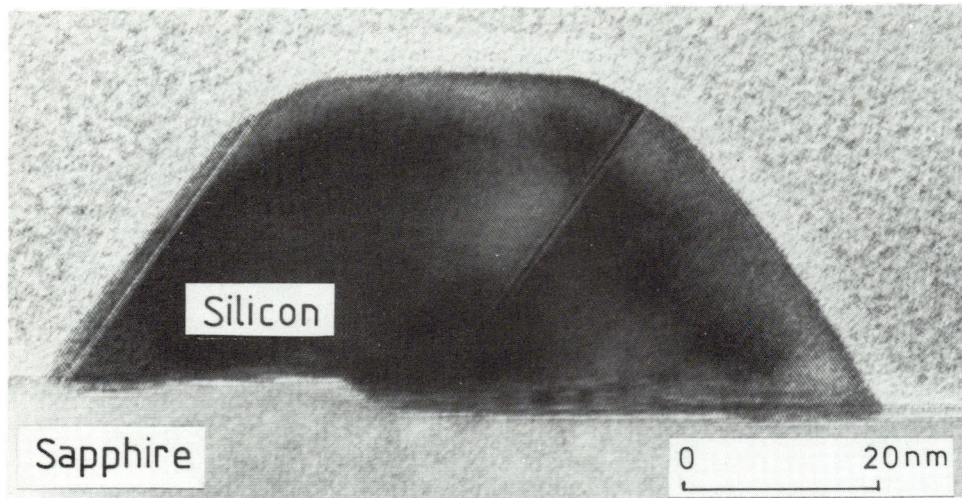

Silicon

Sapphire

0                    20nm

Fig.9 High temperature (1050°C) growth centre.   Note erosion of substrate
and pronounced facets.

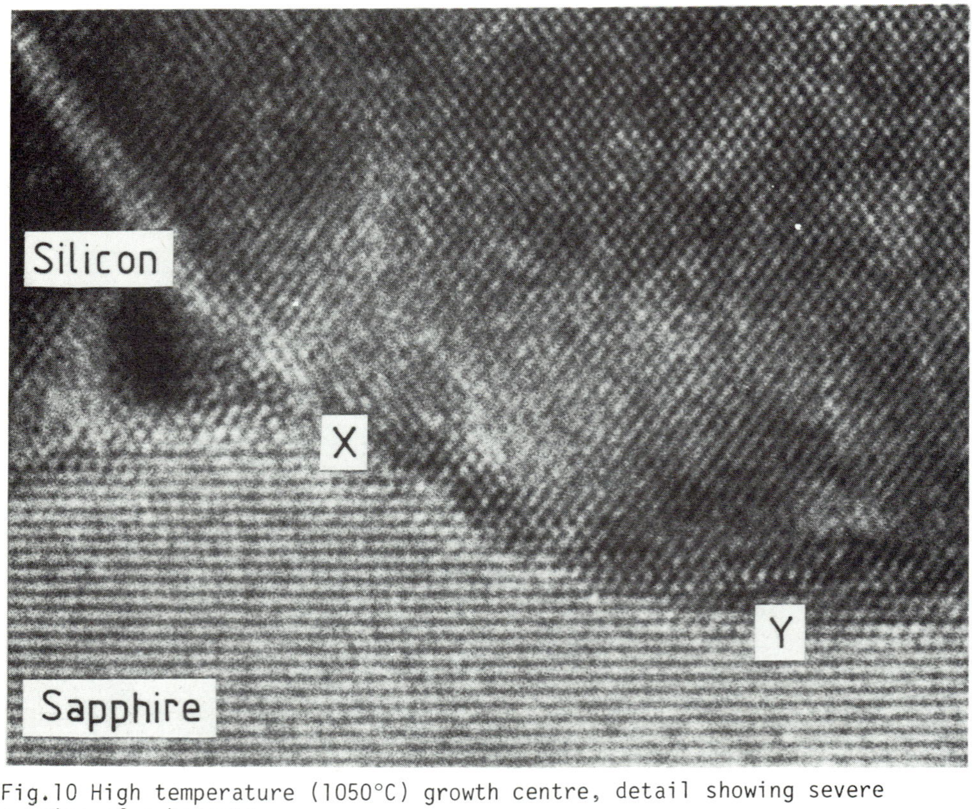

Silicon

X

Y

Sapphire

Fig.10 High temperature (1050°C) growth centre, detail showing severe
erosion of substrate.

*Inst. Phys. Conf. Ser. No. 76: Section 2*
*Paper presented at Microsc. Semicond. Mater. Conf., Oxford, 25–27 March 1985*

# Screw dislocation networks produced by stage IV compression of Ge and Si

P Haasen and H G Brion

Institut für Metallphysik und Kristall-Labor/SFB 126, Universität Göttingen, West Germany

Abstract   Screw dislocation networks are produced by high temperature compression of Si and Ge. The mechanism of their formation was studied by TEM and found to involve climb of 60° dislocations. Climb shows up as the dominant work softening mechanism in these materials at high temperature. An attempt is made to describe the work hardening stages of Ge and Si in a model of dislocation cells.

## I.  Introduction

Screw dislocations form stable hexagonal networks in diamond structure crystals after torsion about a slip plane normal and subsequent annealing (Wagner et al. 1971). It was a great surprise to find such networks also after uniaxial compression at high temperatures, $T>0.8T_m$, (Brion et al. 1985). The mechanism of formation of such networks from arrays of parallel 60° dislocation in different slip systems involves climb as will be discussed below. Climb of edge dislocations is evidently sometimes a kinetically easier process in semiconductor crystals than cross slip of screw dislocations. This has been concluded by Siethoff et al. (1984) from an analysis of work hardening and creep of d.s. crystals at high temperatures and small strain rates. Not only is the SFE $\gamma$ of these crystals relatively small but also its scaling factor ($\mu b$) is relatively large. Therefore, cross slip appears only late in deformation as a work softening stage V while in these materials stage III is often characterized by climb (Siethoff et al. 1984). The intermediate work hardening stage IV of Ge and Si has been investigated by TEM (Brion et al. 1985) as will be discussed in section II. In section III these results will be related to macroscopic work hardening and creep data. Finally, in section IV we will adapt a well known 2-phase dislocation cell model to the explanation of these microscopic and macroscopic observations.

## II.   TEM observations of high T compressed Ge and Si

We have studied <123> oriented Ge crystals compressed to a shear strain of 16%, shear stress $\tau$ = 5MPa at 890° C in nominally single slip ($\dot{\varepsilon}$ = 7·10$^{-4}$ s$^{-1}$). The corresponding data for Si are $T_{def}$=1200° C, $\dot{\varepsilon}$ = 1.7 x 10$^{-3}$s$^{-1}$, $\varepsilon$ = 29%, $\tau$ = 20MPa. Details on specimen preparation

for TEM are given elsewhere (Brion et al. 1985).  Dislocations were
studied and their Burgers vectors analysed at 120 keV in a Philips EM 400
at Stanford and at 1.5 MeV with the Kratos microscope at Berkeley. (We
thank Prof. R. Sinclair, Prof. J. Bravman and Dr. K. Westmacott for their
generous help with the microscopes). For the $\underline{gb}$ = 0 analysis in the main
slip plane (MSP) (422) reflections were used, in the other slip planes
b's were determined by (220) reflections.  Occasionally (660) (220) weak
beam pictures were taken.

One of the hexagonal nets in the MSP typical for high T compressed
Ge and Si is shown in Fig. 1.  The spacing of the nets is about 6 µm in
Ge.  The mesh size is about L = 0.2 - 0.5 µm for Ge, 0.1 - 0.2 µm for Si
which was deformed to a 4 times higher stress.  The L values scale with
the elastic interaction stress of screws $\tau = \mu b/2\pi L$.  The misorientation
provided by such a net is a twist of 0.1°.  Burgers vector analysis of the
hexagonal nets proved that they consisted of three sets of screw dislocations in the MSP.  The meshes become finer with increasing flow stress
as the comparison between Si and Ge shows.  This seems to be the work
hardening mechanism in stage IV.  The question is, however, how one obtains the second (and by superposition with the first, the third) Burgers
vector of the net on deforming a crystal of single slip orientation.

The key observation to answer this question is that of an
occasional array of parallel dislocations in the MSP (Fig. 2).  Of the
nine dislocations;  two are 60° dislocations in the MSP,  three have
one out-of-plane $\underline{b}$ and four have another one in the same set
of secondary planes that intersect MSP parallel to the dislocation lines
shown.  We now imagine a climb process of the two out-of-plane 60° dislocations towards each other.  In reacting with each other they form a
screw dislocation in MSP and, therefore, provide the second $\underline{b}$ there
needed for the formation of hexagonal networks.  One can estimate from
diffusion data that there is enough time during deformation to complete
this climb process.  The 60° dislocations seem to be the dominant feature
that stage IV inherits from stages II and III.  This can be illustrated
by Fig. 3 showing a superimposed cell structure typical for stage II and
a stage IV net.  This Ge crystal has been cooled under load while it continued to deform into the previous stages at the lower temperatures.
Another interesting aspect discussed more fully by Brion et al (1985) are
the square nets observed which could be shown to consist of one set of
MSP screws and one secondary 60° dislocation set oscillating in two
other planes.  This constitutes in our opinion a  futile attempt to form
a hexagonal net by lack of the second out-of-plane dislocation set.


III.  Work hardening rates of Si and Ge at high T


At temperatures above about 0.8 $T_m$, the stress strain curves of Si
and Ge single crystals with a single-slip orientation are characterized
by five deformation stages including two recovery stages (Brion et al.
1981).  Fig. 4 shows these recovery stages (III and V) and the stage of
linear hardening (IV) for Si in a plot of the strain-hardening rate
$\theta = d\tau/d\varepsilon$ vs. shear stress at different temperatures and strain
rates $\dot{\varepsilon}$ (Siethoff et al. 1984).  Steady-state conditions for stage III,
though not reached in the $\tau (\varepsilon)$ -curves can be obtained by extrapolating
the straight-line segment of the $\theta (\tau)$ - curves towards zero,

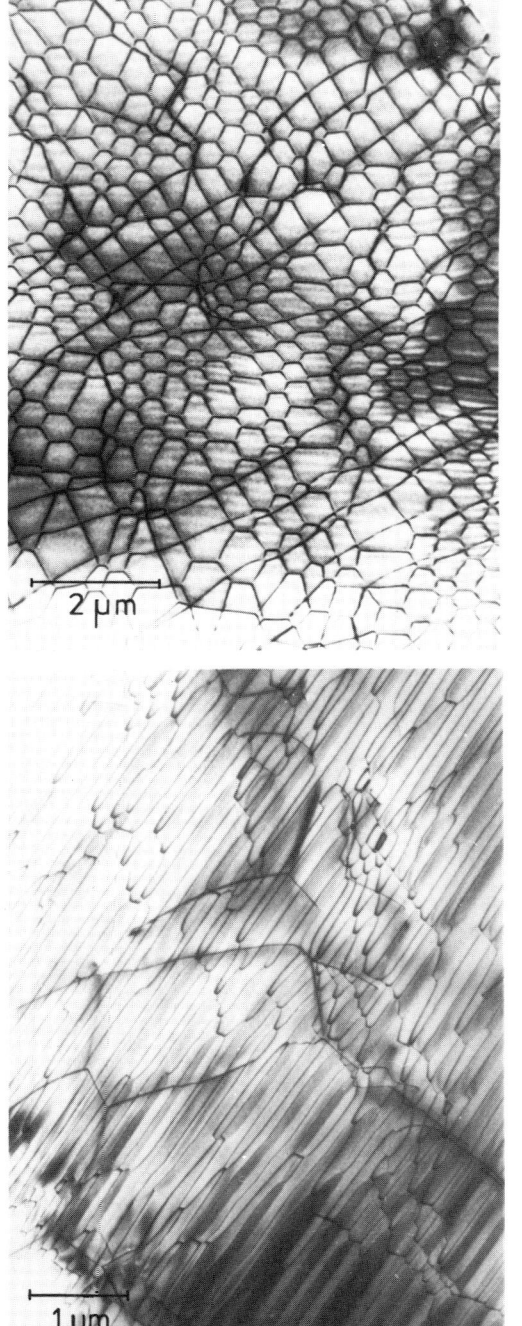

Fig. 1
HVEM survey of dislocation
network in the MSP of Ge
after high T compression

Fig. 2
HVEM survey of parallel
dislocation array in Ge

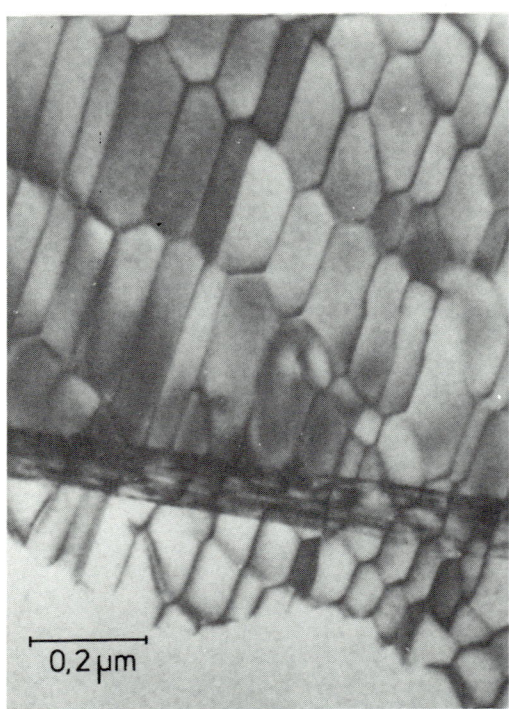

Fig. 3
Edge dislocation walls
after cooling under load
of high T compressed Ge

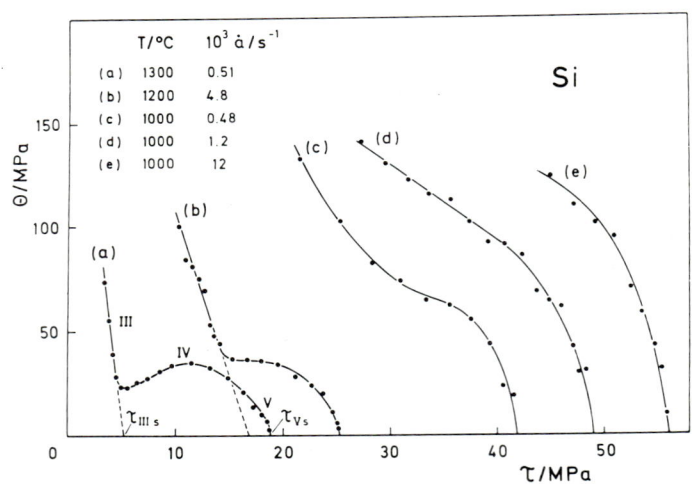

Fig. 4    Work hardening rate of Si vs. stress
at high temperatures (Siethoff et al. 1984)

thus defining the stedy-state stress $\tau_{IIIs}$. Steady-state deformation occurs at $\tau_{vs}$, the relation $\theta(\tau)$ is $_{IIIs}$ non-linear, however, in stage V.

In Si and Ge the steady-state stress $\tau_{IIIs}$ of the first recovery stage obeys a creep equation of the power-law type with activation energies close to those of vacancy self-diffusion. The steady-state stress $\tau_{vs}$ of the second recovery stage, on the other hand, could not be unambiguously analysed on account of the strong scatter of the data. It was shown, however, that the recovery mechanism responsible for stage V is already working just from its beginning. The appropriate stress $\tau_V$ was interpreted to be governed by cross-slip, and it was concluded that cross-slip becomes slower than climb at high temperatures. A cross-over of both mechanisms occurs near 1000° C. This, finally, led to an explanation of the recovery phenomena in the high-temperature plasticity of Si and Ge by two independent mechanisms (climb and cross-slip)both of which show up in the range accessible to measurements (Siethoff, Ahlborn, Schröter, 1984).

An interesting feature of Fig. 4 is the minimum work hardening rate occuring at the transition between stages III and IV, at least for the highest T and lowest $\dot{\varepsilon}$. This must be the point of formation of the screw networks by a climb-controlled reaction of the non-MSP 60° dislocations discussed above. From there on then the screw nets become finer and work hardening increases. At smaller stresses climb is not rapid enough to enable the screws necessary for the network to be formed. It is this minimum $\theta$ point we now focus upon in a theoretical analysis. The model must treat near-edge and screw dislocations separately in their abilities to be stored and to recover and still connect them as they are produced as common loops from the same dislocation sources.

## IV. A two phase-work hardening and softening model

Fig. 5 shows the well known cell model used by Mughrabi, Nix et al. (1984) in fatigue and creep studies. We follow the latter authors in their basic derivation of dislocation storage and annihilation kinetics with the following basic modifications:

(1)  Climb of edges rather than cross slip of screws be the first recovery mechanism;

(2)  Glide be athermal (applied stress equals glide resistance);

(3)  The dimensions of the cells $L_c$ be large compared to the rather sharp cell walls (constant width $L_w$)

Then the following work hardening contributions can be derived for cell interior (c) and cell wall (w)

(1)  $\left. \dfrac{d\tau_c}{d\varepsilon_c} \right|^+ = \dfrac{\mu}{x}$ , x = 300 by adjustment, follows

from a dislocation free path $\Lambda_c \approx 150\,\rho_c^{-1/2}$ ;

$\rho_c$ = dislocation density in the cell, mostly screws;

**CELL STRUCTURE MODEL**

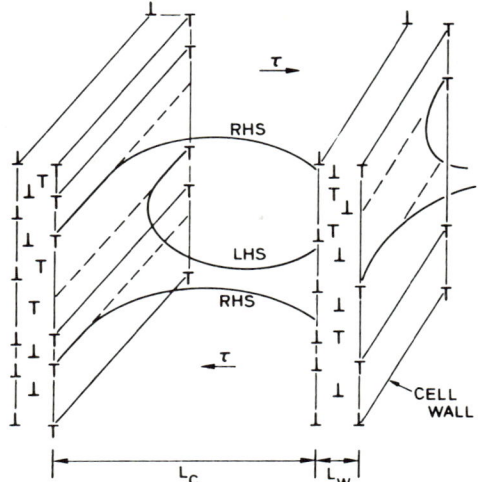

Fig. 5  Model of slip in dislocation cells and the deposition of
edge dislocations in their walls
(Nix et al 1984).

No cross slip is admitted in the cell, $\dfrac{d\tau_C}{d\epsilon_C}\Big|^{-} = 0$, as we do not attempt
to model stage V here.

The edge components of the cellular loops collect in the walls

(2)    $\dfrac{d\tau_W}{d\epsilon_C}\Big|^{+} = \dfrac{\mu}{L_W\sqrt{\rho_W}} = \dfrac{b\mu^2}{L_W\tau_W}$ ,   $\mu$ = shear modulus

where they annihilate by climb as often described.

(3)    $\dfrac{d\tau_W}{d\epsilon_C}\Big|^{-} = -\dfrac{E\tau_W^3 D}{\dot{\epsilon}_C}$ ,   D = diffusion coefficient;
E is a known constant.

The plastic strain in the cell is practically the total one

(4)    $d\epsilon_C \approx d\epsilon$ ,   $\dot{\epsilon}_C = \dot{\epsilon}$

The load distributes itself according to

(5)    $\tau = \tau_C + \dfrac{L_W}{L_C}\tau_W$   for   $L_C \gg L_W$

From (4, 5) it follows with   $L_C = 10\dfrac{\mu b}{\tau_C}$                                    (6)

$\theta = \dfrac{d\tau}{d\epsilon} = \dfrac{\mu}{x} + \dfrac{b\mu^2}{L_C\tau_W} - \dfrac{ED\tau_W^3 L_W}{L_C\dot{\epsilon}} + \dfrac{\tau_W L_W}{10bx}$

$= \dfrac{\mu}{x} + \dfrac{\mu\tau_C}{10\tau_W} - \dfrac{L_W ED\tau_W^3\tau_C}{\dot{\epsilon}\,10\mu b} + \dfrac{\tau_W L_W}{10bx}$            (7)

By (5, 6) $\tau_c$ can be eliminated from (7)

$$\tau = \tau_c \left( 1 + \frac{\tau_w L_w}{10\mu b} \right) \approx \tau_c$$

if $\tau_w \ll \mu$ and $L_w \approx 4nm$ as observed.

Then we obtain

$$\Theta = \frac{\mu}{x} + \frac{\tau\mu}{10\tau_w} - \frac{L_w ED\tau_w^3 \tau}{\dot{\epsilon}\ 10\mu b} \qquad (8)$$

By integration of eq's (1-3) one derives for the interesting range of T and $\dot{\epsilon}$

$$\tau_w^2 = \frac{2xb}{L_w}\ \tau \cdot \mu \qquad (9)$$

and, therefore, finally

$$\Theta = \frac{\mu}{x} + \sqrt{\frac{\tau\mu L_w}{200xb}} - \frac{ED}{10\dot{\epsilon}} \sqrt{\frac{\mu b\tau^5 (2x)^3}{L_w}} \qquad (10)$$

Nix et al (1984) evaluate their situation only numerically. Here we do not fully discuss our analytical solution but just the case of the minimum $\Theta$, where the first and third term in eq. (10) about compensate each other.

Then $\Theta$ min $\approx \sqrt{\frac{\mu L_w}{200xb}} \sqrt{\tau_{min}}$ which is experimentally well fulfilled in

a plot of $\Theta_{min}$ $(\tau_{min}^{1/2})$ obtained for Ge deformed at various temperatures.

The measured slope 11 $N^{1/2}mm^{-1}$ is compatible with $L_w$=4 nm; a very reasonable result. We have not included the emergence of the screw network in our model yet which will lead to the observed $\Theta_{IV}$ but just the preceding climb process. This according to TEM is the prerequisite of the screw network formation from the dissolving cell walls.

## Acknowledgement

Dr H Siethoff has been very helpful in the evaluation of his experimental data. P Haasen is grateful to the Stanford University for the award of the W Schottky Professorship.

## References

Brion H-G and Haasen P 1985 Phil. Mag. in press
Brion H-G, Siethoff H and Schröter W 1981 Phil. Mag. A43 1505
Nix W D, Gibeling J C and Hughes D A 1984 AIME Sympos. on the 50th Annivers. of the Dislocation, in press
Siethoff H, Ahlborn K and Schröter W 1984 Phil. Mag. A50 L 1
Siethoff H and Schröter W 1984 Z. Metallk. 75 475 482
Wagner R, Wöhler F D and Haasen P 1971 Phys. Stat. Sol.(b) 44 381

*Inst. Phys. Conf. Ser. No. 76: Section 2*
*Paper presented at Microsc. Semicond. Mater. Conf., Oxford, 25–27 March 1985*

# A TEM investigation of the plastic zone, dislocation rosettes and cracks around Vickers indentations in silicon

J Samuels, P Pirouz, S G Roberts, P D Warren and P B Hirsch

Department of Metallurgy and Science of Materials, University of Oxford, Parks Road, Oxford OX1 3PH

Abstract  The arrangement of dislocations close to microhardness indentations, and their associated cracks, made in silicon at 350–400°C were investigated by TEM. In the rosette arms, partial dislocations were seen, associated with extended stacking faults (or possibly twins). No dislocations were seen at crack tips. These results are discussed with respect to a model of slip patterns around indentations.

## 1.Introduction

In general, in an indentation test, three parameters can be measured from which information on the plastic and fracture behaviour of the test piece can be obtained:
1)	The hardness, equal to the pressure on the indentation area, which is related to the ease of plastic flow beneath the indenter.
2)	Dislocation rosette sizes, measured after appropriate etching of the specimen. These have been shown to be related to dislocation velocities in several semiconductors (Roberts, Pirouz and Hirsch 1983, 1985).
3)	Radial crack sizes. The extent of these cracks around the indentation may be related to the fracture toughness of the test material (e.g. Lawn, Evans and Marshall 1980).

A proper understanding of all these parameters depends essentially on a knowledge of the details of the plastic zone which forms beneath the indentation. Such an understanding would enable one to clarify:
1)	The relation between the hardness and the more fundamental plastic parameters of the test material (e.g. the yield strength).
2)	The nature of the dislocations forming the rosette and the stresses under which they move.
3)	The nucleation and propagation of the indentation crack systems.

Semiconductors are ideally suited to the understanding of the plastic zone for a number of reasons:
1)	The slip systems in these materials are very well-defined.
2)	There is a wealth of data available on the nature of dislocations and their velocities under various stress and temperature conditions.
3)	Dislocation velocities also depend on the position of the Fermi level in the band gap (i.e. doping).
4)	In compound semiconductors, two types of non-screw dislocations

exist. This last effect can manifest itself in various anisotropies in hardness, dislocation rosette structures and crack behaviour (Warren, Pirouz and Roberts 1984, Hirsch, Pirouz, Roberts and Warren, 1985a,b).

These anisotropies together with the known dependencies of dislocation velocities on doping provide a sensitive tool to investigate the nature of the plastic zone in semiconductors and to follow the micromechanisms acting beneath the indenter in considerable detail.

In this paper, we report briefly some TEM investigations of the plastic zones, rosette dislocations and cracks in Si.

## 2.Observations

Fig. 1 shows a TEM micrograph of a {111} indentation in Si, made with a 10g load at 400°C. Note the heavily dislocated region in the plastic zone forming a triangle at the centre of the micrograph. There are cracks on {110} planes emanating from the apices of the indentation and also shorter fringed bands which correspond to cracks on other inclined planes. Also visible is a hexagonal net of in-plane dislocations which connect some of the perfect dislocations in adjacent arms of the rosette. The rosette dislocations appear to be of two types: normal glide dislocations, and partial dislocations with trailing stacking faults. Note that one side of the rosette arm consists of both whole dislocations and extended stacking faults, whereas the other side contains only whole dislocations. Fig. 2 shows a portion of the rosette arm at a higher magnification, using two different diffraction vectors; in fig.2b, it can be seen that this part of the rosette arm consists of overlapping stacking faults, implying that the rosette dislocations (fig.2a ) are partials. Fig. 3 shows part of a crack tip at higher magnification, using two different diffraction vectors. Each micrograph shows Moire fringes, characteristic of a small rotation across the crack interface. Very few dislocations were observed in the vicinity of the cracks. Similar results have been shown for Si by Hockey and Lawn (1975) and Lawn, Hockey and Wiederhorn (1980).

Fig. 4 shows a low magnification many-beam micrograph of an indentation made at 350°C on an {001} face of Si. In this case, all the rosette dislocations have associated stacking faults, with no sign of any perfect dislocations. Details of the dislocation configuration in the rosette arm are shown at higher magnification in fig. 5. The dislocations in fig. 5a are all partials, and the rosette in fig. 5b consists of stacking faults; preliminary analysis indicates that all the partials are of the same type.

## 3.Discussion

A detailed investigation of the rosette dislocations, slip steps and the median/radial cracks has provided us with a model for the plastic zone formed during the indentation of {111} surfaces of cubic semiconductors. The results have been described by Hirsch et al. (1985a,b). Here, we apply this type of model to the results for indentations made on an {001} surface.

Such a model of the plastic zone under an indentation made on an {001} surface is shown in fig. 6. Basically the same considerations apply as in the indentation of a {111} surface (Hirsch et al. 1985b). Slip close to the indenter would be generated on inclined {111} planes, with dislocations with their extra half-planes on the side of the slip plane nearest the indenter moving from the surface into the crystal. Two types

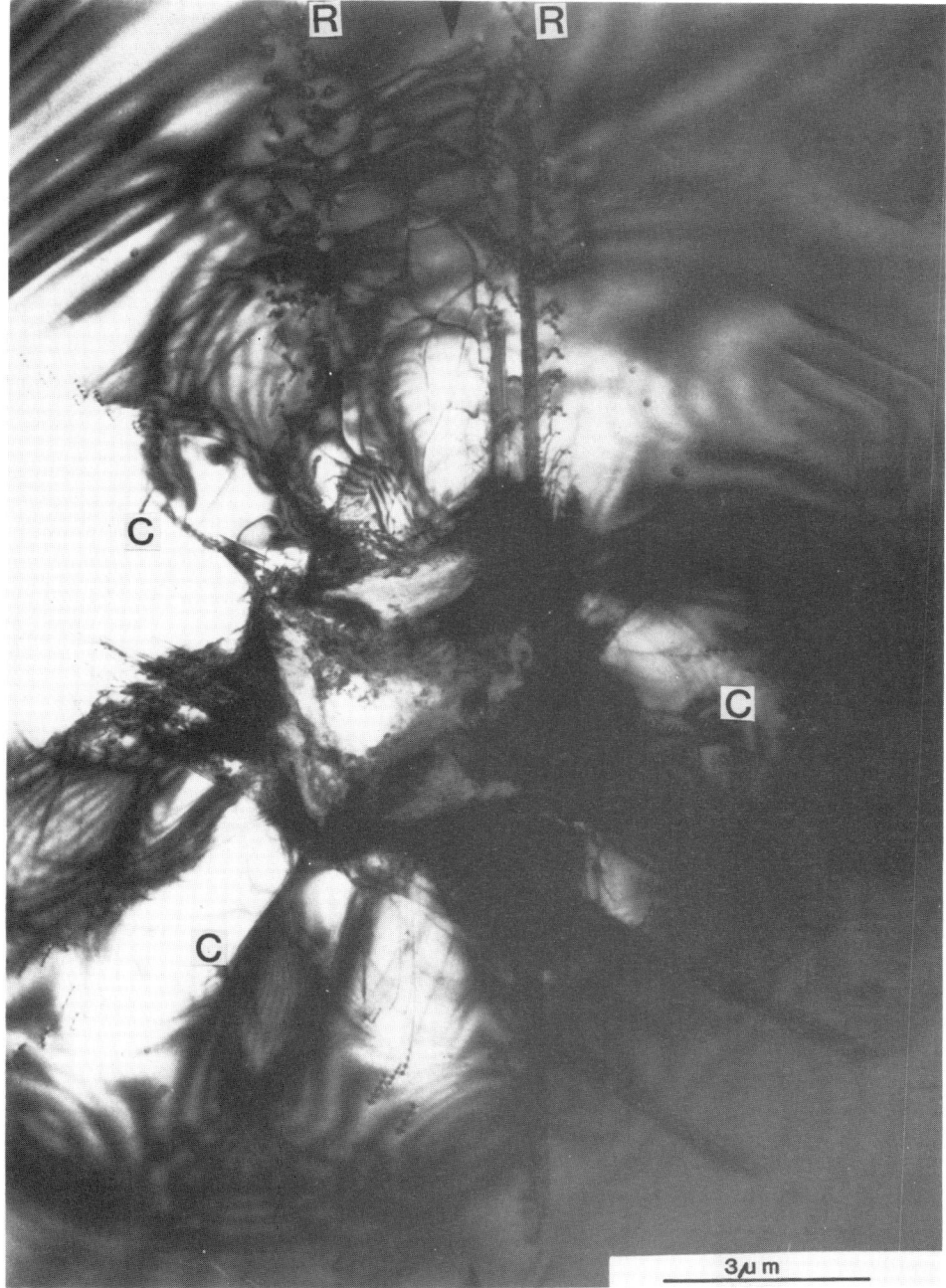

Fig. 1 Multibeam bright-field TEM image of a 10g indentation made at 400°C
on a {111} surface of silicon. Cracks are labelled 'C' and rosette
dislocations 'R'.

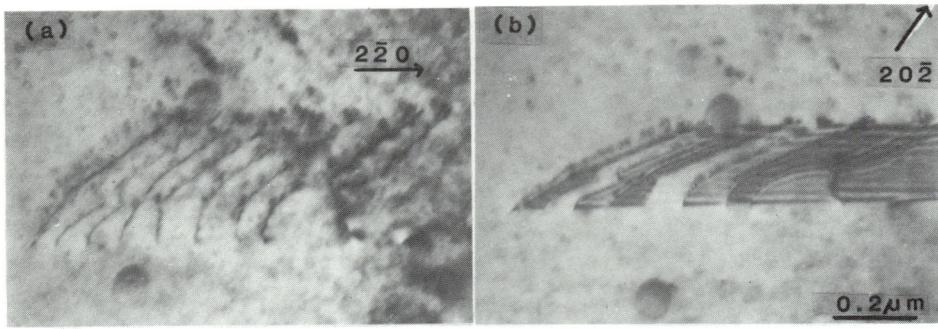

Fig. 2  TEM images of the end of the rosette arm in fig. 1; diffraction conditions to show (a) dislocations, (b) stacking faults.

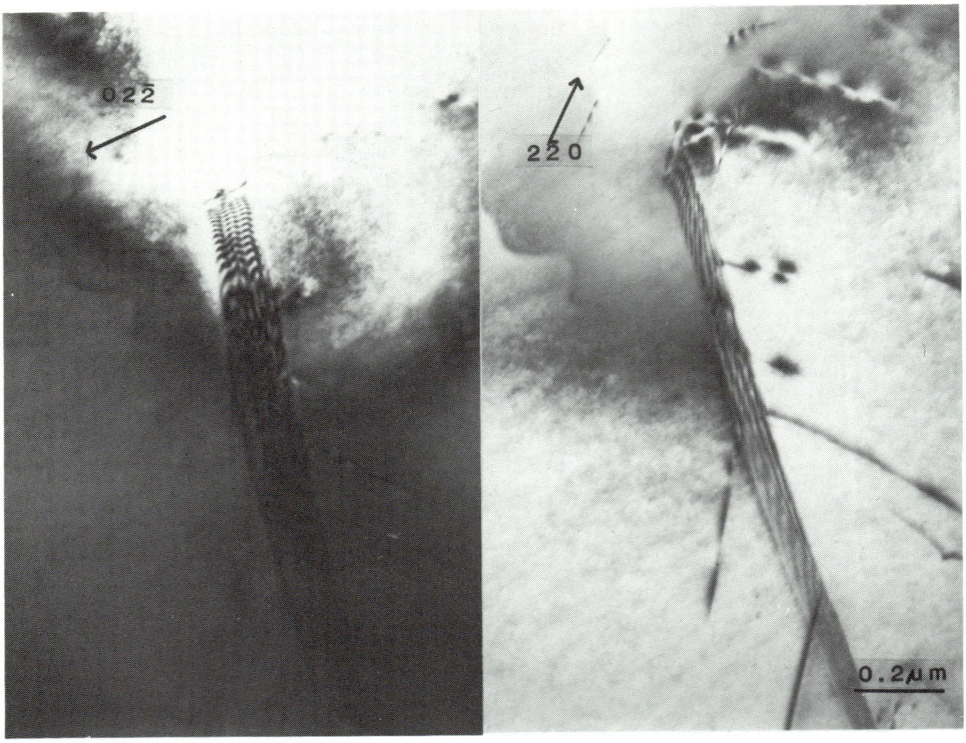

Fig. 3 TEM images of a crack tip, showing Moire fringes under two different diffraction vectors.

of tetrahedron should be considered (fig. 6a). Fig. 6b shows a
cross-section parallel to a <110> direction, showing the planes on which
slip occurs. The plastic zone will therefore consist of dislocations with
Burgers vectors inclined to the surface, whereas rosette dislocations far
from the indentation centre are prismatic half-loops with Burgers vectors
parallel to the surface (Warren et al. 1984) and/or could be partial
dislocations of only one type on each plane forming the rosette (see
below). Dislocations with Burgers vectors inclined to the surface and
gliding on planes ABC' and ABD' (fig. 6a) react to form Lomer-Cottrell
locks along the intersection of the slip planes ('X' in fig. 6b), parallel
to AB. Cracks are nucleated at the Lomer-Cottrell locks. Note that unlike
the case for indentations on {111} surfaces (Hirsch et al. 1985b), the
Burgers vectors of the perfect dislocations AB and C'D' in the two sets of
rosettes along these two directions are mutually orthogonal and no
Lomer-Cottrell locks will form; i.e. no cracks are associated with the
intersection of rosette dislocations. The slip steps formed during the
recovery process, after the unloading of the indenter, are square-shaped
for Si and Ge, and rectangular-shaped for GaAs.

The fact that the rosette dislocations in the {001} case (350°C) seem to be
all partials whereas for the {111} indentation (400°C) both partials and
perfect dislocations are observed may be due to the difference in
indentation temperature; the temperature range 350-450°C) is the critical
range above which indentation hardness decreases rapidly, and deformation
modes may be changing in this range. At low temperatures, partial
dislocations may be more easily nucleated than perfect dislocations. If
the partials are on adjacent {111} planes, such an arrangement would give
rise to a twin with {111} as the composition plane. If the partials are
not on adjacent planes, a set of overlapping stacking faults would result.
It is difficult to distiguish between these possibilities; we have looked
for twin spots in diffraction patterns from the rosette arms, but results
so far have been inconclusive, though Eremenko and Nikitenko (1972) have
shown similar micrographs and interpret these as showing twins. It is
interesting to note that, at the end of some rosette arms, perfect
dislocations of type 1/2<110> can be observed (e.g. dislocations 'E' in
fig. 5a); these may be emissary dislocations, relieving the stress
concentration at the ends of twin lamellae.

## References

Eremenko V G and Nikitenko V I 1972 Phys. Stat. Sol. 14 317
Hirsch P B, Pirouz P, Roberts S G and Warren P D 1985a Proc 2nd Int. Conf.
  on Science of Hard Materials, Rhodes 1984, in press
Hirsch P B, Pirouz P, Roberts S G and Warren P D 1985b Phil. Mag., in
  press
Hockey B J and Lawn B R 1975 J. Mater. Sci. 10 1275
Lawn B R, Evans A G and Marshall D B 1980 J. Am. Ceram. Soc. 63 575
Lawn B R, Hockey B J and Wiederhorn S M 1980 J. Mater. Sci. 15 1207
Roberts S G, Pirouz P and Hirsch P B 1983 J. Physique 44 C4-75
Roberts S G, Pirouz P and Hirsch P B 1985 J. Mater. Sci., in press
Warren P D, Pirouz P and Roberts S G 1984 Phil. Mag. 50A L23

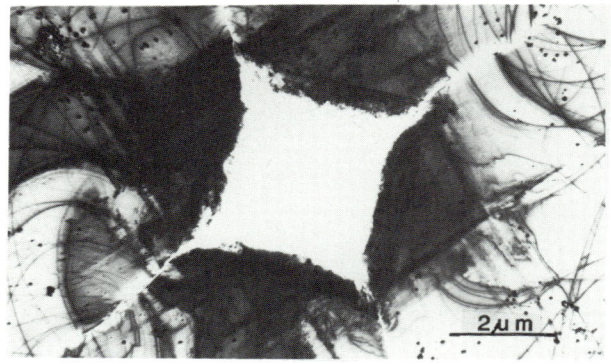

Fig. 4  Multibeam bright-field TEM image of a 50g indentation made at 350°C on {001} silicon.

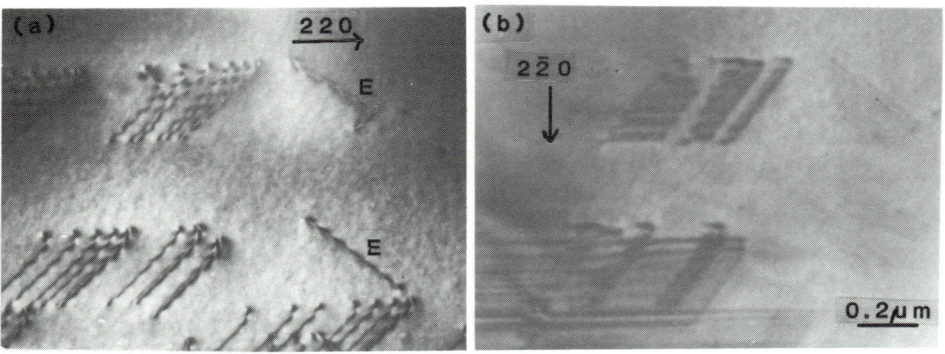

Fig. 5  TEM images of the end of the rosette arm in fig. 4; diffraction conditions to show (a) dislocations, (b) stacking faults.

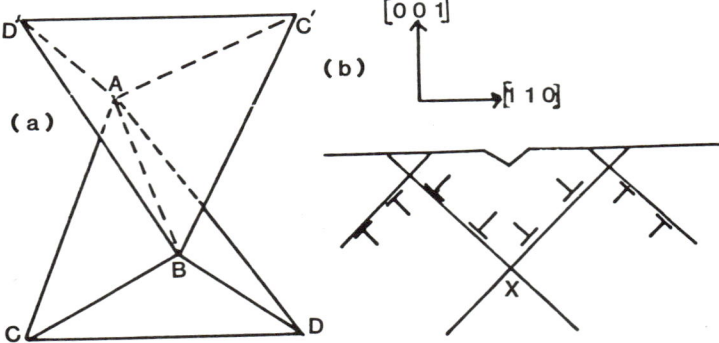

Fig. 6 (a) Orientation of Thompson's tetrahedra of slip planes beneath indentation on an {001} plane; (b) Cross-section parallel to a <110> direction, showing the planes on which slip occurs.

*Inst. Phys. Conf. Ser. No. 76: Section 2*
*Paper presented at Microsc. Semicond. Mater. Conf., Oxford, 25–27 March 1985*

# Dislocations in GaAs

B C De Cooman, K -H Kuesters and C B Carter

Department of Materials Science and Engineering,
Bard Hall, Cornell University, Ithaca, NY 14853, USA

Abstract    Bright-field and weak-beam dark-field microscopy techniques
have been used to study straight dislocations and faulted dipoles in
semi-insulating GaAs and in both MBE- and OMVPE-grown $Al_xGa_{1-x}As/GaAs$
heterojunctions.  The low temperature, high-stress deformation results
in a dislocation configuration where the dislocations lie along their
Peierls valleys.  Using TEM specimens prepared from such deformed
material, the mobility of $\alpha$ and $\beta$ dislocations, the asymmetry of motion
of 60°-$\beta$-dislocations and screw dislocations could be observed directly.
High temperature deformation (425°C) results in the formation of less
well-defined dislocation and many dislocation dipoles, some of which are
found to be faulted dipoles.  A cross-sectional TEM technique allowing
the direct study of screw dislocation motion through epilayers is presented.  This technique showed that dislocation dipoles can be formed at
heterojunctions as a result of dislocation glide.

## 1.  Introduction

It has been shown [1,2,3] that when Si, Ge and GaAs are deformed plastically at high stress (i.e. resolved shear stress > 100 MPa) so as to produce single slip, the dislocations are oriented along the <110> Peierls
valleys of the slip plane.  Such a dislocation configuration, consisting
predominantly of dissociated screw and 60° dislocations, is ideally suited
for the study of electronic [4] and structural properties [5] and for the
measurement  of dislocation velocities [6].  Dislocation glide in compound
semiconductors has been shown [7] to be enhanced by electron-hole recombination.  Both dislocation glide and climb occur during the degradation of
GaAs-based devices such as lasers and LEDs [8].  Dislocation climb can lead
to the development of dislocation networks which are believed to originate
at heterojunction interfaces in GaAs-based devices.

The mechanism of the interaction of dislocations with interfaces and with
point defects, will only be fully understood when the following features
have been characterized [9].

1.  The detailed core structure of well-defined straight dislocations must
be analyzed and the dependence of the dislocated glide velocity on the sign
of the dislocation ($\alpha$ versus $\beta$-dislocation) must be measured.

2.  Because dislocation dipoles are always present in degraded devices, the
exact nature of unfaulted and faulted dipoles in well-defined configurations must be determined.

3. The interaction of well-defined dislocations with interfaces will show the extent to which point defects at the interfaces influence dislocation climb. Again, possible polarity effects must be examined in detail.

For these reasons, procedures have been designed which allow single crystals of GaAs and $Al_xGa_{1-x}As$/GaAs heterojunctions to be deformed in such a way that stright dislocation segments would develop in one slip plane. In the case of devices containing heterojunctions a new cross-sectional technique allowing the observation of dislocations in their slip plane over large distances, has also been developed. The experimental details of this study have been given elsewhere [3,10]. This paper will therefore concentrate on illustrating several new aspects of this research on dislocations in GaAs.

## 2.  Dislocation Motion in Semi-Insulating GaAs

A dislocation configuration consisting almost exclusively of straight segments was obtained by deforming [213]-oriented GaAs crystals in compression under a high stress 150 MPa at 150°C [11]. Only screw dislocations and 60° dislocations with Burgers vector $a/2[0\bar{1}1]$ are observed in the Ga $(\bar{1}1\bar{1})$ slip plane. A method for determining the polarity dependence of the Burgers vector has been proposed by Kuesters et al. [3]. Most of the long, straight dislocation segments as observed in the TEM are in the screw orientation, 60° dislocation segments are usually much shorter and lie close to the $[\bar{1}\bar{1}0]$ and $[10\bar{1}]$ directions. Very few loops with α-segments could be observed. In Fig. 1 a dislocation loop which consists of two 60°-β-segments and two long screw segments is seen to relax from its initial configuration as a result of electron irradiation. Only the lower screw dislocation has moved, with the dislocation developing a large 60°-α-kink moving in the [001] direction; the other screw did not develop such kinks. Eventually the two screw segments annihilated. The dislocation motion was invariably uneven occurring in irregular jumps rather than by continuous glide and only a lower limit for the dislocation velocity can therefore be obtained from sequences of TEM images. For α-dislocations, velocities of greater than $10^{-4}$ cm/sec were observed. This value is considerably larger than previously reported [12]. The high quality of the specimens makes it possible to observe end-on dissociated screw and 60°

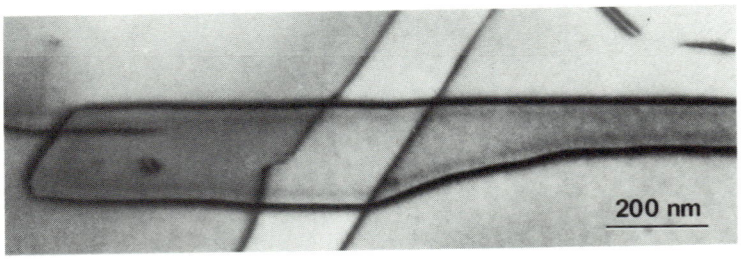

200 nm

Fig. 1. Dislocation loop in GaAs. (022) bright field image. Ga $(\bar{1}1\bar{1})$ up in microscope.

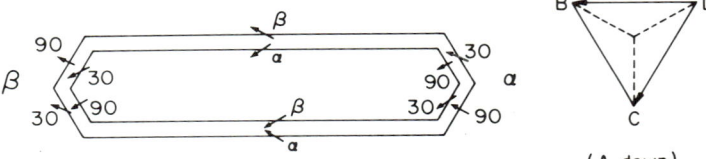

(A down)

dislocations for the determination of the core (glide or shuffle) structure.  This research is still actively in progress.

## 3.  Dipoles in GaAs Deformed at Higher Temperatures

The weak-beam technique has been used to study dislocation dipoles in GaAs
which had been deformed at ~425°C.  The active slip system gave disloca-
tions with a Burgers vector of $\vec{b}_\alpha$ = a/2[011] and $\vec{b}_\beta$ = a/2[0$\bar{1}\bar{1}$], which lay
on the ($\bar{1}1\bar{1}$) glide plane.  In these samples numerous faulted dipoles were
formed and these were always found to be parallel to either the [$\bar{1}\bar{1}$0] or
the [10$\bar{1}$] direction.  The weak-beam images of these dipoles display a clear
reversed inside-outside contrast as is illustrated in Fig. 2.  It can be
shown from geometrical considerations that faulted dipoles which are
parallel to [011] cannot be formed during single slip since none of the
four partials in the dipoles could then have pure edge character.  From the
dislocation loop configuration and the knowledge of the polarity of the
foil surface, it is possible to determine the nature (90$_\alpha$ or 90$_\beta$) of the
edge Shockley partials in weak-beam images: no assumption regarding Z- or
S-configurations need be made.

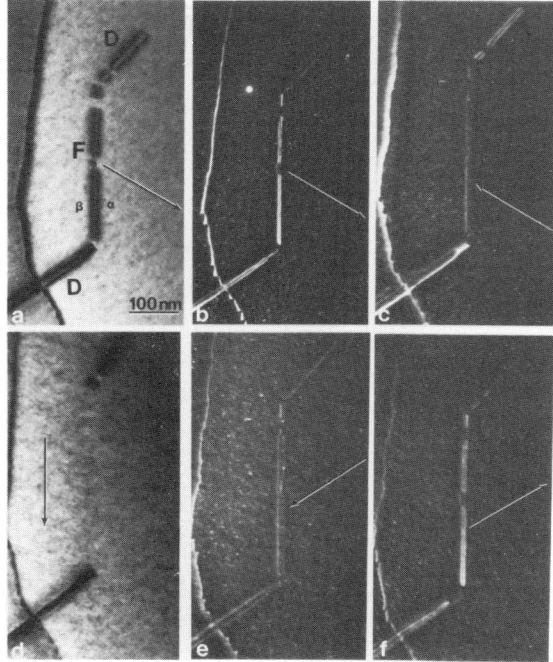

Fig. 2.  {220} contrast
from a faulted dipole in
GaAs. (a) and (d) are
bright field images. (b),
(c), (e) and (f) are weak
beam images (g,3g). As
(1$\bar{1}$1) up in microscope.

The specimen also contains a large number of rows of small loops (Fig. 3).
These loops run closely parallel to <110> type directions and although
their contrast in weak-beam images is faint, it does not vanish for any of

the {220} reflections. This observation suggests that these small loops
result from the break-up of unfaulted dipoles i.e. they are perfect,
a/2<011> dislocation loops. The formation of these rows of dislocation
loops can be explained by a self-climb process caused by the high mobility
of point defects present at the deformation temperature.

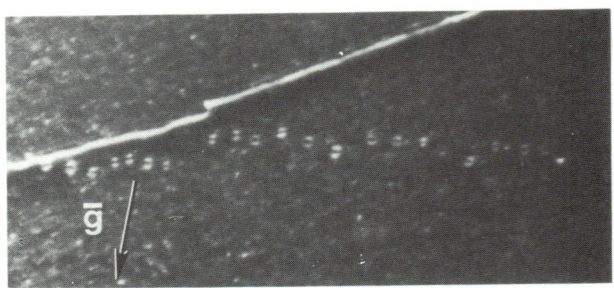

Fig. 3. Row of small
loops in GaAs (g,3g)
weak beam images
showing inside-
outside contrast.

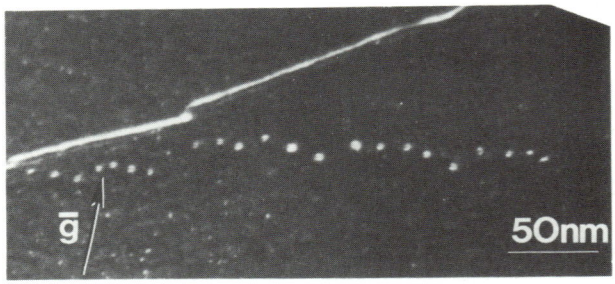

## 4.  Dislocation Motion in $Al_xGa_{1-x}As$/GaAs Heterojunctions

Epilayer structures of alternating layers of GaAs and $Al_xGa_{1-x}As$ grown on
(001) GaAs were deformed in compression at 325°C with the compression axis
oriented along [$\bar{1}$10]. Four slip systems, two for each ((1$\bar{1}$1) and (1$\bar{1}\bar{1}$))
slip plane, are activated in this case (Fig. 4). A new variation on the
cross-sectional TEM technique has been developed in which the deformed
crystals are cut parallel to one of the slip planes, rather than normal to
the (001) epilayer growth surface (Fig. 5). Deformation of OMVPE-grown
material leads to the formation of dislocation dipoles originating at jogs
pinned within the $Al_xGa_{1-x}As$ epilayer; the structure of the dipoles is
apparently not affected by the presence of the layers. Deformation of MBE-
grown material results in dislocations which only interact with the heterojunction. In the latter case, dislocation dipoles are parallel to the
heterojunction and the jogs which give rise to the dipoles appear to be
pinned at the interface itself [10].

In Fig. 6 a segment of a dislocation aligned along the interface (MBE-grown
AlAs/GaAs heterojunction) is shown. Figure 6a shows both edge-on layers
and a dislocation crossing an AlAs layer at A; at B the dislocation is
aligned along the GaAs/AlAs interface. At D the dislocation crosses the
AlAs layer and begins to bend close to the pure screw orientation. In Fig.
6b the specimen was tilted towards [001] to detect significant climb. The

partials were found to be straight over a long distance and nowhere except at D, where the dislocation crosses the AlAs layer, is there a constriction which may show evidence for possible climb.

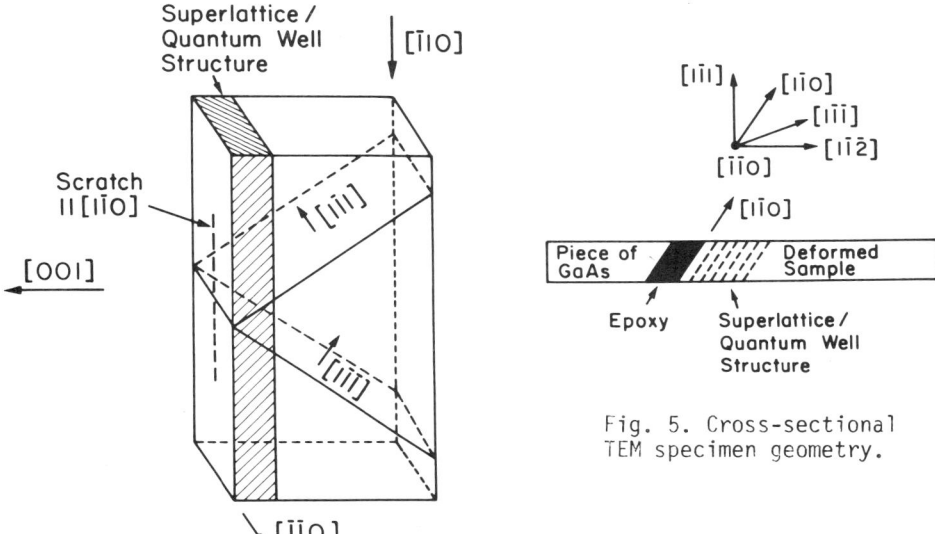

Fig. 4. Deformation geometry for heterojunctions.

Fig. 5. Cross-sectional TEM specimen geometry.

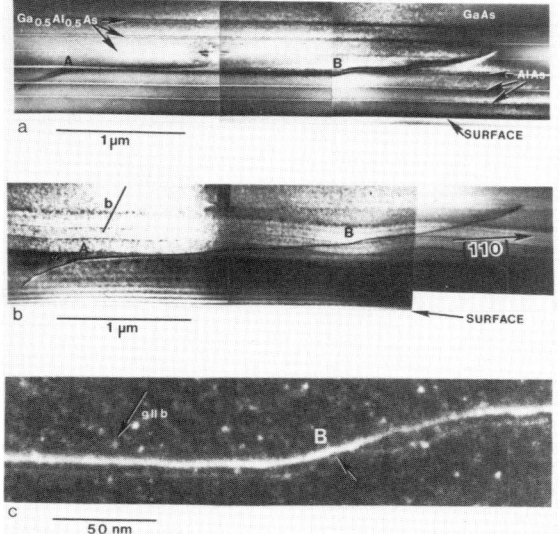

Fig. 6. Example of dislocation aligned along the GaAs/AlAs interface.
(a) View along [1$\bar{1}$0]
(b) View along [1$\bar{1}$1]
(c) Constriction (at B) where the dislocation crosses the wide AlAs layer.

## 5. Conclusions

High stress deformation of GaAs produces very straight dissociated dislocations of well-defined type and polarity which are ideally suited for the
in situ study of dislocation motion during electron irradiation in the
electron microscope. It has been confirmed, by direct observation, that $\alpha$-
dislocations move at much higher velocities than either $\beta$-dislocations or
screw dislocations: the highest dislocation velocities are considerably
greater than previously reported values. The movement of dislocations
appears to be controlled by bowed out segments in the screw dislocations--
which can be thought of as an array of kinks. These observations support
Kimerling's model [13] for dislocation glide in which kink migration is
strongly affected by the electron-hole recombination. Dislocations in
alternating GaAs and $Al_xGa_{1-x}As$ epilayers interact strongly with the heterojunctions. This interaction can either create long dipoles which are
usually oriented along <110> directions or can result in the bending of the
screw dislocations so that they lie parallel to the plane of the interface.

## References

[1]  H. Alexander, C. Kiesielowski-Kemmerich, and E. Weber, Physica 116B
     (1983), 583.
[2]  H. Gottschalk, Proc. Int. Cong. on Electron-Microsc. (Hamburg),
     (1983), p. 527; J. Phys. 40, C6, p. 127.
[3]  K.-M. Kuesters, B.C. De Cooman, C.B. Carter, submitted to Phil. Mag.
     (1985).
[4]  A. Ourmazd, D. Phil. Thesis, University of Oxford (1979).
[5]  A. Olsen and J.C.H. Spence, Phil. Mag. A43 (1981), p. 945.
[6]  G. Feuillet and D. Cherns, Proceedings of the 13th Int. Conf. on
     Defect in Semiconductors (Aug. 12-17, 1984), L.C. Kimerling and J.M.
     Parsey, Jr. eds. (1985), p. 343.
[7]  K. Maeda, M. Sato, A. Kubo, and S. Takeuchi, J. Appl. Phys. 54
     (1983), p. 161.
[8]  P.M. Petroff, J. Semiconductors and Insulators 5 (1983), p. 307.
[9]  P.M. Petroff and L.C. Kimerling, Appl. Phys. Lett. 25 (1976), p. 461.
[10] K.-U. Kuesters, B.C. De Cooman, C.B. Carter, Proc. of 13th Int. Conf.
     on Defects in Semiconductors (August 12-17, 1984), L.C. Kimerling and
     J.M. Parsey, Jr. eds. (1985), p. 351.
[11] J. Klaer, Diploma Thesis, University of Köhn (FRG) (1983).
[12] K. Maeda and S. Takeuchi, Jap. J. Appl. Phys. 20 (198 ), L165.
[13] L.C. Kimerling, Solid State Electronics 21 (1978), 139.

# Electrical activity of dislocations in silicon—a reexamination of Hall effect data

M I Heggie

Department of Physics, University of Exeter, Devon EX4 4QL

   Abstract   Several authors have presented data from the Hall effect in
   deformed silicon and they have their own interpretations of their
   results.  The mutual incompatibility of these interpretations will be
   emphasized in this contribution and a possible resolution of the
   inconsistencies will be proposed.  The suggestion is that there are at
   least two different defects responsible for the electrical activity of
   dislocations and one of these defects may combine with dopant atoms
   when the dopant concentration is high relative to the defect
   concentration to give yet a third level.

## 1. Introduction

The Hall effect in a unipolar semiconductor provides a relatively simple
way of obtaining the charge carrier concentration as a function of
temperature, particularly in the temperature interval between the total
ionization of the dopant and the onset of intrinsic behaviour.  Since
most levels associated with dislocations in silicon are in the lower half
of the gap, most studies have focussed on p–type deformed silicon, notable
amongst them being Labusch and Schröter (1980), Grazhulis, Kveder and
Mukhina (1977) and Ono and Sumino (1980, 1983) – to be referred to
respectively as LS, GKM, OS1 and OS2.

In order to review and compare these works the following notation will be
introduced:  $\mu$ = chemical potential, $N_{Ca}$ = concentration of chemical
acceptors, $N_d$ = formal concentration of dislocation dangling bond sites
assuming a 1:1 correspondence between etch pits and perfect $60^{\circ}$ shuffle
dislocations, f = fractional occupation of these $N_d$ sites by electrons
and p = concentration of holes.  For a given p–type sample, raising the
temperature raises the chemical potential according to $\mu = kT\ell n(C_v T^{3/2}/p)$
where $C_v T^{3/2}$ is the density of accessible states in the valence band
($C_v = 1.95 \times 10^{15}$ $cm^{-3}$ $K^{-3/2}$).  This changes the occupation of the
dislocation states, $f = (p - N_{Ca})/N_d$, in a fashion that depends on the
nature of the gap states to be discussed in Section 2.

## 2. Analysis of Gap States

A concentration of c states per dislocation dangling bond giving rise to
a half-filled band centred on $E_0$ at $T_0$ would trap $\mu$ near $E_0(T)$ until the
band was completely full ($f_{max} \simeq c$), when $\mu$ would rise steeply, or empty
($f_{min} \simeq -c$), when $\mu$ would fall sharply.  However, the concentration c is
normally so high that p = o, ie $f_{min} = -N_{Ca}/N_d$, is reached before $-c$.
Neither of these cases were true for the samples S3, S4, S5, W2, W3, W4
of OS2, for which $N_{Ca}/N_d > -f_{min} < f_{max}$.  This implies the presence of an

extra acceptor below $E_O$ draining off some electrons from the half filled band, reducing its donor character. One consequence of this imbalance between the concentration of dislocation acceptors $N_{da}$ and of dislocation donors $N_{dd}$ ($N_{da} \sim 7$ to $10$ $N_{dd}$) is a temperature dependence in the dislocation level, $E_O(T) = E(OK) + kT\ell n(N_{dd}/N_{da})$ due to the configurational entropy of electrons in this band (see LS). In higher dislocation density samples (S8, W5), where there is not this direct evidence for $N_{dd} \ll N_{da}$, a temperature dependence in $\mu$ of $-2$ to $-3k$ can be discerned at low temperatures (at these temperatures $df/dT$ is small and the chemical potential clings to $E_O(T)$). OS infer that this is due to the configurational entropy term, but this does not seem likely because the same temperature dependence can be seen in GKM in samples for which they were able to show $N_{dd} \gg N_{da}$. Hence, it seems more reasonable to assume that the temperature dependence arises from the entropy of ionization for the dislocation level or band and that the property $N_{dd} \ll N_{da}$ is a property of low dislocation density samples.

The possibility of extra acceptors arising during deformation leads to uncertainty in the position of $f = o$ for the dislocation band and hence in the values of $E_O(T_O)$. However, if several samples differ only in their dislocation density and one common defect is dominant in all of them, then their plots of $\mu$ vs T should all intersect at one unique point $E_O$, $T_O$. The points so derived have been plotted in figure 1, where lines are due to samples for which there are no other suitable sample lines to give an intersection. Included in these latter are samples P2-1 and P3-2 from LS, the sample shown in figure 7 of Schröter, Scheibe and Schön (1980 — to be referred to as SSS) and S8 from OS2. (There was no clear unique intersection between the group S samples of OS2 — the intersection of S6 and S7 was taken for all but S8.)

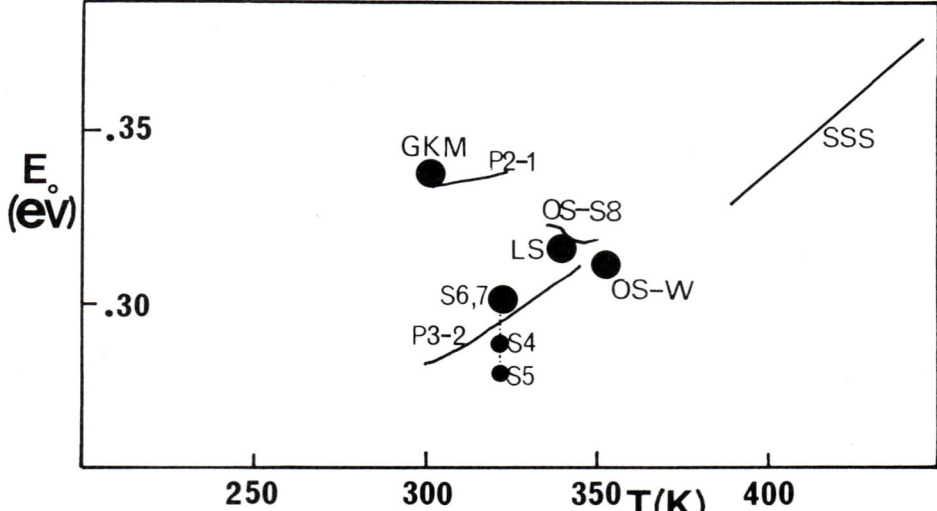

Figure 1    Neutral Levels, $E_O(T_O)$ for p type silicon.

With the exception of two cases the correction in the origin of fractional occupation is less than the concentration of dopant atoms. This suggests the possibility of the dopant, for example boron, combining with the dislocation to form an extra acceptor lying well below $E_O$. If boron were to become three-fold coordinate inside the dislocation core then it might be capable of acting as a double acceptor. This could be the case if boron replaced the threefold coordinated silicon atom in the reconstruction soliton (antiphase defect) described in Heggie and Jones (1983), $B_3^{2-}$ being the analogue of a negatively charged soliton. Such a binding of a soliton by a boron atom would be a natural explanation of the immobilising power of boron (and maybe by analogy phosphorus as $P_3^{2+}$) on dislocations observed by Sumino and Imai (1983). A reservation to this postulate is that the samples S4 and S5 of OS2 apparently require more extra acceptors than dopant atoms – this may be possible if the material used is compensated.

The results for $E_0(T_0)$ are tabulated below.

|        |   |                        | $E_0$(eV) | $T_0$(K) |
|--------|---|------------------------|-----------|----------|
| OS2    | – | S4 S5 S6 S7            | .29–.30   | 325      |
|        | – | S8                     | .32       | 341      |
|        | – | W2, W3, W4, W5         | .31       | 350      |
| LS     | – | P1–1, P1–2, P1–3, P1–5 | .316      | 341      |
|        | – | P2–1                   | .337      | 323      |
|        | – | P3–2                   | .31       | 347      |
| SSS    | – | figure 7               | .33       | 390      |
| GKM    | – | figures 4 and 2        | .337      | 302      |

To examine the defect states more closely the gradients of $\mu$ with respect to T and f must be studied. Consider the general case of a number of defects giving rise to a donor band at $E_d$ and an acceptor band at $E_a$, their separation being U, the Hubbard energy, which could be positive or negative. The approximate occupation of this system is given by

$$f = c \tanh(\mu - E_o)/kT \qquad U < 0 \qquad (1)$$

$$f = 2c \exp(-U/2kT) \sinh (\mu - E_o)/kT \qquad U > 0 \qquad (2)$$

where c is the proportion of possible $60^o$ shuffle dangling bond sites occupied by these defects and

$$E_o = 1/2(E_d + E_a) \qquad \text{and} \qquad U = E_a - E_d.$$

In both cases the rate of change of occupation with temperature is approximately

$$\frac{df}{dT} = \frac{1}{bkT}\left[\frac{d\mu}{dT} - u\right] \qquad \text{near } T_o \qquad (3)$$

where $$E_o \equiv E_o(0K) + uT + \alpha f$$

and $$b = \frac{1}{cw} + \frac{\alpha}{kT_o}, \qquad w \begin{cases} = 1 \text{ for } U < 0 \\ = 2\exp(-U/2kT) \text{ for } U > 0 \end{cases} \qquad (4)$$

Taking $r = N_d/N_{Ca}$ and letting $g_2 = -k[(d\ln p)/d(1/T)]_{TO}$ and
$g_3 = (d\mu/dT)_{TO}$, it can be shown that $g_2/T_0 r = (g_3 - u)/b$. Figure 2 is a
plot of $g_2/T_0 r$ against $g_3$ whose aim was to separate the relative
contributions of u, the temperature dependence of $E_0$, and of b, a
combination of factors: c, the defect concentration, w, dependent on the
effective U and $\alpha$, which takes into account the finite density of states
in the band at $E_0$ and the electrostatic interaction arising from charging
the dislocation line (see LS).

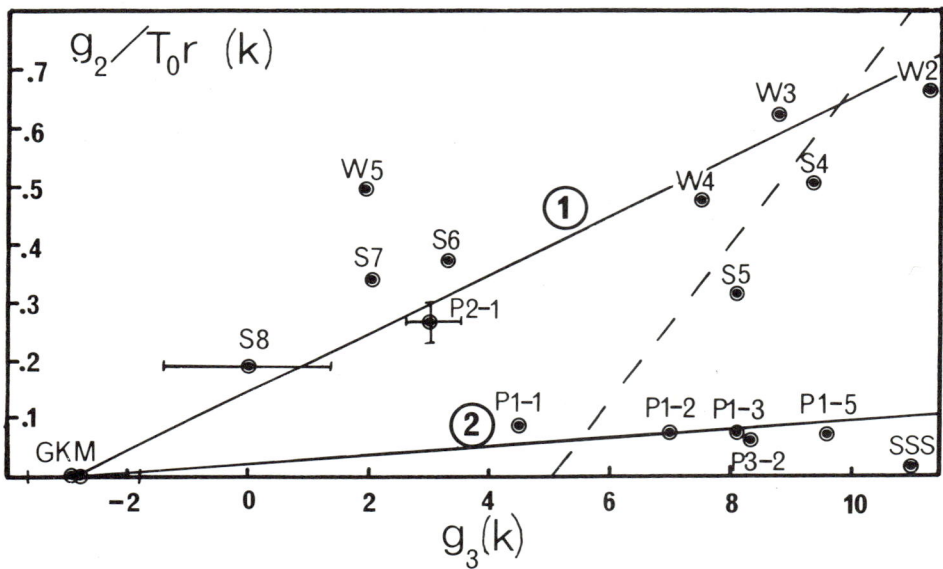

Figure 2    Plot of $g_2/T_0 r$ vs $g_3$ to find u and b

This figure contains the essence of the disparity between the results of
OS2 and LS.  There should be a straight line of positive gradient
connecting each point with the appropriate intercept, u, on the abscissa,
but this plot reveals that the data are broadly split into at least two
groups.  One group includes the samples of OS2 plus P2-1 of LS and the
other contains the other samples of LS.  If both groups include the
samples of GKM, which is almost certain for the first group, they would
be roughly described by u = -3k and 1/b = 5% for the first group (slope
1) and 0.7% (slope 2) for the second group.  In none of the samples is
there a pronounced step in $\mu$ vs f indicative of positive U behaviour,
although this was postulated by SSS with U = 0.28 eV to be a cause of the
low gradient of 0.7%.  In addition there must be some contribution from $\alpha$
to the gradient of $g_2/T_0 r$ vs $g_3$ for the samples of OS2, contrary to Ono
and Sumino's assumptions, because the effect of c on b would not be
sufficient to reduce it to 5% (the concentration, c, being higher than
40% if S4 and W2 are included in this group).  At very low temperatures
the samples S7, S8 and W5 of OS2 and H906 of OS1 give $d\mu/dT$ of -1 to
-3k.  This low temperature tail, when $\mu$ sticks to the defect level as it
moves with T, can give the u characteristic of the defect.  It seems the
main defect in these samples is related to that in GKM samples, which
give u $\simeq$ -3k.

The value of u for the samples of LS is difficult to fix from figure 2 and their low temperature tails seem to be dominated by a dramatic drop in $\mu$ caused by the reduction in screening between dislocation line charges as the hole density becomes very small (see LS). The low temperature tails of the samples H903, H904, H905, M4 and S6 of OS1 and OS2 give a temperature dependence of +5 to +7 k — the negative ionization entropy that this implies is not impossible, but the positive $\mu$ vs T gradient could equally well arise from the descreening of line charges seen in LS samples. Further low temperature measurements on these samples might resolve this ambiguity.

The value of u for the samples in LS must be uncertain, but if it is around zero or negative then the high value of b necessary to give gradient 2 of figure 2 might arise purely from the electrostatic effect (as fitted to the data in LS) or additionally from a large bandwidth for the band (ignored in LS). On the other hand a positive value for u would make such a low gradient unnecessary.

Ono and Sumino dismissed the electrostatic effect originally because of the series of initially n-type samples H903, H904, H905, which gave increasing values of $-k \ dlnp/d(1/T)$, ie 0.31 to 0.39 eV, with increasing density of dislocations. This progression at first sight contradicts the electrostatic effect, which would cause this gradient to decrease with greater dislocation density. This contradiction extends to the low temperature tails of gradient 5 to 7 k, which are displaced towards higher chemical potentials with increasing dislocation density. These observations might be resolved by the following non-unique and tentative model, postulating:

1)   A donor band characterised by u $\simeq$ −3k, increasing non-linearly in concentration with increasing dislocation density,

2)   A half-filled band characterised by u between 0 and 7k well below the donor band at O.K, being predominant in low dislocation density samples — including W2, W3, S4, W4, and S5 of OS2, ie following the broken line for figure 2.

In low dislocation density samples at low temperature $\mu$ stays in the half-filled band, but this band accepts more and more electrons from the increasingly dense donor band as the dislocation density increases. Thus $\mu$ increases through the electrostatic effect or through the finite density of states in the band. At high dislocation densities $\mu$ resides in the donor band, part emptied by the presence of the half-filled band. The behaviour around the temperature of the point of intersection of these two bands depends critically on the density of states that each band contributes near $\mu$, leading to large variations in the effective value of u. Figure 3 illustrates the situation before (a and c) and after (b and d) charge equilibration at low temperature for low (a,b) and high (c,d) dislocation density samples. The band 3 is due to the additional acceptors mentioned in the first part of this section, band 1 is the donor band and band 2 is a half filled band.

This model is one of many possibilities and it should be easy to confirm or deny experimentally, based as it is on the uncertain foundation that u > 0 for band 2. The separation of these effects into properties characteristic of different dislocation densities is a crude one — temperature, rate and mode of deformation must also modify the relative

proportion of each defect.   Typically GKM samples were deformed at 650°C
in compression, OS samples 750°C in tension and LS samples 770-780°C in
compression.

low T, low $N_d$                          low T, high $N_d$

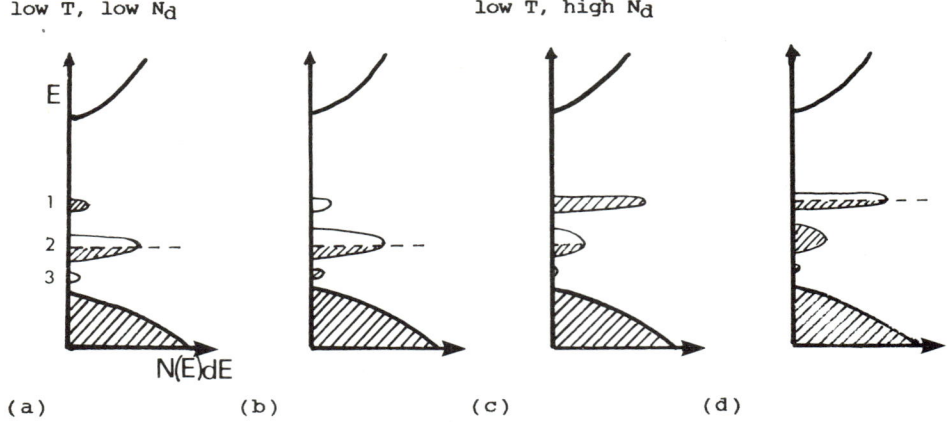

(a)                    (b)                    (c)                    (d)

Figure 3   Densities of states in a tentative model.

## 3. Conclusions

The Hall effect in deformed p-type silicon requires at least two
part-filled defect bands for its explanation - one around .34 eV above Ev
at 300 K, and the other around .31 eV at 350 K, with a third acceptor
state, which becomes important at low dislocation densities.   This state
may arise from the interaction of dopant atoms with the dislocation cores
and in this case may cause problems with DLTS measurements (Kveder et al
1982), where samples have much higher dopant concentrations.

## Acknowledgements

MH acknowledges useful conversations with R Jones, F Louchet and
A Ourmazd, the financial help from CNRS and British Council (Paris) and a
private communication of results from K Sumino.   Most of this work was
carried out at LTPCM-ENSEEG (LA29 de CNRS) Domaine Universitaire,
St Martin d'Hères, France.

## References

Grazhulis V A, Kveder V V and Mukhina V Yu.   1977 Phys. Stat. Sol. (a) **43**
    407
Heggie, M I and Jones R 1983 Phil. Mag. B**48** 365 and 379.
Kveder V V, Osipyan Yu A, Schröter W and Zoth G 1982 Phys. Stat. Sol. (a)
    **72** (1982)
Labusch R and Schröter W 1980 "Dislocations in Solids", ed FRN Nabarro
    (Amsterdam:North Holland) **5** 127
Ono H and Sumino K 1980 Jap. J. Appl. Phys. **19** L629
Ono H and Sumino K 1983 J. Appl. Phys. **54** 4426
Schröter W, Scheibe E and Schön H 1980 J. Micros. **118** 23
Sumino K and Imai M 1983 J. de Physique **44** C4-195

*Inst. Phys. Conf. Ser. No. 76: Section 2*
*Paper presented at Microsc. Semicond. Mater. Conf., Oxford, 25–27 March 1985*

# Interfacial line-defects in heterojunctions

R C  Pond, N A  McAuley and R W  Devenish

Dept. of Metallurgy & Materials Science, The University of Liverpool,
P.O. Box 147, Liverpool L69 3BX

Abstract  The nature of line-defects which can arise in interfaces
depends on the symmetry of the adjacent crystals and their orientation
relationship.  Defects are characterised geometrically by admissible
combinations of symmetry operations, one from each of the crystals. It
is shown that defects can therefore arise in interfaces between lattice-
matched crystals if their point symmetries are not identical.  Observations of dislocations with Burgers vectors equal to $\frac{1}{4}$<111> in $NiSi_2$:Si,
and antisite domain disorder in GaAs:Ge are explained using this theory.
Possible interfacial dislocations in silicon on sapphire are discussed
briefly, and preliminary observations of features consistent with being
interfacial disclinations in $Pd_2Si$:Si are presented.

## 1. Introduction

In crystalline materials the atoms are arranged in an orderly fashion
which exhibits symmetry.  Line-defects in the bulk of a single crystal are
perturbations of this order, and the nature of admissible defects depends
on the crystal's symmetry.  The different types of line-defects which can
arise are characterised geometrically by symmetry operations.  Dislocations, for example, have been studied extensively and are characterised by
their Burgers vectors which must be translation vectors in the case of
perfect defects.  Disclinations and dispirations are also admissible line
defects which can, in principle, exist in crystals that exhibit proper and
screw-rotation axes respectively.  However, except in the case of liquid
crystals, these defects are not generally observed as their elastic strain
energies are prohibitive.  It has been shown in a recent theory, Pond
1985a, that interfacial line-defects are characterised geometrically by
admissible combinations of symmetry operations, one from each of the
adjacent crystals. The object of the present paper is to illustrate that
the variety of interfacial line-defects that can exist is wider than had
previously been recognised.  Interfacial dislocations, disclinations and
dispirations, having relatively modest elastic strain energies (particularly in thin films), can arise in interfaces between crystals which have
either different symmetries or orientations or both.  The discussion is
illustrated by examples of defects observed in interfaces occurring in
semiconductor devices.

## 2. Crystal Symmetry, Surfaces and Interfaces

It is convenient to begin the discussion by considering briefly the
symmetry of single crystals.  We use the notation of Seitz (1936) which
represents a general symmetry operation by the matrix formulation

$[D_i | \underline{\tau}_j]$, which means that the $i^{th}$ point symmetry operation is carried out through a chosen origin, and this is followed by a translation $\underline{\tau}_j$, the $j^{th}$ translation vector. Symmetry operations in nonsymmorphic crystals, i.e. crystals exhibiting mirror-glide planes and/or screw-rotation axes, can be readily represented by including a supplementary displacement, $\underline{\alpha}_i$. Thus, a general symmetry operation is represented by $[D_i | \underline{\alpha}_i + \underline{\tau}_j]$ in such a crystal, where the vector $\underline{\alpha}_i$ depends on the choice of origin and will be zero in the case of symmorphic operations which pass through the chosen origin (Pond and Vlachavas, 1983). Another important class of crystals is that where a crystal's symmetry is lower than that of the lattice on which it is based, and such crystals are designated nonholosymmetric. For example, the sphalerite structure is nonholosymmetric, and can be regarded as being derived from the diamond structure. It is a symmorphic structure so that none of its symmetry operations have supplementary displacements. On the other hand, all of the diamond structure symmetry operations that have non zero supplementary displacements relate crystallographically equivalent sphalerite crystals in which the anion and cation species are interchanged. We refer to sphalerite crystals related in this way as anti-site domains, and refer to the symmetry operations relating them as exchange operations.

The simple fcc, diamond and sphalerite structures exhibit the same set of twentyfour point symmetry operations which have no supplementary displacements. The simple fcc structure exhibits a further twentyfour acting through the chosen origin, and the diamond structure exhibits the same except that all of these have supplementary displacements, $\alpha = \frac{1}{4}<111>$. These latter operations are the exchange operations in sphalerite crystals. The different types of crystal symmetries outlined above have important consequences regarding surface features. Let the unit normal to a given surface be $\underline{n}$. In symmorphic crystals steps having heights equal to $\underline{n}.\underline{\tau}_j$ can exist on any surface, and these are designated complete steps since they separate surfaces which are identical except for their location in space. In the case of nonsymmorphic crystals complete steps can also arise on any surface, but an additional type can also arise on surfaces where the orientation of $\underline{n}$ is left invariant by the operation $[D_i | \underline{\alpha}_i]$.

Such steps are designated demi-steps and have heights equal to $\underline{n}.(\underline{\alpha}_i+\underline{\tau}_j)$.

Thus, for example, demi-steps of height $\frac{1}{4}<001>$ can arise on {001} surfaces of diamond crystals, whereas the minimum complete step height on corresponding surfaces in simple fcc crystals is $\frac{1}{2}<001>$. In the case of the sphalerite structure complete steps can exist since the structure is symmorphic. However, demi-steps can arise on surfaces where an antisite domain boundary emerges onto the surface from the bulk, i.e. where the surfaces separated by the step are related by an exchange operation. It has been shown, (Pond, 1985b), that demi-steps can only exist on {hko} surfaces of diamond, i.e. where at least one Miller index is zero, and where h and k (taken as having no common factors) are a mixture of odd and even integers. Correspondingly, antisite domain boundaries in sphalerite crystals separate crystallographically equivalent (energetically degenerate) surfaces only when they emerge onto {hko} surfaces, and have associated demi-steps when h and k are mixed integers.

We now consider Weingarten-Volterra like line discontinuities in interfaces using the approach of Pond, 1985a. Initially we consider a bi-

crystal to be fabricated by bringing together a white and black crystal
having the appropriate relative orientation and prepared surfaces, as
shown in fig.1(a). Let the downward pointing unit normals be $\underline{n}(w)$ and $\underline{n}(b)$
in the white and black crystals respectively. These surfaces can be
bonded subsequently to form the bicrystal. White and black surfaces which
are crystallographically equivalent to the initial ones can be exposed in
the two crystals, and these new surfaces can be forced together and may
bond to form an interface equivalent to the initial one. Fig.1(b) shows
new surfaces related to the initial ones by translation operations, and by
rotation operations in fig.1(c). The region of new interface is separated
from the initial one by a line-defect which is characterised geometrically
by the operation which brings the new black surface onto the new white one.
In the general case, where the new and original white surfaces are related
by $[D(w)_i | \underline{\alpha}(w)_i + \underline{\tau}(w)_j]$, and the new and initial black by $[D(b)_k | \underline{\alpha}(b)_k +$
$\underline{\tau}(b)_l]$, the compound operation, expressed using the coordinate system of
the white crystal, is given by

$$[D(w)_i | \underline{\alpha}(w)_i + \underline{\tau}(w)_j] \; T \; [D(b)_k | \underline{\alpha}(b)_k + \underline{\tau}(b)_l]^{-1} T^{-1} \qquad (1)$$

where T is the matrix transformation relating the coordinate systems of
the black and white crystals. When the compound operation is a pure translation, the defect is an interfacial dislocation. When it is a proper
rotation the defect is an interfacial disclination, and an interfacial
dispiration when it is a combination of proper rotation and translation.

## 3. Lattice Matched Crystals

When the adjacent crystals are lattice-matched, the transformation matrix
T is the identity matrix, E, and the sets of white and black translation
vectors are identical. However, it can be seen by inspection of equation
1 that interfacial dislocations can arise when symmetry operations, for
which the orthogonal parts are identical but for which the supplementary
displacements are different, exist in the adjacent crystals. For example,
in the case of interfaces between a nonsymmorphic (white) and a symmorphic
(black) crystal, the compound operation becomes

$$[D(w)_i | \underline{\alpha}(w)_i] \; [D(b)_j | 0]^{-1} = [E | \underline{\alpha}(w)_i] \qquad (2)$$

This operation characterises a dislocation with Burgers vector $\underline{\alpha}(w)_i$
(modulo $\underline{\tau}(w)_l$), which separates energetically degenerate interfacial
regions, and can exist on any interfacial plane the orientation of which
is left invariant by any of the set of white symmetry operations having
the supplementary displacement $\underline{\alpha}(w)_i$. Such defects can exist in interfaces between diamond structure and fluorite structure crystals for
example, as occur in certain silicide-silicon contacts and in 'silicon on
insulator' devices. Dislocations of this type have been studied in detail
using weak beam microscopy (Cherns et al. 1984) in $NiSi_2$:Si interfaces.

The Burgers vectors, $\underline{b}$, were found to be $\frac{1}{4}<111>$ (i.e. $\underline{b} = \underline{\alpha}(mod.\tau)$) and
the defects were only observed on {hko} facets, in complete agreement with
the present theory. These observations have been considered in detail in
terms of the different surface structures of the symmorphic silicide overlayer and the nonsymmorphic silicon substrate (Pond and Cherns, 1985). A
schematic representation of a $\frac{1}{4}[1\bar{1}1]$ edge dislocation in a $(1\bar{1}0)$
fluorite: silicon lattice matched interface is shown in fig.2.

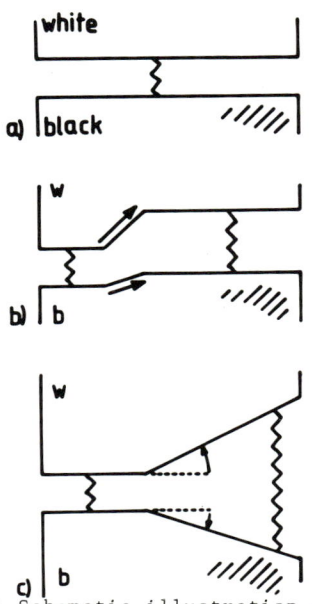

Fig. 1 Schematic illustration of the formation of an interfacial defect; initial white and black surfaces (a), new surfaces related to the initial ones by translation (b) and rotation symmetry operations (c).

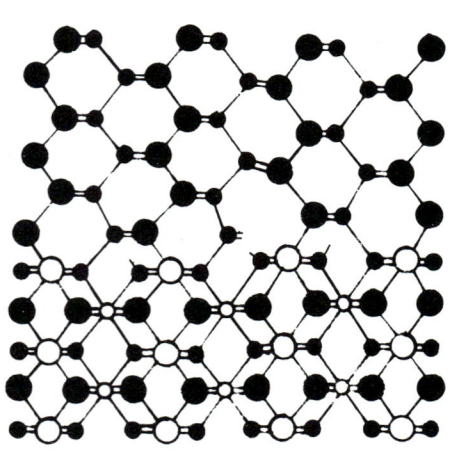

Fig. 2 Schematic [110] projection of a $\frac{1}{4}[1\bar{1}1]$ dislocation in a $(1\bar{1}0)$ interface between lattice matched crystals having the fluorite and diamond structures.

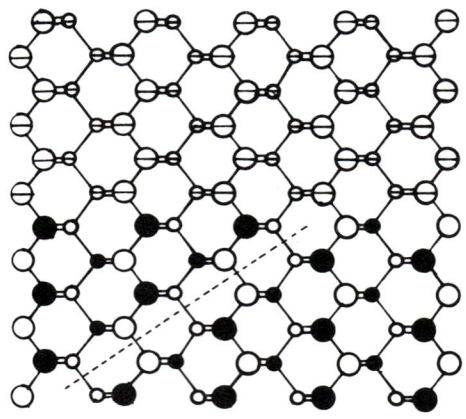

Fig. 3 Schematic [110] projection of a $(1\bar{1}0)$ interface between lattice matched crystals having the sphalerite and diamond structures; an antisite domain boundary in the sphalerite crystal separates energetically degenerate interfacial regions.

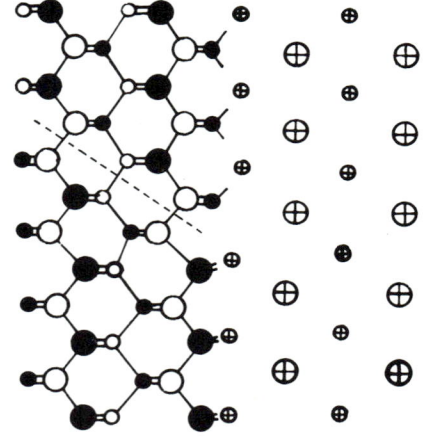

Fig.4 Schematic [110] project of an (001) interface between lattice matched crystals having the sphalerite and simple fcc structures; the line defect has dislocation character with b = $\frac{1}{4}[1\bar{1}1]$.

The epitaxial growth of GaAs on Ge, which closely approximates to a
lattice matched system, is similar to fluorite:silicon interfaces in some
respects.  GaAs has the sphalerite structure and is therefore symmorphic,
but the set of exchange operations for this material is identical to the
set of operations for Ge which have supplementary displacements, $\alpha$.  It
follows that black and white operations can be substituted into equation 1,
yielding the compound operation [E|O], i.e. a defect without dislocation
character.  However, since the GaAs operations employed are exchange operations, it follows that an antisite domain boundary impinges on the boundary
and separates degenerate regions of interface.  As for the previous case,
this can only occur for {hko} interfaces.  Thus, in the epitaxial growth of
GaAs on Ge substrates, antisite disorder can not arise if the substrate
orientation is {hkl}, i.e. where none of the Miller indices is zero. These
predictions are in complete agreement with experimental observations, and
have been discussed in detail by Pond, Gowers and Joyce, 1985. Fig. 3 is a
schematic diagram of an antisite domain boundary separating degenerate
regions of interface in a $(1\bar{1}0)$ GaAs:Ge bicrystal.  Fig. 4 is a schematic
diagram of a (001) GaAs: simple fcc crystal interface.  In this case the
interfacial defect corresponds to the intersection with the interface of
an antisite domain boundary from the GaAs crystal, but the defect now also
has dislocation character with Burgers vector $\underline{b} = \underline{\alpha}$.

## 4. Lattice Mismatched Crystals

The character and role of misfit dislocations in interfaces has been
studied extensively by, for example, Matthews (1975). However, additional
types of dislocations, which can also accommodate misfit, are predicted
according to equation 1.  For example, in the case of silicon on sapphire
devices, the c-mirror-glide plane of the rhombohedral substrate is
oriented parallel to the diamond glide plane of the cubic epitaxial layer.
These glide planes have different supplementary displacements, $\frac{1}{2}c$ for the
sapphire and $\alpha = \frac{1}{4}<111>$ for the silicon, and hence dislocations character-
ised by $[m|\frac{1}{2}c^-][m|\underline{\alpha}]-1 = [E|\frac{1}{2}c-\alpha]$can exist.  Such dislocations have
Burgers vectors equal to $(\frac{1}{2}c-\bar{\alpha})^-$ and are associated with interfacial steps,
as will be discussed in more detail in a subsequent publication.

Consider next the case of two crystals which both exhibit three fold axes,
and where the relative orientation is such that these are slightly mis-
aligned.  According to equation 1, line defects characterised by
$[3^+\mathrm{T}3^-\mathrm{T}^{-1}|0]$ can exist, i.e. disclinations corresponding to small rotations about an axis approximately perpendicular to the three-fold axes.
Preliminary investigations, using transmission electron microscopy, of
$Pd_2Si$:Si interfaces has indicated that such disclinations may be present.
The nominal orientation relationship is (0001) parallel to (111), and
$[11\bar{2}0]$ parallel to $[\bar{2}11]$, with the three-fold axes in the two materials
aligned.  However, bright field images, such as fig.5, reveal mottled contrast resembling that of polycrystalline films, but selected area diffraction shows that the silicide film is basically a single crystal having, on
average, the orientation relationship quoted above.  On the other hand,
the fine structure of the diffraction spots due to $Pd_2Si$, and variations
of the contrast levels and the visibility of lattice planes and moiré
fringes, in individual mottles as the specimen is tilted indicates that
the mottles are three distinct regions of $Pd_2Si$ having slightly different
orientations.  Moreover, the orientations of the mottles are symmetrically
disposed about the nominal orientation relationship.  In addition, it
appears that individual mottles are separated by interfacial line defects,
feature A in fig.5b for example, rather than by planar defects.  These

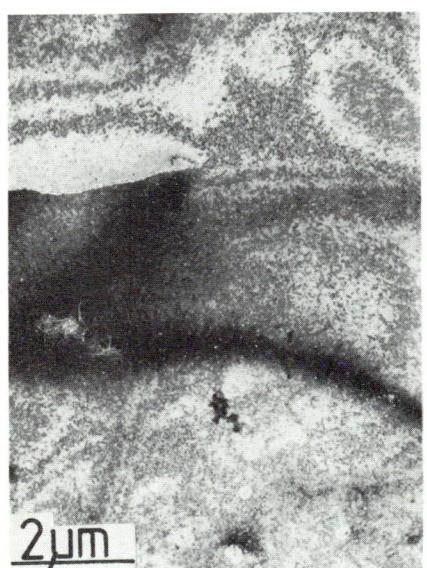 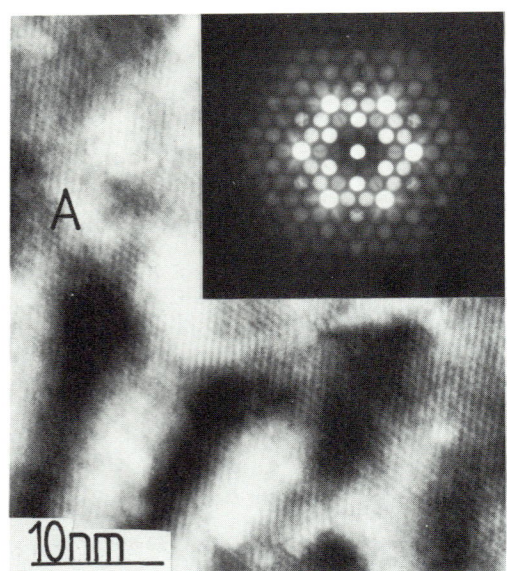

(a)                                                           (b)

Fig. 5 Bright field images of Pd₂Si:Si specimen; (a) low mag. image showing
mottled contrast, (b) high mag. image showing (10Ī0) lattice planes,
moire fringes and line defects between mottles, as at A.   The
incident beam is perpendicular to the interface, and the inset is a
convergent beam diffraction pattern·

observations are consistent with the mottles being regions of silicide
separated from each other by interfacial disclinations having rotations of
the order $0.5^{\circ}$.   The precise orientation relationship of individual
mottles with the substrate and the character of the interfacial defects is
being investigated currently using convergent beam diffraction.

Acknowledgement.

The authors would like to thank Dr. D. Cherns for helpful discussions and
provision of the Pd₂Si:Si specimen.

References

Cherns D, Hetherington CJD, Humphreys C J 1984 Phil. Mag. A49 165
Matthews J W 1975 Epitaxial Growth, Mat. Sci. Series, Academic Press,
    New York
Pond R C 1985(a) Dislocations and the Properties of Real Materials,
    The Institute of Metals
Pond R C 1985(b) Polycrystalline Semiconductors, Ed G Harbeke, Springer-
    Verlag, Berlin
Pond R C and Cherns D 1985 Surf Sci in press
Pond R C Gowers J P and Joyce B A 1985 Surf Sci in press
Pond R C and Vlachavas D S 1983 Proc. R. Soc. Lond. 386A 95
Seitz F 1936 Ann. Math. Statist. 37 17

*Inst. Phys. Conf. Ser. No. 76: Section 3*
*Paper presented at Microsc. Semicond. Mater. Conf., Oxford, 25–27 March 1985*

73

# Characterization of silicon on insulator films realized via zone melting recrystallization

D Bensahel, M Haond and D Dutartre

Centre National d'Etudes des Télécommunications, BP 98, 38243 Meylan-Cedex, France

Abstract    The macroscopical defects encountered during the recrystallization of poly-Si films on insulator by zone melting are described. The behaviour of microscopic defects and their possible localization are then reviewed.

## 1. Introduction

Silicon-On-Insulator (SOI) structures have been studied intensively for several years. Many more or less advanced techniques have been tested, for example, SIMOX (Hemment), ELO (Jastrzebski ), and porous-Si (Imai). So far, the greatest effort has been directed towards SOI obtained by recrystallization via the liquid phase of deposited poly-Si films-on-insulator, mainly an oxidized Si bulk wafer. These techniques utilize c.w. lasers (Colinge 1984) or electron beams (Davis) for localized treatments (roughly limited to the spot diameter), and strip heaters or lamps for the treatment of a whole wafer in one time, via a large linear molten zone (Chen).

We review here the defects that are usually present in these recrystallized films. The techniques used for localization of defects are also discussed.

## 2. Experimental

In **c.w. laser** recrystallization, a focused spot (typically 100 μm in diameter) is scanned across the SOI structure. Dwelling times are in the millisecond range. Laser recrystallization can affect the substrate. Pinizzotto 1983 has observed that slip planes are generated in the substrate. Moreover, he has also reported the presence of dislocations which can extend up to 15 μm into the substrate. However, the slip planes do not affect further proccessing at high temperature. The slip planes are due to thermal stress present in the substrate, especially when seeding windows are opened in it. By reduction of the vertical temperature gradient, i.e., when the substrate temperature is increased (400-600°C), these defects tend to disappear.

In recrystallization induced by **graphite strip heaters** or **lamps**, the molten zone is long (5-10 cm) and wide (2 to 4 mm). In this case, a slow scan is used (0.1 to 1 mm/sec.). Great care has to be taken in the homogeneity and stability of the substrate temperature, which is already high (1000 to 1200°C). Moreover, vertical as well as lateral gradients are present, which can, in the worst case, cause warpage or shrinkage of the whole wafer.

Once recrystallized, the SOI structures are **characterized** in several ways, as follows:

- Optical microscopy in the Nomarski mode gives information on the surface roughness especially when the defects have been chemically decorated (Secco, Schimmel). If the SOI has not been capped, the surface after crystallization will be very smooth and shiny (Colinge 1984). When a cap is present and does not flow, as for Si3N4, the original poly-Si roughness is retained. By contrast, when a SiO2 cap which can flow is used, any change in surface roughness can be observed (Fig. 1a). When investigated in a bright field with an interference filter, the samples can present interference fringes (Fig. 1b) due to interference between the upper and lower surfaces of the film. They appear when mass transport occurs: counting the fringes gives the shape and thickness variations of the recrystallized film which often flows when melting.

- In order to obtain crystallographic information in thin recrystallized films on a large scale, Bezjian et al. have developed the so-called **etch pit grid** technique. A grid-array of etch pits is patterned in Si using an anisotropic chemical etch. The geometry of an etch pit is indicative of the local orientation: a (100) orientation will give a square, the diagonals of which are in the <100> direction. Thus, in-plane misorientation between two (100) grains can be determined by the relative rotation from one square to another, as will be shown below.

- The information given by the above technique is complementary to that obtained by **Transmission Electron Microscopy** (TEM), a powerful, but time consuming technique.


## 3. Macroscopic Defects

- In **laser** recrystallization, depending on the shape of the trailing edge of the spot, we can obtain either a chevron-like structure (round spot (Gat)) or large grains. Stultz et al. 1981 have shown that a liq./sol. interface of 35° with respect to the scan direction produces unidirectional grain growth following the interface at close to 90°. Further grain enlargement is achieved by subsequent overlapping scans. If the sample is not capped, the orientation of the grains will be random.

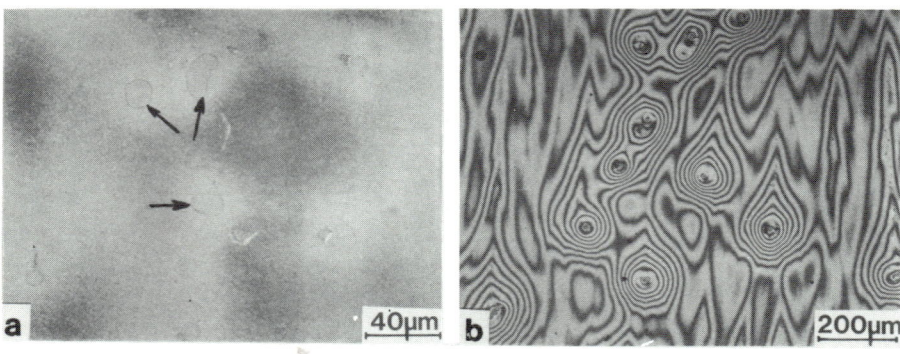

Fig. 1:    (a) Liquid and solid crystallites at the end of the transition zone of a lamp recrystallized sample, (b) Thickness variations around a void as observed through an interferential filter.

When a SiO2 cap is used, a (100) texture is found (Hodé, Stultz 1983). This is interpreted to result from strain due to the cap and the slanted beam.

When a seed opening is etched in the insulator, the (100) orientation of the substrate can be transferred to the SOI structure. However, the orientation is maintained only over 20-100 μm. Further, it turns. There is not yet a clear answer to this change, but the stability of the annealing parameters as well as the thermal gradient as we cross the SOI region might be involved.

Since molten Si on SiO2 does not wet, the laser power interval between "good" annealing and ball-up of the film is slight. 60 nm of Si3N4 on the poly-Si increases the power interval, owing to better wetting between Si and Si3N4 (Kamins). However, Si3N4 is highly resistant and crystallizes instead of flowing at the melting temperature of Si. This often results in cracks in the film due to strains, or in agglomerates created between the Si and Si3N4 which are very difficult to dissolve (Paelinck). Last but not least, electrical measurements on processed recrystallized films capped in this way consistently show surface states in the lower SiO2 interface. These are due to dissolution of some nitrogen during recrystallization. All of these undesirable effects are eliminated when a thin SiO2 layer is present between the poly-Si and the Si3N4 (Colinge 1984).

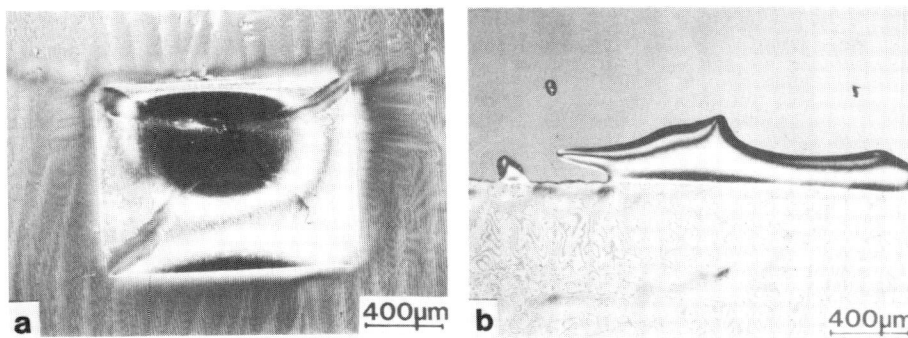

Fig. 2: (a) Melting of the substrate, (b) Agglomeration in the film

- In **lamp** or **strip-heater** recrystallization, where a long and wide molten zone is present, the wetting effects are dramatic.

First of all, the SOI structure needs a cap to prevent agglomeration of the molten Si. The deposited surface layer (typically 1-2 μm of SiO2) can be visualized as performing two functions: reducing the total free energy of the molten layer by providing a second surface for the Si to wet, and mechanically restraining the film from balling up or agglomerating. This effect of de-wetting has been found to be minimized when a second cap of Si3N4 (30 nm thick) is deposited onto the SiO2 cap. Interpretation of this so-called "magic cap" has been based on with the presence of nitrogen which incorporates upon melting of the film (Fan 1984, Weinberg). But this cap is still a subject of controversy. In fact, it has been found that films which are not prone to agglomeration are relatively insensitive to the encapsulant.

Recrystallization is considered bad when one of the following problems or macroscopic defects is encountered. These defects are gross imperfections related to either recrystallization conditions or to properties of the deposited films and/or encapsulant. They are:

- Substrate slip and warpage. For example, we have observed 20 μm shrinkage on a fully recrystallized 4-in wafer correlated with the appearance of slip planes. This shrinkage is even greater when seeding windows are present in the structure, but it can be minimized by the reduction of lateral and vertical gradients.

- Substrate melt through (Fig. 2a). This occurs when the molten zone is too hot and melts the underlying substrate. This results in large, square-shaped areas as discussed by Celler et al. 1984a.

- Isolated thinned regions or voids are observed in the films (Fig. 1b). Their exact origin is not clear and localized impurities or technological defects may be involved. These voids can also be explained by the 10% volumetric contraction of Si upon melting. It is thought that these small de-wetting areas can, if they coalesce, be responsible for film agglomeration.

- Film agglomeration or balling up (Fig. 2b). This is by far the most serious problem. It consists of large areas of Si surrounded by the exposed thermal oxide. Once initiated, the balling up extends all along the molten region. In this case, the solidification restarts mostly in a dendritic way, due to a large supercooling (Dutartre).

## 4. Microscopic Defects

In lamp or strip-heater recrystallization, at the beginning of the recrystallized zone which is (100) textured perpendicular to the surface for a thickness of 0.5 μm, we find a transition zone (Geis 1982a, Cline), containing both partially and completely melted crystallites (Fig. 1a). In films over several microns thick, no predominant texture is observed (Atwater 1984).
Among the several hypothesis published, i.e., anisotropic stress-strain characteristics of Si (Geis 1982a), formation and mobility of dislocations (Cahn), match between planar atomic density of (100) Si and amorphous or microcrystalline SiO2 (Tiller), we favor that of Biegelsen et al. 1984 which shows that (100) crystallites are responsible for the final texture, due to a minimized interfacial energy between Si and SiO2 for the (100) planes.
The zone recrystallized films not only have a strong <100> texture; they also tend to grow with in-plane <100> direction parallel to the scan direction and the <100> orientation (Vu 1983). Generally, the in-plane misorientation can be as high as 15° around <100> parallel to the scan (Geis 1982b).
Inside these grains, which can be several millimeters wide, we find subgrains limited by subgrain boundaries (SGB) in a network which always presents the same characteristics. They run nearly parallel to the scan direction, have a "wishbone" pattern, new SGBs are nucleated to maintain a constant average lateral density, and once nucleated, an SGB does not terminate except by merging. SIMS mapping of recrystallized samples shows a distribution of oxygen along the SGBs (Fan et al. 1984). If one looks at the solidification front, the main results published favor the existence of it which is faceted rather

than planar. A qualitative explanation showing that SGBs are formed at the interior corners of a front with facets defined by alternating (111) and (11$\bar{1}$) planes has been proposed by Geis et al. 1983. The time evolution of the SGB network has been simulated by Pfeiffer et al. 1985. The main assumption of their calculations is that long facets tend to grow faster than short ones because they have more nucleation sites. Thus, short facets shrink in size by being overtaken by longer ones. Fig. 3 shows some results equivalent to their simulations.

- TEM analysis (Haond 1983) gives more particular details of the SGBs (Fig. 4): one, two or several sets of dislocations can be present. All types of low angle boundaries can be found (flexion or twist), but the misorientation between each side of the boundary is usually less than 1°. If the SGBs described previously are dominant, we can find, however, other kinds of defects called secondary-SGBs (S.SGB). As shown in Fig. 4, a S.SGB1 is usually associated with a precipitate. It is made by pure edge character dislocations, the dislocation line being either (101) or ($\bar{1}$01). On the other side, a S.SGB2 always originates along a SG$\bar{B}$ and parallel to the (110) direction. Analysis of the diffraction patterns shows that a large amount of elastic deformation is present in this region, and explains there the strong undulation of the film.
Beneficial effect of these defects could be that they can act as getters. Too great distance between two SGBs is not necessarily desirable. Indeed, we have found that, by lowering the scan speed (<0.05 mm/sec.), the spacing between SGBs is increased, but the grains are no longer defect-free, i.e., they contain stacking faults (Haond 1985).
Generally, according to the purity of the film, we can find along the SBGs precipitates, either coherent or incoherent and identified as hexagonal α or β-SiC (Haond 1983, Biegelsen 1981a, Pinizzotto 1982).
    In the case of thick films (10-80 μm) obtained by the LEGO process (Celler 1984b), large areas free of defects have been obtained. However, the material does contain dislocations and defects in other configurations at a density of 1.E8/cm3 (Pfeiffer 1984).

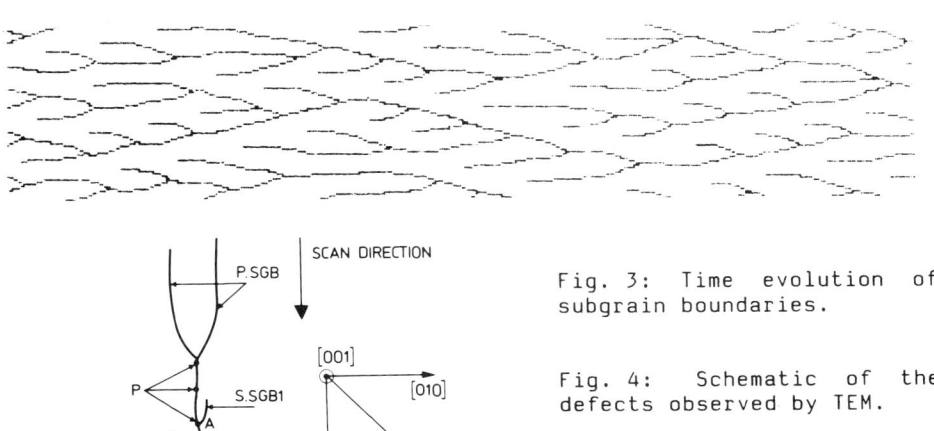

Fig. 3: Time evolution of subgrain boundaries.

Fig. 4: Schematic of the defects observed by TEM.

## 5. Defect Localization

It has become obvious that the SOI emerging technologies will have to live with the defects described above. Localizing them would seem to be the easiest way to get around the problem.

- In the **laser** case, several techniques using differences in thermal conductivity (Biegelsen 1981b, Fennell, Possin, Kawamura, Mukai) or reflectivity (Colinge 1982, 1983a) have been tried. The main idea of these techniques is to initiate solidification in the center of the SOI region and end it in specific tailored regions where the defects will be located. Satisfactory results have been obtained with antireflecting coatings deposited selectively onto the poly-Si film. With a seed and a raster scan of the spot, we find the <100> direction in the SOI region. Mass transport due to the melting and freezing at each scan of the structure by the spot sometimes occurs. However, after roughly 200 µm, the orientation turns, as shown in Fig. 5a. This rotation is accomodated by the generation of isolated dislocations, merely due to the alternative melting and freezing (Fig. 5b).

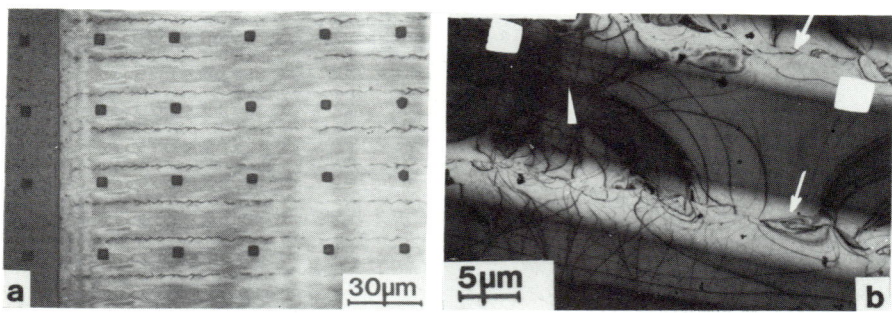

Fig. 5:  (a) Etch pit grid and decoration on laser recrystallized sample with seeding after removal of the cap layers, (b) TEM:Note the localization of defects and correlation between the Kikuchi lines and etch pit.

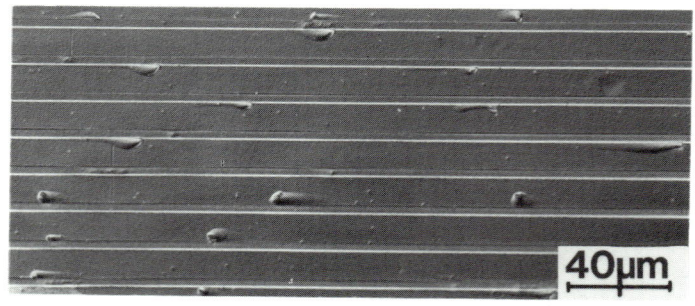

Fig. 6:  Droplets of Si in a periodically seeded sample

- In the **strip heater** or **lamp** case, the reflective properties have less effect than in the laser case. With stripes of carbon resist, some success has been obtained in strip heater recrystallization (Geis 1983). Nearly the same result has been obtained with nitride stripes in the lamp case (Bensahel).

As mentioned above, one remaining problem is the in-plane <100> misorientation. Two techniques have been tested to try and force the growth of this kind of grains only.
As a first technique, we find the work of the MIT group on orientation by means of "hourglass filtering" or growth-velocity competition (Atwater 1983, Smith). Recently, this group has obtained SGBs exhibiting very low angular discontinuities, consisting of isolated in-line dislocations, by reducting the thermal fluctuations in the SOI structure (Geis 1984).
In the second technique, periodic seeding has been used (Fan 1981, Davis). Although it seems very attractive, it is lacking in two main points: first, defects such as SGBs and droplets are left behind within the seeded SOI regions (Fig. 6), also shown in Fan et al. 1981. These droplets are attributed to some constitutionally supercooled areas which lead to trapped liquid surrounded by solidified Si resulting in localized bumping. Second, it is difficult to adjust the power both in the seed and outside it, i.e., in the SOI region (Fig. 7). Insufficient transferred power results in poor epitaxy in the seeding and generates "rungs" which appear as dislocations originating from the seed region and extending into the SOI region (Fig. 8). By contrast, Haond et al. 1984 have found that too much power transferred leads to deep melting in the substrate, and, under certain conditions, even to a regular **sinking** of the oxide stripes to a depth between 1 to 100 µm (Fig. 9).

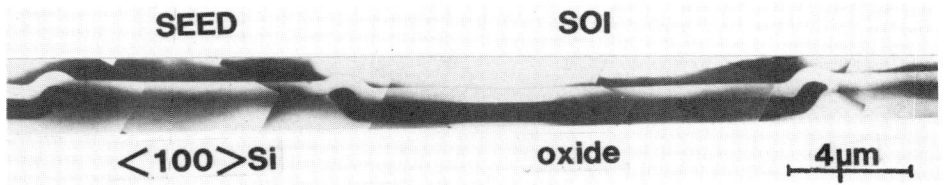

Fig. 7: TEM cross-section of a lamp recrystallized sample periodically seeded.

## 6. Conclusion

A good deal of work has been devoted to obtaining SOI structures by recrystallization of poly-Si films. The greatest efforts have been directed towards the machines necessary to bring recrystallization about. Now, 4-in wafers can be recrystallized with either lasers, strip heaters, or lamp systems. The defects encountered in the films have been investigated.
Since it seems evident that we have to live with them, the main focus is now directed towards their localization. This will define a new kind of circuit design which will minimize the place "lost" due to defects.
Since the grains themselves are free of extended defects, test

circuits made on the films show good performances (Tsaur 1983, Vu 1984). However, we must note that only little work has been directed towards point defects, the study of which will improve circuit performances. For example, studies on the residual N-doping level in slow recrystallization (Ronzani), the existence of surface states, or the origin of leakage currents are still in their infancy. However, we do know that only grain boundaries have a catastrophic effect on circuits (Ng, Colinge 1983b). Up to now, sub-grain boundaries do not seem to affect electrical properties (Tsaur 1982). According to the ultimate use of these films, more or less care will be necessary in the localization of the remaining defects.

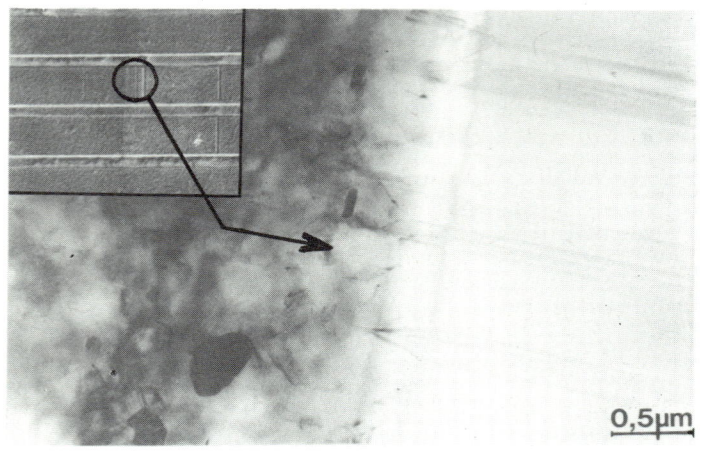

Fig. 8: Optical and electronic micrographs of a periodic seeded sample poorly recrystallized

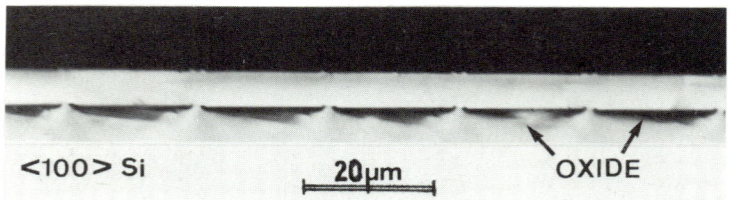

Fig. 9: Sinking of oxide stripes observed in a lamp recrystallized sample periodically seeded.

R. Carre is acknowledged for technical help, as is M. Dupuy for the TEM work. Ms. J. Dargent is also aknowledged for the critical reading of the manuscript.

## References

Atwater H A, Thompson C V, Smith H I and Geis M W 1983 Appl. Phys. Lett.
    43 1126
Atwater H A, Smith H I and Thompson C V 1984 Matt. Lett. 2 269
Bensahel D, Haond M, Vu D-P and Colinge J P 1983 Electron. Lett. 19 464
Bezjian K A, Smith H I, Carter J M and Geis M W 1982 J. Electrochem. Soc.
    129-8 1848
Biegelsen D K, Johnson N M, Bartelink D J and Moyer M D 1981a in Laser
    and Electron Beam-Solid Interactions and Material Processing eds.
    Gibbons J F, Hess L D and Sigmon T W (New York: North-Holland) 487
Biegelsen D K, Johnson N M, Bartelink D J and Moyer M D 1981b Appl. Phys.
    Lett. 38 150
Biegelsen D K, Fennell L E and Zesch J C 1984 Appl. Phys. Lett. 45 546
Cahn R W 1970 in Physical Metallurgy (New York: North-Holland) Ch. 19
Celler G K, Jackson K A, Trimble L E, Robinson McD and Lischner D J 1984a
    in Energy Beam-Solid Interactions and Transient Thermal Processing eds.
    Fan J C C and Johnson N M (New York: North Holland) 409
Celler G K and Trimble L E 1984b in Energy Beam-Solid Interactions and
    Transient Thermal Processing eds. Fan J C C and Johnson N M (New York:
    North-Holland) 567
Chen C K, Geis M W, Choi H K, Tsaur B-Y and Fan J C C Nov. 84, Boston Mat.
    Res. Soc. Meeting paper A7.6
Cline H E 1984 J. Appl. Phys. 55 2910
Colinge J P, Demoulin E, Bensahel D and Auvert G 1982 Appl. Phys. Lett.
    41 346
Colinge J P, Bensahel D, Alamome M, Haond M and Pfister J C 1983a Electron.
    Lett. 19 985 Jap. J. Appl. Phys. Suppl. 22-1 205
Colinge J P, Morel H and Chante J P 1983b IEEE Trans. Electron. Dev. 30 197
Colinge J P Nov. 84 Mat. Res. Soc. Meeting, Boston paper A7.11
Davis J R, McMahon R A and Ahmed H 1983 in Laser Solid Interactions and
    Transient Thermal Processing of Materials eds. Narayan J, Brown W L and
    Lemons R A (New York: North-Holland) 563
Dutartre D 1985 to be published
Fan J C C, Geis M W and Tsaur B-Y 1981 Appl. Phys. Lett. 38 365
Fan J C C, Tsaur B-Y, Chen C K, Dick J R and Kazmerski L L 1984 in Energy
    Beam Solid Interactions and Transient Thermal Processing eds. Fan J C C
    and Johnson N M (New York: North-Holland) 477
Gat A, Gerzberg L, Gibbons J F, Magee T J, Peng P and Hong J D 1978 Appl.
    Phys. Lett. 33 775
Geis M W, Smith H I, Tsaur B-Y, Fan J C C, Silversmith D J and Mountain R W
    1982a J. Electrochem. Soc. 129 2812
Geis M W, Smith H I, Tsaur B-Y, Fan J C C, Maby E W and Antoniadis D A
    1982b Appl. Phys. Lett. 40 158
Geis M W, Smith H I, Silversmith D J, Mountain R W and Thompson C V 1983
    J. Electrochem. Soc. 130 1178
Geis M W, Chen C K, Smith H I, Mountain R W and Doherty C L Nov. 84 Boston,
    Mat. Res. Soc. Meeting paper A7.2
Haond M, Vu D-P, Bensahel D and Dupuy M 1983 J. Appl. Phys. 54 3892
Haond M, Bensahel D and Dutartre D 1984 Electron. Lett. 20 991
Haond M 1985 unpublished results
Hemment P L F Jan. 85 SPIE-85 California, paper review
Hodé J M 1984 Thesis Univ. Grenoble
Imai K, Unno H and Takaola H 1983 J. Cryst. Growth 63 547
Jastrzebski L and Kokkas A G in Energy Beam-Solid Interactions and Transient
    Thermal Processing eds. Fan J C C and Johnson N M 1984 (New York:
    North-Holland) 417

Kamins T I 1981 J. Electrochem. Soc. 128 1825
Kawamura S, Sasaki N, Nakano M and Tagaki M 1984 J. Appl. Phys. 55 1607
Mukai R, Sasaki N, Iwai T, Kawamura S and Nakano M 1983 Proc. IEDM 360
Ng K K, Celler G K, Povilonis E I, Frye R C, Leamy H J and Sze S M 1981
    IEEE Electron. Dev. Lett. 2 361
Paelinck P and Werkman J 1984 Thesis Cath. Univ. Louvain
Pfeiffer L, Kovacs T and West K W Nov. 84 Mat. Res. Soc. Boston, paper A7.3
Pfeiffer L, Paine S, Gilmer G H, Van Saarloos W and West K W 1985 to be
    published
Pinizzotto R F, Lam H W, Malhi S D S and Vaandrager B L 1982 Appl. Phys.
    Lett. 40 388
Pinizzotto R F 1983 J. Cryst. Growth 63 559
Possin G E, Parks H G, Chiang S W and Liu Y S 1982 Proc. IEDM 424
Ronzani A, Vu D P, Haond M and Chantre A May 1985 European Mat. Res. Soc.
    Meeting, Strasbourg
Schimmel D G 1979 J. Electrochem. Soc. 126 479
Secco D'Aragona F 1972 J. Electrochem. Soc. 119 948
Smith H I, Geis M W, Thompson C V and Atwater H A 1983 J. Cryst. Growth
    63 527
Stultz T J and Gibbons J F 1981 Appl. Phys. Lett. 39 498
Stultz T J 1983 Thesis Stanford CA
Tiller W A 1981 J. Electrochem. Soc. 128 689
Tsaur B-Y, Fan J C C, Chapman R L, Geis M W, Silversmith D J and Mountain R W
    1982 IEEE Electron. Dev. Lett. EDL-3 398
Tsaur B-Y, Fan J C C and Geis M W 1983 Appl. Phys. Lett. 41 83
Vu D P, Haond M, Bensahel D and Dupuy M 1983 Appl. Phys. Lett. 54 437
Vu D P, Leguet C, Haond M, Bensahel D and Colinge J P 1984 Electron. Lett.
    20 298
Weinberg Z A, Deline V R, Sedgwick T O, Cohen S A, Aliotta C F, Clark G J
    and Lanford W A 1983 Appl. Phys. Lett. 43 1105

*Inst. Phys. Conf. Ser. No. 76: Section 3*
*Paper presented at Microsc. Semicond. Mater. Conf., Oxford, 25–27 March 1985*

83

# A TEM study of silicon-on-insulator films produced by the dual electron beam recrystallisation of polycrystalline silicon

M Hockly and J R Davis

British Telecom Research Laboratories, Martlesham Heath, Ipswich, IP5 7RE, UK.

Abstract    Cross-sectional TEM studies have been carried out which show that low defect density single-crystal silicon-on-oxide films have been produced using the dual electron beam recrystallisation technique, which employs the seeded lateral regrowth of deposited polysilicon layers.   Control of the movement of the melting and solidification fronts is shown to be crucial to the success of the technique.   A localised characteristic defect has been identified  and its mechanism of formation determined.

## 1. Introduction

Thin single-crystal silicon-on-insulator (SOI) films with low defect densities are desirable for a number of device applications, including VLSI.   One of the methods for producing such films involves the melting and seeded lateral regrowth of polycrystalline silicon deposited on oxide (on a silicon substrate).   The dual electron beam recrystallisation technique has been shown to be capable of producing, by this method, SOI films with excellent structural and electrical characteristics (Davis et al 1983).   In particular, carrier mobilities within a few per cent of bulk silicon values have been measured (Hopper et al 1984).   The present paper reports on a transmission electron microscope (TEM) investigation of such films, using specimens prepared in cross section.

## 2. Experimental

Typical configurations used for the formation of SOI films by the present technique are shown diagrammatically in plan-view in Figure 1 and in

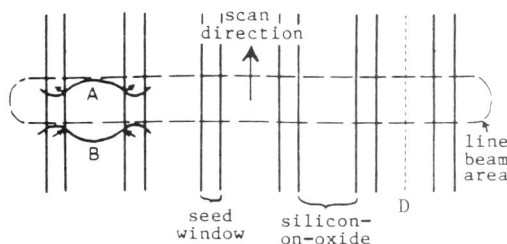

Fig. 1.   Schematic top view of part of specimen during recrystallisation. A – melt front (has component of movement from SOI region to seed window).   B – solidification front (has component of movement from seed window to SOI region).   D – characteristic defect.

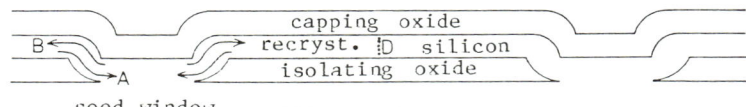

seed window              silicon substrate

Fig. 2.  Schematic cross-sectional view of structure.  A - lateral
         movement of melt front.  B - lateral movement of solidification
         front.  D - characteristic defect (low-angle tilt sub-boundary).

cross-section in Figure 2.  The layers are usually formed as follows:
thermal growth for the isolating oxide layer, LPCVD for the polysilicon
layer, and CVD for the capping oxide layer.  The oxide and silicon layers
are typically 0.5 - 1.0 microns thick, the seed windows are 1 - 4 microns
wide, and the oxide regions can be up to 70 microns wide.  Figure 2 shows
a raised oxide geometry, but geometries involving recessed isolating
oxides and coplanar silicon films are also possible and have been
investigated.  The silicon substrates had (001) surfaces and the seed
windows in this study were aligned in the <110> directions.

Specimens were thinned in {110} cross section for TEM by a combination of
mechanical polishing and ion-beam thinning, and were examined at 200 kV in
a JEOL 200CX TEMSCAN electron microscope with a LINK Systems Energy
Dispersive X-ray (EDX) attachment.

## 3. The Dual Electron Beam Technique

Details of the dual electron beam recrystallisation technique have been
published elsewhere (Davis et al 1983).  One electron beam, rapidly
scanned over the entire wafer area, uniformly heats the wafer from the
back side to about 1000 degrees C.  The second beam, incident on the front
surface, is rapidly scanned over a restricted area to form a so-called
line beam which heats a long thin zone (about 0.1 mm wide by several mm
long).  This heated zone is then swept at approximately 30 cm/s along the
direction parallel to the long thin seed windows in the isolating oxide
(see Fig. 1).  The beam parameters are chosen so that the polysilicon in
the heated zone melts, leaving the substrate silicon and the oxides solid.

Because silicon dioxide has a smaller thermal conductivity than silicon,
that part of the polysilicon layer which is on the oxide is the first to
melt under the influence of the electron beam (Davis 1983).  The melt
front then moves down into the seed-window area and, ideally, melts just
into the substrate silicon.  Upon cooling, the solidification front
follows the reverse path.  Solidification fronts from adjacent seed
windows meet at the centre of each silicon-on-oxide (SOI) region.  As the
zone heated by the line beam is swept across the surface, the melt and
solidification fronts assume forms of the type indicated schematically in
Figure 2.

## 4. Results

The TEM examinations of the seed window areas showed that, by using
appropriate recrystallisation conditions, excellent epitaxial regrowth was
achieved, with no interface being visible between the original substrate
material and the silicon layer.  Although there are defects present at the
seed-window edges (Figure 3), these do not extend into the oxide-isolated
region.  The otherwise excellent  initiation of regrowth has thus led to a

very low density of defects (chiefly
dislocations and twins) in the SOI
region, as shown in Figure 4. (The
centrally positioned defect in Figure
4 is discussed below.)

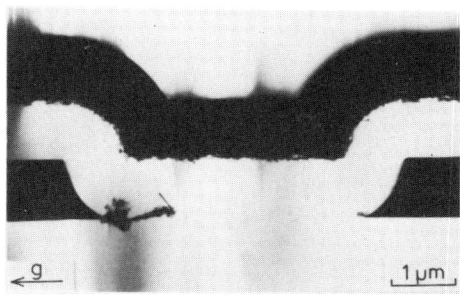

As can be seen in Figs. 3, 4 and 5,
all the interfaces involving the
isolating oxide appear to be smooth
and sharp. This includes the
important so-called "back" interface
between the isolating oxide and the
silicon layer. The etched geometries
of the seed windows are also seen to
be well preserved. The "front"
interface (between the silicon layer

Fig. 3. Cross-sectional TEM micrograph of typical seed-
window region. g = 220.

and the capping oxide) was found to
be of comparable quality, provided that all procedures were carried out in
suitably clean environments. This was not the case for certain of the
polysilicon layers, because of carbon contamination in the system used for
layer deposition. Profiles produced by secondary ion mass spectrometry
(SIMS) revealed the presence of high levels of carbon in such layers, and
also showed that, after dual electron beam recrystallisation, a build-up
of excess carbon was present at the front interface. This correlated with
the observation by TEM of irregularity at this interface, caused by the
presence of crystallites and twins. The poor front interfaces in Figs. 3,
4 and 5, which are examples of this effect, are therefore not
characteristic of the e-beam recrystallisation process, but result from
the excessive levels of carbon. The sweeping of carbon to the front
interface during the recrystallisation indicates that the resolidification
front has a vertical component of motion in the direction away from the
substrate.

EDX analysis was performed on a cross-sectional TEM specimen prepared from
a sample with recessed oxide geometry, in which the polysilicon layer was
implanted with a shallow layer of arsenic ($1 \times 10^{16}$ $cm^{-2}$ at 40 kV)
prior to deposition of the capping oxide. The analysis showed that, after
recrystallisation, the arsenic in the seed-window regions was
redistributed to a greater depth than the bottom of the original
polysilicon layer. In view of the timescales involved, this
redistribution is only possible by liquid phase diffusion, which confirms
that the melt front penetrated into the substrate in the seed window.
Other experiments (Davis 1983) have shown that the substrate below the
isolating oxide remains unmelted.

Earlier etching studies (Davis 1983) had revealed the presence of a
characteristic defect lying along the centre line of each SOI region

Fig. 4.   Cross-sectional TEM montage showing a silicon-on-oxide (SOI)
region with adjacent seed windows. g = 220.

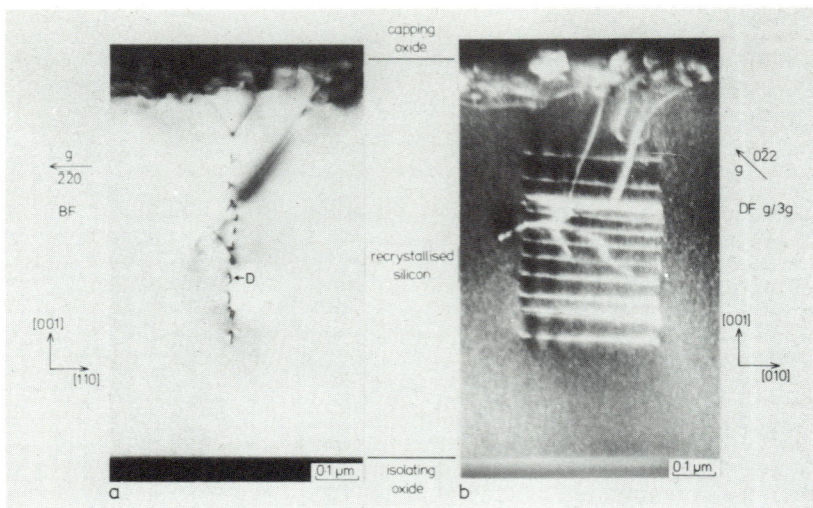

Fig. 5.   Cross-sectional TEM micrographs of low-angle tilt sub-boundary.
(a) At (110) pole.   Dislocations (e.g. D) seen end-on.
(b) After tilting relative to (a) about the [001] direction.

(see Figs. 1, 2,and 4).   Cross-sectional TEM showed the defect to be an
array of edge dislocations, lying one above another, which constitute a
low-angle tilt boundary (sub-boundary).   In Fig. 5(a) the dislocations are
seen almost end-on (110 pole).   They are shown more clearly in Fig. 5(b)
where the specimen has been tilted by approximately 45 degrees about the
[001] direction.

The relative tilts across these sub-boundaries were measured from
diffraction patterns taken from selected areas straddling the
sub-boundaries, using, in some cases, the doubling of Kikuchi lines
visible after tilting along the 004 band.   The measured values of 0.3 −
0.6 degrees were in good agreement with values calculated from the
observed linear dislocation densities in the sub-boundaries.

The existence of the tilt sub-boundaries implies that small
crystallographic misorientations between the silicon layer and the silicon
substrate must arise during the recrystallisation process.   The
crystallographic orientation of the recrystallised layer was measured as a
function of position by monitoring Kikuchi line positions in selected-area
diffraction patterns taken after tilting the specimen along the 004 band
to a suitable orientation near the 010 pole.   The results of this
investigation are depicted schematically in Fig. 6, with the angles

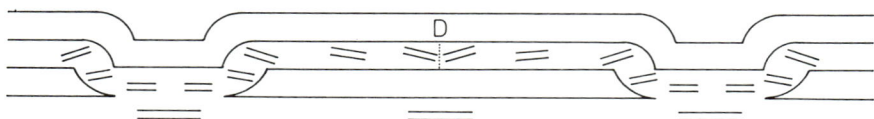

Fig. 6.   Schematic cross-sectional diagram showing misorientations of the
(001) planes in the recrystallised silicon layer (exaggerated for
clarity).   D - position of low-angle tilt sub-boundary.

greatly exaggerated for the sake of clarity (actual values are generally less than 0.5 degrees). As the figure shows, misorientations were observed to arise in the sharply-angled step-ups onto the oxide at the edges of the seed windows. In the material on the oxide the misorientation tends to decrease slightly in the region close to the seed window and then to increase again in the region very close to the central sub-boundary.

Earlier investigations using a bevelling technique (Davis 1983) had found that some net mass transport of material in the layer was occurring, from the SOI regions to the seed windows. This finding was confirmed by the present cross-sectional TEM studies, in which the variations in recrystallised layer thickness could be measured directly. The results are depicted schematically in Fig. 7, where the vertical scale has been greatly exaggerated for the sake of clarity. The measured thickness variations in the important SOI regions were small (less than 10%). A similar difference in thickness was measured between the seed windows and the edges of the SOI regions, but this is not of relevance to device formation.

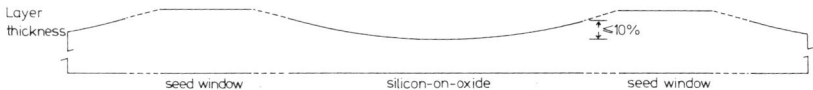

Fig. 7.   Schematic cross-sectional diagram showing the variations in the thickness of the recrystallised silicon layer as a function of position (exaggerated for clarity).

## 5. Discussion

It is clear from the present work that excellent epitaxial regrowth of the polysilicon layer is being obtained, and that partial melting into the substrate is the mechanism whereby this is achieved. The extent of melt penetration into the substrate is apparently small in depth (less than 0.5 microns) and confined to the seed-window regions. The specimen geometry is consequently well-preserved.

For the purposes of device formation, the lowest possible defect densities in the SOI regions are sought. In this respect the defects often observed at the edges of seed windows, which are believed to be caused by contamination from the wet etching process, are not significant as they do not extend into the SOI material. The tilt sub-boundaries, although they are in the SOI material, can be tolerated because their position is so specific that they can be avoided in circuit design.

A model has been put forward describing the movement of the melt and solidification fronts in the layer during the recrystallisation (Davis 1983). This model is supported by results from various other techniques and by calculations of heat flows. The present work confirms and adds detail to this model. In Section 3 above, a description is given of the movement of the melt and solidification fronts in terms of components of motion, both longitudinal (along the direction of sweep of the line beam) and lateral (between the SOI and window regions). Vertical components of motion also exist; the observations on carbon contaminated layers show that the solidification front has a vertical component of motion, away from the substrate.

The model explains the formation of the low-angle tilt sub-boundaries in the following way. As the solidification front moves through the sharply-angled step-up from a seed window onto the isolating oxide, severe stresses develop because of the significant volume change associated with the change of phase of silicon (9%), and the restriction to flow of the molten material imposed by the geometry. These stresses give rise to the observed development of small misorientations in those regions. These misorientations are opposite in sense for the material behind the solidification fronts from neighbouring seed windows, so that where these fronts meet (at the centre of the intervening SOI region) a small relative misorientation exists (generally less than 0.5 degrees), which is accommodated by the formation of a low-angle tilt sub-boundary.

At the position of a sub-boundary, the scope for accommodation of the volume change associated with the change of phase, by gradual distortion of the capping oxide, becomes exhausted. This causes stresses which give rise to the increased misorientations and additional defects which are observed close to the sub-boundaries.

The net transport of layer material from the SOI regions to the seed windows is believed to occur because the capping oxide in the SOI regions, which is flat, sags more easily to accommodate the volume contraction which occurs at melting than does the capping oxide in the seed window regions, which is ridged and therefore mechanically stiffer.

## 6. Conclusions

The present work has confirmed by cross-sectional TEM that dual electron beam recrystallisation can produce low defect density, single-crystal SOI material by a process of melting and seeded epitaxial regrowth. Additionally it has shown that very good interfaces are generally achieved with etched window geometries being well preserved. In specimens with a raised-oxide configuration, the defects running down the centres of the silicon-on-oxide regions were shown to be low-angle tilt sub-boundaries. The work has provided further evidence for a model of the melting/ solidification process which explains the formation of these sub-boundaries and the net transport of material between the SOI regions and the seed windows.

## 7. Acknowledgements

We would like to thank H Ahmed and R A McMahon (Cambridge University Microcircuits Laboratory) for provision of the e-beam recrystallisation facilities, and G F Hopper (GEC Hirst Research Centre) for the polysilicon deposition. Acknowledgement is made to the Director of Research, British Telecom, for permission to publish this paper.

## References

Davis J R 1984 Ph.D. Thesis, Cambridge University
Davis J R, McMahon R A and Ahmed H 1983 J. Phys. C5 337
McMahon R A, Davis J R and Ahmed H 1982 "Laser and Electron-Beam
    Interactions with Solids" Eds Appleton and Celler (North Holland) 783
Hopper G F, Davis J R, McMahon R A and Ahmed H (1984) Electronics Lett.
    20 500

*Inst. Phys. Conf. Ser. No. 76: Section 3*
*Paper presented at Microsc. Semicond. Mater. Conf., Oxford, 25–27 March 1985*

89

# Low angle grain boundaries in zone-melting grown silicon films

H Baumgart and F Phillipp*

Philips Laboratories, Briarcliff Manor, New York 10510, U.S.A.
*Max Planck Institute for Metals Research, D-7000 Stuttgart 80,
Heisenbergstr. 1, West Germany

Abstract   Laser induced zone-melting has been used to recrystallize polycrystalline Si films deposited on amorphous $SiO_2$ substrates. A comprehensive structural analysis was performed by transmission electron microscopy (TEM) to study the lattice defects in the regrown Si films. In all (100) textured recrystallized Si films a characteristic network of low angle grain boundaries has been found to be the dominant defect. These sub-boundaries originate as slip dislocation half loops at internal stress concentrations, which indicate plastic deformation. In those regions where twins were the major defects, the (100) orientation was not preserved.

## 1. Introduction

The nonconventional thin film crystal growth technique of zone-melting was described as early as 1953 in a patent by Leitz as a method of growing single crystalline films. In recent years this approach has been applied to silicon-on-insulator (SOI) structures (Fan et al. 1981) with the result that today device quality large monocrystalline Si films on amorphous insulating substrates are routinely grown in many laboratories. All films produced by zone-melt recrystallization have been shown to contain characteristic low angle grain boundaries, which constitute the major lattice defect in (100) oriented Si films. In zone-melted and $SiO_2$ capped Si films the (100) texture predominates because the interfacial free energy between Si and $SiO_2$ is minimum for {100} planes. In those cases where laser induced recrystallization produced surface textures other than (100) we have observed extensive twin formation of the {111} <112> mode besides the low angle grain boundaries in the Si films. In this paper we present a detailed analysis of the dislocation arrays by TEM and discuss the dependence of their characteristics on the thin film growth conditions. The origin of the individual low angle grain boundaries has been traced to dense tangles of randomly dispersed dislocations which rearrange themselves into configurations of lower energy as the growth front proceeds. The resulting boundary accommodates those dislocations which are not annihilated by other dislocations or at the free surfaces of the thin Si film.

## 2. Results and Discussion

The zone-melting recrystallization of the LPCVD polycrystalline Si films was performed with $CO_2$ laser or $Ar^+$ ion laser irradiation. The sample was heated to a background temperature of 1100°C by a bottom graphite

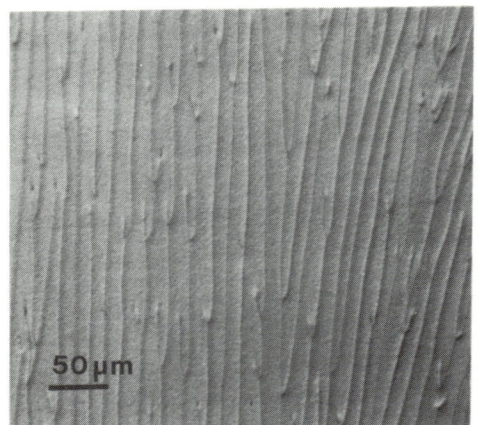

Fig. 1. Optical micrograph of (100) recrystallized Si film containing characteristic low angle grain boundaries

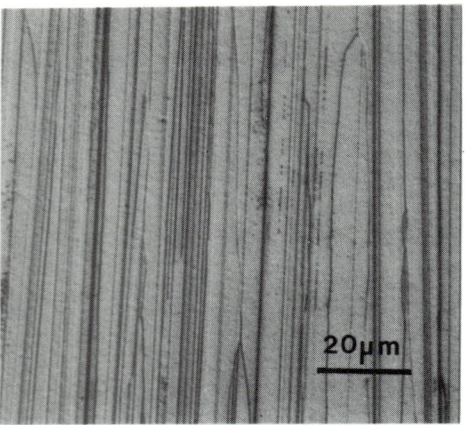

Fig. 2. Surface morphology of (110) zone-melt recrystallized Si showing extensive twinning after Schimmel etching.

heater, while the laser beam was focussed with a cylindrical lens onto the sample to form a several mm wide molten zone in the Si film. We used thermally oxidized Si wafers as substrates to deposit a 0.5 μm film of fine grain polycrystalline Si by low pressure chemical vapor deposition (LPCVD), which was subsequently encapsulated by 2.0 μm of CVD $SiO_2$. In general $CO_2$ laser induced zone-melting at slow scan speeds of 0.5-2.00 mm/sec produces high quality (100) textured monocrystalline Si films with low angle grain boundaries as the only defect as shown in the optical micrograph of Fig. 1.

The (100) texture is predominant in the material solidified from the melt zone. However, the final crystallographic quality of the recrystallized Si film depends also on the speed of the melt zone propagation and the local temperature gradient. Higher scanning speeds (>> 2 mm/sec) usually lead to Si films which are heavily twinned, as demonstrated in Fig. 2, and do not exhibit (100) texture. Similar twinning structures have been observed by Geis et al. (1982) and by Komem and Weinberg (1984) in graphite strip heater processed Si and by Sedgwick et al (1982) in electron beam recrystallized Si films. We have examined the microstructure of regrown Si and Fig. 3 shows a representative TEM micrograph of a twinned region. The surface orientation of the matrix was parallel to the {110} planes. The

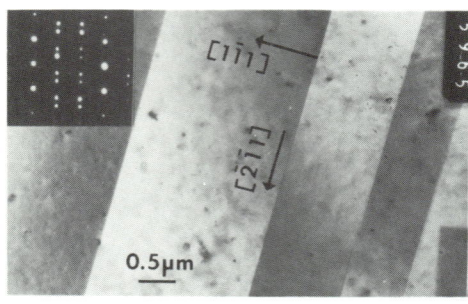

Fig. 3. TEM micrograph of twin structure in (110) oriented recrystallized Si matrix.

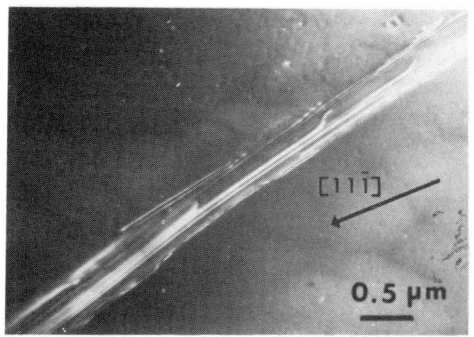

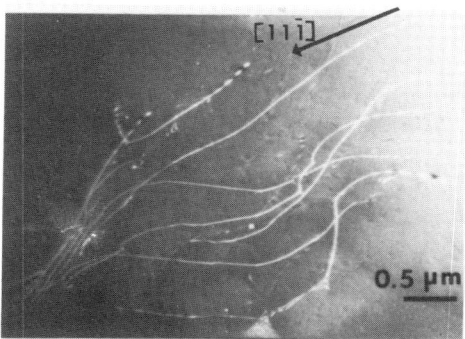

Fig. 4. Dark field TEM image of dislocation array forming a low angle grain boundary.

Fig. 5. TEM micrograph of the source of a low angle grain boundary.

twin planes are $\{1\bar{1}1\}$ planes and the twin edge is parallel to the $[\bar{2}11]$ direction. This is the typical $\{111\}$ <112> twinning mode for f.c.c. crystals and those with diamond structure. We attribute the twinning process to growth twins. Due to the small interfacial free energy of a coherent twin boundary the stacking sequence of crystal planes in the $\gamma_t$ solidifying Si film is relatively easily reversed and a growth twin is produced. Generally higher melt-zone speeds accompanied by local breakdown of the advancing growth front causes extensive twinning by stacking sequence reversal. Once a stable crystal growth mode is established the energetically more favorable (100) surface orientation remains locked-in by continuous self-seeding. In that solidification mode twin formation is suppressed and the regrown Si films exhibit exclusively low angle grain boundaries (Haond et al., 1983 and Komem et al., 1984 ). Figure 1 shows the typical surface morphology of a Si film containing low angle boundaries that were delineated by a chemical defect etch. The subsurface microstructure of these defects is revealed in the TEM micrograph of Fig. 4. In order to resolve the dislocation configuration the specimen had to be tilted by 45°. This darkfield micrograph shows the inner structure of the low angle grain boundary to consist of long parallel arrays of slip dislocations. The misorientation introduced into the Si film by the tilt boundaries varied between 0.8° and 1.2° according to diffraction patterns taken on each side of the various boundaries. We have been able to trace individual low angle grain boundaries to their origin. Figure 5 shows a TEM darkfield image of the source of a low angle boundary. Sub-boundaries in zone-melting grown Si films originate at such random tangles of slip dislocations. The nucleation of these slip dislocations takes place at

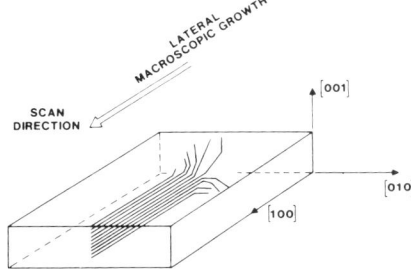

Fig. 6. Schematic diagram of dislocation configuration illustrating the origin and formation of a low angle grain boundary by polygonization.

discrete surface steps which are believed to be uniformly distributed over the surfaces of both the front and back side interface of the recrystallized Si film. It is well established that such surfaces are not planar on an atomic scale but do contain a multitude of surface steps of varying height. These surface steps act as dislocation nucleation sites whenever local stress concentrations exceed a critical value of about G/30, where G is the shear modulus of the material. In such a case local plastic deformation of the solidifying crystal occurs by nucleation of a small dislocation half loop at an activated surface nucleation site. One end of the dislocation half loop gets pinned at the surface step while the other free segment propagates into the crystalline Si film by the process of thermally activated glide. A schematic presentation of this process is depicted in Fig. 6. The pinning points occur at both front and backside interfaces. Under the influence of a local stress field the glide segment of the half loop glides away from the surface and propagates with the trailing edge of the moving solid-liquid interface front through the Si film. During the zone-melting process of the thin film specific growth front instabilities are responsible for the periodic formation of sub-boundaries. Faceting of the solid-liquid growth front (Leamy et al., 1982) causes high local internal stress concentrations. These periodic stress concentrations occur at the inner corners of the growth facets because it is there that impingement of growth steps and impurity incorporation takes place. Since zone-melting crystal growth occurs at high background bias temperatures ($\approx 1100°C$-$1200°C$) the dislocations in the random array rearrange themselves by glide and climb into a more stable vertical dislocation wall as schematically depicted in Fig. 6. (Baumgart et al., 1985). This process of polygonization leads to the characteristic dislocation configuration of the low angle boundary.

## References

Baumgart H and Phillipp F 1985 Energy Beam-Solid Interactions and Transient Thermal Processing, eds D K Biegelsen, G A Rozgonyi and C V Shank (New York: North Holland)
Fan J C C, Geis M W and Tsaur B Y 1981 Appl. Phys. Lett. 38 365
Geis M W, Smith H I, Tsaur B Y, Fan J C C, Silversmith D J and Mountain R W 1982 J. Electrochem. Soc. 129 2812
Haond M, Vu D P and Bensahel D 1983 J. Appl. Phys. 54 3892
Komem Y and Weinberg Z A 1984 J. Appl. Phys. 56 2213
Leamy H J, Chang C C, Baumgart H, Lemons R A and Chen J 1982 Materials Letters 1 33
Leitz E Brit. Pat. 691 335 (1953)
Sedgwick T O, Geiss R H, Depp S W, Hanchett V E, Huth B G and Graf V 1982 J. Electrochem. Soc. 129 2802

*Inst. Phys. Conf. Ser. No. 76: Section 3*
*Paper presented at Microsc. Semicond. Mater. Conf., Oxford, 25–27 March 1985*

# Transmission electron microscopy of preamorphized, shallow implanted and rapid thermally annealed silicon

D K Sadana,* T Sands,** W Maszara*** and G A Rozgonyi***

*Philips Research Laboratories, Signetics Corporation, Sunnyvale, CA 94086; **Lawrence Berkeley Laboratory, University of California, Berkeley, CA 94720; ***Materials Engineering Department, North Carolina State University, Raleigh, NC 27650.

Abstract    Conventional as well as high resolution transmission electron microscopy has been performed on amorphous (a)/crystalline (c) interfaces in pre-amorphized surface layers produced by $Si^+$ or $Ge^+$ implants into Si.  It is found that $Si^+$ implants invariably create a broad transition region at the a/c interface.  Misoriented (3-4°) microcrystallites present within the transition region act as nucleation sites for "hairpin" dislocations in the rapid thermally regrown implanted region.  Uniform amorphization and sharpness of the a/c interface are found to be a prerequisite for defect-free regrowth of the preamorphized layer.

## 1. Introduction

Implantation of dopant into an amorphous surface region circumvents dopant channeling and thereby prevents the formation of a channeling tail in the dopant profile.  The amorphous surface can be obtained either by deposition of amorphous Si by LPCVD, evaporation and sputtering or by implanting the surface region with $Si^+$ or $Ge^+$.    However, annealing of the deposited layers on a single crystal Si substrate [(100) as well as (111)] tends to create stacking faults, microtwins and misoriented cyrstallites in the regrown region (Queisser et al 1962).    Therefore, implantation by $Si^+$ or $Ge^+$ into Si is the preferred method to obtain an amorphous surface region.  (Tsai and Streetman 1979).  Amorphization by Si is typically achieved by three consecutive high dose implantations with decreasing energies (e.g., 300, 150 and 70 keV) to ensure homogeneity of the amorphized layer (Carter et al 1984, Maszara et al 1984).

In this paper, shortcomings of Si amorphization prior to shallow (<0.2 μm) dopant implantation will be illustrated.  The effects of in-situ annealing and of the presence of a wide transition region at the amorphous (a)/crystalline (c) interface on amorphization and recrystallization processes will be discussed.  The superiority of $Ge^+$ implantation over $Si^+$ implantation in achieving uniform amorphization and regrown region of high structural perfection will be demonstrated.

## 2. Experimental

Surface amorphization of (100) Si wafers was accomplished by either single (300 keV), double (300+150 keV or 150+70 keV) or triple energy $Si^+$ implantation (350+150+70 keV) in the dose range $2 \times 10^{15}$ – $10^{16}$ $cm^{-2}$

**Now at Bell Communication Research Inc., Murray Hill, NJ 07974.

at nominal liquid nitrogen (LN) or room temperature (RT). Following pre-amorphization, the shallow boron distribution was obtained by implanta-tion of 42 keV $BF^+_2$ to a dose of $2 \times 10^{15}$ $cm^{-2}$, again at either RT or LNT. Solid-phase epitaxial regrowth of the amorphous layers was induced by RTA with incoherent light at temperatures of 950-1150° for 10 seconds. The distributions of boron and fluorine before and after RTA were obtained by SIMS (results not included here). Plan-view TEM and XTEM, both in conventional as well as high resolution modes, allowed the determination of defect distributions.

## 3. Results

The analysis by XTEM revealed that only under certain $Si^+$ implantation conditions does a completely amorphized surface layer result. However, a single 300 keV $Si^+$ RT implant ($1 \times 10^{16}$ $cm^{-2}$) was found to produce a buried layer of small dislocation loops (40 nm) and defect clusters beginning at 280 nm below the surface. No evidence of any amorphous material in the implanted region was present. Implantations with 150 keV and 70 keV $Si^+$ ions at RT ($5 \times 10^{15}$ $cm^{-2}$) produced a 230 nm buried amorphous layer beginning at 45 nm below the surface and a 150 nm wide continuous amorphous layer (extending to the surface), respectively. If the standard triple implantation scheme with the typical sequence of 300 keV, 150 keV and 70 keV is employed, the resulting structure is shown in Fig. 2b. Having determined the damage structures present for each energy/dose combination, the damage distribution observed in Fig. 2b is not surprising. When this experiment was initially carried out with the triple RT implants, these defect structures were difficult to interpret. The damage structures after a double implant (300+150 keV and 150+70) Fig. 2a and 2c. The results of Fig. 2 are consistent with those in Fig 1. The presence of small buried crystallites (c) was also observed near the surface (Fig. 2a).

Triple LNT implants created a uniform amorphous surface region of ≈700 nm width. Subsequent 42 keV $BF_2^+$ or $B^+$ implantations to a dose of $2 \times 10^{15}$ $cm^{-2}$ at RT or LNT were carried out into these wafers. The amorphous layer (without any dopants) regrew perfectly by solid phase epitaxy on the underlying Si substrate leaving no secondary defects. However, a layer of dislocation loops (type I damage) remained at a depth (640 nm) which corresponded to the original a/c interface. (Maszara et al 1984, Carter et al 1984).

The subsequent $BF_2^+$ implantation at nominal room temperature into the preamorphized Si substrate caused sufficient beam heating to increase the width of the transition region at the a/c interface from 100nm to 180nm. These changes at the interface were found responsible for the nucleation of "hairpin" dislocations on RTA. The $BF_2^+$ related damage near the surface showed the presence of fine F clusters (1.5-4 nm), $B_2O_3$ microcrystallites (5 nm) and stacking faults bounded by Schockley partial dislocations after RTA. (Carter et al 1984, Sands et al 1984).

When triple RT implanted samples or LN samples (Fig. 3a) that had bad thermal contact during implantation were subsequently rapid thermally annealed (without any dopants), hairpin dislocations were again seen to nucleate at the a/c interface (Fig. 3b). It was also possible to create hairpins on RTA even in the samples with uniform amorphous layers if these samples were preheated at 200-500°C (thus simulating the effect of beam heating) (Maszara et al 1985).

In order to achieve better control over the a/c interface and thus suppress the nucleation of hairpins, $Ge^+$ implantation (300 keV, $10^{15}$ or
$10^{16}$ $cm^{-2}$) into Si has been performed (Sadana et al 1984, Seidel et
al 1984). A continuous amorphous layer (340 nm for $10^{15}$ $cm^{-2}$ or
420 nm for $10^{16}$ $cm^{-2}$) has been shown to consistently form by $Ge^+$
implantation irrespective of the implantation temperature (LNT or RT).
Futhermore, the nature of the a/c interface was found to remain
unperturbed by the subsequent $BF_2^+$ implantation under similar
conditions to those which caused observable changes at the a/c interface
in the $Si^+$ amorphized case (Fig. 4a). Upon RTA at 1100°C/10 sec., no
hairpin dislocations were found to be present in the observed area
(indicating that the density of hairpins is $\leq 10^7$ $cm^{-2}$); only a layer
of dislocation loops corresponding to the original a/c interface was
present. Furthermore, surface damage was also less severe than that of
the Si preamorphized case in that only small F clusters ($\approx 20$ Å) with no
specific crystallographic structures were detected by high resolution TEM
in the $BF_2^+$ implanted region. The most important finding was the
fact that RTA at either 1150°C/10 sec or 1100°C/30 sec eliminated
even the layer of dislocation loops at the a/c interface which is always
present in ion implanted and annealed semiconductors (including compound
semicondutor) (Sands et al 1985). These differences can be understood by
closely examining the amorphous crystalline transition regions in the
$Si^+$ and $Ge^+$ implanted Si (Fig. 4a and b). Silicon implanted at RT
with $Ge^+$ (300 keV, $10^{16}$ $cm^{-2}$) contains an undulating but fairly
abrupt interface (peak-to-valley depth difference $\approx$ 7-10 nm) (Fig. 4b)
whereas silicon self-implanted at LNT can contain an a/c transition
region on the order of 100 nm wide (Fig. 4a).

## 4  Discussion

### 4.1  Dynamic Annealing During RT Implants

The size of the amorphous (or damage) zone created at the end of an ion
track depends on the mass of the ion for a given host matrix and implant
temperature. Small a-zones in crystalline matrix can be efficiently
annihilated or reduced in dimensions if point defects (interstitials and
vacancies) in the surrounding region are mobile (Washburn et al 1983).
Such a situation probably exists at room temperature or above. In the
case of high energy (300 keV) RT $Si^+$ implantation, the higher power in
the beam results in substrate ·heating. Therefore, the a zones that may
have formed directly at the end of the $Si^+$ tracks shrink faster than
new ones that are created. Coalescence of the mobile excess interstitials creates the observed small dislocation loops in the implanted
region (Fig. 1a). At lower energies (150 or 70 keV), the substrate temperature probably does not rise high enough to completely annihilate the
a zones. Therefore, at sufficiently high doses ($5\times10^{15}$ in the present
case), a continuous or a buried "a" layer can be created.

### 4.2  Interfacial Dislocation Loops and Conditions for Their Annihilation

Details of the origins of the different types of defects in preamorphized
and RTA samples have been discussed in our earlier publications (Maszara
et al 1982, Carter et al 1984, Sands et al 1985). It is obvious that
excess self-interstitials are created below $R_p$ during ion implantation. In the case where an amorphous layer is formed, the excess interstitials immediately below the $R_p$ typically reside within the amorphous
material. This is because the typical widths of the observed amorphous

layers are found to be $R_p + n \Delta R_p$ where n could be an integer or a fractional number. On subsequent annealing, the excess interstitials within the amorphous material add extra atomic layers during epitaxial regrowth. However, excess interstitials below the a/c interface will either diffuse to nearby sinks (e.g., an a/c interface) or they will cluster to form the nuclei of interstitial loops. Consequently, the density of interstitial dislocation loops (either Frank-type, $b=a/_3\langle 111 \rangle$, or perfect $b=a/2\langle 110 \rangle$) will be determined by the competition between diffusion to interstitial sinks and the nucleation rate of interstitial loops.

For the rough a/c interfaces such as those in the $Si^+$-implanted samples, the efficiency of the amorphous-crystalline interface as an interstitial sink is increased because of the greatly increased interface area. While the crystalline material grows from all directions in the initial stages of annealing, the excess silicon atoms are rejected to the amorphous material. As the amorphous zones between growing islands shrink and disappear, the excess silicon atoms coalesce to form interstitial dislocation loops. The result is a spatial distribution of interstitial loops which reflects the width of the original amorphous-crystalline transition region. Since the width of the transition region is much narrower in the case of $Ge^+$ than $Si^+$, the spatial distribution of the interstitial loops is also more well defined than in the $Si^+$ case. Further annealing coarsens the interstitial loop distribution. Along with the possible annihilation of interstitial loops by vacancy indiffusion, the lack of dislocation loops in certain $Ge^+$ implanted and RTA samples can be qualitatively explained by nucleation theory considering the temperature/time conditions for which the nucleation of loops may be inhibited during annealing (Sands et al 1985).

### 4.3 Hairpin Dislocations

Figure 5a shows a cross-section of the a/c transition region in a $Si^+$-preamorphized sample subsequently implanted at (nominal) room temperature with $BF_2^+$. The a/c transition region is very broad ($\approx 120$ nm). The high resolution image (Fig. 5b) reveals the presence of misoriented ($\approx 3-4^o$) crystallites surrounded by amorphous material near the upper portion of the transition region at a depth of 700 nm. The misorientation is thought to be a result of stresses in the transition region which may be caused by the difference in density between amorphous and crystalline silicon.

Hairpin dislocations are created when, during RTA, the advancing and recrystallizing a/c growth front intersects a small misoriented crystallite in the upper portion of the a/c transition region. A perfect dislocation segment is formed at the intersection to accommodate the misorientation. As the combined growth front continues to move, the dislocation segment wraps around the misoriented material to form a half loop. The growth front emanating from the misoriented material remains slightly ahead of the surrounding primary growth front. Verification of this growth front morphology has been provided by the XTEM image of the partially regrown sample ($750^oC$, 10 sec) (Sands et al 1985).

It is clear from this model that a broad transition region containing a high density of misoriented crystallites results in the nucleation of a high density of hairpin dislocations. Thus, the above model also explains the observation that hairpin dislocations are also formed during

RTA of RT preamorphized Si without the subsequent $BF_2^+$ implant (Fig. 3). The $Ge^+$ preamorphized samples with very narrow a/c transition regions contain a much lower density of hairpin dislocations following RTA which is again consistent with the above model because fewer mis-oriented cyrstallites are expected in this case.

## 5. Conclusions

1. Extreme care should be taken to prevent substrate heating during triple $Si^+$ implant.

2. A number of different types of defects at different depth levels are usually created when $BF_2^+$ implanted $Si^+$ preamorphized wafers are rapid thermally annealed.

3. Defect-free regrowth can be achieved in $Ge^+$ preamorphized Si by RTA.

## 6. Acknowledgments

The authors gratefully recognize the important contributions of Professor Jack Washburn (U. C. Berkeley), Professor J. Wortman (N. C. State University) and Maurice Norcott (Philips Labs, Sunnyvale) to this paper.

## References

Carter C, Maszara W, Sadana D K, Rozgonyi G A, Liu J and Wortman J 1984 Appl. Phys. Lett. 44 459
Maszara W, Carter C, Sadana D K, Liu J, Ozguz V, Wortman J and Rozgonyi G A 1984 Proc. Mat. Res. Soc. (Boston) 23 285
Maszara W, Sadana D K, Rozgonyi G A, Sands T, Washburn J and Wortman J 1985 Proc. Mat. Res. Soc. (Boston) in Press
Queisser H J, Finch R H and Washburn J 1962 J. Appl. Phys. 33 1536
Sadana D K, Maszara W, Wortman J, Rozgonyi G A and Chu W K 1984 J. Electrochem. Soc. 131 943
Sands T, Washburn J, Gronsky R, Maszara W and Sadana D K 1984 Appl. Phys. Lett. 45 982
Sands T, Washburn J, Gronsky R, Maszara W, Sadana D K and Rozgonyi G A 1985 Proc. 13th Int'l. Conf. on Defects in Semi-conductors eds L C Kimerling and J M Parsey Jr AIME, NY 9 531
Sands T, Washburn J, Myers E and Sadana D K 1985 Nucl. Inst. and Meth. B7/8 337
Seidel T, Knoell R, Stevie F A, Poli G and Schwartz B 1984 Proc. Elec-trochem. Soc. 84/1 in Press
Tsai M Y and Streetman B G 1979 J. Appl. Phys. 50 183
Washburn J, Murty C, Sadana D K, Byrne P, Gronsky R, Cheung N and Kilaas R 1983 Nucl. Inst. and Meth. 209/210 345

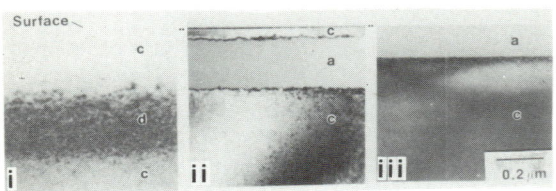

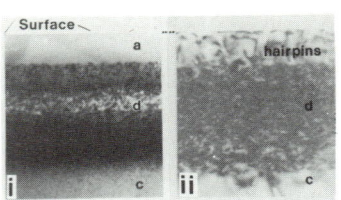

Fig. 1 Damage distributions vs. implant energy for $Si^+$ implants into (100) Si (same beam current 0.25 uA cm$^{-2}$) (i) 300 keV, $10^{16}$ cm$^{-2}$, (ii) 150 keV, $2 \times 10^{15}$ cm$^{-2}$ and (iii) 70 keV, $2 \times 10^{15}$ cm$^{-2}$.

Fig. 3 Damage distributions in (i) triple energy implants of Si same as Fig.2 (ii) but the Si wafer had a bad thermal contact with the mount during the implant and (ii) RTA 1150°C for 10 sec.

Fig.2 Damage distributions in double and triple implants ;(i) (300, $10^{16}$ + 150, $2 \times 10^{15}$ + 70, $2 \times 10^{15}$) keV, cm$^{-2}$ and (iii) (150, $2 \times 10^{15}$ + 70, $2 \times 10^{15}$) keV, cm$^{-2}$.

Note :    In Figs. 1-3,
          a - amorphous
          c - crystal
          d - damage

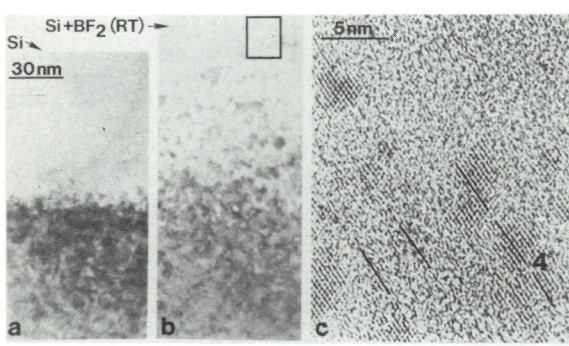

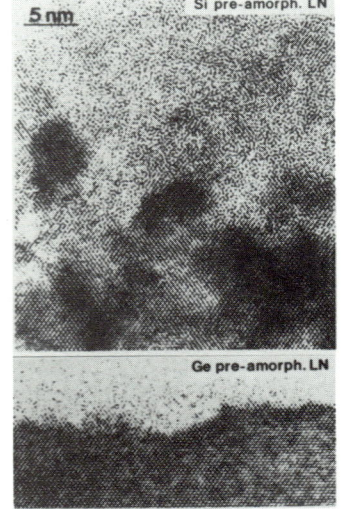

Fig. 5 Transition regions in (a) triple energy LN $Si^+$ implanted Si, (b) subsequently RT $BF_2^+$ (42 keV, $2 \times 10^{15}$ cm$^{-2}$) Si and (c) high resolution image of the indicated area in (b). Microcrystallites imbedded in amorphous surrounding show small misorientation (3-4°).

Fig. 4   HRTEM images of the transition regions in LNT $Si^+$ (triple energy) and $Ge^+$ (300 keV, $10^{16}$ cm$^{-2}$) implanted Si.

*Inst. Phys. Conf. Ser. No. 76: Section 3*
*Paper presented at Microsc. Semicond. Mater. Conf., Oxford, 25–27 March 1985*

# Residual defects following rapid annealing of boron implanted silicon with and without preamorphisation by silicon implantation

D G Hasko, R A McMahon, H Ahmed, W M Stobbs[*] and D J Godfrey[**]

Microelectronics Research Laboratory, Cambridge University, Cambridge Science Park, Milton Road, Cambridge CB4 4BH, U.K.   [*]Department of Metallurgy and Materials Science, Cambridge University, Pembroke Street, Cambridge CB2 2QN, U.K.;   [**]GEC, Hirst Research Centre, East Lane, Wembley, Middlesex, HA9 7PP, U.K.

Abstract    Implants of boron into silicon which has been made amorphous by silicon implantation have a shallower depth profile than the same implants into silicon.   This results in higher activation and restricted diffusion of the B implants after annealing, and there are also significant differences in the microstructure after annealing compared with B implants into silicon.   Rapid isothermal heating with an electron beam and furnace treatments are used to characterize the defect structure as a function of time and temperature.   Defects are seen to influence the diffusion of non-substitutional boron.

## 1. Introduction

The lateral dimensions of semiconductor devices for VLSI circuits continue to be reduced.   In CMOS circuits scaling requires reductions in the junction depths of ion implanted sources and drains, together with higher carrier concentrations, El-mansey (1982).   Boron has the highest solid solubility of p-type dopants, but has the disadvantages of high diffusion rates and an extended range due to channelling.   The alternative of implanting the boron as $BF_2$ gives a reduced range, but leaves unwanted fluorine in the silicon lattice.   Amorphisation of a silicon wafer prior to boron implantation has been seen to give a reduced implant range as the ion channelling is eliminated, Tsai and Streetman (1979).   Furthermore, preamorphisation alters the defect structure remaining in the annealed wafer, Carter et al. (1984), resulting under some conditions, in higher activation and different diffusion behaviour than for similar implants into crystalline silicon, Godfrey et al. (1984).   The nature and density of defects determined by cross-sectional TEM, are described as a function of temperature and time following rapid isothermal heating or furnace treatments, for both preamorphised and non-preamorphised boron implants. Rapid isothermal heating using an electron beam is a well-characterised technique for the annealing of semiconductor wafers, McMahon et al. (1983). In this method, both the annealing temperature and time may be closely and independently controlled by employing feedback from an optical pyrometer directed at the back of the silicon wafer.   The minimum processing time for a given peak temperature is limited by the heating and cooling rates. The annealing schedules used here are described by a peak annealing temperature and a time for which the wafer was held at that temperature.   All samples used for e-beam annealing were capped with 0.5μm of deposited oxide to eliminate dopant loss.   The furnace annealing treatment was 1hr.

at 950°C in a $N_2$ ambient.    The density of defects after annealing was compared to the non-substitutional component of the boron concentration
profile.    This component was determined by comparing the boron atomic concentration profile, derived from SIMS, with the carrier concentration
profile, derived from spreading resistance measurements.    The annealing
of boron, implanted with doses in excess of $10^{15}/cm^2$ at 25keV, at temperatures below 1000°C or for times shorter than a few minutes, results in a
well defined excess peak, Godfrey et al. (1984).    The defect density profile was assessed for small loops by counting the number of them within
successive layers of equal thickness, while the density of more extended
defects was assessed by counting the number of defects crossing lines
drawn at the centres of the layers.    The defect density profiles obtained
in this way may only be used to show the variation of density with depth
in a particular sample.    Defect densities in samples annealed with different treatments may not be compared due to the different nature of the
defects in each case.

## 2. Implants without Preamorphisation

Furnace annealing of a silicon wafer, which has been implanted  with a
high dose of B ions, leaves a characteristic, dense tangle of dislocation
loops.    A cross-sectional TEM micrograph of a furnace annealed specimen
with a $1.10^{16}$ $B/cm^2$ at 25keV implant is shown in Fig. 1(a), and the relative defect density is shown compared to the non-substitutional boron concentration in Fig. 2(a).    Rapid isothermal annealing at 1000°C for 30s
also leaves a dense tangle of loops, see Fig. 1(b), but after annealing at
1100°C for 30s, see Fig. 1(c), the loop density was reduced and the damaged
region extends deeper into the crystal.    This behaviour is qualitatively
matched by changes in the non-substitutional boron concentration, see
Figs. 2(b) and 2(c).

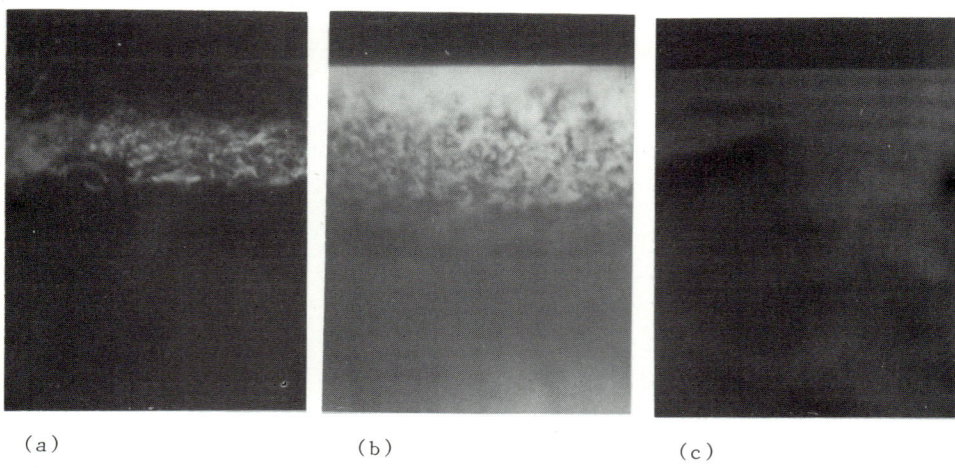

   (a)                          (b)                          (c)

Fig.1    Cross-section TEM of silicon with a 25keV B
implant (a) after furnace annealing at 950°C for
1hr in $N_2$, (b) after rapid annealing at 1000°C
for 30s, and (c) after rapid annealing at 1100°C          010          0.1µm
for 30s.

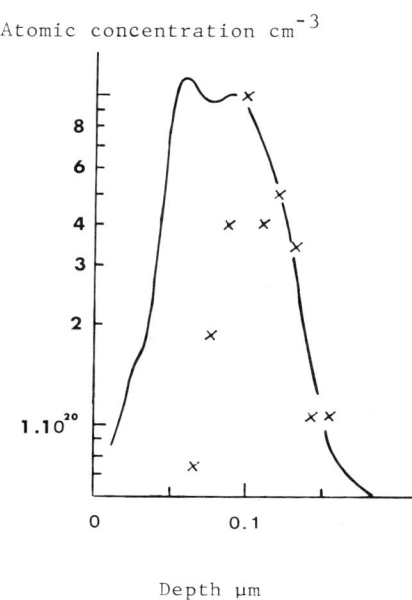

(a)

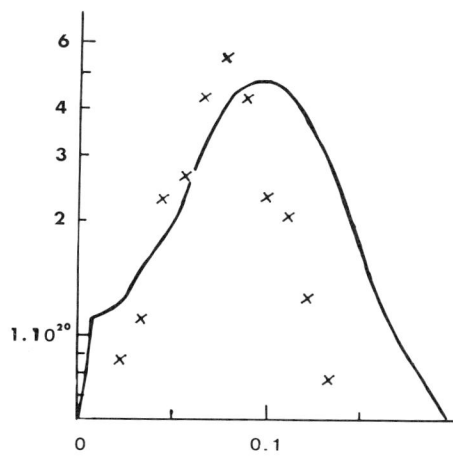

(b)

Atomic concentration cm$^{-3}$

(c)

Depth µm

Fig. 2 Boron concentration
profiles at the highest concentration part of the implant, (a) after
furnace annealing at 950°C for 1hr,
(b) after rapid annealing at 1000°C
for 30s and (c) after rapid annealing at 1100°C for 30s. Also plotted
on the same depth scale, but with
an arbitrary linear density scale
are the relative defect concentrations (X) obtained by inspection
of the micrographs, as explained in
the text.

## 3. Implants with Preamorphisation

The preamorphisation was accomplished by a single silicon implant of
$5.10^{15}$ Si$^{28}$/cm$^2$ at 180keV, with the wafer temperature not exceeding 150°C.
This formed an amorphous layer extending 0.35µm from the top surface, see
Fig. 3(a).   A dense band of small, principally perfect, <110> loops,
20-60Å in diameter lies at the bottom of the amorphous layer.  Some small
crystallites, approximately 150Å in diameter, were also observed at the
surface where the amorphisation was not complete.  Furnace annealing of a
preamorphised sample results in a very different microstructure, see Fig.
3(b) compared with the behaviour of boron implanted into crystalline

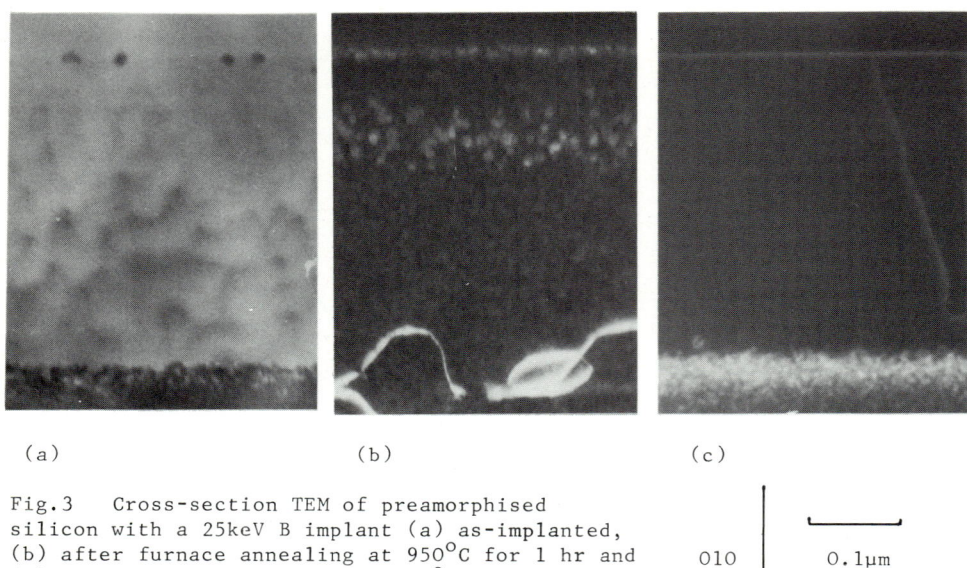

(a)                          (b)                          (c)

Fig.3   Cross-section TEM of preamorphised
silicon with a 25keV B implant (a) as-implanted,
(b) after furnace annealing at 950°C for 1 hr and          010    0.1µm
(c) after rapid annealing at 1000°C for 30s.

material.   There is a band of small defects, approximately 60Å in diameter,
at the peak of the boron concentration profile, ∿ 0.1µm below the surface.
At the depth of the original amorphous-crystalline interface of the as-
implanted sample, 0.35µm, there is now a band of large loops, 2000-3000Å
in diameter, which are well below the deepest part of the boron implant.
The relative defect density is compared to the non-substitutional boron
concentration in Fig. 5(a).   Rapid isothermal annealing at 1000°C for 30s,
see Fig. 3(a), or at 1100°C for 30s, leaves few defects in the boron implanted region.   Loops are present at the original amorphous-crystalline interface but remain relatively unchanged in size for this treatment.   The main
feature of the recrystallized layers is the presence of a few large
'hairpin' defects generally lying in the trace of the 010 plane on 111
planes with <110> type burgers vectors.   The 'hairpins' appear to owe their
origin to shear processes initiated at the original interface between the
crystalline and amorphous zones, presumably to accommodate stresses associated with the volume change due to the implant.   There is evidence that
these defects move by climb; note, for example, the denudation of small
loop damage at the apparent initiation point in Fig.4.

Fig. 4   Detail of a 'hairpin'
close to the band of defects
situated at the original
amorphous crystalline interface

Atomic concentration cm$^{-3}$

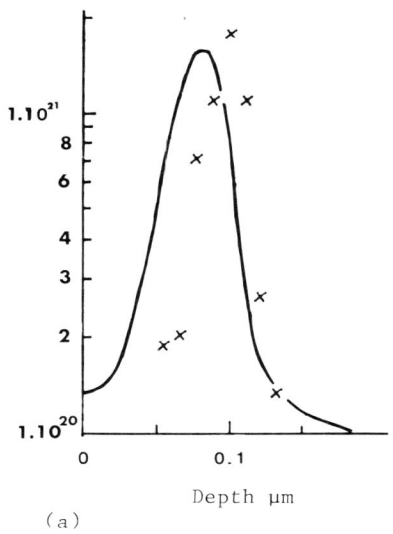

(a)

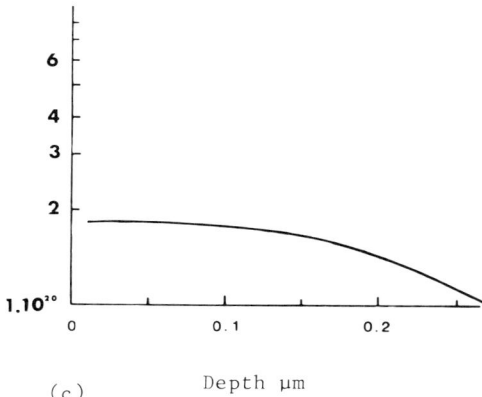

(c)          Depth μm

Atomic concentration cm$^{-3}$

(b)

Fig.5   Boron concentration profile at the highest concentration part of the implant, (a) after furnace annealing at 950°C for 1hr, (b) after rapid annealing at 1000°C for 30s and (c) after rapid annealing at 1100°C for 30s. The relative defect concentration is shown as in Fig.2.

## 4. Discussion

Residual defects after the annealing of boron implants into crystalline silicon may be associated with the non-substitutional part of the boron profile. The defects and the excess boron do not, however, always coincide with depth as the anneal proceeds, suggesting that the evolution of the loops which form immediately upon heating, is not associated directly with the diffusion of the excess boron. The interaction between the defects and the excess boron influences the diffusion of the excess boron, since the non-gaussian excess boron peaks were only observed in the presence of a high density of defects, Figs 2,5. Reduced defects remain in furnace annealed samples where boron has been implanted into amorphous silicon instead of implantation into crystalline silicon. This density of defects is too low to influence the diffusion of the excess boron. With a rapid anneal, the defects result only from the creation and regrowth of

the amorphous layer.  The boron appears to cause no additional defects.
The density of 'hairpin' defects found was much less than has been observed
in other studies, Carter et al (1984).  This may be associated with differences in the pre-amorphisation schedules.  For example, Carter et al.(1984),
used three silicon implants at energies of 70, 150 and 300keV, with doses
of $5.10^{15}$, $5.10^{15}$ and $10^{16}$ Si/cm$^2$ instead of the single implant used in
this study.  The density of 'hairpins' depends on the total boron and silicon dose, but annealed specimens given only the pre-amorphisation described
here were found not to contain this type of defect.  Although the preamorphisation was not complete, with some small crystallites present at the
surface after silicon implantation, no residual damage results.  This may
be due to the greatly inhibited regrowth rate which is experimentally observed close to surfaces, Olsen et al (1980).  A combination of rapid annealing and a pre-amorphisation by a single implant is effective in producing shallow boron junctions with improved dopant activation.  Pre-amorphisation also produces a marked reduction in the number of defects in the
boron implanted region.

## Acknowledgments

D. G. Hasko acknowledges the SERC, U.K. for an Information Technology
Fellowship, and R. A. McMahon acknowledges British Telecom for a research
Fellowship at Corpus Christi College, Cambridge.

## References

El-mansy Y IEEE Trans. Electron. Dev. 1982 ED-29 567
Tsai M Y and Streetman B G 1979 J. Appl. Phys. 50 183
Carter C , Maszara W, Sadana D K, Rozgonyi G A, Liu J and Wortman J
   1984 Appl. Phys. Lett. 44 (4) 459
Godfrey D J, McMahon R A, Hasko D G, Ahmed H and Dowsett M G
   1984 Proc. of the Materials Research Society: Energy beam-solid
   Interactions and Transient Thermal Processing   Boston, Mass.
   (to be published)
McMahon R A, Ahmed H, Godfrey D J and Yallup K J 1983 IEEE Trans ED30, 1550
Olson G L, Kokorowski S A, McFarlane R A and Hess L D 1980 Appl. Phys.
   Lett 37 (11) 1019

*Inst. Phys. Conf. Ser. No. 76: Section 3*
*Paper presented at Microsc. Semicond. Mater. Conf., Oxford, 25–27 March 1985*

105

# Effect of transient annealing on grain growth and structure of polycrystalline silicon films

S J Krause, S R Wilson*, W M Paulson* and R B Gregory*

Dept. of Mechanical and Aerospace Engineering, Arizona State University, Tempe, AZ 85287, USA
*Semiconductor Research and Development Laboratory, Motorola Inc., 5005 E. McDowell Road, Phoenix, AZ 85008, USA

Abstract   Transient annealing effects on the grain size, structure and morphology of B and As implanted films of polycrystalline silicon were determined by transmission electron microscopy.  Originally, grains had a bimodal distribution of smaller equiaxed grains and larger columnar grains.  The growth of both types of grains was described with a modified model for interfacial energy driven grain growth.  Bend contours in annealed unimplanted films were not present in annealed ion implanted films due to formation of a cellular network structure in the grains.

## 1. Introduction

Polycrystalline silicon is commonly used as an interconnect and as a gate material for most metal-oxide-semiconductor (MOS) devices.  To reduce resistivity of polycrystalline silicon, it is ion implanted and annealed during the same steps as are the source and drain of the MOS device. Depending on the temperature of deposition, the silicon may be completely amorphous, partially amorphous and partially crystalline, or entirely crystalline (Kamins et al 1978).  The polycrystalline silicon itself may have a single or bimodal distribution of grain sizes (Duffy et al 1983) which have equiaxed and/or columnar morphologies and which may or may not be heavily textured (Anderson 1973).  Residual stresses may be present in as-deposited or annealed films depending on conditions of deposition or annealing (Choi and Hearn 1984).  During annealing the presence of As or P can cause dopant enhanced diffusion of Si which increases the grain growth rate.  This process has been modeled with an interfacial energy driven grain growth model by Wada and Nishimatsu (1978) for furnace annealing of P doped polycrystalline silicon and by Krause et al (1984) for transient annealing of As doped polycrystalline silicon.  In this paper we have characterized grain growth and changes in grain size, structure and morphology of ion implanted polycrystalline silicon during the short times and high temperatures characteristic of transient annealing.

## 2. Experimental

The substrates used in these experiments were 75 mm (001) Si wafers on which a 0.1 micron $SiO_2$ film was grown.  Undoped polysilicon, about 300 nm thick, was deposited on the oxidized wafer surface at 625°C by low pressure chemical vapor deposition of silane.  The films were then implanted with As or B (60 keV, $5.0 \times 10^{15}/cm^2$) and capped with sputtered

$SiO_2$. A few films were left unimplanted. Wafers were transient annealed for up to 25 seconds total time to a maximum temperature of 1250°C using a Varian IA 200 rapid isothermal annealer with a resistively heated graphite sheet. The annealer has been described elsewhere (Wilson et al 1984). The nominal heater set point temperature of 1150°C was measured by a thermocouple located 2 mm behind the heater. The true wafer temperature was measured by an optical pyrometer located directly behind the wafer. The pyrometer was calibrated from temperatures measured by a thermocouple located on a silicon wafer which had reached thermal equilibrium.

The grain size and structure in the polycrystalline films were determined by transmission electron microscopy (TEM). The films were bright and dark field imaged at 200 kV in a JEOL 200CX by standard techniques. Grain size distributions were determined by measuring between 30 and 100 grains from a micrograph. Average grain sizes were specified from the peaks in a given distribution.

## 3.  Grain Morphology and Distribution of Sizes

The average grain sizes for all samples are tabulated as a function of annealing condition in Table I. The as-deposited films, as shown for the plan and cross section views in Figures 1A and 1B, display a bimodal distribution of grain sizes. The first population in the as-deposited film is composed of small equiaxed grains about 5 nm in diameter which are located chiefly at the bottom of the film and form early in the deposition process. The second population is composed of elongated grains of a columnar morphology with an average diameter of 20 nm which arise from preferential growth of some grains during deposition. Duffy et al (1983), in a study of crystallite sizes by X-ray line broadening measurements, also reported a bimodal distribution of grains for polycrystalline silicon films deposited at 622°C. The effect of annealing on a typical film, one which was As implanted and annealed at a set point of 1150°C for 12.5 seconds, is shown in plan and cross section views in Figures 2A and 2B. Two different sets of grains have grown to average diameters of 31 nm and 103 nm and both have equiaxed morphologies. Other micrographs of both As and B implanted films show similar features. Thus, the bimodal distribution of grain sizes in as-deposited films persists during transient annealing, although the two populations develop equiaxed morphologies as growth progresses.

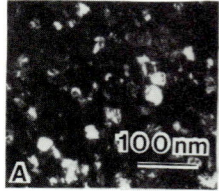 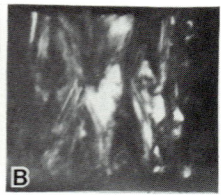 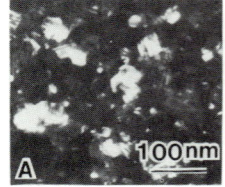 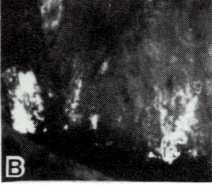

Figure 1. A) Plan and B) cross-section views of as-deposited film.

Figure 2.  A) Plan and B) cross-section views of As implanted film annealed at 1150°C for 12.5 seconds.

## 4.  Effect of Dopant on Grain Growth

Figure 3 shows micrographs of films annealed at a set point temperature of 1150°C. The As implanted films shown have been exposed for A) 10, B) 12.5

C) 15, and D) 17.5 seconds. The B implanted films have been held for
E) 10, F) 12.5, G) 15, and H) 17.5 seconds. In both As and B implanted
films grain sizes increase with increasing annealing time. However, the
grains grow faster for the As implanted films compared to the B implanted
films. At 17.5 seconds the sets of grain sizes are 195 nm and 310 nm for
As implanted films and are 110 nm and 75 nm for B implanted films.

TABLE I. Grain size as a function of annealing conditions

| Set Point Temp.($^{0}$C) | Peak Wafer Temp.($^{0}$C) | Exposure Time(sec.) | Effective Time(sec.) | $\sqrt{Dt^{*}/T}$ (cm/K$^{1/2}$x10$^{-6}$) | Grain size(nm) 1st | 2nd |
|---|---|---|---|---|---|---|
| 25 | - | - | - | - | 20 | 5 |
| 1150, As | 855 | 7.5 | 0.6 | 0.1 | 30 | 5 |
| 1150, As | 910 | 10.0 | 1.0 | 0.2 | 47 | 5 |
| 1150, As | 1075 | 12.5 | 1.65 | 1.1 | 103 | 31 |
| 1150, As | 1177 | 15.0 | 2.0 | 2.5 | 178 | 89 |
| 1150, As | 1210 | 17.5 | 2.82 | 3.7 | 275 | 156 |
| 1150, As | 1271 | 20.0 | 6.75 | 8.0 | 295 | 168 |
| 1150, As | 1276 | 25.0 | 11.5 | 10.2 | 310 | 195 |
| 1150, B | 909 | 7.5 | 0.75 | 0.2 | 21 | 7 |
| 1150, B | 1050 | 10.0 | 0.85 | 0.7 | 36 | 15 |
| 1150, B | 1147 | 12.5 | 1.25 | 1.6 | 62 | 25 |
| 1150, B | 1207 | 15.0 | 1.5 | 2.6 | 145 | 75 |
| 1150, B | 1244 | 17.5 | 2.4 | 4.1 | 175 | 100 |

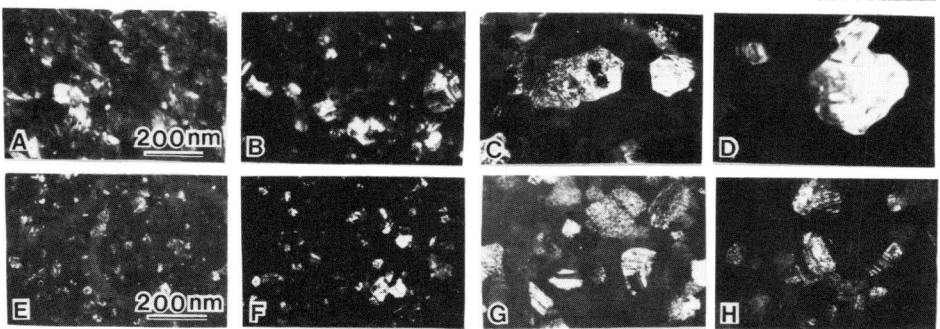

Figure 3.  Dark field micrographs of transient annealed As implanted films
exposed for A)10, B)12.5, C)15, and D)17.5 sec. and B implanted films
exposed for E)10, F)12.5, G)15, and H)17.5 sec. at set point temperature.

As discussed in an earlier paper, this is due to enhancement of the
diffusion rate of Si by the As dopant (Krause et al 1984). A similar
effect is observed for a P dopant, but a B dopant does not increase the
diffusivity of Si (Wada and Nishimatsu 1978). Under similar annealing
conditions, the relative ratio of grain sizes of As implanted films versus
B implanted films should be equal to the square root of the ratio of Si
diffusivity in As versus B doped films. Here, grain sizes in a given
population were compared for similar film annealing conditions. The ratio
of average grain sizes in As doped versus B doped films is about a factor
of 1.8. For doping concentrations used here, the ratio of the square root
of the diffusivity of the As doped versus the B doped Si is predicted to
be about 1.4 (Wada and Nishimatsu 1978). Although the trend is correct,

the ratio of grain sizes is higher than predicted.

## 5. Modeling Grain Growth During Transient Annealing

In transient annealing significant diffusion occurs during the rapid
temperature rise and fall and must be considered in modeling grain growth.
We have done this by modifying an interfacial energy model used by Wada
and Nishimatsu (1978). First, the effect of changing temperature is
compensated for by calculating the diffusion coefficient (D) of Si in
polycrystalline Si based on the peak temperature of the wafer. Second, an
effective time (t\*), as described by Shewmon (1963), is calculated based
on the time-temperature profile, the peak temperature (T), and the
appropriate activation energy $(E_a)$. For $E_a$ we chose the value of 2.4 ev
used by Wada and Nishimatsu (1978). The modified model for interfacially
driven grain growth can then be given by the relationship: $r \propto \sqrt{Dt^*}/T$. D
may be calculated from $D = D_0 \exp(E_a/kT)$ where $D_0$ is the proportionality
constant, (and is assumed here to be 1 $cm^2$/sec. for simplicity) and k is
Boltzman's constant.

In Figures 4 and 5 grain size is plotted as a function of $\sqrt{Dt^*}/T$ for As
implanted samples and B implanted samples, respectively, annealed at
various time-temperature conditions. In both sets of samples the two
populations of smaller and larger grain sizes from the bimodal
distributions have been plotted as open and closed circles, respectively.
For the plot of As implanted samples some earlier data from annealing at
a 1200°C set point have been included. For both As implanted and B
implanted films lower values of $\sqrt{Dt^*}/T$ show a linear relationship with
grain size, indicating that the modified model for growth is appropriate
for transient annealing. The slope of the line is higher for As implanted
films than for B implanted films due to As dopant enhanced diffusion of Si.

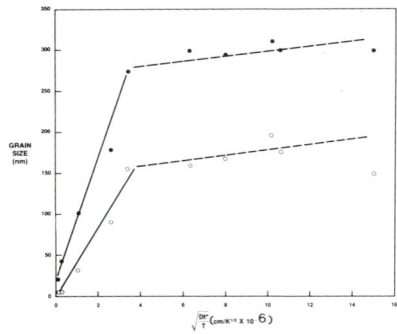

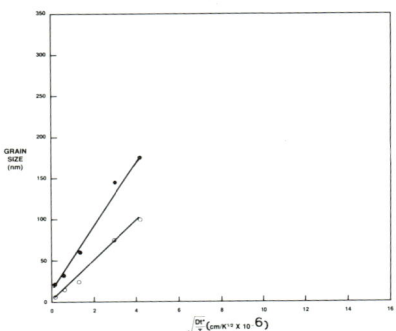

Figure 4. Grain size as a function    Figure 5. Grain size as a function
of $\sqrt{Dt^*}/T$ for As implanted films.    of $\sqrt{Dt^*}/T$ for B implanted films.

In the As implanted films the slope of the plot is reduced at higher
values of $\sqrt{Dt^*}/T$, indicating that grain growth rate is lower. This is
due to geometrical constraints on growth. The regions for reduced growth
rates are represented by dashed lines on the plot because the scatter of
the data is high. For the population with larger grains, the growth rate

is lowered when the grain size reaches the thickness of the film. A similar effect has been observed for furnace annealing of polysilicon films by Jain and Overstraeten (1975) and in transient annealing of polysilicon films by Pinizzotto (1984). For the population of smaller grains, the growth rate is reduced when the grains impinge upon one another. Since only a fraction of the smaller grains actually participate in grain growth, they do not impinge upon one another until they reach a diameter of about 150 nm.

## 6. Grain structure

Shown in Figure 6 are pairs of bright field (BF) and higher magnification dark field (DF) micrographs of annealed films which are A) umimplanted, B) As implanted, and C) B implanted. In the umimplanted film in Figure 6A bend contours are seen in all grains in the BF image and there is little, if any, fine structure in the DF image. In the As implanted films the BF image shows bend contours in some of the grains while the DF image shows some fine structure at grain boundaries and within some grains. In the B implanted films BF image shows virtually no bend contours while the DF image shows fine structure throughout entire grains. Upon closer inspection the fine structure in the B or As doped films is seen to consist of a cellular network structure with cells 4 to 6 nm in diameter. Similar cellular network structures have been observed by Cullis et al (1981) in laser annealing of implanted silicon on sapphire, except that the size of the cells is about 50 to 150 nm. The micrographs in Figure 6 show that the amount of bend contouring in the grains is reduced as the amount of cellular network formation is increased. This is probably due to relief of residual strains causing variations in the diffracting conditions which cause bend contours.

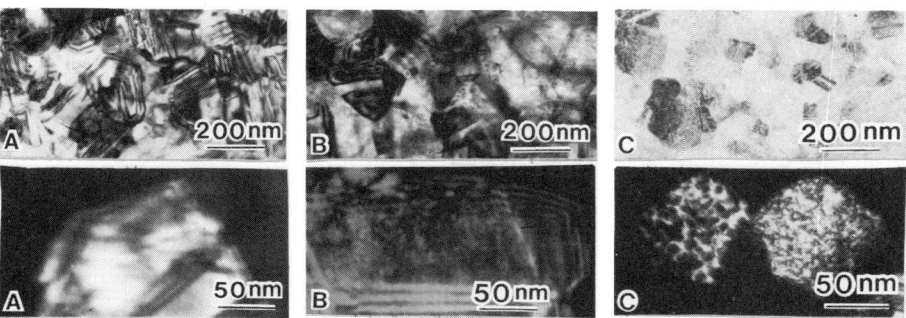

Figure 6. Bright field-dark field pairs of annealed films A) unimplanted and held 20 sec., B) As implanted and held 17.5 sec., and C) B implanted and held 17.5 sec.

## 7. Conclusions

In transient annealing of B or As implanted polycrystalline silicon films an original bimodal distribution of grain sizes persists during annealing. However, the original populations of smaller equiaxed grains and larger columnar grains evolves into two populations of equiaxed grains.

Grain growth of As implanted films is more rapid than that of B implanted films due to enhanced diffusion of Si in the presence of As.

Grain growth during early stages of transient annealing can be described with a modified model for interfacial energy driven grain growth. However, at later stages of annealing the grain growth rate of As implanted films is eventually lowered by geometrical constraints.

Cellular network structures form in the grains of B implanted films and form outward from grain boundary regions of As implanted films during transient annealing. These cellular network structures reduce residual strains and associated bend contours in B implanted and As implanted films compared to unimplanted films without the cellular structure.

## Acknowledgements

The authors gratefully acknowledge discussions with Peter Fejes and also use of the facilities at the High Resolution Electron Microscopy Center supported by a grant from the National Science Foundation and Arizona State University.

## References

Anderson R R 1973 J. Electrochem. Soc. 120 1540
Choi M S and Hearn E W 1984  J. Electrochem. Soc. 131 2443
Cullis A G, Webber H C, Chew N G, Hill C, and Godfrey D J 1981 Inst. Phys. Conf. Ser. 60 95
Duffy M T, McGinn J T, Shaw J M, Smith R T, and Soltis R A 1983 RCA Review 44 287
Fairfield J M and Masters B J 1967 J. Appl. Phys. 38 3148
Jain R K and Overstraeten R J 1975 J. Electrochem. Soc. 122 552
Kamins T I, Mandurah M M, and Saraswat K C 1978 J. Electrochem. Soc. 125 927
Krause S J, Wilson S R, Paulson W M, and Gregory R B 1984 Appl. Phys. Lett. 45 778
McLean D 1957 Grain Boundaries in Metals Oxford University Press London
Pinizotto R F 1984 Proc. Mat. Res. Soc. in press
Seto J Y W 1976 J. Appl. Phys. 47 5168
Shewmon P G 1963 Diffusion in Solids McGraw-Hill New York
Tannenbaum E T 1961 Solid-State Electronics 2 123
Wada Y and Nishimatsu S 1978 J. Electrochem. Soc. 125 927
Wilson S R, Gregory R B, Paulson W M, Hamdi A H, and McDaniel F D 1984 J. Appl. Phys. 55 4162

*Inst. Phys. Conf. Ser. No. 76: Section 3*
*Paper presented at Microsc. Semicond. Mater. Conf., Oxford, 25–27 March 1985*

# Regrowth of deposited amorphous silicon films by IR light annealing

F S Huang and J C Guo

Institute of Electrical Engineering, National Tsing Hua University, Hsinchu
Taiwan 300, Republic of China

Abstract   IR light has been used to produce crystal growth in deposited
amorphous silicon layers.  The microstructure of the annealed films was
observed.  The high annealing temperature and long annealing time give
silicon nitride formation, which was verified by TEM and Auger analyses.
The pulsed IR light annealing gives the largest grain size among the
various annealing times. The electrical properties are also measured
from a simple n(a-Si)/p(c-Si) junction.  We correlate the TEM results
with I-V measurements.

## 1. Introduction

In recent years, many people have investigated the epitaxial regrowth of
deposited amorphous silicon (a-Si) and ion-implanted a-Si on crystalline
silicon substrates.   Roth and Anderson (1977) reported the solid-phase
growth of epitaxial Si thin films in an ultra high vacuum chamber at temperatures of 500-600°C.  Hung et al (1980) also used a vacuum quartz-tube
furnace to obtain epitaxial growth. For adiabatic heating, the liquid-phase
epitaxial growth occurs during resolidification of the flash melted surface.
Pulsed-electron-beam annealing (Greenwald et al 1979), pulsed-laser-anneal-
ing (McMahon and Ahmed 1980), and CW laser annealing (Celler 1979) of ion-
implanted silicon have been studied.  The recrystallization of amorphous Si
films is a vast area of scientific and practical interest.

This work focuses on the annealing of deposited amorphous silicon with IR
light and also gives the correlation between the electrical properties and
grain size.  The a-Si films prepared by plasma enhanced chemical vapor deposition and by r.f. sputtering were studied.  After various heating
processes, TEM micrographs, Auger spectra and I-V data were obtained and
the comparisons were made.

## 2. Experiments

We prepared the a-Si films using an r.f. sputtering system and plasma
enhanced chemical vapor deposition (PECVD) system.  The deposition conditions of r.f. sputtering are described in Table 1(a).  The forward r.f.
power applied to the target was 180W.  The mixed gas pressure (Ar+H$_2$) was
about 20m torr, and the substrate temperature was 250°C during deposition.

In the plasma enhanced CVD system, amorphous Si was produced by chemical
vapor deposition of the mixed gas SiH$_4$+H$_2$ assisted by the glow discharge
induced by the r.f. power.  The substrate temperature was set at 250°C.
Film depositions for various conditions are listed in Table 1(b).

P type (100) Si with resistivity in the range 15-20Ω-cm was used for substrates. We deposited the n-type a-Si films onto the c-Si substrates to form a p-n junction. After the annealing process, Mo was sputter-deposited onto the n-type polycrystalline Si films to make the ohmic contact. Then, we implanted $B^+$ into the other side of the substrate to render the surface $P^+$ (the dose was about $10^{15}cm^{-2}$ and energy about 25keV) and evaporated Al to make an ohmic contact. Thus, the Mo/poly-Si/c-Si/Al structure was fabricated. The I-V measurements were performed by using an HP 4140B.

An IR gold image furnace was used for IR light annealing. Nitrogen was introduced as carrier gas during annealing.

The microstructure of annealed films was studied using a TEM (JEOL-100B) operated at 100kV. Auger electron spectra were taken using an Anelva AES-350 for several samples.

3. Results and Discussion

Various annealing processes were chosen and the results are listed in Tables 2-6. Table 2 collects the observations from TEM for long time annealing. The thermal nitridation was identified from

| Sample | target | Ar:H$_2$ | thickness |
|--------|--------|----------|-----------|
| HN1 | HN | pure Ar | 4250Å |
| HN91 | HN | 9:1 | 2000Å |
| HN2 | HN | pure Ar | 9600Å |
| LN11 | LN | 1:1 | 6000Å |

Table 1(a) Sputtered a-Si deposition conditions.

| Sample | gas ratio (SiH$_4$:H$_2$) | thickness |
|--------|---------------------------|-----------|
| PECVD1 | 1:4 | 3000Å |
| PECVD2 | 1:4 | 10000Å |
| PECVD3 | 1:4 | 5000Å |

PECVD*3    PECVD3 implanted with $P^+$
$N_{dose}=10^{16}cm^{-2}$, E=150keV

Table 1(b) PECVD a-Si deposition conditions.

| Sample | annealing condition | | grain size |
|--------|---------------------|---|------------|
| HN1 | 1300°C | 4 min | nitridation |
| | 550°C | 30 min | |
| HN91A | 1200°C | pulse | nitridation |
| | 550°C | 30 min | |
| HN91B | 1300°C | 1 min | nitridation |
| | 550°C | 30 min | 250Å |
| HN91C | 1300°C | pulse | nitridation |
| | 550°C | 30 min | 1/3μm |
| LN11A | 1250°C | 10 min | nitridation |

Table 2  Long time annealing of various samples.

the TEM diffraction pattern as shown in Fig. 1. This abnormal diffraction pattern even existed after pulsed processing at 1300°C followed with 550°C, 30 minutes heating. Figure 2 shows a TEM diffraction pattern of sample HN91C. This nitridation is also confirmed by Auger spectrometry. We observe the existence of N(380eV) in Fig. 3. Table 3 presents evidence that short annealing time can get rid of the nitridation and 1-2 second pulse annealing is enough for the recrystallization.

We also show results for various a-Si films taken through the same annealing process in Table 4. The PECVD a-Si always has larger grains than sputtered a-Si. We believe that the higher deposition rate of HN2 prevents oxygen contamination and yields a larger grain size than that of LN11. The PECVD system gives a-Si films with good quality and clean interfaces which favour grain growth. The Argon present in sputtered films does not favour crystal growth.

The grain size of samples PECVD3 after various temperature pulse annealing cycles are listed in Table 5. The PECVD*3 samples are those PECVD3 implanted with $P^+$ at $10^{16}\,cm^{-2}$ with energy 150keV before the annealing processes. This implanted $P^+$ only penetrates of the order of a thousand Å. It cannot change the interface condition. But the heat generated from implanting $P^+$ had an annealing effect on a-Si films, and reduced some trap states. So we expect that electrical properties can be improved, but not the grain size. From the data in Table 5, we conclude that high temperature favours the regrowth and $P^+$ ion implantation has no effect on grain size. The electrical properties will be discussed later. Then, we studied the grain size as a function of temperature. From Table 5, the grain size increases as annealing temperature rises.

From the slope of log L vs 1/T as shown in Fig. 4, we can get the activation energy for samples PECVD3 and PECVD*3. The secondary recrystallization activation energy is about 1.08eV for PECVD*3 and 1.52eV for PECVD3. Wada and Nishimatsu (1978) reported 1.0eV for the secondary recrystallization activation energy. Fig. 5(a) is a TEM diffraction pattern of sample PECVD3F. The micrograph is illustrated in Fig. 5(b). The grain size of ~1μm can be estimated.

Now, we show in Table 6 that PECVD a-Si samples with various thicknesses under pulsed IR light annealing at 1250°C have almost the same grain size. We conclude that thickness has no sig-

Fig. 1   TEM diffraction pattern of sample HN1.

Fig. 2   TEM diffraction pattern of sample HN91C.

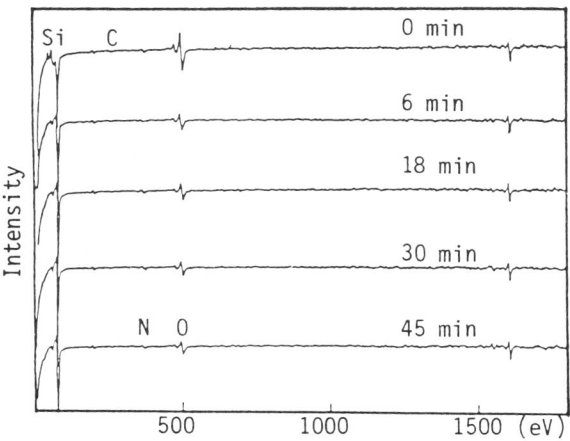

Fig. 3   AES of sample HN91C.

nificant effect on the grain growth in the thin film, contrary to the relative thick film (about 100 μm). Thick films after annealing always have a final grain size of about the same as film thickness.

The current-voltage characteristic curves of n(poly-Si)/p(c-Si) junctions are also measured. The ideality factor (n) and reverse leakage current of various samples are listed in Table 7. It shows that larger grain poly-Si generally has smaller n and reverse leakage current.

An n value close to 1 and a small leakage current imply a perfect interface, which also favours grain growth. So, the interface cleanness is an important factor for regrowth and rectification performance. Meanwhile, we have made a comparison between the unimplanted and P$^+$ implanted films. The P$^+$ implantation had no significant effect on the grain growth, but it improved the electrical properties by making the n value close to 1 and yielding smaller reverse leakage current. This can be explained by the annealing effect of ion implantation.

| Sample | Annealing | Condition | Grain size |
|--------|-----------|-----------|------------|
| LN11A | 1250°C | 10 min | thermal nitridatic |
| LN11B | 1250°C | 5 min | 0.1μm |
| LN11C | 1250°C | pulse | 0.1μm |

Table 3  Various annealing times at 1250°C on LN11 samples.

| Sample | Annealing | Condition | Grain size |
|--------|-----------|-----------|------------|
| LN11C | 1250°C | pulse | 0.1μm |
| HN2B | 1250°C | pulse | 0.6μm |
| PECVD1B | 1250°C | pulse | 0.8μm |
| PECVD2B | 1250°C | pulse | 0.9μm |

Table 4  1250°C pulse annealing for various samples.

| Sample | Annealing | Condition | Grain size |
|--------|-----------|-----------|------------|
| PECVD3C | 1000°C | pulse | 0.15μm |
| PECVD3D | 1100°C | pulse | 1/6μm |
| PECVD3E | 1250°C | pulse | 0.9μm |
| PECVD3F | 1300°C | pulse | 1.0μm |
| PECVD*3C | 1000°C | pulse | 0.15μm |
| PECVD*3D | 1100°C | pulse | 0.3μm |
| PECVD*3E | 1250°C | pulse | 0.8μm |
| PECVD*3F | 1300°C | pulse | 1.0μm |

Table 5  Various temperature pulse annealing of PECVD3 and PECVD*3 samples.

## 4. Conclusions

We have investigated the regrowth of deposited a-Si films. Among various annealing times, the pulsed IR light annealing on PECVD a-Si at 1300°C gives the best result. In this process, oxygen and nitrogen are not found to be incorporated into the a-Si, and that favours grain growth. The diode parameter n of this n(a-Si)/p(c-Si) junction is 1.27.

The grain size depends on interface cleanness, annealing temperature and film quality. A clean interface, high annealing temperature and PECVD a-Si film favour grain growth. The thickness of film does not have a significant influence on the result.

The rectification of the n-p junction also strongly depends on the microstructure of the recrystallized poly-Si film. Usually, a larger grain size gives a smaller n value and reverse leakage current. In addition, implantation into PECVD films improves rectification performance.

References

Celler G K 1979 J. Appl. Phys. 50 7264
Greenwald A C, Kirkpatrick A R, Little R G and Minnucci J A 1979 J. Appl. Phys. 50 783
Hung L S, Lau S S, Von Allmen M, Mayer J W and Ullrich B M 1980 Appl. Phys. Lett. 37 909
McMahon R A and Ahmed H 1980 Appl. Phys. Lett. 37 1016
Roth J A and Anderson C L 1977 Appl. Phys. Lett. 31 689
Wada Y and Nishimatsu S 1978 J. Electrochem. Soc. : Solid-State Science and Tech. 125 1499

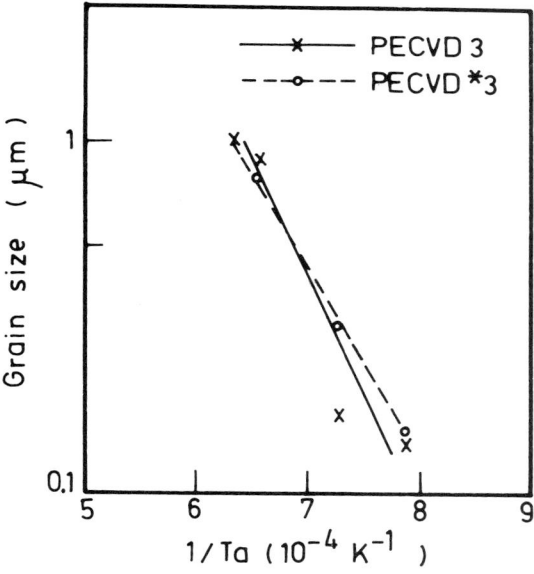

Fig. 4  Temperature dependence of grain size for samples PECVD3 and PECVD*3.

Fig. 5(a)  TEM diffraction pattern of sample PECVD3F.

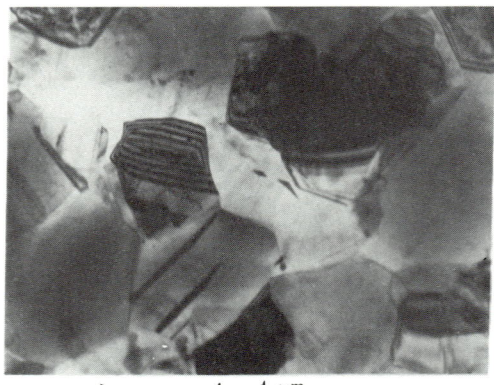

⊢————————⊣ 1 μm

Fig.5(b). TEM bright field image of sample
          PECVD 3F.

Table 6. Relation between grain growth & film
         thickness annealing condition:
         pulsed IR light annealing at 1250°C.

| Sample | PECVD1 | PECVD2 | PECVD3 |
|--------|--------|--------|--------|
| thickness | 3000 Å | 10000 Å | 5000 Å |
| grain size | 0.8 μm | 0.9 μm | 0.9 μm |

Table 7. TEM & I-V measurement of various
         samples.

| Sample | grain size | I-V n(0-0.2v) | I$_r$(-0.2v) |
|--------|-----------|--------------|-------------|
| HN91A | thermal-nitridation | 3.2 | 12.25 μA |
| LN11B | 0.1 μm | 2.6 | 6 μA |
| HN91C | 1/3 μm | 1.4 | 1 μA |
| PECVD3E | 0.9 μm | 2.7 | 2 μA |
| PECVD*3E | 0.8 μm | 1.66 | 1 μA |
| PECVD3F | $\geq$ 1 μm | 2 | 2 μA |
| PECVD*3F | 1 μm | 1.27 | 0.1 μA |

*Inst. Phys. Conf. Ser. No. 76: Section 3*
*Paper presented at Microsc. Semicond. Mater. Conf., Oxford, 25–27 March 1985*

# The microstructure of transiently annealed GaAs

M A Shahid,  K K Patel and B J Sealy

Department of Electronic & Electrical Engineering,
University of Surrey, Guildford, Surrey GU2 5XH, U.K.

Abstract   $Mg^+$ ion implanted GaAs has been transiently annealed and studied by transmission electron microscopy and electrical measurements. High percentage electrical activities of 80% and 100% were measured for doses of $1 \times 10^{14}$ and $5 \times 10^{13}$ respectively following annealing for 30 seconds. Only interstitial perfect dislocation loops with $\frac{1}{2}<110>$ Burgers vectors were observed and their density was found to decrease with increasing temperature.

## 1. Introduction

It is well known that p-type layers in GaAs can be produced relatively easily by ion-implantation. However, the broadening of the impurity profile is difficult to control due to the high diffusivity of p-type dopants in GaAs during post-implantation annealing. $Be^+$, $Mg^+$, $Zn^+$ and $Cd^+$ all produce p-type activity in GaAs and have been the subject of many publications. $Zn^+$ and $Cd^+$ being rather heavy ions require relatively high ion energies for a given projected range and because of their heavy mass induce relatively high levels of damage in the substrate. $Mg^+$ on the other hand besides being a light ion is non-toxic (unlike $Be^+$) and can be implanted at a suitable depth in GaAs at moderate ion energies creating far less damage than $Zn^+$ and $Cd^+$. Despite these important advantages of this ion over others only a small amount of work has been reported in the literature (Choe et al. 1980, Yeo et al. 1981,82). Moreover, most of this work deals with long time furnace annealing where redistribution of Mg can be appreciable especially for high doses. Transient or short time annealing may well prevent diffusion of Mg but this technique has not yet been applied extensively to acceptor ions and only one paper has been published so far (Blunt et al. 1984) which describes the electrical properties of $Mg^+$ ion implanted GaAs. In this paper we will present the results of electrical measurements and transmission electron microscopy for transiently annealed $Mg^+$ implants in GaAs.

## 2. Experimental

$Mg^+$ ions were implanted at room temperature into undoped, semi-insulating (100) GaAs in a non-channelling direction. Ion doses of $5 \times 10^{13}$ and $1 \times 10^{14}$ $Mg^+cm^{-2}$ were used at an ion energy of 100 keV. After depositing CVD $Si_3N_4$ at a temperature of 635°C for 30 seconds annealing was carried out either on a graphite strip heater or in an incoherent light furnace (Sealy et al. 1985). A total time of 30 seconds at temperature was selected for all samples using temperatures of up to 950°C. After annealing, the samples were cut into clover leaf shapes and the sheet hole concentration and sheet resistivity were measured. The hole concentration and mobility profiles were also determined from differential Hall effect measurements using the Van der Pauw method. The samples were also studied by transmission electron microscopy (TEM). For this, specimens were cleaved into 2 x 2mm squares. After protecting the implanted side with Lacomit, the samples were thinned for electron transmission from the back surface only. A JEOL 200CX scanning transmission electron microscope was used to record bright field and (g, 3g) dark field images and selected area diffraction patterns from individual sample areas.

## 3. Results

### 3.1 Electrical

Figure 1 shows the plots of sheet hole concentration and sheet resistivity as a function of annealing temperature. At the lower temperatures, the $1 \times 10^{14}$ $Mg^+cm^{-2}$ implant shows the highest percentage electrical activity. However, at 900°C and above the sheet hole concentration is not different from that of the $5 \times 10^{13}$ $Mg^+cm^{-2}$ implant. This figure also shows that the best sheet resistivity for the two doses are obtained for samples annealed at 900/950°C.

Typical hole concentration and mobility profiles are shown in Figure 2,

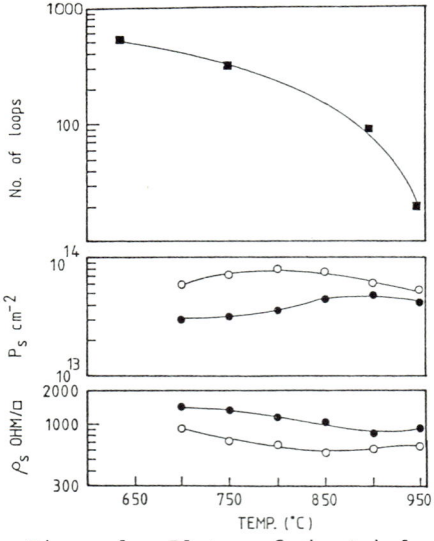

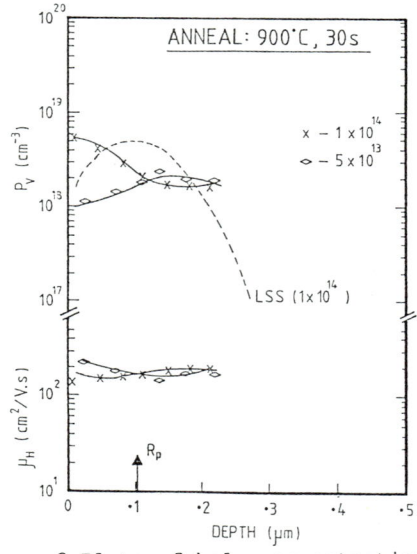

Figure 1. Plots of sheet hole concentration ($P_S$), sheet resitivity ($\rho_S$) and loop density (per 3.5μm²) as a function of annealing temperature.
●,■ = $5 \times 10^{13}$ $Mg^+cm^+$,
O = $1 \times 10^{14}$ $Mg^+cm^{-2}$

Figure 2. Plots of hole concentration ($P_V$) and hole mobility ($\mu_H$) as a functioin of depth.
◆= $5 \times 10^{13}$ $Mg^+cm^{-2}$,
X − $1 \times 10^{14}$ $Mg^+cm^{-2}$

for both doses annealed at 900°C for 30 seconds. For the dose of
$1 \times 10^{14}$ Mg$^+$cm$^{-2}$, a hole concentration of about $5.5 \times 10^{18}$ cm$^{-3}$ is
recorded near the surface. This drops to a value of about $1.5 \times 10^{18}$
cm$^{-3}$ at a depth of about 0.1 µm and stays at this value up to 0.2 µm. On
the other hand the $5 \times 10^{13}$ Mg$^+$cm$^{-2}$ implant shows a gradual increase in
hole concentration with depth and shows a peak of about $2 \times 10^{18}$cm$^{-3}$ at
a depth of about 0.1 µm. In both cases, the hole mobilities have
similar values.

## 3.2 Transmission Electron Microscopy

Figure 3 shows a set of micrographs for the samples implanted with
$5 \times 10^{13}$ Mg$^+$cm$^{-2}$. The micrograph of Figure 3a is produced from the

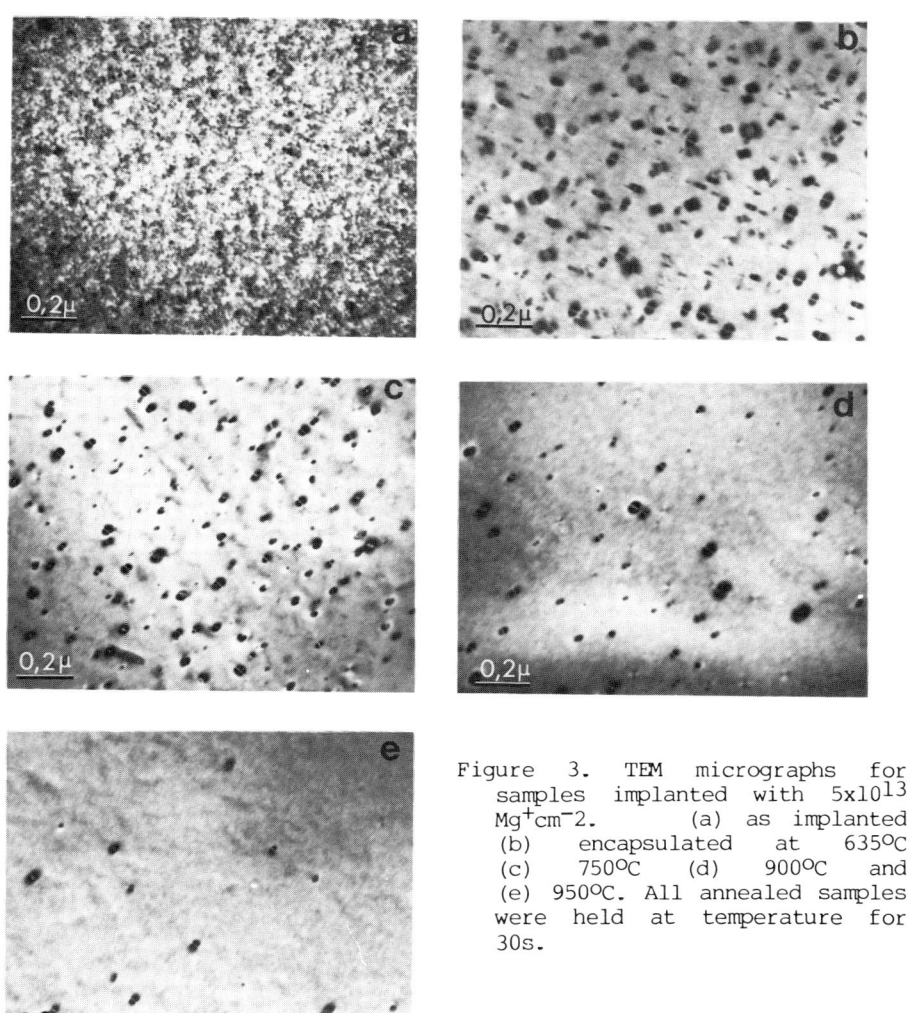

Figure 3. TEM micrographs for
samples implanted with $5 \times 10^{13}$
Mg$^+$cm$^-$2. (a) as implanted
(b) encapsulated at 635°C
(c) 750°C (d) 900°C and
(e) 950°C. All annealed samples
were held at temperature for
30s.

as-implanted sample and shows a structure typical of material implanted at this dose. Some annealing does take place during encapsulation by CVD $Si_3N_4$ at 635°C for 30 seconds (Figure 3b). However, the density of the dislocation loops is very high ( ~1.6 x $10^{10} cm^{-2}$). As the annealing temperature is raised from 700°C to 950°C the loop density reduces (Figures 3c, 3d, 3e). Included in Figure 1 is a plot of defect density as a function of annealing temperature, illustrating the correlation between sheet resistivity and loop density. Figure 4a shows a micrograph from an encapsulated sample implanted with a dose of 1 x $10^{14}$ $Mg^+ cm^{-2}$. Here the microstructure is different from that observed in

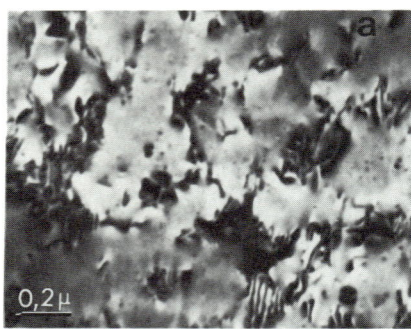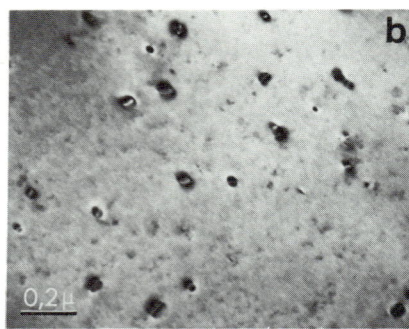

Figure 4. Transmission electron micrographs from samples implanted with $1x10^{14}$ $Mg^+$ $cm^{-2}$. (a) encapsulated at 635°C (b) annealed at 900°C for 30 seconds.

Figure 3b in that instead of a high density of dislocation loops, highly strained material together with dislocation lines and loops is observed. However, annealing at higher temperatures produces similar results to the 5 x $10^{13}$ $Mg^+ cm^{-2}$ implant. For instance, the micrograph of Figure 4b shows a sample implanted with 1 x $10^{14}$ $Mg^+ cm^{-2}$ and annealed at 900°C for 30 seconds. The loop density in this micrograph is similar to that of Figure 3d produced from 5 x $10^{13}$ $Mg^+ cm^{-2}$ implants annealed at 900°C for 30 seconds. A detailed analysis (Shahid et al. 1983) carried out on these samples shows that the dislocation loops are perfect interstitials possessing $\frac{1}{2}<110>$ Burgers vectors. These loops are distributed equally on the {110} planes both perpendicular and inclined to the (100) surface plane of GaAs.

## 4. Discussion

The published results on $Mg^+$ implants into GaAs have mainly been concerned with conventional furnace annealing (Choe et al. 1980, Yeo et al 1981,82). As far as the authors are aware, the microstructure of $Mg^+$ implanted GaAs has not been reported. Our present results show that there is generally a correlation between the electrical properties and the defect density as a function of annealing temperature. This is particularly true for the low dose of 5 x $10^{13}$ $Mg^+ cm^{-2}$.

Yeo et al.(1982) have shown from SIMS analysis of $Mg^+$ implanted GaAs that the outdiffusion of Mg depends strongly on the implanted ion dose in that high doses result in a high level of Mg outdiffusion towards the surface and into the encapsulant ($Si_3N_4$ in our case). Our results agree with this, in that the 1 x $10^{14}$ $Mg^+ cm^{-2}$ implants show a relatively high hole concentration near the surface, higher than predicted by the theoretical profile. We suggest that the redistribution of Mg has occurred during annealing. This effect is not observed in the samples implanted with a dose of 5 x $10^{13}$ Mg$^+$ cm$^{-2}$. A difference between the two doses is also evident in the microstructure after encapsulating with 635°C for 30 seconds. The higher damage density in the as-implanted material for a dose of 1 x $10^{14}$ Mg$^+$cm$^{-2}$ results in streaking in the diffraction spots in the selected area diffraction pattern and dislocation lines and loops in the micrographs after encapsulation.

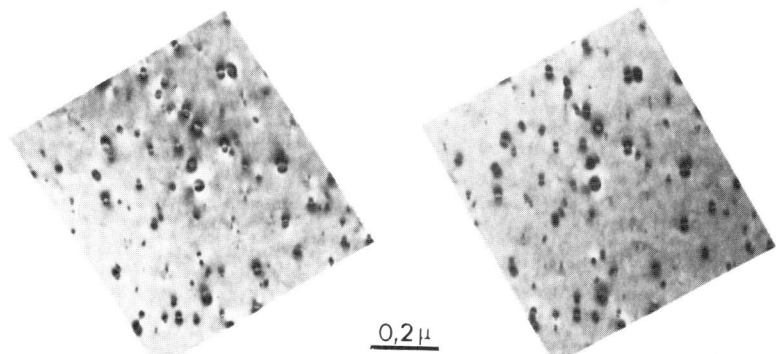

0,2 µ

Figure 5. A stereo pair of TEM micrographs from a sample implanted with 5x$10^{13}$ Mg$^+$ cm$^{-2}$ and encapsulated at 635°C showing depth distribution of loops.

After annealing in the temperature range 700 to 950°C, samples contain only one type of defect. These are in the form of perfect interstitial dislocation loops possessing $\frac{1}{2}$<110> Burgers vectors. Stereo microscopy shows that these defects are contained within a depth of about 0.25 m below the surface within which most of the carriers are measured. The density of these defects decreases with increasing temperature. These findings are similar to those for selenium implanted GaAs (Shahid et al. 1983). In a recent publication we have shown (Shahid et al. 1985) that for a dose of 1 x $10^{14}$ ions cm$^{-2}$ at an ion energy of 300 keV for Se$^+$ or Sn$^+$, the results of annealing up to a temperature of up to 950°C are different from those concerning Mg$^+$ implanted GaAs. Both Se$^+$ and Sn$^+$ implanted samples have a relatively high background density of dislocation loops and lines. In addition, Sn$^+$ implanted samples have a large density of $\frac{1}{3}$<111> faulted loops and these loops are decorated. In order to produce a residual defect density comparable to Mg$^+$ implanted GaAs a high temperature of 1000°C or higher is required. At such a high temperature precipitates both in Se$^+$ and Sn$^+$ implanted samples are observed. We have not annealed Mg$^+$ implanted samples at such high temperatures. However, from the published results and present studies it can be inferred that for doses higher than 5 x $10^{13}$ Mg$^+$cm$^{-2}$, outdiffusion will be a major problem.

## 5. Summary

A correlation between the electrical properties and dislocation loop density has been found for Mg$^+$ implanted GaAs after transient annealing. Throughout the temperature range of 700 to 950°C, only perfect interstitial dislocation loops with $\frac{1}{2}$<110> Burgers vectors are seen and their density is found to decrease with increasing annealing temperature. The loop density also correlates with a decrease in resistivity with temperature. For a high dose of $1 \times 10^{14}$ $Mg^{+}cm^{-2}$ a relatively high hole concentration is measured for a 30 second anneal at 900°C indicating possible outdiffusion of Mg towards the surface.

Acknowledgements

The authors wish to thank the SERC for financial support, R. Bensalem for encapsulating the samples, J. Mynard and the staff of the Accelerator Laboratory and the Microstructural Studies Unit, University of Surrey, for technical assistance in ion-implantation and transmission electron microscopy.

References

Blunt R T, Szweda R, Lamb M S M and Cullis A G 1984
    Elec. Lett. 20 444
Choe B D, Yeo Y K and Park Y S 1980 J. Appl. Phys. 51 4742
Sealy B J, Bensalem R and Patel K K 1985 Nucl. Instr. B in press
Shahid M A, Moffatt S, BNarrett N J, Sealy B J and Puttick K E 1983
Rad.Eff. 70 291
Shahid M A, Bensalem R and Sealy B J 1985 Proc. Mat. Res. Soc. Symp. in
    press
Yeo Y K, Park Y S and Choe B D 1981 Nucl. Instr. Meth. 182/183 609
Yeo Y K, Park Y S, Pedrotti F L and Choe B D 1982 J. Appl. Phys. 53 6148

*Inst. Phys. Conf. Ser. No. 76: Section 4*
*Paper presented at Microsc. Semicond. Mater. Conf., Oxford, 25–27 March 1985*

# Characterization of MBE grown Si on (001) GaAs by transmission electron microscopy

C W T Bulle-Lieuwma, P C Zalm and M P A Viegers

Philips Research Laboratories, P O Box 80.000, 5600 JA Eindhoven, The Netherlands

Abstract  Epitaxial layers of silicon, grown on (001) GaAs substrates by Molecular Beam Epitaxy (MBE) at 600°C were investigated by TEM. Samples with overlayers of 250 Å and 2000 Å were studied in planar and cross-sectional modes. In the planar specimens Moiré patterns are observed with a periodicity in accordance with the 4% lattice mismatch, but heavily dislocated. After complete removal of the substrate, these dislocations were also seen in the bare Si layers. They appeared to be threading from the surface to the interface and back with only short segments (0-200 Å) along the interface. In HREM images of cross-sectional specimens, dislocations were found to be either perfect or extended on (111) planes. Their geometry indicates that they were formed by a glide process on the inclined (111) planes, probably disturbed by mutual interactions between the dislocations because of the high dislocation densities involved. This is a very promising feature, because glide only starts after the growth of a critical thickness below which the Si layers on GaAs are expected to be pseudomorphic.

## 1. Introduction

Since MBE has proven to be capable of realizing very abrupt transitions for a wide variety of heterostructures, this technique is very suitable for growing epitaxial Si on GaAs. In such a structure it would be possible to combine silicon and GaAs device technologies on a single wafer. The lattice constants of Si (a=5.43 Å) and GaAs (a=5.65 Å) give rise to a relatively large mismatch of about 4 %. Due to the lattice mismatch, nucleation difficulties in the initial stages of growth can occur and may result in the formation of lattice defects. It has been suggested that these difficulties can be avoided to a great extent by first growing a buffer layer, for instance germanium (Sheldon et al. (1984), Fletcher et al. (1984)). Layers of $Ge_xSi_{1-x}$ films on Si(100) have also been reported (Bean et al. (1984a), (1984b), Fiory et al. (1984), Kasper et al.(1975)) which can be used for this purpose. The growth of epitaxial GaAs layers directly on single crystal Si(001) substrates by MBE has also been succesfully performed (Wang (1984), Tsaur and Metze (1984)). However, antiphase disorder can occur and may influence the bulk material properties away from the hetero-structure interface. Wang (1984) has shown that the antiphase disorder in GaAs directly grown on (001)Si can be suppressed when the Si surface is first exposed to an As beam. The first layer at the interface then consists of predominantly As-Si bonds, and contains steps of even numbers of atomic layer height. Very recently we reported a

first attempt to grow Si directly on (001) oriented GaAs substrates using
MBE (Zalm et al. (1985). In-situ Low Energy Electron Diffraction (LEED)
measurements showed that, up to a layer thickness of about 10 monolayers
($\sim 14$ Å) the unit mesh of the Si has the same size as the unit mesh of
GaAs, which indicates pseudomorphic growth. For a layer thickness of 100
monolayers the unit cell has shrunk to its normal size. Rutherford Backscattering (RBS) measurements of 5000 Å thick layers, using ion channelling, revealed high disorder at the GaAs-Si interface, but the Si crystallinity improved markedly towards the Si-surface. With Medium Energy
Ion Scattering (MEIS) it was established that no island formation occurs
during the initial stages of growth. Here we report the first results of
TEM investigations in cross-sectional and planar modes of lattice defects
in samples with overlayers of 250 Å and 2000 Å, grown at 600°C.

## 2. Experimental

Polished (001) oriented GaAs wafers of 7x25 mm$^2$ and 0.4 mm thick, cut
from Cz grown crystals, were mounted in a Si:MBE apparatus described in
detail elsewhere (de Jong et al. (1983)). Clean surfaces were prepared by
bombardment with 600 eV Ar-ions to a total dose of about $1\times10^{16}$cm$^{-2}$. The
disorder introduced by sputtering was removed by thermal annealing for 90
minutes at 600°C. Si layers of different thicknesses were deposited at
600°C. Part of the in-situ analysis equipment consisted of a four-grid
LEED system. After removal from the UHV chamber completed samples were
investigated by RBS, ion channelling, MEIS and by TEM. Samples with overlayers of 250 Å and 2000 Å were studied by TEM in cross-sectional and
planar modes. The plan view specimens were prepared by drilling ultra-
sonically 3 mm discs from the wafer. They were thinned from the rear side
by jet-etching with chlorinated methanol as etchant.The etching was conducted until the GaAs substrate was completely removed from a small
region (0.5 mm diameter) in the middle of the specimen. There, the Si
layer can be examined without interference from the substrate, whereas at
the edges both the Si-layer and the GaAs substrate can be studied on top
of each other.
Cross-sectional specimens of (110) orientation were made by first cutting
two bars out of the wafer along (110) directions and gluing them together
with the epilayers facing each other. Then slices were cut off the sand-
wich, mechanically polished and further thinned by Ar-ion milling at
about 6 kV on a rotating sample stage.
Ion milling is carried out at a relatively large angle of incidence of
25° with the sample surface and at 6 kV, in order to obtain a high
sputter rate. The final milling is carried out at an angle of incidence
of 10°, in order to smooth the surface of the specimen (Bulle-Lieuwma
(1985)) and at reduced voltage to remove to a great extent the amorphous
top layer caused by the Ar$^+$ bombardment. In spite of the difference in
the sputter etch rates of GaAs and Si by a factor of 1.5, thin smooth interface regions can be obtained by milling at grazing incidence.
TEM micrographs were obtained with a Philips EM 420 ST microscope,
equipped with a double tilt goniometer stage and LaB$_6$ filament, operated
at 120 kV. The high resolution structure images were obtained along the
<110> direction by axial illumination with 9 beams contributing to the
image (of (111), (002), and (220)-type).

## 3. Results and Discussion

Planar sections of Si layers of 250 Å and 2000 Å revealed Moiré fringes

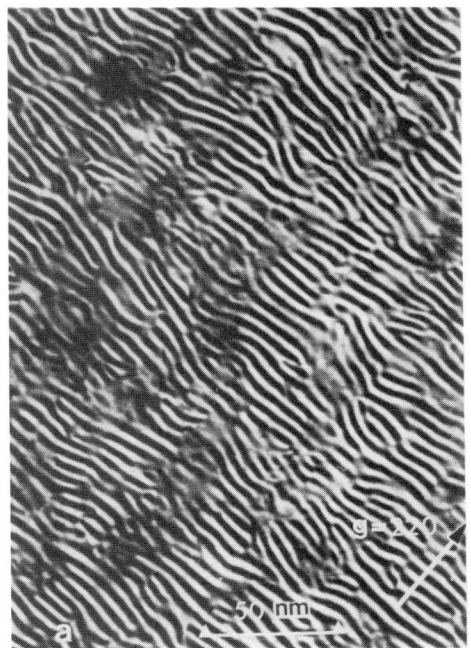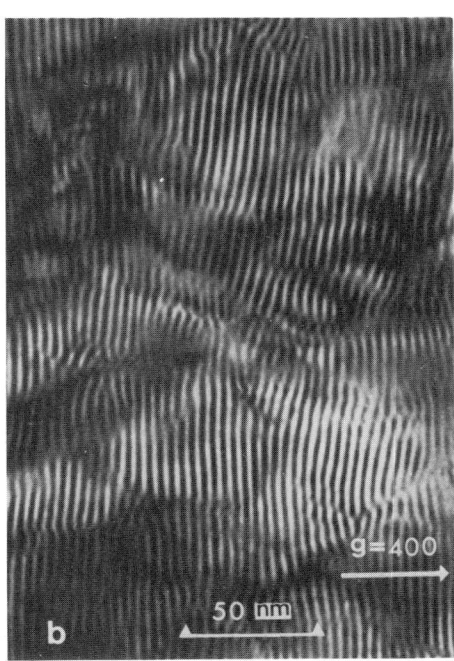

Fig. 1. Bright Field images of a planar section of a Si layer of 2000 Å
on (001) GaAs, showing Moiré fringes and dislocations for a)
g=(?20) and b) g=(400).

perpendicular to the chosen diffraction vectors g, respectively g=220,
g=2$\bar{2}$0,g=400 and g=040 (fig.1). This means, that there is a small difference in lattice spacing between the epitaxial layer and the substrate
and that on the average no rotational misorientation of the layers has
been produced. The periodicity is in accordance with the 4 % lattice mismatch between GaAs and Si.
It can also be seen in fig. 1 that the fringes are locally bent.    At
some places they even fail to continue.Although we have not yet analyzed
these phenomena in detail, most of the distortions of the Moiré fringes
in Fig. 1 can be attributed to the presence of dislocations threading
through the Si-epilayer. The average dislocation distance is about 200 Å.
The dislocations are most pronounced in the Moiré fringes obtained with
g=220. It means that they are preferably located on (110) planes. When
they are also glissile and their habit planes are of (111) type, they
have to be threading through the silicon layer preferably along <112> directions.
In those regions of the specimens where the GaAs substrate had been removed completely, a high density of dislocations was observed (fig. 2a).
With weak beam imaging, tangled dislocations could be resolved and seem
to be dissociated (fig. 2b). Stereo microscopy revealed that these dislocations are threading through the Si layer from the surface to the
interface and back with only short segments along the interface.In some
regions a square network of dislocations has partly been formed. In
cross-sectional specimens, imaged under dynamical two-beam conditions,
the threading dislocations could also clearly be seen (fig. 3). Their
directions are very close to the <112> -direction, proposed above.

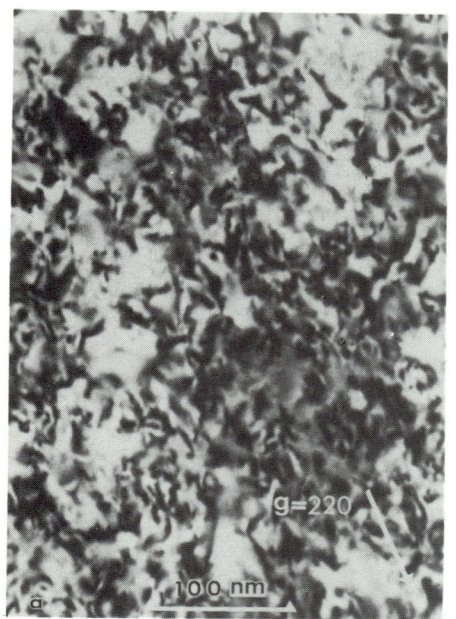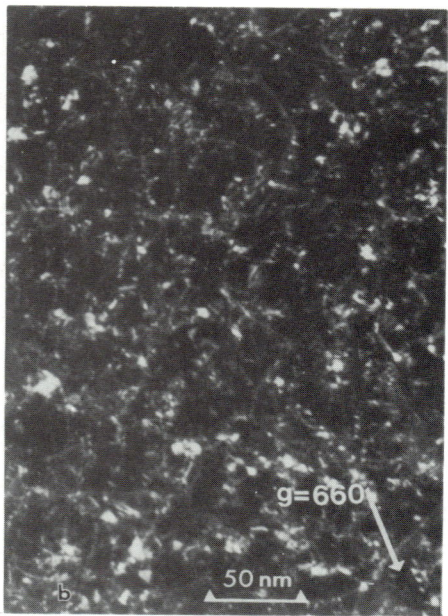

Fig. 2 a) Transmission Bright Field plan view image of 2000 Å Si(001).
The GaAs substrate has been completely removed in this region.
b) Weak Beam micrograph of Si(001), showing threading dislocations
with short segments along the interface.

The crystallographic quality of the interface region was further investigated by High Resolution Electron Microscopy (HREM). Although there is a
relatively large mismatch between GaAs and Si it is evident from fig. 4
that Si has been grown coherently on GaAs. At the interface region there
is a gradual transition from GaAs to Si over a distance of about 5 monolayers. Such a transition can be described by an intermixing of the
elements or by an abrupt transition on an interface, which contains atomic steps. The roughness is possibly due to the previously described
cleaning procedure of the GaAs surface prior to MBE growth.
The defect structure, as revealed in fig. 4, consists of perfect dislocations located close to the interface and small stacking faults on (111)
planes. Their number is largest near the interface decreasing towards the
top of the silicon layer. Because LEED measurements indicated that
pseudomorphic growth occurs up to a film thickness of only 14 Å, we
attribute the observed defect structure to the relaxation of the misfit
strain in layers of increased thickness. Below the critical thickness
there is a planar tensile strain in the Si film equal to the misfit defined as $\Delta a/a$, where $\Delta a$ is the difference between the lattice parameters
of GaAs and Si. The misfit strain can first be accomodated in a homogeneous tetragonal distortion, so that pseudomorphic growth occurs. The
most common way in which the misfit strain may relax is by the formation
of 60-degree dislocations, running along the interface    (Matthews
(1975)). These dislocations are generally dissociated into a 30-degree
and a 90-degree Shockley partial dislocation. Each of them may glide on
the (111) planes towards the interface.
In a previous study of Si grown on GaP (Viegers et al. (1984)) we have

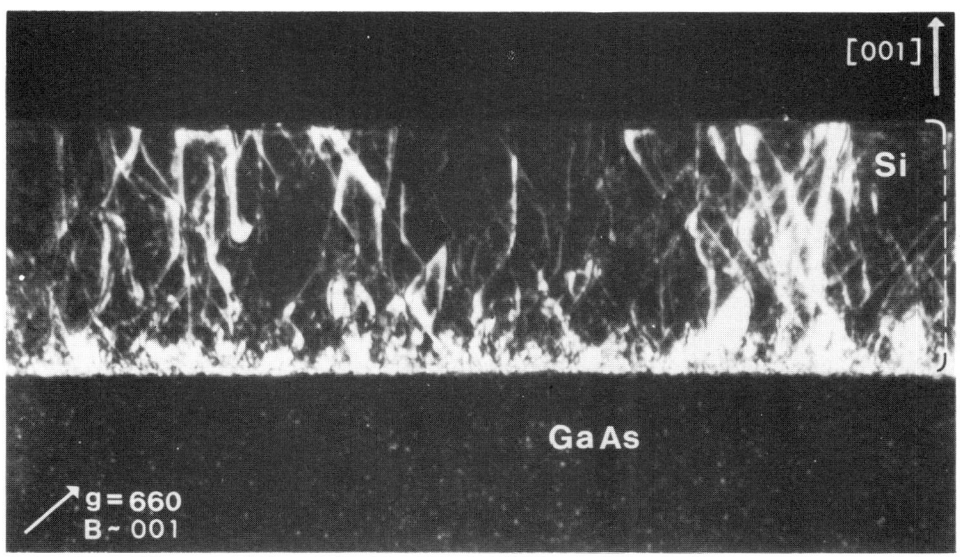

Fig. 3  Dark Field cross-sectional image along <110> of 2000 Å Si on GaAs(001). In the Si layer a high density of dislocations is present. In the GaAs substrate no defects are observed.

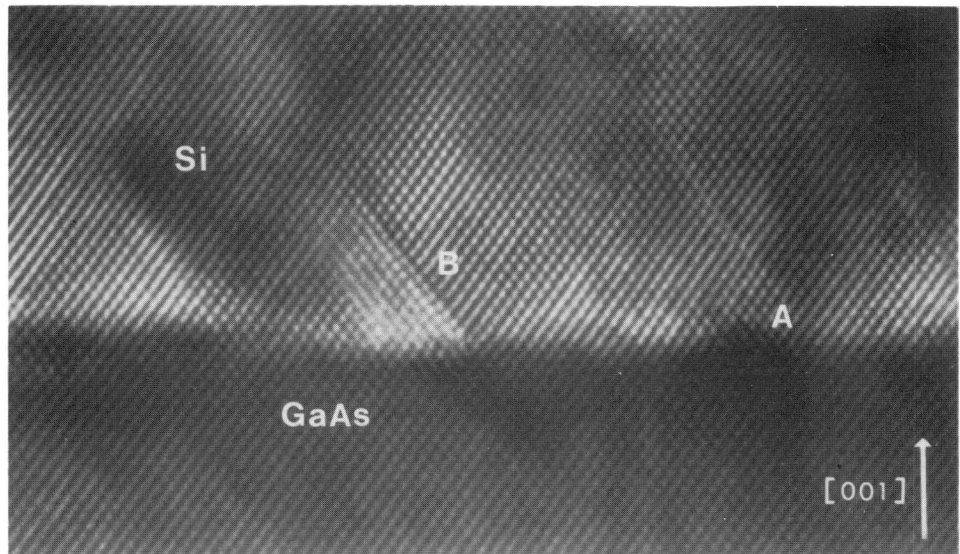

Fig. 4  HREM-image along <110> of the GaAs/Si interface region, showing perfect dislocations (A) close to the interface and small stacking faults (B) on (111) planes.

argued that the unit displacement b, which results in a 60-degree dislocation, is achieved by two movements, first by b1, yielding a 90-degree Shockley partial dislocation, and then by b2, yielding a 30-degree Shockley partial dislocation. The geometry of Si on (001) GaAs is the same and therefore causes the 90-degree partial dislocation to nucleate first. Along these lines the stacking faults in fig. 4 have to be interpreted as dissociated 60-degree dislocations. However, in comparison with Si grown on GaP with a mismatch of only 0.4 %, we now have a much larger strain field and a correspondingly increased density of dislocations. In plan view specimens we only saw short segments in the interfacial plane, so it is obvious that the glide process has been disturbed by mutual interactions between the dislocations.
From the Si on (001) GaP studies (Viegers et al. (1984), we concluded that there must be a substantial barrier to dislocation formation preventing thermodynamic equilibrium. To investigate the existence of a similar barrier in the case of Si on GaAs we compared the defect density in layers of different thicknesses. We found that in layers of 2000 Å grown at 600°C and in layers of 250 Å grown at 500, 550 and 600°C the defect structures were essentially similar. This indicates that there is no thermodynamic equilibrium between strain and dislocations. It means that either the misfit strain has already been relaxed completely at a thickness smaller than 250 Å, or there indeed exists a barrier for dislocation formation. In the latter case it promises that extended defects can be prevented, for instance by preventing the occurrence of nucleation sites or by reduction of the growth temperature, which reduces the mobility of dislocations.

## Acknowledgements

All samples were grown at the "FOM Instituut voor Atoom en Molecuul Fysica" in Amsterdam. We are indebted to Prof.Dr. F.W. Saris for his generous hospitality and Mr. P.M.J. Marée for assistance during sample growth. We also like to thank Dr. J. Haisma for stimulating interest.

## References

Bean J C, Sheng T T, Feldman L C, Fiory A T and Lynch R T 1984a Appl. Phys. Lett. 44 102
Bean J C, Feldman L C, Fiory A T, Nakahara S and Robinson I K 1984b J. Vac. Sci. Tech. A2 436
Bulle-Lieuwma C W T 1985, to be published
Fiory A T, Bean J C, Feldman L C and Robinson I K 1984 J. Appl. Phys. 56 1227
Fletcher R M, Ken Wagner D and Ballantyne J M 1984 Proc. MRS Conf. Boston Vol. 25 (New York: North Holland) 417
de Jong T, Douma W A S, Smit L, Korablev V V and Saris F W 1983 J. Vac. Sci. Tech . B1 888
Kasper E, Herzog H J and Kibbel H 1975 Appl. Phys. 8 199
Matthews J W 1975 Epitaxial Growth, Part B, ed J W Matthews (New York: Academic Press) 559
Sheldon P, Jones J M, Hayes R E, Tsaur B-Y and Fan J C C 1984 Appl. Phys. Lett. 45 274
Tsaur B-Y and Metze G M 1984 Appl. Phys. Lett. 45 535
Viegers M P A, Bulle-Lieuwma C W T, Zalm P C and Marée P M J 1984 Proc. MRS Conf. Boston
Wang I J 1984 Appl. Phys. Lett. 44 114
Zalm P C, Marée P M J and Olthof R I J 1985 Appl. Phys. Lett. 46

*Inst. Phys. Conf. Ser. No. 76: Section 4*
*Paper presented at Microsc. Semicond. Mater. Conf., Oxford, 25–27 March 1985*

# Electron microscope characterisation of silicon layers grown by molecular beam epitaxy

N G Chew, A G Cullis, C A Warwick, D J Robbins, R W Hardeman and D B Gasson

Royal Signals and Radar Establishment, St Andrews Road, Malvern, Worcs. WR14 3PS

Abstract    Transmission electron microscopy has been used to study epitaxial Si layers deposited onto single crystal Si substrates by molecular beam epitaxy.   The structural results obtained have been correlated with spectrally resolved luminescence data acquired using an SEM based low temperature cathodoluminescence system.  The dependence of layer defect structure upon surface cleaning procedures and deposition temperature have been characterised and the effects of particle inclusion during growth determined.

## 1. Introduction

The growth of thin semiconductor layers by molecular beam epitaxy (MBE) is currently attracting great scientific and technological interest.  This ultra-high-vacuum-based process enables the growth of epitaxial layers with very precise control of individual layer thickness, interface sharpness and doping profiles.  To realise the full potential of this growth technique it is necessary to characterise the growth process and the resulting layers as fully as possible.  Hence, in the present work transmission electron microscopy and scanning cathodoluminescence (CL) have been used to characterise the crystallographic structure arising from growth of MBE Si layers on bulk Si substrates.  The results obtained have been correlated with photoluminescence and secondary ion mass spectrometry (SIMS) data and attributed as far as possible to specific characteristics of the growth process.

## 2. Experimental

The undoped Si layers were deposited onto 2" Si wafers in a VG366 MBE system.  Wafers were cleaned prior to deposition either by simple heating or by sputtering with 5-6keV $Ne^+$ ions. Si evaporation was from an electron-beam heated high purity source, with substrate temperatures generally being maintained in the range 750-850°C.

Specimens for structural characterisation were thinned to electron transparency in either plan or cross-sectional configuration by sequential mechanical polishing and low voltage $Ar^+$ ion-milling.  Thinned samples were examined either in a JEOL 120C transmission electron microscope (TEM) using 120keV electrons or in a JEOL 4000EX (resolution at Scherzer defocus ~1.8Å). High resolution images were recorded on <110> projection using axial illumination at 400keV.  Spectrally resolved cathodoluminescence assessment was carried out on samples cooled to ~10K in a scanning electron microscope operated at 10-20keV beam energy.  Beam currents employed were typically

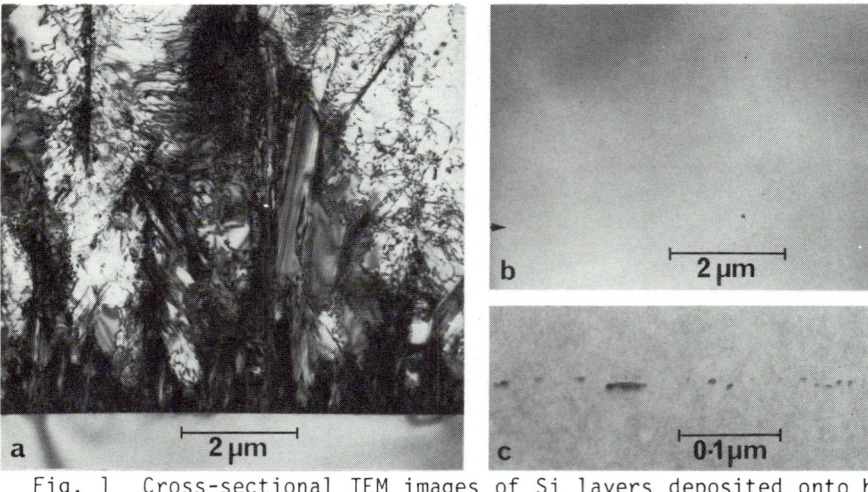

Fig. 1 Cross-sectional TEM images of Si layers deposited onto heat cleaned (001) Si substrates at: (a) 750°C; (b) 850°C (interface arrowed); (c) interface region of layer shown in (b).

~50nA. Wavelength dispersive optics gave a spectral resolution of ~10nm and the detector used was a liquid nitrogen cooled Ge p-i-n diode (noise equivalent power $10^{-15}W(Hz)^{-1/2}$).

## 3. Results and Discussion

### 3.1 Si Layers Grown on Heat Cleaned Substrates

The structure of layers grown onto heat cleaned substrates was found to be very dependent on the substrate temperature during deposition. This is illustrated in Fig. 1 for layers grown at 750°C and 850°C. It is clear from Fig. 1(a) that the layer grown at the lower temperature is highly defective with many inclined microtwin lamellae emanating from the interface region and a very high density of dislocations present throughout the layer. Photoluminescence spectra from this sample exhibited characteristic D1 to D4 peaks which have been previously attributed to recombination at lattice dislocations (Drozdov et al 1976). CL mapping with the D4 luminescence peak gave uniform image contrast which is consistent with the very high dislocation density observed. Layers grown at the higher substrate temperature were substantially defect-free as shown in Fig. 1(b). There was, however, a distribution of small inclusions on the interfacial plane in these layers, as is revealed in the higher magnification image of Fig. 1(c). The photoluminescence spectra from these high quality layers showed no features attributable to crystallographic defects except when layers were deposited at very high growth rates - layers grown under these conditions are discussed in section 3.3.

The observed change of structure with substrate temperature is attributed to the removal of native oxide from the substrate surface by a reactive etching process with the impinging Si beam. It has previously been shown that the heating procedures employed in these studies do not completely remove oxide from the substrate surface (Hardeman et al 1985). At low growth temperatures the proposed reactive etching of the oxide does not proceed sufficiently quickly for the oxide to be removed before Si over-

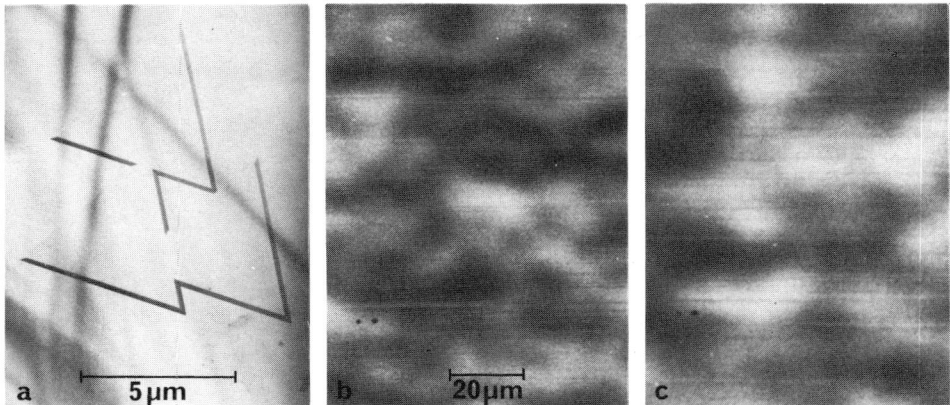

Fig. 2 Si layer deposited onto imperfectly prepared (111) Si substrate.
(a) Plan TEM image; (b) 1.08eV (band-edge) CL image and (c) 1.04eV CL
image.   Note, dark areas of (b) correspond to bright areas of (c).

growth occurs and the highly defective epitaxy of Fig. 1(a) results.   At
higher growth temperatures however the oxide is substantially removed before
deposition takes place and layers of high structural quality are produced.

The interfacial inclusions illustrated in Fig. 1(c) are likely to be
remnants of the surface oxide (see, for example, Ponce (1985));   these
inclusions may be responsible for the nucleation of layer defects observed
occasionally in layers deposited at the higher growth temperature.    An
example of a defective layer grown on a heat cleaned (111) substrate is
shown in the plan view TEM image of Fig. 2(a).   It can be seen that many
inclined stacking faults are present, forming arrays of truncated inverted
tetrahedra.   By measuring the length of the lines of intersection of these
faults with the specimen surface (along <110> directions) it was established that they were, indeed, all nucleated at a depth corresponding to the
substrate/layer interface.   Photoluminescence spectra from this layer showed
weak peaks corresponding to the D1-D4 lines and also bands at 1.036eV and
1.008eV, the latter having previously been attributed to stacking faults in
bulk Si (Sauer et al 1985).   The CL micrographs of Fig. 2 show images
obtained from this sample using 1.08eV band-edge (b) and 1.04eV (c) luminescence.   The excellent correlation between the CL images (dark regions of
(b) corresponding to bright regions of (c)) and the TEM observations demonstrate unambiguously that the 1.05eV luminescence is due to the presence
of grown-in stacking faults.

3.2 Si Layers Grown on Ne$^+$ Sputter Cleaned Substrates

Layers grown on Ne$^+$ sputter cleaned substrates, even at the lower growth
temperatures of ∿750°C, generally contained few extended defects.   However,
for layers deposited at these low temperatures there was, in all cases, a
band of crystallographic damage in the interfacial region.   This is illustrated in the cross-sectional micrograph of Fig. 3(a) where a line of
small dislocation loops is seen to be present just below the layer/substrate
interface.   The interfacial plane itself contains a dispersion of small
precipitates.   The sample of Fig. 3(a) was Ne$^+$ sputter cleaned for 2½mins,
increasing the sputtering time to ∿30mins yielded the structure shown in
Fig. 3(b).   The crystallographic damage is similar to that in (a) but the

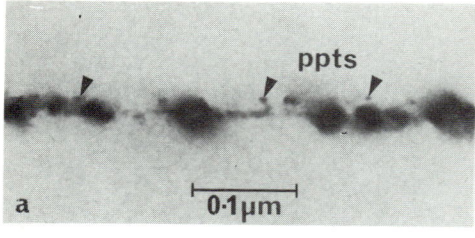

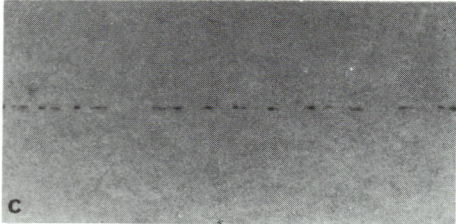

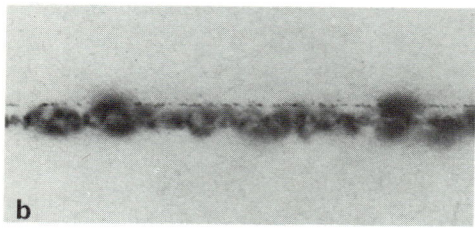

Fig. 3 Cross-sectional TEM images of interface regions of Si layers deposited onto (001) Si substrates after $Ne^+$ sputter cleaning: (a) $2\frac{1}{2}$mins sputtering, 750°C anneal and deposition; (b) 30mins sputtering, 750°C anneal and deposition; (c) $2\frac{1}{2}$mins sputtering, 900°C anneal and 750°C deposition.

number of interfacial precipitates has increased significantly. SIMS data for these specimens showed a heavy metal "spike" (particularly for Mo) at the interface, the magnitude of which increased with $Ne^+$ sputtering time. It is thought likely that these heavy metal impurities, which may result from inadvertent sputtering of surrounding metal within the growth chamber, are aggregated on the interface resulting in the observed localised precipitation. The damage band at the interface is attributed to incomplete annealing of ion damage produced during the $Ne^+$ sputtering. This was confirmed by annealing specimens at higher temperatures (∼900°C for 5mins) before layer deposition commenced; in this case the damage band was completely eliminated although the interfacial precipitation remained.

Although the disordered interface region present in these samples did not, in general, give rise to extended layer defects occasional defects nucleated at the interface were observed - an example present in a layer grown on (011) Si is shown in the high resolution micrograph of Fig. 4.

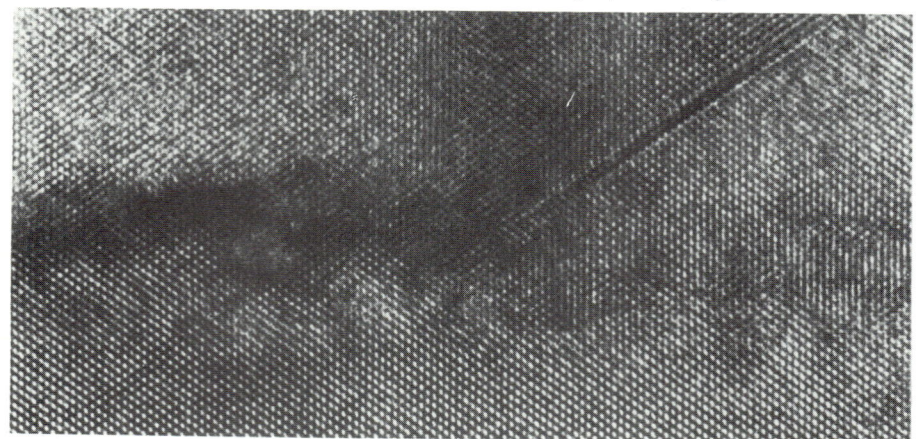

Fig. 4 High resolution cross-sectional TEM image of Si layer deposited onto $Ne^+$ sputter cleaned (011) substrate following 750°C anneal, showing defects present in disordered interface region.

3.3 Layer Defects Arising from Particle Inclusion During Growth

It was noted that the MBE Si layers, particularly when grown under high growth rate conditions, contained many (typically $10^2$-$10^4$cm$^{-2}$) topographical imperfections. These surface defects, which have been observed by other workers (Kubiak et al), are illustrated in the secondary electron image of Fig. 5(a). Photoluminescence spectra from these layers exhibited the characteristic D1-D4 bands and subsequent CL mapping with both band edge (1.10eV) and the D4 (0.99eV) luminescence peaks proved unambiguously that the D4 luminescence was localised and associated with these topographical defects (see Fig. 5).

TEM analysis of plan view specimens for both (001) and (111) orientations showed these defects to consist of a "core" of misoriented polycrystals and microtwin lamellae surrounded by arrays of stacking faults on the inclined {111} planes, as shown in Fig. 6(a) and (b). The origin of these defects is elucidated in the cross-sectional micrograph of Fig. 6(c). In this sample the interface was delineated by an ion-cleaning damage band and it can be seen that during layer growth a polycrystalline particle (believed to be Si) became attached to the sample surface and locally inhibited layer growth, resulting in the formation of a cavity. Misoriented material was nucleated at this perturbation and defect formation led to the structure previously described. It is thought likely that these gross defects, which are apparently present throughout the layer, may all be attributed to the inclusion of particles which originate within the apparatus during growth.

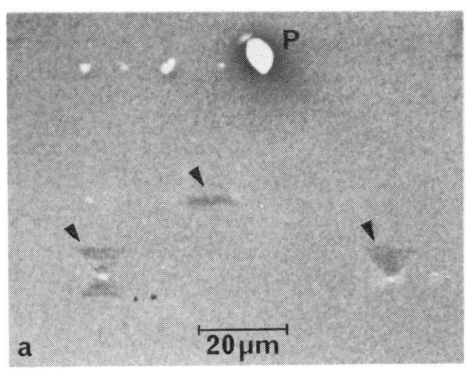

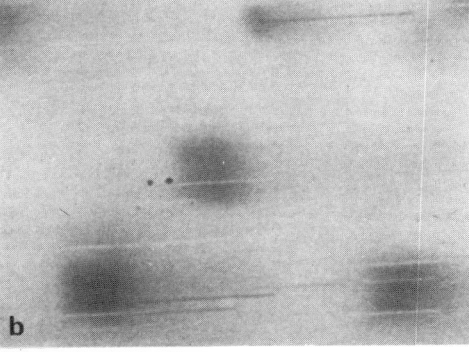

Fig. 5 Si layer deposited onto (001) Si substrate under high growth rate conditions: (a) secondary electron image; (b) 1.10eV (band-edge) CL image and (c) 0.99eV (D4) CL image. Note particle on surface (P) shadowing band-edge luminescence, also the correspondence of dark spots in (b) and bright spots in (c) with the array of surface defects arrowed in (a).

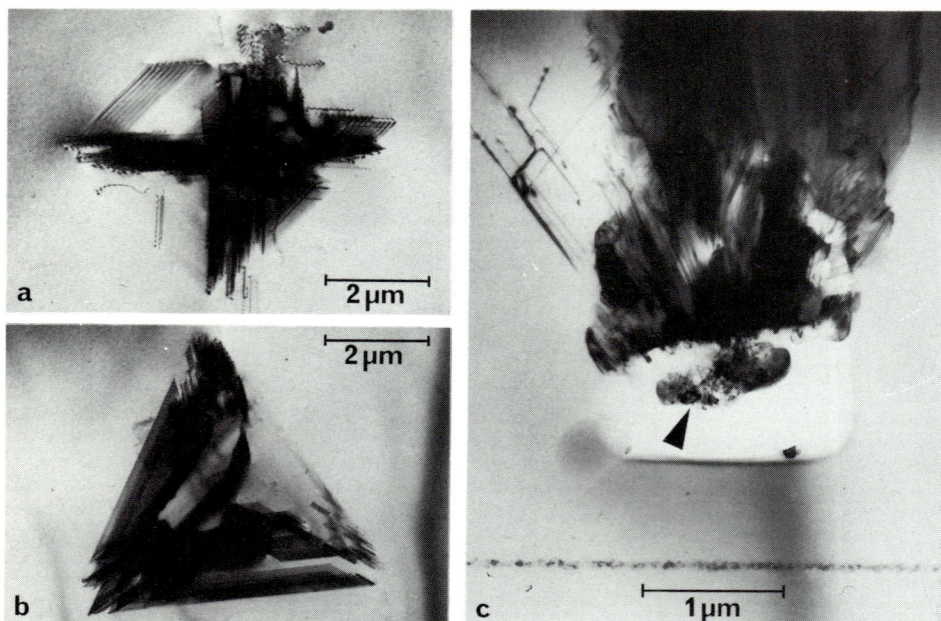

Fig. 6   TEM images of defects present in Si layers grown under high growth rate conditions: (a) Plan image of (001) layer; (b) Plan image of (111) layer and (c) cross-sectional image of (001) layer deposited onto Ne⁺ sputter cleaned substrate (particulate inclusion arrowed).

## 4. Conclusions

The nature of Si substrate cleaning procedures critically affects the structural quality of homoepitaxial MBE Si layers.  The removal of native oxide from heat cleaned substrates by a temperature dependent reactive etching process has been demonstrated.  In addition, the possible incorporation of crystallographic damage and heavy metal impurities during ion cleaning procedures is described.  The nature of topographical surface defects in MBE layers has been characterised and their origin attributed to the inclusion of particles of polycrystalline Si during growth.  This work has also illustrated the important role of scanning cathodoluminescence in correlating TEM microstructural studies and luminescence data.

## References

Drozdov N A, Patrin A A and Tkachev V D 1976 JETP Lett. $\underline{23}$ 597
Hardeman R W, Robbins D J, Gasson D B and Daw A 1985 Proc. 1st Int. Symp.
   on Si MBE (Pennington: Electrochem. Soc.)
Kubiak R A A, Leong W Y and Parker E H C Private Communication
Ponce F A 1985 This proceedings volume p 1
Sauer R, Weber J, Stolz J, Weber E R, Kusters K-H and Alexander H 1985
   Appl. Phys. $\underline{A36}$ 1

*Inst. Phys. Conf. Ser. No. 76: Section 4*
*Paper presented at Microsc. Semicond. Mater. Conf., Oxford, 25–27 March 1985*

# STEM characterization of precipitates in Czochralski-grown silicon

D M Dlamini

Cavendish Laboratory, Madingley Road, Cambridge CB3 OHE

Abstract    Observations of Czochralski-grown (CZ) silicon by STEM have revealed the presence of precipitates of tin about 4 nm in diameter. They appear in the form of clusters approximately 1 µm in size. In transmission EBIC the particles produce an enhanced bright contrast associated with Z-contrast effects, whereas in bulk EBIC in the SEM the clusters appear dark because they act mainly as recombination centres. The total amount of tin is surprisingly high (5 ppm) and is thought to be introduced during crystal growth but can be put into solution by ageing at elevated temperatures.

## 1. Introduction

Metallic impurities (e.g. precipitates) are known to degrade the characteristics of semiconductor devices and circuits by acting as shorts or high-field regions when they are located in the vicinity of p-n junctions (Cullis et al (1974) and Goetzberger et al (1960)). Due to practical demands in microelectronic device fabrication, their effects have been viewed with increased concern.

Electrical or electronic characterization of impurities has been frequently performed on bulk materials and devices with the aid of a scanning electron microscope (SEM) in the EBIC (electron beam induced current) mode; such an instrument is also capable of providing information about the chemical composition of imperfections. However, in spite of its wide application it is limited by its effective probe size, particularly where particles with nanometre dimensions are to be examined.

For this reason, use can be made of a Vacuum Generators HB5 STEM (scanning transmission electron microscope) for observing the structural, chemical and electrical properties of precipitates in commercially-available silicon single crystals. Previously, Fathy and Pennycook (1981) have applied a similar technique and reported microanalysis of nickel and copper precipitates in silicon epitaxial layers.

## 2. Experimental

The CZ silicon materials were supplied in wafer form by Siltronix (Switzerland) and are n-type (P-doped), {100} oriented, with a resistivity of 10 ohm-cm, and are 0.25 mm thick.

Specimen preparation was carried out by initially cutting 3 mm discs from the wafers with an ultrasonic drill, followed by chemical polishing to a final thickness less than 400 nm in a solution consisting of 1 part HF (48%) to 8 parts $HNO_3$ (70%). For purposes of EBIC studies, a layer of 5N aluminium, less than 50 nm thick, was deposited on one side of the thin sample to introduce a Schottky barrier. The resultant specimen was mounted on a double-tilt EBIC cartridge for analysis.

3.  Results and Analysis

Structural examination of the specimens revealed the presence of precipitate clusters less than 1 μm in size. Fig. 1(a) is an example of a micrograph showing the clusters and fig. 1(b) is a higher magnification image of the cluster marked H. The clusters do not show any contrast due to strain in the matrix and their density (about $2 \times 10^9$ m$^{-2}$) is observed to increase with foil thickness, indicating that they are uniformly dispersed throughout the volume of the specimen.

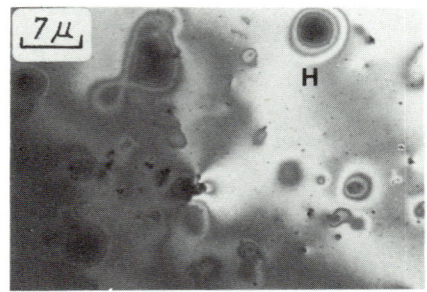

Fig. 1. (a) Illustration of precipitate          (b) Magnified image of cluster
            clusters in CZ silicon.                           marked H.

The clusters are composed of minute, spherical precipitates approximately 4 nm in size. Figs. 2(a) and (b) are respectively bright field and annular dark field images of the precipitates. (The particles marked A and B were contaminated by the electron beam probe during X-ray microanalysis.)

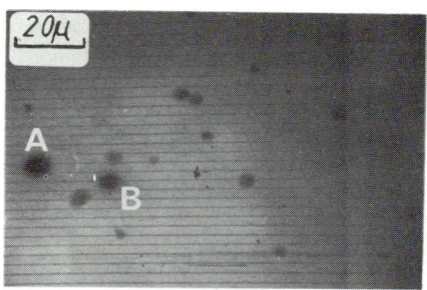

(a)  Bright field.                                  (b)  Annular dark field.

Fig. 2.  Images of individual precipitates within a cluster.

Notice the Z-contrast effect in the ADF image, suggesting that the atomic number of the precipitates is higher than for silicon. The precipitates are crystalline in nature in accordance with the observation of electron diffraction spots on CBED patterns as well as the appearance of Moiré fringes.

Chemical identification of several particles from different batches of wafers leads to the conclusion that they are composed of tin. Figs. 3(a) and (b) are typical X-ray spectra from a precipitate and silicon matrix respectively. Two strong peaks appear in the former: (i) a silicon $K_\alpha$ peak at 1.74 keV and (ii) a tin $L_\beta$ peak at about 3.66 keV. There is at least one other peak at approximately 4 keV, as expected for tin.

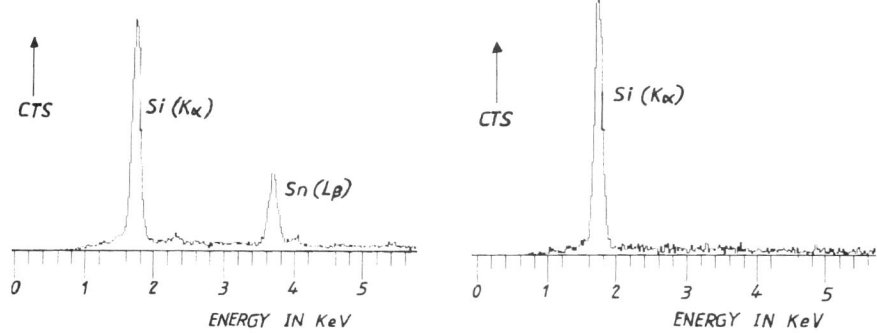

Fig. 3:  X-ray spectra from  (a) a precipitate    (b) silicon matrix

The estimated area density of the (individual) precipitates within a cluster is 2 x $10^{15}$ m$^{-2}$, whilst the number of tin atoms per cubic metre is 5 x $10^{25}$, so that the overall atomic fraction becomes approximately 5 ppm. Trumbore's solid solubility data (Trumbore, 1960) gives a comparable tin concentration of the order of $10^{25}$ atoms/m$^3$.

Electrical analysis by EBIC has shown that the tin precipitates produce a strong bright contrast as illustrated in the EBIC image (fig. 4(b)). Fig. 4(a) is a corresponding bright field micrograph. Several authors (e.g. Chynoweth et al (1959), Ravi et al (1973) and Kittler (1980)) have previously interpreted similar bright contrast seen in the SEM in terms of depletion region effects, charge multiplication, microplasma formation, etc. However, it seems no mention has been made so far about the atomic nature of the defect in question relative to the surrounding matrix.

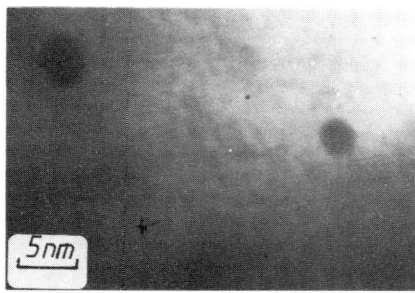

Fig. 4: (a) Bright field micrographs showing individual tin precipitates.

(b) EBIC micrograph revealing bright contrast from precipitates.

In the present report, the observation is accounted for by two contributions, bearing in mind that the region of the (thin) Schottky diode in which the precipitates are situated will essentially be depleted of carriers: (i) Z-contrast and (ii) field concentration effects - the atomic number of tin, Z(Sn) is about four times Z(Si), so that carrier production in the former is expected to be effectively higher. On the other hand, the maximum field at the precipitate is calculated to be three times the applied field (i.e. the depletion-region field). Therefore, a combination of the two processes is believed to explain the comparatively pronounced collection of carriers excited in the region of the precipitate, hence the observed bright contrast. Note that the image shows very high spatial resolution as expected for these mechanisms.

Results based on heat-treatment experiments suggest that the tin can be dissolved by ageing for one hour at temperatures around 800°C and then cooling to room temperature through radiation. This information is in good agreement with data published by Trumbore, Isenberg and Porbansky (1958).

It can be concluded that the presence of tin precipitates (with such high concentrations) can easily affect the performance of microcircuits by lowering their breakdown voltages as a consequence of the enhanced fields caused by the metallic precipitates. Further, a cluster crossing a junction can be assumed to cause leakage currents through short-circuiting. This problem could be reduced by annealing the initial wafers prior to circuit fabrication, provided the electrical effects of the dispersed solute are negligible.

## Acknowledgements

The author wishes to thank Dr. L.M. Brown for his guidance and critical discussions during the course of this work. Thanks are also due to Professor U. Valdrè (University of Bologna, Italy) for advice on experimental work and for useful comments. Finally, the author acknowledges contributions from many colleagues, particularly present and former members of the Microstructural Physics group.

## References

Chynoweth A G and Mckay K G 1959 J. Appl. Phys. 30 1811
Cullis A G and Katz L E 1974 Phil. Mag. 30 1419
Fathy D and Pennycook S J 1981 Inst. Phys. Conf. Ser. No. 60 243
Goetzberger A and Shockley W 1960 J. Appl. Phys. 31 1821
Kittler M 1980 Kristall und Technik 15 575
Trumbore F A., Isenberg C R and Porbansky E M 1958 J. Phys. Chem. Solids 9 60
Trumbore F A 1960 Bell System Tech. J. 39 205

*Inst. Phys. Conf. Ser. No. 76: Section 4*
*Paper presented at Microsc. Semicond. Mater. Conf., Oxford, 25–27 March 1985*

139

# TEM and RBS studies of hydrogen implanted silicon and silicon carbide

S Wood,* J Greggi, Jr,* W J Choyke,▲ J A Spitznagel,▲  N J Doyle,+
and R B Irwin+

*Westinghouse R&D Center, Pittsburgh, PA 15235
+University of Pittsburgh, PA  15260
▲Westinghouse R&D Center and University of Pittsburgh

Abstract    Single crystal silicon and silicon carbide were implanted at
temperatures of 96°K, 300°K and 800°K with 50 keV to 80 keV $H^+$ ions to
fluences of 2 x $10^{17}$ $H^+/cm^2$ or 8 x $10^{17}$ $H^+/cm^2$. Post implantation
annealing was conducted at $\geq 800°K$. Microstructural changes were
investigated with cross-sectional TEM and RBS/channeling. Results show
that for silicon, a highly hydrogen doped amorphous layer is formed at
96°K whereas at 300°K, crystallinity is maintained despite hydrogen
trapping. Annealing behavior depends strongly upon the as-implanted
microstructure and the stability of the Si-H layer. Silicon carbide
develops an amorphous damage layer at temperatures $\leq 1000°K$ when
implanted with 8 x $10^{17}$ $H^+/cm^2$.

## 1.  Introduction

Hydrogenated amorphous silicon and silicon based alloys are of interest for
their use in solar cell applications (Roedern, et al, 1984). The
deliberate introduction of hydrogen by direct implantation is of increasing
importance for the minimization of leakage currents and the introduction
of shallow donors in devices and for the passivation of silicon solar
cells. Thus, it is important to understand the nature of the damage
introduced by hydrogen doping and the interaction of hydrogen with the
implantation produced defects. We report here the continuation of a
study on single crystal FZ <111> silicon and present some initial results
on silicon carbide.

Previous cross-section transmission electron microscopy (XTEM) and
Rutherford Backscattering (RBS) data by Wood, et al (1983) and Choyke,
et al (1983) have shown that implantation of monoenergetic (50 or 75 keV)
hydrogen into single crystal FZ silicon at 95°K and 300°K produces a well
defined damage band for fluences $\geq$ 1 x $10^{16}$ $H^+/cm^2$ initially centered near
$x_m$, the peak of the damage distribution calculated from the energy
deposited in elastic scattering. Increasing the hydrogen fluence at 95°K
and 300°K produces an asymmetric broadening of the damage band towards the
front (implanted) surface. At 95°K and 2 x $10^{17}$ $H^+/cm^2$, Choyke, et al
(1985) observed the formation of an amorphous zone whereas at 300°K, the
entire damage zone retained its crystallinity.

●Supported in part by NSF Grant DMR-84-03596.

a)   96°K

b)   96°K + 800°K/12800s
     anneal

c)   800°K

d)   300°K

e)   300°K + 800°K/12800s
     anneal

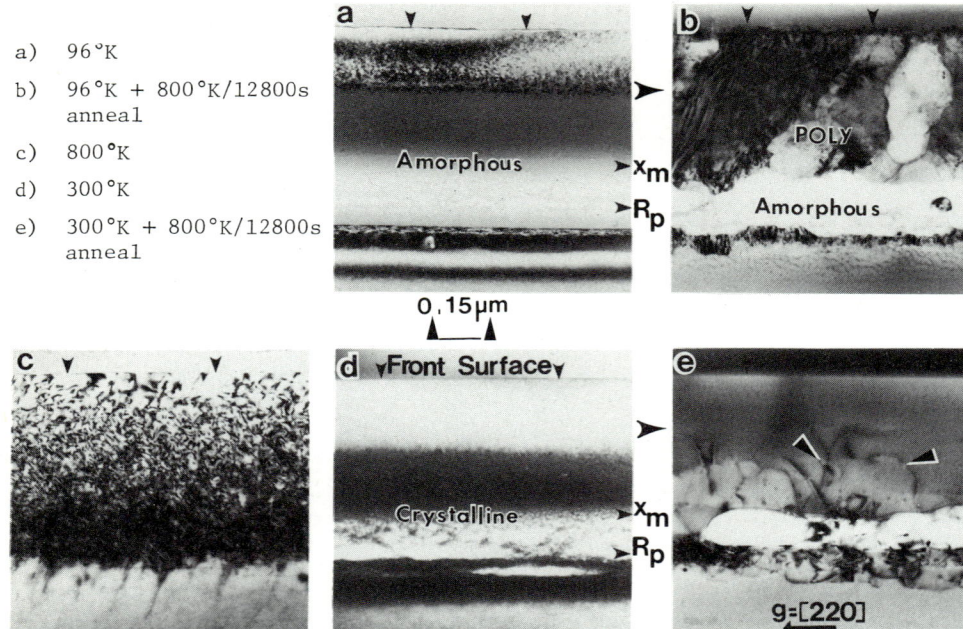

Figure 1.    XTEM BF microstructures obtained from as-implanted and implanted
             and annealed silicon.  All specimens were implanted to 8 X 10$^{17}$
             H$^+$/cm$^2$ at 50 keV.

The present study concentrates on a comparison of as-implanted and implanted
plus annealed silicon and silicon carbide.  The 800°K anneal temperature
selected for silicon was based on observations by Chu, et al (1977) which
suggested a precipitous drop in the direct backscattering yield and the
onset of pronounced dechanneling in the temperature range 723°K - 823°K for
<100>CZ silicon implanted with hydrogen at elevated temperature.  For
silicon carbide, the anneal temperature was scaled with respect to that
utilized for silicon based on anticipated vacancy mobility and the Debye
temperature.

2.    Experiment

Samples used were either Monsanto N-type float zone <111> Si with a
conductivity of 30-60 Ωcm or single crystal <0001> SiC, predominantly 6H.
All Si samples were cleaned using a 16 step procedure used for integrated
circuit device fabrication.  SiC crystals were heated at 900°C in air for
∿5 minutes, then immersed in HF for ∿20 minutes, followed by water, acetone
and alcohol rinses.  They were finally wiped dry with lint free paper.
TEM specimens were either cut from 1 X 1 cm squares used for RBS/channeling
analyses or directly implanted 2 mm X 2 mm squares.  Hydrogen implants
were generally made at 50 keV for silicon and 80 keV for SiC with a 200 keV
implanter equipped with post acceleration magnetic mass separation.  All
samples were implanted at a beam current of 10 μA/cm$^2$ at pressures of
10$^{-7}$ Torr.  Hydrocarbon build-up was minimized by cryogenic shielding.
Post-implantation annealing of Si was also conducted in vacuum at pressures
∿10$^{-7}$ Torr, whereas SiC was annealed in argon.  Annealing and implantation
times were equivalent.  Channeling measurements were conducted using a

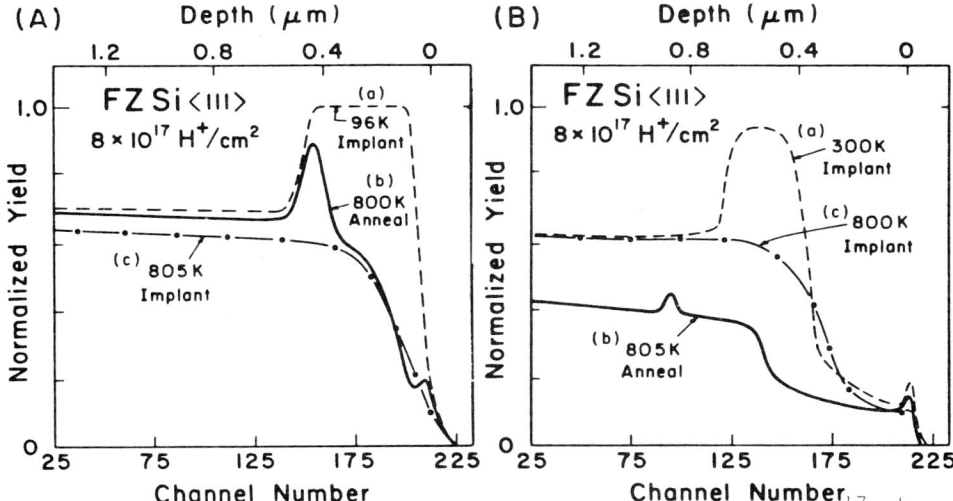

Figure 2.   Channeling spectra from FZ <111> Si implanted to 8 X 10$^{17}$ H$^+$/ cm$^2$, comparing low temperature implants with annealed and high temperature (800°K) implants.

1.5 MeV beam from a 2 MeV Van de Graaff.  Beam currents were normally 15 μA and the spot size was 0.8 mm.  The beam divergence was less than 0.03° and backscattering measurements were made at 168°.  Preparation of cross-section TEM specimens was described previously by Wood, et al (1983). An additional dimpling step has been added to reduce ion milling time. All TEM studies were performed on a Philips 400T electron microscope operating at 120 keV.

3.   Results and Discussion

Most emphasis was placed upon specimens implanted to 8 X 10$^{17}$ H$^+$/cm$^2$, some of which were annealed for 12,800s to correspond to their implantation time at a beam current of 10 μA/cm$^2$.  Microstructures imaged using a 2-beam dynamical diffraction condition  generally near a [112] zone and with g = [220] are presented in Figure 1 for silicon.  Corresponding RBS/ channeling curves are shown in Figure 2.

3.1  Comparison of 96°K and 300°K Silicon Implanted to 8 X 10$^{17}$ H$^+$/cm$^2$

As observed previously at lower fluences, implantation at 96°K produces an amorphous zone whereas at 300°K, crystallinity is retained, despite the formation of a highly doped layer whose imaging characteristics are very different from those of damaged single crystal silicon.  The amorphous zone formed at 96°K now has two distinctly different layers - one located approximately between $x_m$ and $R_p$ (end of projected range for the implanted H) in which most of the trapped hydrogen is located, and a second zone closer to the surface.  Microdiffraction shows that the ring patterns are somewhat different for these two regions, indicating variations in short range order.  The amorphous zone is bounded by two crystalline damage layers and the front one extends to the implant surface.  Higher magnification images indicate that this layer is composed of a high number density of small (<5 nm) defect clusters, similar to the structure of the damage layer at 300°K.  The sharp interface between the amorphous and crystalline zones at

96°K is consistent with the concept of a damage threshold for amorphicity, as predicted by Christel, Gibbons and Sigmon (1981).

Direct comparison of TEM and RBS data can be made for the 96°K specimens since all implants were made at 50 keV. At 300°K, however, we are confined to a qualitative comparison because the RBS channeling samples were implanted at 75 keV. The direct backscattering peak (Figure 2A) at 96°K correlates with the location of the amorphous zone delineated by the TEM. At 300°K, the normalized yield does not approach that of the random, which is consistent with the formation of a highly damaged but crystalline region. The total hydrogen retained in the damage zone (calculated from the dip in the random RBS curve) was $\sim$60% of that implanted at 96°K and $\sim$40% at 300°K. Its location corresponded to the rear portion of the direct backscattering peak and its concentration was saturated at $\sim$3.7 X $10^{17}$ $H^+$/ $cm^2$. This indicates an average stoichiometry of $SiH_n$ with n = 0.8 $\pm$ 0.2, in agreement with previous IR measurements by Stein and Peercy (1980) suggesting monohydride formation.

## 3.2  800°K, 8 X $10^{17}$ $H^+$/$cm^2$ Implant into Silicon

Implantation at 800°K (Figure 1c) produces a damage zone comprised of better defined dislocations and loops which extends from $R_p$ to the implant surface. Weak beam dark field images suggest the linear defects extending beyond $R_p$ are narrow twins. The total hydrogen retained in the damage zone was zero.

## 3.3  Implantation to 8 X $10^{17}$ $H^+$/$cm^2$ Followed by an 800°K Anneal

Annealed microstructures are presented in Figures 1b and 1e. The 300°K implant yields a predominantly single crystal layer with a high residual dislocation density and some polycrystalline silicon near $R_p$. Line dislocations have grown beyond the as-implanted damage layer towards the front surface. Many are pinned by bubbles or voids. Since no hydrogen dip was observed (Irwin, 1984), either no hydrogen is present in the voids, or the hydrogen concentration varies sufficiently slowly that it cannot be detected by this technique. Voids are present throughout the entire region between $R_p$ and the front surface. There is also a "line" of large, elongated, low atomic density features at $x_m$ which correspond to the region of high hydrogen doping in the as-implanted crystal. In the 96°K implanted and annealed specimen (Figure 1b), it is polycrystalline silicon growth which has dominated the recovery process. A residual hydrogen doped layer remains, located approximately between $x_m$ and $R_p$. A narrow layer of residual damage, with some polycrystalline silicon also exists beyond $R_p$. It is clear that the $SiH_n$ phase formed by a 96°K hydrogen implant is more stable than that developed at 300°K presumably because of as-implanted amorphicity. Comparison of the XTEM microstructures with the RBS/ channeling curves indicates the following. In the 96°K annealed specimen, the steep dechanneling which begins at the front surface is due to the polycrystalline silicon. The direct backscattering peak is from the residual amorphous $SiH_n$ layer. In the 300°K annealed specimen, the steep dechanneling portion of the curve is due primarily to the high dislocation density near $R_p$ (the small amount of polycrystalline silicon also contributes to this).

## 3.4  Implantation to 2 X $10^{17}$ $H^+$/$cm^2$ Followed by an 800°K Anneal

This yields microstructures (Figure 3) very similar to those already discussed in Figure 1e. The retained hydrogen concentration is less than half of the saturated value of $\sim$3 X $10^{17}$ $H/cm^2$ (Choyke, et al, 1985). Thus,

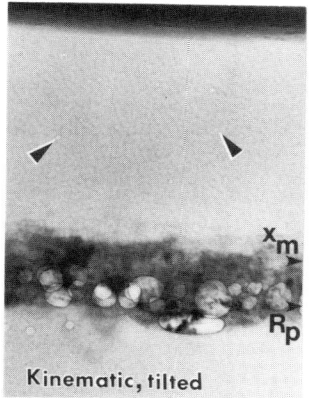
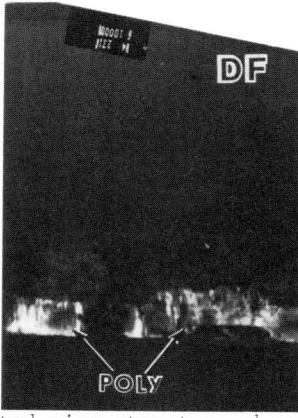

Figure 3. Microstructure of FZ <111> Si Implanted to 2 X $10^{17}$ $H^+/cm^2$ at 50 keV and 96°K after Annealing at 800°K for 3300s

even though the as-implanted microstructure shows that a narrow layer (∼0.1 μm wide) centered at $x_m$ was amorphous, the hydrogen concentration was apparently too low to stabilize the $SiH_n$ phase for these annealing conditions. Polycrystalline silicon formation is more pronounced than in the 300°K/annealed 8 X $10^{17}$ $H^+/cm^2$ implant specimen and is usually associated with large (<1 μm diameter) voids. Smaller voids are present throughout the entire region between $x_m$ and the front surface.

3.5  Implanted and Implanted plus Annealed SiC

Microstructures of as-implanted and annealed 6H <0001> SiC crystals are presented in Figure 4. Surprisingly, implantation of 8 X $10^{17}$ $H^+/cm^2$ at 1000°K produces an amorphous zone bounded by two highly damaged regions. At 300°K, the amorphous zone shows a dual structure similar to that discussed previously for silicon. Its boundaries are clearly defined by the disappearance of the 6H fringes in the lattice image (not shown). The hydrogen dip indicates that most of the retained hydrogen is present at depths between $x_m$ and the back damage zone. At a fluence of 8 X $10^{17}$ $H^+/cm^2$, Irwin (1984) calculated that the average stoichiometry of the hydrogen region has saturated at $SiCH_{n'}$ with n' = 1.0 ± 0.2. SAD ring patterns from the two zones indicate differences in short range order.

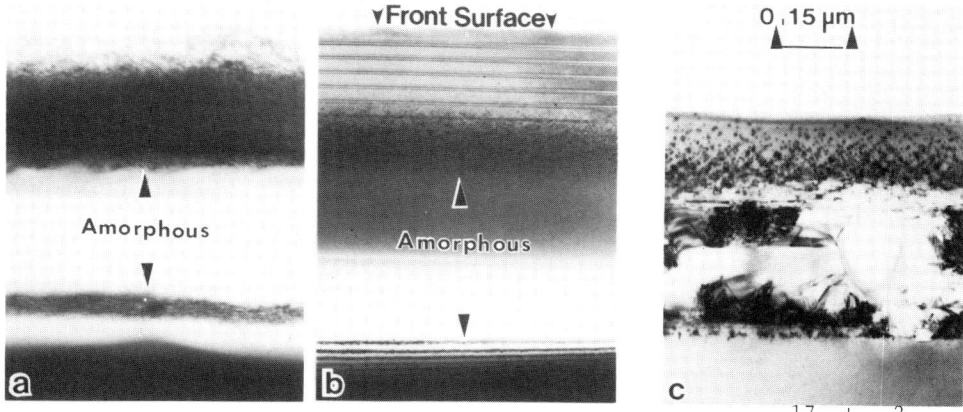

Figure 4. XTEM of 6H <0001> SiC after Implantation to 8 X $10^{17}$ $H^+/cm^2$ at 80 keV: a) at 1000°K; b) at 300°K: c) at 300°K after Annealing at 1600°K for 12,800s

Annealing produces a complex microstructure. In the layers corresponding
to the as-implanted damage regions in front of and behind the amorphous
layer, the high density of "black spot" damage zones have coarsened to form
a relatively high number density of resolvable ($\leq$15 nm diameter) dislocation loops. In the front layer, the loop density decreases progressively
such that no line defects are observed between the front surface and a
depth of $\sim$0.1 μm. Annealing of the as-implanted amorphous layers produces
a highly faulted structure with polycrystals and large void formation.
It is clear that the $SiCH_n$ phase was not stable at 1600°K. This is
consistent with luminescence data indicating that the C-H bond breaks
$\sim$1500°K since it is probably stronger than the Si-H bond (Choyke, 1985).

4.  Summary

Implantation of hydrogen in single crystal Si produces an amorphous zone
at 96°K but not at $>$300°K. In SiC, the damage zone was amorphous at
$T \leq 1000°K$, at $8 \times 10^{17}$ $H^+/cm^2$. At this fluence, the amorphous zone
exhibits a dual layer character in both materials, with most of the
retained hydrogen trapped in the deeper layer. Silicon recrystallization
behavior and stability of the Si-H phase depend strongly upon implant
temperature and fluence (i.e. hydrogen concentration). An amorphous Si-H
layer, with a saturated hydrogen concentration $\sim$3.7 $\times$ $10^{17}$ $H/cm^2$ remains
stable during an 800°K anneal, whereas a crystalline layer with the same
concentration does not. Polycrystalline silicon formation dominates
recovery in the former (96°K implant) while single crystal regrowth
dominates after 300°K implantation. Amorphous layers with <3.7 $\times$ $10^{17}$ $H/cm^2$
also decompose upon annealing and only a small region of polycrystalline
silicon is produced, near $R_p$. Annealing SiC at 1600°K also causes
decomposition of the hydrogen rich layer (even though the saturation
hydrogen concentration was exceeded) and a highly faulted structure with
some polycrystal formation results.

References

Choyke W J, Irwin R B, McGruer J N, Townsend J R, Doyle N J, Hall B O,
    Spitznagel J A and Wood S, 1983 Nucl. Instr. and Methods, 209/210 407

Choyke W J, Spitznagel J A, Doyle N J, Wood S, and Irwin R B, 1985 MRS
    Proceedings

Choyke W J, 1985 Unpublished research

Christel L A, Gibbons J F and Sigmon T W, 1981, J. Appl. Phys. 52 7143

Chu W K, Kastl R H, Lever R F, Mader S and Masters B J, 1977 "Ion Implanta-
    tion in Semiconductors 1976" ed. Chernow, Borders and Brice, Plenum
    Press, 483

Irwin  R B, "RBS/Channeling Analysis of Hydrogen Implanted Single Crystals
    of FZ Silicon and 6H Silicon Carbide", 1984  Ph.D Thesis, U. of Pittsburgh

Roedern B von, Mahan A H, Williamson D L and Madan A, 1984 5th Symposium on
    "Materials and New Processing Technologies for Photovoltaics"

Wood S, Greggi Jr J, Spitznagel J A, Doyle N J, Irwin R B, Townsend J R,
    and Choyke W J, 1983 Inst. Phys. Conf. Ser. No. 67, 247

*Inst. Phys. Conf. Ser. No. 76: Section 4*
*Paper presented at Microsc. Semicond. Mater. Conf., Oxford, 25–27 March 1985*

145

# The morphology of LPCVD polycrystalline Si

R E  Mallard+, D C  Houghton+*, G J C  Carpenter** and F R  Shepherd+

+ Bell-Northern Research, Ottawa, Canada
* Present address:  National Research Council, Ottawa, Canada
**PMRL, CANMET, Energy, Mines and Resources, Ottawa, Canada

Abstract   The abnormally rapid growth of isolated grains in LPCVD polysilicon which can result in a rough nodular morphology has been investigated.   Comparison of as deposited nodular polysilicon layers bearing a high density of large silicon grains with smooth polysilicon layers using XTEM shows that accelerated growth of favorably oriented grains begins at the columnar growth stage in the deposition.   No impurities in the nodular regions could be detected by STEM-EDX, EEL imaging in TEM, SIMS or Auger microanalysis.   The nodular growth may be suppressed by the two stage deposition of an amorphous Si layer followed by a regular polysilicon layer.

## 1.  Introduction

Low pressure chemical vapour deposition (LPCVD) at temperatures $\sim620°C$ is the standard deposition process to produce high quality polysilicon films for MOS devices.   The growth, structural and electrical properties of such films have been extensively studied (Anderson 1973, Kinsbron et al 1983, Kamins et al 1973, Kamins et al 1978, Kamins 1980, Harbeke et al 1984, Hendriks et al 1984).   Although films with a smooth surface morphology and reproducible grain structure are routinely obtained, occasionally abnormally rapid grain growth occurs in localised areas, resulting in a rough nodular or "hazy" surface which is unsuitable for device fabrication.   Optical inspection of hazed wafers, because of the distribution of the hazy nodules on the surface, shows that in many cases the occurance of haze is related to the presence of particulates or impurities which modify the growth behaviour of the films.   This paper discusses the role of impurities and deposition temperature on the grain growth and morphology of LPCVD poly-Si.

## 2.  Experimental

LPCVD Si films were deposited onto n-type $4-6\Omega$cm Si wafers which had a 100 nm thermal oxide.   Poly-Si layers were deposited from decomposition of $SiH_4$ at a pressure of 200 mtorr and a substrate temperature of $\sim 625°C$, whereas amorphous $\alpha$-Si layers were deposited at substrate temperatures below $\sim 580°C$.   The morphology and structure of both hazy and regular poly-Si films was studied by SEM and by both plan view and cross-section TEM.   Surface impurity segregation was assessed using AES and bulk impurity concentrations were determined by SIMS and Spark Source Mass Spectroscopy (SSMS). The possibility of segregation of impurities to grain boundaries and regions of high fault density was investigated using a dedicated STEM (VGHB5) and a new analytical microscope, the EEL imaging TEM (ZEISS EM902).

## 3. Results and Discussion

The SEM micrographs in figure 1 show the typical surface morphology of (a) regular polysilicon and (b) polysilicon with $\sim10^6$ cm$^{-2}$ of nodules produced by localised abnormal grain growth.

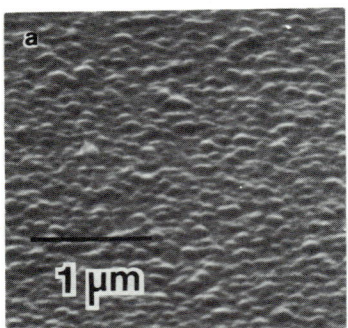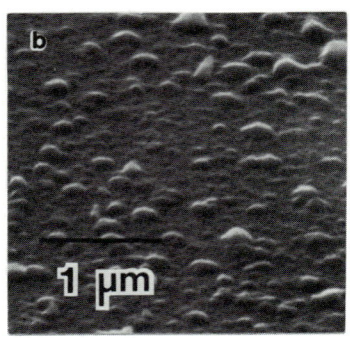

Fig. 1 SEM micrographs of surface of LPCVD polysilicon layers deposited at 625°C. (a) "Regular" polysilicon (b) Hazy polysilicon exhibiting abnormal grain growth.

The growth morphology of these films which were deposited under nominally identical conditons is shown in cross section in figure 2. The larger grains giving rise to the haze radiate from some growth centre which occurs in the region where the fine equiaxed grain structure is being overtaken by a columnar structure. A qualitative indication of the strong <110> texture of these films is seen in fig. 2c, which is a dark field image of a hazy "nodule" in a more advanced growth state, formed from the (220) spot highlighted in the corresponding SADP in (d). It was clear that the relative intensity of the {111} ring was considerably reduced in comparison with plan view specimens, as expected from the texture. The large grains, like the neighbouring predominantly <110> oriented columnar "regular" grains, are heavily twinned with many faults along {111} planes which lie roughly perpendicular to the substrate and parallel to the growth direction. The textured columnar structure develops because of granular epitaxial growth which occurs preferentially on {110} planes over the temperature region of this deposition ($\sim0.5T_m$).

To determine the bulk impurity concentration of polysilicon layers with and without a high incidence of nodules, SIMS, WDX-ray and SSMS were carried out. No significant differences in impurity levels were observed. In addition, surface analysis by AES and Auger sputter depth profiling revealed no evidence of impurities.

In view of the possibility that the nuclei or impurity rich regions of hazy nodules may be highly localised, two microanalytical techniques

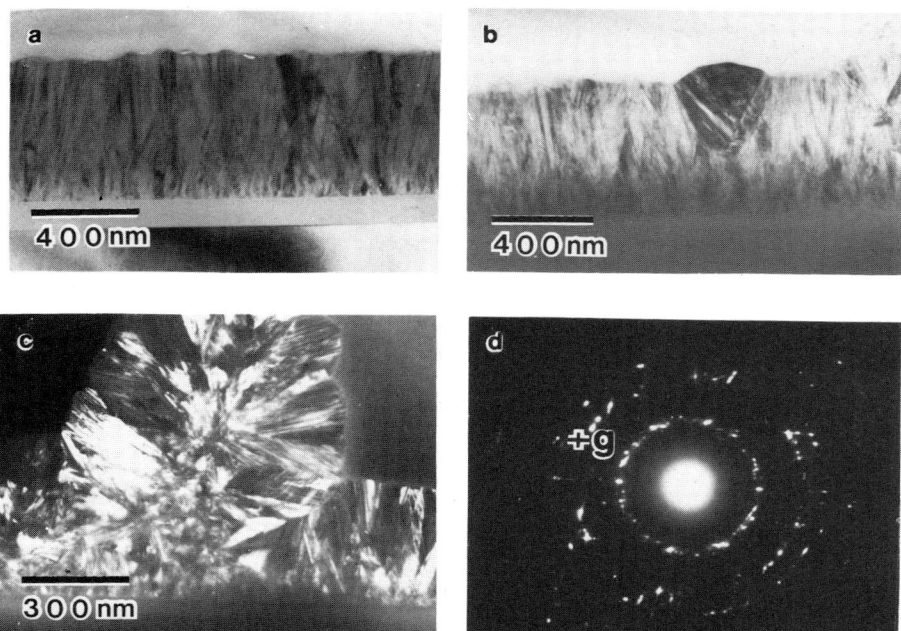

Fig. 2 XTEM micrographs of LPCVD polysilicon deposited at 625°C. (a)
regular polysilicon. (b) Hazy "nodules" as in fig. 1(b). (c) Dark
field XTEM micrograph from a (220) reflection of a large hazy nodule
(d) SADP from nodule shown in (c).

with high spatial resolution were employed:  STEM-EDX and EEL imaging
TEM.  For elements where $Z \geqslant 11$, FEG STEM-EDX allows the use of a 1 nm
probe at 1 nA and 100 keV, with a minimum detectable mass of $\sim 10^{-21}$g.
This corresponds, in a $\sim$ 50 nm thick specimen, to a minimum
detectability limit of $\sim$ 200 impurity atoms or 1/10 of a monolayer in
a sub-boundary oriented parallel to the probe.  In spite of this high
spatial resolution, STEM microanalysis did not identify regions of
high impurity concentration.

The EEL imaging microscope was used to investigate the surface or
boundary segregation of light elements B, C, N and O which would not
be detectable by STEM-EDX.  In biological test specimens a minimum
detectable mass of $2 \times 10^{-21}$g and a spatial resolution of 0.3 - 0.5 nm
have been demonstrated (Adamson-Sharpe and Ottensmeyer 1981).  To
calibrate the sensitivity and spatial resolution of this technique for
semiconductor applications a 3 nm oxide present on the polysilicon
surface prior to cross sectioning for XTEM microanalysis was examined.
This oxide layer is clearly visible by EEL contrast in figure 3.
However, identical analytical procedures  for B, C, N and O detection
in hazy polysilicon indicated that the concentrations of these
elements were below the sensitivity of the EEL imaging technique.

Although no impurities were detected by any of the analytical
techniques tried, there is nevertheless strong evidence to support the

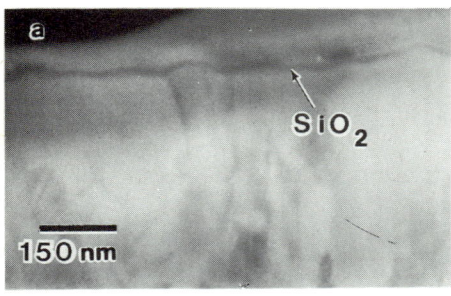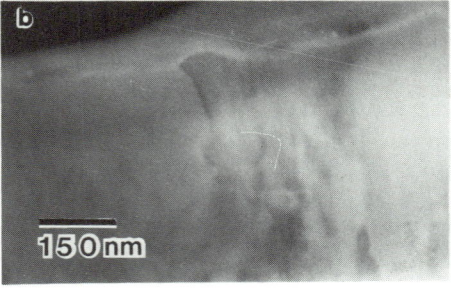

Fig. 3   EEL image micrographs of x-section through annealed LPCVD
polysilicon showing a 3nm surface oxide (a) taken at 510±7.5eV and (b)
taken at 550±7.5eV.

suggestion that the haze is contamination related because of the
spatial distribution of the large haze grains.  This contamination is
likely therefore in small localised concentrations below our detection
limits.    It is useful to consider several possible impurity related
origins of haze which include:

a) wafer borne particulates as a possible source of nuclei
   initiating rapid grain growth during early stages of
   deposition.
b) gas borne particulates
c) vapour phase nucleation of silicon due to instabilities in
   $SiH_4$.  Solid silicon particles condense in the vapour and form
   a "soot" on the growing polysilicon surface.   The
   microstructure of these nodules would not resemble the
   columnar morphology as exhibited in figure 2.
d) a localised source of contamination in the $SiH_4$ source gas or
   purging gas.

Impurities from any of the above sources may alter or catalyse the
local growth rate.

The rate controlling step for the growth of LPCVD silicon layers from
kinetic considerations appears to be the adsorption of $SiH_4$ on the
growing interface (Kinsbron et al 1983).   Consequently an increase in
the number of active sites through an increase in fault density or the
presence of a catalyst may drastically alter the local growth rate.
The mobility of an impurtiy within the polysilicon layer during growth
may be considered by assuming grain boundary diffusion to dominate and
estimating the supply of impurity atoms from the bulk to the surface.
The rate of incorporation of an impurity in the growing layer can be
compared with this out diffusion flux.   Accelerated growth will be
maintained when

$$C_{is} \cdot V \leqslant Dgb(dCi/dx) \cdot \partial/l \cdot K,$$

where   $C_{is}$   is   the   surface   impurity   concentration   necessary   to
enhance the growth rate, V the polysilicon growth rate, Dgb is grain
boundary diffusivity, (dCi/dx) the impurity concentration gradient
through the thickness of polysilicon, $\partial$ and l are the grain boundary
width and spacing, and k is a geometrical factor ~2.    Inserting
reasonable values for these parameters at 600°C:   $C_{is}$ ~ 0.01; V,
$10^{-8}$ $cmS^{-1}$; Dgb (Burger and Donovan 1967), $10^{-8}cm^2S^{-1}$; dCi/dx, $10^4cm^{-1}$
and $\partial/l \cdot k$, 0.01 reveals that the grain boundary flux is at least 4

orders of magnitude greater than that required to maintain a surface concentration of ~1% of a potential catalyst (e.g., Fe (Houghton and Carpenter 1982)) during growth at normal rates. An impurity element at that surface concentration would have been detected by the analytical techniques tried (eg. SIMS), even if it were homogeneously distributed in the film. It is likely therefore that the impurity, after participating in the accelerated growth mechanism, would have had to again diffuse to the new surface or have been volatilized.

Although the exact role of impurities in influencing the growth rate of crystals during LPCVD deposition is not fully understood, one method in controlling the formation of haze would be to alter the deposition scheme to confine impurity elements or to modify the growth process in such a way that the nucleation of fast growing <110> oriented haze grains is supressed.

It has been proposed (Gardiner et al 1982) that a two stage deposition process, involving the deposition of a thin amorphous $\alpha$-Si layer at Ts $\leqslant 600°C$ followed by a thick conventional polysilicon layer at Ts$\geqslant 600°C$, will improve the specular uniformity and decrease the incidence of haze in polysilicon films. A XTEM micrograph through one of these so called "low-high" films is shown in fig. 4. The 50nm thick amorphous layer crystallizes during the high temperature deposition. The grain size of the crystallised $\alpha$-Si is large and the density of faults low compared to the adjacent "high" temperature Si. This will potentially limit the flux of a diffusing impurity species to the deposition surface. Crystallised amorphous layers usually exhibit textures unlike the <110> preferred orientation of conventional polysilicon (Anderson, 1973, Kinsbron et al 1983, Kamins 1980, Harbeke et al 1984) and would therefore tend to delay the development of a <110> texture in the "high" films. The detailed structure of "low-high" films and their role in preventing the occurence of haze will be the subject of future investigations.

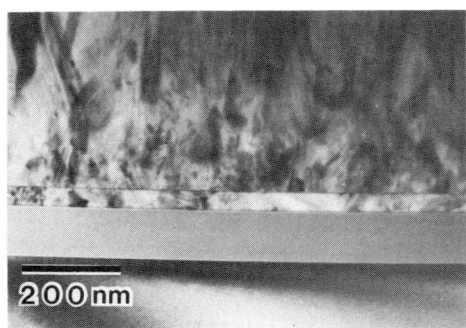

Fig. 4 XTEM micrograph showing microstructure of annealed "low-high" composite polysilicon.

4. Summary

A comprehensive microstructural and microanalytical investigation of the haze phenonmenon in polysilicon deposition has failed to reveal the source of nuclei or identify any impurity elements. There is strong evidence from the distribution across a wafer that the haze is caused by contamination. Although we have detected impurities in other samples (Houghton and Carpenter, 1983), the impurity levels in these samples are below the detection limits of the analytical techniques used. Localised impurities on rapidly growing surfaces of very small concentrations (say ~1 atomic%) might account for the enhanced growth rate observed in hazy polysilicon and impurities in this concentration and distribution might not be detected by any of the techniques tried. Since the standard polysilicon films exhibit a very high density of rapid diffusion paths along <111> oriented faults, these impurities may easily travel to the surface to induce the enhanced deposition rate. The occurrence of abnormal grain growth appears to be suppressed by utilizing a low temperture α-Si film as a layer prior to conventional polysilicon deposition. The crystallised α-Si is structurally different from polysilicon deposited at higher temperatures and might inhibit the transport of impurity atoms to the growth surface, while also influencing the development of the texture of the high temperature film.

## 5.  Acknowledgements

This work was supported by Northern Telecom Electronics Limited. The technical assistance and suggestions of several colleagues, particularly J. Powell, J. Boyd and J. Ellul at NTEL and Marc Charest at CANMET are gratefully acknowledged. P. Ottensmeyer (University of Toronto) and G.R. Piercy (McMaster University) made available EEL, TEM and STEM-EDX facilities respectively.

## References

Adamson-Sharpe K.M. and Ottensmeyer P. (1981), J. of Microscopy
    122 309
Anderson R.M. (1973), J. Electrochem. Soc. 120 1540
Burger R.M., Donovan R.P. (1967), in Physics of Semiconductor Devices,
    ed. S.M. Sze (New York: Prentice Hall) p.31. Value quoted is for
    Fe diffusion through bulk Si. Grain boundary value is likely to be
    even higher.
Gardiner J.R., Makarewicz S.R. and Pliskin W. (1982), IBM Technical
    Disclosure Bulletin 24 11A
Harbeke G., Krausbauer L., Steigmeler E.F., Widmer A.E., Kappert H.F.
    and Neugebauer G. (1984), J. Electrochem. Soc. 131 675
Hendriks M., Radelaar S., Beers A.M. and Bloem J. (1984), Thin Solid
    Films 113 59
Houghton D.C., Carpenter G.J.C. (1983), MRS Proc. Electron Mic. Mat.
    Vol 31, eds. Krakow, W., Smith, D.A., Hobbs, L.W. (New York:
    Elsevier) p.255
Kamins T.I. (1980), J. Electrochem. Soc. 127 686
Kamins T.I., Mandurah M.M. and Saraswat K.C.(1978) J. Electrochem.
    Soc. 125 927
Kamins T.I. and Cass T.R. (1973), Thin Solid Films 16 147
Kinsbron E., Sternhem M. and Knoell R. (1983), Appl. Phys. Lett.
    42 835

*Inst. Phys. Conf. Ser. No. 76: Section 4*
*Paper presented at Microsc. Semicond. Mater. Conf., Oxford, 25–27 March 1985*

# Grain growth in heavily implanted LPCVD polysilicon layers during annealing

A H Reader, F W Schapink* and S Radelaar*

Philips Research Laboratories
P.O. Box 80.000, 5600 JA Eindhoven, The Netherlands.
* Laboratory of Metallurgy, Delft University of Technology,
Rotterdamseweg 137, 2628 AL Delft, The Netherlands

Abstract   Polycrystalline silicon layers were heavily
implanted (doses >5x10$^{14}$ atoms cm$^{-2}$) with either
phosphorus or arsenic ions. Direct heating in a
transmission electron microscope revealed that a rapid
increase in the average grain size of the layers
occurs within the first few minutes of an anneal above
700°C. The larger grains form by the crystallization of
the amorphous implanted film produced by the
implantation. The crystallization mechanism observed
favours lateral growth of grains in particular
orientations and produces the noted ⟨110⟩ texture.

## 1. Introduction

Polycrystalline silicon (polysilicon) layers are widely
used in the semiconductor industry as interconnect material
between devices and as gate electrodes for MOS transistors
and capacitors. The layers are normally deposited by the
chemical vapour deposition (CVD) technique on oxide covered
subtrates. For the above applications layers must be highly
doped as low resistivity material is required. Doping by ion
implantation is frequently carried out because good control
over the impurity profiles can be obtained. Wada and
Nishimatsu (1978) found that considerable grain growth can
occur during the anneals required for further wafer
processing. In this paper, we present our investigations of
the mechanism of grain growth in implanted polysilicon
layers. These investigations have been confined to the growth
that occurs below 1000°C, as it is thought that different
mechanisms operate at higher temperatures (Wada and
Nishimatsu 1978). Both cross-sectional and plan-view
transmission electron microscope (TEM) specimens have been
prepared to study the morphology changes within the layers
that occur during annealing. The growth mechanism has been
investigated by in situ TEM heating experiments.

## 2. Experimental

Low pressure (LP) CVD polysilicon layers, approx. 0.46 μm
thick, were grown on silicon coated ⟨100⟩ silicon substrates
at Philips Research Laboratories. The layers were then ion
implanted with either phosphorus or arsenic at 150 keV with
doses of greater than $5 \times 10^{14}$ atoms cm$^{-2}$. An implantation
voltage of 150 keV was chosen as the majority of the
resulting dopant profile (Gibbons 1980) is contained within
the top half of the layer. The implanted wafers were cut
into smaller samples, some of which were isothermally
furnace-annealed at 950°C (± 10°C) in an argon gas stream.
Specimens suitable for electron microscopy were produced
from implanted and from unimplanted and annealed samples,
respectively, by mechanical polishing and ion-beam thinning
from the substrate side (plan-view specimens). Specimens of
implanted samples therefore consisted of a thin electron
transparent amorphous silicon film (∝-film) around the
central hole, supported by a ring of unimplanted CVD layer
and substrate (fig. 1). These implanted specimens were used
in the in situ TEM heating experiment to determine the grain
growth mechanism. It should be noted that the electron
transparent area of annealed specimens contains the majority
of the implanted part of the polysilicon layer. The specimen
preparation technique removes the substrate and the deeper,
unimplanted regions of the layer. Cross-sectional TEM
specimens were manufactured using the method of Pettit and
Booker (1971). Specimens were examined in a Philips EM400T
operating at 120 keV.

## 3. Results

Figures 2 and 3 show in situ TEM annealing experiments
carried out on implanted plan-view specimens containing both
an ∝-film and some of the underlying polysilicon layer. The
micrographs in these figures were obtained from specimens
implanted with arsenic or phosphorus with doses of $5 \times 10^{14}$
and $2 \times 10^{16}$ atoms cm$^{-2}$ respectively. Note that the average
grain size increased very rapidly while the temperature of
specimen was still rising to 950°C (figure 2). Simultaneously, the electron diffraction patterns indicated an increase
in the crystallinity in the layer. These observations
suggest that the rapid growth phenomenon is associated with
crystallization of the ∝-film produced by ion-implantation.
Heavily phosphorus implanted layers also exhibited this
behaviour when subjected to similar in-situ anneals.
Figure 3 shows that the rapid increase in grain size
which is associated with the crystallization of the ∝-film
occurs during anneals above 700°C. The specimen used in this
case was isothermally in situ annealed for 4 hours at 20°
intervals from 650°C. No observable change in grain morphology was noted during anneals to 730°C. Crystallization of
the   ∝ -film occurred during the anneal at 750°C which
resulted in an increased grain size as seen in figure 3b.
Subsequent heat treatments at temperatures up to 790°C did
not produce any further growth.

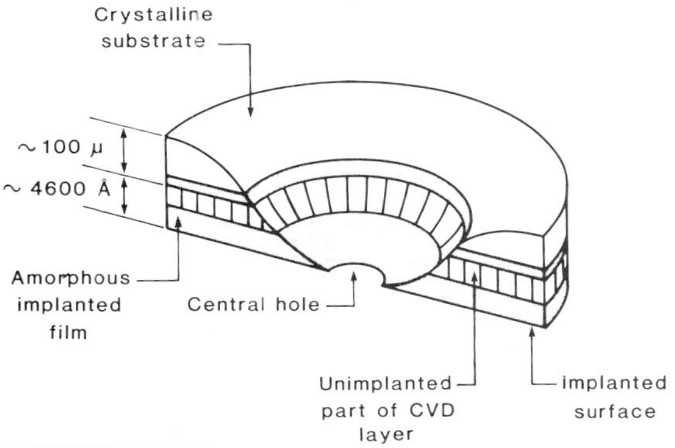

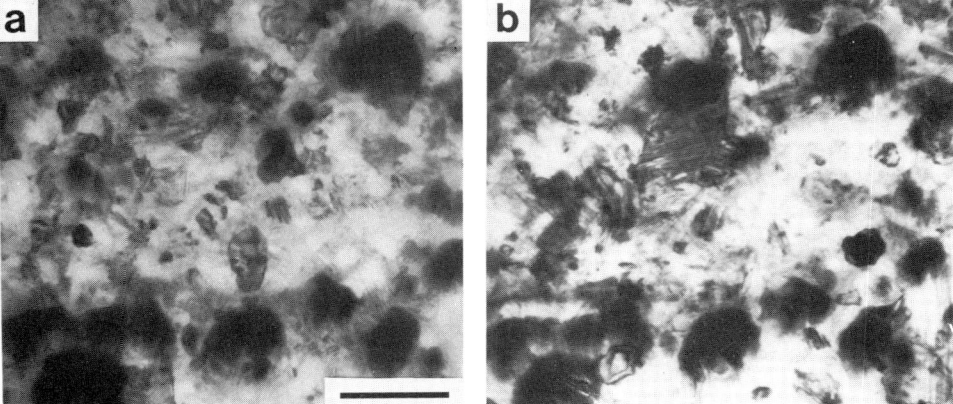

Crystalline
substrate

~100 µ

~4600 Å

Amorphous
implanted
film

Central hole

Unimplanted
part of CVD
layer

Implanted
surface

Fig. 1 Schematic draw-
ing of a plan-view TEM
specimen used in the
investigation of the
grain growth mechanism.

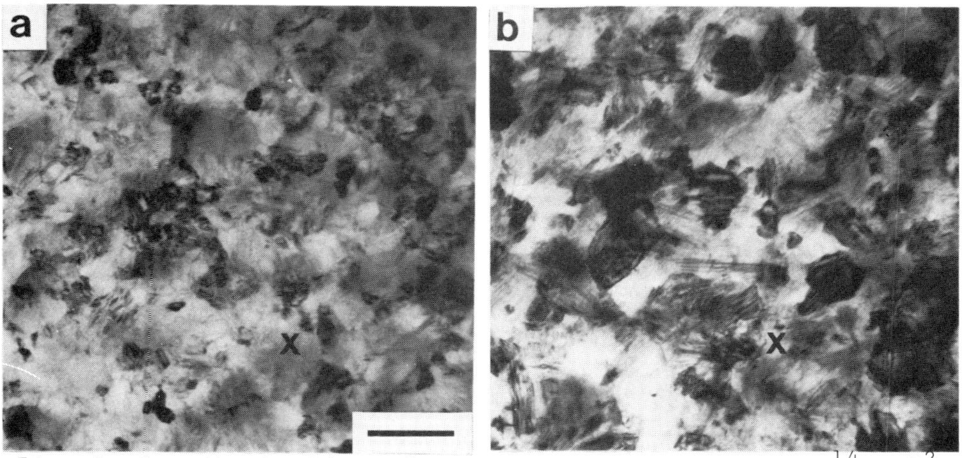

Fig. 2 In situ heating of an arsenic implanted ($5\times10^{14}$ cm$^{-2}$, 150 KeV) specimen: a) before heating, b) same area immediate-ly after heating to 950$^{\circ}$C. Marker 0.2 um.

Fig. 3 In situ heating of a phosphorus implanted ($2\times10^{16}$ cm$^{-2}$ 150 KeV) specimen: a) after 4 hrs. at 730$^{\circ}$C b) same area after 4 hrs. at 750$^{\circ}$C. The "islands" of dark contrast are artefacts on the specimen's surface. Marker 0.2 um.

Plan-view TEM specimens obtained from annealed samples indicated that the observations described above also occur in furnace-treated bulk specimens (Reader et al 1984). A large increase in the average grain size occurred during the first anneal after implantation and the only effect of prolonged heat treatment at 950°C was to reduce the number of lattice defects within the grains. The lattice defects were in the form of stacking faults and microtwins, most of which were perpendicular to the surface (figure 4). Electron microdiffraction indicated that about 85% of the grains had surface normals within 15° of the $\langle 110 \rangle$ crystallographic direction.

The effect on the grain morphology of the implantation and anneal can be seen by examining cross-sectional TEM specimens. Figure 5 shows that after annealing the grains in the part of the layers which had been made amorphous by implantation were much larger than those in the as-grown material. In the unimplanted part of the layers (nearer the interface with the substrate) the original small grains appear to be unaffected by the implantation. A subsequent 1/2 hour anneal at 950°C caused little change in this part of the layer. A small amount of grain growth was, however, noted after longer anneals at 950°C.

In situ TEM heating of a cross-sectional specimen of an implanted layer (figure 6) showed that epitaxial regrowth starts at many points on the interface between the amorphous region and the polysilicon underlying the $\alpha$-film. No movement of the amorphous/polysilicon interface was observed during a 20 minute heat treatment at 630°C. However, crystallization of the $\alpha$-film occurred rapidly during a subsequent anneal at 700°C.

The crystallization and grain growth behaviour of $\alpha$ -silicon was investigated by in situ heating locally unsupported $\alpha$-films (prepared as described earlier and illustrated in figure 1). When $\alpha$-films crystallize, many dendrite-like grains form and grow (figure 7). These grains contain a series of "spinal" stacking faults or micro-twins on vertical (11$\bar{1}$) planes and possess [011] surface normals. Longitudinal growth of the grains occurs by the extension of the spinal faults/micro-twins into the $\alpha$-silicon in the [2$\bar{1}$1] direction, as has been noted by Drosd and Washburn (1982). The interface planes between the grains and the $\alpha$ -silicon on either side of the spinal fault(s) were determined as (1$\bar{1}$1) for either the matrix or the twin when the spinal planes were (11$\bar{1}$). Further results on the crystallization of $\alpha$-silicon are to be published.

## 4. Discussion

The mechanism by which rapid grain growth occurs in heavily implanted polysilicon layers annealed between 700°C and 1000°C can be understood with reference to the in situ experiments described above. During the first few minutes of an anneal as the temperature of a specimen passes approximately 700°C, crystallization of the amorphous implanted films begins on the underlying polysilicon by epitaxial regrowth. As the underlying polysilicon possesses some $\langle 110 \rangle$ texture

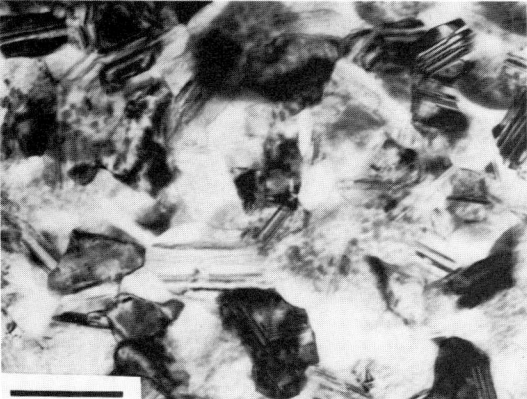

Fig. 4  Example of grain structure observed in implanted specimens after furnace annealing at 950°C (+ 10°C). Note stacking faults and microtwins within grains. Marker 0.2 um.

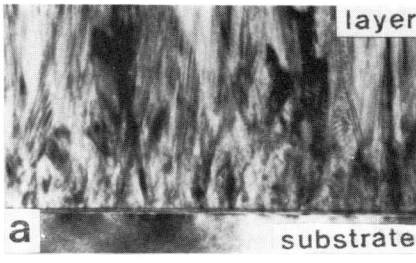

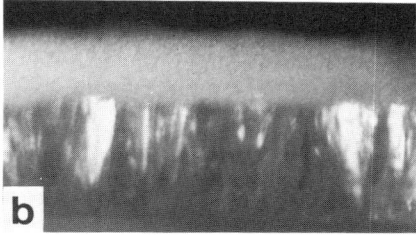

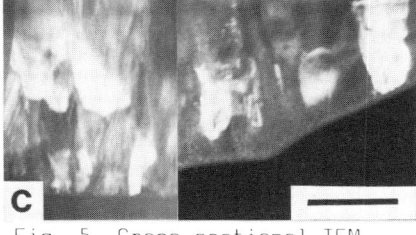

Fig. 5  Cross sectional TEM micrographs showing microstructure of layer: a) as-grown layer (bright field micrograph); b) implanted with 2.5x10$^{15}$ As$^+$ cm$^{-2}$, note amorphous region at top of layer; c) annealed for 1/2 hr at 950°C, note large grains in previously amorphous region. Marker 0.2 um.

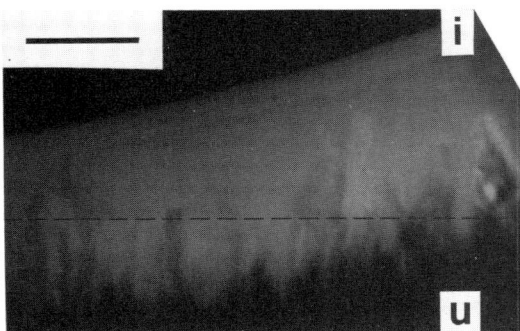

Fig. 6  In situ heating of a phosphorus implanted (2x10$^{16}$ cm$^{-2}$, 150 KeV) cross-sectional specimen. Crystallization of the **α**-film begins, by epitaxial re-growth, at the interface (dashed line) between the implanted (i) and unimplanted (u) parts of the layer. Marker 0.2 um.

Fig. 7  Typical grain observed during crystallization of a locally unsupported **α**-film - phosphorus implantation (2x10$^{16}$ cm$^{-2}$, 150 KeV). Note the vertical "spinal" stacking faults/microtwin. Crystallization occurs by the extension of faults/microtwins. The "islands" of dark contrast are artefacts on the specimen's surface. Marker 0.5 um.

(Hendriks et al 1984), some of the grains growing (epitaxial
ly) into the $\alpha$-film will    also have $\langle 110 \rangle$
orientations. $\langle 110 \rangle$ oriented grains have a relatively fast
lateral expansion as they are assisted by the stacking
fault/microtwin growth mechanism outlined above.
Crystallization of the $\alpha$-film is complete when the growing
grains reach the surface of the layer. By this time, the
110   oriented crystals will have expanded laterally more
than grains of other orientations. This mechanism thus
creates the morpology observed in the TEM specimens of
annealed layers. Cross-sectional specimens (figure 5) show
that certain grains have undergone a large amount of lateral
growth and that this growth occurs in the previously
amorphous part of the layer. Many of the grains seen in
plan-view specimens (figure 4) contain stacking faults and
microtwins, and possess $\langle 110 \rangle$ surface normals. As pointed
out in the experimental section, microstructure in the top
part of the layer is only observed in plan-view specimens.
Thus, the noted lattice faults and texture are a consequence
of the mechanism of rapid growth at the beginning of the
anneal.

5. Conclusions

     When a heavily phosphorus or arsenic implanted
polysilicon layer is introduced into a furnace,
crystallization occurs as the temperature of the wafer
increases above about 700°C. The concomitant increase in
grain size is thought to be achieved by a crystallization
mechanism that occurs in the amorphous implanted region of
the layer. It is, however, not achieved by a lateral
movement of the grain boundaries that existed before
implantation and it does not occur throughout the entire
thickness of the layer.

Acknowledgements

     This work is part of the research programme of the
Foundation for Fundamental Research on Matter (FOM-Utrecht)
and has been made possible by financial support from the
Netherlands Organization for the Advancement of Pure
Research (ZWO-The Hague).

References

Drosd R and Washburn J 1982 J. Appl. Phys. 53 397
Gibbons J F  1980 Handbook on Semiconductors ed T S Moss
   (Amsterdam: North Holland) vol. 3 pp 601-640
Hendriks M, Delhez R, de Keijser Th H, Radelaar S, Habraken
   F H, Kuiper A E and Boudewijn P R 1984 J. Appl. Phys. 56
   2751
Pettit H and Booker G 1971 EMAG ed W Nixon (Bristol: Inst.of
   Phys.) Conf. Series 10 pp. 290-3
Reader A H, Schapink F W and Radelaar S 1984 MRS-Europe
   Conf. (Les Ulis:  Les Edition de Physique) to be published
Wada Y and Nishimatsu S 1978 J. Electrochem. Soc. 125 1499

*Inst. Phys. Conf. Ser. No. 76: Section 4*
*Paper presented at Microsc. Semicond. Mater. Conf., Oxford, 25–27 March 1985*

# The observation and significance of amorphous regions in implanted and annealed polysilicon films

D L Black, J P Lavine, S-T Lee, and D L Losee

Research Laboratories, Eastman Kodak Company
Rochester, New York 14650, USA

Abstract    Cross-sectional transmission electron microscopy is used
to examine silicon films deposited at 580 and 620°C in a low-pressure
chemical vapor deposition system. The films are implanted with
phosphorus and annealed at 800°C. The diffused phosphorus profiles
are measured with secondary ion mass spectrometry and show very
little enhanced grain boundary diffusion. The transmission electron
microscopy shows amorphous regions remain in the films deposited at
580°C. These regions appear to be due to the extremely slow velocity
for grain boundary migration during primary recrystallization. It
is possible these amorphous regions are connected with the lack of a
diffusion tail. The TEM results are also used to explain the
electrical properties of the silicon films.

## 1.    Introduction

Polycrystalline silicon layers are used as electrodes in integrated
circuits. The drive to fabricate smaller devices produces a need for
smoother polysilicon films with lower resistivities, more controlled
doping profiles, and finer grains. The last two film properties help
control the profile of the polysilicon during oxidation and etching
(Irene et al. 1981). Anderson (1973) showed that lower deposition
temperatures lead to smoother films, which allow finer patterning. Ion
implantation permits a better control of the doping profile than the more
usual diffusion doping of polysilicon. Thus, an investigation was
started to characterize polysilicon films deposited by low-pressure
chemical vapor deposition (LPCVD).

LPCVD silicon films deposited at 580°C consist of amorphous silicon,
although a few grains may be observed (Becker et al. 1984, Duffy et al.
1983, Kinsbron et al. 1983, Losee et al. 1984, McGinn et al. 1983).
These studies also show that the films crystallize upon annealing and
that the crystallites are equiaxed with a wide distribution in size. In
fact, many of the crystallites are larger in diameter than the columnar
grains found in the polycrystalline films deposited at 620°C (Duffy
et al. 1983, Falckenberg et al. 1979, Lu et al. 1984, McGinn et al.
1983). The lower-deposition-temperature films develop grains with
extensive microstructure upon annealing (Becker et al. 1984, Kinsbron
et al. 1983). This makes it difficult to detect the grain boundaries.
The investigation reported here shows how transmission electron microscopy is helping to elucidate the differences observed in impurity

diffusion and in electrical properties between the 580 and 620°C films. In addition, amorphous regions are seen in the 580°C films after ion implantation and high-temperature annealing. This is the first obser- vation of such regions in annealed LPCVD polysilicon films.

## 2.   Experimental Techniques

Silicon films ~0.4 µm thick were deposited on a 0.1 µm thermally grown oxide on p-type <100> silicon substrates. The silicon films were deposited in a conventional LPCVD system at 580 or 620°C with deposition rates of 38 and 100 A/min, respectively. The films deposited at 580°C are amorphous, and the 620°C films are polycrystalline. Phosphorus was then implanted at 50 keV, and a cap layer of 0.5 µm of silicon dioxide was deposited at 425°C. The films were then annealed at 800°C in $N_2$ for various times.

The phosphorus profiles were measured with a Cameca IMS-3f secondary ion mass spectrometer (SIMS), with both $Cs^+$ and $O_2^+$ used for the sputtering beam. Hall-effect measurements were made to determine the effective carrier concentration and the effective carrier mobility of the films. A four-point probe was used to measure the sheet resistivity of the poly- silicon films.

Several film samples were prepared for cross-sectional transmission electron microscopy by using lapping and ion-milling techniques. A Philips 420T electron microscope was used for morphological and structural investigations of thin regions of the samples. Convergent- beam electron diffraction (CBED) was used to identify amorphous and crystalline regions in the silicon films. A 5.0 nm diameter probe beam was used for the CBED.

## 3.   Results

The as-implanted phosphorus profiles are shown in Fig. 1. The columnar nature of the 620°C deposited film appears to give rise to a slight tail, which may be due to channeling. The phosphorus profiles in Fig. 2 demonstrate the diffusion that occurs during a 30-min anneal at 800°C in $N_2$. The 620°C deposited films show more diffusion and a grain boundary diffusion tail. The profiles for the three lowest doses were fit to the solutions of a two-dimensional grain and grain boundary diffusion model previously described (Losee et al. 1984). The grain diameters are assumed to be 0.08 and 0.02 µm for the 580 and the 620°C deposited films, respectively. These values are based on the TEM results presented below. The ratio of grain boundary to grain diffusion coefficients is less than 6 for the 580°C deposited films. The 620°C deposited films yield a grain boundary diffusion coefficient that is about 10 times larger. Figure 3 shows the sheet resistances of a set of samples annealed for 30, 120, or 360 min at 800°C in $N_2$. In all cases, the 580°C films are less resistive than the 620°C films. This result agrees with those for boron and arsenic of Becker et al. (1984) but differs from the single point for phosphorus of Wu et al. (1984). The Hall-effect measurements of Fig. 4 show larger effective carrier concentrations and higher Hall mobilities in the 580°C deposited films for identical phosphorus implant doses. The differences in electrical properties are consistent with finer grains in the 620°C deposited films and, hence, with more sites for precipitation

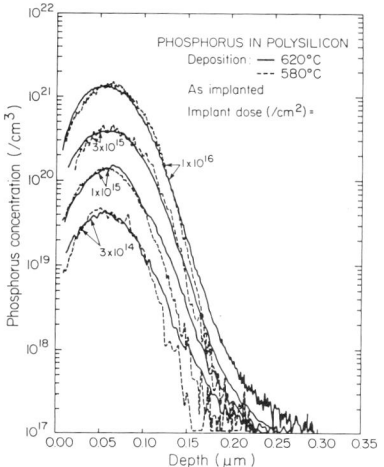

Fig. 1  SIMS profiles of the
as-implanted phosphorus.
(---) 580°C deposited film;
(——) 620°C deposited film.

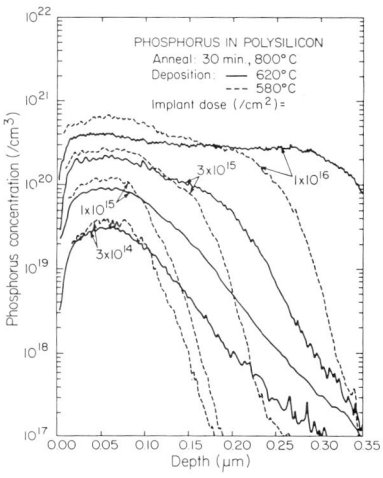

Fig. 2  SIMS profiles of the
phosphorus after annealing.
(---) 580°C deposited film;
(——) 620°C deposited film.

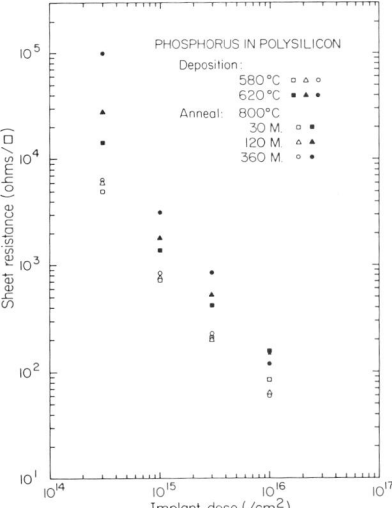

Fig. 3  Sheet resistance from
four-point probe measurements.

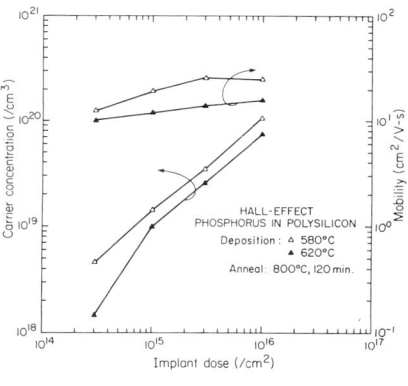

Fig. 4  Carrier concentration
and mobility measured by the
Hall effect.

in these films.  The present Hall-effect measurements show both films to have grain boundaries with similar electrical properties.

Transmission electron microscopy was performed on four samples annealed for 30 min at 800°C in $N_2$.  The micrographs for the 620°C deposited films appear in Figs. 5 and 6.  The grain structure is columnar with very fine grains.  Some grain branching appears within the films.  In addition, the heavy implant sample of Fig. 6 shows wider grains within the top third of the film.  The phosphorus dose of $1.0 \times 10^{16}/cm^2$ converts the top 0.12 μm or so of the polysilicon to amorphous silicon.  This approximate depth agrees with calculations of energy deposition during ion implantation that are performed with the Monte Carlo computer program TRIM (Biersack and Haggmark 1980).  A TEM cross section of the silicon film would then resemble Fig. 3b of Chew et al. (1983).  The amorphous silicon is reconverted to polysilicon during the 800°C anneal, and it appears that some new grains nucleate and grow along with the columnar grains.  This makes the implant damage boundary visible.

Figures 7 and 8 are the micrographs for the 580°C deposited films.  The grains are equiaxed and random in size and orientation.  These figures show that the annealed 580°C deposited films do have many grains that are larger than those in the annealed 620°C deposited films.  In addition, the 580°C deposited films are seen to have multiple layers of grains.  These observations on grain size and orientation are in agreement with previous work (Becker et al. 1984, Duffy et al. 1983, Kinsbron et al. 1983, McGinn et al. 1983).

Phosphorus diffusion was simulated for a film with an array of grains that resembles Figs. 7 and 8.  A Monte Carlo method (Lavine 1984) was used to show that the diffusion profile has a tail when the grain boundary diffusion is more than ten times faster than the grain diffusion.  The lack of an observed diffusion tail led to further investigation with convergent-beam electron diffraction.  The diffraction patterns for the 620°C deposited films show only crystalline regions.  However, the 580°C deposited films have regions that lack crystalline diffraction patterns as the inset in Fig. 8 demonstrates.  These amorphous regions are composed of silicon and they occur between grains and within what appears to be large grains.

Amorphous regions between grains have been seen in ceramics (Williams and Newbury 1984).  The occurrence of the amorphous regions in the present silicon films appears to be due to the extreme slowness of grain boundary migration during primary recrystallization (Schins et al. 1980, Tsaur and Hung, 1980, Wada and Nishimatsu, 1978).  These studies lead to an estimate of at most 3 A/min for the rate of grain boundary migration at 800°C.

4.    Conclusions

The annealed silicon films have grain structures similar to those of previous observations.  The 580°C deposited films lead to smoother surfaces and lower resistivities.  The 580°C deposited films show shallower diffusion profiles than the 620°C deposited films.  The amorphous regions observed in the 580°C deposited films are suspected to be related to the slow diffusion.

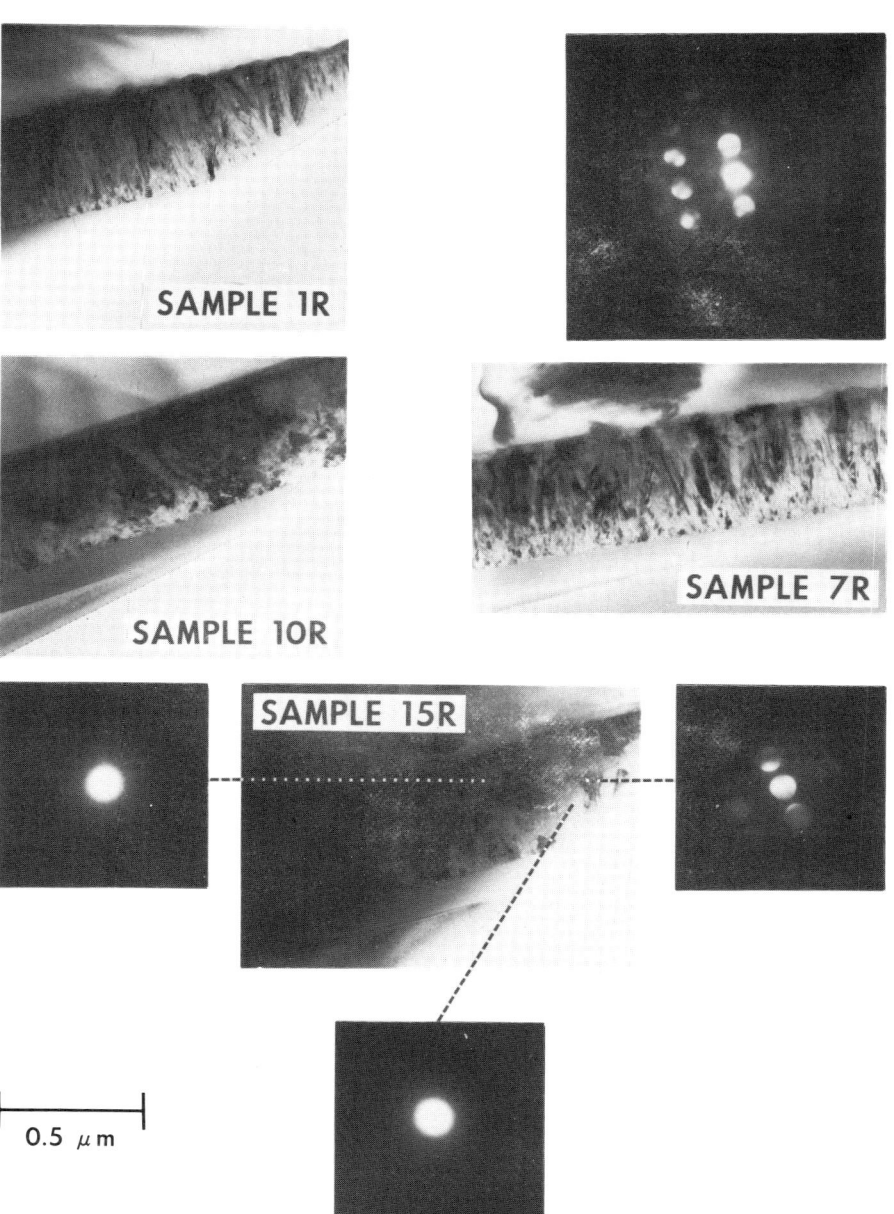

SAMPLE 1R

SAMPLE 10R

SAMPLE 15R

SAMPLE 7R

0.5 μm

Fig. 5   Micrograph of the 620° C deposited film.
The implant dose is 3 x 10¹⁴/cm². Sample 1R.

Fig. 6   Micrograph of the 620° C deposited film.
The implant dose is 1 x 10¹⁶/cm². Sample 7R.

Fig. 7   Micrograph of the 580° C deposited film.
The implant dose is 3 x 10¹⁴/cm². Sample 10R.

Fig. 8   Micrograph of the 580° C deposited film.
The implant dose is 1 x 10¹⁶/cm². Sample 15R.

Acknowledgement

We thank J W LeClair for the Hall-effect measurements and C M Jarman for
his assistance with the SIMS measurements.

References

Anderson R M 1973 J. Electrochem. Soc. 120 1540
Becker F S, Oppolzer H, Weitzel I, Eichermüller H and Schaber H 1984
    J. Appl. Phys. 56 1233
Biersack J P and Haggmark L G 1980 Nucl. Inst. Meth. 174 257
Chew N G, Cullis A G, White J C and Cox T I 1983 Inst. Phys. Conf. Ser.
    No. 67 473
Duffy M T, McGinn J T, Shaw J M, Smith R T, Soltis R A and Harbeke G 1983
    RCA Review 44 313
Falckenberg R, Doering E and Oppolzer H 1979 Proc. Electrochem. Soc. Fall
    Meeting 1429
Irene E A, Tierney E, Blum J M, Aliotta C F, Lamberti A C and Ginsberg B
    J 1981 J. Electrochem. Soc. 128 1971
Kinsbron E, Sternheim M and Knoell R 1983 Appl. Phys. Lett. 42 835
Lavine J P 1984 Simulation of Semiconductor Devices and Processes
    eds Board K and Owen D R J (Swansea: Pineridge Press) 467
Losee D L, Lavine J P, Trabka E A, Lee S-T, and Jarman C M 1984 J. Appl.
    Phys. 55 1218
Lu N C C, Lu C Y, Lee M K, Shih C C, Wang C S, Reuter W and Sheng T T
    1984 J. Electrochem. Soc. 131 897
McGinn J T, Kappert H F, Krausbauer L, Shaw J M and Widmer A E 1983 Proc.
    Electrochem. Soc. Spring Meeting 647
Schins W J H, Bezemer J, Holtrop H and Radelaar S 1980 J. Electrochem.
    Soc. 127 1193
Tsaur B Y and Hung L S 1980 Appl. Phys. Lett. 37 648
Wada Y and Nishimatsu S 1978 J. Electrochem. Soc. 125 1499
Williams D B and Newbury D E 1984 Adv. Electron. Electron Phys. 62 161
Wu C P, Schnable G L, Lee B W and Stricker R 1984 J. Electrochem. Soc.
    131 216

*Inst. Phys. Conf. Ser. No. 76: Section 5*
*Paper presented at Microsc. Semicond. Mater. Conf., Oxford, 25–27 March 1985*

163

# Silicides and lateral diffusion couples

J C Barbour, J W Mayer, and L R Zheng[a]

Dept. of Materials Science and Engineering, Cornell University, Ithaca, New York, USA 14853.
[a] Permanent address: Shangai Institute of Metallurgy, Shangai PRC

Abstract    Thin-film    lateral-diffusion    couples,    formed    by overlapping deposited layers of metal and Si, form silicides whose growth kinetics and phase sequences provide the bridge between conventional    thin-film    and    bulk-diffusion-couple    silicide formation. The lateral extent of the silicide is up to 100 microns which allows measurement of composition gradients within growing phases and across the interface between phases. Self-supporting lateral Ni-silicide couples, 500-600 Å thick, have been analyzed in a scanning transmission electron microscope (STEM) by microdiffraction and energy dispersive X-ray spectroscopy (EDS) to identify phases and compositions.

## 1. Introduction

Silicide formation has conventionally been studied using thin metal films deposited on single crystal silicon (Tu and Mayer 1978) or with bulk diffusion couples (Tu et al 1983), and the two approaches have yielded different results. In order to understand the microscopic mechanisms of silicide formation and relate the kinetics of thin-film silicide formation to bulk-phase silicide formation, electron microscopy has been used to study silicide growth in thin-film lateral diffusion couples. The Ni-Si equilibrium phase diagram (Hansen 1959) in fig. 1 is representative of silicide forming systems in that congruent and non-congruent phases are present and the lowest

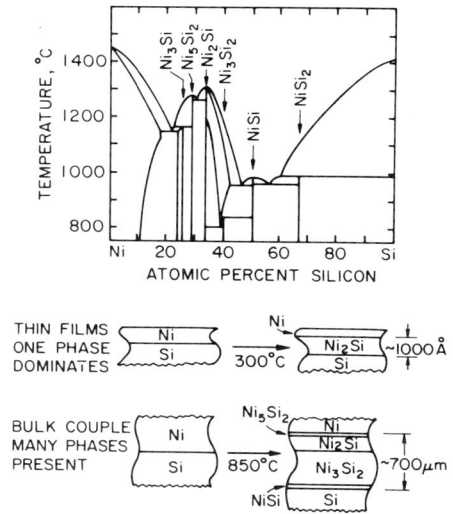

Fig. 1    Equilibrium    phase diagram    for    Ni-Si,    the formation    of    1000 Å thick $Ni_2Si$ layer with a deposited $Ni$ thin film on Si and the formation    of    multiple phases in    a    bulk    Ni-Si    diffusion couple.

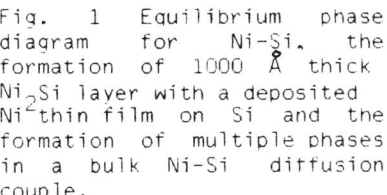

eutectic is approximately 1000°C. Thin metal films deposited on single crystal silicon substrates form silicides well below the eutectic temperature, typically 200 to 300°C for near noble metals. Only one silicide phase grows, e.g. $Ni_2Si$, (Canali et al 1979) until the metal film is consumed and then the next more Si-rich congruent phase (NiSi) is formed and grows without the formation of the intermediate, non-congruent phase ($Ni_3Si_2$). In contrast, bulk couples (Tu et al 1983) are analyzed after high temperature processing (700-850°C) and exhibit many phases simultaneously. The compound $Ni_3Si_2$ exhibits the most extensive growth in these bulk Ni-Si couples, whereas $Ni_3Si_2$ has not been found in thin-film silicide formation.

## 2. Formation of Conventional Thin-film Silicides

Thin-film silicides have been investigated using Rutherford backscattering spectrometry (RBS) to establish silicide composition and glancing angle X-ray diffraction (XRD) to identify phases (Canali et al 1979). The depth resolution of RBS is approximately 100 Å, but laterally the area analyzed is about 1 $mm^2$. The RBS spectrum in fig. 2 shows that a uniform layer of $Ni_2Si$ is formed between Ni and Si. This layer grows with $(time)^{1/2}$ kinetics and an activation energy of about 1.5 eV (Tu and Mayer 1978).

Implanted marker studies show that Ni is the dominant diffusing species in $Ni_2Si$.

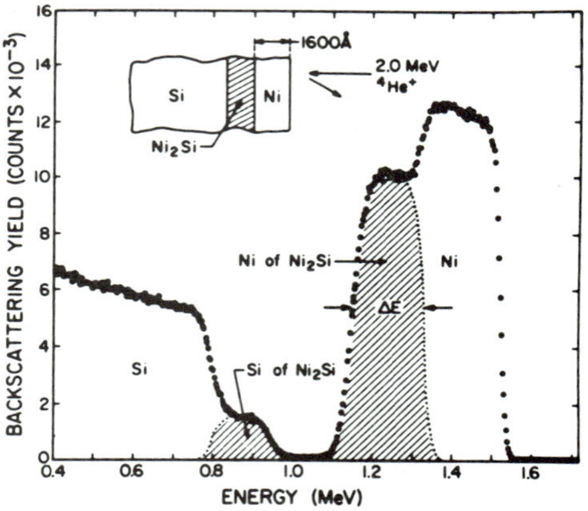

Fig. 2 Rutherford back-scattering spectrum showing the formation of a layer of $Ni_2Si$ between Ni and Si.

## 3. Formation of Lateral Silicide Structures on Inert Substrates

Lateral diffusion couples were formed by depositing 1200 Å of Ni through a mask (prepared by anisotropic etching of a Si wafer) which is in contact with single crystal Si (500 Å thick) on a sapphire substrate. The samples were then vacuum annealed in order to study the lateral growth of silicides. Scanning electron microscopy (SEM) was used to observe the spread of the Ni-silicide at the initial boundary. In the temperature range of 400 to 800°C, this silicide layer grew laterally (from right to left in fig. 3) with behavior which is characteristic of a diffusion limited process (Zheng 1983 and 1984).

The silicide composition was determined from electron microprobe measurements of the relative Si/Ni X-ray yield as compared with the X-ray yields from silicide standards. Figure 3 shows the composition profile (the relative X-ray yield) measured at approximately 4 micron

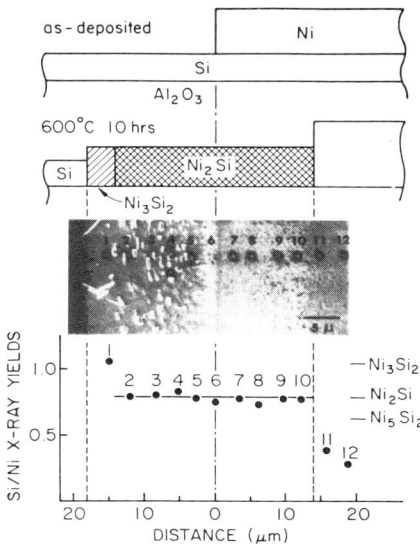

Fig. 3 Schematic diagrams and SEM micrographs showing the predominant growth of the $Ni_2Si$ phase after annealing a lateral diffusion couple at 600 °C for 10 hours. The composition profile for the Ni-silicide was determined from relative X-ray yields as compared with standards.

intervals. The composition in the silicide layer, after annealing for 10 hours at 600°C, is predominantly that of $Ni_2Si$ with an uncertainty of less than ±6%. The dominant growth of one phase is a common feature in the initial stages of all thin film silicide reactions. A small area near the pure Si region has a composition close to that of $Ni_3Si_2$. However, the extent and composition of the $Ni_3Si_2$ area, shown as a single point on the composition profile, has a large uncertainty as a result of the limited lateral resolution in the scanning electron microscope.

## 4. Growth Rates in Ni-, Pd-, and Pt-Si Lateral Couples

Lateral silicide couples were made for three near-noble metals (Ni, Pd, and Pt) by depositing metal islands on single crystal Si as previously described. These samples were annealed in a vacuum of $1 \times 10^{-7}$ torr at temperatures between 350 and 750°C. The silicide growth kinetics were determined by using an SEM to measure the lateral growth length, $L_D$, and to determine the composition from relative Si/metal X-ray yields. For the temperatures shown in

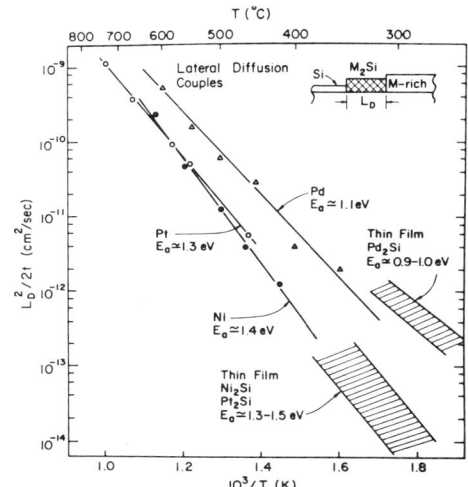

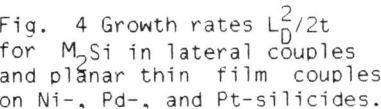
Fig. 4 Growth rates $L_D^2/2t$ for $M_2Si$ in lateral couples and planar thin film couples on Ni-, Pd-, and Pt-silicides.

fig. 4, each metal-silicon couple initially formed a predominant growth phase which was the metal-rich silicide, $M_2Si$, and had a growth length which was between 1 and 20 microns. Also, $L_D$ was found to be proportional to the square root of annealing time. This type of growth kinetics is consistent with the diffusion limited kinetics exhibited in conventional thin film growth of the $M_2Si$ phase.

The growth rates, $L_D^2/2t$ in terms of $cm^2/sec$, for the Ni, Pd, and Pt lateral silicide formation are plotted in fig. 4. Growth rates for conventional planar thin-film silicide formation (shaded regions) are shown for comparison. This figure indicates that the rates as well as the activation energies for lateral silicide formation are in close agreement with those rates and activation energies for planar thin-film silicide formation. The rates for the lateral silicides extend to higher temperatures due to the large diffusion lengths ($10\mu m$) in the lateral couples as compared to those lengths ($0.1\mu m$) in thin film couples. Further, the growth rates for the lateral couples at the highest temperatures are comparable to the growth rates for bulk couples. Therefore, the kinetic behavior of lateral couples provides a bridge between bulk and thin film behavior.

## 5. Formation of Multiple Phases in Lateral Couples

Initially, one phase ($Ni_2Si$), dominates the lateral growth in Ni-silicide couples, similar to the behavior of planar thin-film reactions. This result is in contrast with the behavior of bulk couples annealed at temperatures between 750 and 850°C, where many phases are found. Lateral diffusion couples annealed at and above 600°C for times sufficiently long such that the growth length exceeds 20 microns exhibit a sequence of phases as found in bulk diffusion couples. Electron microprobe measurements (fig. 5) on a sample annealed for 60 hours at 600°C show the phase sequence $Ni_5Si_2$, $Ni_2Si$, $Ni_3Si2$, and NiSi. Annealing for different times at 600°C showed that the phase $Ni_2Si$ grew to about 25 microns at which time other phases began to appear

Fig. 5 Schematic diagram and composition profile showing multiple phase formation in a lateral silicide couple annealed at 600°C for 60 hours. The phase sequence $Ni_5Si_2$, $Ni_2Si$, $Ni_3Si_2$ and NiSi agrees with that in bulk couples (fig. 1) and follows the progression on the equilibrium phase diagram.

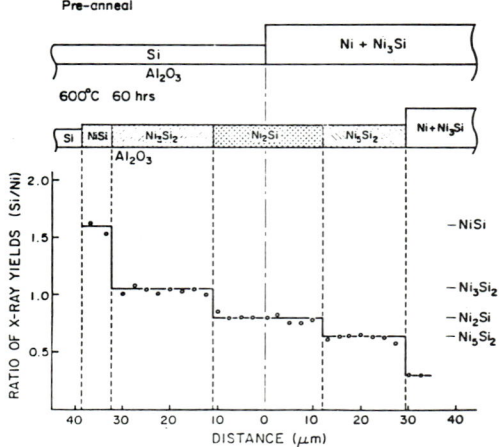

and grow. We believe the lateral diffusion couples annealed at high temperature reflect the behavior that would be found in bulk couples in the early stages of silicide formation: growth of one phase dominates until a critical length is reached where upon additional phases form.

## 6. Thin-film Lateral Silicide Couples

At temperatures below 500°C a high lateral resolution is needed to look for the presence of $Ni_3Si_2$ and to measure composition gradients. Energy dispersive spectroscopy (EDS) in a scanning transmission electron microscope (STEM) allows lateral analysis of areas separated by only 10-20 A in a thin film. Therefore, thin self-supporting lateral diffusion couples were developed (Chen et al 1984) to gain the needed spatial resolution. Figure 6 shows the results (Barbour et al 1985) of a study to determine the accuracy to which compositions can be measured using EDS. Thin film Ni-Si standards on NaCl substrates were analyzed using RBS to determine compositions. The standards were placed on TEM grids after dissolving the substrate in water. An EDS analysis of several samples of varying

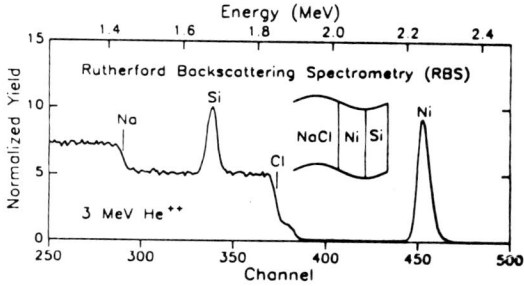

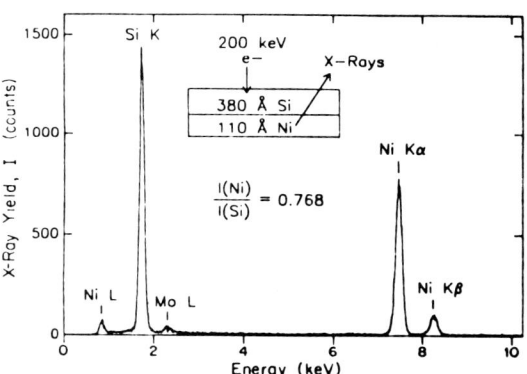

Fig. 6 Compositions in thin film lateral diffusion couples can be measured using EDS to an accuracy of about ±4% when thin film standards are available. RBS can be used to measure the composition and thickness of the standards prepared by electron beam deposition onto dissolvable substrates. The proportionality factor between the composition ratio and the X-ray intensity ratio obtained by this method has a standard deviation of 3.5%. Therefore, only composition gradients of 4% or larger are statistically measurable.

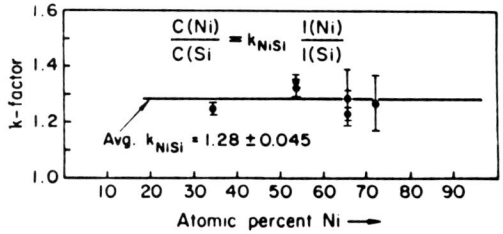

composition showed that the proportionality factor (k-factor) between the Ni-Si composition ratio and the Ni-Si X-ray yield had a standard deviation of 4-5%. Thus the smallest composition gradient detectable in the lateral diffusion couple is 4-5%.

Figure 7 schematically shows a self-supporting lateral diffusion couple annealed at 450°C for 12 hours. This sample was prepared by an electron beam evaporation of silicon (amorphous) onto a NaCl substrate followed by evaporation of a thick Ni island through a shadow mask in contact with the sample. The substrate is dissolved in water and the self-supporting lateral couple is placed on a molybdenum TEM grid before vacuum annealing. The techniques which can be used in a STEM to examine lateral silicide formation and probe the region of the growth front near the silicon include: EDS, microdiffraction, and electron energy loss spectroscopy (EELS). This paper gives the results from an EDS analysis used to identify compositions and from the microdiffraction used to identify phases.

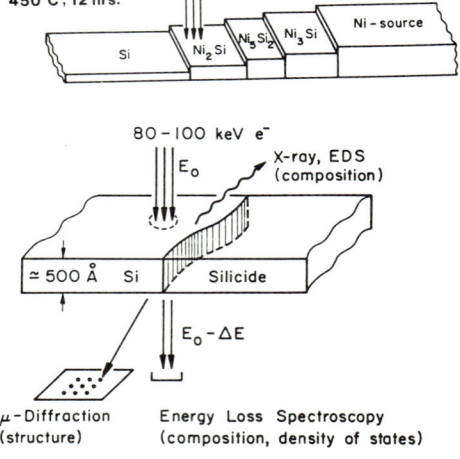

Fig. 7 Scematic diagram showing the techniques compatible with a STEM which can be used to probe the growth front in self-supporting lateral silicides.

A transmission electron microscopy (TEM) image of the growth front after a 12 hour anneal at 450°C is shown in fig. 8. The grains in the right portion of the figure are crystals of $Ni_5Si_2$ and the long grains in the middle to left portion of the figure are single crystals of $Ni_2Si$. Both phases form long grains in the direction of growth outward from the nickel source region (far to the right of the figure). These phases were identified by microdiffraction patterns as illustrated below the micrograph, and the compositions were confirmed using EDS. Also, microdiffraction (using a 150-200 Å electron beam) helped identify small discontinuous regions of NiSi in the initial stages of formation and growth between the $Ni_2Si$ and the Si. Samples annealed for longer times at 450°C showed a more continuous NiSi phase but no evidence was found for the presence of $Ni_3Si_2$. The presence of $Ni_3Si_2$ between NiSi and $Ni_2Si$ would indicate that $Ni_3Si_2$ is a stable equilibrium phase below 500°C and can nucleate in thin films but does not grow because of slower growth kinetics. Therefore, the samples

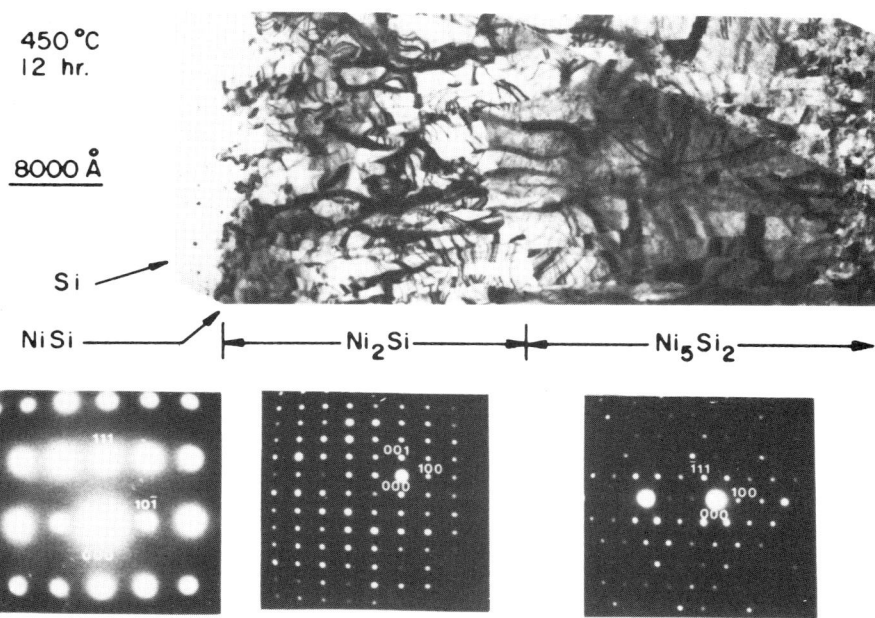

450 °C
12 hr.

8000 Å

Si

NiSi ————— Ni₂Si ————— Ni₅Si₂ ➤

Fig. 8 (top) Micrograph of the growing phases in a self-supporting lateral diffusion couple near the growth front. The phases from the various regions were identified using microdiffraction.

Fig. 9 (right) shows the results using a 500 Å electron probe to collect EDS spectra along the grains shown above. No gradient was detected along the Ni₅Si₂ and the Ni₂Si, to within ±4%. The lower plot, (b), shows preliminary results from using a 10 Å probe to look for the presence of Ni₃Si₂ between NiSi and Ni₂Si.

EDS ANALYSIS OF LATERAL SILICIDES
450°C, 12 hrs.

(a) Composition Gradient

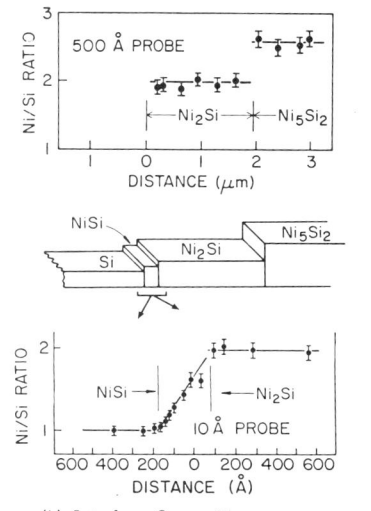

(b) Interface Composition

annealed at 450°C for 12 hours were examined with a 10 Å electron  probe
to look for evidence of $Ni_3Si_2$.

Figure 9 shows the results of an EDS analysis  to  examine  the  lateral
growth region pictured in fig.  8.  A large area (500 Å) probe was used
to measure compositions at several points along the growth direction  in
the    $Ni_2Si$  and  $Ni_5Si_2$  phases.   The   error   bars   indicate   an
uncertainty of ±5%  and  the  graph  shows  the  phases  are  uniform  in
composition  to  better  than  4%.   No composition gradients were detected
within these  phases  thereby  giving  an  upper  limit  on  composition
gradients  of  about  4%.    The  preliminary  results  (Barbour and Batson
1985) of the analysis using the 10 Å electron probe in a  STEM  show  a
linear  gradient  between  the  $Ni_2Si$ and NiSi phases and do not show a
strong step indicative of $Ni_3Si_2$.  The  large distance over which the
gradient  spans  is  due  to  a  tilted  interface.   We  speculate that
$Ni_3Si_2$ is formed only at the higher temperatures where both  Ni  and
Si  readily diffuse whereas at the lower temperatures Ni is the dominant
diffusing species.

Figure 10 shows an enlargement of the long grains of $Ni_2Si$ which  can
be analyzed not only for gradients

**450 °C  12 hr.        Ni Silicide Growth Front**

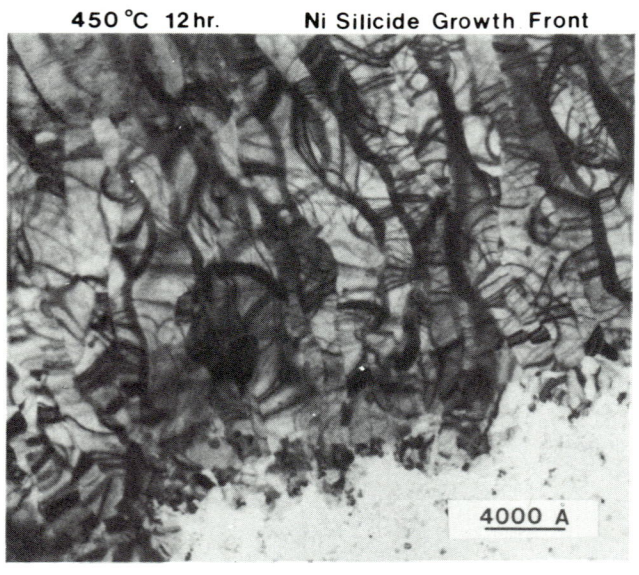

4000 A

Fig.  10. A transmission  electron  micrograph  showing  long  grains  of
$Ni_2Si$  at  the  growth  front  of  a self-supporting lateral diffusion
couple annealed at 450°C for 12 hours.

along the grains in the direction of growth (down, in fig. 10) but also for gradients within the phase perpendicular to the growth direction. e.g. at grain boundaries. In summary, the use of lateral diffusion couples in conjunction with electron microscopy allows one to gain increased lateral resolution for composition analyses (up to 10 Å between areas) which can then be used to model the formation of silicides.

## 7. Acknowledgements

The authors thank L S Hung for his advice and suggestions. The work was supported in part by NSF through the Materials Science Center (electron microscopy facility and technical operations laboratory) at Cornell University, Ithaca, New York, USA. We thank P E Batson for the use of the STEM at IBM Research. The techniques for creating lateral diffusion couples were developed at NRRFSS, Cornell University. Work on self-supporting lateral silicides has been funded by the ONR (L Cooper).

## References

Barbour J C, Sickafus K and Nastasi M 1985 J. Vac. Sci. Technol. submitted
Barbour J C and Batson P E 1985 unpublished work
Canali C, Majni G, Ottaviani G and Celotti G 1979 J. Appl. Phys. 50 255
Chen S H, Zheng L R, Barbour J C, Zingu E C, Hung L S, Carter C B and Mayer J W 1984 Mater. Lett. 2 469
Hansen M 1959 Constitution of Binary Alloys (New York:McGraw-Hill) pp 1039-1042
Tu K N and Mayer J W 1978 Thin Films-Interdiffusion and Reactions ed J M Poate, K N Tu, and J W Mayer (New York:Wiley-Interscience) Chap. 10
Tu K N, Ottaviani G, Gosele U, and Foll H 1983 J. Appl. Phys. 54 758
Zheng L R, Hung L S and Mayer J W 1983 J. Vac. Sci. Technol. A1 758
Zheng L R, Zingu E C and J W Mayer 1984 Mat. Res. Soc. Symp. Proc. ed J E E Baglin, D R Campbell and W K Chu (New York:North-Holland) pp 75-85

*Inst. Phys. Conf. Ser. No. 76: Section 5*
*Paper presented at Microsc. Semicond. Mater. Conf., Oxford, 25–27 March 1985*

173

# Interfacial atomic structure and Schottky barrier height

J.M. Gibson, R.T. Tung, C. A. Pimentel[*] and D.C. Joy
AT&T Bell Laboratories, 600 Mountain Avenue, Murray Hill, NJ 07974, USA
* on leave from Instituto de Fisica, University of Sao Paulo, Brazil

**Abstract** The epitaxial growth of metals on silicon has led to new phenomena in Schottky barrier formation. Furthermore, the uniformity of epitaxial layers permits unique correlation of atomic structure and electrical properties. We demonstrate this for the $NiSi_2$/Si system. A new method for the microscopic determination of Schottky barrier height, involving electron-beam induced current in the SEM, is explored. Results suggest that the local barrier height in non-epitaxial $NiSi_2$ films displays an orientation dependence which is consistent with our single-crystal measurements. Such microscopic techniques show potential for similar structure/property correlation in non-uniform systems.

## 1. Introduction

Although the Schottky barrier was the first solid-state rectifying junction to be discovered, the factors determining the barrier height between a particular metal and semiconductor still remain controversial. At least two different mechanisms can dominate. The barrier height may simply be determined by the difference between the work function of the metal and the electron affinity of the semiconductor, as originally suggested by Schottky. Alternatively, the Fermi-level at the metal-semiconductor interface may be pinned by defects according to the model of Bardeen. The metal-silicide compounds are particularly stable metal contacts on silicon, and thus have practical applications in devices. The fact that inter-reaction is almost invariably involved in silicide formation leads to separation of the interface from the original Si surface and thus to relatively clean interfaces. Nonetheless, studies of Schottky barrier height for silicides have not revealed consistent agreement with either the Schottky or Bardeen models[1].

A variable in such studies which is difficult to isolate is non-uniformity at the silicide-silicon interface, and possibly of the barrier height. By the use of epitaxial silicides and ultra-high vacuum preparation conditions this can be eliminated and some very novel results on Schottky barrier formation have resulted. For example, a systematic orientation dependence of barrier height is revealed in the $NiSi_2$/Si system for the first time. With such uniform single-crystal interfaces it is also possible to relate atomic structure information obtained from high-resolution electron microscopy studies to macroscopic physical properties. Results will be summarized in this paper.

Having observed that in such controlled systems there exists an intriguing orientation dependence of Schottky barrier height, which has technological and scientific relevance, we have explored a microscopic measuring technique for Schottky barrier height using electron beam induced conductivity (EBIC). This method may allow extension of experiments to non-uniform systems, which are much more common.

## 2. Results

Nickel and cobalt disilicides are cubic materials with the $CaF_2$ structure and lattice constants 0.5% and 1.2% less than silicon (at room temperature). These are formed as the end phase when reacting a deposited metal film with silicon and exhibit epitaxy on (111) and (100) substrates. Both silicides contain two orientations with $[111]MSi_2\|[111]Si$, when grown on (111) Si under normal conditions. These orientations are doubly-positioned and we refer to them as A, where $[110]MSi_2\|[110]Si$, and B, where $[110]MSi_2\|[\overline{1}\overline{1}0]Si$. The interface structure of such films was first examined by Cherns et.al.[2] who found remarkably flat interfaces with structure which was in good agreement with a simple model known as the 7-fold model (from the co-ordination of Ni atoms at the interface). The model is shown in figure 1 for the A and B $NiSi_2$/Si interfaces. On (100) Si, interfaces are non-planar, with a tendency to facet onto (111) planes, but the (100) facets show good agreement with a simple 6-fold model[3] [4]. However, the Schottky barrier height of all these films seems to be constant at about 0.66eV[5] independent of orientation. In fact a similar barrier height is also found for non-epitaxial $Ni_2Si$ and NiSi on silicon[5].

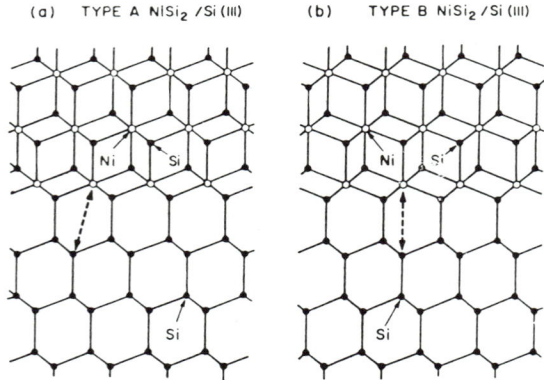

Figure 1: The 7-fold co-ordinated model for a) A orientation and b) B orientation $NiSi_2$ films on (111) Si. The dotted lines join Ni and Si atom pairs discussed in the text.

Using ultra-high vacuum growth techniques we have reported that it is possible to grow true single-crystals of $NiSi_2$ on silicon with uniform interfaces for type A *or* type B and (100)[6]. Cobalt disilicide can also be grown as a single crystal on (111) silicon with such techniques but only with the B orientation[7]. The basis of these methods is in-situ cleaning of the silicon surface prior to deposition to the point where no contamination is seen by Auger electron spectroscopy and the surface reconstructs, e.g. to the (7x7) structure for Si (111). Deposition of the metal, by electron-beam evaporation, is followed by in-situ annealing to form the disilicide phase. For the case of $CoSi_2$ on (111) silicon, single-crystal growth is obtained simply by following the same deposition/temperature cycles in-situ UHV as would be normally used ex-situ, i.e. 100-1000 Å Co deposited at room temperature and a 850-950 °C anneal for about 30 minutes[8]. Single-crystal growth of $NiSi_2$ requires somewhat unusual conditions. In order to avoid double-positioning on (111) Si and for flat interface formation on (100) Si, a growth prescription known as the template method was developed[6] [9]. In this method a very thin layer of Ni ( < 30 Å) is deposited on the atomically-clean Si substrate at room temperature. This layer is then annealed at 400-500°C. At this low temperature, which is insufficient for $NiSi_2$ formation in thicker films, $NiSi_2$ layers form whose orientation is a function of the deposited Ni thickness. For Ni thickness less than 8Å the layer orientation is type-B. For Ni thickness in the range 16-20 Å the layer orientation is type-A. At other thicknesses A+B orientation is found[6]. These layers are known as "templates" for the reason that *after* formation excess Ni can be deposited on their surfaces at higher temperatures (>650 °C) to form thick films, which have an orientation (either A or B or A+B) determined by the template layer. Figure 2 shows high-resolution transmission electron micrographs of (a) a type-A film and (b) a type B film less than 100 Å thick. These layers are psuedomorphic, i.e. contain no misfit dislocations, when formed as templates, although thicker layers grown at higher temperatures have regular networks of misfit dislocations[3]. High-resolution images such as figure 2 demonstrate the uniformity of these interfaces. Following the models proposed by Cherns et. al.[2] for A+B films grown under less clean conditions, we have studied interface structure primarily by measurement of the relative rigid lattice shift. This technique should be accurate to about 0.2Å[10] and gives excellent agreement with the 7-fold model for both A and B interfaces[3]. On the (100) Si surface, the template method is also uniquely able to stabilize a (100) interface and provide uniform films, whose interface structure is well-fitted by a 6-fold model[3][4] More

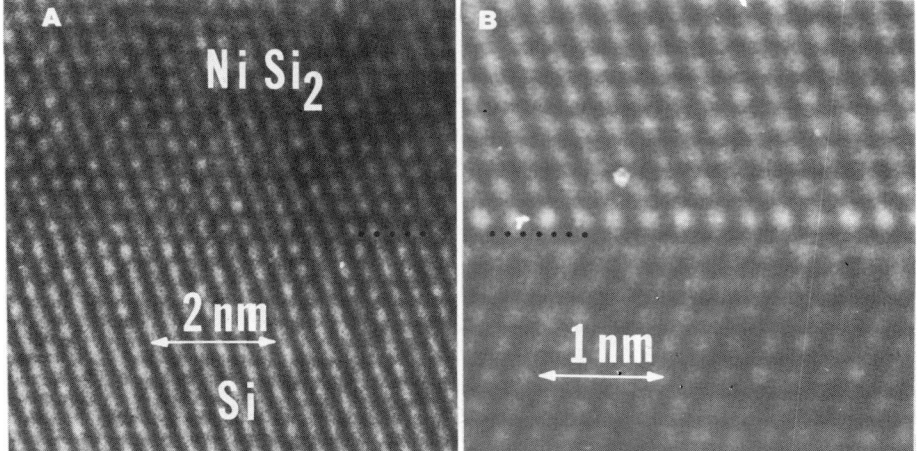

Figure 2: Axial bright-field transmission electron micrographs of a) an A orientation template and b) a B template. Images were taken near the Scherzer defocus on a JEOL 200CX instrument with point-to-point resolution 2.5Å. Specimen thickness was less than 100Å.

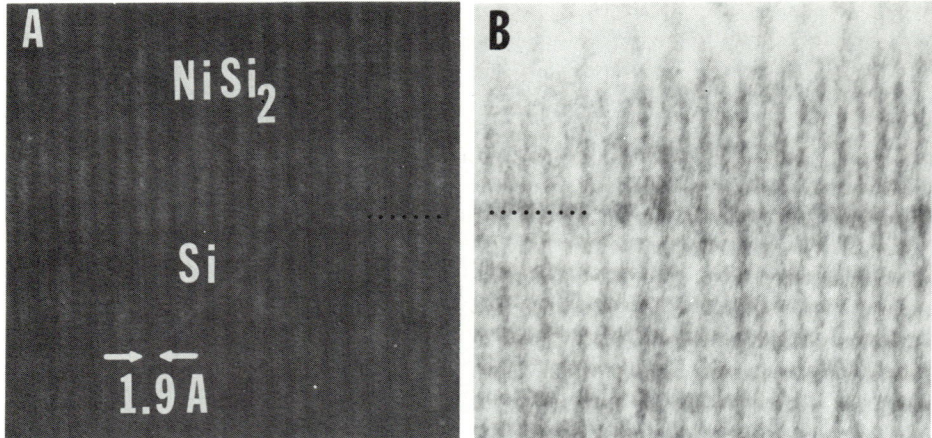

Figure 3: Axial bright-field electron micrographs from the Berkeley Atomic Resolution Microscope of $NiSi_2$ (111) templates on Si: a) of A orientation; b) of B orientation. Both images are taken in the <112> direction where the first transfer function ring includes the 1.9 Å (220) periodicity.

recent results on interface structure determination in this system are shown in figure 3 which comprises high-resolution axial bright field electron micrographs of $NiSi_2$ templates taken with the Atomic Resolution Microscope (ARM) in Berkeley, California. On this instrument, the familiar <110> projection of these interfaces reveals little more detail than is normally obtained with an instrument such as the 200CX. However figure 3 shows <112> projections of (a) an A and (b) a B $NiSi_2$/Si (111) (template) interface demonstrating 1.9Å resolution. The tilt capability of the ARM ($\pm40^\circ$) also allows visualization of the same interface area in different zone axes. The point resolution of this instrument, operating at 1MV, is better than 1.6Å. (Although a vibration problem has in these images allowed such resolution only in one direction. This problem has recently been overcome). From such images we can measure new components of the rigid shift between lattices which gives further accuracy to our interface structure determination. The measurements are in agreement with the seven-fold model which would predict no lattice shift of the (220) planes across these interfaces. The ARM tilt capability also permits reconstruction of structure at non-flat interfaces.

It might be appropriate at this point to discuss other methods for determining interface structure which have recently been used on $NiSi_2$/Si (111) interfaces. An elegant recent report of ion-channeling and blocking at the $NiSi_2$/Si interface (using a B-template) found very good quantitative agreement with the seven-fold model[11]. In fact better quantitative accuracy was achieved than could be expected with high resolution electron microscopy (at present), so that a subtle contraction of interfacial bonds of about 0.06 Å could be detected. On the other hand using X-ray interferometry Akimoto et. al.[12] deduced that a 5-fold model applied for (111) $NiSi_2$ on Si. Unfortunately, these authors misunderstood the nature of the interface and thus their results should be viewed sceptically. Their technique, which could possess similar accuracy to ion-channeling, was misinterpreted because the interface that they studied was certainly not regular. A thick $NiSi_2$ film was used which undoubtedly contained misfit dislocations, and was not single-crystalline, so that the local lattice registry was not regular. Such effects seriously limit the sensitivity of techniques like these with no spatial resolution and which are sensitive only to shifts between the entire lattice of a film and the substrate. Only by studying very thin uniform and psuedomorphic films can useful results be obtained. In other cases high-resolution electron microscopy has unique capabilities in studying structure and is in any case an important measure of interface regularity and local structure.

The Schottky barrier heights ($\phi_b$) of these template grown single-crystalline layers have been measured[13]. These and more recent results are summarized in Table 1.

Table 1: Schottky barrier heights for epitaxial silicides grown under UHV conditions.
Measurements were made on n-type substrates.

| Composition | Orientation | $\phi_b$ (I-V) | $\phi_b$ (C-V) |
|---|---|---|---|
| $NiSi_2$ | (111) A | 0.65 eV | 0.65 eV |
| $NiSi_2$ | (111) B | 0.79 eV | 0.79 eV |
| $NiSi_2$ | (111) A+B | 0.66eV | 0.7-0.8 eV |
| $NiSi_2$ | (100) | - | 0.48 eV |
| $CoSi_2$ | (111) B | 0.64 eV | - |

Measurements were made either by current-voltage (I-V) and/or capacitance-voltage (C-V) characteristics. Ideality factors were found to be very close to unity.

The results shown in table 1 for UHV grown single-crystalline A and B films are quite surprising in comparison with previous data for the $NiSi_x$ system which showed insensitivity of $\phi_b$ to most experimental variables[5]. The difference between A and B barrier heights cannot be simply explained by the Schottky model, since the work function of the metal and electron affinity of the semiconductor should be unaffected by a $180^0$ rotation about (111).

Another model involving defect Fermi-level pinning at the $NiSi_2$ interface would predict the same barrier height for A and B[14]. Work by McGill[15] has indicated that defect densities necessary for Fermi-level pinning must be very high ( several atomic %) because of the existence of metal-induced gap states which can very effectively screen defects. Therefore it is likely that the observed difference of barrier height is intrinsic and is related to bonding at the metal-semiconductor interface. It seems unlikely that this could be the case, given the similarity in the 7-fold model for the A and B cases, unless there was some rearrangement of electrons from Ni to Si atoms. We speculate that at the B interface, the Ni atoms at the interface might share electrons with the indicated Si atoms in figure 1 (dotted line). This atom pair is considerably closer at the B interface. The existence of a small ($\approx 0.06$Å) contraction deduced by van Loenen et. al.[11] is consistent with some interaction between this Ni-Si pair. Tersoff[16] has proposed another, simple, model, based on the apparent similarity of the B interface to hexagonal Si, which may explain the barrier height difference between A and B, but would not address other results such as (100) and CoSi$_2$. It should be noted that all $\phi_b$ measurements were made both in thick and thin films so that any effect of misfit dislocations (absent in the templates) can be discounted.

In any case the experimental data on interface structure is highly refined and should provide sufficient input for the theoretical modelling of barrier heights atomistically. This could lead to progress in the physical understanding, and level of technological control, of metal-semiconductor interfaces.

How can these measurements be reconciled with the measurement of a low barrier height for the mixed A+B NiSi$_2$ films? Since I-V characteristics are commonly used to measure $\phi_b$, it would be expected that the lowest barrier height would always dominate since $I=Ae^{q(V-\phi_b)/kT}$. In fact the C-V measurements, which are linearly proportional to the barrier height, do indeed show a higher barrier height for A+B films than do I-V measurements. This is consistent with a mixed barrier system. To clarify this issue, and allow the extension of these experiments to many other non-uniform systems, it would be very desirable to have a method of making barrier height determinations with some spatial resolution. Huang et. al.[17] proposed that electron-beam induced voltage is a suitable technique, in which the fine probe of a scanning electron microscope is used to generate carriers and effectively forward bias a Schottky barrier locally until the generation current equals the diode current. This voltage and its dependence on beam current allows simple measurement of barrier height. However, it is not clear that this technique can be easily extended to non-uniform systems since the diode current is not constrained to flow across the local Schottky barrier.

Electron beam induced current (EBIC) appears to have potential for local barrier height determination. Although diffusion current is a contribution to EBIC, there is a considerable effect of Schottky barrier height on the EBIC signal, particularly if the beam voltage is optimized so that the majority of carriers are generated in the depletion region. One might expect this effect to arise from differences in the depletion layer width which is approximately proportional to the square root of the local barrier height. Thus greater barrier height should produce more EBIC and thus appear brighter in EBIC images. We are studying this dependence quantitatively with Monte Carlo modelling of the generation process[18] and the exact mode of dependence is still unclear. However qualitative results are persuasive. EBIC images from very thin A,B and A+B templates taken at 5.5kV are shown in figure 4. Secondary electron images from the same areas show no contrast, confirming the uniformity of the films. EBIV measurements on the A and B films agree well with the abovementioned barrier heights. The uniformity of the image contrast confirms the uniformity of the barrier heights for the thin A and B films. In comparison the A+B film shows mottled contrast where dark areas appear to correlate with A and bright areas with B grains in the films. In fact the EBIC signal in these grains, when they are sufficiently large, displays the same dependence on beam voltage and current as do the single-crystal films and we therefore conclude that the local barrier height in A+B films is the same as the appropriate single-crystal value.

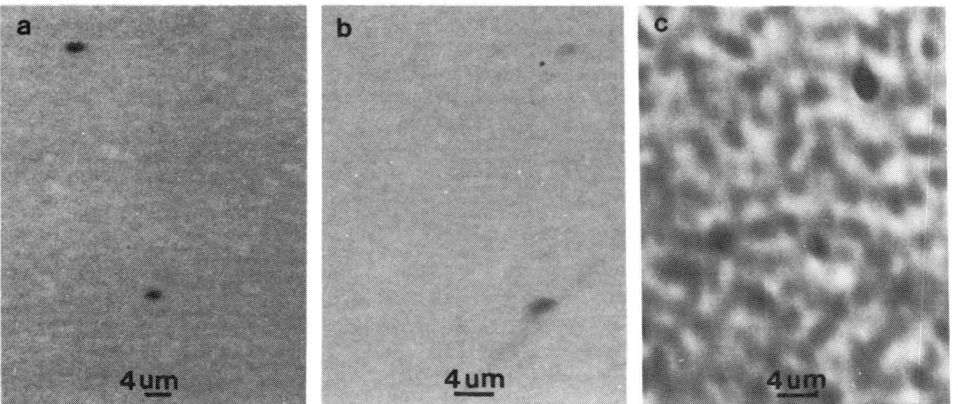

Figure 4: EBIC images at 5.5kV accelerating voltage of a) an A template; b) a B template and c) an A+B template; all are about 60 Å thick.

These measurements may resolve the apparent anomaly between our results for single-crystalline silicides and others. The EBIC technique, in combination with microscopic structural determination of interface structure, allows extension of these experiments to many other systems. Indeed it may prove that many other Schottky barrier systems reveal orientation dependence of the barrier height when probed on a microscopic scale.

Cobalt silicide, when grown by UHV techniques, displays only the B orientation. Therefore the dependence of Schottky barrier height on crystalline orientation has not been investigated. Interfaces in the $CoSi_2$/Si system are also atomically abrupt, but on (111) the B interface appears to best fit the 5-fold co-ordinated model[3]. One intriguing consequence of our UHV growth of $CoSi_2$ is its exceptional electron transport properties parallel to the interface. Hensel et. al.[19] have reported that the resistivity of $CoSi_2$ (111) films of 100 Å thickness is within a few percent of the bulk resistivity at all temperatures. Hall effect and magnetoresistance measurements indicate a carrier mean-free path of $\approx$ 200 Å at room temperature. These results imply that electrons carrying current in such thin $CoSi_2$ films are not losing energy on scattering from the interfaces i.e. that electron scattering must be specular. This result has implications for unique devices fabricated with Si/$CoSi_2$/Si heterostructures and for new regimes of physical measurements.

In other epitaxial thin film systems we have also attempted to correlate interface structure with physical properties. The alkaline-earth fluorides are epitaxial insulators on Si and other semiconductors[20] which permit correlation of well-behaved interfaces[21] with electron mobility and interface state density in field-effect-transistor structures[22]. $Ge_xSi_{1-x}$/Si superlattices have also been the subject of study[23], where the potential exists for correlation between interface electronic transport in a 2-dimensional electron gas and interface roughness and alloy structure.

## Conclusions

Through the use of epitaxial growth and clean interfaces, new phenomena in Schottky barrier formation have been revealed. In particular an orientation dependence has been found in $NiSi_2/Si$ barriers. Perfect interfaces are seen by high-resolution electron microscopy and the atomic structure is well-modelled. This structure/property relationship potentially allows a better understanding of the physics of the metal/semiconductor interface as well as better control of Schottky barriers in technology.

It is suggested that electron beam induced conductivity in the scanning electron microscope is a viable method for locally studying barrier height and it is shown that the local barrier height in thin mixed orientation $NiSi_2$ films is like that in single-crystal films of appropriate orientation. Other unique properties of epitaxial thin films on semiconductors related to their structural properties are described.

### Acknowledgements

The authors acknowledge invaluable discussions with P.M. Petroff, D. Fathy, S.J. Pennycook and R. Hull. The assistance of M.L. McDonald was also invaluable. One of us (CAP) thanks CNPq, Brazil, for provision of a visiting fellowship.

## References

1. see for example, S.P. Murarka, "Silicides for VLSI Applications", Academic Press (New York), 1983.

2. D.C. Cherns, G.R. Anstis J.L. Hutchison and J.C.H. Spence, Phil. Mag. **A46** , 849 (1982).

3. J.M. Gibson, R.T. Tung and J.M. Poate, Mat. Res. Soc. Proc. **14** , 395 (1983).

4. D.C. Cherns, C. Hetherington and C.J. Humphries, Phil. Mag. (1984)

5. P.E. Schmid, P.S. Ho, H. Foll and T.Y. Tan, Phys. Rev. B. **28** , 4593 (1983).

6. R.T. Tung, J.M. Gibson and J.M. Poate, Phys. Rev. Lett. **50** , 429 (1983).

7. R.T. Tung, J.C. Bean, J.M. Gibson, J.M. Poate and D.C. Jacobson, Appl. Phys. Lett. **40** , 684 (1982).

8. H. Ishiwara, M. Nagatumo and S. Furukawa, Nucl. Inst. Meth. **149** , 417 (1978).

9. R.T. Tung, J.M. Gibson and J.M. Poate, Appl. Phys. Lett. **42** 888 (1983).

10. J.M. Gibson, Ultramic. **14** , 1 (1984)

11. E.J. van L Loenen, J.W.M. Frenken, J.F. van der Veen and S. Valeri, Phys. Rev. Lett. **827** , (1985).

12. K. Akimoto, T. Ishikawa, T. Takahashi and S. Kikuta, Jap. J. of Appl. Phys. **22** , L798 (1983).

13. R.T. Tung, Phys. Rev. Lett. **52** 461 (1984).

14. O.F. Sankey, R.E. Allen and J.D. Dow, Sol. Stat. Comm. **49** , 1 (1984).

15. A. Zur, T.C. McGill and D.L. Smith, Phys. Rev. B **28** , 2060 (1983)

16. J. Tersoff, Phys. Rev. Lett. **52** , 465 (1984).

17. H.C.W. Huang, C.F. Aliotta and P.S. Ho, Appl. Phys. Lett. **41** , 54 (1982).

18. D.C. Joy and C. Pimentel, this volume

19. J.C. Hensel, R.T. Tung, J.M. Poate and F.C. Unterwald, Appl. Phys. Lett. **44** , 913 (1984).

20. J.M. Phillips, L.C. Feldman, J.M. Gibson and M.L. McDonald, J. Vac. Sci. Tech. **A1** , 563 (1983).

21. J.M. Gibson and J.M. Phillips, Appl. Phys. Lett. **43** , 828 (1983).

22. T.P. Smith, III, J.M. Phillips, W.M. Augustyniak and P.J. Stiles, Appl. Phys. Lett. **45** , 907 (1984).

23. R. Hull, J.M. Gibson and J.C. Bean, Appl. Phys. Lett. **46** , 179 (1985).

*Inst. Phys. Conf. Ser. No. 76: Section 5*
*Paper presented at Microsc. Semicond. Mater. Conf., Oxford, 25–27 March 1985*

183

# Studies of misfit strains in NiSi$_2$/Si bicrystals

C J Kiely and D Cherns

H.H. Wills Physics Laboratory, University of Bristol, Tyndall Avenue,
Bristol BS8 1TL, UK

Abstract    The accommodation of misfit in epitaxial NiSi$_2$/Si films has
been examined by TEM.  A combination of lattice imaging and diffraction
contrast techniques has shown new defects in both NiSi$_2$/(111)Si and
NiSi$_2$/(001)Si bicrystals.  Misfit strains are examined by observing the
behaviour of bend contours crossing discontinuous deposits of NiSi$_2$ on
(001)Si and by convergent beam diffraction.

## 1. Introduction

The microstructure of NiSi$_2$/Si films has attracted considerable interest
over the last few years.  The fact that both NiSi$_2$/(111)Si and NiSi$_2$/(001)
Si films are epitaxial having a small natural mismatch (0.4%), and possess
interfaces which are abrupt and atomically smooth over large distances
(often > 0.1 μm), has enabled interfacial structure to be studied by
lattice imaging techniques.  Atomic models have thereby been derived for
the NiSi$_2$/(111)Si orientations A and B interfaces (Cherns et al 1982), and
the NiSi$_2$/(001)Si interface (Cherns et al 1984).  Recent work by Tung
(1984) has shown that these three NiSi$_2$/Si interfaces can display markedly
different Schottky barrier heights confirming that studies of NiSi$_2$/Si
films may prove to be of considerable use in understanding the electrical
properties of metal/semiconductor contacts.

However, it is clear from further studies, particularly in NiSi$_2$/(001)Si
films (Cherns et al  1984, Cherns and Pond 1984), that the defect structure of NiSi$_2$/Si films is complex.  In this paper we examine some new defects both in NiSi$_2$/(111)Si and NiSi$_2$/(001)Si films by combined lattice
imaging and diffraction contrast techniques and use diffraction techniques
to examine the strains present in the films.

## 2. Experimental

NiSi$_2$/Si films were prepared by depositing about 500Å of Ni onto freshly
cleaned (111) and (001) Si substrates held at 200°C in evaporators operating at about 10$^{-6}$ torr.  The bicrystals were subsequently annealed in
vacuo for 1 hour at 800°C to produce NiSi$_2$ deposits.  (Some of the samples
were transferred to a second vacuum chamber for the annealing process).
Electron transparent specimens were prepared by backthinning plan-view
specimens with HF/HNO$_3$ and by ion-thinning cross-sectional samples using
3-5 keV Ar ions at 15° incidence.  Specimens were examined mostly in a
Philips EM430 transmission electron microscope operating at 300 kV.

## 3.1. NiSi$_2$/(111)Si films

NiSi$_2$/(111)Si films showed mostly epitaxial islands oriented with NiSi$_2$
either parallel to the (111)Si substrate (orientation A) or rotated by 180°
about the [111] substrate normal (orientation B).  In plan-view specimens,
orientation A islands sometimes showed misfit dislocations with Burgers
vectors $\frac{a}{2}<1\bar{1}0>$ lying in the film plane, as previously observed (Foll et al
1982).  Misfit dislocations with partial Burgers vectors $\frac{a}{6}<112>$ were
often observed in orientation B islands as also expected;  such dislocations denote an atomic step in the (111) interface (Cherns et al 1982).  In
addition long straight defects were sometimes seen in orientation A interfaces (Fig. 1).  These defects lie accurately along (110) directions and
were found to be invisible in the g = 220 reflection parallel to the line
direction when viewed close to the [111] film normal implying an edge displacement vector.  A preliminary analysis showed the defect to be visible
in g = 311 reached by tilting about an axis perpendicular to the line direction, which suggests an inclined Burgers vector.  Images in various reflections obtained by tilting parallel to the line direction suggested
the defect may be of the same type as those observed in orientation B
regions (Fig. 2), or possibly a Frank dislocation of $\frac{a}{3}<11\bar{1}>$ type.

Examination of lattice images from orientation B sections also showed new
defects which may be faults on (100) planes in the silicide layer (Fig.2).
The displacement of the fringes in the direction arrowed indicates a fault
vector with a projected component $\frac{a}{4}<110>$.

## 3.2. NiSi$_2$/(001)Si Films

NiSi$_2$/(001)Si films also showed epitaxial islands all oriented with NiSi$_2$

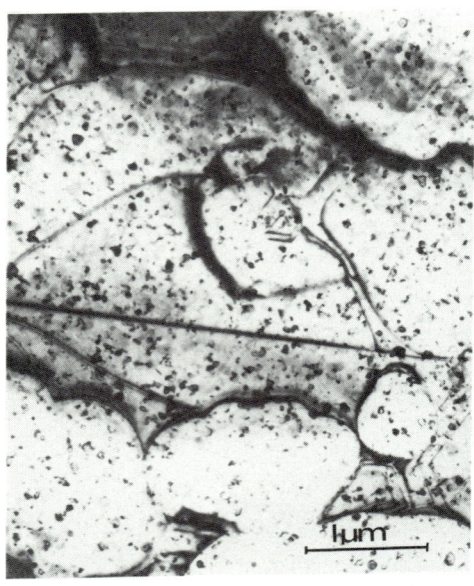

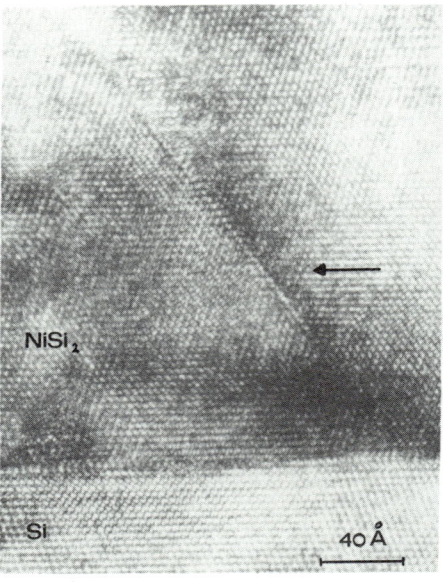

Fig. 1  Typical long straight
dislocation observed in orientation
A NiSi$_2$/Si(111).

Fig. 2  (110) lattice image of
faults in orientation B NiSi$_2$/Si(111)

parallel to the underlying substrate as expected.  Some of the samples
showed regions which were misoriented suggesting some contamination of the
substrate during growth or incompleted conversion of the deposit to the disilicide.  Well oriented regions in plan-view samples showed dislocations
with Burgers vectors b = $\frac{a}{4}$<111> as previously identified (Cherns and Pond
1984).  These are misfit dislocations separating areas of interface which
are crystallographically related by a rotation of 90° about the [001] film
normal.  In addition to these dislocations faint striations accurately
parallel to <110> directions in the film plane were often observed (Fig.3).
The contrast from these striations was much less intense than that displayed by the $\frac{a}{4}$<111> dislocations and was least when g lay parallel to the
line direction, implying an edge-type displacement vector.  Within one
domain of the NiSi$_2$/(001)Si interface the striations lay all in the same
direction, changing direction by 90° on crossing a $\frac{a}{4}$<111> dislocation into
a neighbouring domain as indicated.  Striations were often closely spaced
down to about 60Å apart or less and were sometimes observed to terminate
abruptly as arrowed.

The nature of these striations is not fully understood at present.  The
unidirectional nature of the striations within one domain and the behaviour at the domain boundary (the $\frac{a}{4}$<111> dislocation) suggests that the
striations are associated with the NiSi$_2$/(001)Si interface, where the [110]
and [1$\bar{1}$0] directions are not equivalent.  Microfacetting is one possible
explanation and this is illustrated in Fig. 4.  The nature of (100)/(111)
facet intersections in NiSi$_2$/(001)Si is such that the facet intersections
in Fig. 4(a) viewed down [110] say, in the interfacial plane require
$\frac{a}{4}$<111> dislocations.  Facet intersections in the same domain viewed down
[1$\bar{1}$0] (Fig. 4(b)) require no displacements in the limit of the model used
(that is, no interface relaxation).  The low contrast of the striations
in Fig. 3 appears consistent with the model in Fig. 4(b).

The low contrast of the striations in Fig. 3 might also suggest the intriguing possibility of a reconstructed interface.  However, the question of
microfacetting requires more detailed studies.  Indeed the examination of
cross-sectional samples shows that closely spaced interfacial facets do

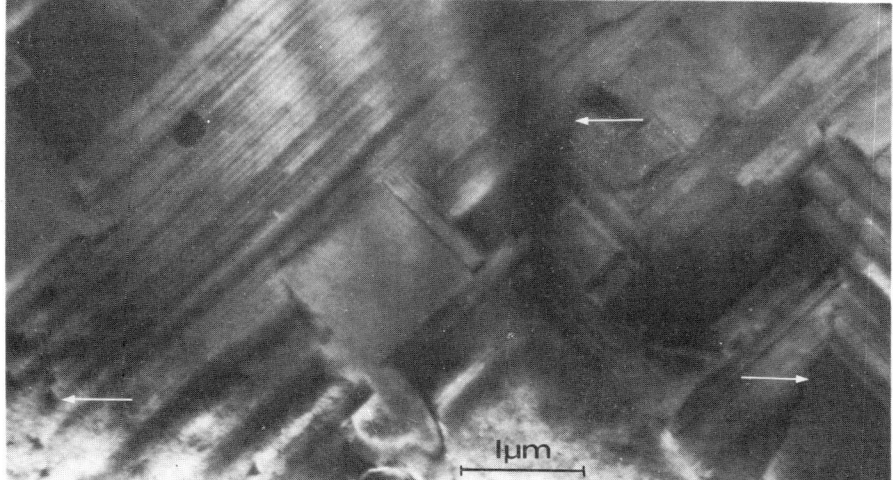

Fig. 3  Striations in domains of NiSi$_2$ on Si(001).  Domain boundaries
correspond to $\frac{a}{4}$<111> dislocations (arrowed)

Fig. 4  Schematic diagram of (100)/(111) facet intersections along
(a) [110] and (b) [1$\bar{1}$0] direction.

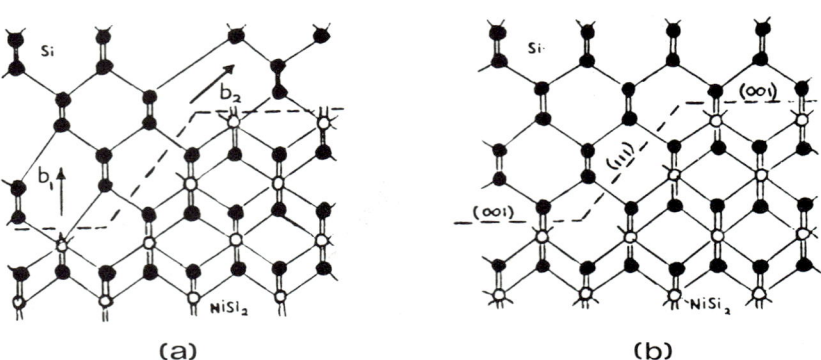

(a)                                                    (b)

exist in these films.  The observation that striations can apparently terminate within a domain also suggests that facet propagation in the NiSi$_2$/
(001)Si interface requires examination.

### 3.3. Misfit Strains

The islands of NiSi$_2$ in both NiSi$_2$/(111)Si and NiSi$_2$/(001)Si films were
either pseudomorphic or contained too few dislocations or other defects to
fully accommodate the 0.4% mismatch between NiSi$_2$(a = 5.406Å) and Si
(a = 5.428Å).  The NiSi$_2$ and Si films are thus expected to be in biaxial
tension and compression respectively, leading to tetragonal distortions of
opposite sign in samples where the layers are of comparable thickness.
Fig. 5 compares two bright-field images of the same area of a NiSi$_2$/(001)Si
film viewed approximately down the [00̄1] film normal and down a <110> axis
inclined at 45° to the interface.  The NiSi$_2$ deposit is on the upper side
of the foil.  In Fig. 5(a) the 200, 020, 220, and 2$\bar{2}$0 bend contours pass
smoothly between regions of bicrystal A and regions B where only the Si
substrate is present;  200 and 020 appear only weakly in the Si being kine-
matically forbidden.  In Fig. 5(b) the bend contour from the same 200
planes again passes smoothly between Si and bicrystal regions.  However,
the bend contours from inclined planes such as the 111 contour indicated
are displaced on crossing between Si and bicrystal regions; in this case
the displacement is about 0.2°.

A complete understanding of the bend contours in Fig. 5 has yet to be established.  Assuming for simplicity that the silicide and silicon layers in
region A are of comparable thickness and constrained to be flat, we expect
the tetragonal distortions to cause rotations of inclined planes, like the
(111) planes, in opposite senses in the two layers. (see Fig. 6).  On the
other hand, planes perpendicular and parallel to the interface remain un-
rotated.  These  rotations can be up to 0.3° for the planes inclined at 45°
to the interface.  Assuming tentatively that the bend contour is situated
at the Bragg condition for the bottom layer, (a reasonable assumption in
the limit as the layer thickness increases), then the displacements observed in Fig. 5(b) are in the expected sense.  Turning the foil over again

Fig. 5  Bright field image of an island g NiSi$_2$ on (001)Si (a) down [001]
film normal, (b) down a <110> axis inclined at 45° to the interface.

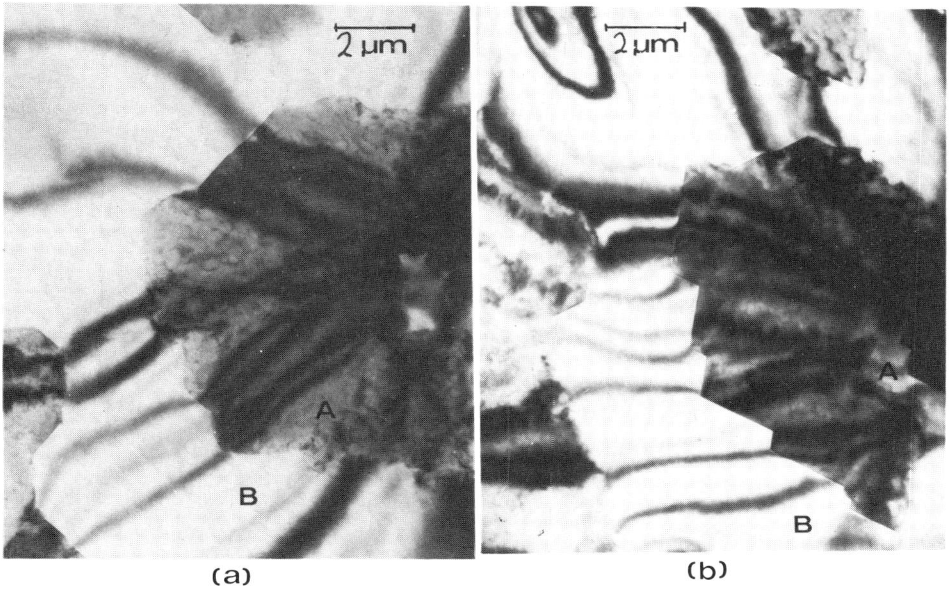

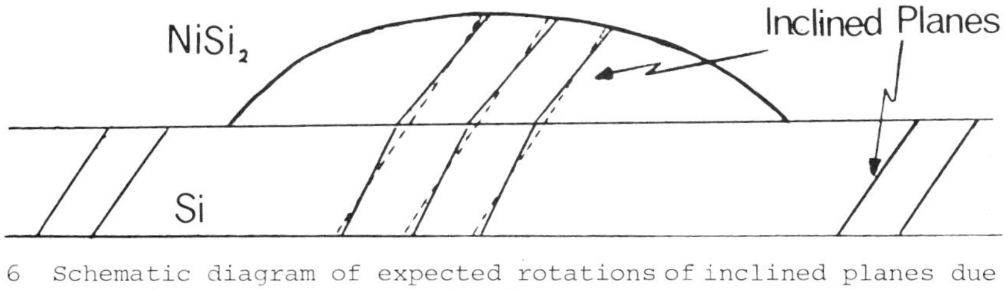

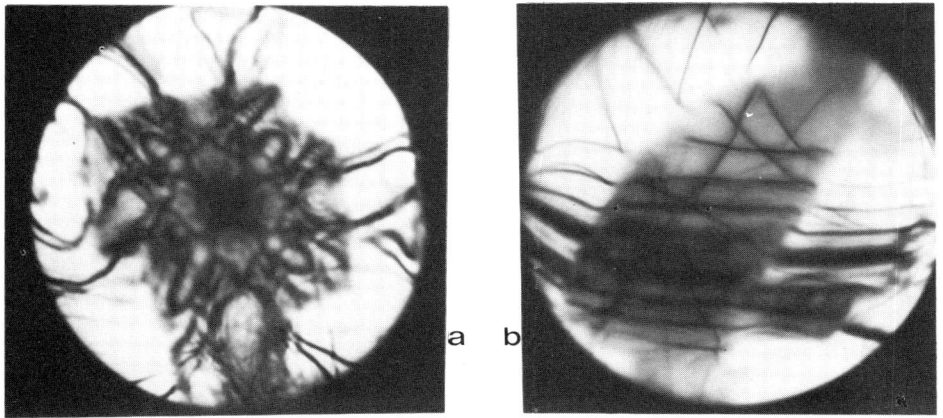

(a)                                                          (b)

Fig. 6  Schematic diagram of expected rotations of inclined planes due to
unrelaxed misfit strains in NiSi$_2$/Si films.

Fig. 7  Wide angle (000) diffraction disc from a film of NiSi$_2$/(001);
(a)with [001] nearly parallel to the beam direction (b) from inclined planes

produced displacements of bend contours from inclined planes in the same region consistent with that in Fig. 6.

The fact that a heavily bent region is present in Fig. 5 makes a quantitative analysis of strains formidable. However, bend contour patterns may also be generated on a flat region of sample using the diffraction technique described by Tanaka (1980). In this technique a wide angle defocussed probe illuminates an extended region of the sample and the diffraction mode is selected. A selected area aperture can be used to eliminate all but a single diffraction disc which displays an image of the illuminated area. The spatial resolution achieved is of the order of the smallest incident probe size. Fig. 7 illustrates the wide angle (000) disc taken at 120 kV from an $NiSi_2/(001)Si$ film. Fig. 7(a) is taken from a flat region with [001] nearly parallel to the beam direction, and the bend contours are seen to be continuous through the bicrystal region (cf. Fig. 5(a)). Fig. 7(b) shows a bend contour from inclined planes and is seen to be displaced on crossing the bicrystal region. (cf. Fig. 5(b)).

Conclusions

The presence of new defects in both $NiSi_2/(111)Si$ and $NiSi/(001)Si$ films has been illustrated. An analysis of strains in $NiSi_2/(111)Si$ and $NiSi_2/(001)Si$ films using the methods illustrated in Figs. 5 and 7 is being carried out. Comparative studies of strains are also being carried out using convergent beam electron diffraction patterns taken on plan and cross-sectional view specimens.

Acknowledgements

We are grateful to C.J.D. Hetherington for provision of some of the $NiSi_2/(001)Si$ samples used in this work, and to D.J. Eaglesham for helpful discussions. We would also like to acknowledge the support of the SERC.

References

Cherns D, Anstis GR, Hutchison JL, Spence JCH, 1982 Phil.Mag.A. <u>46</u> 849.
Cherns D, Hetherington CJD, Humphreys CJ, 1984 Phil.Mag.A. <u>49</u> 165.
Cherns D, Pond RC, 1984 Materials Research Society Symposium Proceedings,
    Vol. 25, 423.
Foll H, Ho PS, Tu KN 1982 Phil.Mag.A. <u>45</u> 31.
Tanaka M, Saito R, Ueno K, Harada Y, 1980 J.Electron Microscopy <u>29</u> 408.
Tung RT 1984 J.Vac.Sci.Technol. B2(3) 465.

*Inst. Phys. Conf. Ser. No. 76: Section 5*
*Paper presented at Microsc. Semicond. Mater. Conf., Oxford, 25–27 March 1985*

# Cross-section TEM study of the simultaneous formation of shallow junctions and TiSi$_2$ contacts by rapid thermal annealing

J Vanhellemont[1], H Bender[2], J Van Landuyt[1], C Claeys[3] and G Declerck[3]

[1]Universiteit Antwerpen, RUCA, Groenenborgerlaan 171, B-2020 Antwerpen, Belgium
[2]IMEC, c/o Universiteit Antwerpen, Groenenborgerlaan 171, B-2020 Antwerpen, Belgium
[3]IMEC, Kapeldreef 75, B-3030 Heverlee, Belgium

Abstract   The results are presented of a detailed HREM and HVEM study of the simultaneous formation of a n$^+$p junction and its TiSi$_2$ contact on (001) p-type silicon substrates. A Sb$^+$ or As$^+$ ion implantation is performed before or after (mixing) the Ti sputter deposition and is followed by a 5 s thermal anneal in a N$_2$ ambient. After the anneal no defects are found for the As$^+$ implantation. For the Sb$^+$ implantation without mixing small dislocation loops are observed at the original a-Si/Si interface. In the mixing case small crystalline precipitates which are semi-coherent with the silicon matrix are revealed.

## 1. Introduction

Titanium disilicide is a promising candidate for the replacement of poly-silicon as the contact material in VLSI circuits due to its low resistivity and its compatibility with high temperature processing steps. The silicide contacts can be formed by several techniques such as cosputtering of Ti and Si, composite target sputtering or sputtering from a sintered single compound target (McLachlan and Avins 1984). These sputtered and cosputtered films are all amorphous so that an additional thermal anneal step is required to obtain the crystalline silicide contacts. Conventional heat treatments can cause considerable redistribution and out-diffusion of the dopant atoms. Recently, atomic mixing by ion implantation of dopant species through thin metal films on silicon substrates has been proposed to be combined with rapid thermal annealing (Maex et al 1984, Maex and De Keersmaecker 1984) because of the possibility to simultaneously activate the ion implanted dopants, to regrow the amorphized silicon with little redistribution of the dopant atoms and to form silicide contacts by the same short time annealing step. The simultaneous formation of shallow n$^+$p junctions and TiSi$_2$ contacts by a rapid thermal annealing was studied by Maex and De Keersmaecker (1984). The present paper reports the results of a cross-section TEM study of samples prepared by this technique.

## 2. Experimental techniques

In this work (001) oriented p-type Czochralski silicon wafers with a resistivity of 15-30Ω cm are used. After a HF dip to remove the native oxide layer the wafers are covered with a 50 nm thin Ti layer by means of DC magnetron sputtering. During the As$^+$ (5.10$^{15}$/cm$^2$ at 200 keV) and the Sb$^+$ (1.10$^{15}$/cm$^2$ at 450 keV) ion implantation special care is taken to keep the

wafer temperature below 120°C by conductive cooling. The 5s thermal anneal
at 1000°C is carried out in a N$_2$ ambient with a bank of tungsten halogen
lamps.

The (110) oriented cross-section specimens for TEM are prepared by ion
milling. They are investigated by means of high voltage electron microscopy (HVEM) at an acceleration voltage of 1250 kV. The high resolution
work is performed in a JEM 200 CX at 200 kV.

3. Results and discussion

As is illustrated on figures 1, 2 and 3a, the implantation with Sb$^+$ or As$^+$
ions creates an amorphous layer in the silicon substrate. For the Sb$^+$-
implantations the a-Si layer extends to about 320 nm without mixing and to
280 nm when mixing is used. In all cases the a-Si/Si interface is very
rough in agreement with the reports in the literature (Fletcher et al 1981).
When the ion implantation is performed after the Ti deposition an additional thin amorphous layer is observed between the polycrystalline Ti and

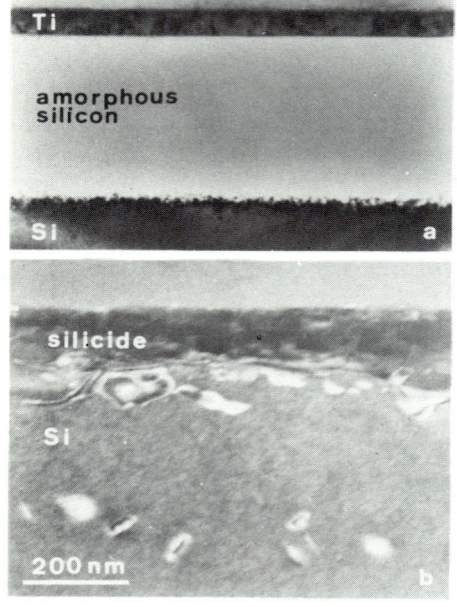
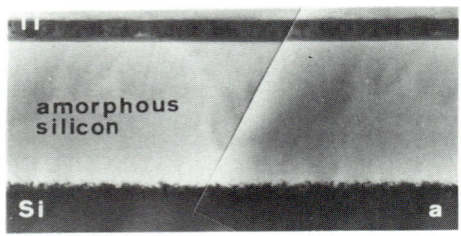
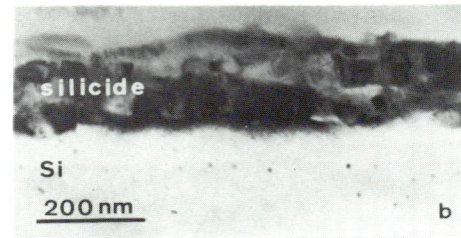

Fig. 1 a) Cross-section HVEM
micrograph showing the amorphized
silicon substrate after the Sb$^+$
implantation without mixing.

b) After the thermal anneal step
small dislocation loops are observed at the previously a-Si/Si
interface.

Fig. 2a) The configuration after the
Sb$^+$ implantation with interface mixing.

b) After the rapid annealing tiny
precipitates are observed below the
polycrystalline silicide layer in the
regrown Si-substrate.

the a-Si layer (fig. 2b and 3b). This layer shows a different absorption
contrast than the a-Si one and is probably due to a mixing of Ti and Si.

After the thermal anneal step no defects are observed for the As[+] implantation (with mixing). For a Sb[+] implantation however, the mixing and the
non-mixing case show a remarkable difference. Without mixing a large number of small dislocation loops is observed near the previously a-Si/Si
interface (fig. 1b). The loops are about 80 nm large and have a Burgers
vector of the $\frac{a}{2}<011>$ type (fig. 4).

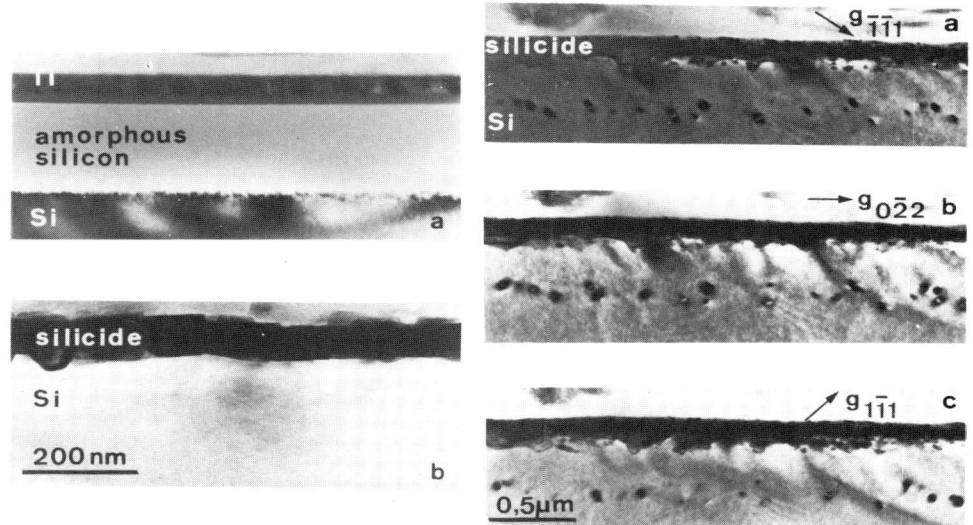

Fig. 3 a) Cross-section image af-    Fig. 4  The small defects can be
ter an As[+] implantation with inter-  identified as dislocation loops
face mixing.                          with $\bar{b} = \frac{a}{2}<011>$
b) After the thermal annealing no
defects are observed.

The situation changes drastically when the Sb[+] implantation is performed
through the Ti film, i.e. with mixing. In that case tiny defects are observed scattered all over the previously a-Si region. The contrast of
these defects changes only very little for different two beam diffraction
conditions suggesting that they are not dislocation loops but small precipitates embedded in the Si matrix without high stresses. When investigating the same samples in the high resolution microscope the precipitates
seem at first to be amorphous. Indeed, when the specimen is oriented to
obtain optimal resolution for the Si matrix, no lattice fringes can be observed in the precipitate and the image is similar to the one obtained for
an amorphous particle embedded in the Si matrix. It was however observed
that the contrast of the particles increases when tilting the specimen away
from the exact $(110)_{Si}$ pole. Orientation of the specimen by trial and error
allows us to obtain Moiré fringes in the precipitate indicating a crystalline
nature. Due to the rather low density of the defects (for HREM work) only

in a few cases the two conditions to obtain high resolution images, i.e. a precipitate in a very thin region and an orientation suitable for lattice imaging, were fullfilled (Fig. 5 and 6). As can be seen on the optical diffractograms of figures 5 and 6 the diffraction spots corresponding with one set of (111) planes of silicon coincide with the diffraction spots of a set of lattice planes of the precipitate, which indicates that the precipitates are semi-coherent with the silicon matrix. The possible compositions for the precipitates are : Sb, Sb/Ti or Ti/Si. From the compounds with these compositions, which are described in the X-ray powder diffraction files, the one which fits best with the optical diffractogram of figure 5 (i.e. $d_{hkl}$ and angles) is the orthorhombic ($TiSi_2$) 12Q with the zirconium disilicide type of structure (ASTM 10-225). Both the (001) and the (100) section of this compound correspond with the diffractogram. However, for both these orientations HREM image simulations performed with the real space method (Van Dyck and Coene 1984, Coene and Van Dyck 1984) as a function of the specimen thickness and defocus value,show no correspondence with the experimental image. Moreover the proposed structure cannot explain the diffraction pattern shown in figure 6. It is therefore concluded that the precipitates cannot have the ($TiSi_2$) 12Q structure.

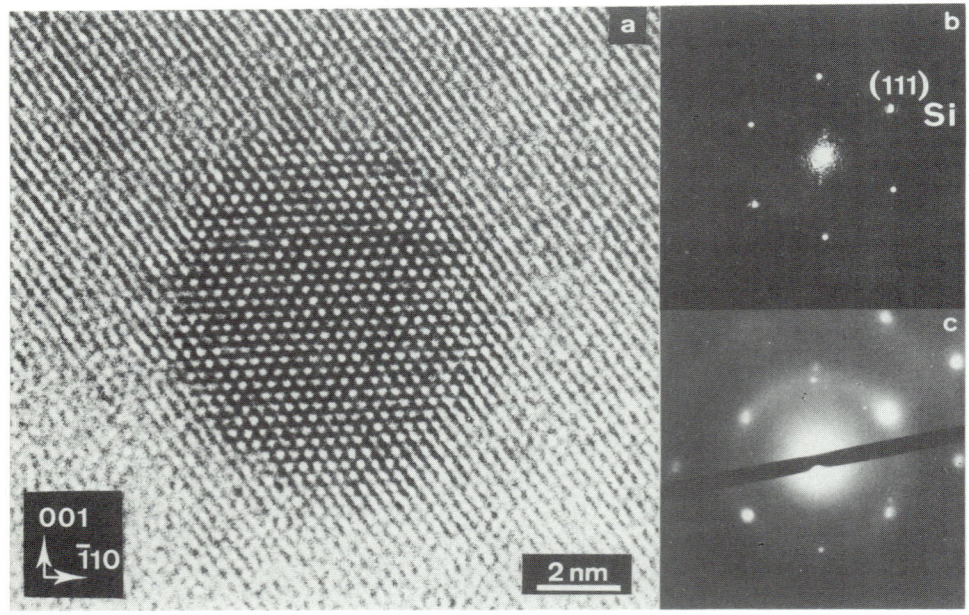

Fig. 5 a) HREM image of a precipitate. One set of (111) planes of the silicon matrix coincides with one set of planes of the precipitate.
b) Optical diffractogram of the precipitate.
c) Electron diffraction pattern of the precipitate region.

X-ray micro-analysis of the precipitate of figure 5,using a JEOL 100C scanning transmission electron microscope, shows no difference in Sb concentration when compared with the surrounding Si matrix but clearly indicates an increased Ti content in the particle.  It is therefore concluded that the precipitate is most likely composed of a $TiSi_x$ compound with x close to two in agreement with the fact that for the  refractory silicides the most stable phase is the silicon rich one, i.e. M(etal)$Si_2$(Mc Lachlan and Avins 1984).  The precipitate of figure 6, for which no X-ray microanalysis was performed, gives an optical diffraction pattern which can also be attributed to Sb (Bender 1985) so that it might be possible that there exists a mixture of two different types of precipitates (Sb and $TiSi_x$).  Anyhow these results indicate that a fraction of the Ti layer is driven in by the $Sb^+$ ions during the ion implantation.  During the subsequent short thermal anneal step these Ti atoms migrate and form silicon rich $TiSi_x$ precipitates in the regrown Si layer.

Etching the thinned TEM specimens with 3-1 stain (3 $HNO_3$, 1 HF, 10 HAc) for a few seconds allows us to visualize the junction depth.  The etching technique is calibrated for As implants by comparing calculated dopant profiles with the delineated junction depths observed in the TEM.  For this purpose p-type silicon is implanted with $As^+$ ($1.10^{16}/cm^2$, 40 keV) followed by a drive-in step at 950°C in $N_2$ for different times.  Computer simulations of the As profiles were obtained by using SUPREM-3.  The comparison of the calculated profiles and the experimentally delineated junction depths shows that the 3-1 etch delineates doping levels of about $5.10^{18}$ to  $1.10^{19}$ atoms/ $cm^3$.  The simulated junction depth lies about 25 nm deeper in the bulk. For the considered 200 keV As implantation this technique results in a delineated junction depth of about 0.13 μm which corresponds very well with electrical measurements giving an As concentration of $1.10^{19}/cm^3$ at 0.13 μm (Maex and De Keersmaecker 1984). Applying the same  etchant for the $Sb^+$ implanted samples gives a junction depth of about 0.4 μm which according to the electrical measurements should correspond with a doping concentra-

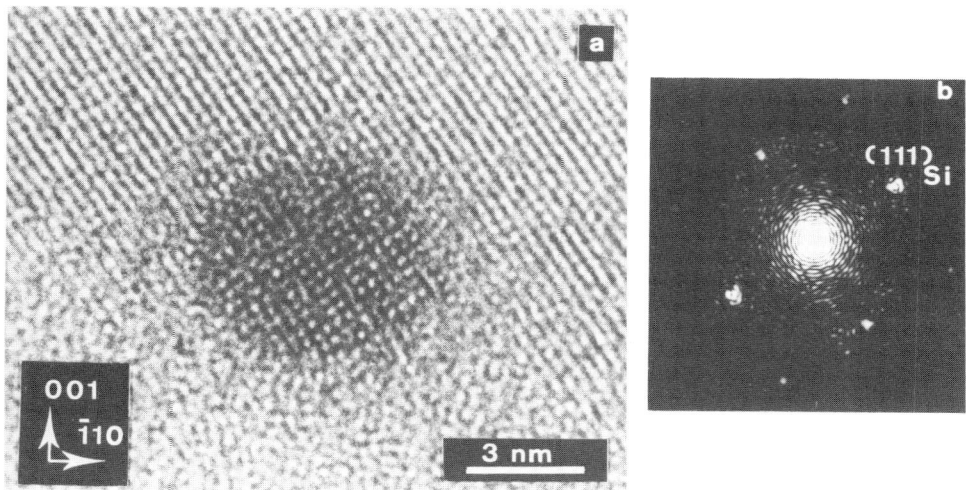

Fig. 6 a) HREM image of a precipitate lying close to the edge of the specimen, which is amorphized by the ion thinning.
b) Optical diffractogram of the precipitate.

tion of $1.10^{19}/cm^3$. A detailed analysis of the different doping profiles
is performed by Maex and De Keersmaecker (1985).

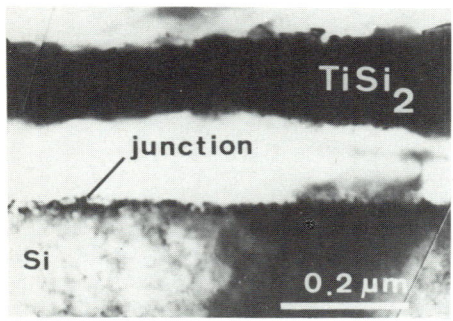

Fig. 7  HVEM image after a 5 s etch
in 3-1 stain (As implantation with
mixing).

## 4. Conclusion

The simultaneous formation of shallow junctions and $TiSi_2$ contacts by using
ion implantation either before or after Ti deposition combined with rapid
thermal annealing is very promising for implementation in VLSI processing.
The reported TEM observations show that after an $As^+$ implantation through
a Ti film a defect free regrowth of the amorphous Si layer can be obtained.  In the case of a $Sb^+$ implantation however, both the mixing and the
non-mixing cases give rise to the formation of defects in the recrystallized a-Si region.

## Acknowledgement

The authors wish to thank the ion implantation and silicide groups of the
ESAT laboratory, KU Leuven, for providing the silicide material for TEM
investigation.  We are particulary grateful to Ir $^K$ Maex and Dr R De Keers-
maecker for many helpful discussions on the subject.
Drs E Van Cappellen, Universiteit Antwerpen, UIA, is acknowledged for the
STEM X-ray microanalysis.
J Vanhellemont is indebted to the Belgian Science Foundation (IIKW) for
his fellowship.

## References

Bender H  1985  these Proceedings
Coene W and Van Dyck D  1984  Ultramicroscopy 15 41
Maex K, De Keersmaecker R, Alkemabe P F A, van der Wey WF and F H P M
    1984  Proc. MRS Europe Conference  eds Pinard P and Kalbitzer S (Les
    Editions de Physique) pp 315-322
Maex K and De Keersmaecker R F 1984  Proc 14th European Solid State Device
    Research Conference  eds Noblane J P and Zimmerman J  to be published
    in Physica B
Maex K and De Keersmaecker R  1985 Proc MRS Spring Meeting San Francisco
    to be published
McLachlan D R and Avins J B  1984 Semiconductor International 129
Van Dyck D and Coene W  1984 Ultramicroscopy 15  29
Fletcher J Narayan J and Holland O W  1981 Inst. Phys. Conf. Ser. 60  295

*Inst. Phys. Conf. Ser. No. 76: Section 5*
*Paper presented at Microsc. Semicond. Mater. Conf., Oxford, 25–27 March 1985*

# HREM study of nitrogen-annealed silicon-rich tungsten silicide

J Vanhellemont[*], J Van Landuyt[*], C Claeys[☆], G Declerck[☆], H L Babbar[♦],
D N Nichols[♦] and C Anagnostopoulos[♦]

[*]Univ. Antwerpen, RUCA, Groenenborgerlaan 171, B-2020 Antwerpen, Belgium.
[☆]IMEC, Kapeldreef 75, B-3030 Heverlee, Belgium.
[♦]Research Laboratories, Eastman Kodak Company, Rochester, NY 14650, USA.

Abstract   The results are presented of a high-resolution electron microscopy (HREM) study of the annealing in a nitrogen ambient of a silicon-
rich tungsten silicide layer, sputter deposited on (001) p-type silicon
substrates. The as-deposited amorphous $WSi_{2-x}$ film is separated from
the silicon substrate by a 1.5 nm oxide layer. After the annealing at
950°C in $N_2$, islands of epitaxially grown silicon are observed underneath the silicide layer. At the original interface a high density of
4-10 nm large amorphous precipitates are formed. They act as nuclei for
defects in the epi-layer such as stacking faults and twin lamellae.

## 1. Introduction

Refractory metal disilicides are the subject of extensive study in various
laboratories because of their low resistivity, which makes them very attractive for replacing polysilicon as the interconnection medium in VLSI circuits (Murarka 1981, Tsai et al. 1981, Inoue et al. 1983).
Another advantage is their resistance against chemical reagents and high
temperatures used in VLSI processing, so that their implementation will not
require drastic changes in the processing techniques that are well established for silicon gate technology. $WSi_2$ has a low resistivity, is resistant to chemical reagents currently used for cleaning, can withstand rather
high process temperatures, and allows the growth of stable $SiO_2$ layers. The
properties of the $WSi_2$ layer strongly depend on the annealing conditions.
This paper presents the results of a HREM cross-section study of the annealing  in a nitrogen ambient of a silicon-rich amorphous tungsten silicide
layer.

## 2. Experimental Techniques

In this work (001) p-type silicon wafers were used.  After mechanical scrub-
bing, followed by a standard RCA chemical cleaning, $WSi_{2-x}$ was sputter deposited from a hot-pressed composite $WSi_3$ target. The annealing was done in
a diffusion furnace at 950°C for 30 min in a $N_2$ ambient.

Plan-view TEM specimens were prepared by ultrasonic drilling followed by
back-side chemical thinning in 1(HF):8($HNO_3$).  The (110) oriented cross-
section specimens were prepared by a sawing technique followed by ion
milling (Vanhellemont et al. 1983). The HREM work was done at 200 kV on a
JEM 200 CX electron microscope.

3. Results and Discussion

Cross-section HREM showed that the as-deposited $WSi_{2.x}$ film forms a 375 nm thick amorphous layer which is separated from the silicon substrate by another 1.5 nm thin amorphous layer revealed by a different absorption contrast (Fig. 1). Most likely the thin interfacial layer is related to a native silicon oxide layer already present before sputtering. The silicon surface was very flat with fluctuations of less than 1 nm.

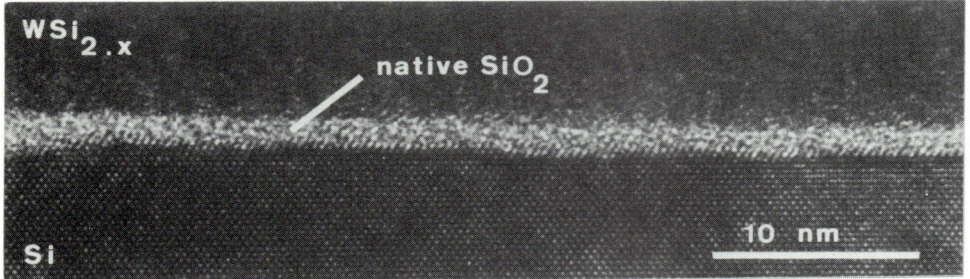

Figure 1 HREM micrograph showing the native oxide between $a-WSi_{2.x}$ and Si.

After the annealing, plan-view transmission electron microscopy (TEM) revealed the presence of a polycrystalline silicide layer with a grain size of about 100 nm (Fig. 2). On the electron diffraction pattern in Figure 2b, diffraction rings corresponding with the lattice planes (002), (101), (110) and (103) of tetragonal ($WSi_2$) 6B can clearly be observed (ASTM 11-195). The (002) spot of the (100) section of Si is used for calibration.

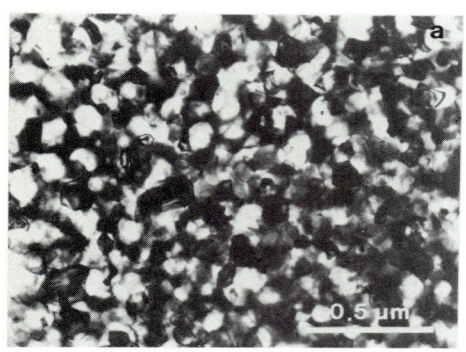 

Figure 2 a) Plan-view of the annealed $WSi_2$ film.
        b) Diffraction pattern corresponding with polycrystalline $WSi_2$.

Cross-section investigation showed that islands of epitaxially grown silicon are formed underneath the WSi$_2$ layer. They may extend over several micrometers, causing a considerable lifting of the silicide layer from the silicon surface (Fig. 3). The formation of these islands can be explained by
a migration of the excess silicon in the silicide layer towards the interface under the influence of the large stresses in that area, and subsequent
precipitation on the Si-surface.

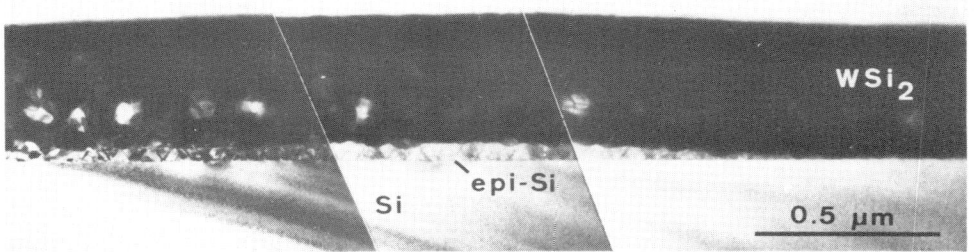

Figure 3 Cross-section image illustrating the formation of an epitaxial
        Si layer lifting the WSi$_2$ layer from the original Si surface.

This epitaxial layer contains a large number of defects. Higher magnifications show that these faults originate at small precipitates lying at
the original a-WSi$_{2-x}$/Si interface (Fig. 4). Figure 4b gives a selected
area diffraction pattern of the highly defective area. Between the bright
diffraction spots of the (110) section of Si, two additional weak spots
occur in each <111> direction, lying respectively at 1/3 and 2/3 of the
distance between the silicon spots, thus corresponding to a 3-fold period
in the <1$\bar{1}$1> directions. The occurrence of these additional spots can be

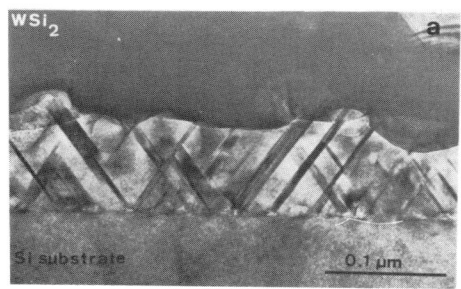

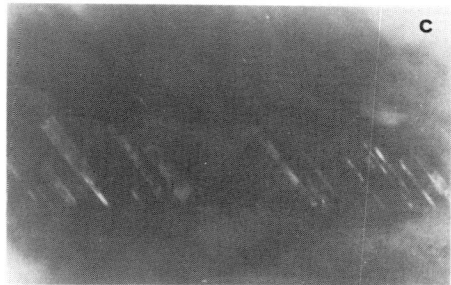

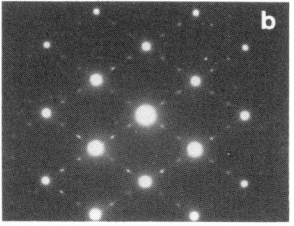

Figure 4a) The highly defective
        epi-layer.
      b) Selected area diffrac-
         tion pattern of the
         epi-layer.
      c) Dark-field image
         showing twins of one
         family.

explained by the presence of extensive twinning in the epi-layer combined with double diffraction. The dark-field micrograph of Figure 4c, which is the result of selecting two extra spots from one <111> direction, clearly shows the large number of corresponding twin bands in the epi-layer.

HREM investigation of the precipitates in very thin sections yielded images typical of an amorphous phase (Fig. 5). The location of these amorphous particles at the original interface suggests that the native oxide layer, present in the as-deposited material, transforms into amorphous $SiO_x$ precipitates simultaneously with the epitaxial growth of the Si between the oxide particles. This process  is similar to that described for the annealing  of a polysilicon layer deposited on a silicon substrate, giving rise to epitaxial growth and balling up of the native oxide to form precipitates at the original Si surface (Albu-Yaron et al. 1984). The precipitates act as nuclei for the extensive twinning in the epi-layer, which probably removes some of the stresses caused by both the silicide layer and the precipitates. The same figure shows part of a $WSi_2$ grain of which one set of lattice planes is resolved. The inset gives an optical diffraction pattern obtained at the $WSi_2$/Si interface. From this it can be concluded that the silicide lattice planes are of the {101} type of tetragonal ($WSi_2$) 6B. In the epi-layer stacking faults originating at the $WSi_2$ grain are also observed.

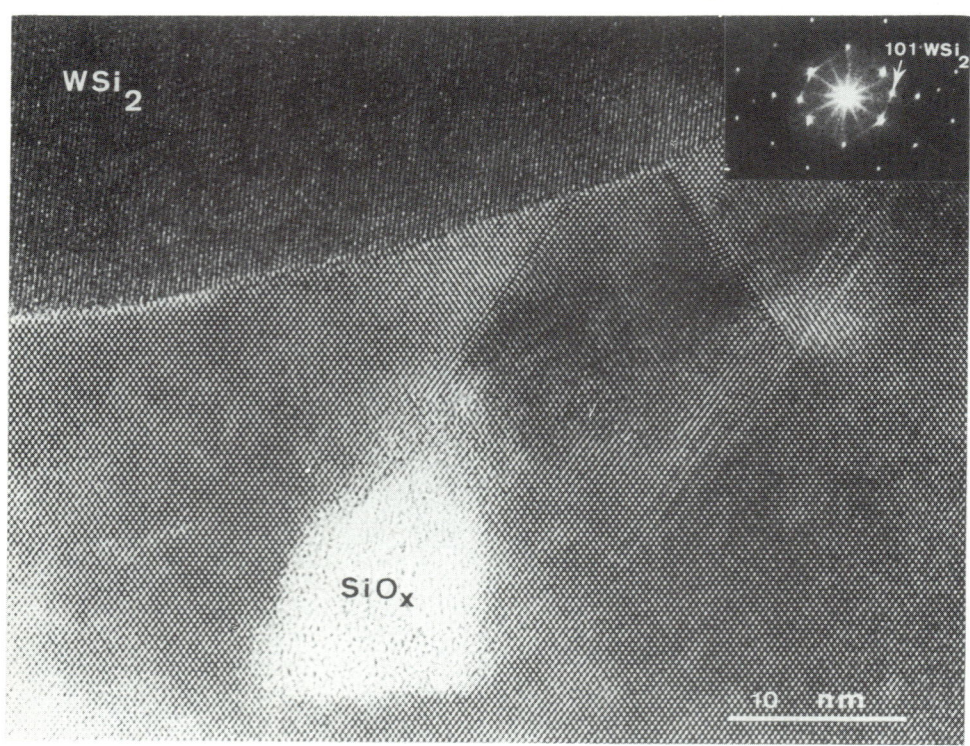

Figure 5 HREM image of an a-$SiO_x$ precipitate close to a $WSi_2$ grain. The inset shows an optical diffraction of the $WSi_2$/Si interface .

Figure 6 shows a somewhat thicker region where twins originating at the
interface precipitates are visible. In this case the twins extend throughout the whole foil thickness so that no interference with the perfect Si
matrix is observed. Optical diffraction revealed only one additional diffraction spot resulting from the twinning. No double diffraction can occur
for this configuration.

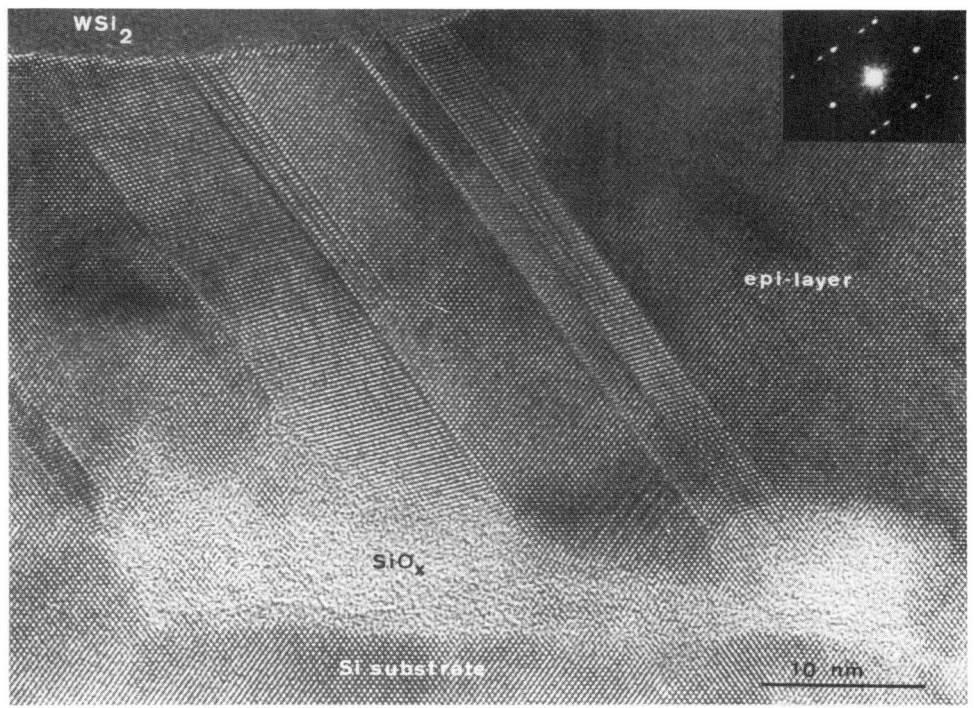

Figure 6 Twins originating at the interface precipitates.

As is illustrated in Figure 7, in thicker regions lamellae are observed
which show a 3-fold period in the <111> directions. Optical diffraction
of such regions leads to a diffraction pattern similar to the one obtained
in the electron microscope (cf. Fig. 4b). One can explain this 3-fold
period by the overlap of a twin lamella and the silicon matrix, which by
double diffraction would also produce the observed diffraction pattern.
An alternative explanation, however, is also possible. Recently Hendriks
et al. (1984) and Lereah and Grünbaum (1984) reported the existence of polytypic silicon modifications in polycrystalline silicon films. High local
stresses during epitaxial growth can induce the formation of a high density of stacking faults and/or microtwins, giving rise to a new silicon
polytypic modification. Work is going on to distinguish between the two
possible explanations.

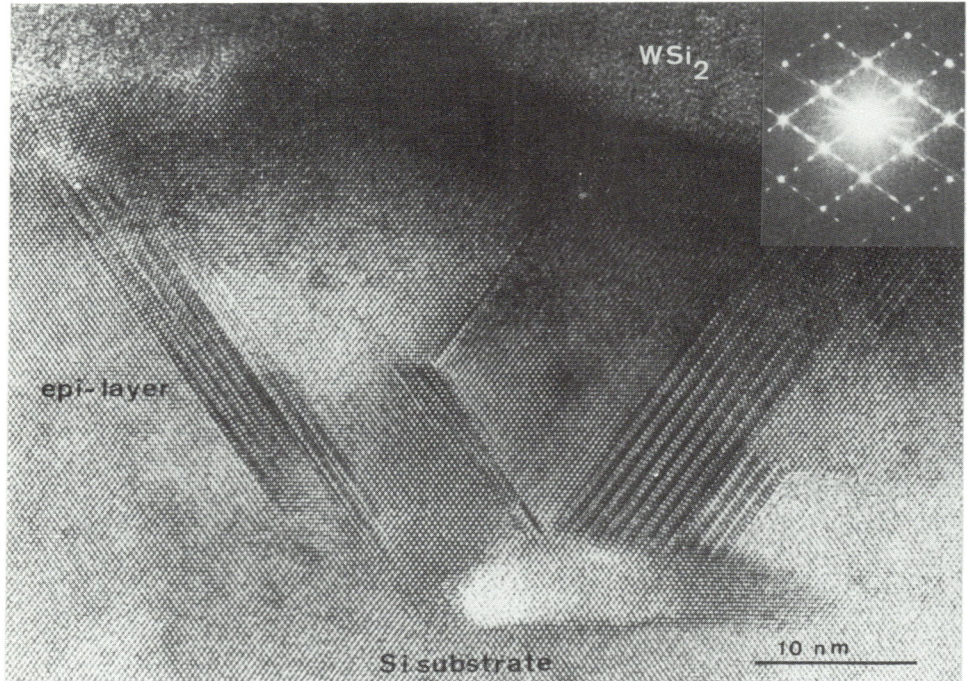

Figure 7 HREM image of a twinned region with overlapping silicon matrix.

Acknowledgement

J Vanhellemont is indebted to the Belgian Science Foundation (IIKW) for his fellowship.

References

Albu-Yaron A, Barry J C and Booker G R  1984  Proc. 8th European Congress on Electron Microscopy  eds Csanady A, Röhlich P and Szabo D (Budapest) pp 521-522
Hendriks M, Radelaar S, Beers A M and Bloem J  1984  Thin Solid Films  113 59
Inoue S, Toyokura N, Nakamura T and Ishikawa H  1983  JARECT Vol 8  Semi-conductor Technologies  ed Nishizawa J  pp 45-54
Lereah Y and Grünbaum E  1984  Phil. Mag. A 50  1
Murarka S P  1981  Semiconductor Silicon  eds Huff H R, Kriegler R J and Takeishi Y (Pennington : The Electrochem. Soc. Softbound Ser.) pp 551-561
Tsai M Y, Chao H H, Ephrath L M, Crowder B L, Cramer A, Benneth R S, Lucchese C J and Wordeman M R  1981 Semiconductor Silicon  eds Huff H R Kriegler R J and Takeishi Y (Pennington : The Electrochem. Soc. Softbound Ser.) pp 573-587
Vanhellemont J, Bender H, Claeys C, Van Landuyt J, Declerck G, Amelinckx S and Van Overstraeten R  1983  Ultramicroscopy  11  303

*Inst. Phys. Conf. Ser. No. 76: Section 6*
*Paper presented at Microsc. Semicond. Mater. Conf., Oxford, 25–27 March 1985*

201

# Developments in the TEM study of compound semiconductors

G R Booker

Department of Metallurgy & Science of Materials,
University of Oxford, Parks Road, Oxford OX1 3PH

Abstract    This paper describes some of the trends presently taking
place related to TEM studies of compound semiconductors.  Specimen
thinning procedures are initially discussed.    TEM examinations of
compound semiconductors are being widely performed to investigate the
inhomogeneities present, these including crystallographic defects,
interfaces, compositional variations, clustering, impurity segregation,
precipitation, etc.  Results are described relating to melt-grown GaAs,
ternary and quaternary epitaxial layers grown on binary substrates such
as GaAlAs/GaAs, GaInAs/InP, GaInAsP/InP, CdHgTe/CdTe, etc.

## 1. Thinning and Examination of TEM Specimens

For TEM studies of crystalline semiconductor specimens, the main
examination methods used are two-beam, weak-beam and lattice-imaging,
these corresponding to progressively better spatial resolution.  For 100
and 200kV TEMs and for materials of average atomic number, the thickness
of the thinned specimens used for these examination methods are typically
~100, ~30 and ~10nm respectively.  Consequently, on going to progressively
better resolution, it becomes increasingly more difficult to prepare
thinned specimens of the quality required.

For plan-view and angle-lap specimens, chemical thinning is often used
because it is rapid and does not usually introduce damage into the
specimen.  However, it is not precisely controllable.  For cross-section
specimens, ion-beam thinning is almost always used because it has the good
control that is necessary.    However, it introduces damage into the
specimen.

For many semiconductor materials, standard $Ar^+$ ion-beam thinning
produces a thin amorphous layer on the surface of the specimen, and
immediately beneath this a thin damage region containing clusters and
loops up to a maximum size of typically ~5nm across.  When such thinning
is performed from both sides of the specimen, the undamaged material of
interest is sandwiched between two such sets of damage.  The cluster/loop
damage is most clearly revealed by TEM weak-beam examinations, as shown
in Fig.1a for Si and Fig.1b for GaAs.  Care in the initial preparation of
the surface and in optimising the thinning conditions, e.g. low beam
currents and voltages, and low incident angles for the beam, can minimise
this damage but not eliminate it.

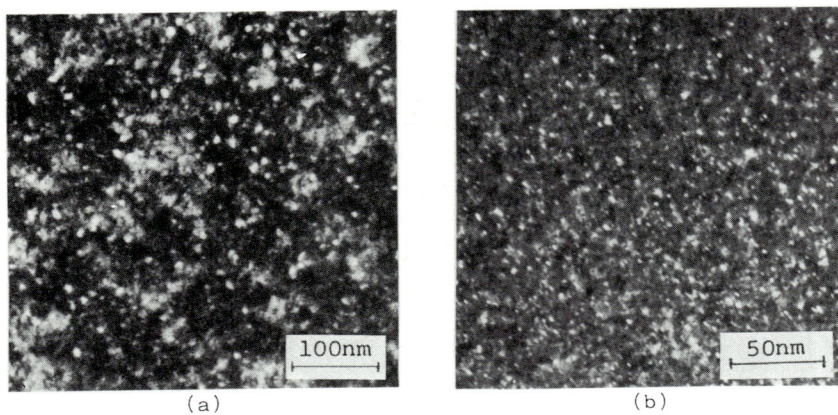

(a)                                    (b)

Fig.1 Specimens Ar$^+$ ion-beam thinned and examined by TEM using the weak-
beam technique. a) Si, b) GaAs. The bright spots are cluster/loop
damage introduced by the ion-beam thinning.

For example, in a TEM weak-beam investigation of LEC GaAs slices
implanted with 240keV Si$^+$ ions in the dose range $3 \times 10^{12}$ to $10^{14} \mathrm{cm}^{-2}$
(Stewart, Blunt, Booker and Sanders 1983, Stewart 1985), it was not
possible to use Ar$^+$ ion-beam thinned specimens to observe the defects
arising from the Si$^+$ ion implantation when these defects were smaller than
~5 to 10nm across. This is because such defects were obscured by the
cluster/loop damage introduced by the Ar$^+$ ion-beam thinning. For the
examination of Si$^+$ ion implantation defects of this size, these authors
had to use chemically thinned specimens. The same difficulty arises when
TEM examinations are made of small defects present in Si specimens if Ar$^+$
ion-beam thinning is used.

For some compound semiconductors, Ar$^+$ ion-beam thinning can give
additional problems. For example, with InP specimens, P is
preferentially lost from the surface and In globules are formed during the
thinning. When such specimens are subsequently examined by TEM, the In
globules often obscure the defects that were initially present in the InP.
This effect can also be minimised but not eliminated. With CdTe and
especially CdHgTe specimens, extremely pronounced damage arises from Ar$^+$
ion-beam thinning.

When semiconductors are ion-beam thinned, the damage which is
introduced modifies the electrical properties of the material to a certain
depth below the surface. For example, when Davidson (1972) used 4keV
N$^+$ ions and an incident angle of 25° to thin Si specimens, combined
electrical conductivity and mobility measurements showed that the material
within ~40nm of the surface was electrically inactive, but the material
beneath this was not significantly electrically changed. Conversely, when
Lidbury (1974) used 1keV N$^+$ ions and an incident angle of 20° to thin GaAs
specimens, analogous measurements showed that the electrical activity was
significantly changed down to depths of ~0.5μm, i.e. depths much greater
than the ion range. Lidbury attributed this to the creation by the ion-
beam of deeply diffusing, electrically compensating defects in the GaAs.

There is now much experience in the use of reactive ion sputtering as a dry-etching process in the fabrication of semiconductor devices. In this process the incident ions are chosen so that they chemically combine with the semiconductor atoms to form molecules, and these molecules are readily ejected from the semiconductor specimen surface. The process allows low-energy ions to be used, and so the damage introduced can be small. Reactive ion-beam methods are now being used to thin semiconductor specimens for TEM examinations. For example, Cullis, Chew, Hutchison, Irving and Geiss (1985) and Cullis (1985) used $I^+$ ions in a modified ion-beam thinning apparatus and obtained encouraging results. Thus, InP specimens were thinned without significant formation of In globules, and CdTe specimens were thinned without the introduction of pronounced damage. This represents an important advance in TEM specimen preparation procedures for semiconductor specimens. It is possible that reactive ion-beam thinning may, with further work, enable Si and GaAs specimens to be produced with less damage than that which occurs at present using $Ar^+$ ion-beam thinning.

## 2. Particles in GaAs Ingot Material

Liquid-encapsulated Czochralski (LEC), semi-insulating (SI) GaAs ingot material is of considerable importance because, for example, of its potential use as substrate slices for the fabrication of high-performance integrated circuits. The semi-insulating properties of this material arise mainly from the presence of the EL2 deep-level, which considerable evidence now shows to be due to either $As_{Ga}$ anti-site defects or complexes involving these defects. In order to produce such material, the GaAs ingots are generally grown from slightly As-rich melts. The distribution of the EL2 centres across the GaAs slices is not uniform, the variation often correlating with grown-in dislocations present as low-angle boundaries. TEM examinations of such GaAs material often reveal numerous particles, mainly on the dislocations.

Cullis, Augustus and Stirland (1980) examined such particles ~50nm across in LEC Cr-doped SI GaAs using TEM/TED methods and showed that the individual particles consisted of single-crystal As. Although this result was consistent with the material being As-rich, it initially seemed perhaps surprising that such relatively large particles could have formed by conventional nucleation and diffusion processes. Stirland, Augustus, Brozel and Foulkes (1984) examined such particles ~70nm across in LEC undoped SI GaAs using similar methods and also showed that the particles were single-crystal As. Ponce, Wang and Hiskes (1984) examined particles up to ~20nm across in LEC undoped SI GaAs using the TEM lattice-imaging method. The particles tended to be mainly amorphous and were not identified. Lodge, Booker, Warwick and Brown (1985) examined particles up to ~100nm across in LEC undoped SI GaAs using the TEM two-beam and weak-beam methods. The particles exhibited moiré fringes (Fig.2), the spacings and directions of which were consistent with the particles being single-crystal As. Smaller particles exhibiting similar moiré fringes were also present on the dislocations, the fringes being clearly discernable for particles down to ~15nm across. These TEM results show that As precipitates on dislocations in LEC SI GaAs specimens, and that the precipitation can be pronounced. Such precipitation may play an important role in determining the electrical and luminescent properties of the material.

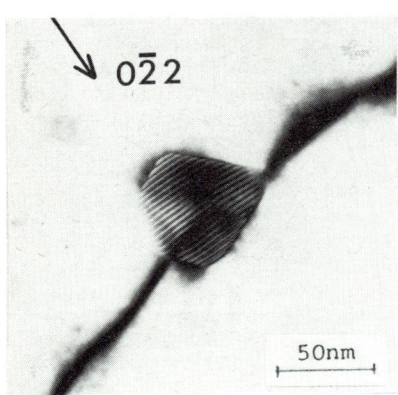

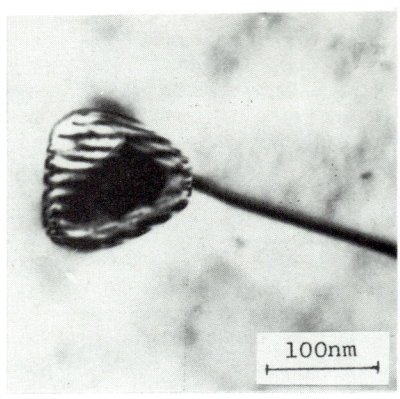

Fig.2 LEC undoped SI GaAs with
particle on dislocation
(Lodge et al. 1985).

Fig.3 LPE CdHgTe layer grown
on CdTe substrate with
particle on dislocation
(Lyster et al. 1985).

Particles similar to those described above for GaAs are often present
in other semiconductor materials. For example, Lyster, Chew, Booker and
Astles (1985) examined such particles in LPE CdHgTe layers grown on CdTe
substrates using the TEM two-beam and weak-beam methods. The particles
were up to ~150nm across and were present on dislocations forming an
interface network between the two materials. The particles exhibited
moiré fringes (Fig.3) and these indicated that the individual particles
consisted sometimes of one single-crystal, and sometimes of 2 or 3 single-
crystals together occasionally with what seemed to be a void. These
particles were not identified. However, they occur in the interface
region where there are large Cd and Hg concentration gradients, i.e. where
significant diffusion of these elements took place during the layer growth
and subsequent cooling down, and so it is possible that these particles
also arise directly from the alloying elements.

3. Inhomogeneities in Hetero-epitaxial Layers

Group III-V hetero-epitaxial layers are being used for a variety of
devices, e.g. emitters and detectors for fibre-optic communications.
Ternary or quaternary alloy layers are often grown on binary substrates,
combinations such as GaAlAs/GaAs, GaInAs/InP, GaInAsP/InP commonly being
used. The layers are grown by a variety of methods embracing liquid phase
epitaxy (LPE), vapour phase epitaxy (VPE), metallo-organic chemical vapour
deposition (MOCVD), molecular beam epitaxy (MBE), etc. For the
GaAlAs/GaAs system, the layers are closely crystallographically matched
for all layer compositions. For the GaInAs/InP system, the layer
composition used is generally $Ga_{.47}In_{.53}As$ because this gives exact
lattice-match with the InP substrate. For the GaInAsP/InP system, a
variety of layer compositions can be used which give exact lattice-match
but also allow the band-gap energy to be selected.

When such layers are grown, a number of different types of
inhomogeneity can occur. For example, an interface dislocation network
can arise if the composition of the layer does not correspond to exact

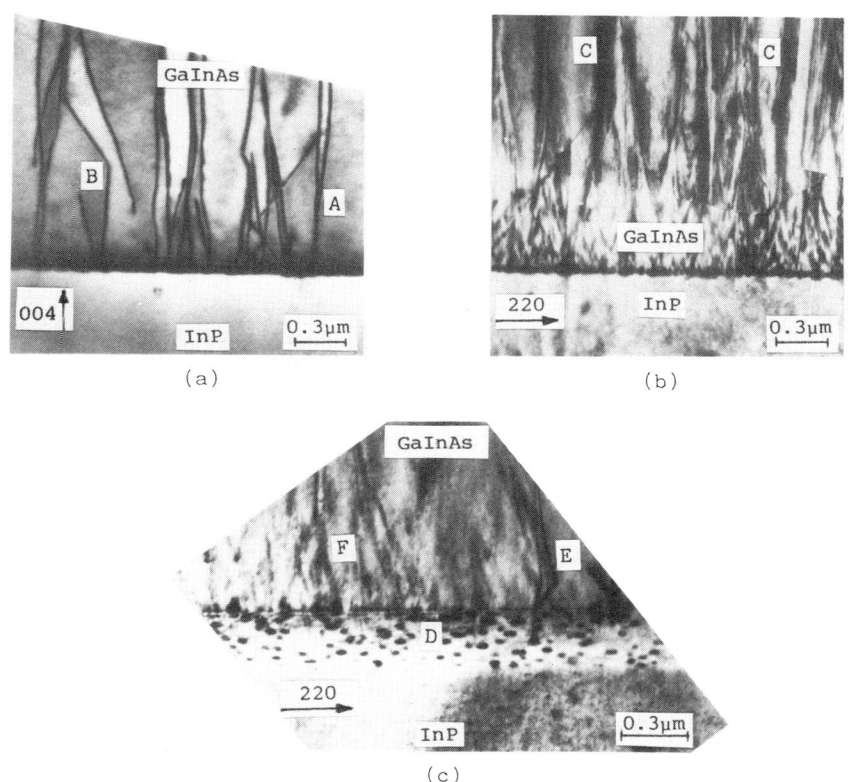

Fig.4 VPE GaInAs layer grown on a (001) InP substrate. TEM cross-sections
of typical areas.
a) 004 reflection.   A - dislocation dipole, B - bent dislocation.
b) 220 reflection.   C - bend contour contrast.
c) 220 reflection after tilting about [110].   D - In globules,
   E - dislocation dipole, F - bend contour contrast.
(Al-Jassim et al. 1984).


lattice match.  Dislocations and stacking faults can be generated at the
interface and run through the layer if the condition of the substrate
surface is not good or the initial growth of the layer is poor.   A
progressive change in the composition of the layer in the direction
perpendicular to the interface can occur, known as grading, due to a
progressive change in the source reactants.   Local changes in the
composition can also occur in the direction parallel to the interface.
When all of these gross inhomogeneities are eliminated, there remain the
threading dislocations in the layer arising from the grown-in dislocations
in the melt-grown substrate slices, and variations in the composition of
the layer on a fine scale due to alloy clustering and possibly also alloy
ordering.

    As an example of some of the gross inhomogeneities that can arise, a
TEM examination of a VPE GaInAs layer grown on an (001) InP substrate is

described (Al-Jassim, Norman and Booker 1984). TEM cross-section micrographs obtained using a 004 reflection (Fig.4a) showed both dislocations and stacking faults generated at the GaInAs/InP interface and running through the layer, the density being $\sim 10^8$ to $\sim 10^9$ cm$^{-2}$ when referred to the original layer surface. Similar micrographs obtained using a 220 reflection (Fig.4b) showed the dislocations and stacking faults, but in addition a pronounced wavy dark structure running through the layer was present. This latter structure behaved like bend contours, but did not appear to be associated with buckling of the foil because it was often more pronounced when the foil was thicker. When such specimens were tilted in the TEM through a large angle about the $[110]$ direction and a 220 reflection was used (Fig.4c), small dark blobs could be seen in the interface region. These blobs correspond to particles up to $\sim 50$nm across and of density $\sim 10^9$ to $\sim 10^{10}$ cm$^{-2}$. The micrographs also showed that dislocations were generated at $\sim 1$ to $\sim 10\%$ of these particles and ran through the layer, and that the wavy bend contour type of contrast also often commenced at such particles.

A possible explanation for this behaviour is that the heating of the InP substrate slice in the growth apparatus prior to deposition causes loss of P from the surface, and large numbers of small In globules form which are molten when the layer begins to grow. A small fraction of the globules each generate a pair of dislocations, but the majority give rise to local irregular columns in the layer with an alloy composition different from the remainder of the GaInAs layer. Associated with such columns are lattice planes that have different spacings and are bent, and these give rise to the wavy bend contours. Clearly such columns will give contrast effects that are a maximum when using a 220 reflection, and a minimum when using a 004 reflection, as is observed.

The growth methods initially used for such hetero-epitaxial layers were the LPE and VPE methods. However, there is a move towards the use of the MOCVD and MBE methods because of their potentially better control of layer thickness and composition. TEM examinations by Hockly and White (1984) of early MOCVD GaInAs layers grown on (001) InP substrates showed that the initially grown portions of the layers corresponded to a very disturbed region which gave rise to numerous dislocations running through the layer. The distrubed region exhibited moiré fringes and these were interpreted as being due to extremely large lattice-mismatches, the magnitude of which approached the difference between the InP and InAs lattices. Modifications to the growth reactor to give faster switching times, and the use of an InP buffer layer, significantly reduced the number of such dislocations. TEM examinations of the best quality MOCVD GaInAs layers now being grown show virtually no such dislocations.

TEM cross-section examinations of MOCVD GaInAs layers grown on (001) InP substrates sometimes show a series of fine bands running parallel to the interface (Norman 1985) (Fig.5). The bands are irregularly spaced and of widths down to typically $\sim 5$nm, and are generally most pronounced when a 002 reflection is used. These bands correspond to variations in layer composition and occur during layer growth if the gases are not under precise control and fluctuations arise. The composition variations are analogous to those that occur when superlattices are grown. However, in the present case the variations arise inadvertently and in an irregular manner.

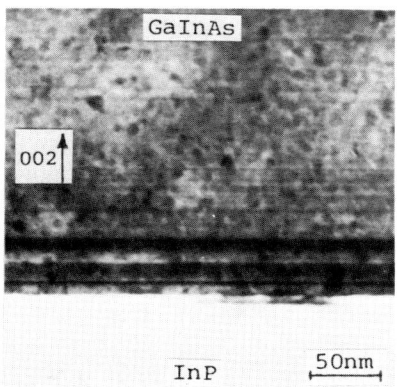

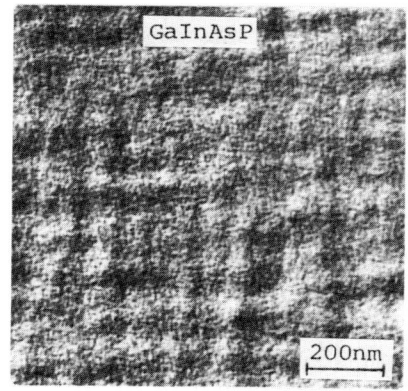

Fig.5 MOCVD GaInAs layer grown on
(001) InP substrate. TEM cross-
section specimen, 002 reflection.
Bands parallel to interface are
due to variations in alloy layer
composition (Norman 1985).

Fig.6 LPE GaInAsP layer grown on
(001) InP substrate. TEM
plan-view of layer, 220
reflection. Coarse tweed
structure and fine granular
structure are due to alloy
clustering (Norman and
Booker 1985a).

## 4. Alloy Clustering

Many Group III-V ternary and quaternary alloy systems exhibit a
miscibility gap.    Hence, some of these alloys, when below particular
temperatures, should be unstable and alloy clustering, i.e. local
variations in chemical composition, can then occur.      Henoc, Izrael,
Quillec and Launois (1982) investigated a series of LPE GaInAsP layers
grown on (001) InP substrates at ~640°C.    TEM examinations showed that
those specimens with compositions lying within the unstable region of the
spinodal decomposition diagram as calculated by de Cremoux, Hirtz and
Ricciardi (1981) exhibited a coarse tweed structure with bands of spacing
~200nm lying along the $[100]$ and $[010]$ cube directions (similar to those
shown in Fig.6).    Conversely, those specimens with compositions lying
outside the unstable region showed no such bands.      Combined scanning
transmission electron microscopy (STEM) and energy dispersive X-ray (EDX)
chemical analysis showed that the bands corresponded to periodic
variations in composition ranging from typically $Ga_{.30}In_{.70}As_{.66}P_{.34}$ to
$Ga_{.39}In_{.61}As_{.57}P_{.43}$ .    Such layers also exhibited a fine granular
structure on the ~10nm scale.

Considerable further work has been performed on such structures in LPE
GaInAsP layers.    For example, Treacy, Gibson and Howie (1985) concluded
that the band contrast observed by TEM mainly arose because of local
elastic relaxations of the surface of the thin foil specimen, this being
associated with a periodic variation in lattice parameter present in the
initially grown layer.    Norman and Booker (1985a and b) observed satellite
spots in TED patterns obtained from such specimens, the satellites being
interpreted in terms of a periodic variation in lattice parameter on the
scale of the fine granular structure.    It was concluded that the fine
structure probably also arose by spinodal decomposition.    Mukai (1983)

showed that the carrier mobility and photoluminescence were significantly reduced when the layers were grown under conditions corresponding to the unstable region of the spinodal decomposition diagram. TEM examinations showed that such alloy clustering also occurred in Group III-V ternary and quaternary layers grown by other methods, e.g. VPE, MOCVD and MBE (Norman and Booker 1985b).

5. Dislocation Generation and Movement

When semiconductor epitaxial layers are grown on single-crystal semiconductor substrates, stresses that are present during growth or subsequent cooling down can cause dislocations to be generated and/or move in the specimens. For example, the interface dislocation networks that often arise can form by a number of methods such as a) threading dislocations bending around at interfaces, b) dislocations being generated at the layer surface and moving down to the interface, and c) dislocation sources operating in the interface region. Whether a network occurs or not depends on the lattice-mismatch, temperature, time at temperature, layer thickness and the particular semiconductor materials. The experimental results show that there is a major difference in behaviour for Group III-V binary layers and Group III-V ternary and quaternary layers. For example, for homo-epitaxial binary layers in which the mismatch arises from the presence of dopants, the threshold lattice-mismatch value that needs to be exceeded to produce interface dislocation networks is typically $\sim 10^{-4}$. However, for hetero-epitaxial ternary and quaternary layers in which the mismatch arises mainly from the alloy compositions, the threshold is typically $\sim 10^{-3}$ to $\sim 10^{-2}$. A possible explanation for this behaviour is that alloy clustering is present in the ternary and quaternary layers, this makes the material harder, and so it is more difficult to generate and move dislocations.

Watts and Willoughby (1984a and b) made microhardness indentation measurements at room temperature for a series of LPE $Ga_xIn_{1-x}As_yP_{1-y}$ layers that were lattice-matched to InP substrates and embraced the complete composition range from InP to $Ga_{.47}In_{.53}As$. The hardness progressively increased as y increased, reached a maximum for y in the range 0.6 to 0.7, and thereafter decreased. This result was explained by Watts and Willoughby in terms of solid solution hardening effects together with possibly a change in the active slip system on going from InP to the alloys. However, it is pointed out here that the range of values of y obtained for maximum hardness correspond closely to the middle of the unstable region of the spinodal decomposition diagram for this alloy system as calculated by de Cremoux et al.(1981), suggesting that alloy clustering may contribute to the hardness occurring. Watts and Willoughby (1984a) also measured the hardness variation for analogous GaInAs alloys and obtained a maximum at the composition $Ga_{.70}In_{.30}As$. This may be compared with the middle of the unstable region of the calculated spinodal decomposition diagram for this alloy system which occurs at $Ga_{.45}In_{.55}As$, and so the agreement is less good for this alloy system. Nevertheless, it seems that alloy clustering may again play a role. It would be advantageous if such hardness measurements could be made at higher temperatures, especially corresponding to the layer growth temperatures, because more direct correlations could then be made.

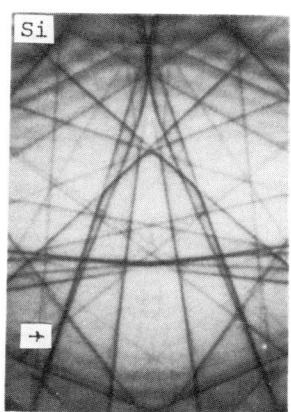 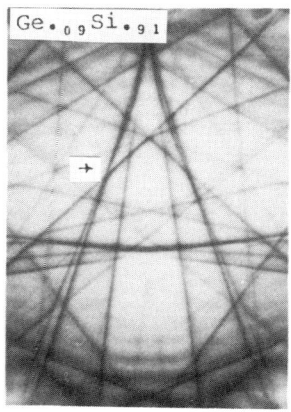

Fig.7 CBED patterns from Si substrate and MBE Ge$_{.09}$Si$_{.91}$ layer in TEM
cross-section specimen. Note change in position of intersections of
(14 $\bar{4}$ 2) and ($\bar{13}$ 5 $\bar{1}$) lines shown arrowed due to small differences
in lattice plane spacings (Fraser et al. 1985).

6. Measurement of Lattice-strains at Interfaces

When semiconductor hetero-epitaxial layers are grown, any lattice
mismatches that are present can give rise to elastic strains at the
interface. One method for investigating such strains is to obtain TED
patterns from TEM cross-section specimens and to select a number of
convenient high-index (hkl) Kikuchi lines that intersect one another.
Such TED patterns are then obtained under identical conditions from the
substrate and layer, and small differences in lattice parameter are
revealed as small differences in the positions of the Kikuchi line
intersections. The lattice parameter difference can be deduced either
from calculations of the changes in the Bragg angles for the particular
(hkl) lines, or by using an experimental calibration procedure.

Simmonds and Booker (1982) and Simmonds (1982) applied this method
using conventional selected-area TED patterns to Ga$_{.35}$Al$_{.65}$As layers grown
on (001) GaAs substrates, and considered the intersections of four pairs
of lines of types {14 6 0} {9 5 1}, {12 4 0} and {13 3 1}. (The method
is essentially that of Walker and Booker (1982) who measured small changes
in lattice parameters of bulk semiconductor specimens by using SEM
electron channelling patterns). The differences in the positions of the
lines from GaAs and GaAlAs regions were interpreted in terms of
calculations based on cubic lattices on the assumption that the tetragonal
distortion of the GaAlAs lattice in the bulk specimen would to a first
approximation be relaxed to give a cubic lattice in the thin foil
specimen. Some of the analyses gave a mismatch value of $\sim 9 \times 10^{-4}$, which
was close to the calculated value assuming a fully relaxed GaAlAs lattice,
but other analyses gave markedly incorrect values. It was concluded that
the method was probably sound and capable of a potential accuracy of
$\sim 3 \times 10^{-4}$, but that computer simulated patterns would need to be obtained
taking into account a certain amount of tetragonal distortion together
with surface relaxation effects.

Fraser, Maher, Humphreys, Hetherington, Knoell and Bean (1985) applied
this method using convergent-beam electron diffraction (CBED) patterns to
MBE $Ge_{.090}Si_{.910}$ and $Ge_{.045}Si_{.955}$ layers grown on (100) Si substrates
(Fig.7).  Patterns obtained from Si regions were matched against computer
simulated patterns based on the cubic Si lattice to establish the
'reference state'.  Patterns obtained from GeSi regions were then matched
against computer simulated patterns based on appropriate tetragonal
lattices until the best fit was obtained.  The tetragonality deduced from
the patterns proved to be approximately half that calculated for the fully
unrelaxed bulk specimen, e.g. for the $Si/Ge_{.090}Si_{.910}$ specimens, $2.6\times10^{-3}$
rather than $5.5\times10^{-3}$.  This and other considerations suggested that
additional surface relaxations had probably taken place, and these are now
being taken into account in the computer simulations.  When this work of
Fraser et al is complete and the method is firmly established, it will be
an extremely powerful technique because it will have high spatial
resolution, e.g. ~4nm, as well as good accuracy, and will be eminently
suitable for investigating strains at hetero-epitaxial interfaces,
including superlattice structures.

The author wishes to thank the many colleagues who have kindly
contributed to the paper by providing specimens, photographs, information
and useful discussions.

## References

Al-Jassim M M, Norman A G and Booker G R 1984 unpublished results
Cremoux B de, Hirtz P and Ricciardi J 1981 Inst. Phys. Conf. Ser. No.67
    p115
Cullis A G, Augustus P D and Stirland D J 1980  J. Appl. Phys. 51 2556
Cullis A G, Chew N G, Hutchison J L, Irving S J C and Geiss J 1985 (this
    Conference)
Cullis A G 1985 private communication
Davidson S M 1972 J. Physics E : Scientific Instruments 5 23
Fraser H L, Maher D M, Humphreys C J, Hetherington C J D, Knoell R V and
    Bean J C 1985 (this Conference)
Henoc P, Izrael A, Quillec M and Launois H 1982 Appl. Phys. Lett. 40 963
Hockly M and White E A D 1984 J. Crystal Growth 68 334
Lidbury D P G 1974 D. Phil thesis, University of Oxford
Lodge E A, Booker G R, Warwick C A and Brown G T 1985 (this Conference)
Lyster M, Chew N G, Booker G R and Astles M G 1985 unpublished results
Mukai S 1983 J. Appl. Phys. 54 2635
Norman A G 1985 unpublished work
Norman A G and Booker G R 1985a J Appl. Phys. 57 4715
Norman A G and Booker G R 1985b (this Conference)
Ponce F A, Wang F C and Hiskes R 1984 Semi-Insulating III-V Materials ed
    D C Look and J S Blakemore (Shiva) p.68
Simmonds M J and Booker G R 1982 unpublished results
Simmons M J 1982 Part II thesis, University of Oxford
Stewart C P, Blunt R T, Booker G R and Sanders I R 1983 Physica 116B 635
Stewart C P 1985 D. Phil thesis, University of Oxford
Stirland D J, Augustus P D, Brozel M R and Foulkes E J 1984 Semi-
    Insulating III-V Materials ed D C Look and J S Blakemore (Shiva) p.91
Treacy M M J, Gibson J M and Howie A 1985 Phil. Mag. (in press)
Walker A R and Booker G R 1982 Proc. 10th International Congress on
    Electron Microscopy, Hamburg vol.1 p.651
Watts D Y and Willoughby A F W 1984a Materials Letters 2 355
Watts D Y and Willoughby A F W 1984b J. Appl. Phys. 56 1869

*Inst. Phys. Conf. Ser. No. 76: Section 6*
*Paper presented at Microsc. Semicond. Mater. Conf., Oxford, 25–27 March 1985*

# The association of EL2° with single dislocations in GaAs

D J  Stirland, M R  Brozel* and I  Grant**

Plessey Research (Caswell) Ltd., Caswell, Towcester, Northants. NN12 8EQ UK

* Dept. of Electrical and Electronic Engineering, Trent Polytechnic,
  Burton Street, Nottingham NG1 4BU.

**ICI Wafer Technology Ltd, Maryland Ave, Tongwell, Milton Keynes MK15 8HF

Abstract    2" diameter wafers cut from indium doped, liquid encapsulated Czochralski (LEC) grown GaAs ingots have been examined using
calibrated etchants to reveal dislocations, and transmission infrared
microscopy to detect the deep donor level EL2°. In regions of low
dislocation density ($\sim5 \times 10^2$ cm$^{-2}$) single isolated dislocations were
identified by their etching behaviour. At the identical sites, infra-
red micrographs showed increased absorption due to increased EL2°
concentration. The results thus demonstrate directly that enhancement
of $\left[EL2^\circ\right]$ occurs at single dislocations.

## 1. Introduction

Martin et al (1981) originally reported that characteristic "W" and "U"
shaped variations in $\left[EL2^\circ\right]$, the concentration of the deep donor level EL2°,
were to be found across $<110>$ diameters of $\{001\}$, undoped, LEC, semi-
insulating GaAs wafers. They also noted similarly shaped fluctuations in
dislocation densities. Subsequently, Brozel et al (1983) showed that fine
scale ($\sim100\mu m$ size) fluctuations of $\left[EL2^\circ\right]$ were superimposed on the long
range "W" and "U" variations, and that these could be directly related to
fine scale inhomogeneities in dislocation densities. Near infrared
absorption measurements at wavelengths of $1\mu m$ and $2\mu m$ were used to determine $\left[EL2^\circ\right]$, transmission infrared microscopy was employed to display the
absorbing EL2° regions as a qualitative distribution across entire 2"
diameter wafers, and A/B etching (Abrahams and Buiocchi 1965) was performed
to reveal dislocation distributions across the specimen surfaces. EL2°
related absorption observations and measurements require specimen
thicknesses $\geqslant2$ mm, whereas typical A/B etch depths are usually only
$\sim10-20\mu m$, so it is surprising that correlation between the two distributions was found. The correlation was achieved because the fine scale
inhomogeneities in the distributions lay along $<001>$ in $\{001\}$ specimens,
and along $<110>$ in $\{110\}$ specimens. However, it is difficult to achieve a
one-to-one correspondence between localised enhanced $1\mu m$ absorption and
dislocations because the average dislocation density encountered in
typical, commercially available, undoped, semi-insulating GaAs is $\sim5-10 \times$
$10^4$ cm$^{-2}$. The fine scale inhomogeneities indicate that this high
dislocation concentration is not uniformly distributed. Indeed no
correlation would have been possible had the dislocation density been
uniform.

We have now been able to demonstrate directly that there is a one-to-one correspondence between single dislocations and sites of enhanced [EL2°] by utilising very low dislocation density, LEC, semi-insulating GaAs containing ~0.1% indium.

## 2. Experimental Details

Four indium doped GaAs ingots were grown, using in situ synthesis, in a Metals Research MSR6/R high pressure puller. The indium was added either as the metal or as high purity InAs. After growth, each crystal was cut into {001} wafers, some having standard device type thicknesses (~350μm) and some having thicknesses of 1 mm or 3mm. The 1 mm thick wafers were etched in molten KOH. The easily visible etch pits thus generated afforded a rapid assessment of the dislocation distributions both across the wafer areas and at specific positions down the ingot length. The 3 mm thick wafers were doubly polished and initially examined using the transmission infrared absorption method previously described (Brozel et al 1983). This preliminary examination showed that many of the wafers contained central and peripheral regions exhibiting the complex absorption structures which we have previously associated, in undoped GaAs, with high concentrations of dislocations (Stirland et al 1983). However, a number of wafers also exhibited an annular region at approximately half radius which contained very little absorbing structure. In addition, some wafers showed low absorption in the central region also (Foulkes et al 1984). The work described in the following sections was carried out on two wafers, 1 mm and 3 mm thick, from ingot A1008 which exhibited regions of low dislocation density and low absorption respectively in the central 3-4 mm diameter area. The 1 mm thick wafer had been cut from the 'shoulder' region at the seed end, and the 3 mm wafer had been cut from approximately half way down the ingot length (melt fraction g~0.4). After detailed infrared examinations of the central region of the 3 mm wafer had been made this specimen was etched for different intervals in the A/B etchant, at room temperature. Additional removal of material was performed using a bromine-methanol chemico-mechanical polish between two of the etch stages. A/B etched surfaces were observed by Nomarski interference contrast.

## 3. Experimental Results

Fig. 1(a) shows part of the central region of the 1 mm seed end wafer after an etch in molten KOH. The central core, exhibiting an etch pit density (e.p.d.) of ~5 x $10^3$ cm$^{-2}$, is sharply delineated by an abrupt drop in e.p.d. to a value essentially zero in the surrounding regions. Fig. 1(b) shows a higher magnification view of several typical etch pits within the central region.

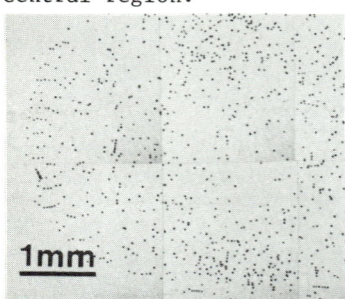

Fig. 1(a) Central region of KOH
etched wafer

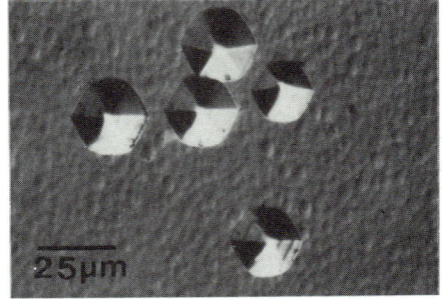

Fig. 1(b) Individual KOH etch pits
from Fig. 1(a)

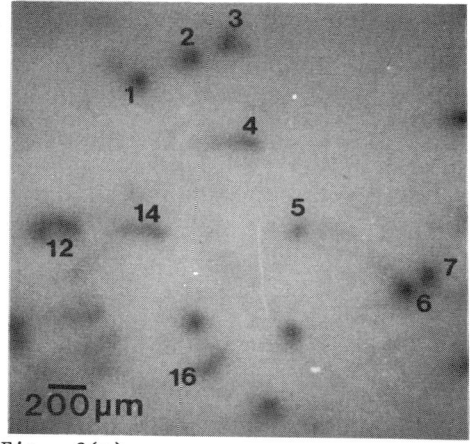

Fig. 2(a)

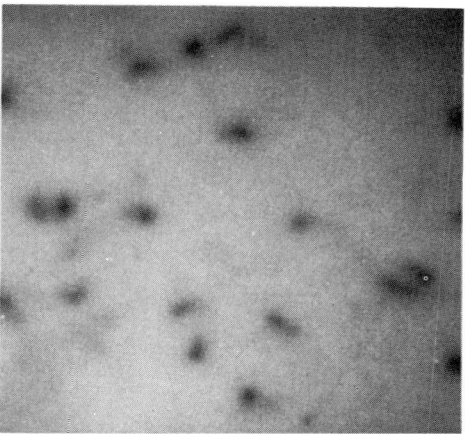

Fig. 2(b)

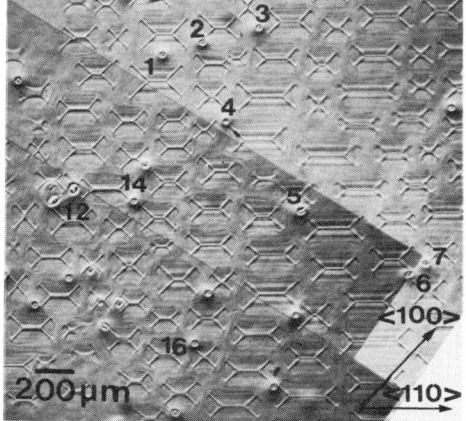

Fig. 2(a) Transmission infrared micrograph of central region of 3mm thick wafer. 0° beam incident angle

Fig. 2(b) 10° beam incident angle

Fig. 3 Central region of 3 mm thick wafer after 7.5 min A/B etch

Fig. 2(a) is an infrared transmission micrograph of the central region of the 3 mm thick wafer cut from approximately half way down crystal A1008. The sample is reasonably transparent in the 1μm wavelength region, with the exception of isolated areas which are strongly absorbing. The density of these absorbing areas is ~5 x $10^2$ cm$^{-2}$. We have previously shown that similar effects can usually be associated with the absorption of light by the deep donor level EL2° (Brozel et al 1983). Moreover, high $\left[\text{EL2°}\right]$ is found where dislocation density is high.

Recently we have indicated that infrared microscopic information can be augmented by the use of a pseudo-3D technique (Brozel et al 1985). This method makes use of two relatively tilted transmission images, produced by rotating the specimen through ± 5°, which are then viewed in an appropriate 3D-viewer. Fig. 2(a) and 2(b) represent such a 3D stereo-pair, with beam incident angles of 0° and 10° respectively. The 3D image shows clearly that the majority of the dark absorbing regions are rod-like, and extend from top to bottom surfaces. In order to determine whether the absorbing regions can be associated with the presence of one, or several, dislocations the specimen was etched for 7.5 min at room temperature in the A/B etchant. This treatment removes ~18μm thickness of GaAs, but the memory effect of the etchant preserves an image of the dislocation content of the removed material on the final surface (Stirland and Ogden 1973). Fig. 3 shows a micrograph of the same region as that of Fig. 2 after this treatment. The most obvious feature of the area is the oriented cell-like array of linear markings along <100> and <110>, which have previously been associated with the effects of constitutional supercooling (Bardsley et al 1962). This arises from high concentrations of impurity (indium) in the melt coupled with relatively rapid growth rates. Because repeated etching experiments have shown that the majority of these features extend unchanged down the <001> growth axis for considerable distances (hundreds of microns) they provide useful positional reference markers for other etch features. In addition to these structures individual etch features can be seen at the numbered locations. Comparisons of these with Figs. 2(a) and 2(b) show that there is an etch feature corresponding with every dark (absorbing) rod. Indeed, comparison of these micrographs and others we have studied shows that there is a one-to-one correspondence between areas of high $[EL2^o]$ and single etch features. It should be noted that the core defect density is an order of magnitude lower, at ~5 x $10^2$ cm$^{-2}$, for the specimen at position g~0.4 than at the seed end.

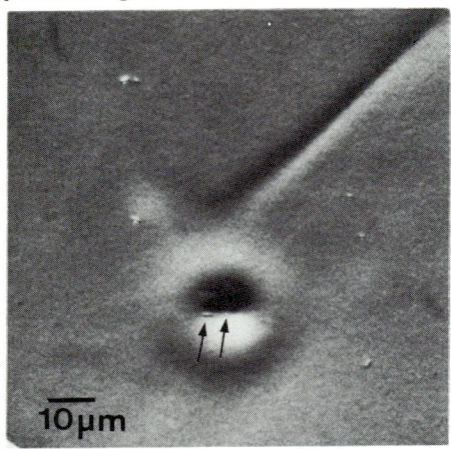

Fig. 4(a) Etch feature #2 from Fig. 3; 7.5 min A/B etch

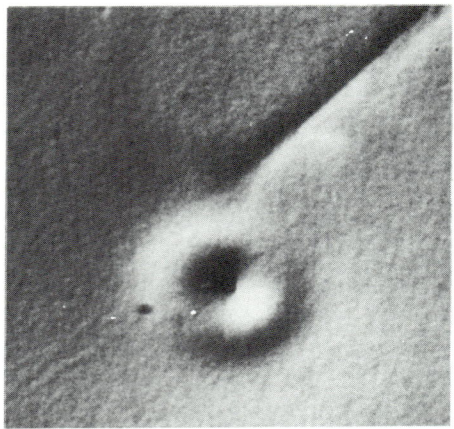

Fig. 4(b) Etch feature #2 from Fig. 3; 15 min A/B etch

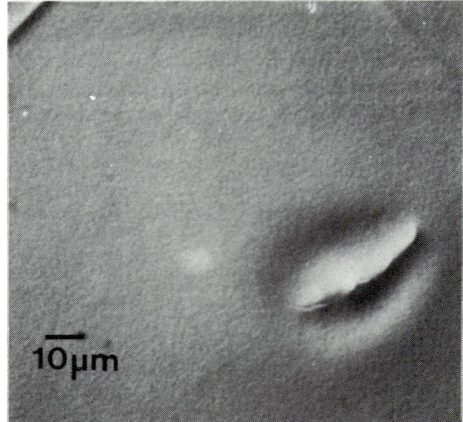

Fig. 5(a) Etch feature #5 from Fig. 3; 7.5 min A/B etch

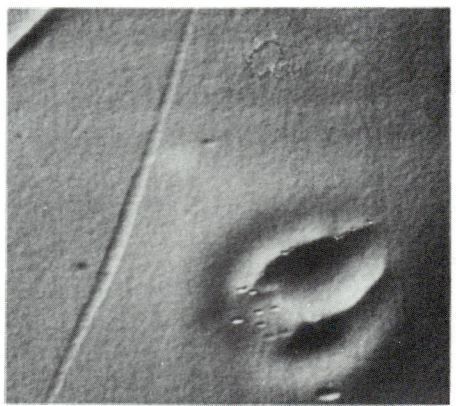

Fig. 5(b) Etch feature #5 from Fig. 3; 15 min A/B etch

Fig. 4(a) shows a single etch feature (#2) on Fig. 3 at high magnification. From the various Nomarski contrast changes it can be deduced that the central part of this feature has a raised conical configuration, and that two etch pits (arrowed) lie close to the tip of the cone. Surrounding the central part is a shallow annular mound whose diameter is ~50-70 μm. The same feature #2 was examined after a further 7.5 min A/B etch. This treatment results in removal of an additional ~18 μm thickness of GaAs. Fig. 4(b) indicates that the etch feature is essentially similar to that shown in Fig. 4(a), with the exception of the etch pits, which enlarge from 2.4 μm and 1.7 μm lengths along <110> to 3.5 μm and 2.6 μm respectively. The position of the central cone relative to the cellular boundary markings of the constitutional supercooling structure for Fig. 4(a) and 4(b) is unchanged within measurement accuracy.

Fig. 5(a) shows another etch feature (#5) on Fig. 3 at a similar magnification to that of Figs. 4(a) and 4(b). The central region enclosed by a slightly elongated shallow mound consists of a curved ridge, containing several small etch pits with ~1.8-2.8 μm lengths along <110>. Fig. 5(b) shows the same feature after a second (7½ min) etch. The memory effect of the A/B etch (Stirland and Ogden 1973) ensures that the initial configuration of the ridge is preserved; in addition a second ridge practically "reflecting" the first is now apparent. The second ridge contains many etch pits (0.5-1.0 μm) along its length. Several larger fresh etch pits have appeared in the region between the ends of the ridges.

4. Discussion

In principle, it would have been possible to achieve the correlation between enhanced [EL2°] regions and single dislocations by using the molten KOH etch to indicate the sites of individual dislocations as in Fig. 1. However, information obtained from KOH etch micrographs is restricted to one parameter (dislocation density), whereas additional features of individual dislocations can be deduced by careful interpretation of A/B etch sequence micrographs.

Although only two of the etch features shown in Fig. 3 have been described in detail, they are representative of many of the 25 individual features which have been examined. The appearances of the central regions of all of the features examined to date are consistent with a previous interpretation (Stirland 1977) that they result from the A/B etch attack at single dislocations. In addition, the presence of etch pits in close proximity to the dislocations indicates that precipitates, possibly of hexagonal arsenic (Cullis et al 1980), are attached to the dislocations. The conical centre to defect #2 does not alter with repeated etching because the dislocation is essentially lying exactly along <001>, at least over the ~36 μm depth exposed by the two A/B etch treatments.

The curved ridge at the centre of defect #5 (Fig. 5(a)) can be interpreted as the projected image of part of a helical dislocation with axis along <001>. The second A/B etch exposes a further half turn of the helix (Fig. 5(b)). This particular defect was followed through 4 different A/B etch sequences, including a chemico-mechanical polish to remove 125 μm between the second and third etches. Figs. 5(a) and 5(b) represent the third and fourth stages: the first two stages gave essentially similar curved ridges. Again, the presence of many etch pits at each stage of etching indicates that precipitates are attached to the helical dislocation.

The shallow annular mounds surrounding the central dislocations in all of the 25 examined defects have not previously been encountered. However, comparison of Fig. 2 and Fig. 3 suggests that the diameters of the infrared absorbing rods are similar to the diameters of the shallow mounds ($\sim$50-100$\mu$m). Because the rods represent regions of enhanced $[EL2^o]$, as confirmed by photoquenching experiments (Foulkes et al 1984), it is possible that the shallow mounds are due to the effects of increased $[EL2^o]$ on the A/B etch rate. Alternatively, however, these raised etch features could be due to segregation of indium to the dislocations. Experiments are underway to resolve these alternatives.

5. Conclusions

We have shown that regions of enhanced $[EL2^o]$ can be correlated with single dislocations at the centre of <001> axis GaAs. In addition, we have indicated that $EL2^o$ centres are distributed in cylinders $\sim$50-100$\mu$m diameter about these dislocations. It is possible that the A/B etchant is capable of revealing these regions of enhanced $[EL2^o]$ by a differential attack rate compared with the attack rate at the GaAs matrix.

Acknowledgements

The authors are grateful to E.J. Foulkes for important preliminary measurements, and to H.A. Bethell for skilled experimental assistance. It is a pleasure to acknowledge the support and encouragement of D.T.J. Hurle. This work has been performed with the support of Procurement Executive, Ministry of Defence, sponsored by DCVD.

References

Abrahams M S and Buiocchi C J  1965 J. Appl. Phys. 36 2855
Bardsley W, Boulton J S and Hurle D T J  1962 Solid State Electronics 5 395
Brozel M R, Foulkes E J and Stirland D J  Proc. 11th Int. Symp. GaAs and Related Compounds, Biarritz 1984 To be published in 1985
Brozel M R, Grant I, Ware R M and Stirland D J 1983 Appl. Phys. Lett. 42 610.
Cullis A G, Augustus P D and Stirland D J 1980 J. Appl. Phys. 51 2556
Foulkes E J, Brozel M R, Grant I, Singer P, Waldock B and Ware R M Proc. 1984 3rd Int. Conf. on Semi-Insulating III-V Compounds, Kah-Nee-Ta Ed. D C Look and J S Blakemore (Shiva) p160
Martin G M, Jacob G, Goltzene A, Schwab C and Poiblaud G 1981 Inst. Phys. Conf. Ser. 59 281
Stirland D J 1977 GaAs and Related Compounds (Edinburgh 1976) Inst. Phys. Conf. Ser. 33a 150
Stirland D J, Grant I, Brozel M R and Ware R M 1983 Proc. 3rd Oxford Conference on Microscopy of Semiconductor Materials Inst. Phys. Conf. Ser. 67 285
Stirland D J and Ogden R 1973 Phys. Stat. Sol. (a) 17 K1.

*Inst. Phys. Conf. Ser. No. 76: Section 6*
*Paper presented at Microsc. Semicond. Mater. Conf., Oxford, 25–27 March 1985*

# TEM study of grown-in dislocations in semi-insulating, undoped LEC GaAs

E A Lodge[1], G R Booker[1], C A Warwick[2] and G T Brown[2]

[1]Department of Metallurgy & Science of Materials,
University of Oxford, Parks Road, Oxford OX1 3PH
[2]Royal Signals & Radar Establishment, St Andrews Road,
Great Malvern, Worcs WR14 3PS

Abstract    The grown-in dislocations and associated point defects in SI, undoped LEC GaAs are being characterised with the complementary techniques of X-ray topography, infra-red imaging, cathodoluminescence and TEM. Conventional two-beam and weak-beam TEM images are presented of dislocations in different crystals which are decorated with precipitate particles of 15–90nm diameter. A Moiré fringe analysis of one such particle is consistent with it being elemental As.

## 1. Introduction

In order for GaAs to realize its potential for use as fast logic devices in the computer industry and as analog microwave devices, substrate slices of uniform electrical properties are necessary for integrated circuit device fabrication. At present there are local variations in electrical properties across substrate slices. These are associated with deep levels, in particular EL2, which is generally attributed to antisite defect complexes and considered to be related to grown-in dislocations formed by thermal stresses.

The LEC GaAs ingots are grown from an As-rich melt and are subsequently annealed at 950°C for ~ 5 hours. Slices cut from the ingots have a resistivity ~ $10^7$ ohm cm and a low $10^{15}$ cm$^{-3}$ carbon content. The EL2 density is ~ $10^{16}$ cm$^{-3}$ and that of the grown-in dislocations is $10^4$ – $10^5$ cm$^{-2}$, being highest at the edge and in the centre of $[001]$ slices.

In order to establish a representative correlation of the macroscopic defect distribution with the defect microstructure the following procedure is used. Firstly, a reflection X-ray topograph (Brown et al 1984) is taken of the slice in order to image the dislocations in about the top 10μm of material. This shows cell structures composed of polygonized dislocations forming cell walls ~ 40μm wide surrounding dislocation-free areas ~ 500μm in diamter. Secondly, a 1μm wavelength infra-red absorption image is taken using slices ~ 2.5mm thick to show the distribution of neutral EL2 in the cell walls and lineage features across the slice. However, since this technique averages through the complete slice thickness, the fine detail does not correlate directly with that of the X-ray topograph. Thirdly, both plan-view $[001]$ and cross-section $[011]$ TEM samples are prepared by chemical, $Cl_2$/methanol, and/or Ar-ion beam thinning techniques from selected areas of the slices corresponding to high dislocation density. Fourthly, SEM CL studies (Warwick and Brown

1985) are performed at 10K on the TEM samples with a spatial resolution
~ 3µm. The 1.51eV luminescence band is due to recombination of free holes
at neutral shallow donors and the dark spots in the luminescence correspond to the intersection of the surface with individual dislocations.
High spatial resolution TEM is then performed to determine the nature of
the dislocations observed in the SEM CL map. A JEOL 200CX TEM is used and
it is therefore necessary that dislocations are in a ≤ 100nm thick region
of the sample for the weak-beam study.

2. Results

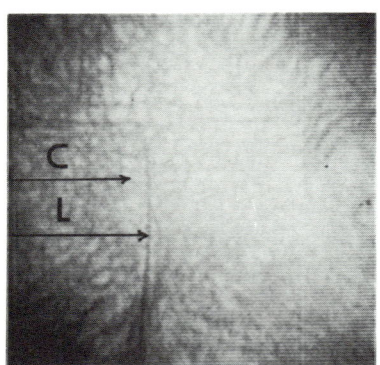

Fig.1 is a 1µm wavelength
absorption image of the central
30mm of a 50mm diameter [001]
slice of MMT 78A (Cambridge
Instruments ingot code) in which
C denotes a cell wall and L a
lineage feature.

1

Figs.2 and 3 are SEM CL total light images of an [001] TEM sample of
MMT63A in which a, b, c and d are areas in which cell walls are close to
the thin edge around the hole in the sample. A bright lineage feature is
seen to run across the top of the sample which can subsequently be given
further ion-beam thinning to render the lineage feature visible for TEM
study. Fig.3 is a higher magnification image of area b showing networks
of dislocations in the cell wall.

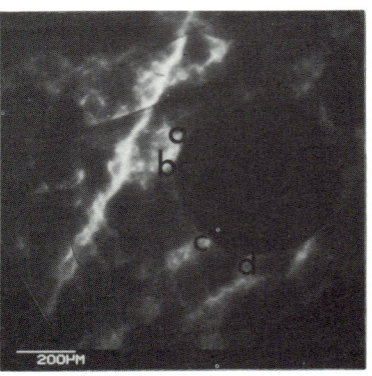

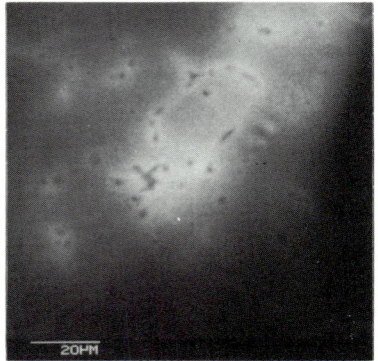

2

3

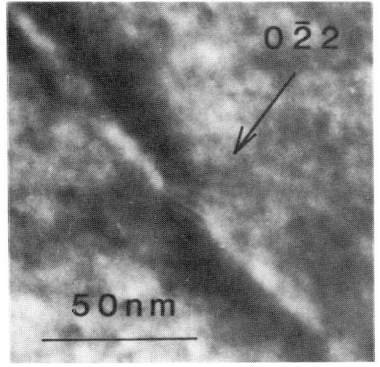

Fig.4 is a g = 0$\bar{2}$2 bright-field TEM image of a dislocation in MMT78A [011]. It is decorated with a particle ~ 15nm diameter showing 3.1nm Moiré fringes.

4

Fig.5 is a g = $\bar{2}$20 bright-field TEM image of a dislocation in the MMT63A [001] sample of Fig.2. It is decorated with a particle ~ 90nm diameter which shows 3.1nm Moiré fringes. Fig.6 is the corresponding dark-field weak-beam g,5g $\bar{2}$20 image of the particle showing additional topographical information.

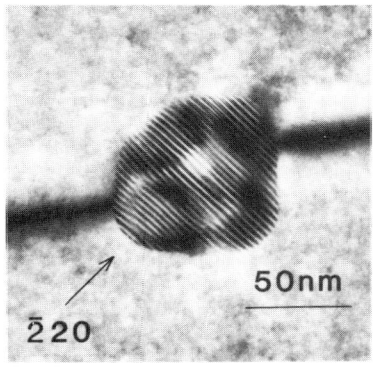

5

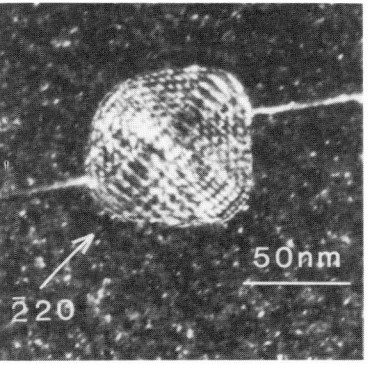

6

Fig.7 is a g = 0$\bar{2}$2 bright-field TEM image of a dislocation in MMT78A [011]. It is decorated with a particle ~ 70nm diameter which shows strong 3.1nm Moiré fringes. Fig. 8 is the corresponding g = 400 bright-field image which shows weak 1.5nm Moiré fringes. In both cases the Moiré fringes are orthogonal to the g diffraction vector, indicating that they are of the 'parallel' type.

The following Moiré fringe analysis is consistent with an assignment of the particle as elemental hexagonal As, $a_o$ = 0.3760nm, $c_o$ = 1.0548nm (Swanson and Fuyat 1953). The spacing between Moiré fringes due to parallel lattice planes with spacings $d_1$ and $d_2$ is $D = d_1 d_2 / |d_1 - d_2|$. Now GaAs $d_{0\bar{2}2}$ = 0.1998nm with As $d_{120}$ = 0.1879nm gives D = 3.1542nm, and GaAs $d_{400}$ = 0.1415nm with As $d_{2\bar{2}\bar{5}}$ = 0.1289nm gives D = 1.4476nm. Furthermore the angle between GaAs(0$\bar{2}$2) and GaAs(400) = 90° while that between As(120) and As(2$\bar{2}$5) = 98.6°. (Also note As($\bar{1}\bar{2}$0) $\Lambda$ As($\bar{2}$25), As(120) $\Lambda$ As(2$\bar{2}$5) and As($\bar{1}\bar{2}$0) $\Lambda$ As($\bar{2}\bar{2}$5) = 98.6°, while As(120) $\Lambda$ As($\bar{2}$25), As($\bar{1}\bar{2}$0) $\Lambda$ As($\bar{2}\bar{2}$5), As(120) $\Lambda$ As($\bar{2}\bar{2}$5) and As($\bar{1}\bar{2}$0) $\Lambda$ As(2$\bar{2}$5) = 81.4°). The difference of ± 8.6° between the matrix angle and the particle angle could explain why

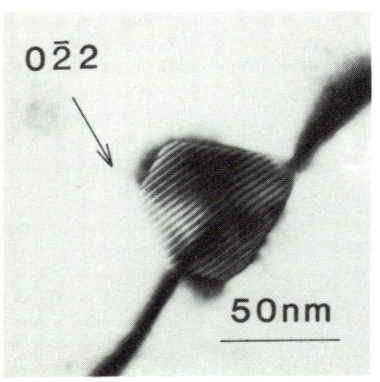

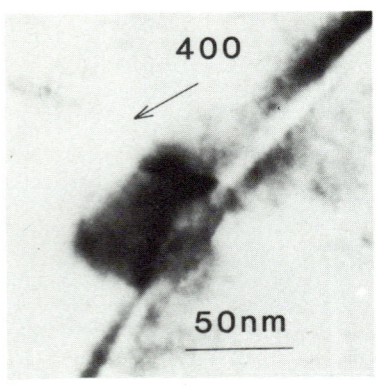

7
8

the Moiré fringe contrast for $g$ = 0$\bar{2}$2 is strong and that for $g$ = 400 is weak. This assignment is in agreement with the detailed transmission electron diffraction analysis of Cullis et al (1980) and that of Stirland et al (1984) but is in disagreement with that of Cornier et al (1984).

## 3. Conclusion

Preliminary results have been presented of a TEM study to characterise the grown-in dislocations in a series of SI, undoped LEC GaAs slices by analysis of type and detailed structure, to determine the nature of any precipitate particles and to look for preferential decoration of different dislocation types. Such information is important as the growth process is being refined in an effort to produce dislocation-free, electrically uniform material. The existence of As particles in the GaAs lattice is relevant to the question of the nature of EL2 and its interaction with the dislocation structure. High resolution TEM studies could potentially reveal the atomic arrangement at dislocation cores and the initial growth of such particles.

## Acknowledgements

The authors wish to thank N G Chew and D J Stirland for useful discussions and Cambridge Instruments for supplying the GaAs slices. The work is supported by the Ministry of Defence (CVD).

## References

Brown G T, Skolnick M S, Jones G R, Tanner B K and Barnett S J 1984 Semi-Insulating III-V Materials ed D C Look and J S Blakemore (Shiva) pp76-82
Cornier J P, Duseaux M and Chevalier J P 1984 Appl. Phys. Lett. 45 (10) 1105
Cullis A G, Augustus P D and Stirland D J 1980 J. Appl. Phys. 51 (5) 2556
Stirland D J, Augustus P D, Brozel M R and Foulkes E J 1984 Semi-Insulating III-V Materials ed D C Look and J S Blakemore (Shiva) pp91-4
Swanson H E and Fuyat R K 1953 NBS Circular 3 539
Warwick C A and Brown G T 1985 Appl. Phys. Lett. 46 (6) 574

*Inst. Phys. Conf. Ser. No. 76: Section 6*
*Paper presented at Microsc. Semicond. Mater. Conf., Oxford, 25–27 March 1985*

221

# Grain boundaries in GaAs

C B Carter, N H Cho, Z Elgat, R Fletcher and D K Wagner*

Department of Materials Science and Engineering, Bard Hall
*School of Electrical Engineering, Phillips Hall,
Cornell University, Ithaca, NY 14853, USA

Abstract    GaAs bicrystals have been produced by growing the GaAs directly on Ge bicrystals using the organo-metallic vapor phase epitaxy
technique.  The use of this new approach is illustrated for the first-
order, $\Sigma=3$, twin and for $\Sigma=5$, [001] tilt boundary.  Both interfaces are
invariably faceted.  Growth on a {110} substrate also produces numerous
microtwins which can lead to the formation of higher-order twin interfaces.  This effect is illustrated by the observation of the second-
order $\Sigma=9$ interface.

## 1.  Introduction

There have been several theoretical studies of the structure of grain
boundaries in GaAs [1-3] but very few experimental observations have been
reported.  In comparison, recent studies of grain boundaries in Si and Ge
[4-8] have not only led to a greater understanding of these interfaces but
have also resulted in a much greater understanding of the structure of lattice dislocations in these materials.  The reason for this rapid advance in
understanding defects in Si and Ge was a direct result of the development
of techniques for forming bicrystals either by hot-pressing together two
single crystals or by growing bicrystals by the Czochralski technique.  The
former technique has so far proved to be unsuitable for GaAs partly because
it is a compound but primarily because if dissociates incongruently at temperatures above 500°C.  Grain boundaries have been examined in melt-grown
GaAs [9] but there is then no control over the geometry of the interface.

Two techniques have therefore been developed to overcome these difficulties.  In the first [10], two single crystals of GaAs are cut and held together so that the plane separating them would correspond to the desired
grain boundary if they were in intimate contact.  A layer of amorphous $SiO_2$
is deposited over the junction of two grains and a thin film of GaAs is
then grown on the two GaAs single crystals.  These epilayers then grow over
the $SiO_2$ and join to form a grain boundary.  This technique has the disadvantage that the interface plane is not easily controlled.

A new technique [11] has therefore now been developed whereby GaAs grain
boundaries are grown directly on Ge bicrystal substrates.  This paper illustrates the type of grain boundary which can be grown using this new
technique.  The technique will be demonstrated by the growth of two types
of grain boundary: the first-order twin ($\Sigma=3$) grown on a {110} substrate
and a $\Sigma=5$ tilt boundary grown on a {001} substrate ($\Sigma$ represents the in-
verse of the fraction of lattice sites which are common to the two grains

[12,13]). This new approach has the further advantage that a grain boundary in GaAs can then be compared directly to the corresponding inter- face in the polar material.

The first-order twin boundary in compound semiconductors has been observed for example in InSb, GaSb and ZnS. Holt [1] noted that the structure of such twins could take on two forms which he termed the paratwin and the orthotwin. The two forms differ in that the bond across the coherent {111} plane would be either all Ga-As bonds or all anti-site bonds (i.e. Ga-Ga or As-As bonds). A recent study of antiphase boundaries (APBs) in GaAs sug- gests that interfaces which locally involve a large number of antisite bonds are unlikely to occur [14]. The first-order twin boundary was chosen for detailed study because the same interface has been studied extensively in Si and Ge. Such interfaces in the elemental semiconductors are almost invariably faceted with $(111)_M$, $\{112\}_M\{112\}_T$, $\{111\}_M\{115\}_T$ and $\{002\}_M\{122\}_T$ facets being most common [15]: the planes are indexed for a twin with a common (111) plane and $[1\bar{1}0]_M$ parallel to $[\bar{1}10]_T$; M and T rep- resent matrix and twin and are interchangeable. The GaAs has been found actually to grow on the {110} substrate in different twin related configu- rations. When these different twins grow together, they can form second- and third-order twins; these twins are also referred to as $\Sigma=9$ and $\Sigma=27$ grain boundaries.

The $\Sigma=5$ boundary is particularly interesting because it has the highest density of coincident sites after the first-order twin, but unlike the $\Sigma=3$ boundary, it involves extensive bond distortion at the interface. It can be constructed by rotating one grain through an angle of 38°57' about the [001] axis, with respect to the other grain. This type of interface has been studied in Ge and Si. It has recently been pointed out [16] that un- like the case of fcc metals, a rotation of 38°57' in GaAs actually produces a different boundary from the 51°3' rotation. The Ge bicrystal has been chosen so as to produce $\Sigma=5$ tilt boundaries in GaAs. By analogy with re- sults of studies of similar interfaces in fcc metals, it might be antici- pated that the $\Sigma=5$ boundary would also tend to facet along special planes which actually correspond to planes with a high density of coincident lat- tice sites.

## 2. Experimental Details

Germanium bicrystals containing $\Sigma=3$, <110> and $\Sigma=5$, <001> tilt boundaries were grown using the Czochralski method with two carefully oriented seeds. The Ge bicrystals were mechanically cut normal to the tilt axis and polished. TEM examination of specimens prepared from the Ge bicrystals showed that the grain boundaries were very close to the desired geometry and that few subgrain boundaries were present in the Ge grains. The Ge substrate was cleaned by boiling in acetone and then methanol for 6 minutes each, rinsing with deionized water, etching in HF and drying in $N_2$ gas. An epilayer of GaAs was then grown on the Ge by high pressure (76 torr), organo-metallic vapor phase epitaxy (OMVPE). The substrate temperature was held at 650°C during growth and the final thickness of the GaAs epilayers was shown to be 1.5 μm. The TEM samples were ion-thinned almost exclu- sively from the Ge side and were examined in a Siemens Elmiskop 102 operat- ing at 125 kV.

## 3. Experimental Results and Discussion

Figure 1 shows a dark-field image of a grain boundary (G) grown on a $\Sigma=3$, first-order twin in Ge. The corresponding diffraction pattern confirmed that the majority of the boundary lies parallel to the common (111) plane; there is a small deviation from the exact coincident orientation which is accommodated by the dislocations labelled D. The situation can be confused by the presence of a large number of microtwins (e.g. M) which were usually present in these specimens. It is interesting to note that the microtwins did not only form by a 70°53' rotation about the ±[1$\bar{1}$0] growth axis but also formed with similar rotations about other, inclined <110> axes. Microtwins similar to that shown in Fig. 1 are now used to examine the facet-

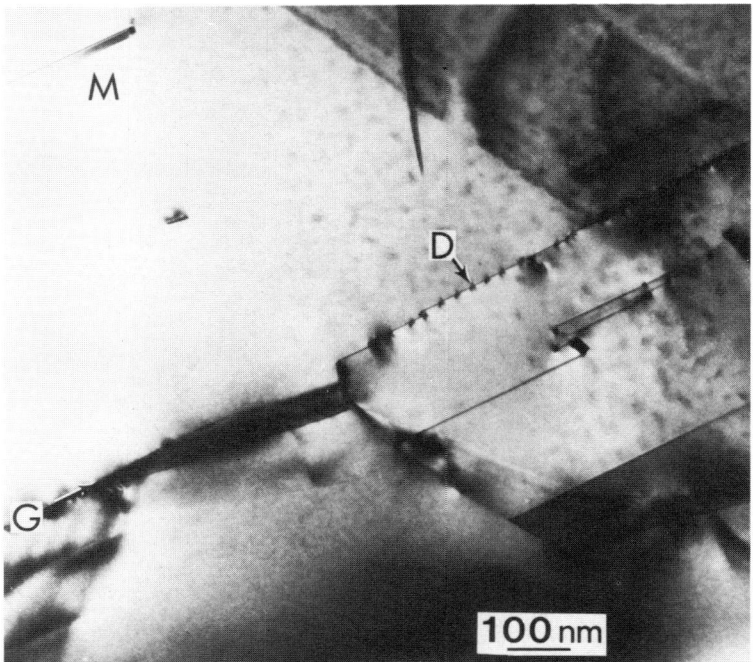

Fig. 1 A first-order twin boundary (G) in GaAs grown on a Ge bicrystal with a {110} surface. D indicates an intrinsic dislocation; M is a microtwin.

ing of the Σ=3 boundary. The fact that such microtwins are present can be appreciated more clearly from the image shown in Fig. 2. In this study, it has been found that the first-order twin interface in GaAs is always faceted with facets parallel to the common {111} plane being particularly common although other facet planes are common. When a pair of first-order twins which have a different twin plane coalesce, the Σ=9, second-order twin interface is formed as occurs at D in Fig. 2c. This interface is also invariably faceted.

The presence of a large number of microtwins in these samples also ensures that suitable twin interfaces will be present in the thin area of the specimen. Figure 3 is a high-resolution image from such an area and reveals the presence of Σ=3 interfaces (AC, AB, CD), Σ=9 interfaces (AD) and Σ=27 interfaces (BD). Images have been simulated for these twin boundaries

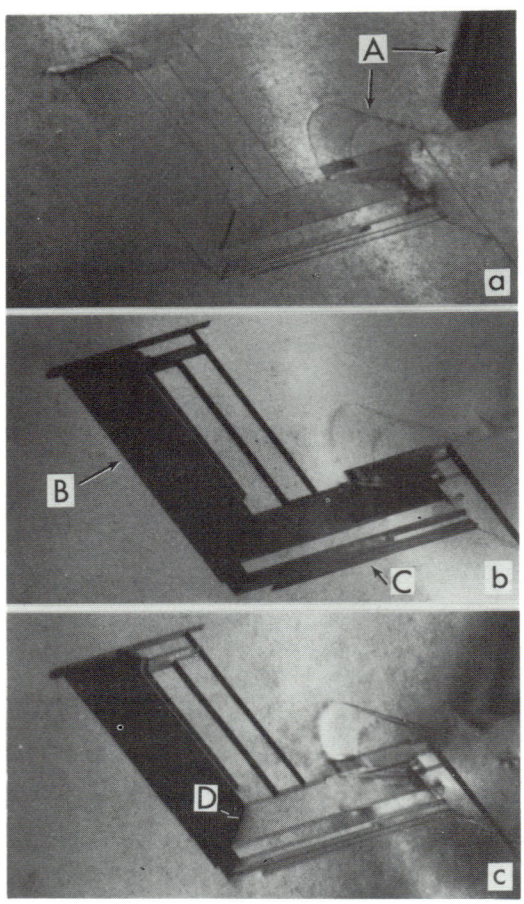

Fig. 2  Microtwins in GaAs grown on a {110} Ge substrate.  A is an antiphase boundary, B and C are the {111} twin planes and D is a Σ=9 interface.
All are bright-field images a) at {110} pole; b) common {111} reflection
excited; c) {111} reflection for one twin orientation satisfied.

using the multislice program developed at ASU and show excellent agreement.
With the microscope available at present it is not possible to differen-
tiate between the paratwin and the orthotwin [1].  Although it will still
not be possible to distinguish between Ga and As with the availability of
the new generation of 400 kV high-resolution microscopes, it will be possible to distinguish between the two types of interfaces: this result can
be recognized more easily for InP where the difference in atomic scattering
factors is further increased [3].

The Σ=5 boundary in GaAs was grown on a Σ=5, [001], bicrystal with a
nominal (130) boundary plane.  It is illustrated in Fig. 4 which shows a
short, heavily faceted segment imaged in bright-field.  It is believed that
the inclined section consists of three facets: one is the common {130}
facet (B) and lies parallel to the beam direction, the second is an edge-on
facet parallel to the (100) plane (A) in one grain, and the third facet is

shallowly inclined to the surface and may be parallel to the common (001) plane in which case it would be a pure twist segment. The formation of straight {130} facets is particularly interesting because this facet contains a high density of coincident lattice points. Further analysis including the use of high-resolution electron microscopy, is in progress to analyze the detailed configuration of this interface.

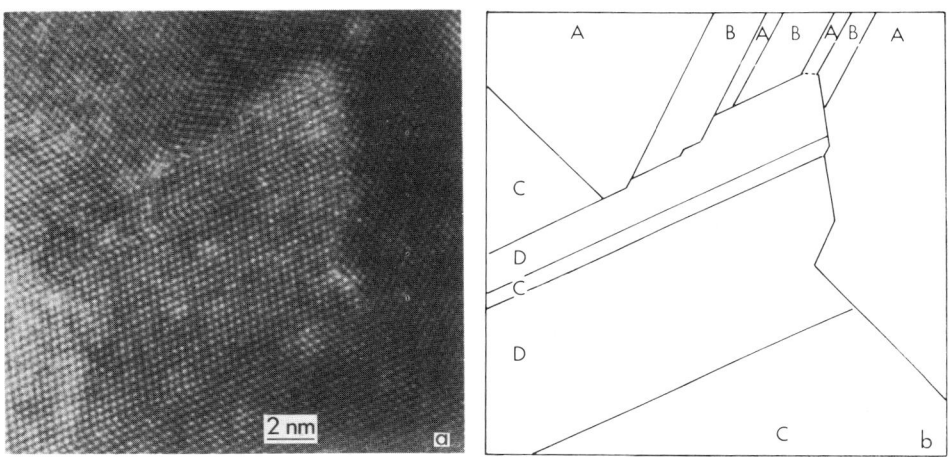

Fig. 3  High-resolution image of coalescing microtwins resulting in the formation of Σ=9 (AD) and Σ=27 (BD) twin boundaries by the coalescence of first-order twins.

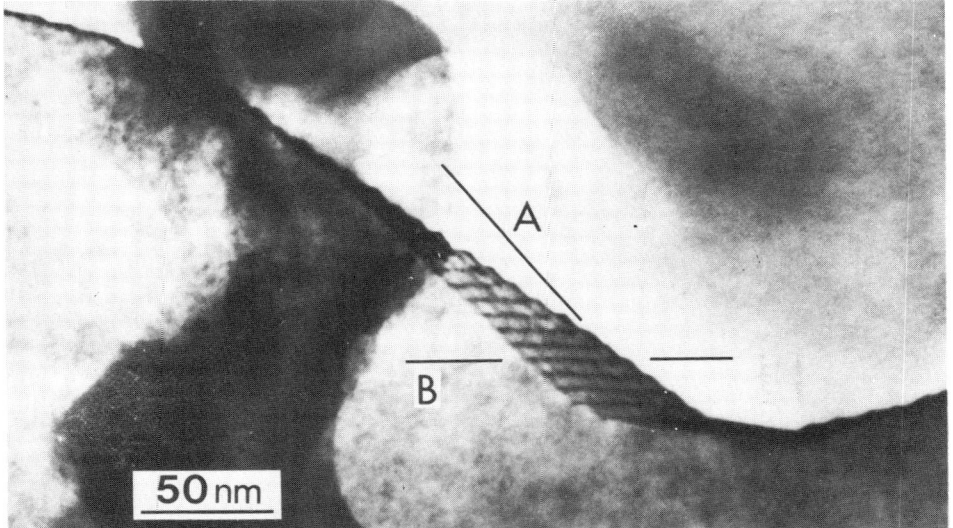

Fig. 4  A Σ=5 grain boundary in GaAs.  This segment has been selected to illustrate the strong tendency of the interface to facet.

## 4. Conclusion

Well-controlled bicrystals of GaAs can be grown on Ge bicrystal substrates with either a {110} or {001} surface. The technique has been illustrated for the first-order, Σ=3 twin and the Σ=5 high-angle boundary grown on {110} and {001} surfaces respectively. Growth on the {110} substrate also produces many microtwins. The Σ=3, Σ=5, and the accidentally occurring Σ=9, boundaries were all invariably faceted in GaAs. This result may be associated with the growth mechanism or it may be, at least, in part related to the non-polar nature of the material.

## Acknowledgments

The authors would like to thank Mr. B. C. De Cooman and Dr. W. B. Schaff for helpful discussions and assistance and Mr. Ray Coles for maintaining the microscope. The microscope is a Materials Science Center Facility and is supported in part by NSF. This research is supported by the U.S. Army Research Office under contract no. DAAG29-82-K-0148.

## References

[1] D.B. Holt, J. Phys. Chem. Solids, 25, 1385-1395 (1964).
[2] J.W. McPherson, G. Filatovs, E. Stefanakos and W. Couis, J. Phys. Chem. Solids, 41, 747-756 (1980).
[3] B.C. De Cooman, N.-H. Cho, Z. Elgat and C.B. Carter, Submitted to Ultramicroscopy (1985).
[4] B. Cunningham, H. Strunk, and D.G. Ast, Appl. Phys. Lett., 40, 237-239 (1982).
[5] C.B. Carter, in Grain Boundaries in Semiconductors (Elsevier), p. 33 (1982).
[6] M.D. Vaudin, B. Cunningham and D.G. Ast, Scripta Metallurgica, 17, 191-198 (1983).
[7] A.-M. Papon, M. Petit, and J.-J. Bacmann, Phil. Mag., 49, 573-589 (1984).
[8] C. D'Anterroches and A. Bourret, Phil. Mag., 49, 783-807 (1984).
[9] M.G. Spence, W.J. Schaff and D.K. Wagner, in Grain Boundaries in Semiconductors (Elsevier), p. 125 (1982).
[10] J.P. Salerno, B.W. McClelland, P. Vohl, J.C.C. Fan, W. Macropoulos and C.O. Bozler, in Grain Boundaries in Semiconductors (Elsevier), p. 77 (1982).
[11] N.H. Cho, C.B. Carter, Z. Elgat and D.K. Wagner, MSC Report No. 5525, Cornell University, submitted for publication (1985).
[12] D.A. Smith and R.C. Pond, International Metals Reviews, 61-74, June (1976).
[13] R.W. Balluffi, A. Brokman and A.H. King, Acta. Metall., 30, 1453-1470 (1982).
[14] N.H. Cho, B.C. De Cooman, D.K. Wagner and C.B. Carter, MSC Report No. 5524, Cornell University, submitted for publication (1985).
[15] J. R. Kohn, Am. Miner. 43, 263 (1958).
[16] R.C. Pond and D.B. Holt, J. Phys. 43, C157 (1982).

*Inst. Phys. Conf. Ser. No. 76: Section 6*
*Paper presented at Microsc. Semicond. Mater. Conf., Oxford, 25–27 March 1985*

# In-situ observations of the development of heavy-ion damage in semiconductors

M L Jenkins*[+], T J Chandler*, I M Robertson[+] and M A Kirk**

* Department of Metallurgy & Science of Materials,
  University of Oxford, Parks Road, Oxford OX1 3PH
[+] Materials Research Laboratory, University of Illinois,
  Urbana, IL 61801, USA
**Materials Science and Technology Division,
  Argonne National Laboratory, Argonne, IL 60439, USA

Abstract    In-situ observations on ion-beam induced amorphisation of
GaAs, GaP and Si are reported.  Direct-impact amorphisation was found
to occur in GaAs irradiated with 100keV $Xe^+$ ions to low doses at low
temperature (~40K) in contrast to previous room temperature irradiations.  In GaP and in silicon, where heavy projectiles do cause direct
impact amorphisation at room temperature, the evolution of the damage
structure with ion dose was studied.  The defect yield both in GaP
irradiated with 100keV $Kr^+$ ions and in Si irradiated with 100keV $Xe^+$
ions was found to decrease monotonically with increasing dose over the
dose range $10^{15}$–$10^{17}$ ions $m^{-2}$.

## 1. Introduction

Ion implantation has been used successfully to dope silicon for several
years, and there has been considerable progress in the use of this technique to dope compound semiconductors such as GaAs and GaP.  An unavoid-
able drawback of ion irradiation, especially for compound semiconductors
where annealing can pose problems, is radiation damage of the substrate,
which can lead at high ion implant doses to the formation of a buried or
surface amorphous layer.  Two mechanisms have been suggested for the
mechanism by which such amorphous layers are produced.  In the first, the
amorphous layer is built up by the overlap of small amorphous regions
created within individual displacement cascades.  'Direct impact amorphisation' is known to occur in Si and in GaP provided that the energy
density within cascades is sufficiently high (see e.g.  Chandler and
Jenkins 1983, Howe et al 1980).  In the second the crystalline to amorphous transition is triggered when the cumulative defect density, arising
from the spatial overlap of two or more cascades, reaches some critical
value.  Both mechanisms may occur simultaneously, and a theory has been
developed to take account of this (Webb and Carter 1979).  We report here
some preliminary results on experiments which explore the mechanisms of
ion-beam induced amorphisation of semiconductors further.  We have carried
out in-situ heavy-ion irradiations of GaAs, GaP and Si to investigate:

(i)  Direct impact amorphisation at low temperatures in GaAs.   In room
     temperature irradiations amorphous zones are not observed at low
     doses (Chandler and Jenkins 1983).

(ii) The evolution of damage with dose and the transition to an amorphous layer in GaP and in Si under irradiation with heavy projectiles (Xe$^+$ and Kr$^+$ ions respectively) known to cause direct impact amorphisation at low doses.

## 2. Experimental Procedures

The in-situ experiments were performed on the High-Voltage Electron Microscope (HVEM) - Tandem Accelerator Facility at Argonne National Laboratory. This facility consists of two accelerators, a 300keV Texas Nuclear accelerator and a NEC 2MeV Tandem accelerator, either of which produce ion beams that can be directed onto the HVEM sample position by an ion-beam interface. Only the first of these accelerators was used in the present experiments. Details of the facility are given by Taylor et al (1981). The low-temperature experiment in GaAs was performed using a single-tilt, liquid-helium cooled sample stage with the specimen maintained at about 40K. A (100) foil was irradiated with 100keV Xe$^+$ ions to a dose of $5.10^{15}$ ions m$^{-2}$. The experiments in GaP and Si were performed at room temperature using a standard double-tilt holder. (100) foils were irradiated with 100keV Kr$^+$ and Xe$^+$ ions respectively to total doses $\geqslant 10^{17}$ ions m$^{-2}$ in dose increments of typically $5.10^{15}$ ions m$^{-2}$. Typical dose rates were about $0.7.10^{14}$ ions m$^{-2}$ s$^{-1}$ ($10\mu$A m$^{-2}$). The Kratos EM7 HVEM was operated at 200kV. Foils were viewed flat-on under well defined diffraction conditions. The angle of incidence of the ion beam was about 60° to the specimen surface. During irradiation of the Si specimens the foils were viewed continuously using a high-resolution video recorder. After each irradiation step diffraction conditions were re-set if necessary and micrographs were recorded.

## 3. Results and Discussion

### 3.1 The Low Temperature Experiment on GaAs

Figure 1 shows an area of a GaAs foil just following irradiation at 40K with 100keV Xe$^+$ ions to a dose of $5.10^{15}$ ions m$^{-2}$ and imaged using a diffraction vector $\underline{g}$=400. The dark-dot contrast features in this thin area are characteristic of amorphous zones. The defect yield (taken to mean in this paper the ratio of visible distinct contrast features to the number of incident ions) is about 0.4. This relatively high value is consistent with the formation of amorphous zones by a process of direct impact amorphisation within individual displacement cascades.

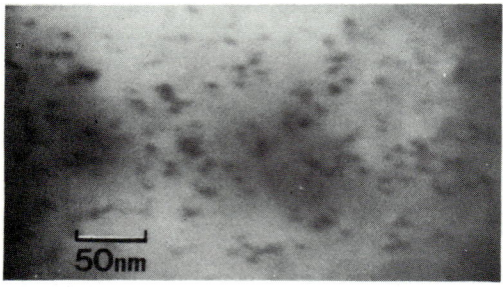

Fig.1. Amorphous zones in GaAs produced by low-temperature (40K)irradiation with 100keV Xe$^+$ ions to a dose $5.10^{15}$ m$^{-2}$.

Direct impact amorphisation has been found to occur with high efficiency in GaP and Si in room temperature irradiations with heavy projectiles, but in GaAs room temperature irradiation with 100keV $Xe^+$ ions was found to produce at low doses only a very low yield of weak contrast features which could not be identified unambiguously (Chandler and Jenkins 1983). Only at doses $\geqslant 3.10^{16}$ ions $m^{-2}$ were amorphous regions produced (Chandler 1984). The above results suggest that the point defect production processes and the initial creation of amorphous zones are similar in GaAs, GaP and Si, but thereafter annealing processes may occur in GaAs in irradiations at room temperature. This conclusion has also been reached by authors reporting Rutherford back-scattering experiments (e.g. Ahmed et al 1980, Williams and Austin 1980).

Unfortunately, immediately after the micrograph of Fig.1 was recorded the specimen shattered, terminating the experiment, so the annealing behaviour of the defect regions could not be investigated.

3.2 Damage Evolution in GaP and in Si

Damage evolution seemed to follow a similar pattern in both materials. Figures 2 and 3 show stages in damage development in GaP irradiated with 100keV $Kr^+$ ions and in Si irradiated with 100keV $Xe^+$ ions respectively. The lower dose behaviour is shown in more detail in Fig.2 whilst damage development over the whole dose range studied is shown in Fig.3. In each case only a few of the irradiation steps are illustrated. A careful analysis of these and other micrographs of the series and of video recordings revealed the following:

(a) Some dark-dot contrast regions appeared at an early stage in the irradiation sequence and persisted in a more or less unchanged form through many of the remainder of the series. Examples of such regions are arrowed and labelled 1 in Figs.2 and 3. The observation in these thin areas of foil of dark contrast dots at low dose is consistent with direct impact amorphization.

(b) Some defect regions faded in contrast from one irradiation to the next, and over the course of several irradiations disappeared altogether. Neighbouring defect regions were not similarly affected, and the same change was found in consecutive micrographs obtained under slightly differing contrast conditions, so the possibility that this is a contrast artefact can be discounted. Examples of such

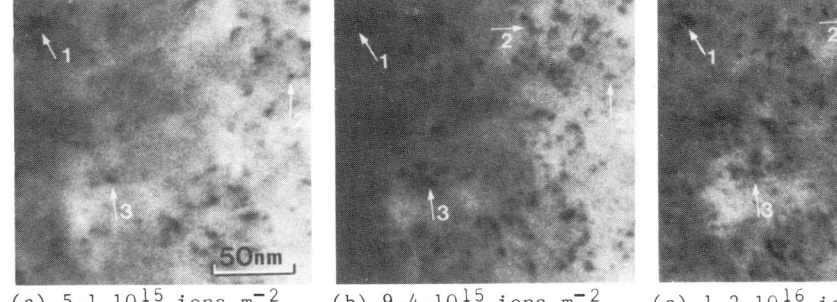

(a) $5.1.10^{15}$ ions $m^{-2}$     (b) $9.4.10^{15}$ ions $m^{-2}$     (c) $1.2.10^{16}$ ions $m^{-2}$

Fig.2. In-situ irradiations of GaP with 100keV $Kr^+$ ions. The same area is shown in each micrograph, imaged in dark-field using $\underline{g}$=220.

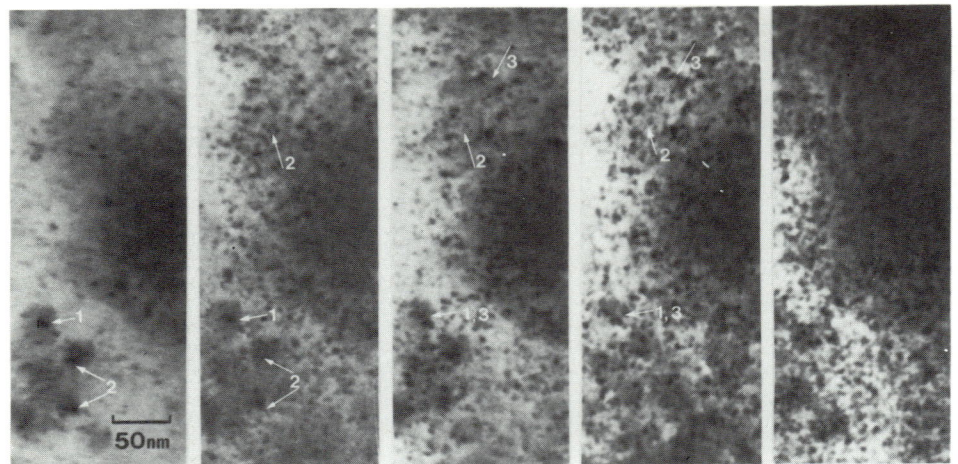

(a)0.73     (b)3.1         (c)7.0        (d)8.8        (e)17.4.10$^{16}$ m$^{-2}$
Fig.3. In-situ irradiations  of Si with 100keV Xe$^+$ ions.  The same area is
shown in each micrograph, imaged in dark-field using $\underline{g}$=220.

regions are arrowed and labelled 2 in Figs.2 and 3.

(c) When  new  contrast  dots  appeared  they  did  so  suddenly,  in Si between
    consecutive frames of the video recording.

(d) Occasionally,  defect  regions  became  stronger  in  contrast  from  one
    irradiation to the next, whilst not seeming to change much in overall
    size.   More often, when new contrast centres formed close to existing
    features the region between the dark centres showed a light grey con-
    trast.   As the dose increased to ~10$^{17}$ ions m$^{-2}$ larger black/grey
    contrast areas formed, encompassing regions previously occupied by one
    or several dark dots.   Regions showing these features are arrowed and
    labelled 3 in Figs.2 and 3.   At a dose of ~2.10$^{17}$ ions m$^{-2}$ (Fig.3e)
    partially connected dark contrast areas covered much of the field of
    view, but these could not be correlated on a one-to-one basis with the
    dark dots visible in the same foil area at lower doses.

The  consequences  of  the  various  effects  described  above  on  the  overall
density  of  visible  contrast  features  is  shown  in  Fig.4.   At low doses
($\leqslant$10$^{16}$ ions m$^{-2}$) defect yield values for both Kr$^+$$\rightarrow$GaP and Xe$^+$$\rightarrow$Si were high
($\geqslant$0.3), consistent with a direct impact amorphization mechanism.   At doses
$\geqslant$2.10$^{16}$ ions m$^{-2}$, however, the yields dropped steeply to much lower values
($\leqslant$0.1) and then showed a continuing steady decline for doses up to the
maximum employed.   The areal density of defect regions showed a tendency
to saturate at a value of 5-7.10$^{15}$ m$^{-2}$.   Even at doses approaching 10$^{17}$
ions m$^{-2}$, about ten times the cascade overlap dose, separate contrast dots
could usually be distinguished (Fig3d).   By a dose of 1.7.10$^{17}$ ions m$^{-2}$,
however,  such  a  distinction  was  less  meaningful  and  the  concept  of  a
'defect yield' was no longer valid (Fig.3e).

The  in-situ  experiments  at  Argonne  were  complemented  by  more  extensive
room-temperature  irradiations  of  GaP  and  GaAs  carried  out  in  an  ion
implanter  at  AERE  Harwell.    In  general  observations  made  in  these

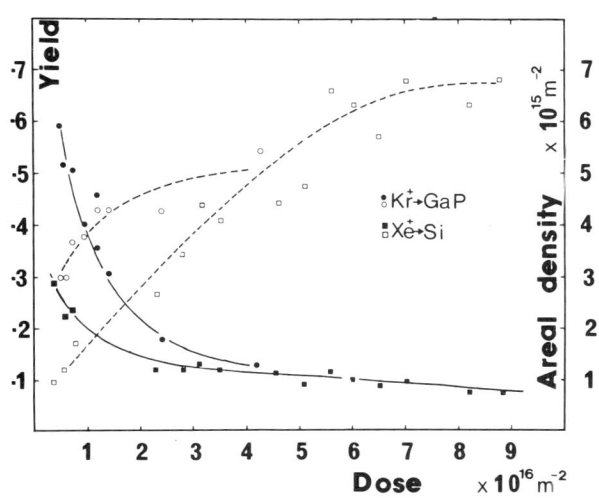

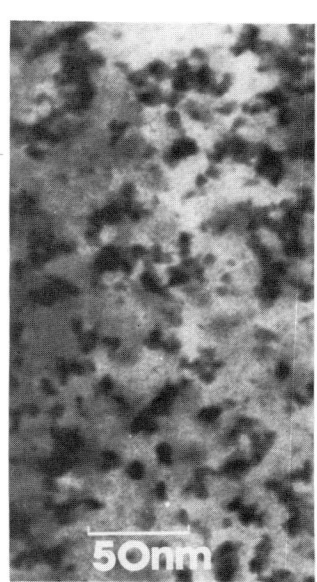

Fig.4. Defect yields (solid curves and symbols) and areal densities (dashed curves and open symbols) for 100keV $Kr^+\to GaP$ (circles) and 100keV $Xe^+\to Si$ (squares).

Fig.5. Amorphous areas in GaAs after 100keV $Kr^+$ ion irradiation to a dose $3.10^{17}$ ions $m^{-2}$.

experiments parallelled those of the in-situ experiments described above. In particular, in specimens of both GaP and GaAs irregular partially-connected 'amorphous areas' were found at doses $1-3.10^{17}$ $Kr^+$ ions $m^{-2}$ (Fig.5). This structure would appear to be somewhat more developed but essentially similar to that seen in Fig.3e. Complete surface amorphisation was observed in GaP at doses of $3.10^{17}$ $W^+$ ions $m^{-2}$ and at $1.10^{18}$ $Kr^+$ ions $m^{-2}$, but was not found in GaAs at doses up to these levels. This is consistent with more extensive annealing in GaAs in room temperature irradiations.

The observations described above may be compared with experiments in Si carried out by Ruault et al (1983). These authors studied the formation and evolution of radiation damage in Si due to room-temperature Bi implantation at 50keV and 200keV using the in-situ ion irradiation facility at Orsay. The chief conclusion of this work was that amorphous layers are not formed by the overlap of amorphous zones produced in the cores of displacement cascades. Rather, so called 'grey zones' formed in the vicinity of neighbouring dark contrast features and extended towards the ion-entry surface, and these were believed to be the precursors of amorphous layers. Grey zones were believed to form by the overlap of lightly damaged regions in the wings of cascades. Our observations would tend to support this hypothesis.

A second conclusion reached by Ruault et al (1983) was not however substantiated in the present work. They reported a time evolution of individual damage features. In their geometry ions were incident at a shallow angle (~30°) to the specimen surface, and damage formed along elongated damage tracks. The contrast along such tracks was found to develop into 'subclusters' over a period of about ten seconds. Ruault et al concluded from this that these small subclusters were not formed directly in regions of high deposited energy density along the track, but were a result of defect motion.   In our experiments no such time evolution of new defect regions was observed:   new contrast spots appeared suddenly.   In addition our observation of the formation of defect regions in GaAs at 40K also supports the idea that amorphous zones form spontaneously at cascade cores without the need for long-range defect migration.

The observation (b) above points to a dynamic mechanism of radiation-induced annealing, which has not been reported in previous TEM studies. Beam annealing effects during implantation of GaAs and GaP have been reported previously by several authors in Rutherford back-scattering studies, albeit at higher doses and dose rates (e.g. Ahmed et al, 1980, Williams and Austin 1980).   An interesting feature of our observations was that some defect regions appeared more resistant to annealing than others. A possibly    related effect was reported by Williams and Austin, who found that in GaAs disorder partially annealed during room- and higher-temperature irradiation was difficult to remove completely below an anneal temperature of 870K, whereas defect regions 'frozen-in' at liquid nitrogen temperature could be recrystallised epitaxially at temperatures below 520K.   This suggested the presence of defect complexes with higher activation energies in partially annealed layers.   Our observations here on the different   susceptibilities   to   annealing   of   apparently   similar   defect regions, and our observations reported in  3.1 that high-dose irradiations of GaAs lead to defect regions stable at room temperature, whereas amorphous zone produced by direct impact amorphization were not observed at room temperature, both suggest that the atomic structures of 'amorphous' zones are not uniform, but may differ from zone to zone and depend on the material and the irradiation conditions.

The mechanism of radiation-induced annealing is not yet understood in detail.   One interesting possibility is local annealing due to thermal spikes at displacement cascades.   The cascade 'temperatures' are sufficiently high and thermal spike lifetimes sufficiently long for appreciable thermal annealing to occur in cascade peripheries.   This has been discussed in more detail by Chandler (1984).

## Acknowledgements

This work was supported by the US Department of Energy, Division of Materials Science, through contracts DE-AC02-76ER01198 (U of I) and W-31-109-ENG-38 (ANL).   We thank E A Ryan and A Philippides for assistance with operation of the Argonne HVEM-Tandem.

## References

Ahmed N A G, Christodoulides C E and Carter G 1980 Rad. eff. $\underline{52}$, 211
Chandler T J 1984 D Phil Thesis, University of Oxford
Chandler T J and Jenkins M L 1983 Inst. Phys. Conf. Ser. No.67, 297
Howe L M, Rainville M H, Haugen H K and Thompson D A 1980 Nucl. Inst. and
    Meth. $\underline{170}$, 419
Ruault M O, Chaumond J and Bernas H 1983 Nucl.Inst. and Meth. $\underline{209/210}$, 351
Taylor A, Wallace J R, Ryan E A, Philippides A and Wrobel J R 1981 Nucl.
    Inst. and Meth. $\underline{189}$, 211
Webb R and Carter G 1979 Rad. eff. $\underline{42}$, 159
Williams J S and Austin M W 1980 Nucl. Inst. and Meth. $\underline{168}$, 307

*Inst. Phys. Conf. Ser. No. 76: Section 6*
*Paper presented at Microsc. Semicond. Mater. Conf., Oxford, 25–27 March 1985*

# Structural defects in cadmium telluride grown from the vapour phase

K Durose, G J Russell and J Woods

Department of Applied Physics & Electronics,
University of Durham, South Road, Durham, DH1 3LE,U.K.

Abstract The structure of crystals of CdTe grown by a vapour phase
technique has been investigated by CTEM and etching techniques. The
main defects studied were twins, sub-grain boundaries and precipitates.
Etching studies showed that the twins were of the ortho-type but CTEM
examination revealed the presence of dislocation arrays along their
boundaries. The sub-grains ranged in size from 5 to 500 μm with an
average size of 150 μm. Misorientations between adjacent sub-grains
were generally less than 30'. Tellurium precipitates up to 10 μm were
also observed to decorate some defects, especially grain boundaries.

## 1. Introduction

The band gap of $Cd_xHg_{1-x}Te$ (where $x = 0.2$) is in the infra-red region of
the electromagnetic spectrum making this material suitable for the
fabrication of thermal imaging devices. However, difficulties in preparing it in suitable bulk crystal form have led to the development of
epitaxial techniques for its deposition onto foreign substrates such as
CdTe (Mroczkowski and Vydyanath, 1981), GaAs (Geiss et al, 1985) and
sapphire (Tennant et al, 1985). Initially liquid phase epitaxial methods
were used (Mroczkowski and Vydyanath, 1981) and more recently metal-
organic chemical vapour deposition techniques have been put to this
application (Irvine et al, 1984). CdTe is a suitable substrate material
since it has a good lattice match with $Cd_xHg_{1-x}Te$ and does not lead to the
introduction of foreign impurities into the epilayers by interdiffusion.
However, defects in the substrate material may have an influence on the
defect content and compositional homogeneity of the epilayers deposited
upon them. Since the quality of the epilayers must be high for thermal
imaging applications (Irvine et al, 1984) the defect content of the CdTe
substrates is of paramount importance. The majority of work on crystals
of CdTe concerns material grown from the melt. However, in this laboratory
large single crystals of CdTe up to 29 mm in diameter have been grown from
the vapour phase in sealed silica ampoules (Durose et al, 1985). This has
the advantages over melt growth techniques of enabling growth to take
place at lower temperatures and providing some control over the stoichiometry of the crystals grown. This paper reports the structural
characterisation of vapour grown CdTe by CTEM and etching techniques. In
particular the observations of twins and sub-grain boundaries are
described.

## 2. Experimental

The growth of large crystals of CdTe from the vapour phase in sealed silica capsules and some preliminary transmission electron microscope observations have been described earlier (Durose et al, 1985). Thin sections for study in the JEM 120 transmission electron microscope were prepared by chemical jet polishing using a solution of 2% bromine in methanol. Etching studies of the same material were conducted using three different reagents. These were:-

(1) 10 cc $HNO_3$ + 20 cc $H_2O$ + 1.5 mg $AgNO_3$-E Ag-1(see Inoue et al,1962).
(2) 3HF + $2H_2O_2$ + $2H_2O$ -²Nakagawa's etch³(see Nakagawa et al, 1979).
(3) 0.5% $Br_2$ in $CH_3OH$ with photostimulation (Williams, 1984).

Surfaces were prepared for etching studies by mechanically polishing oriented slices of material firstly with 12.5 µm and then with 1 µm alumina. Finally they were chemically polished with a solution of 2% bromine in methanol.

## 3. Results and Discussion

Twins have been found to be an important type of defect in these CdTe single crystals and consequently they have been studied in detail by both etching and CTEM techniques. Of the three etchants investigated, that termed E Ag-1 was found to be the most useful as it produced etch pits having symmetry which related to the orientation of the surfaces being etched. It is a known polar etch for CdTe and has been studied recently by Iwanaga and Yoshiie (1983) and Lu et al (1985). In the present work the observation by Fewster and Whiffin (1983) that the Nakagawa etch produces pits on Cd {111}A but not on Te {$\overline{1}\overline{1}\overline{1}$} B faces was used to determine the polar action of the E Ag-1 etch. From this it was found that E Ag-1 produces triangular pits on both polar faces, those on Cd {111} A having apices while those on Te {$\overline{1}\overline{1}\overline{1}$} B were flat bottomed. This implies that the facets of the etch pits are comprised of Te {$\overline{1}\overline{1}\overline{1}$} B planes.

In order to establish whether the twins were of the 'ortho-' (rotation or <110> tilt with θ = 250° 32') or the 'para-' (reflection or <110> tilt with θ = 70° 32') type (Holt, 1964), the following experiment was performed. A boule of CdTe was oriented by the Laue back reflection technique and a slice was cut from it such that the sawn face intersected a {111} twinning plane at an angle of ≃ 3°. The surface was polished and then etched with Inoue's E Ag-1 reagent.

An optical micrograph of the etched surface in the region of a twin boundary is shown in figure 1. The pits on either side of the boundary are of the same polar type and correspond to those seen on a Te ($\overline{1}\overline{1}\overline{1}$) surface. The two sets of pits are also rotated relative to one another by 180° indicating that the twin is of the ortho-type. No para-twins have been observed which is consistent with predictions based on energetic considerations.

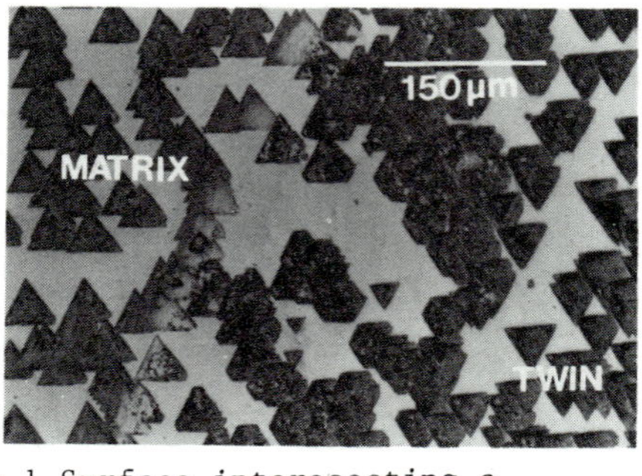

Fig.1.Surface interesecting a twinning plane at an angle of 3° after etching with E Ag-1

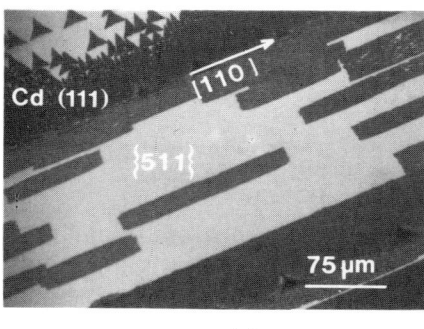
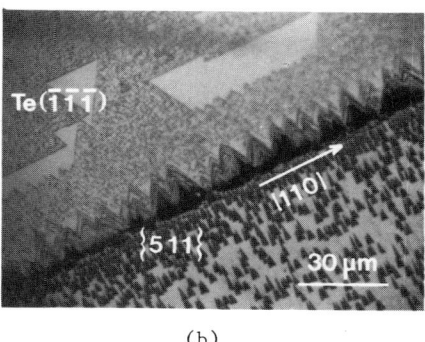

(a)　　　　　Fig 2　　　　　(b)

Fig 2. Effect of E Ag-1 on twin bands on opposite {111} faces
        (a) Cd {111}, (b) Te {$\bar{1}\bar{1}\bar{1}$} .

Twin bands intersecting accurately aligned Cd and Te {111} planes were
also investigated by etching with E Ag-1. The exposed surfaces of such
twin bands are {511} and these are also polar faces. The differences
in the etching action of E Ag-1 on {511} Cd and Te faces are shown in
the optical micrographs in figures 2a and b. These micrographs reveal
two particularly important features. Firstly, the shapes of the pits on
{511} and {$\bar{5}\bar{1}\bar{1}$} are consistent with the intersections of tetrahedra of
Te{$\bar{1}\bar{1}\bar{1}$} planes with these faces. This model accurately predicts the apex
angle of the isosceles triangular pits in figure 2b to be 22° 36' and can
account for the trough-like nature of the larger rectangular pits in figure
2a. Secondly, arrays of etch pits are observed at the twin boundaries on
both faces. These pits are not of the types observed either on {511} or
{111} surfaces and are clearly related to some interfacial defect. Pits at
the twin boundaries were also observed after etching with the Nakagawa
etchant. Kumagawa (1983) has reported the presence of pits at twin
boundaries in GaSb after etching with a reagent which does not produce
dislocation etch pits. The pitting in that case was attributed to steps
(i.e. short sections of lateral twin boundaries) on the twin boundaries.
When boundaries such as those seen in figure 2 were observed by CTEM,
arrays of dislocations lying coincident with the composition planes of
the twin boundaries were observed as is shown in figure 3. Both the

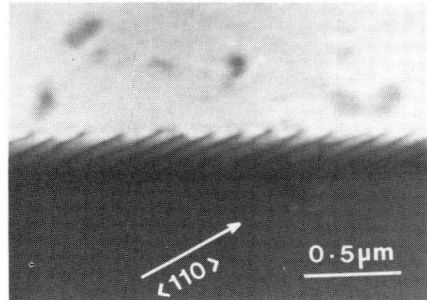
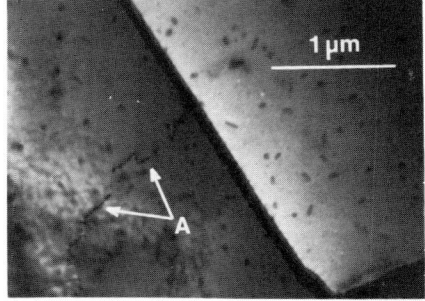

Fig 3.Transmission electron micro-　Fig 4.Transmission electron micrograph
      graph of twin boundary showing         of a crystallographic grain
      the presence of a dislocation          boundary.
      array.

regularity and close spacing of the dislocations in this micrograph
suggest that the etch pits observed along twin boundaries are not
necessarily related to such dislocation arrays. This suggests that other
defects, such as steps, are present in addition to dislocation arrays at
these twin boundaries.  The origin of these dislocation arrays may be
related either to the formation of the twins or to the dislocation
polygonisation process discussed below.

In addition to the twins, other crystallographic boundaries have been
observed. One such example is the crystallographic grain boundary shown
in the electron micrograph of figure 4.  Analysis of the electron
diffraction patterns taken from either side of this boundary indicated that
it is of a rotation/tilt type rather than having a simple tilt character.
Close examination of figure 4 reveals the presence of dislocations
along  the  boundary  as  is  evidenced  by  the  interruption  of  the
fringe contrast.  The other important feature of this micrograph is the
row of dislocations which can be seen leading away from the grain boundary
at A.  This constitutes a very small angle sub-grain boundary. Most of the
sub-grain boundaries which were observed corresponded to a larger misorientation than this and a typical example is provided by the electron
micrograph shown in figure 5.  This boundary is comprised of an array of
dislocations which are predominantly of the same type.  The difference in
orientation between the two sub-grains
was determined by measuring the relative
displacement of the Kikuchi lines from
spots in diffraction patterns taken
from either side of the sub-grain
boundary.  In this way the misorientation of the sub-grains in figure 5 was
calculated to be $\sim$ 18'.

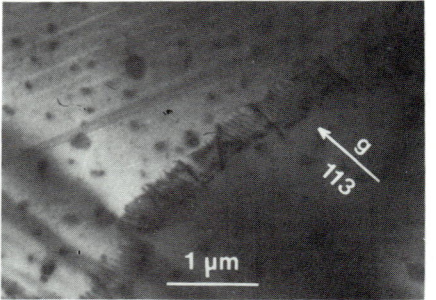

The larger scale features of these
sub-grain boundaries were studied
in more detail by etching techniques
and optical microscopy. The Nakagawa
and E Ag-1 reagents reveal them as
arrays of etch pits whereas a solution
of 0.5% bromine in methanol exposes        Fig 5.Transmission electron
them as raised linear features. Since            micrograph of a sub-
E Ag-1 tends to be inhomogeneous in its          grain boundary
action and as the Nakagawa etchant is
strongly orientation dependent, the solution of 0.5% bromine in methanol
was considered to be the preferred sub-grain boundary etchant.  The sub-
grains ranged from 5 to 500 μm in size with their average width being about
150 μm.  Their size and morphology has been seen to vary both from grain to
grain and within individual grains.  Of particular importance is the fact
that the pattern of sub-grain boundaries appears to be influenced by the
presence of other defects.  For example, they are often seen to be
associated with the junctions between lateral twin boundaries and twin
boundaries on {.111} planes as at B in figure 6a.  While many sub-grain
boundaries terminate at twin boundaries, a few have been observed to cross
them.  Regarding the origin of sub-grain boundaries, it is probable that
they arise from the migration of grown-in or stress induced dislocations
by a polygonisation process (see Hall and Vandersande, 1978).  This could
be a direct consequence of the fact that the crystals are maintained at a
relatively high temperature for several days during the growth process.

The final observation to be reported regarding defects in these crystals concerns large precipitates ∿ 10 μm in diameter. These were found to be present at grain boundaries, grain boundary junctions, grain boundary-twin boundary intersections and at lateral twin boundaries. Figure 6b records an optical micrograph of a grain boundary on a surface which has been etched with a solution of 0.5% bromine in methanol. Several precipitates are clearly present on this grain boundary and those labelled A in the figure are associated with sub-grain boundaries. These precipitates were identified as tellurium by energy dispersive x-ray analysis in a scanning

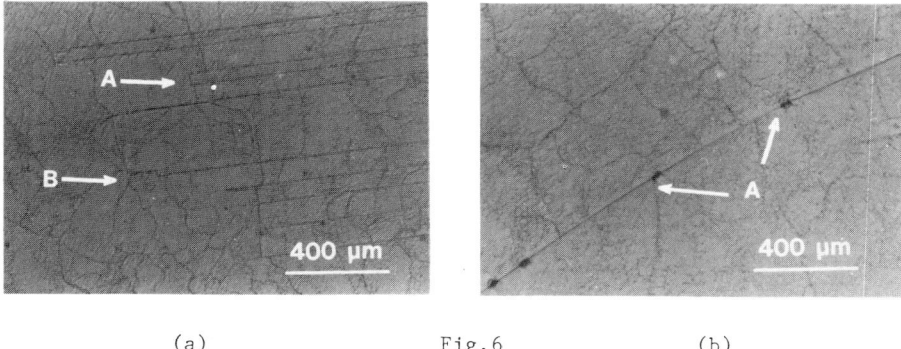

(a)                    Fig.6                    (b)

Fig 6.    Surfaces etched with a solution of 0.5% bromine in methanol
          showing (a) a lateral twin boundary at A and (b) tellurium
          precipitates at a grain boundary.

electron microscope. Much smaller precipitates of sub-micron dimensions were observed in an earlier CTEM study of this material (Durose et al, 1985). The origin of the tellurium precipitates may be explained by the fact that these crystals were grown from a tellurium rich vapour and that this element has a retrograde solid solubility in CdTe (Anderson et al, 1982).

4. Summary

This work has shown that the twins which are frequently encountered in CdTe grown from the vapour phase are of the ortho-type. In addition CTEM studies have revealed the presence of dislocation arrays at the boundaries of these defects. However etching investigations suggest that short sections of lateral twin boundaries may also be present. Sub-grain boundaries are another important defect in these crystals and these have been studied both by CTEM and etching techniques. The average sub-grain size was approximately 150 μm and the misorientations of up to 30' between adjacent sub-grains were observed. It is proposed that these sub-grain boundaries may arise as a result of a dislocation polygonisation process. The interaction between twins and sub-grain boundaries has been observed together with the identification of tellurium precipitates located at energetically favourable sites in the lattice.

Acknowledgements

The authors wish to express their thanks to Mr N F Thompson for his invaluable assistance in growing the crystals and to Mr T Harcourt for the cutting and polishing of all of the crystal slices studied.

References

Anderson P L, Schaake H F and Tregligas J H 1982 J. Vac. Sci. Technol. 21 125
Durose K, Russell G J and Woods J 1985 J.Crystal Growth, in the press
Fewster P F and Whiffin P A C 1983 J. Appl. Phys. 54 4668
Geiss J, Gough J S, Irvine S J C, Blackmore G W and Mullin J B 1985 J.Crystal Growth, in the press
Hall E L and Vandersande J B 1978 J. Am. Ceram. Soc. 61 417
Holt D B 1964 J. Phys. Chem. Solids 25 1385
Inoue M, Teramoto I and Takayanagi S 1962 J. Appl. Phys. 33 2578
Irvine S J C, Tunnicliffe J and Mullin J B 1984 Materials Letters 2 305
Iwanaga H, Yoshiie T, Takeuchi S and Mochizuki K 1983 J.Crystal Growth 61 691
Kumagawa M 1982 Jpn. J. Appl. Phys. 21 804
Lu Y C, Route R K, Elwell D and Feigelson R S 1985 J. Vac. Sci. Technol. A3 264
Mroczkowski J A and Vydyanath H R 1981 J. Electrochem. Soc. 128 655
Nakagawa K, Maeda K and Takeuchi S 1979 Appl. Phys. Lett. 34 574
Tennant W E, Gertner E R, Blackwell J D and Rode J P 1985 J. Crystal Growth, in the press
Williams D J, Private communication

*Inst. Phys. Conf. Ser. No. 76: Section 6*
*Paper presented at Microsc. Semicond. Mater. Conf., Oxford, 25–27 March 1985*

239

# Sphalerite to Wurtzite phase transformations in single-crystal CdS films

C Feldman, G Deutscher, Y Lereah* and E Grünbaum*

Department of Physics and Astronomy, Faculty of Exact Sciences,
*Department of Electronic Materials and Devices and Electromagnetic
Radiation, Faculty of Engineering,
Tel-Aviv University, Ramat-Aviv, 69978 Israel

Abstract  Transmission electron microscopy was used for in-situ observations of transformations in single-crystal (100) sphalerite CdS films to
the wurtzite phase.  The nucleation was induced by a pulse of a concentrated electron beam, and the lateral growth by partial concentration of
the electron beam and/or by general heating of the film by means of wire
resistance.  A relation was found between the defect nature of transformed wurtzite areas and their relative orientation to the original
sphalerite lattice.  A discussion of the transformation mechanisms is
given.

## 1. Introduction

CdS is one of the II-VI semiconductor compounds that occurs in two crystalline phases:  The wurtzite (hexagonal) type (the common phase), and the
sphalerite (cubic) phase, which can be obtained by epitaxial growth (Holt
1974).  In this work, the transformations to the wurtzite phase in thin
films of single-crystal (100) sphalerite CdS were performed in-situ and
examined in the electron microscope.

## 2. Experimental Method

The sample, single-crystal (100) sphalerite CdS films (Fig.1), 800-2500 Å
thick, were prepared by epitaxial growth in high-vacuum on (100) NaCl substrates, and possessed a density of $10^{12}$-$10^{13}$ defects (dislocations and
stacking-faults) per $cm^2$.  They were floated off the substrates in water
and collected on grids, which were mounted on a heating holder in a
Philips EM-300 transmission electron microscope.  This holder was provided
with a wire resistance heater and the temperature was measured with a thermocouple.  Phase transformations were performed in-situ by two methods:
1. Localized heat pulse obtained by concentrating on an area of 1-2 μm a
   100 keV electron beam during a short period.
2. General heating by means of the heating holder.

## 3. Results

For nucleation, the local heating by the electron beam was carried out on
the sphalerite CdS films in 2 different ways:
1. The electron beam was concentrated suddenly and dispersed immediately
   afterwards.  A transformed wurtzite single-crystal area with a diameter
   of a few microns appeared (Fig.2 and 3).

2. The electron-beam was concentrated suddenly and left on for a period of a few seconds before being dispersed.  A transformed area consisting of a few wurtzite grains (Fig.4), instead of  a single one, was observed. As before, they appeared immediately after dispersing the beam, and not while it was concentrated. (If the concentrated electron beam was incident on the film during a longer period, no  transformation but evaporation of the film was observed).

By general heating of the films by means of the heating holder up to 550°C, no transformation was observed.  At this temperature, parts of the films began to evaporate (as revealed by small holes created in the film), and domains with a "black-white" contrast, due to faulting on the {111} planes, appeared (Holt et al. 1976).

Fig. 1:   TED pattern of the original single-crystal (001)
          sphalerite CdS film (left). $\vec{B}$ = [001].
Fig. 2:   TED pattern of transformed single-crystal (0001)
          wurtzite area (right). $\vec{B}$ = [0001].

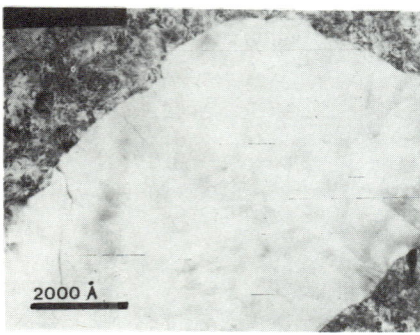 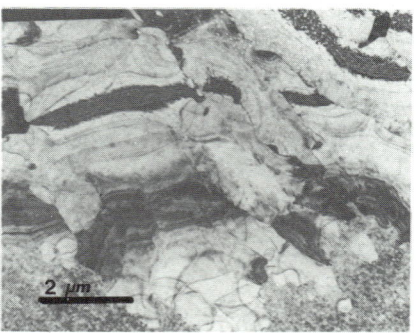

2000 Å

2 µm

Fig. 3:   TEM micrograph of one wurtzite grain, induced by concentrating a
          100keV electron  beam during a short period.
Fig. 4:   TEM micrograph of several wurtzite grains, induced by concentrating a 100keV electron beam during a few seconds.

The size of the wurtzite nuclei could be enlarged if the electron beam was partially concentrated at their boundary (Fig. 5). The same result could be obtained by heating the whole film to 380 - 550°C, using the heating holder only. The growth rate was 0.1-4Å/s, and could be increased up to a few microns per second, if in addition to the general heating the electron beam was partially concentrated at the wurtzite grain boundary. Increasing the temperature to more than 550°C, the lateral growth was almost completely stopped, and did not continue even when the temperature was lowered subsequently. As mentioned before, the "black-white" domains appeared in the sphalerite region at 550°C (Fig. 6).

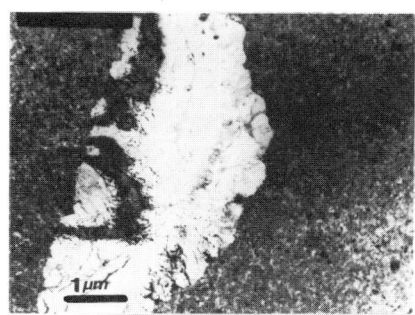 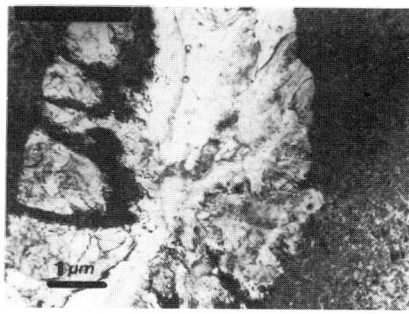

Fig. 5: TEM micrographs showing lateral growth of a wurtzite grain induced by partial concentration of the electron beam on its boundary.

Fig. 6: TEM micrograph of black and white domains contrast in the sphalerite region around the wurtzite area, in a film which was heated above 550°C.

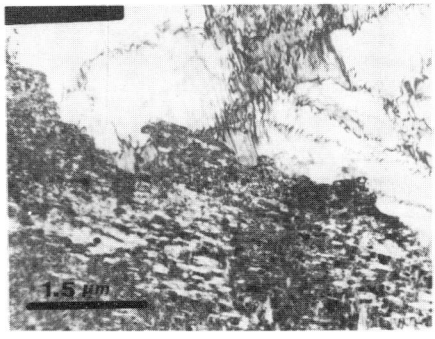

Two different types of microstructures were present in the transformed areas (unrelated to the local heating period, i.e. the number of the wurtzite grains present):
1. Areas that were completely free of stacking faults, and included long and straight dislocations, with a density of $5\text{-}7{\cdot}10^9$ per cm$^2$ (Fig. 7). A definite crystallographic relation between the lattices of the wurtzite phase and the previous sphalerite phase existed: the wurtzite basal plane was parallel to the film surface, i.e. $(0001)_w \parallel (001)_s$, and one of the wurtzite prism-planes made an angle of about 3° with one of the other {100} sphalerite planes (Fig. 8).

2. Areas (most frequently present) that included parallel stripes, due to stacking-faults in the (0001) planes, and small isolated dislocations with a density of $1-3 \cdot 10^{10}$ per $cm^2$ (Fig. 9). For these areas, no crystallographic relation between the lattices of both phases was found.

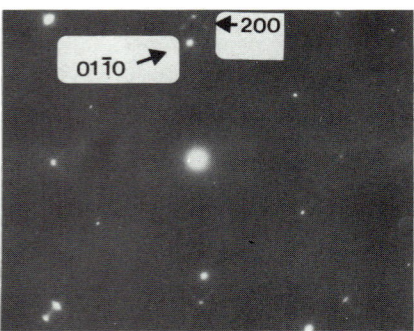

Fig. 7:    TEM micrograph of 1st type wurtzite area showing long dislocations only (left).

Fig. 8:    TED pattern (corresponding to Fig. 7) of both sphalerite region and wurtzite area, showing their relative crystallographic orientation (right). $\vec{B} = [001]$.

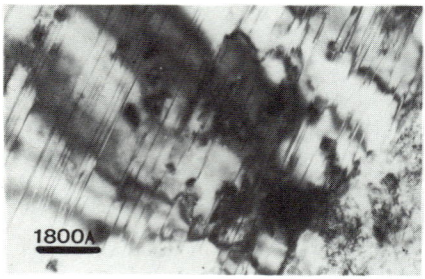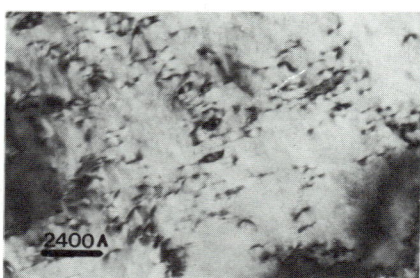

Fig. 9:    TEM micrographs of 2nd type wurtzite area, showing stacking-faults and small dislocations.

## 4. Discussion

The results provided a clear distinction between the two consecutive processes of the phase transformation: nucleation and lateral growth. The creation of the wurtzite nuclei appears to be induced by the strains introduced in the films by the concentrated electron beam. The results showed that the chance for their creation increased with the duration of the concentrated beam. However, they could not be detected until the beam was dispersed and the strains which were released allowed their lateral growth.

## 4.1 Nucleation Mechanism

The nucleation mechanism is basically a mechanism of a local distortion due to the strains. For the wurtzite areas with $(0001)_w \parallel (001)_s$ we consider a model, where the unit cell of sphalerite CdS is represented as one central octahedron and 8 tetrahedrons (composed of cadmium atoms, and 4 sulphur atoms in the centers of 4 tetrahedrons). A distortion of a half-octahedron is considered, in order to create a semi-hexagonal unit-cell, which will serve as a wurtzite nucleus.

The distortion process is related to the displacement of 5 Cd atoms: 4 of them remain in the plane of the half-octahedron after the distortion to the hexagonal basal-plane, and the fifth atom (at the top of the octahedron) is also displaced out of the (100) plane, in order to give a distance of c/2 from the plane of the 4 atoms. Thus, the semi-hexagonal cell includes the same tetrahedron as in the original FCC unit-cell, and the energy of a sulphur atom in this new tetrahedron will be equal to that in its previous tetrahedron. Hopping of a sulphur atom into this new tetrahedron results in the formation of a wurtzite nucleus. This jump is energy-favoured, due to the higher energy of the S atom in the old (now deformed) tetrahedron.

The calculations for the distortion of the half-octahedron to create a semi-hexagonal unit-cell are based on the assumption that the energy for the (small) displacement of each Cd atom to its new site is proportional to $R^2$, where R is the distance between the old-site in the FCC cell and the new-site in the HCP cell. All the displacements can be expressed in terms of the parameters of the original and final unit-cells, and the distortion angle, $\Phi$, in the distortion plane (see Fig. 10). The value of $\Phi$ is obtained by minimizing the sum of the squared displacements with the only variable $\Phi$. The result is $\Phi = 3.16°$, which agrees with the measured one (within the limit of measurement accuracy).

Fig. 10: Schematic representation of the displacements of the Cd atoms: The original sites are marked by medium circles (the upper octahedral site includes a white ring). The new sites are marked by large circles. The displacements of the 5 atoms are marked with arrows.

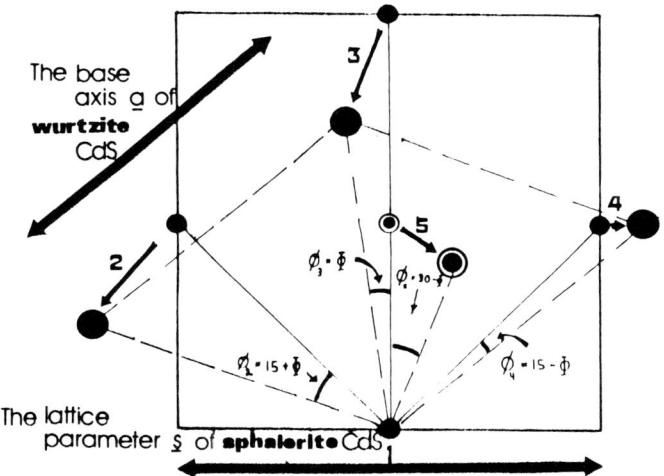

## 4.2  Lateral-growth Mechanism

This is essentially a mechanism of phase transformation of the sphalerite film at the boundary of the wurtzite nuclei; the necessary heat energy was provided by the electron beam and/or the wire resistance.  As the temperature exceeded 550°C and caused faulting on the {111} planes, the available energy was not sufficient for further lateral growth.

Further information about the lateral growth mechanism is provided by the micro-structure of the  transformed areas:

1.  For initial nuclei, which had their c axis perpendicular to the (001) film surface, the transformation process began along the c axis, and only then the initial  wurtzite nuclei began to grow laterally, by phase transformation of successive prism-planes, as  was confirmed by the appearance of straight dislocations.  No stacking-faults of the (0001) planes were introduced because of the small film thickness, since the two free film surfaces prevent large strains during the transformation across the film thickness.

2.  For initial nuclei, which their c axis forming various angles with the normal to the (001) film surface, lateral growth is not isotropic, and the thickness of the film plays no role.  Therefore, the (0001) planes have shown stacking-faults, which play the role of re-adjusting the large mismatch between the two lattices. This in turn gave rise to short dislocations, because they were interrupted by the stacking-faults.

## 5.  Summary

The sphalerite to wurtzite transformation in CdS films is composed of two consecutive processes:  the nucleation, due to the local distortion which is obtained by a localized heat pulse; and the lateral growth in the temperature range 380 - 550°C.

The main features of the transformations are:  the possibility to obtain a transformed single-crystal wurtzite CdS area from a single-crystal sphalerite CdS film by enlarging the first nucleus; and the reduced defect concentration ($10^9$-$10^{10}$ per cm$^2$), which is 3 orders of magnitude less than in the original sphalerite films.

## Acknowledgement

This research was partially supported by Southern California Edison.

## References

Holt D B  1974 Thin Solid Films 24 1.
Holt D B, Abdalla M I, Gejji F H and Wilcox D M  1976 Thin Solid Films 37 91

*Inst. Phys. Conf. Ser. No. 76: Section 6*
*Paper presented at Microsc. Semicond. Mater. Conf., Oxford, 25–27 March 1985*

# InP growth around mesas defined by {111}In and (001) planes

S Mahajan*!, R A Logan**, S N G Chu**, H M Cox*, V G Keramidas* and M A Koza*
*Bell Communications Research, Inc. Murray Hill, NJ 07974
**AT&T Bell Laboratories Murray Hill, NJ 07974

Abstract    InP has been grown around mesas etched in (001) InP substrates by liquid phase and vapor phase epitaxy. The resulting {111}In - and (001) - growth interfaces have been examined by spatially resolved photoluminescence and cross-sectional transmission electron microscopy. In each of the two cases, the radiative quality of different interfaces is comparable and good, and does not appear to be affected by the presence of strain centers along the interfaces grown by liquid phase epitaxy.

## 1. Introduction

Many buried heterostructure laser designs appear promising for light wave communication systems (Li 1983). In these device structures, the high dielectric, low bandgap active region in which the optical field and injected carriers are confined is imbedded in a low dielectric, high bandgap medium to form a low current threshold, high performance laser. This structure is usually achieved by a two-step liquid phase epitaxial (LPE) growth. In the first LPE cycle, a series of epitaxial layers of different compositions and conductivity-types are grown. This is followed by etching to delineate mesas and subsequent regrowth in the second cycle. The mesas are delineated by {111}In stop-etch side wall planes and have a narrow lateral dimension (~1μm) to ensure lowest order mode operation of the laser. While these procedures reproducibly yield high quality lasers in the (Al,Ga,As) system (Logan 1982), their extension to the (In,Ga,As,P) system results in a low laser yield with many short circuited devices.

In order to understand the origin of short-circuited devices, Logan et al. (1983) have simulated the device situation by growing InP layers around mesas delineated in the (001) InP substrates. They have shown that when the standard cleaning procedures are used, interfaces between the {111}In planes of the mesas and regrowths are non-radiative as revealed by spatially resolved photoluminescence (SRPL). However, isotropic etching of ~1000A InP, achieved by anodization and stripping of the grown oxide prior to regrowth, promotes good uniform growth resulting in radiative interfaces (Logan, Temkin, Merritt and Mahajan 1984). This surface cleaning procedure also permits the formation of high performance double channel buried heterostructure lasers (Logan, Temkin, Merritt and Mahajan 1984).

---

!Permanent Address: Department of Metallurgical Engineering and Materials Science, Carnegie-Mellon University, Pittsburgh, PA 15213

In the present study, interfaces between the {111}In stop-etch planes
after anodization and oxide stripping and InP growths obtained by LPE
have been examined by transmission electron microscopy (TEM). In
addition, the characteristics of these interfaces have been compared
with those obtained by the vapor phase epitaxy (VPE) of InP using the
chloride process (Koza and Cox 1984).

## 2. Experimental Details

Narrow - as well as wide-waisted mesas, defined by the {111}In stop-etch
plane side-walls, were formed by masking polished (001) InP substrates
along the [110] and [1$\bar{1}$0] directions with thermally deposited SiO2 and
etching in 1% Br-methanol solution. For the growth of InP by LPE, as-
etched mesas were anodized (Studna and Gualtieri 1981) and the resulting
oxide layers were removed prior to growth. Subsequently, InP was grown
in a hydrogen ambient in a graphite boat using conventional two-phase
epitaxy. The melt was equilibrated at 655°C, and cooled at 0.7°/min,
with the etched substrate placed under the melt from 650 to 640°C so as
to grow several micrometers of InP around the mesas with a minimal
melt-back of the original surface.

During the preheating prior to growth, the substrate was protected by a
GaAs cover piece (Kinoshita, Okuda and Uematsu 1983) which was cleaned
by anodization (Logan, Schwartz and Sundburg 1973). While this may not
be as effective as a phosphorus overpressure to minimize substrate
degradation it was found to be adequate and the simple cleaning procedure
between usage assured reproducibility.

For the growth of InP by VPE, mesas were in-situ etched for about 3 mins
and subsequently InP was deposited. During the deposition, substrates
were maintained at 680°C and the In to P flux ratio was ~ 3.

To ascertain the radiative quality of the resulting {111}In and (001)
interfaces, small pieces were cleaved from different wafers and examined
using SRPL. The facet luminescence was excited with a weakly focused
Ar ion laser light and observed with an infrared microscope equipped with
an X100 objective and an S-1 image converter.

For cross-sectional TEM, small pieces were cleaved from the as-grown
wafers and subsequently thinned using the procedure developed by Chu
and Sheng (1984). Thinned samples were examined in a JEM 200 microscope
operating at 200keV.

## 3. Results

Figures 1(a) and (b) show, respectively, SRPL images of InP growths
obtained around the wide-waisted mesas by LPE and VPE. The short inclined
segments are the {111}In - growth interfaces, whereas the horizontal
portions represent the (001) - growth interfaces. It is clear that the
radiative quality of the two types of interfaces is comparable in each
of the two cases. Similar results were obtained for the narrow-waisted
mesas.

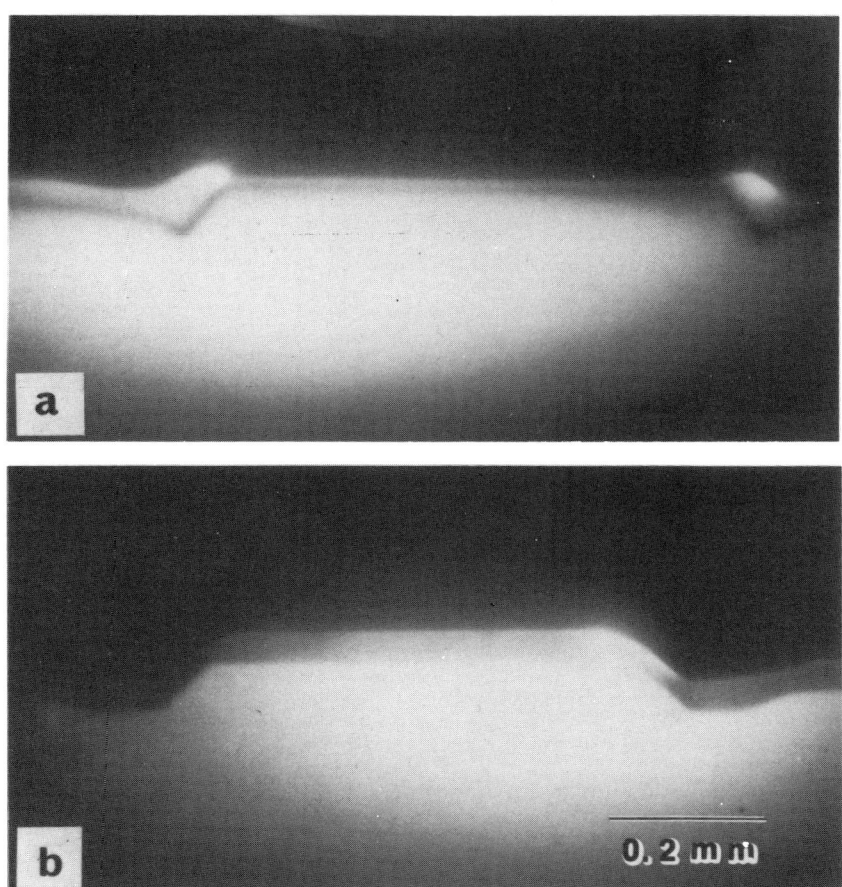

Figure 1.   Spatially resolved photoluminescence images of InP growths
            obtained around wide-waisted mesas by (a) liquid phase epitaxy
            and (b) vapor phase epitaxy.  The short inclined segments and
            horizontal portions are, respectively, {111}In - and (001) -
            growth interfaces.

Shown in Fig.2(a) is an edge-on image of a portion of the {111}In - LPE
growth interface obtained using the 222 reflection which is normal to
the interface.  It is apparent that structural features exhibiting
black-white contrast are present in the boundary similar features are
seen along the (001) - LPE growth interface imaged using the 004
reflection which is also normal to the interface, Fig. 2(b).  When these
structural features were examined with reflections lying in the interface,
the black-white contrast was not observed, implying that the strain
associated with the microstructural features is, in each of the two cases,
perpendicular to the interface.

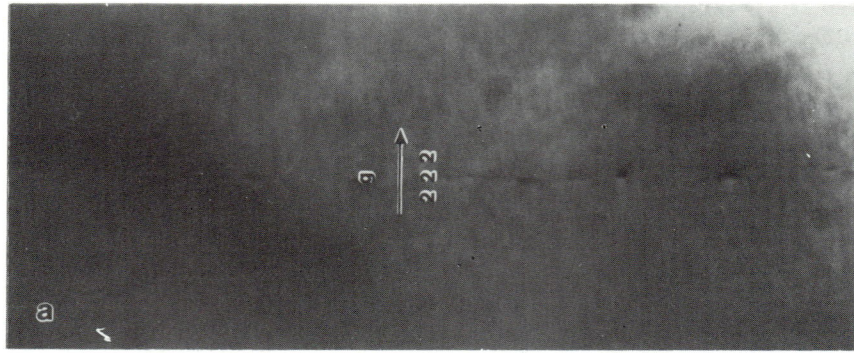

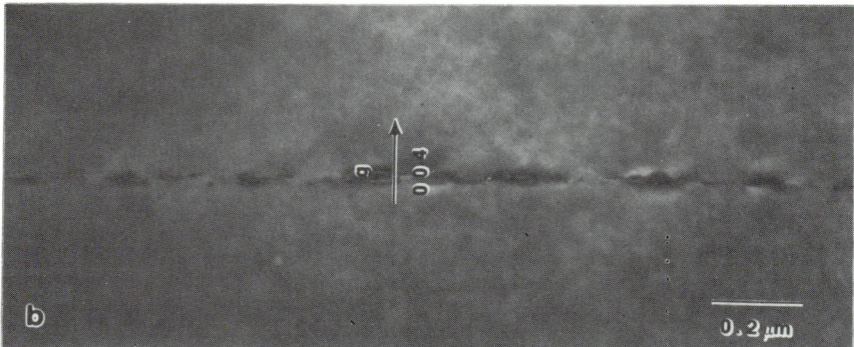

Figure 2.   An electron micrograph showing (a) {111}In - and (b) (001) -
            growth interfaces obtained by liquid phase epitaxy.  Note
            the presence of strain centers along the two interfaces.


Figure 3(a) shows the {111}In - and (001) - VPE growth interfaces,
whereas Fig. 3(b) is an enlargement of a portion of the {111}In-growth
interface.  These interfaces are quite diffuse and features exhibiting
the black-white contrast are not seen along the boundaries.

## 4. Discussion

Two interesting observations emerge from the preceding study.  First, the
radiative interfaces between the {111}In planes of mesas etched in (001)
InP substrates and InP growth can be obtained by LPE and VPE.  An
additional step in the cleaning procedure is involved in the case of LPE.
This entails anodization of the as-etched mesas, followed by stripping
of the resulting oxide prior to growth.  On the other hand, the VPE growth
involving in-situ etching yields radiative interfaces without using
anodization.  Furthermore, it is observed that, in the absence of
anodization, it is extremely difficult to grow InP on the {111}In surfaces
of the mesas by LPE (Logan, Temkin, Merritt and Mahajan 1984).  These
results suggest that the as-etched {111}In surfaces of the mesas are not
suitable for the growth of InP by LPE.

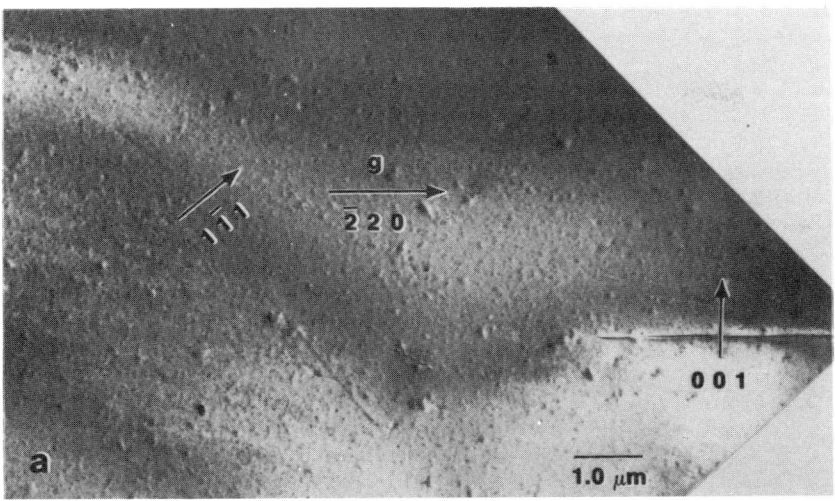

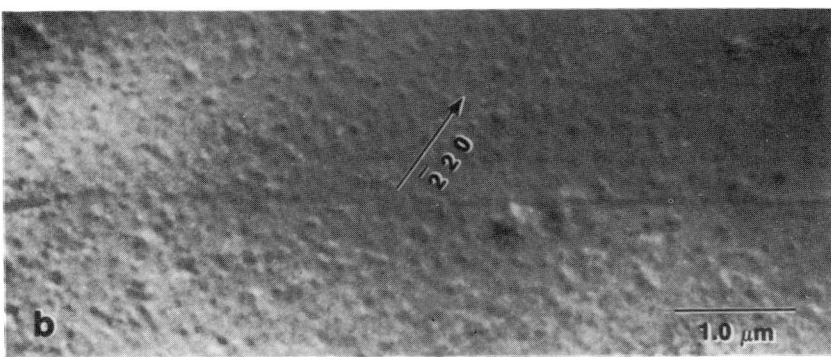

Figure 3.   An electron micrograph showing (a) {111}In - and (001) - growth
interfaces obtained by vapor phase epitaxy, whereas (b) shows
an enlargement of a portion of the {111}In-growth interface.

The difficulty associated with the growth of InP on the {111}In surfaces
of etched mesas by LPE may be due to either non-stoichiometry of the
surface or some ubiquitous contaminant.  For the growth to occur on the
{111}In plane, a layer of phosphorus atoms should deposit-first,
followed by a layer of In atoms.  Since the layers are grown from an
In-rich melt containing a very small amount of P, the activity of P at
the {111}In surface should be extremely small.  It is then conceivable
that a slight amount of the stoichiometric disturbance or contamination
is sufficient to prevent the uniform deposition of the phosphorus layer.
It should be noted that when regrowth is preceded by a severe melt-etch
with an unsaturated In-melt, one obtains a clean radiative regrowth
interface.

In the case of VPE, the situation is quite different. Not only the flux of P in the incoming gas stream is fairly high (In:P::3:1), but also the ability to in-situ etch in a controlled manner facilitates growth.

Second, the strain centers are observed along the {111}In - and (001) - LPE growth interfaces. An obvious question concerns their origin. Since anodized surfaces are rough (Logan 1984), it is easy to visualize that In-rich melt droplets could be trapped at the interfaces, thus giving rise to the observed contrast.

Acknowledgments

The authors would like to thank J.H. Wernick for fruitful discussions and gratefully acknowledge the technical support of H.G. White and F.R. Merritt.

References

Chu S N G and Sheng T T 1984 J. Electrochem. Soc., 131 2663.
Kinoshita J, Okuda H and Uematsuy 1983 Elect. Lett. 19 215.
Koza M A and Cox H M 1984 Unpublished Research, Bell Communications
    Research Inc., Murray Hill, NJ 07974.
Li Tingyi 1983 IEEE J. Selected Areas in Communication 1 356.
Logan R A 1982 Proceedings of SPIE 323 150.
Logan R A 1984 Unpublished Research, AT&T Bell Laboratories, Murray
    Hill, NJ 07974.
Logan R A, Schwartz B and Sundburg W J 1973 J. Electrochem. Soc. 120
    1385.
Logan R A, Henry C H, Merritt F R and Mahajan S 1983 J. Appl. Phys.
    54 5462.
Logan R A, Temkin H, Merritt F R and Mahajan S 1984 Appl. Phys. Letts.
    45 1275.
Studna A A and Gualtieri G J 1981 Appl. Phys. Lett. 39 965.

*Inst. Phys. Conf. Ser. No. 76: Section 6*
*Paper presented at Microsc. Semicond. Mater. Conf., Oxford, 25–27 March 1985*

# TEM and STEM study of the microstructure of ternary and quaternary III−V epitaxial layers

F Glas*, P Hénoc* and H Launois**

\* CNET Laboratoire de Bagneux
\*\* Laboratoire de Microstructures et de Microélectronique - CNRS
196 rue de Paris - 92220 Bagneux - FRANCE

Abstract    We report a TEM and STEM study of ternary and quaternary III-V semiconductors epitaxial layers. The fine granular contrast (scale ~ 15 nm) characterizing micrographs of such samples (InGaAsP, InGaAs, InGaP) is shown to be strain-induced rather than structure factor-induced. Electron diffraction reveals planes of diffuse intensity in the reciprocal space, indicating correlated atomic displacements along the [110] directions ; no evidence of a three-dimensional chalcopyrite-type ordering of the III atoms in InGaAs is found.

## 1. Introduction

It has been predicted that ternary and quaternary bulk III-V semiconductor alloys should be unstable with respect to composition fluctuations, provided the constituent binaries (e.g. GaAs and InAs for the InGaAs ternary) have different lattice parameters (de Crémoux et al 1981, Stringfellow 1982a). Subsequently Stringfellow (1982b) argued that the strains between regions differing in composition arising from this lattice mismatch would inhibit the decomposition. However, it has been established by STEM microanalysis that, in a certain composition and temperature range, InGaAsP and InGaAs layers grown by Liquid Phase Epitaxy (LPE) lattice-matched on (001) InP, exhibit composition fluctuations (Hénoc et al 1981, Glas et al 1982). These are associated with contrast elements parallel to [100] and [010] directions, visible mainly in dark field (DF) images. The fluctuations and contrast have a quasiperiod of 100 to 150 nm. Another prominent feature of DF micrographs of such layers is a finer granular contrast of size ~ 15 nm. Following the predictions of alloy instability and its experimental discovery in LPE layers, this contrast has been attributed by several authors (Gowers 1983, Booker 1984, Mahajan et al 1984, Mahajan 1984) to composition fluctuations, but without any direct proof. In this paper, we report a detailed analysis of this fine contrast and discuss its possible origin. We then consider the information brought by electron diffraction to the knowledge of local order in such samples.

## 2. The Fine Scale Contrast

Using a VG HB5 STEM, we observed the fine scale contrast (Fig. 1) in InGaAsP and InGaAs epitaxial layers grown on (001) InP, and in InGaP grown on (001) GaAs. We checked that the existence of this contrast does not depend on the method of epitaxy used : it is present in InGaAsP sam-

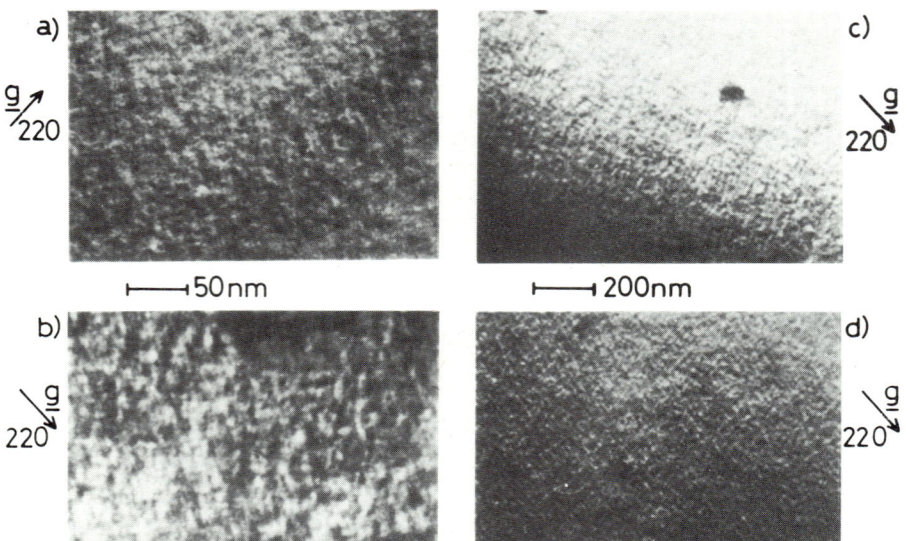

Fig. 1 220 dark field STEM images of (a) LPE and (b) chlorides VPE InGaAsP ($\lambda$ = 1.3 $\mu$m) and of (c) LPE and (d) MBE $In_{.53}Ga_{.47}As$.

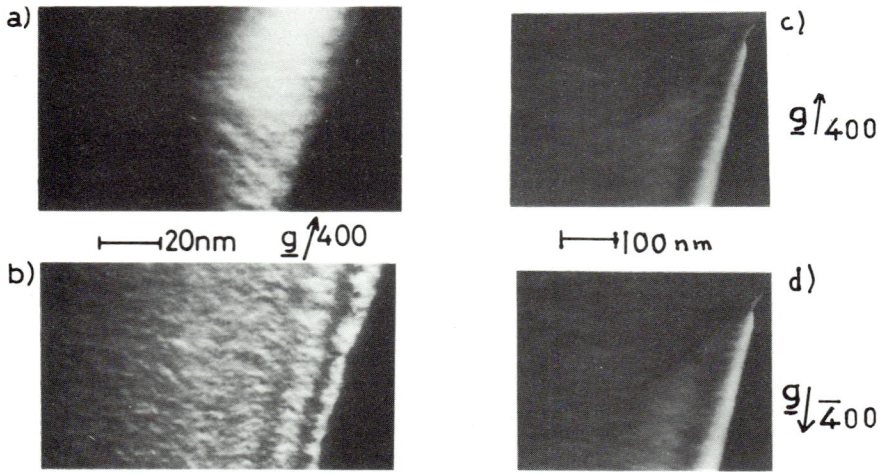

Fig. 2 (a,b,c,d) Dark field STEM images of a thin 45° wedge in $In_{.53}Ga_{.47}As$ (the thickness at a given distance of the edge is equal to that distance). (a) $\underline{g}$ = 400, s=o (b) $\underline{g}$ = 400, $s \sim 4/\xi_{400}$ (c) $\underline{g}$ = $400^{-}$, s=o (d) $\underline{g}$ = $400^{-}$, s=o.

(e) Variation with reduced thickness $t/\xi_{400}$ of the fine contrast measured in 400 dark field images of $In_{.53}Ga_{.47}As$. $Ig^{M}$ and $Ig^{m}$ are the maximum and minimum diffracted intensities.

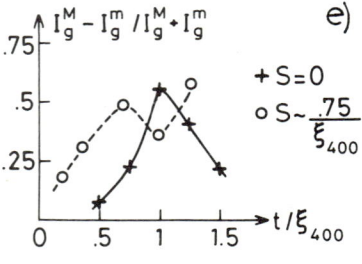

ples grown by LPE (Fig. 1(a)) (with emission wavelengths $\lambda$ ranging from 1.15 to 1.38 $\mu$m), by chlorides Vapor Phase Epitaxy (VPE) (Fig. 1(b)), and by Organo-Metallic Chemical Vapor Deposition, and in $In_{0.53}Ga_{0.47}As$ grown by LPE (Fig. 1(c)) or Molecular Beam Epitaxy (MBE) (Fig. 1(d)). For LPE samples, it exists whether or not the composition and temperature lie in the immiscibility domain defined by the presence of the $\sim$ 150 nm composition fluctuation (Glas et al 1982). However, this contrast does not appear in binaries (GaAs, InP), or in GaAlAs, whose constituent binaries GaAs and AlAs are nearly lattice-matched.

We studied qualitatively and quantitatively this contrast in two-beam DF micrographs of epitaxial $In_{.53}Ga_{.47}As$, varying the diffraction vector $\underline{g}$, the deviation parameter s and the specimen thickness t. These last two quantities are easily monitored using thickness fringes on 45° wedges originating from the accidental cleavage along (101) planes of specimens thinned from the (001) oriented substrate (Fig. 2).

(1) In g=220 images, the $\sim$ 15 nm contrast displays in places a preferential orientation along $\lfloor 100 \rfloor$ and $\lfloor 010 \rfloor$ (Fig. 1(b)) ; in g=400 images, it is elongated perpendicularly to $\underline{g}$, keeping a $\sim$15 nm scale along $\underline{g}$ (Fig. 2(a)) ; it is very weak in g=200 images.

(2) In 400 images taken with s=o (Fig. 2(a)), the contrast is very weak if $t < 0.75\ \xi_{400}$ ($\xi_{400} \sim$ 67 nm is the 400 extinction distance), but is visible up to thicknesses of several 100 nm ; if s is increased, the contrast becomes visible at smaller thicknesses (Fig. 2(a,b,e)).

(3) We could not correlate most contrast features between micrographs taken with different values of s. Indeed, if s is increased, the characteristic scale of the contrast diminishes, and fringe-like contrast of various spacings (observed down to $\sim$2 nm) appear (Fig. 2(b)).

(4) For s=o, the contrast is reversed between 400 and $\overline{4}00$ images (Fig. 2(c,d)).

(5) For the same g=400, the contrast is reversed between s and -s. This is shown in Figure 3, where both sides of the central fringe of a bend contour are successively brought by tilting to the same area of the specimen.

Since the relative changes of the structure factors with In content x(for an alloy where III atoms are randomly distributed on their sublattice)

are $\dfrac{1}{F_{200}}\ \dfrac{dF_{200}}{dx}\bigg|_{x=.53} \approx 1.36$ and $\dfrac{1}{F_{400}}\ \dfrac{dF_{400}}{dx}\bigg|_{x=.53} \approx .36$, the difference in

amplitude between 200 and 400 contrasts (result (1)) implies that these are not images of domains whose difference in composition would directly produce structure factor variations. Mikkelsen Jr et al (1983) considered the possible existence in $In_xGa_{1-x}As$ of pseudo-chalcopyrite domains, where In and Ga atoms would be ordered on their sublattice, as Cu and Fe are in chalcopyrite. However, result (5) excludes that the fine contrast is an image of individual domains of orientation variants of such a superstructure, or any other, since the resulting image would not be altered by changing s to -s. We thus conclude that the granular contrast is not a direct image of three-dimensional microdomains.

Gowers (1983) suggested that such a contrast arises from composition fluctuations : these would induce a periodical modulation $R=bsin(\theta+2\pi z/L)$ of the lattice spacings through the thickness of the crystal, $\theta$ being a phase angle whose variation between neighbouring columns produces contrast. Gowers tested his hypothesis by solving the two-beam dynamical

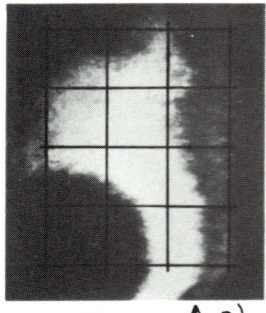 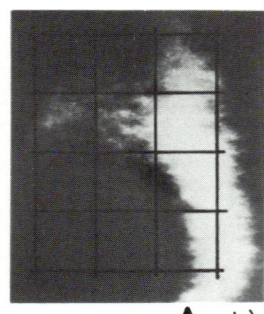

Fig.3 400 dark field STEM images of a bend contour in the same area of an $In._{53}Ga._{47}As$ specimen. The area above the arrow is set at $S\sim1/2\xi_{400}$ in (a) and $S\sim-1/2\xi_{400}$ in (b).The grid has the same position with respect to the specimen in (a) and (b).

⊢——⊣20nm   ↑ a) $\underline{g}$↓ 400            ↑ b)

diffraction equations with a resulting modulation $\Delta s(z)=\underline{g}.dR/dz$ of the deviation parameter s through each column, for a mean value $\overline{s}=o$. We have extended his calculations for $s\neq o$. Our calculations agree with results (2) and (4). They agree with (3) only if we suppose that the modulation has components with different spatial periodicities in the plane of the layer. A simple change through the foil thickness of only the structure factor and extinction distance (which again could be induced either by composition fluctuations or by ordering variations) at constant s contradicts result (5). Hence we conclude that lattice bending (variation of s) is the main cause of the contrast. Moreover, a variation of strains with depth is compatible with our results. However, a fluctuation of composition on the scale of the contrast is no necessarily required. Such a fluctuation can not be justified by the existence of the 150 nm quasiperiodic fluctuation, since the latter has been up to now observed only for precise conditions of one particular epitaxy method (LPE) (Glas et al 1982), whereas we show the former to exist in a much wider range of samples. For instance, it is certain that in such non stoichiometric compounds, the distribution of the III (and V for quaternaries) atoms on the scale of the unit cell will deviate from the average. This could induce atomic displacements, and the interaction of these strain fields could create the fine contrast. Strains could also exist at the interfaces between hypothetic domains, each being a variant of a given superstructure (with atoms ordered on their sublattice), or at interfaces between such domains and a disordered matrix. Hence, the precise origin of the granular contrast remains to be determined.

3. The Search for a Superstructure

The hypothesis of local ordering of III atoms in $In._{53}Ga._{47}As$ was tested by Transmission Electron Diffraction (TED), using a Siemens Elmiskop. We considered two possibilities :
- for a pseudo-chalcopyrite superstructure suggested by Mikkelsen Jr et al (1983) after their EXAFS experiments indicated that As atoms are considerably displaced from their virtual lattice positions, the perio-dicity of the crystal is doubled in one of the three [100] directions ('c-axis'). If this one lies in the plane of the layer, superstructure spots of the type $(0, \overline{3}/2, 1)$ or $(1, \overline{1}/2, 0)$ (indexes refer to the zinc-blende lattice) should be visible in [123] zone axis diffraction patterns.
If the c-axis is parallel to the [001] growth axis, extra spots such as (0,3,1/2) should appear in [116] zone axis patterns. We could not find

any of these extra spots (Fig. 4(a, b)) using a diffracting area of size ~ 1 μm.
- for a pseudo-CuAu I superstructure, defined by a stacking along any of the [100] directions of alternating planes of Ga and In, superstructure spots such as (0,1,1) should appear in [100] type zone axis patterns. We did not observe them in the ⌊001⌋ pattern (Fig. 4(c)).
Since the diffracting area is much larger than the scale of the fine contrast, we considered the possibility of destructive interferences between waves diffracted by antiphase domains of these hypothetic super-structures. We thus performed STEM microdiffraction experiments with a ~1 nm diameter probe on 10 to 100 nm thick areas : it was again impossible to observe any of the expected extra spots. We conclude that there is no three-dimensional pseudo-chalcopyrite or [001] pseudo-CuAu I ordering of III atoms in epitaxial $In_{.53}Ga_{.47}As$. This is however not incompatible with the atomic displacements revealed by EXAFS, provided there is no volume correlations between them.

## 4. The Planes of Diffuse Intensity

However, InGaAsP and $In_{.53}Ga_{.47}As$ lattice-matched to InP show evidence of local interatomic correlations. TED patterns (Figs. 5(a,b), 4(a)) of such samples reveal lines of diffuse intensity. In [001] zone axis patterns, these are segments parallel to (110) and (1$\bar{1}$0) directions, but passing some distance away from all the (2h,2h,0) spots (h integer ≠o). Fig. 5(c) indicates schematically the positions of the most intense of these segments. By tilting the specimen towards other zone axes (e.g. [111] in Fig. 5(b)), we followed these lines in the reciprocal space : they are the traces on the Ewald sphere of portions of planes of diffuse intensity perpendicular to all (220) directions. The diffuse intensity varies along and across the planes, with maxima located near but away from the diffraction spots. We attribute the reported elongation of the (4,0,0) spot (Mahajan 1984) to the intersection in its vicinity of planes perpendicular to (220), (2$\bar{2}$0), (202) and (20$\bar{2}$). This is confirmed by the persistence of a strong diffuse intensity even when the spot itself is very faint (Fig. 4(a)). Such planes have already been observed by X ray diffraction (C. Gors et al).
Microdensitometric traces of these diffuse lines show that :
(i) the distance between the (2h,2h,0) spot and the neighbouring lines is about 3 to 5% of $g_{220}$, but does not increase with the order of the spot (ie between 220, 440, 660,...)

Fig.4 Electron diffraction patterns of an $In_{.53}Ga_{.47}As$ sample. Zone axes are (a) ⌊123⌋ (b) ⌊116⌋ (c) ⌊001⌋. Arrows indicate the expected positions of some (a,b) pseudo-chalopyrite and (c) pseudo CuAu I super-structure spots.

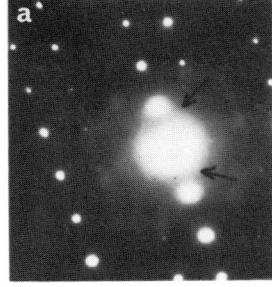 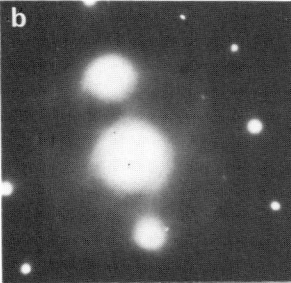 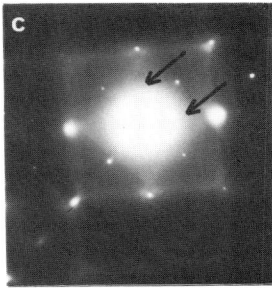

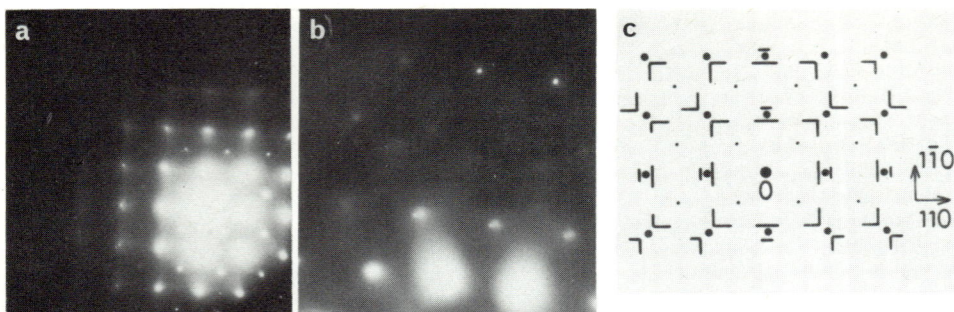

Fig.5 (a) [001] and (b) [111] zone axis electron diffraction patterns of an In.₅₃Ga.₄₇As sample, showing lines of diffuse intensity parallel to (110) and (1̄10) directions. (c) Locus in the [001] reciprocal plane of the most intense portions of these lines (0 is the transmitted beam).

(ii) the FWHM of the lines increases with this order (it is about $|g_{220}|/4$ for 220 and $|g_{220}|/3$ for 440)
(iii) there is more diffuse intensity in the portions of planes located on the low angle side of a diffraction spot than in those located on the high angle side of the same spot
(iv) along a diffuse line, the intensity is maximum close to the diffraction spots
Such planes exist in InGaAsP and In.₅₃Ga.₄₇As matched to InP irrespective of the method of epitaxy used and of the presence of the 150 nm composition fluctuations. They are absent from binaries and GaAlAs, where phonon related planes exist, but pass exactly on the diffraction spots. There is no plane passing through or near the (0,0,0) spot. We thus explain these planes by the correlation of atomic displacements along [110] directions. (i) suggests that the resulting atomic chains should exhibit a superperiodicity and be several nm long, but the large width of the planes (ii) indicates a distribution of such superperiods. (iii) might be related to the larger ionic radius and scattering factor of In compared to Ga. (iv) indicates the existence of some correlations in the directions perpendicular to the chains. Such correlations might result from the growth process, which is believed to proceed preferentially along the [110] oriented atomic steps. However, a detailed model remains to be found.

References

Booker GR 1984 Inst. Phys. Conf. Ser. N°68 417
De Crémoux B, Hirtz and Ricciardi P 1981 Inst. Phys. Conf. Ser. N°56 115
Glas F, Treacy MMJ, Quillec M and Launois H 1982 J. Physique 43 C5-11
Gors C and Comès R Private communication
Gowers JP 1983 Appl. Phys. A 31 23
Hénoc P, Izrael A, Quillec M and Launois H 1982 Appl. Phys. Lett. 40 963
Mahajan 1984 Inst. Phys. Conf. Ser. N°67 259
Mahajan S, Dutt BV, Temkin H, Cava RJ and Bonner WA 1984 J. Crystal Growth 68 589
Mikkelsen Jr JC and Boyce JB 1983 Phys. Rev. B 28 7130
Stringfellow G B 1982a J. Crystal Growth 58 194
Stringfellow G B 1982b J. Electron. Mat. II 903

Acknowledgements : The authors thank S. Slempkes and M. Allovon for the samples, and C. Gors and R. Comès for fruitful discussions.

Inst. Phys. Conf. Ser. No. 76: Section 6
Paper presented at Microsc. Semicond. Mater. Conf., Oxford, 25–27 March 1985

# TEM and TED studies of alloy clustering in GaInAsP, GaInAs and GaInP epitaxial layers

A G Norman and G R Booker

Department of Metallurgy & Science of Materials,
University of Oxford, Parks Road, Oxford OX1 3PH

Abstract    Transmission electron microscopy (TEM) and transmission
electron diffraction (TED) have been used to study GaInAsP alloy semiconductors.  TEM diffraction contrast effects and satellite spots in
TED patterns characteristic of alloy clustering as expected in spinodally decomposed alloys have been observed in LPE,VPE, MOCVD and MBE
grown layers.  Examination of a <100> orientation cross-sectional
sample of a LPE GaInAsP alloy has revealed significant differences in
clustering behaviour between the <100> directions parallel to the
epitaxial layer interface and the [001] growth direction.

## 1. Introduction

GaInAsP alloy semiconductors are of increasing importance for a wide range
of semiconductor devices, e.g. infra-red emitters and detectors, field
effect transistors, etc. (Pearsall 1982).  Thermodynamic analysis of the
phase diagram of this system (Foster and Woods 1971, de Cremoux et al
1980, Stringfellow 1982 and Onabe 1982) has indicated the presence of a
miscibility gap and the possibility of alloy clustering occurring by
spinodal decomposition at temperatures close to those used in epitaxial
growth.  Calculations by de Cremoux et al (1980) predicted the behaviour
shown in Figure 1 for the GaInAsP system.  The continuous curves are iso-
therms representing the boundary of the unstable (inside of curves) and
stable (outside of curves) regions of the phase diagram.  The lines PQ and
RS indicate alloy compositions lattice-matched to InP and GaAs substrates
respectively.

TEM investigations by a number of authors of liquid phase epitaxial (LPE)
grown GaInAsP alloy semiconductors have revealed diffraction contrast
effects similar to those observed in spinodally decomposed alloys.  A
coarse tweed-like structure (~150nm scale) associated with compositional
modulations occurring along the [100] and [010] directions has been
observed in alloys grown inside the unstable region of the phase diagram
at the growth temperature (Henoc et al 1982, Glas et al 1982, Launois et
al 1983, Mahajan et al 1983 and 1984, Ueda et al 1984, Norman and Booker
1985).  This coarse tweed-like structure has been investigated in detail
by Treacy et al (1985) who emphasised that the primary source of TEM contrast is due to near surface elastic relaxation of shear stresses present
in the spinodally decomposed material.  A fine granular structure (~15nm
scale) has been observed in LPE samples grown both inside and outside the
unstable region of the phase diagram at the growth temperature (Henoc et
al 1982, Glas et al 1982, Launois et al 1983, Gowers 1983, Mahajan et al
1983 and 1984, Ueda et al 1984, Norman and Booker 1985) and has also been

observed in molecular beam epitaxial (MBE) material (Gowers 1983). This has been attributed to alloy clustering occurring by spinodal decomposition (Gowers 1983, Mahajan et al 1983 and 1984, Ueda et al 1984, Norman and Booker 1985). Recently Norman and Booker (1985) have reported the presence of satellite spots in [001] TED patterns of LPE GaInAsP layers indicating a periodic variation in lattice parameter along the [100] and [010] directions of a wavelength of ~12nm correlating with the scale of the fine granular structure.

In the present paper new observations are reported on spinodal decomposition related effects in TEM and TED studies of GaInAsP, GaInAs and GaInP layers grown by LPE, MBE, metal organic chemical vapour deposition (MOCVD) and vapour phase epitaxy (VPE). All samples were thinned for TEM examinations by standard techniques and examined at 100kV using a Philips EM300. All TEM micrographs were taken using dark field conditions.

## 2. Results and Discussion

### 2.1 LPE GaInAsP Layers on (001) InP Substrates

TEM and TED examinations have been made on two types of LPE GaInAsP layers grown at temperatures between 550 and 660°C. Type A layers had a composition (point A, Figure 1) lying inside the unstable region of the phase diagram at the growth temperature whilst type B layers had a composition lying outside (point B, Figure 1). A coarse tweed-like structure (~150nm) was visible along the [100] and [010] directions in type A layers using a <220> type reflection, but not in type B layers as shown in Figure 2(a) and (b). When viewed in <400> type reflections, e.g. Figure 2(c), only the contrast bands perpendicular to the g vector were visible indicating strains present along the <100> type directions. A fine granular structure (~15nm scale) was visible in both type A and type B layers and showed a preferred orientation when imaged in <400> type reflections as shown in Figure 2(c) and (d). The fine granular structure was found to be more pronounced in type A layers than in type B layers, e.g. compare Figure 2(a) and (b). TED examinations of type A and type B layers were carried out at the [001] pole. Type A layers, Figure 3, were found to possess satellite spots associated with all the main spots except for the central spot. The occurrence and behaviour of the satellite spots is characteristic of periodic variations in the lattice parameter along the [100] and [010] directions (Norman and Booker 1985). The modulation wavelength of these periodic variations was deduced from the TED patterns to be 10-15nm for both of the <100> directions, thus correlating well with the observed TEM fine granular structure. Analogous TED patterns from type B layers revealed either no satellite spots or only faint satellite spots, e.g. Figure 4.

We believe the differences between type A and type B layers can be explained as follows. The coarse tweed-like structure in type A layers has been attributed (Launois et al 1983) to an interfacial spinodal decomposition mechanism occurring between the liquid and the growing solid. The fine granular structure is believed to arise from spinodal decomposition occurring in the solid phase. Type A layers can decompose during layer growth as well as during cooling down from the growth temperature. Thus in type A layers a coarse tweed-like structure and a fine granular structure can develop during layer growth and subsequent cooling and a significant compositional modulation amplitude can result giving rise to strong TEM contrast and satellite spots in TED patterns. In Type B layers

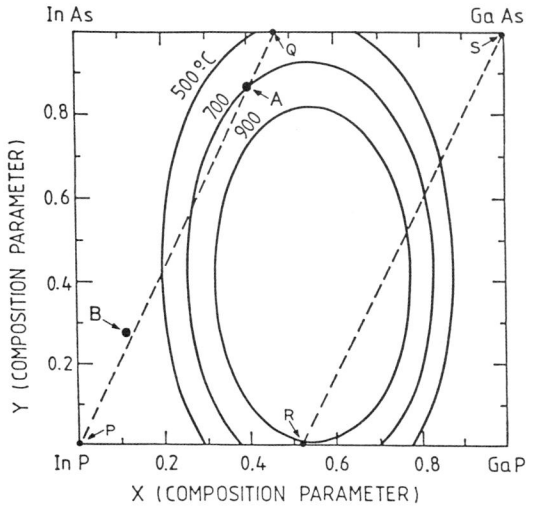

Fig.1 Stable and unstable regions
for GaInAsP calculated by
de Cremoux et al (1980).

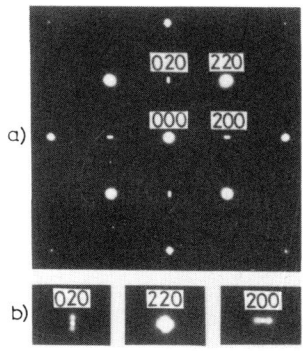

Fig.3 TED pattern of
[001] pole from type A
LPE GaInAsP layer show-
ing satellite spots.
(a) low mag. (b) enlarge-
ment of spots illustrat-
ing form of satellites.

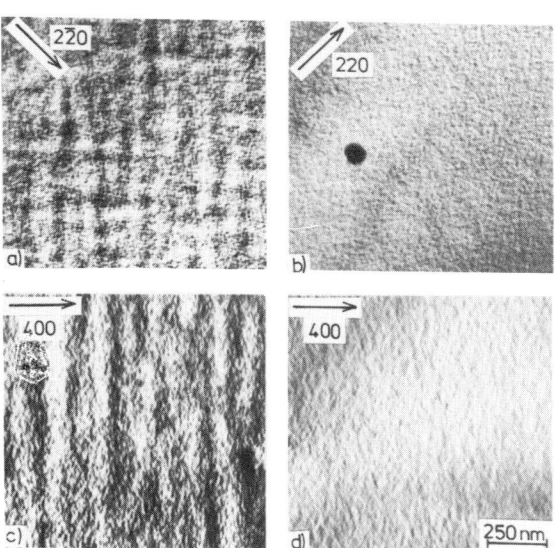

Fig.2 TEM contrast observed in
type A, (a) and (c), and type B,
(b) and (d), LPE GaInAsP layers.

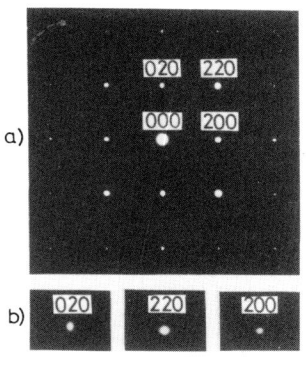

Fig.4 TED pattern of
[001] pole from type B
LPE GaInAsP layer show-
ing no satellite spots.
(a) low mag. (b) enlarge-
ment of spots.

no decomposition can take place during layer growth, hence the absence of a coarse tweed-like structure, but on cooling decomposition can occur although at much lower temperatures. In this case only a small composition modulation amplitude results giving rise to a weak TEM contrast and very faint or no satellite spots.

## 2.2  3-Dimensional Nature of Spinodal Decomposition in LPE GaInAsP Layers

This has been investigated in a type A LPE GaInAsP layer by examining plan-view and <100> cross-sectional specimens. The appearance of the spinodal decomposition in the <100> directions parallel to the epitaxial layer interface is illustrated above in Figure 2(a) and (c). The appearance of the spinodal decomposition in a [010] cross-section of the same layer is shown in Figure 5. The TEM contrast obtained using the [400] and [004] reflections differs markedly as can be seen in Figure 5(a) and (b). On using the [400] reflection parallel to the interface a linear contrast similar to that observed in plan-view samples is visible, orientated perpendicular to the g vector, indicating compositional variations occurring in the <100> directions parallel to the interface as already demonstrated by the plan-view results. However, on using the [004] reflection no such contrast is observed indicating no appreciable variations in composition in the growth direction. Figure 5(c) illustrates the contrast obtained using a <220> reflection which is similar to that observed in the [400] reflection.

TED results, Figure 5(d), show satellite spots associated with the 200 spot but not with the 002 spot, thus supporting the TEM contrast results in that no appreciable decomposition occurs in the [001] growth direction. In bulk samples spinodal decomposition would be expected to occur equally along the three elastically soft directions (Cahn 1962) which in GaInAsP are the <100> directions. In the case of a thin epitaxial layer not perfectly lattice-matched to the substrate, stresses will exist in the <100> directions parallel to the interface but not in the [001] growth direction due to the relaxation of stress in this direction by tetragonal distortion. Stress is known to affect the morphology of spinodal decomposition (Cahn 1968) and the presence of misfit stresses parallel to the epitaxial layer interface could explain the observed morphology of decomposition.

## 2.3  LPE, MBE, MOCVD and VPE GaInAs and VPE GaInP

As shown in Figure 6(a) a fine granular structure (10-15nm), but no coarse tweed-like structure, was observed in MOCVD GaInAs grown at 650°C on InP. This behaviour, similar to type B LPE GaInAsP layers, is as expected since the layer composition, point Q in Figure 1, lies outside the unstable region at the growth temperature of 650°C. Figure 6(b) illustrates the TEM contrast observed in an LPE GaInAs layer grown at 585°C which meant that the layer composition was very close to lying inside the unstable region as indicated in Figure 1. A fine granular structure and a very faint coarse tweed-like structure are visible indicating that during growth the layer may have just started to decompose forming a coarse tweed-like structure similar to that observed in type A LPE GaInAsP layers. Figure 6(c) illustrates the TEM contrast observed in a VPE GaInAs layer grown on InP at 650°C. A fine granular structure is again observed as expected from the high growth temperature. The large strain contrast features are believed to be caused by composition variations in the layer originating from growth on a slightly thermally degraded InP substrate and not from clustering or spinodal decomposition.

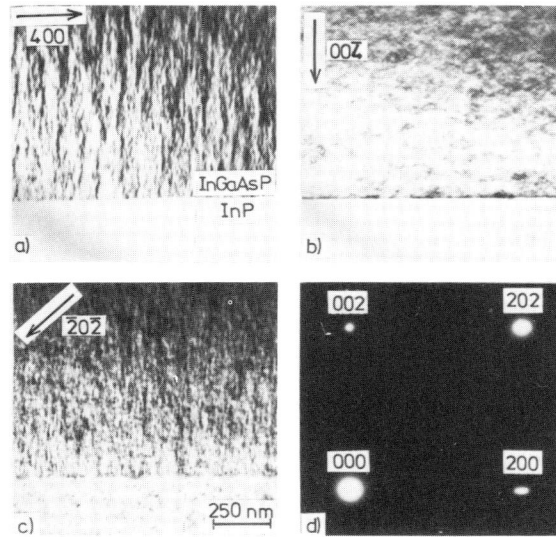

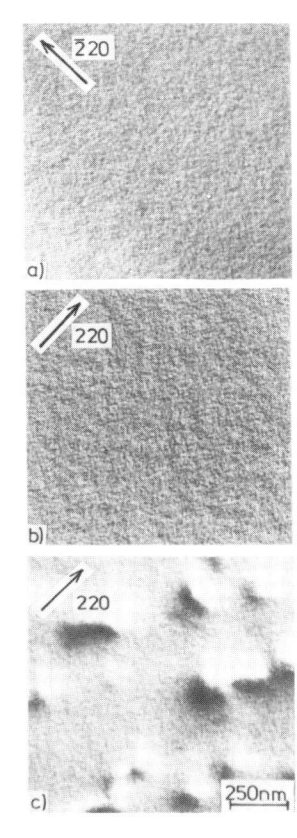

Fig.5 (a), (b) and (c) TEM contrast observed in [010] cross-section of type A LPE GaInAsP layer. (d) TED pattern [010] pole illustrating no satellites associated with 002 spot.

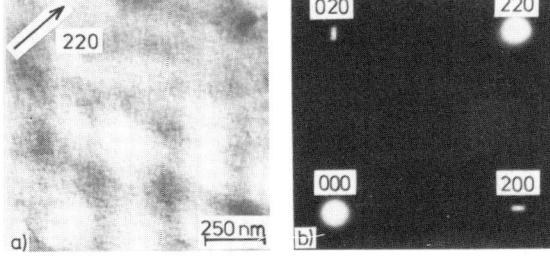

Fig.6 TEM contrast observed in (a) MOCVD GaInAs (Tg=650°C). (b) LPE GaInAs (Tg= 585°C) (c) VPE GaInAs (Tg=650°C)

Fig.7 (a) TEM contrast observed in MBE GaInAs (Tg≈560°C). (b) TED pattern of [001] pole showing satellite spots.

Tg is the growth temperature.

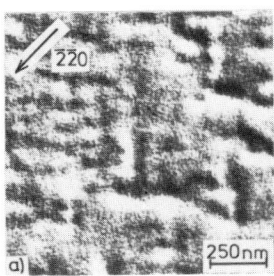

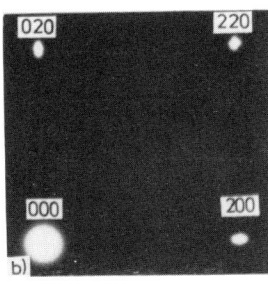

Fig.8 (a) TEM contrast observed in VPE GaInP (Tg=750°C). (b) TED pattern [001] pole show-ingsatellite spots.

Examinations of an MBE GaInAs layer grown on InP at ~560°C revealed the presence of both a fine granular structure (~10-15nm) and a coarse tweed-like structure (~250nm) as illustrated in Figure 7(a). TED patterns of the [001] pole, Figure 7(b), revealed the presence of satellite spots similar to those previously observed in type A LPE GaInAsP layers. The presence of this coarse tweed-like structure and strong satellite spots suggest that at ~560°C the ternary GaInAs lattice-matched to InP (point Q, Figure 1) lies inside the unstable region.

TEM examinations of a VPE GaInP layer grown on GaAs, composition point R in Figure 1, at 750°C also revealed the presence of both a fine granular structure (~10-15nm) and a coarse tweed-like structure (100-200nm), Figure 8(a), similar to that observed recently in LPE GaInP by Ueda et al (1984). Satellite spots were also visible in TED patterns of the [001] pole as shown in Figure 8(b). This behaviour suggests that at 750°C, GaInP lattice-matched to GaAs lies inside the unstable region as predicted by Figure 1.

The coarse tweed-like structure has previously only been reported in LPE grown material and has been interpreted as arising from spinodal decomposition occurring by an interfacial diffusion mechanism at the liquid/growing solid interface (H Launois et al 1983). The present results on MBE GaInAs and VPE GaInP indicate that a coarse tweed-like structure can develop in material grown by other techniques.

## Acknowledgments

The authors wish to thank British Telecom and Plessey Caswell for provision of samples, Drs M Hockly, M R Taylor and J Gowers for useful discussions, and the Science and Engineering Research Council and British Telecom for support.

## References

Cahn J W 1962 Acta. Met. 10 179
Cahn J W 1968 Trans. Met. Soc. AIME 242 166
de Cremoux B, Hirtz P and Ricciardi J 1981 GaAs and Related Compounds, Vienna 1980. Inst. Phys. Conf. Ser. 56 pp115-124
Foster L M and Woods J F 1971 J Electrochem. Soc. 118 1175
Glas F, Treacy M M J, Quillec M and Launois J 1982 J. de Physique 43 Colloq. C5, Supplement to No.12 C5-11
Gowers J P 1983 Appl. Phys. A 31 23
Henoc P, Izrael A, Quillec M and Launois H 1982 Appl. Phys. Lett. 40 963
Launois H, Quillec M, Glas F and Treacy M J 1983 GaAs and Related Compounds, Albuquerque 1982 Inst. Phys. Conf. Ser. 65 pp537-544
Mahajan S 1983 Microsc. Semicond. Mater. Conf., Oxford 1983 Inst. Phys. Conf. Ser.67 pp259-272
Mahajan S, Dutt B V, Temkin H, Cava R J and Bonner W A 1984 J. of Crystal Growth 68 589
Norman A G and Booker G R 1985 J. Appl. Phys. in press
Onabe K 1982 Jpn. J. Appl. Phys. 21 L323
Pearsall T P 1982 GaInAsP Alloy Semiconductors (New York: Wiley)
Stringfellow G B 1982 J. Crystal Growth 58 194
Treacy M M J, Gibson J M and Howie A 1985 submitted to Phil. Mag.
Ueda O, Isozumi S and Komiya S 1984 Jap. J. Appl. Phys. 23 L241

*Inst. Phys. Conf. Ser. No. 76: Section 6*
*Paper presented at Microsc. Semicond. Mater. Conf., Oxford, 25–27 March 1985*

# Precipitate formation in heavily doped GaInAsP alloys grown by liquid phase epitaxy

P Charsley and R S Deol
Department of Physics, University of Surrey, Guildford GU2 5XH

Abstract Transmission electron microscope techniques have been used to study the microstructure of $Ga_xIn_{1-x}As_yP_{1-y}$ (y=2.1x with y=0.07) grown by liquid phase epitaxy on Fe–doped InP substrates. Precipitates of various sizes and morphology have been observed. The origin of some of these second phase particles is attributed to iron–phosphide precipitation resulting from iron diffusion from the substrate.

## 1. Introduction

High resistivity, lattice matched $Ga_xIn_{1-x}As_yP_{1-y}$ with y=2.1x and y=0.07,was grown by liquid phase epitaxy (LPE) on a semi–insulating Fe–doped InP substrate. The high resistivity was achieved by heavily doping the material with Mn and Ge. This alloy was preferred to high resistivity InP because it was easier to grow with good morphology and freedom from surface terracing. Both p- and n–type layers of the quaternary alloy were produced and Hall measurements showed that in all cases the mobility was surprisingly high in view of the large values of ionized impurity scattering to be expected (Shantharama et al. 1985). Since one explanation lies in the possible formation of precipitates involving Mn and Ge an electron microscope study of these layers has been undertaken. A number of different types of precipitate have been observed some of which can be explained by iron diffusion from the substrate into the epilayer. Other precipitates rich in Mn have also been observed. The only previous published work, known to the authors, on precipitation in heavily doped LPE GaInAsP layers is by Ueda et al. (1983).

## 2. Experimental Details

Lattice matched layers ∼ 1.2μm in thickness of $Ga_{0.03}In_{0.97}As_{0.07}P_{0.93}$ were grown by LPE on Fe–doped InP substrates of {100} orientation. Materials with four different levels of Ge doping, but with a fixed Mn doping level, were investigated; these are listed in Table 1 together with details of their electrical characteristics from the work of Shantharama et al. (1985).

Plan–view transmission electron microscope (TEM) studies were primarily carried out on samples of material A, B and D together with cross–sectional studies on material B. The plan–view specimens were thinned, from the substrate side, using mechanical polishing followed by jet–chemical thinning using a bromine–methanol solution; the cross–sectional specimens were thinned by $Ar^+$ ion–milling at 3 keV following mechanical polishing. A JEOL STEM 200cx, operating at 200kv, was used for the microscopy and for some of the energy dispersive X-ray

Table 1.   Details of Specimen Doping and Electrical Characteristics

| Material | Dopant in mg/g in melt | | Majority carrier type | Carrier concentration $cm^{-3}$ | Mobility $cm^2V^{-1}s^{-1}$ |
|---|---|---|---|---|---|
| | Mn | Ge | | | |
| A | 0.00344 | 0 | p | $6.3\times10^{15}$ | 139 |
| B | 0.00344 | 1.91 | p | $2.5\times10^{15}$ | 136 |
| C | 0.00344 | 2.668 | p | $6.0\times10^{15}$ | 121 |
| D | 0.00344 | 3.792 | n | $1.0\times10^{17}$ | 500 |

spectroscopy (EDS) studies.   A Philips 400T was also used in the latter work.

## 3. Results

Precipitates were observed in all of the material types studied, most of the work being done on material types A, B, and D.   Although individual precipitates were small (usually $\leqslant$ 0.1$\mu$m) there were clearly several different types of morphology and, in certain instances, well-defined configurations of precipitates were observed.   These features may be summarized under 5 headings:

(i)     Linear arrays of very small precipitates, closely spaced, often contiguous, and showing weak strain contrast.   They were most commonly observed in material A and occasionally in B and D.

(ii)    Precipitates in well-defined linear arrays but widely spaced and exhibiting strong strain contrast.   They were observed in materials A and B.

(iii)   Isolated small precipitates, which were approximately equiaxed, 0.1 to 0.2$\mu$m in size, and which exhibited only very weak strain contrast.   These precipitates were observed in all materials.

(iv)    Clusters of larger precipitates (up to ~ 0.5$\mu$m), very irregular in form, exhibiting strain contrast.   In some cases dislocations were clearly associated with them.   They were only observed in material B.

(v)     Small platelet precipitates, ~ 10 to 20nm in size, showing weak strain contrast.   They were observed in materials B and D.

Precipitates of types (i) and (ii) are shown in Figs. 1(a) and 1(b), respectively, in a specimen of material A.   In both cases the precipitates are aligned approximately parallel to the epilayer surface; in Fig. 1(a) along a <210> and in Fig. 1(b) a <110> direction.   There would appear to be a clear distinction between these types of precipitates particularly in the presence of pronounced strain field contrast in Fig. 1(b); this has been established as due to a positive dilatation.

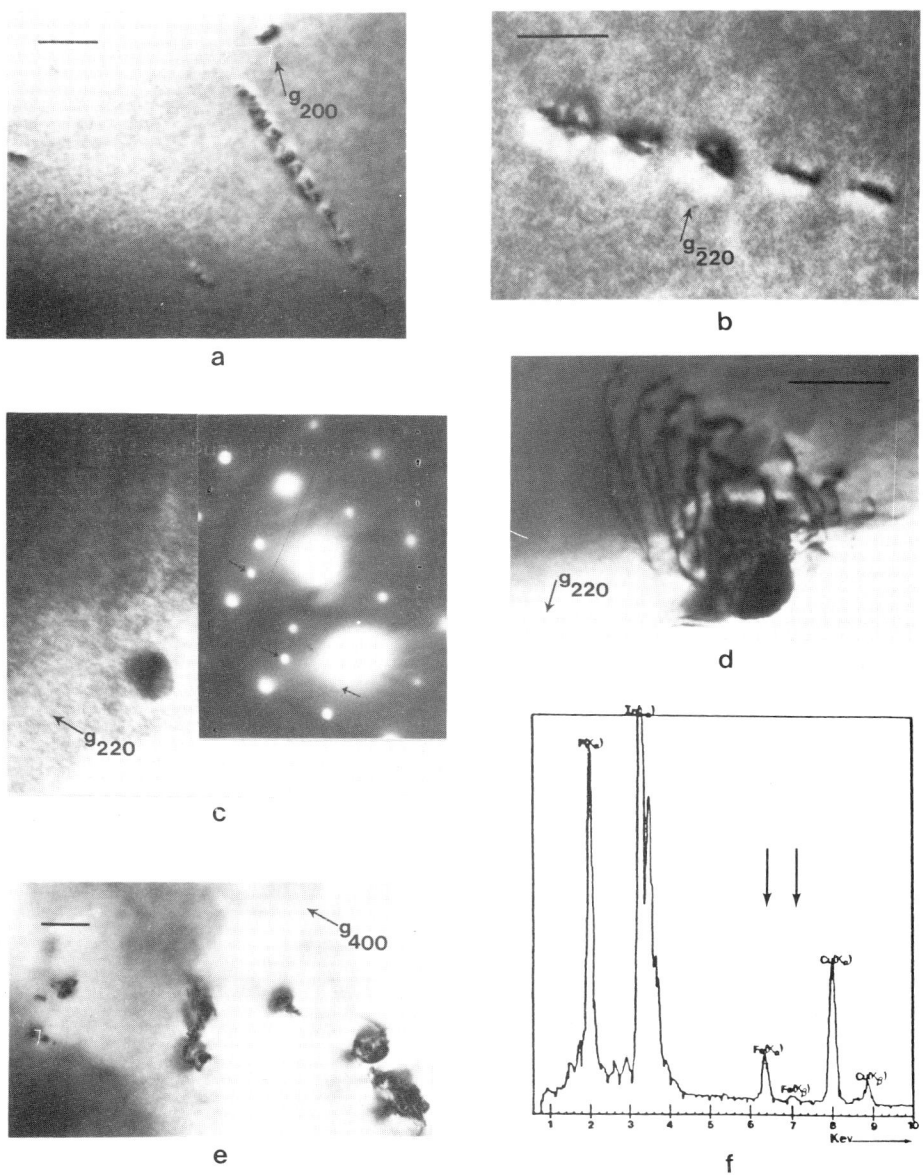

Fig. 1.   All scale markers are 0.2μm.   All micrographs are bright field
TEM except (b) which is a centred dark-field TEM image: (a) and (b)
precipitates in material A of type (i) and (ii) respectively; (c) a
precipitate of type (iii), in material B, together with the associated
SAD pattern.   Some extra-reflections are arrowed; (d) a precipitate of
type (iii) in material B; (e) a typical cluster of precipitates of type
(iv) in material B; (f) an EDS spectrum from a type (iv) precipitate in
material B.   Arrows indicate the Fe peaks.

Figure 1(c) shows a precipitate of type (iii), observed in material B. Its morphology is quite distinct from the types (i) and (ii) precipitate. The diffraction spots in the SAD pattern, which are not associated with the matrix, correspond to a structure with d-spacings in reasonable agreement with the published values for $Fe_2P$ (ASTM index). The precipitate shown in Fig. 1(d), also observed in material B, is thought to be of type (iii). Its retention after thinning can be explained by the difference in its chemical composition from the matrix and its rounded nature can be clearly seen. The dislocations visible in this micrograph can be ascribed to localized stresses associated with the precipitate. The formation of dislocations will be easier when they can be generated at the epilayer surface i.e. when the precipitate itself is close to the surface, as in this example. A typical cluster of precipitates of type (iv) in material B is shown in Fig. 1(e). This precipitate appears to be made up of an irregular arrangement of much smaller precipitates with some associated dislocations. An EDS spectrum from such a precipitate, Fig. 1(f), shows a significant iron concentration. An EDS spectrum from a plate-like precipitate in material-type C, which had dimensions $\sim 1.5\mu m$, was found to have a high concentration of Mn and no detectable concentration of Fe. The smallest type of individual precipitate (type (v)) is shown in Fig. 2(a), occurring in material D.

A cross-sectional TEM specimen of material B, Fig. 2(c), shows several new features. There is extensive precipitation in the interfacial region which extends to a depth of $\sim 0.3\mu m$ in the substrate and $\sim 0.1\mu m$ in the epilayer, there is also some precipitation visible within $\sim 0.1\mu m$ of the epilayer surface. Very little or no precipitation is visible within the bulk of the epilayer. Stereo micrographs have shown that the precipitates in Fig. 2(c) do not lie at the foil surface and cannot therefore be attributed to ion beam damage. Fig. 2(b) shows a SIMS profile for material of closely similar compoisition to material B. There is a peak in the iron concentration at the interface for a thickness of $\sim 0.3\mu m$ which corresponds reasonably well with the observed distribution of precipitates in cross-section.

## 4. Discussion

The studies of precipitate morphology, configurations and associated strain fields suggest that at least 5 types of precipitate occur in the epilayer surface regions of $Ga_{0.03}In_{0.97}As_{0.07}P_{0.93}$ heavily doped with Mn and/or Ge. Types (iii) and (iv) of these precipitates are associated with Fe, type (iii), having been identified as $Fe_2P$, occurs in all materials investigated whereas type (iv) has only been observed in material of type B. It is possible that this is a different iron-phosphide. Several different types of iron-phosphide have been observed by Lee et al. (1977), Rumsby et al. (1981), Miyazawa et al. (1982) and Smith et al. (1984) in Fe-doped InP grown by the LEC technique. However the difference in morphology may be determined by effects of doping on the diffusion rates.

It is likely that iron-phosphide precipitates are formed by the diffusion of iron from the substrate. The rapid diffusion of iron from an Fe-doped InP substrate into an epitaxial layer has been previously reported by Chevrier et al. (1980) using SIMS. DLTS studies by Bremond et al. (1982) have also demonstrated this. Our SIMS and cross-sectional TEM results support this view. These results show that pronounced precipitation

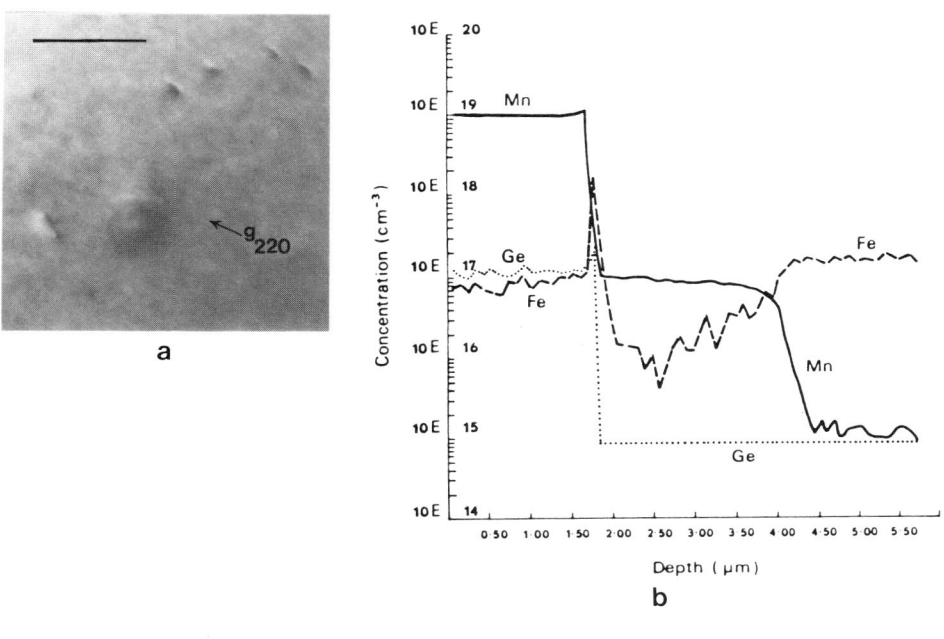

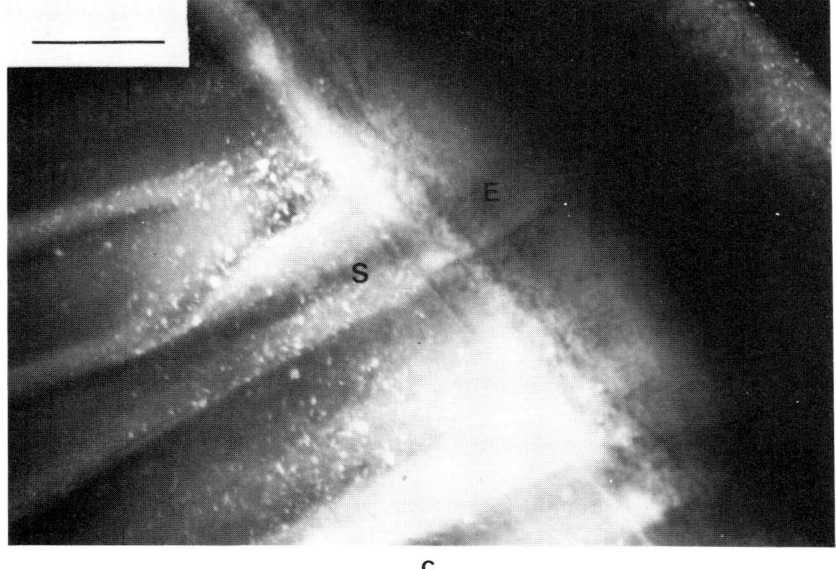

Fig. 2.   All scale markers are 0.2µm.   (a) Bright field TEM micrograph
illustrating precipitates of type (v) observed in material D; (b) a SIMS
profile for Mn, Ge and Fe in a material of similar composition to
material B; (c) cross-section TEM dark field image of material B.   S is
the substrate and E is the epilayer.

occurs near the interface both in the substrate and the epilayer. Although it has not yet been verified it is highly probable that these precipitates are also iron-phosphide in view of the SIMS results. Although these results indicate a high concentration of Fe throughout the epilayer, the apparent absence of precipitates other than at the interface and near the surface can be qualitatively understood by supposing that the precipitation is determined by thermal history. Both the interface and epilayer surface are regions of comparatively rapid cooling. The work quoted above (Chevrier et al. 1980) indicates the occurrence, under certain circumstances, of a high Fe-concentration at the surface as well as at the interface. The present results may provide an explanation of these results since if, during the epilayer growth, iron-phosphide precipitates are formed, a further diffusion of Fe into those regions would be expected. SIMS can not distinguish between iron in solution and present in precipitates. Further cross-sectional TEM combined with EDS is highly desirable.

Clearly the existence of other, very different morphologies, suggests that chemically different precipitate types occur in the epilayer. A large Mn rich precipitate has been observed but it remains undetermined whether the types (i) and (ii) precipitate are Mn rich. The results on specimens B and D have revealed an additonal precipitate morphology (type (v)) which occurs when both Mn and Ge are present. The extremely complex precipitation, in part due to Mn and Ge doping, further complicated by the diffusion of Fe from the substrate, is being studied further by EDS.

## 5. Acknowledgements

One of us (RSD) wishes to thank the SERC for financial support. We would also like to thank Dr. P.D. Greene and Elaine Allen of STL for providing the samples and for valuable discussions and Loughborough Consultants for the SIMS results. Members of the Microstructural Studies Unit, University of Surrey, are thanked for experimental assistance. Dr. U. Bangert is also thanked for her help in preparing the cross-sectional specimen.

## References

Bremond G, Nouailhat A and Guillot G 1982 Inst. Phys. Conf. Ser. 63 239
Chevrier J, Armand M, Huber A M and Linh N T 1980 J. Electron. Mat. 9 745
Lee R N, Norr M K, Henry R L and Swiggard E M 1977 Mater. Res. Bull. 12 651
Miyazawa S and Koizumi H 1982 J. Electrochem. Soc. 129 2335
Rumsby D, Ware R M and Whitaker M 1981 J. Cryst. Growth 54 32
Shantharama L G, Adams A R, Allen E M and Greene P D 1985 Proc. Gallium Arsenide and Related Compounds (Biarritz 1984) – in the press
Smith N A, Harris I R, Cockayne B and MacEwan W R 1984 J. Cryst. Growth 68 517
Ueda O, Umebu I and Kotani T 1983 J. Cryst. Growth 62 329.

*Inst. Phys. Conf. Ser. No. 76: Section 6*
*Paper presented at Microsc. Semicond. Mater. Conf., Oxford, 25–27 March 1985*

269

# Ion backscattering and channelling studies of InGaAs epitaxial layers on InP substrates

J M Cole[*], L G Earwaker[*], N G Chew[+], A G Cullis[+] and S J Bass[+]

[*]Department of Physics, University of Birmingham, P O Box 363, Birmingham B15 2TT.
[+]Royal Signals and Radar Establishment, Malvern, Worcs WR14 3PS.

Abstract The $In_{1-x}Ga_xAs/InP$ system has been studied by transmission electron microscopy combined with Rutherford backscattering and channelling spectrometry. The layer thickness data as determined by Rutherford backscattering and computer simulation were found to be in good agreement with the transmission electron microscopy determinations. Channelling measurements on the lattice mismatched InGaAs-InP system indicate higher dechannelling rates along the <110> directions than along the [001] growth direction. This behaviour is opposite to that observed for single crystal unstrained sphalerite structures and is consistent with the contraction or expansion which occurs along the growth direction to accommodate the lattice mismatch. A quantitative estimate of 0.0126 ± 0.0034 was obtained for the layer strain in the $In_{0.58}Ga_{0.42}As/InP$ system.

## 1. Introduction

Recently, there has been considerable interest in periodic structures of lattice mismatched crystalline materials, usually referred to as Strained Layer Superlattices (Mayer et al 1973, Blakeslee and Alliotta 1970, Picraux et al 1983, Chu et al 1982). Early work in the field was confined to the closely lattice matched systems, such as GaAlAs/GaAs (Mayer et al 1973), as it was believed to be necessary to match the lattice constants in order to maintain crystal perfection. However, it was predicted (Blakeslee and Alliotta 1970), that the generation of defects would be prevented in a system whose average lattice constant, over the layers, matched that of the substrate. In Strained Layer Superlattices (SLS), the lattice mismatch is accommodated by the layer strain. Electronic properties such as the band gap and transport parameters are found to vary according to the degree of lattice strain. Thus by variation of the layer composition it is possible to grow superlattices with specific electronic properties.

The dechannelling of alpha particles in ion channelling experiments is sensitive to the lattice strain. Studies on the InGaAs/GaAs (Picraux et al 1983) and InAs/GaSb (Chu et al 1982, Saris et al 1980) systems have found that channelling along the [001] growth direction is, in general, better than that observed along the <110> directions. This deviation from single crystal behaviour has been shown to be due to the presence of layer strains, by computer simulation (Barrett et al 1971) and by experiment (Chu et al 1982, Saris et al 1980) for the InAs/GaSb system.

In this work, concentration and layer thickness measurements have been made for a number of InGaAs/InP structures, using Rutherford backscattering spectrometry (RBS) combined with computer simulation techniques and transmission electron microscopy (TEM). Ion channelling studies were performed to determine the effects of layer strain on dechannelling rates.

2. Experimental Details

A series of samples consisting of single layers approximately 3000Å thick of $In_{1-x}Ga_xAs$ with various values of the parameter x, were grown on (001) InP substrates by Metal Organic Chemical Vapour Deposition (MOCVD).

Samples for TEM analysis were thinned in cross section using mechanical polishing and reactive $I^+$ ion milling (Chew and Cullis 1984). Thinned specimens were examined by TEM using 120keV electrons.

The principles of RBS and channelling as applied to these superlattices have been discussed in the literature (Mayer et al 1973, Picraux et al 1983, Chu et al 1982). In this study, backscattering and channelling measurements were made using a 2MeV $^4He^+$ beam generated by the 3MV Dynamitron Accelerator at the University of Birmingham. The samples were mounted on a triple axis goniometer and the backscattered ions were collected by a surface barrier detector placed at a backward angle of 150 degrees with respect to the beam direction.

3.   Results and Discussion

3.1  Transmission Electron Microscopy

A number of the $In_{1-x}Ga_xAs$ layers on InP were examined in cross-section by TEM and, often, the ternary semiconductor was substantially free of extended crystallographic defects. Figure 1a shows an image of a 3300Å alloy layer of composition $In_{0.53}Ga_{0.47}As$ and it is clear that, although the material contains few visible defects, strong mottled contrast can be seen on a scale of 100-300Å. This is thought to result from spinodal

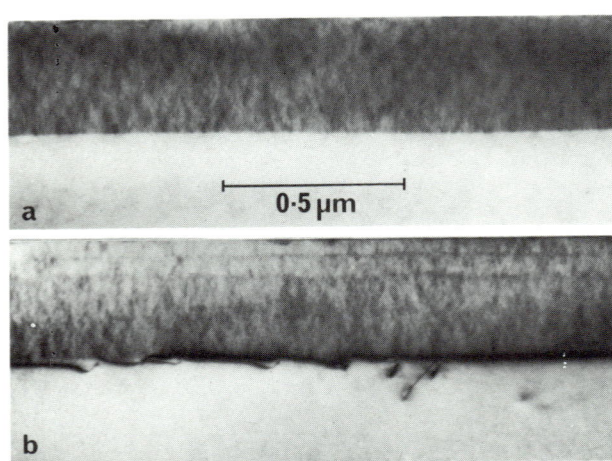

Fig. 1.   Cross-sectional transmission electron images of $In_{1-x}Ga_xAs$/InP structures     a)  x = 0.47;  b)  x = 0.34

decomposition of the alloy in a manner described elsewhere in these proceedings (Norman and Booker 1985). In addition, this particular layer is closely lattice-matched to the InP substrate so that the interface region is also essentially defect-free. For comparison, Fig.1b shows a 3200Å alloy layer of composition $In_{0.66}Ga_{0.34}As$. In this case, misfit dislocations lying in the interfacial region are clearly evident. The layer itself contains relatively small numbers of defects, although some dislocations were seen and alloy clustering decomposition has once again taken place.

## 3.2  Random Alignment RBS Results

Random alignment RBS was performed for each of the samples. The origin of the shape of the RBS spectra, for the $In_{1-x}Ga_xAs/InP$ system, is described in Fig. 2. However, two important points should be emphasized. First, the two edges at the high energy end of the spectra are both caused by surface elements; In and the indistinguishable Ga and As, which have similar masses. Second, the front and back edges of the peak in each spectrum are, in fact, attributed to the same point in the structure; namely the interface between the surface InGaAs and the InP substrate. The lower energy edge is caused by the decrease in Ga and As at the interface. Thus the shape and size of this peak produces useful information concerning the structure in the interface region. A typical spectrum and computer simulation are shown in Fig. 2, for the lattice matched case of 3300Å $In_{0.53}Ga_{0.47}As$ on InP.

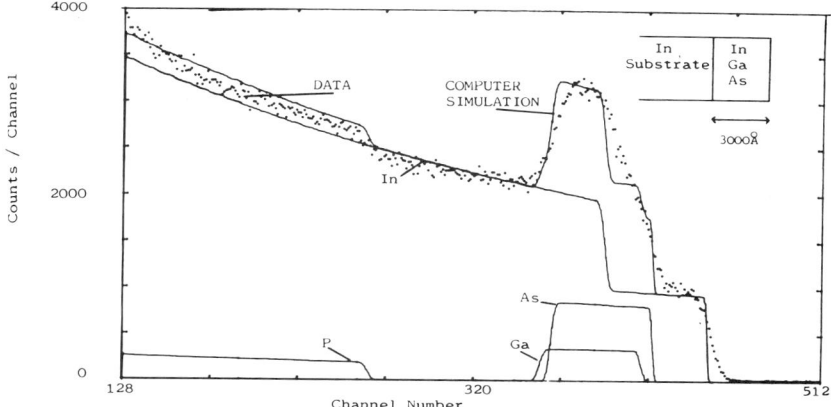

Fig. 2.  Description of the sample structure and the individual elemental contributions to the RBS spectra for $In_{0.53}Ga_{0.47}As$ on InP.

The results of the composition and layer thickness determinations by computer fitting to the RBS data are summarised in Table 1.

The determinations of the compositional data are independent of the layer strains so that using Vegard's Law, measurements of the unstrained lattice parameter are obtained. It should be noted that the thickness of the surface layer rises slowly with an increase in the composition parameter and that the layer thickness data are in excellent agreement with the values obtained by TEM.

| SAMPLE | X-VALUE RBS | LAYER THICKNESS (Å) TEM | RBS | CHI MIN (001) | (110) | CHI (Substrate) (001) | (110) |
|--------|-------------|------|------|------|------|------|------|
| A | 0.530±0.010 |  | 4000±100 | 8.4±0.4 | 6.6±0.3 | 41.7±0.2 | 55.0±0.2 |
| B | 0.470±0.010 | 3300±100 | 3300±100 | 6.6±0.3 | 3.6±0.3 | 18.1±0.1 | 11.7±0.1 |
| C | 0.460±0.010 | 3200±100 | 3300±100 | 8.1±0.3 | 5.8±0.2 | 31.9±0.3 | 56.2±0.2 |
| D | 0.440±0.010 |  | 3300±100 | 9.7±0.3 | 6.4±0.2 | 42.4±0.3 | 62.8±0.4 |
| E | 0.42±0.010 |  | 3100±100 | 11.1±0.2 | 11.6±0.6 | 49.1±0.4 | 64.5±0.7 |
| F | 0.34±0.010 | 3200±100 | 3000±100 | 18.2±0.4 | 11.8±0.4 | 72.1±0.6 | 83.7±0.5 |
| InP |  |  |  | 3.07±0.18 | 2.05±0.12 | 14.13±0.11 | 7.48±0.06 |
| GaAs |  |  |  | 3.70±0.21 | 3.10±0.12 | 11.10±0.10 | 8.53±0.13 |

Table 1.  Summary of the RBS and Channelling results on the $In_{1-x}Ga_xAs$ on. InP structures.

### 3.3  Channelling Alignment RBS

The measurement of the layer strain, present in the lattice mismatched samples, required that ion channelling was performed along the growth direction and compared with that observed along the inclined <110> major crystallographic axes (Picraux et al 1983, Chu et al 1982), see Figure 3.

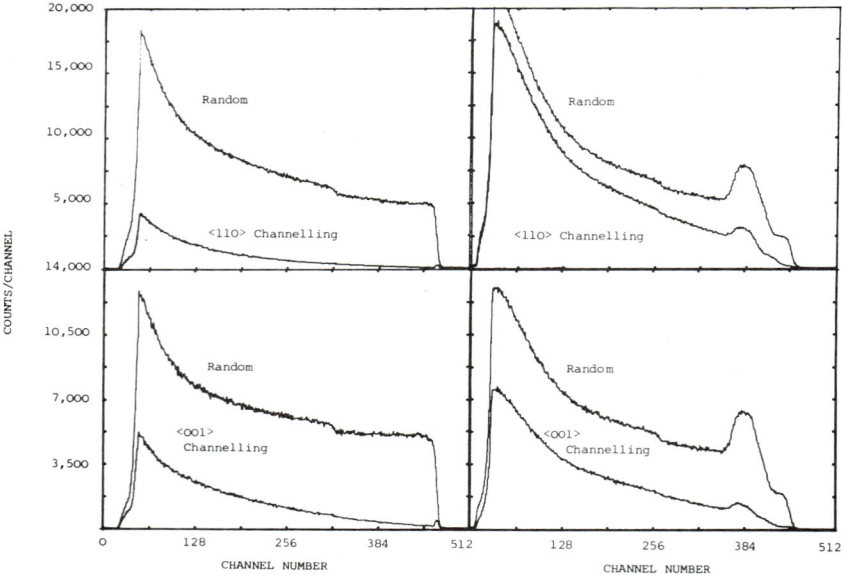

Fig. 3.  Rutherford Backscattering random and channelled spectra for <001> and <110> alignment of InP and of $In_{0.54}Ga_{0.46}As$ on InP.

For comparison, the same measurements were taken for single crystal InP and GaAs samples.

The <110> channelling measurements were performed at an inclination of 45 degrees with respect to the <001> orientation in a direction chosen to minimise the effects of the large angle rotation on the relative depth scales, so that direct comparison of the <001> and <110> channelling spectra was possible. The random yields, for comparison with the <110> channelling were taken at this 45 degree inclination. The minimum yield measurements (Table 1) were taken as the ratio (Chi min), of the channelled to random yields just below the surface peak. This yield ratio calculation was also performed for each sample at depths corresponding to the region of the crystal just behind the surface InGaAs layer and to the deep InP substrate (Table 1).

The $In_{0.58}Ga_{0.42}As/InP$ specimen was subjected to the most exhaustive channelling investigation. In the <001> alignment, reference channelling spectra were obtained. In the <110> alignment the surface and substrate were channelled separately, by the use of an energy window of variable position (Picraux et al 1984). The angular positions at which these two types of channelling behaviour were observed, were noted and used to estimate an effective layer strain.

From the channelled Rutherford backscattering spectra (Fig. 3), and the summary of results (Table 1), for the layer InGaAs/InP structures, there arise two important points.
(a) The degree of ion dechannelling is dependent on the value of the x parameter. The dechannelling is greatest for the case with a value of x which is furthest from the lattice match value. This observation applies for both crystal dechannelling axes.
(b) For any particular value of x other than that for the lattice match condition, the observed <110> dechannelling is greater than that observed for <001> alignment.

These results are visible qualitatively (Fig. 3), but can be seen more quantitatively by a close inspection of Table 1. To obtain a measure of the rates of dechannelling, each of the channelled spectra were divided by the corresponding random spectrum. The results of this operation for the case of 3200Å $In_{0.54}Ga_{0.46}As$ on InP and for the single crystal InP are shown in Fig. 4.

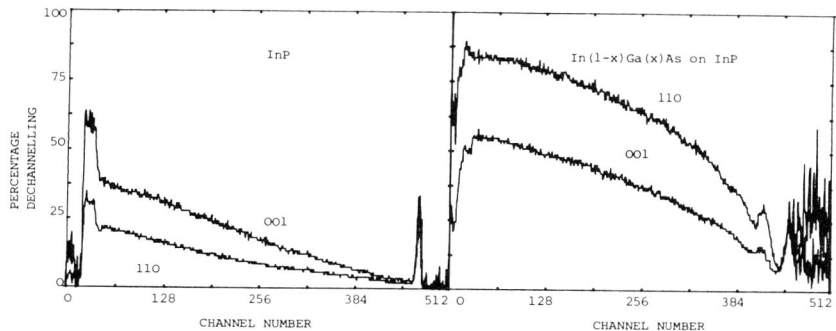

Fig. 4. Dechannelling plots (channelled spectrum divided by random spectrum) for InP and $In_{0.54}Ga_{0.46}As$ on InP.

From this figure it can be seen that for InP the yield measurements indicate higher ratios for the ‹001› with respect to the ‹110› alignments, not only at the surface but throughout the depth of the sample. It is clearly illustrated in Fig. 4 that this behaviour is different from that observed for the InGaAs/InP structures which have similar ‹001› and ‹110› alignment yield ratios at the surface, but dechannelling within the structure in the ‹110› alignment causes the ‹110› yield ratio to rise above the ‹001› alignment value. It is important to note that the crystal quality of sample closest to the lattice match (x = 0.47), as determined by the channelling analysis, was found to be comparable to that determined for the single crystal InP. Thus, the spinodal decomposition of the alloy observed by TEM had little effect upon $^4He^+$ ion dechannelling. In addition, the general increase in both ‹001› and ‹110› dechannelling with decrease in x value below 0.47 is in accord with a general increase in layer defect density and the presence of misfit dislocations at the layer/substrate interface.

The observation of relatively high ‹110› dechannelling with respect to that for ‹001› alignment in each of the layer systems can be used to determine the layer strains. Optimum surface and substrate channelling directions for the $In_{0.58}Ga_{0.42}As$ on InP structure were obtained by performing detailed angular scans. A model of the distortion produced in the surface layer is shown in Fig. 5, and the two channelling axes are indicated. The spectra obtained from the ‹110› channelling of the specimen surface and substrate are shown in Fig. 6.

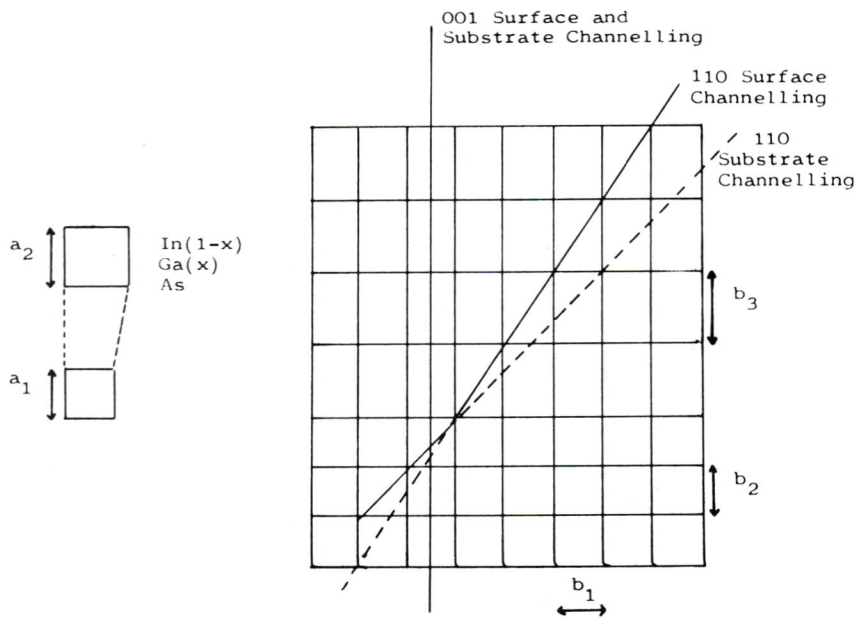

Fig. 5.   The accommodation of lattice mismatch by layer strains and the angular separation of the substrate and surface ‹110› channelling directions.

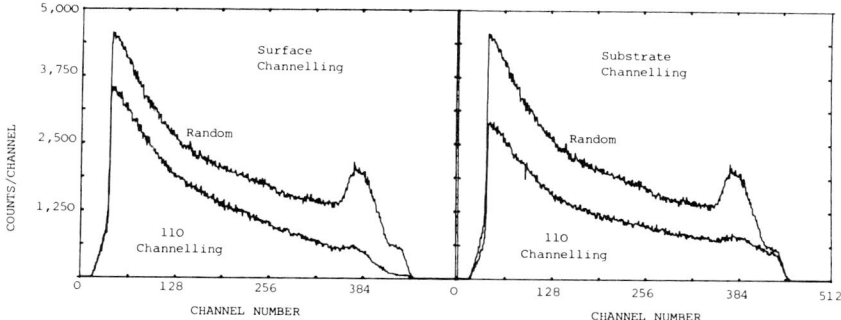

Fig. 6. Random and channelled Rutherford backscattering for surface and substrate <110> alignment in the $In_{0.58}Ga_{0.42}As$ on InP system.

The angular separation between the two alignment positions was found to be $\Delta\emptyset = 0.36 \pm 0.10$ degrees. If it is assumed that the single in-plane lattice constant is that of the substrate then, using the notation discussed in Fig. 5, $\Delta\emptyset = \tan^{-2}(b_3/a_1)$; where $a_1$ is the InP unstrained lattice constant (5.869Å), the new InGaAs strained layer parameters are $b_3 = 5.943 \pm 0.020$Å and $a_1 = b_1 = 5.869$A. Thus the calculated layer strain as defined by $((b_3-b_1)/b_1) = 0.0126 \pm 0.0034$, which is comparable with the value of 0.0091 reported for $GaAs_xP_{1-x}/GaP$ by Picraux et al (1984).

It is interesting to note that, a value for Poissons Ratio of $0.35 \pm 0.12$ can be determined for this surface InGaAs layer. This is obtained by using a value of $5.888 \pm 0.004$Å for the unstrained lattice constant obtained from the Rutherford backscattering compositional determinations and Vegard's law, combined with the values of the strained lattice parameters determined by the channelling measurements. This value of Poissons Ratio is indicative of the consistency of these measurements with other strain measurements. In fact, Blakemore (1982) reported a value of 0.31 for Poissons ratio for GaAs, at 300K.

It should be noted that the consistency can be further illustrated by assuming that the strained parallelopiped (Fig. 5) is formed from an unstrained cube with negligible change in volume. This allows a value for the unstrained lattice parameter of $5.893 \pm 0.006$A, to be calculated from the channelling data without reference to the RBS compositional determinations. This value is consistent with the parameter determined earlier ($a_2 = 5.888$ Å.)

4. Conclusions

The results reported demonstrate the successful combination of RBS and computer simulation techniques with ion channelling and cross sectional TEM in the determinations of composition, thickness and strain state for $In_{1-x}Ga_xAs$ layers grown on (001) InP substrates. The layer thickness data as obtained by RBS and computer simulation were shown to be in excellent agreement with the TEM determinations. The determinations of the unstrained lattice parameters by RBS and computer simulations were shown to be consistent with estimates obtained from the ion channelling results. Qualitative dechannelling effects have proved to be consistent with the generation of tetragonal distortions along the growth direction to accommodate lattice mismatch. The strain produced in a 3200Å layer of $In_{0.58}Ga_{0.42}As$ grown on InP was estimated to be 0.0126 ± 0.0034 by the ion channelling angular scan method. This estimate is comparable to reported strain estimates in similar III-V systems.

References

Barrett J H 1971 Phys. Rev. B3 1527
Blakemore J S 1982 J. Appl. Phys. 53 10 R123
Blakeslee A E and Alliotta C F 1970 IBM J. Res. Develop. 14 686
Chew N G and Cullis A G 1984 Appl. Phys. Lett. 44 142
Chu W K, Saris F W, Chang C A, Ludeke R and Esaki L 1982 Phys. Rev. B26 1999.
Mayer J W, Ziegler J F, Chang L L, Tsu R and Esaki L 1973 J. Appl. Phys. 44 2322
Norman A G and Booker G R 1985  This proceedings volume
Picraux S T, Dawson L R, Osbourn G C and Chu W K 1983 Nucl. Instr. & Meth. 218 57
Picraux S T, Biefield R M, Osbourn G C and Chu W K 1984 Thin Films and Interfaces II Eds. J E E Baglin, D R Campbell and W K Chu (Amsterdam: North Holland) p 477
Saris F W, Chu W K, Chang C A, Ludeke R and Esaki L 1980 Appl. Phys. Lett. 37 1999
Whiteley J W and Ghandhi S K 1983 Interfaces and Contacts, Eds. R Ludeke and K Rose (Amsterdam: North Holland) p 145

*Inst. Phys. Conf. Ser. No. 76: Section 6*
*Paper presented at Microsc. Semicond. Mater. Conf., Oxford, 25–27 March 1985*

# Lattice relaxation in thin compositionally-modulated semiconductor films

J M  Gibson , M M J  Treacy$^*$, J C  Bean and  R Hull

AT&T Bell Laboratories, 600 Mountain Avenue, Murray Hill, NJ 07974, USA
* Exxon Research and Engineering Co., Annandale, NJ 08801

**Abstract** The effects of lattice relaxation in thin samples of compositionally modulated semiconductors are discussed. It is shown that even for very weak strain modulations ($<10^{-3}$) strong diffraction contrast can occur. This is the origin of some "tweed"-like contrast seen in quaternary semiconductor alloys. Relaxation can also affect the lattice parameter in strained-layer superlattices and can explain some anomalous reports. Finally, effects on high-resolution imaging of interfaces between strained-layers are considered.

Although the effects of elastic relaxation of strain fields due to dislocations, point defects and precipitates in the thin foils used for transmission electron microscopy have received due attention[1] there has been surprisingly little awareness revealed in the literature for similar effects in the case of of strained boundaries and periodically-strained systems. Spinodal decomposition and strained-layer superlattices provide two important examples of the latter and cross-section interface studies of epitaxial thin films an example of the former. We have found that relaxation effects in such systems are no less significant than for the case of point and line defects. Our first studies, for the spinodal system $In_xGa_{1-x}As_yP_{1-y}$ showed that relaxation is the dominant source of long-period contrast[2]. For the case of strained-layer superlattices, relaxation-induced contrast can be dominant when the structure factor modulation is weak[3] and in any case is an important source of contrast. In this paper we review these results and introduce considerations of such contrast at a single interface, in particular the effect of relaxation on high-resolution imaging which can be significant.

Interest in this subject was aroused by studies of the liquid-phase epitaxially grown $In_xGa_{1-x}As_yP_{1-y}$ system, since it had been predicted by de Cremoux et.al.[4] and Stringfellow[5] that spinodal decomposition should exist in this system. Experimental results using transmission and analytical electron microscopy from Henoc et. al.[6] provided evidence for long period ($\approx$ 200nm) spinodal decomposition. However, the "tweed"-like contrast with this period, which was also reported by other groups, was never explained in any detail. We have analyzed this contrast[2] and found that it can be reliably attributed to spinodal decomposition but for quite unexpected reasons. In fact, the contrast cannot be explained by any simple effects such as structure factor changes and lattice dilatation, which apart from giving the wrong behaviour with parameters such as operating reflection, deviation parameter and thickness, are generally too weak to be seen in this system.

The contrast can be attributed to relaxation of the stress fields associated with spinodal decomposition in a thin foil. For a bulk material which is spinodally decomposed in one direction (with sinusoidal lattice parameter modulation) the strain in the modulation direction (x)

$$\epsilon_{xx} = \frac{1+\nu}{1-\nu}\epsilon_0 \sin(\alpha x) \tag{1}$$

(where $\alpha = \frac{2\pi}{\Lambda}$ and $\Lambda$ is the spinodal wavelength) is exaggerated by typically two times over the "stress-free" value $\epsilon_0$ due to the suppression of expansion in other directions. In the extreme case where a foil of thickness t $<<$ $\Lambda$ is formed for TEM, one of the bulk stresses which causes the enhancement of the modulation amplitude is completely removed, which leads to an amplitude

$$\epsilon_{xx} = \epsilon_{zz} = (1+\nu)\epsilon_0 \sin(\alpha x) \tag{2}$$

In this case the strain has relaxed by a considerable factor from the bulk value in the direction parallel to the spinodal modulation and has also been introduced in the thin foil direction. This would have a significant effect on lattice parameter measurements in the case of spinodal decomposition and has been ignored in most earlier studies[7] yet can explain some apparent anomalies[3]. For the quaternary semiconductor case, and indeed for most spinodal studies with transmission electron microscopy, the ratio of the specimen thickness to wavelength is of order one and neither the fully relaxed nor bulk case applies. We have solved this problem using the plane strain approximation and linear, isotropic elasticity theory to yield the general result for the strain[2]

$$\epsilon_{zz} = 2\epsilon_0 \frac{1+\nu}{1-\nu} \left[ \frac{(\alpha\frac{t}{2})\cosh(\alpha\frac{t}{2})\cosh(\alpha z) - (\alpha z)\sinh(\alpha z)\sinh(\alpha\frac{t}{2}) - (1-2\nu)\sinh(\alpha\frac{t}{2})\cosh(\alpha z)}{\sinh(\alpha t) + \alpha t} - \frac{1}{2} \right] \sin(\alpha z) \tag{3}$$

The result implies that there is bending of the lattice planes $(\approx \frac{\partial \epsilon_{xx}}{\partial z})$ which can give rise to significant variations in diffraction contrast images even if $\epsilon_0$ is as small as $10^{-4}$. For the quaternary case, this bending is the major source of contrast and can explain the experimental results in detail with the use of the 2-beam Howie-Whelan equations and the column approximation.

Similar effects will exist in any system containing periodic strain fields which is thinned in cross-section for electron microscopy and can be described by a Fourier series of the above solutions. One important example of this sort is the strained-layer superlattice which has received a great deal of attention in semiconductors recently[8]. Attempts have been made to measure coherent tetragonal distortions from thin foils which have ignored this major effect[9] [10]. Apparently anomalous results in these studies can be explained by relaxation and consideration of diffraction by superlattices[3][11].

The theory can be extended to non-periodic systems such as a single strained interface. This case has been solved numerically by Auret et. al. [12] but without recognition of the long distance relaxation. In our studies the single interface case can readily be derived from the superlattice case as shown in figure 1. As a

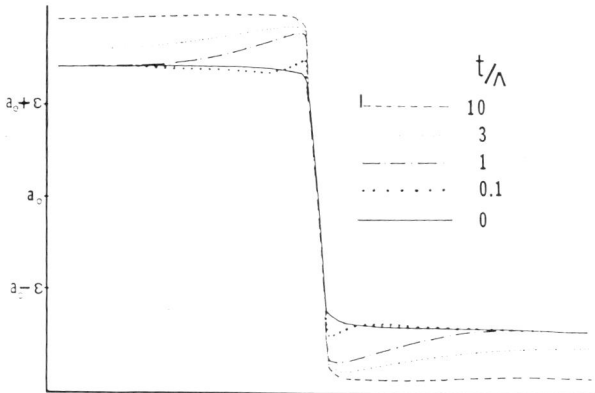

Figure 1: The value of lattice parameter, averaged through the film thickness, for strained-layer superlattices of different periods relative to the foil thickness $(\frac{\Lambda}{t})$. In each case the scales are adjusted so that the superlattice period is apparently constant. As $\frac{t}{\Lambda}$ nears infinity the results asymptotically approach the case of a single interface.

function of $\frac{t}{\Lambda}$, the mean value of the superlattice lattice parameter is seen to relax only near the interfaces. This can be understood in the Fourier representation as follows: as the superlattice period become long compared with the film thickness, the fundamental period and its low multiples can be considered to be fully relaxed as in equation 3 above. Full relaxation does not occur for strain components of frequency higher than $k \approx \frac{1}{t}$ and at the extreme limit of $k \gg \frac{1}{t}$ relaxation is absent. Since high spatial frequencies are "localized" at interfaces in a superlattice, the changes in lattice parameter and bending are localized in distances of order t from a single interface. The projected lattice parameter as plotted in figure 1 is only of limited value. A fuller understanding of image contrast must include image simulations including bending effects and the local value of dilatation. Fortunately, contrast effects in axial bright field images should not be very serious if the thickness is considerably less than the extinction distance.

Apart from contrast effects at interfaces, relaxation can affect measurements of rigid shifts between lattices. Extravagant accuracies have been claimed for this technique, based only on instrumental considerations. However, in many cases strain is involved across the interface and relaxation will cause a significant difference. For example, a change in lattice spacing of 1% induced by relaxation, might cause an error of several percent magnitude in rigid shift measurement, depending on how measurements were made. Nevertheless, the effects are easy to calculate and incorporate in measurements. It is certainly an effect to be aware of in such studies.

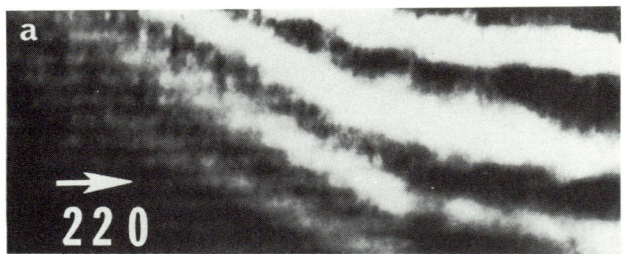

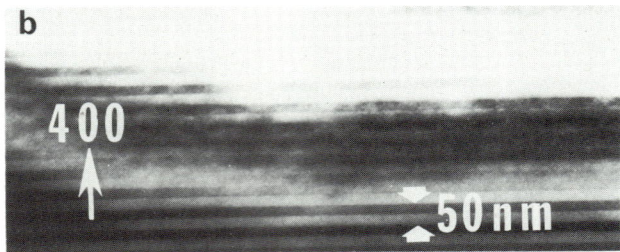

Figure 2: Dark-field 2-beam images taken at the exact Bragg condition from the same area of $Ge_{0.05}Si_{0.95}$/Si 500Å period superlattice. (a) 220 reflection perpendicular to the modulation direction. (b) 400 reflection parallel to the modulation direction. The latter shows relaxation effects from the 0.1% amplitude strain wave.

In diffraction contrast images, such as the dark-field images shown in figure 2, strong relaxation-related contrast can occur even from weak modulations. In figure 2, the strain amplitude is only $10^{-3}$. One common characteristic of relaxation induced contrast is that it is strong only in reflections parallel to the modulation direction. By detailed image simulation we hope to use this technique to study strain fields in very thin layers and superlattices.

In conclusion we have demonstrated that elastic relaxation is a significant source of contrast in thin films of compositionally modulated semiconductors. In particular for the case of spinodal decomposition in quaternary semiconductor alloys it is only through consideration of relaxation that one can correlate diffraction contrast with decomposition. Values of strain amplitudes can thus be deduced. In the case of strained-layer superlattices, errors have occurred in interpretation with neglect of relaxation effects. In the future, elastic relaxation contrast effects in modulated structures may be taken beyond the realm of artefact and exploited as a method of measuring very sensitively the elastic constants of thin layers and superlattices.

Furthermore, it should be emphasized that relaxation may be a serious limitation to interpretation of interface images in strained systems in that it can affect measurement of lattice rigid shifts and may change the appearance of the image at the interface. Image calculations following the simple models we have described here should be made to understand these effects in strained systems.

We gratefully acknowledge discussions with A. Howie, H. Launois, J.M. Brown and D. Maher.

*References*

1. e.g. W.J. Tunstall, P.B. Hirsch and J. Steeds, Phil. Mag. **9** , 99 (1964)

2. M.M.J. Treacy, J.M. Gibson and A. Howie, Phil. Mag., to appear

3. J.M. Gibson and M.M.J. Treacy, Ultramicroscopy **14** , 345 (1984);

4. de Cremoux, B., Hirtz, P. and Ricciardi, J., in "GaAs and related compounds", ed. by H.W. Thim, (Institute of Physics: London), (1981).

5. Stringfellow, G.B. J. of Cryst. Growth, **58** , 194 (1982).

6. Henoc, P., Izrael, A., Quillec, M. and Launois, H., Appl. Phys. Lett. **40** , 963 (1982).

7. e.g. R. Sinclair, R. Gronsky and G. Thomas, Acta. Met. **34** , 789 (1976).

8. G. Osbourn, J. Appl. Phys. **53** , 1587 (1982).

9. J.M. Brown, N. Holonyak, M.J. Ludowise, W.T. Dietze and C.R. Lewis, Appl. Phys. Lett. **43** ,863 (1983).

10. J.M. Brown, N. Holonyak, R.W. Kaliski, M.J. Ludowise, W.T. Dietze and C.R. Lewis, Appl. Phys. Lett. **44** , 1158 (1984).

11. J.M. Gibson, R. Hull, J.C. Bean and M.M.J. Treacy, Appl. Phys. Lett., April 1985

12. F.D. Auret, C.A.B. Ball and H.C. Snyman, Thin Sol. Films **61** , 289 (1979).

Inst. Phys. Conf. Ser. No. 76: Section 6
Paper presented at Microsc. Semicond. Mater. Conf., Oxford, 25–27 March 1985

# TEM studies of defects in MOCVD GaAs/GaAlAs double heterostructures

U Bangert and P Charsley
Department of Physics, University of Surrey, Guildford

Abstract   Crossections of MOCVD grown GaAs/GaAℓAs double heterojunction
structures are investigated by TEM.  The substrate/epilayer interface
is found to be the main source for defects such as stacking fault
"pyramids", dislocations and dislocation loops.  Possible causes for
the defect formation, namely contamination of the substrate, excess Aℓ
and lattice mismatch are discussed.

## 1. Introduction

The MOCVD technique has become attractive for the growth of GaAs/GaAℓAs
heterojunction structures because high quality material with very good
layer uniformity over large areas can be achieved (Nakanisi, 1984). MOCVD
also makes it easier to grow multiple layer structures as required for
double heterojunction pulsed lasers (Whiteaway et al, 1981) and quantum
well lasers (Burnham et al, 1984).

Recent reports on high quality multiple layer structures grown by MOCVD
suggest that interfaces viewed by HREM are dislocation free and smooth
within 1–2 atomic layers.  Leys et al (1984) attribute observed
roughnesses in the substrate/epilayer interface to contamination of the
substrate with carbon and/or oxygen impurities.

Dislocations, stacking faults and faulted loops are commonly observed in
epitaxial layers grown by LPE.  The usual explanations are substrate
contamination and/or segregation of dopants (Kotani et al, 1977).
Augustus et al (1981) observed stacking faults in MOCVD grown GaAℓAs
layers.  It was suggested that their generation mechanism is analogous
to misfit dislocations and that their growth is maintained by
incorporation of impurities or dopants.

In this paper we studied MOCVD as grown double heterjunction pulsed laser
structures by crossectional TEM.  Possible contamination of the substrate
and excess Aℓ in the GaAℓAs epilayer will be discussed as sources for
defects and mismatch in the substrate/epilayer interface.

## 2. Experimental

The TEM samples consisted of MOCVD grown double heterojunction structures
on (001) GaAs substrates.  Details of the structure are given in Fig. 1.
The MOCVD growth procedure is described elsewhere (Thrush et al, 1984).
For planar viewing in the TEM (along <001>) specimens were chemically
jet-etched from the substrate side with a solution of Br:CH$_3$OH in the
proportion of 1:9.  The GaAs contact layer was taken off prior to the jet

etching procedure by etching first in $H_2O_2$ and ammonia solution (1:19) and afterwards in dilute HCℓ.

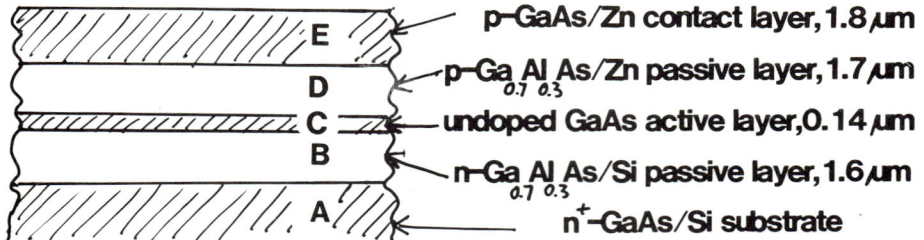

p-GaAs/Zn contact layer, 1.8 μm

p-Ga$_{0.7}$Al$_{0.3}$As/Zn passive layer, 1.7 μm

undoped GaAs active layer, 0.14 μm

n-Ga$_{0.7}$Al$_{0.3}$As/Si passive layer, 1.6 μm

n$^+$-GaAs/Si substrate

Fig.1 Schematic crossection through the double heterostructure

For crossectional viewing (along <110>) the material was cleaved along <110>, mechanically polished down to ~50μm and then ion beam milled with a 3 keV Ar-beam at glancing incidence. Two sets of samples were prepared consisting of "good" laser material (threshold current ~1.5 kA cm$^{-2}$) and "bad" laser material (threshold current ~6.6 kA cm$^{-2}$). TEM was performed with a 200 CX JEOL electron microscope at 200 kV.

3.Results

All samples show stacking faults and dislocations. The stacking faults are intrinsic or extrinsic and bounded by Shockley partials or they are joined by stair rod dislocations to form more complex arrays. Fig.2 shows four stacking faults on adjacent {111} planes forming a "pyramid". They are alternately intrinsic and extrinsic. The origin of the pyramid is calculated to be very close to the substrate interface using the known thicknesses of the epilayers and on the assumption that the stacking faults can penetrate through the layers. This was found to be true from crossectional specimens (see below). Other such truncated pyramids were seen with closely similar dimensions which supports the view that they originate close to a single surface.

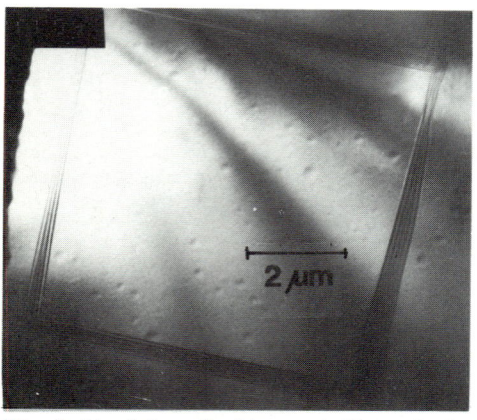

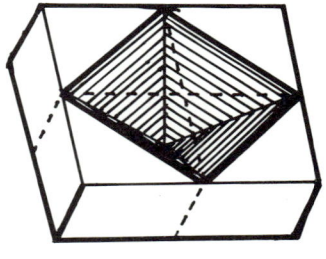

2 μm

Fig.2 Stacking faults on adjacent {111} planes forming a "pyramid". Only the fringes around the base of the pyramid can be seen on the micrograph

Crossectional view of the "good" laser material showed mainly defect free
specimens (Fig.3a). Occasionally stacking faults progressing through the
epilayers were observed. The interfaces between the individual layers
showed narrow light bands which Thrush et al (1984) attributed to an
Aℓ-transient introduced during MOCVD growth. The "bad" laser material
showed a high defect density in the n-passive GaAℓAs layer close to the
substrate/epilayer interface (Fig.3b).

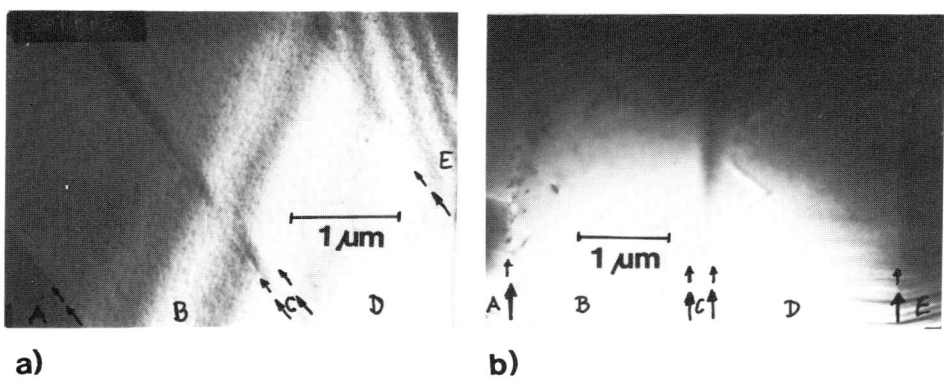

**a)**                                    **b)**

Fig.3 a) Crossection of "good" laser material; threshold current 1.5 kA
cm$^{-2}$; no defects. b) crossection of "bad" laser materials; threshold
current 6.6 kA cm$^{-2}$; accumulation of defects at the substrate/epilayer
interface A/B

The defects were identified as straight dislocations, stacking faults,
large faulted loops (Fig.5) and small loops directly in the interface
(Fig.4). Examination of some of the large loops showed that they are
intrinsic with a Burgers' vector of the type $\frac{1}{6}$ <121>. Closer examination
of the interface itself revealed a fringe structure which had zero
contrast when g was parallel to the boundary and maximum contrast when g
had a large component normal to it (Fig.6). All subsequently grown
interfaces and layers are mainly defect free.

## 4.Discussion

Stacking fault complexes are widely observed in Diamond structures
particularly in Si (Booker, 1964; Ogden et al 1974). They have also been
reported for MOCVD grown GaAℓAs on GaAs (Augustus et al 1981). Ogden et
al found that carbon contamination on (111) substrates could set off the
growth of stacking fault tetrahedra which then propagated with the
epitaxial layer. Augustus et al suggest misfit dislocations as a
possible nucleation agent. In our case the MOCVD reactor is flushed with
trimethyl-Aℓ and trimethyl-Ga vapours when still cold, prior to MOCVD
growth. If there were traces of oxygen in the system the alkyls might
react to form patches of oxide on the substrate. These might then set
off the growth of a stacking fault pyramid. Another nucleation agent
could be islands of Ga left on the substrate after arsenic loss during
heating. The latter process is less likely since the reactor is kept
under an arsine overpressure. Contamination of the substrate could also
account for the accumulation of small loops in the interface.

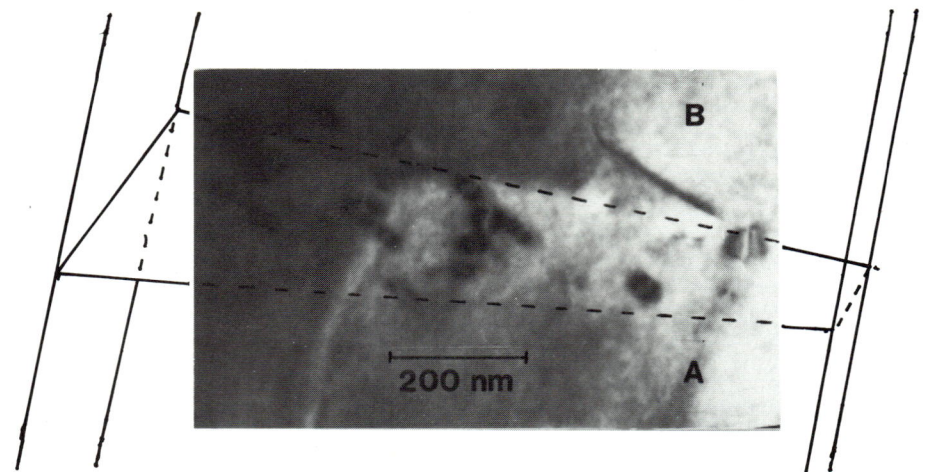

Fig.4 Small loops in substrate/epilayer interface when tilted ~30° with respect to e—beam about <110>

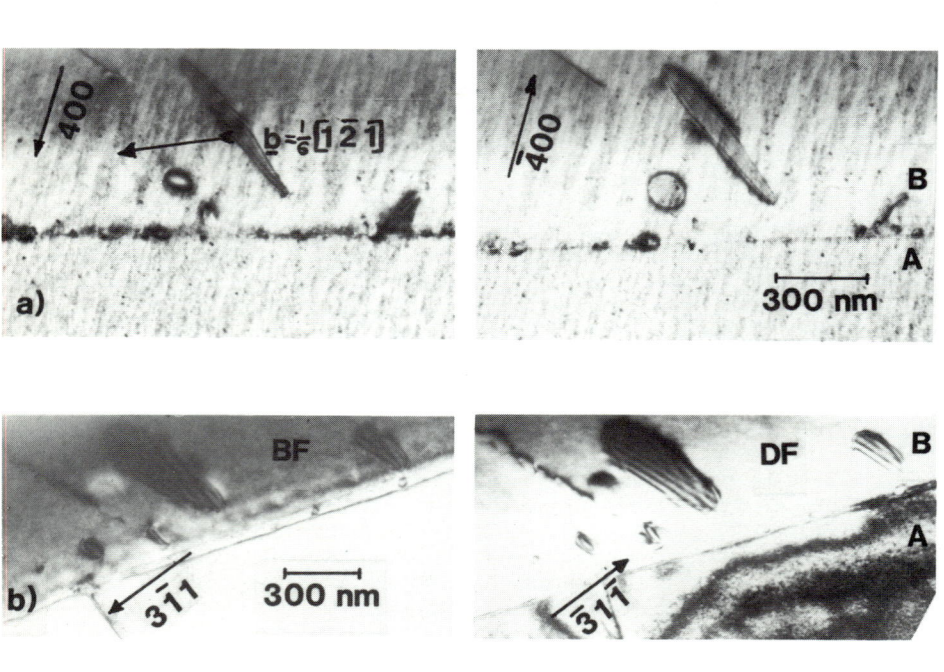

Fig.5 (a)  Faulted intrinsic loop on a ($\bar{1}$11) plane.  Inside/outside contrast for opposite signs of g; s < 0; b = $\frac{1}{6}$ [1$\bar{2}$1]

      (b)  Bright field and centred dark field image for faulted intrinsic loops

Faulted loops were observed by various workers (Stirland et al, 1978). The usual suggestion is that the loop is caused either by microprecipitates or segregation of dopants or contaminants (Ogden et al, 1974), even elemental As (Cullis et al, 1980) on the loop plane. We observe large, faulted, intrinisic loops with bounding Shockley partials in the epilayer next to the substrate. SIMS results as well as our own TEM and some preliminary X-ray microanalysis studies show that there is a very high Aℓ-concentration near the substrate. The Aℓ transient is due to the operation of the valves (Thrush et al, 1984). The incorporation of islands of high concentration of Aℓ or the formation of Aℓ-compounds or Aℓ-precipitates might provide the necessary distortion for a faulted dislocation loop to form. When the final composition in the gas flow is established the fault will not propagate.

It was first thought that the fringes in the interface arise due to the difference in the structure factors for GaAs and GaAℓAs. In order to create the observed number of fringes a composition with a very high Aℓ content is required for several hundred atomic layers.

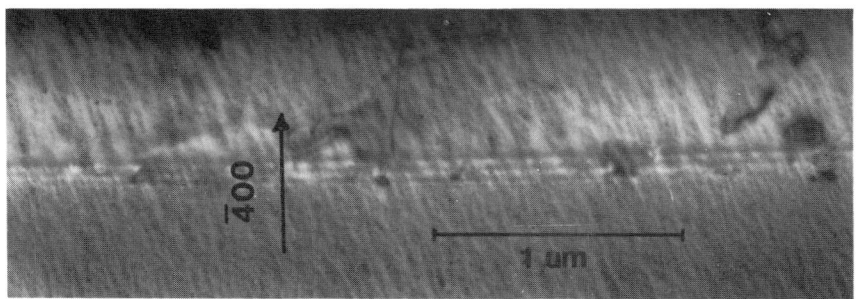

Fig.6 Fringes in substrate/epilayer interface when tilted with respect to e-beam about ‹110›

Structure factor fringes, however, should not show changes in intensity for all our reflections chosen. The fact that the fringes are most intense for g-vectors with a large component normal to the interface and invisible when g is parallel to it make an argument in favour of lattice mismatch more ¯plausible. A mismatch causing displacement only in direction of the epitaxial growth would not be noticeable for reflections parallel to the interface. It could be caused by the condition (contamination, substrate morphology) of the substrate. Regions of mismatch in their turn could contribute to the production of the observed defects near the interface which could relax the localised strain.

### Acknowledgements

We would like to thank SERC for financial support and Dr E J Thrush and Dr J E A Whiteaway for providing the material and for valuable discussion.

References

Augustus P D, Bradley R R, Inst. Phys. Conf. Ser. 60, 325 (1981)
Booker G R, Disc. Faraday Soc. 38, 298 (1964)
Burnham R D, Streifer W, Paoli T L, Holonyak N, Jr, J.Cryst. Growth 68, 370 (1984)
Cullis A G, Augustus P D, Stirland D J, J. Appl. Phys. 51, 2556 (1980)
Kotani T, Ueda O, Akita K, Nishitani Y, Kusonoki T, Ryuzan O, J.Cryst. Growth 38, 85 (1977)
Leys M R, Van Opdorp C, Viegers M P A, Talen-Van Der Mheen H, J.Cryst. Growth 68, 431 (1984)
Nakanisi T, J. Cryst. Growth 68, 282 (1984)
Ogden R, Bradley R R, Watts B E, phys. stat. sol (a) 26, 135 (1974)
Stirland D J, Augustus P D, Straughan B, J. Mater, Sci. 13, 657 (1978)
Thrush E J, Whiteaway J E A, Wale-Evans G, Whight D R, Cullis A G, J. Cryst. Growth 68, 412 (1984)
Whiteaway J E A, Thrush E J, J.Appl. Phys. 52, (3), 1528 (1981)

*Inst. Phys. Conf. Ser. No. 76: Section 7*
*Paper presented at Microsc. Semicond. Mater. Conf., Oxford, 25–27 March 1985*

289

# Investigation of defects in GaAsP superlattices

**M M Al–Jassim and K M Jones**

Solar Energy Research Institute, 1617 Cole Blvd., Golden, Colorado 80401 USA

**Abstract**    GaAsP superlattices have been incorporated in GaAs/GaAsP cascade solar cell structures in an attempt to reduce the dislocation density in the high band gap cell. The structures were grown on GaAs substrates by MOCVD. Defects and irregularities in these structures were investigated by TEM, EBIC, X-ray microanalysis and chemical etching.

## 1.  Introduction

Multijunction cascade solar cells are of increasing importance since they achieve higher conversion efficiency by capturing a larger portion of the solar spectrum.  In a typical multijunction cell, the high and low band gap cells are stacked in such a way that sunlight falls first on the material with the larger band gap.  Lower energy photons not absorbed in the latter are transmitted to the second cell.  The two main problems in cascade cells are:  (i)  the mismatch between the different semiconducting materials making up the cells which often results in a high density of misfit and threading dislocations, (ii)  the difficulty in fabricating a low resistance interconnect between the two cells.  In the present work a novel cascade structure (Wanlass and Blakeslee 1982) incorporating a strained layer superlattice is used to alleviate the above mentioned problems.  The proposed function of the superlattice is to block the propagation of threading dislocations.  The aim of this study is to investigate the structural defects and compositional inhomogeneities and to devise methods to improve the device performance.

## 2.  Experimental

Fig. 1 shows a schematic cross-section of the cascade structure used in this study.  The layers were grown on (100) GaAs substrates by MOCVD using trimethylgallium, $PH_3$, and $AsH_3$.  The growth rate and temperature were 0.1 $\mu m$ $min^{-1}$ and 750°C respectively.  The layer structure consists of:  (i)  a GaAs homolayer containing the low band gap cell, (ii)  a GaAsP graded layer, (iii) a $GaAs_{.7}P_{.3}$ buffer layer, (iv) $GaAs_{1-x}P_x/GaAs_{1-x'}P_{x'}$ superlattice, and (v) a $GaAs_{.7}P_{.3}$ layer matched to the average composition of the superlattice.  This layer embraces the wide band gap cell.

The TEM specimens were prepared by standard techniques and examined at either 100 kV using a JEOL 100cx or 1000 kV using a Kratos EM-1500.  The SEM cleaved cross-sections were first etched in a $KOH/K_4Fe(CN)_6$ solution

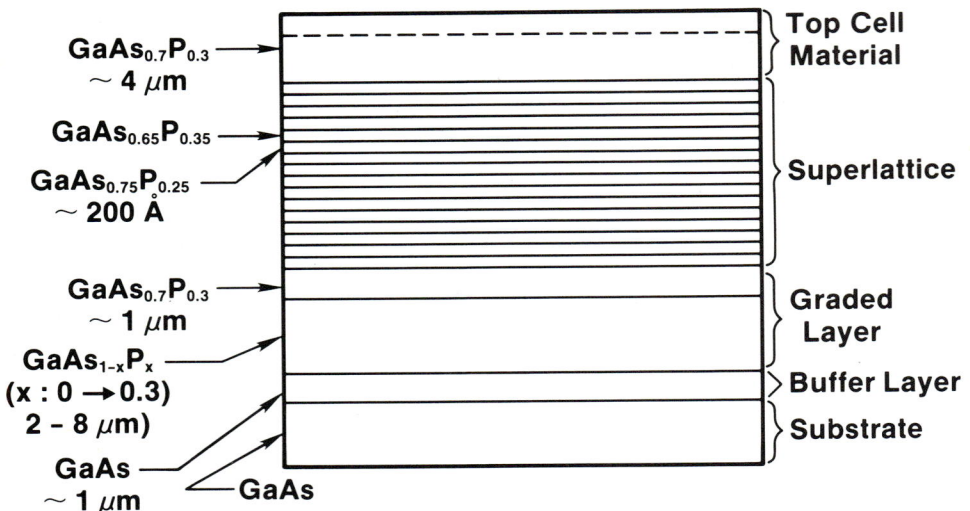

GaAs$_{0.7}$P$_{0.3}$ ~ 4 $\mu$m — Top Cell Material

GaAs$_{0.65}$P$_{0.35}$
GaAs$_{0.75}$P$_{0.25}$ ~ 200 Å — Superlattice

GaAs$_{0.7}$P$_{0.3}$ ~ 1 $\mu$m — Graded Layer

GaAs$_{1-x}$P$_x$ (x : 0 → 0.3) 2 - 8 $\mu$m) — Buffer Layer

GaAs ~ 1 $\mu$m — Substrate

GaAs

Fig. 1  Device Structure

for 3-5 seconds, then examined in the SEM attachment to the 100cx.  The EBIC samples were examined in a JEOL 35c SEM at 25 kV.

## 3.  Results and Discussion

Fig. 2 is a TEM cross-sectional montage revealing the overall dislocation configuration in a typical specimen that contains a GaAs$_{.75}$P$_{.25}$/ GaAs$_{.65}$P$_{.35}$ superlattice.  It shows a region extending from the GaAs substrate to the specimen surface.  It is evident from this figure that the graded layer contains an extensive and dense network of dislocations at each compositional step.  These dislocations lie mainly in the (100) plane along the [011] and [0$\bar{1}$1] perpendicular directions.  Most of them are 60° dislocations with inclined Burgers vector of the type $\frac{a}{2}$ <110>. Additionally, threading dislocations varying in density and orientation tend to originate in the graded layer and continue to propagate through the constant composition region.  Most of these dislocations are inclined to the [011] plane of the cross-section and therefore soon run out of the foil.  They are characterized by their oscillatory contrast.  However, some threading dislocations tend to align themselves along the [100] direction.  TEM cross-sectional examinations revealed that the former type may bend over by slipping on {111} planes to form part of the misfit dislocation networks in the graded layer, while the latter type tends to propagate through the graded layer.

Since the threading dislocation density is of paramount importance in this work, it was studied as a function of grading rate (i.e. the total thickness of the graded layer) and the number of steps in that layer.  A series of samples each consisting of a graded layer and a GaAs$_{.7}$P$_{.3}$ top layer was grown for that purpose.  It was found that the threading dislocation density in the constant composition layer steadily  decreased

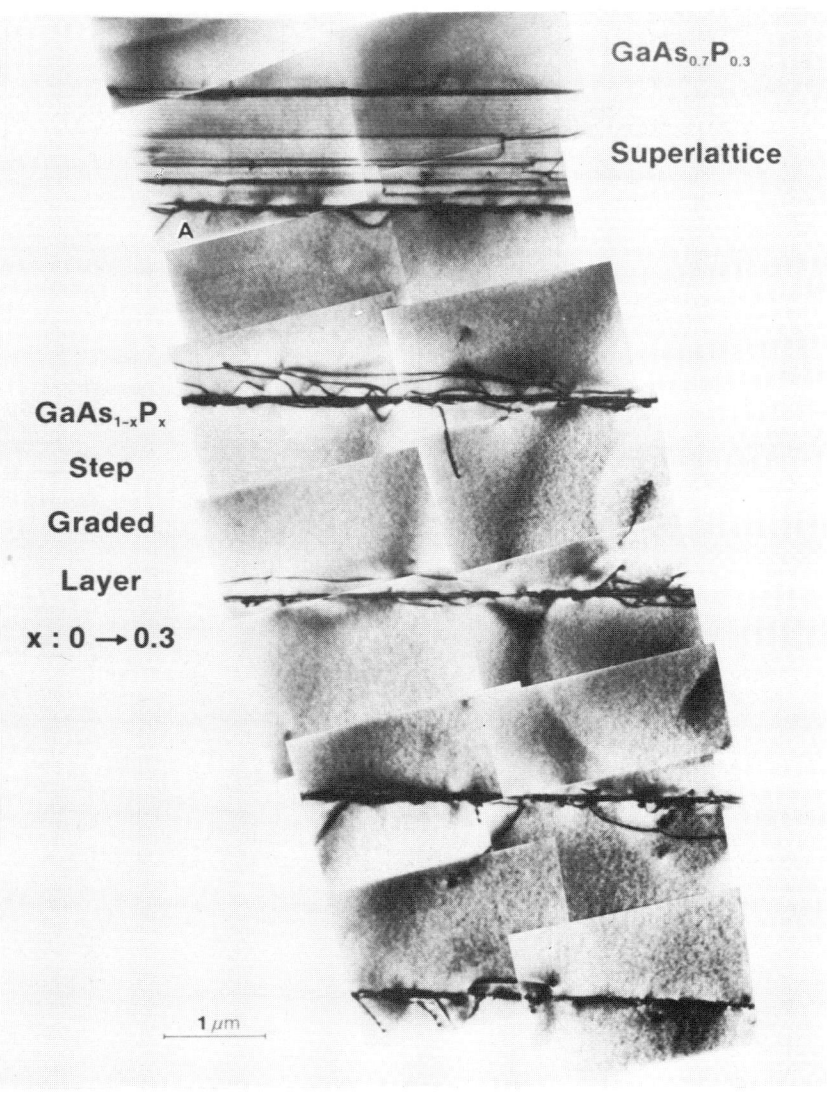

GaAs$_{0.7}$P$_{0.3}$

Superlattice

GaAs$_{1-x}$P$_x$

Step

Graded

Layer

x : 0 → 0.3

A

1 $\mu$m

Fig. 2   TEM cross-sectional montage of a typical specimen
containing a planar superlattice

upon decreasing the grading rate to ~4 at. % $\mu$m$^{-1}$.  Fig. 3 is an EBIC
micrograph of a GaAs$_{.7}$P$_{.3}$ layer grown on a graded layer using the latter
grading rate.  The threading dislocation density in this sample is ~8x10$^5$
cm$^{-2}$.  However, further reduction in the grading rate did not result in
any further reduction in dislocation density.  The number of steps proved
to play an equally crucial role.  Samples having 1,2,3,4,5,7, and 10
compositional steps were examined.  The dislocation density decreased
with increasing the number of  steps  up  to  4  steps,  but  no  further

reduction was achieved by using 5,7,10 steps, or continuous grading. The minimum dislocation density obtained in the present work by using a graded layer alone is ~$7 \times 10^5$ cm$^{-2}$.

The superlattice (SL) in Fig. 2 is planar and uniform. It contains a high density of misfit dislocations, most of which are believed to form by the bending of threading dislocations propagating from the graded layer. The morphology of the SL was extensively studied in this work. Fig. 4 shows a pair of SEM micrographs of two cleaved and

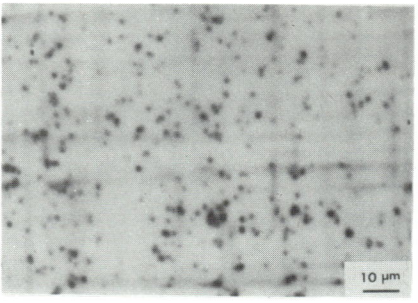

Fig. 3  EBIC micrograph of a GaAs$_{.7}$P$_{.3}$ cell grown on a step graded GaAsP layer

etched SL structures. The morphology of such cross-sections exhibited a wide range of appearance. It varied from flat, parallel, nearly perfect SL layers (Fig. 4a) to extremely distorted layers. The nature of the distortion is clearly revealed in Fig. 4b where it is seen to consist of two components. One is a progressive bending or warping of the sublayers. The other is roughly sinusoidal deviation in and out of the cleavage plane, caused by nonuniform attack of the etchant over the originally planar cleaved surface. The sites of maximum etching are also the centers of dense dislocation clusters, as revealed by TEM. Fig. 5 is a TEM cross-sectional micrograph of a typical distorted SL. The distorted regions are depicted as dark areas with high density of threading dislocations emanating from them and propagating through the top layer to the specimen surface. X-ray microanalysis studies of these regions performed on thin TEM foils revealed that they contain 30-50% higher phosphorus concentration than the adjacent undistorted regions. This indicates that the high localized mismatch caused by the compositional variation is the cause of the high density of threading dislocations generated by distorted SL's.

Many growth parameters were investigated in an attempt to eliminate the distortions. The largest effect was achieved be reducing the interlayer

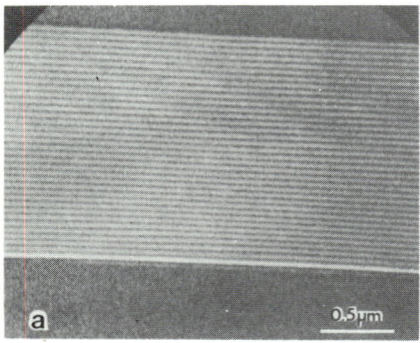 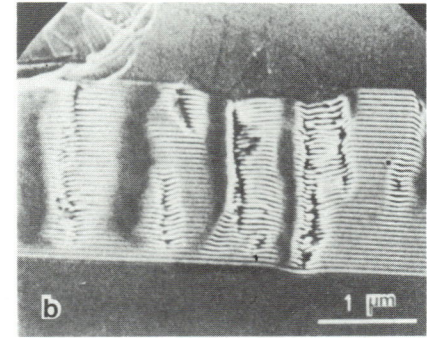

Fig. 4  SEM micrographs of etched cross-sections of:  (a) planar super-lattice (b) distorted superlattice

misfit, i.e., the interlayer compositional difference $\Delta x$, while keeping the average value of x constant and matched to that of the layers above and below the SL. For example, a $\text{GaAs}_{.75}\text{P}_{.25}/\text{GaAs}_{.65}\text{P}_{.35}$ ($\Delta x=0.1$) super-lattice grown at a rate of 0.04 $\mu m$ $min^{-1}$ is planar (Fig. 2 and 4a), whilst a $\text{GaAs}_{.80}\text{P}_{.20}/\text{GaAs}_{.60}\text{P}_{.40}$ ($\Delta x = 0.2$) is severely distorted (Fig. 4b and 5). The second factor determining the amount of distortion was the growth rate of the SL. It was found that faster growth rate favours planar and more uniform superlattices. Lowering the growth temperature also reduced the tendency toward layer distortion. However, it was not as dramatic an effect as that of the above mentioned factors.

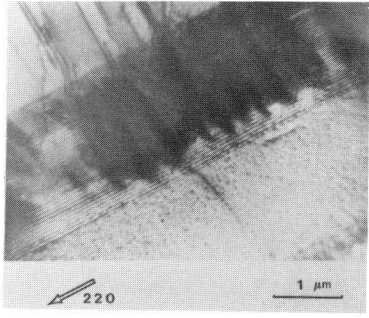

Fig. 5 TEM cross-section of a distorted superlattice

This phenomenon of morphological distortion in GaAs/GaAsP superlattices has been well characterized in the present work, and the growth conditions to avoid it have been unveiled. However, its origin is not very clear at this stage. Several models such as diffusion-induced disorder, constitutional supercooling, and impurity segregation have been suggested (Blakeslee et al 1985).

As mentioned above, the main objective in using a SL in this work is to block the propagation of threading dislocations from the graded layer to the high band gap cell. In order to avoid the morphological distortion described earlier, we have mainly used $\text{GaAs}_{.75}\text{P}_{.25}/\text{GaAs}_{.65}\text{P}_{.35}$ SL's for this purpose. The bending and elimination of threading dislocations was investigated as a function of the thickness of individual layers in the SL and the number of layers. The bending of threading dislocations is energetically favourable if the thickness of the individual layers exceeds a critical thickness $h_c$ given by (Matthews et al 1976):

$$h_c = \frac{b(1-\nu \cos^2 \alpha)}{8\pi\ f(1+\nu)\cos\lambda}\ \ln\ (\frac{h_c}{b})$$

Using the following values: b=4 Å, $\nu$ =0.33, f=0.0036 and cos $\lambda$ =cos $\alpha$=0.5, $h_c$ was calculated to be ~250 Å. A series of samples having layer thicknesses in the range 130 Å–350 Å and a number of SL layers ranging from 10-80 was examined. The density of interface dislocations in the SL was used as a qualitative measure of the bending efficiency of the SL. The density of threading dislocations, measured from EBIC dark spot count, in the $\text{GaAs}_{.7}\text{P}_{.3}$ layers above and below the SL was used as a quantitative measure. Superlattices having layers <200 Å thick exhibited little or no bending, while those with layer thickness ~250 Å contained dense dislocation networks lying in the (100) plane. These are formed by the bending of threading dislocations into misfit orientation (A in Fig. 2). TEM plan-view examination of SL's with 250 Å layers revealed that these dislocations lie on perpendicular <110> directions and have similar characteristics to the misfit dislocations in the graded layer. The threading dislocation density in this type of specimens was $9\times10^4$ cm$^{-2}$ – $2\times10^5$ cm$^{-2}$ and $7$-$9\times10^5$ cm$^{-2}$ in the layers above and below the SL respectively. Superlattices having 350 Å layers did not effect any further reduction nor did they generate "new" dislocations. The number of layers in the SL proved to be less crucial than the thickness of these layers. Superlattices having 20-30 layers proved to be as efficient a dislocation filter as those with 60 layers. TEM cross-sectional examination revealed that most bending occurred in the first 10-20 layers.

The inclination of dislocations played a major role. Most inclined dislocations were observed to bend over to misfit orientation (A in Fig. 2) while dislocations along or close to the [100] direction propagated through the superlattice with little or no change in orientation (Fig. 6). This is thought to be due to the fact that only inclined dislocations can glide on {111} planes.

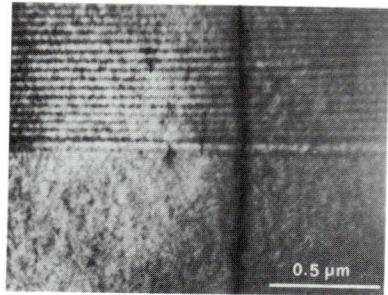

Fig. 6  TEM cross-section of a threading dislocation propagating through the superlattice

The dislocation reduction effected by the SL was only a factor of 5-10. This is believed to be due to: (i) About 10% of the threading dislocations are along the [100] direction and therefore do not bend; (ii) The mismatch between the superlattice as a whole and the layers above and below. Matthews and Blakeslee (1976) pointed out that such a mismatch results in the generation of new dislocations. The cross-sectional montage in Fig. 2 supports this and reveals the generation of extra dislocations at the top superlattice interface.

## Acknowledgments

The authors wish to thank A. Kibbler, J. M. Olson and M. W. Wanlass for growing the samples, and A. E. Blakeslee for valuable discussions, and C. R. Herrington and A. Mason for technical assistance. This work was supported by the U.S. Dept. of Energy under Contract No. DE-AC02-83CH10093.

## References

Blakeslee A E, Kibbler A and Wanlass M W 1985 Superlattices and Microstructures 1 339
Matthews J W and Blakeslee A E 1976 J. Cryst. Growth 32 265
Matthews J W, Blakeslee A E and Mader S 1976 Thin Solid Films 33 253
Wanlass M W and Blakeslee A E 1982 Proc. 16th Photovoltaics Specialists Conf. San Diego CA pp 584

*Inst. Phys. Conf. Ser. No. 76: Section 7*
*Paper presented at Microsc. Semicond. Mater. Conf., Oxford, 25–27 March 1985*

# A TEM examination of AlGaAs/GaAs superlattice structures grown by MBE

M R Taylor, M Hockly, D A Andrews and G J Davies

British Telecom Research Laboratories, Martlesham Heath, Ipswich,
IP5 7RE, UK.

Abstract    A selection of GaAs/ $Al_{.3}Ga_{.7}As$ superlattice structures,
grown both on GaAs and on $Al_{.3}Ga_{.7}As$ buffer layers, has been studied by
TEM using specimens prepared in cross-section. It has been shown that
there is an optimum temperature range for the growth of high quality
AlGaAs layers which is different from the optimum temperature range for
GaAs layers. Appropriate conditions are established for the growth of
good quality multi-quantum well structures with planar interfaces on
AlGaAs buffer layers.

## 1. Introduction

Multi-quantum well/ superlattice structures are finding increasing
importance in the device world both for their optical and their electron
transport properties. They have considerable potential as light sources
for use in optical communications systems, but their efficient operation
depends particularly on the exact configuration and composition of the
constituent layers. In this paper, TEM observations of a range of
AlGaAs/GaAs structures containing superlattices, and grown by molecular
beam epitaxy (MBE), are related to growth conditions. A study of the
conditions under which undulations develop during the growth of AlGaAs
resulted in a method for optimising the control of growth to achieve
planar interfaces, whilst still maintaining desirable optical and
electrical properties. The structures described are not intended
specifically for device applications, but are experimental in nature.
Their chief function is the development of growth techniques and of
theoretical ideas of multi-quantum well (MQW) operation.

## 2. Experimental

The superlattice structures were grown on (100) GaAs substrates by MBE
(Andrews et al 1984 and Davies and Andrews, 1984) using a VG V80H system
incorporating effusion cells and microprocessor controlled shutters.
Temperatures were monitored with a thermocouple mounted behind the
rotating substrate holder block. Calibration with respect to specimen
temperature was by dual wavelength infra-red pyrometry, melting points and
observations by RHEED. The nominal composition of all the AlGaAs layers
was 30% AlAs when grown at 630 deg C. Growth rates were 1 to 1.5 μm per
hour with group V:group III flux ratios between 5:1 and 10:1, as measured
by a moveable ion gauge. The arsenic species supplied was largely $As_4$.

Transmission electron microscopy was carried out at 200 kV using a JEOL
JEM 200CX TEMSCAN system. Cross-sectional samples of (011) orientation

were prepared by argon ion thinning at 5 kV, 20 μA and 15 deg incidence (Fletcher et al 1981). All the micrographs were recorded using 200 dark field imaging to take advantage of the large structure factor (and hence contrast) differences between AlAs (or AlGaAs) and GaAs for this reflection (Petroff et al 1978). In this imaging mode, GaAs appears relatively dark and AlGaAs brighter, the brightness increasing with increasing Al concentration. At the 200 dark field orientation, the projected width of a plane inclined at the Bragg angle in a 100 nm thick specimen is only 0.5 nm. Determinations of superlattice periods were made from measurements on diffraction patterns of superlattice spot spacings.

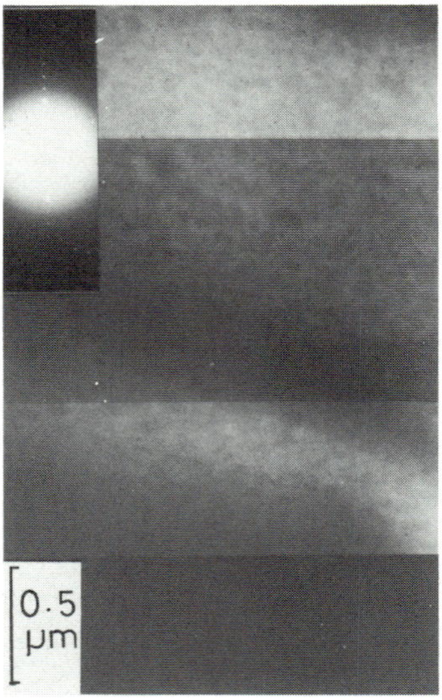

SURFACE

AlGaAs
($710^{\circ}$C)

SUPER-
LATTICE
(GaAs 4 nm/
AlGaAs 7 nm,
$630^{\circ}$C)

AlGaAs
($710^{\circ}$C)

GaAs
BUFFER
($630^{\circ}$C)

SUBSTRATE

### 3. Results and Discussion

Results will be described for a range of AlGaAs/GaAs superlattice structures with

Fig. 1 Typical structure: specimen F, with its specific growth temperatures and dimensions. Diffraction pattern :- 200 matrix spot and superlattice spots.

various periods and width ratios, and whose surrounding layers are in some cases GaAs, and in others AlGaAs. No crystalline defects were found in any of the structures. An example of a typical structure is given in Fig 1.

Factors needing attention when considering superlattice quality include layer flatness and uniformity. Previous workers (eg Petroff et al 1978 and 1984) have discussed the formation of surface undulations in AlGaAs layer growth, and the rapid reversion to planarity brought about by subsequent overgrowth with GaAs. In further characterising this phenomenon, we begin by presenting results for superlattices containing AlGaAs layers which were grown directly on GaAs buffer layers at 630 deg C. Consider first specimen A (Fig 2) grown on a buffer layer 0.4 μm thick. All regions of the superlattice have interfaces which are flat to within approximately 1 nm. Layers are of uniform thickness and constant spacing throughout. Specimens grown under similar conditions with much smaller layer widths behave in the same way (specimen B, Fig 2).

The situation is quite different for superlattice layers grown at the same temperature on an AlGaAs layer. Superlattice structure C (Fig 2) was grown on an AlGaAs confining layer 1 μm thick under the same growth conditions as structures A and B. The interface is undulating with deviations from planarity of up to 10 nm, and an undulation wavelength of typically 300 nm. Growth of 5 to 10 periods of the superlattice structure is sufficient to remove these undulations. Small deviations from planarity, typically up to 2 nm within a lateral distance of 10 nm, and typically spaced 50 to 100

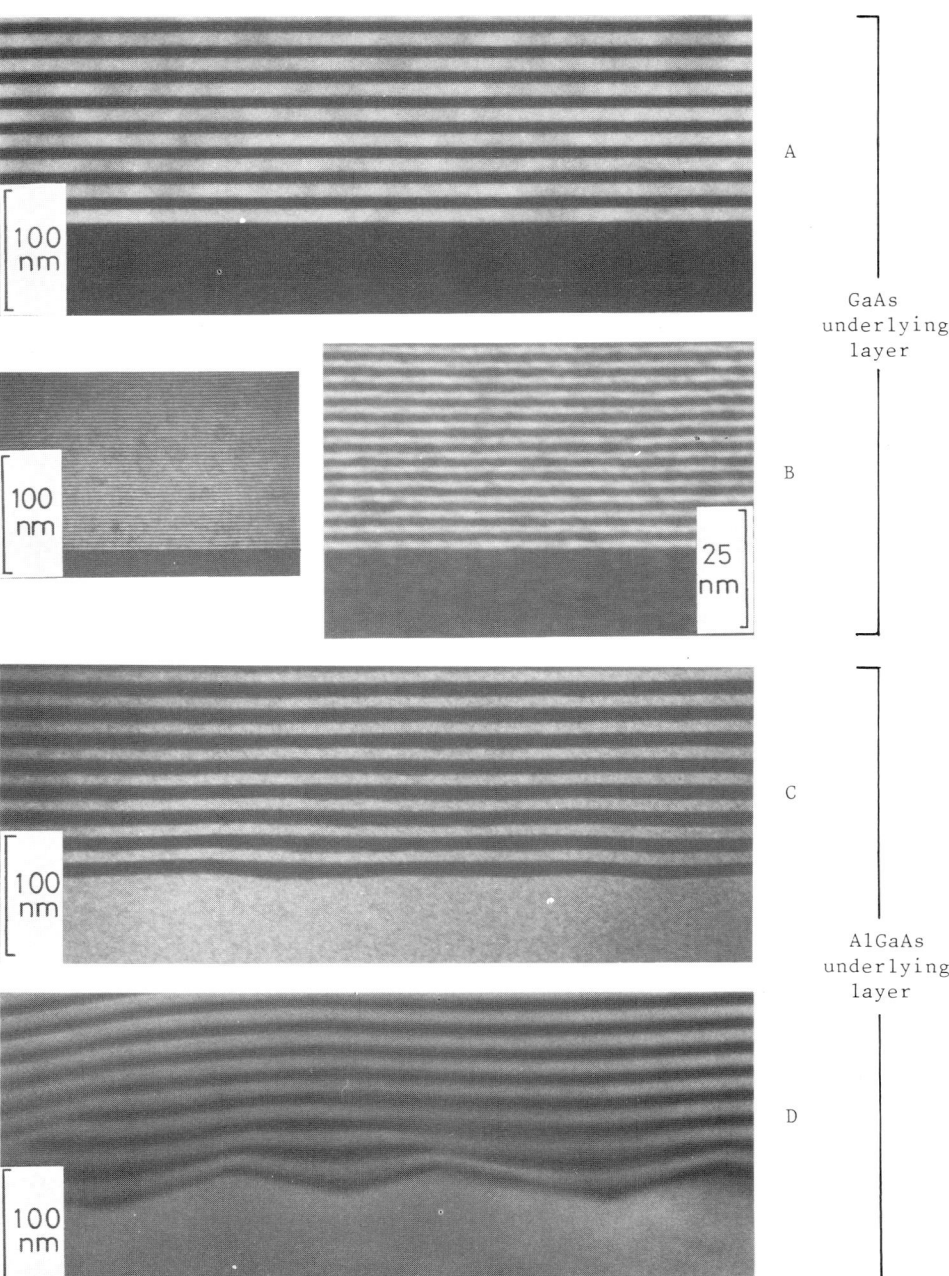

Fig. 2 Specimens A to D, grown at 630 deg C, showing the lower interface (nearest to the substrate) of the superlattice. Layer widths within the superlattice are (GaAs/ AlGaAs) :- A  8 nm/ 11 nm ; B  1.7 nm/ 1.7 nm ; C  11 nm/ 11 nm ; D  8 nm/ 10 nm .

nm apart, may still be observed throughout the superlattice, at the upper surfaces of the constituent AlGaAs layers. Undulations have redeveloped at the top surface of the upper AlGaAs confining layer, even though its interface with the superlattice is planar.

Much worse deviations from planarity are possible as can be seen in specimen D (Fig 2) grown under similar conditions to specimen C. Even though the GaAs buffer layer was planar, the 1 μm thick overlying AlGaAs confining layer had developed facets with deviations from planarity of up to 20 nm. These were rapidly smoothed by overgrowth with the superlattice, though they were so gross initially that they were not totally eliminated. The upper AlGaAs confining layer, grown after the superlattice, again developed a badly faceted surface. Indeed the deviations from planarity at the surface were readily observed by optical microscopy (Nomarski interference contrast) as an uneven linear structure perpendicular to the plane of the TEM cross-section, (011), and approximately parallel to the [01$\bar{1}$] direction. Faceted striations were observed in both of the AlGaAs confining layers in this specimen. Also present are mini-deviations from flatness within the superlattice stack, as described for the previous specimen.

Since useful MQW devices often require superlattices to be sandwiched between thicker AlGaAs confining layers for optical and electrical confinement, better control of interface planarity in such structures is clearly necessary. The next sample to be described shows the effects of changes of growth temperature, at constant incident reactant fluxes (Fig 3). Comparatively thick (~ 100 nm) alternate layers of AlGaAs (nominally 30% AlAs) and GaAs were grown at successively higher temperatures, the temperature changeover taking place when each AlGaAs layer had been grown for half its total growth time. A common growth time was used for all the AlGaAs layers, and a common but different growth time was used for all the GaAs layers. Undulating interfaces were found at the tops of AlGaAs layers grown both at low and high temperatures, but layers grown at 680 deg C and 690 deg C had flat interfaces. The amplitudes of the undulations were mostly less than 10 nm, with wavelengths ranging from 100 to 170 nm. All interfaces at the tops of GaAs layers were planar.

Some of the results may be accounted for on the basis of a consideration of the thermodynamic

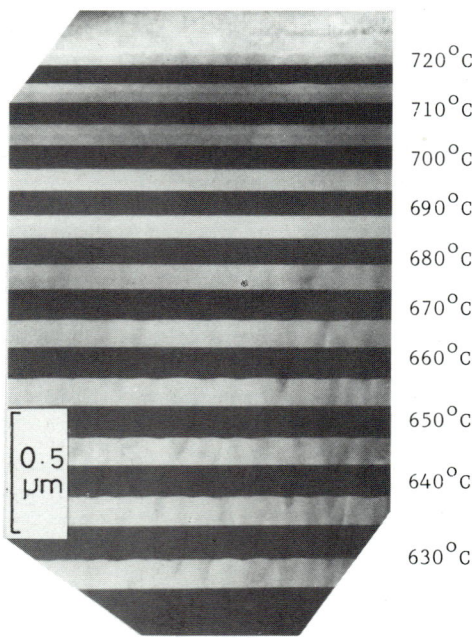

720°C
710°C
700°C
690°C
680°C
670°C
660°C
650°C
640°C
630°C

0.5 μm

Fig. 3 Comparatively thick (~ 100 nm) alternating layers of GaAs and AlGaAs grown, with common growth times, at successively higher temperatures (indicated above).

aspects of MBE growth, (Heckingbottom 1984), and will receive detailed
treatment later (Andrews et al 1985). A brief summary appears below. For
the growth of GaAs and AlGaAs under the normally used conditions of an
arsenic rich surface, and for growth temperatures below 600 deg C, it can
be assumed that all the group III arriving species are incorporated. Under
these conditions, the ratio of Al to Ga in a growing AlGaAs layer is the
same as the ratio of the Al beam flux to the Ga beam flux. If the
temperature is raised, increased evaporation of the group III species from
the surface takes place.  This causes a decrease in growth rate,
manifested in Fig 3 as a decrease in layer thickness. The rates of loss of
Ga and Al are, however, not in the same ratio as the incident beam fluxes,
but are proportionally greater for Ga than for Al. AlGaAs which has been
grown at a sufficiently high temperature therefore has a compositional
ratio of Al to Ga which is higher than the ratio of the incident beam
fluxes, and which increases with increasing temperature. This effect is
manifested in Fig 3 as a brightening in contrast at the positions in the
AlGaAs layers where the temperature was stepped up. It is not, however,
obvious why such phenomena should result in undulating surfaces at the
higher temperatures in AlGaAs, but not in GaAs. At lower temperatures
(below 680 deg C) the undulations are possibly associated with the
hindered dissociation of $As_4$ tetramers to $As_2$, and a consequent
inefficient supply of As to appropriate lattice sites. The presence of Al
is presumed to make control of the surface population of As atoms
relatively more difficult, thus accounting for the difference between GaAs
and AlGaAs surfaces.

For particular growth conditions there was an optimum temperature range (~
680 - 690 deg C) for the growth of AlGaAs layers with flat interfaces.
A different optimum temperature (approximately 630 deg C) was appropriate
for the growth of GaAs, if excessive re-evaporation was to be avoided, and
good optical and electrical properties were to be achieved (Morkoc et al
1982). To achieve the desired structures, it was decided to grow AlGaAs
confining layers near their optimum temperature so that their surfaces
remained flat, and to grow the multi-quantum well superlattice stack near
the optimum temperature for GaAs. Our results demonstrate that the
AlGaAs layers within the superlattice then have little opportunity to
develop undulations before the next GaAs layer is grown. In principle,
good results should also be achievable if the whole structure were to be

Fig. 4    Specimen E, showing the lower interface of the superlattice.
Growth temperatures for the various regions are shown. Layer widths within
the superlattice were :- GaAs 4 nm; AlGaAs 6 nm.

grown at the higher temperature, but this approach was not pursued in these initial experiments because of possible additional difficulties in controlling GaAs growth rates.

Specimen E (Fig 4) shows the results of growing a superlattice on a 0.27 µm thick AlGaAs confining layer by the above two-temperature method. Both the initial interface between the underlying GaAs buffer and the lower AlGaAs confining layer (not shown), and the interface between the lower AlGaAs confining layer and the superlattice are flat to within 2 nm. Close examination of the lower AlGaAs confining layer adjacent to the first layer in the superlattice stack reveals a band of darker contrast approximately 50 nm thick, which may be interpreted as representing the region formed after the temperature had been dropped to 610 deg C in preparation for growth of the superlattice. In specimen F, (Fig 1),the superlattice was grown on a 0.7 µm AlGaAs confining layer using the two-temperature method. Gentle undulations were found at the interface between the bottom AlGaAs confining layer and the superlattice. Their amplitude was approximately 5 nm, and average spacing approximately 1500 nm. Both these structures represent a considerable improvement on layer structures grown at 630 deg C throughout.

4. Conclusions

1. The existence of optimum temperatures for the growth of high quality defect-free AlGaAs and GaAs layers with planar interfaces was established, for a particular set of incident reactant fluxes.

2. It was shown that a two temperature growth technique may be used to achieve high quality AlGaAs/GaAs superlattice structures with planar interfaces in the case where the confining layers are AlGaAs.

5. Acknowledgements

Acknowledgement is made to the Director of Research, British Telecom, for permission to publish this paper, and to the Department of Trade and Industry for partial funding of this work under 'JOERS', the Joint Opto-Electronics Research Scheme.

References

Andrews D A, Heckingbottom R and Davies G J 1985 to be published
Andrews D A, Scott E G, Houghton A G N, Rodgers P H and Davies G J 1984 Proc. 3rd Int. MBE Conf, San Francisco, August 1984, to be published in J. Vac. Sci. Tech.
Davies G J and Andrews D A 1984 Vacuum 34 543
Fletcher J, Titchmarsh J M and Booker G R 1980 Inst. of Phys. Conf. Ser. No. 52 153
Heckingbottom R 1984 Proc. 3rd Int. MBE Conf, San Francisco August 1984, to be published in J. Vac. Sci. Tech.
Morkoc H, Drummond T J, Kopp W and Fischer R 1982 J. Electrochem. Soc. 129 824
Petroff P M, Gossard A C, Wiegmann W and Savage A 1978 J Crystal Growth 44 5
Petroff P M, Miller R C, Gossard A C and Wiegmann W 1984 Appl. Phys. Lett. 44 217

*Inst. Phys. Conf. Ser. No. 76: Section 7*
*Paper presented at Microsc. Semicond. Mater. Conf., Oxford, 25–27 March 1985*

301

# The structure of ion implanted Al$_x$Ga$_{1-x}$As/GaAs superlattices

B C De Cooman, S H Chen, C B Carter, J Ralston*and G D Wicks*

Department of Materials Science and Engineering, Bard Hall,
Cornell University, Ithaca, New York 14853, USA
*School of Electrical Engineering, Phillips Hall,
 Cornell University, Ithaca, New York 14853, USA

Abstract    Cross-sectional Transmission Electron Microscopy (XTEM) and
Raman Spectroscopy were used to study the defect structure of annealed
high dose ion-implanted Al$_x$Ga$_{1-x}$As/GaAs (x = 0.3) superlattices.  The
results show clearly that the amount and depth of superlattice layer
ion-beam mixing depends on the ion mass and fluence used during
implantation.  In superlattices that retain their structure after
implantation and annealing the distribution of the defects is
inhomogeneous: most defects are nucleated in the GaAs layer.

## 1.  Introduction

Localized mixing and doping of superlattices has recently been proposed as
a method for obtaining regions of different bandgap and mobility monolith-
ically integrated on the same substrate [1].  Diffusion or ion-implantation
of photomasked superlattice structures can, in principle, be used to
achieve any type of required integrated circuit.

The results reported show a wide scatter in experimental conditions depending on the choice of ion mass, ion energy and dose, anneal time and temperature used.  Consequently considerable disagreement regarding the extent
of superlattice mixing still seems to exist in the literature.  Doping by
diffusion of Zn in Al$_x$Ga$_{1-x}$As/GaAs [2,3] and In$_x$Ga$_{1-x}$As/GaAs [4] superlattices appears to disorder the superlattice entirely and results in a single
layer of bulk crystal with a composition determined by the superlattice
layer composition x and thickness.  Similar results have been reported for
the diffusion of Si into Al$_x$Ga$_{1-x}$As/GaAs superlattices by Meehan et al.
[5].  Coleman et al. [6] have studied the implantation of 375 keV Si into
AlAs/GaAs superlattices and found, by means of photoluminescence and scanning electron microscopy (SEM), that a low mobility Al$_x$Ga$_{1-x}$As alloy is
formed in the implanted region.  According to the Rutherford Back-Scattering (RBS) study of Myers et al. [7], the structure of a GaAs$_x$P$_{1-x}$/GaP
superlattice is not lost after the implantation of 75 keV Be$^+$ ions and
subsequent anneal at 825°C for 10 min.  Recently Picraux et al. [1] have
reviewed the implantation of Be, N, Si, and Zn into In$_x$Ga$_{1-x}$As/GaAs
strained-layer superlattices and found that the superlattice structure
survives the implantation and annealing treatment in most cases.

In the present work a different experimental approach, has been followed:
cross-sectional TEM and Raman Spectroscopy have been used to assess the

structural aspects of ion-implanted $Al_xGa_{1-x}As/GaAs$ superlattices. Raman scattering can be used to observe the vibrational spectra of each individual layer in a superlattice [8]. For $Al_xGa_{1-x}As$ alloys the Raman spectrum shows essentially 2 peaks, one GaAs-like mode and the other AlAs-like mode [9]. In the backscattering geometry from a {100} type surface only GaAs-like and AlAs-like LO-phonons appear in the Raman spectrum. The position of the peak in the Raman spectrum is determined by x, the ternary allow composition. The main drawback of the technique is the lack of precision in estimating the exact penetration depth of the laser beam.

(200)-Dark-field electron microscopy of superlattices yields cross-sectional images which are sensitive to the composition of the superlattice layers; the contrast in those images can be used to determine the Al concentration within the $Al_xGa_{1-x}As$ ternary [10,11]. High-resolution microscopy and weak-beam electron microscopy were used to study the defects generated by the ion-implantation process.

2. Experimental

In the present work 600 nm thick MBE-grown $Al_xGa_{1-x}As/GaAs$ superlattices (x = 0.3) of 10 nm thick GaAs layers alternating with 10 nm thick $Al_xGa_{1-x}As$ layers or 5 nm thick GaAs layers alternating with 10 nm thick $Al_xGa_{1-x}As$ layers were implanted with Be (80 keV), Mg (100 keV), Si (240 keV) and Se (175 keV) at doses of $1\times10^{15}$ $cm^{-2}$ and then annealed at 640 °C in an inert gas atmosphere for 4 hrs. The specimens were prepared for cross-sectional electron microscopy by mechanical polishing and ion-milling. Implantation damage distribution was calculated using a Monte Carlo computer program [12].

3. Results and Discussion

After Si-implantation and annealing, the Raman Spectra show only GaAs-like and AlAs-like LO-phonons corresponding to an $Al_{0.22}Ga_{0.78}As$ alloy indicative of extensive intermixing. The results of the Raman spectroscopy study have been reported elsewhere [13].

In cross-sectional TEM, three distinct regions can be distinguished in annealed ion-implanted superlattices. In the topmost region the damage has caused the formation of an amorphous layer. This amorphous region is only present in the case of Se implantation: it partially recrystallizes during implantation. It is this dynamical recrystallization which results in the formation of stacking-faults and microtwins. In the second region either total, or at least appreciable, ion-beam mixing of the GaAs and $Al_xGa_{1-x}As$ layers occurs. The extent of the intermixing depends on whether the topmost layers are amorphized, because the interdiffusion of crystalline GaAs and $Al_xGa_{1-x}As$ layers is extremely small (diffusion coefficient, D $\leq 10^{-18}$ $cm^2/sec$) [14]. Crystal-lattice defects are not present in this region. This type of region occurs in Mg, Si, and Se implanted superlattices and is absent in the case of Be implantation. The third region contains many small dislocation loops which have formed by the agglomeration of point defects during the annealing process. Below this defect region the superlattice has essentially retained its as-grown structure. Figure 1 shows a (200) dark-field image of an annealed, Se implanted superlattice in which the three regions discussed above are clearly visible. Figure 2 shows an image of a Si-implanted superlattice before annealing. Except for the presence of regions of micro-twins in the dynamically recrystallized region and a region of dislocation loops below the surface, the superlattice has retained its original structure. Table 1 summarizes the results from the

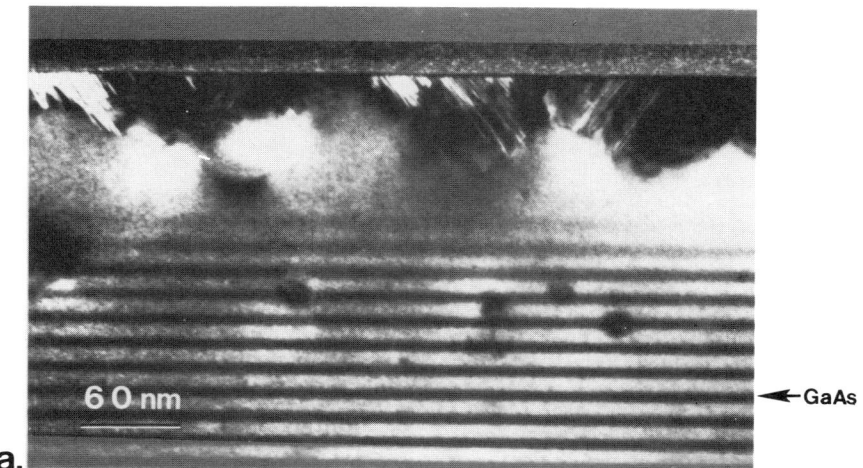

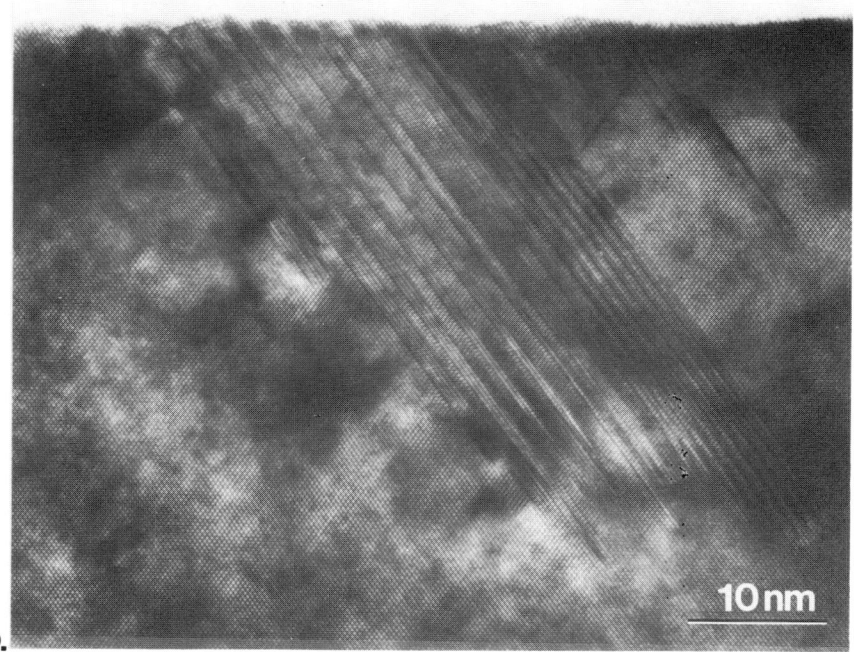

Fig.1. Cross-sectional images of an annealed Se-implanted superlattice
(a) (002) dark-field image.
(b) lattice image of the topmost layers.

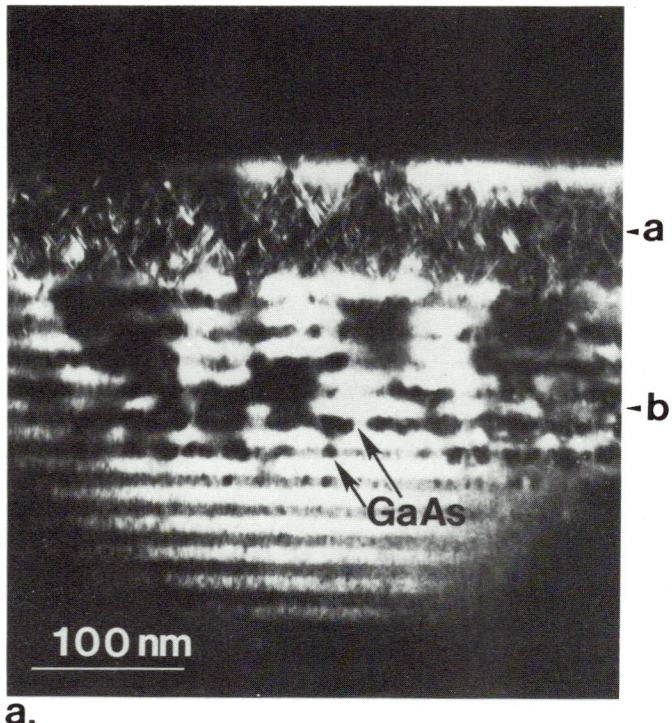

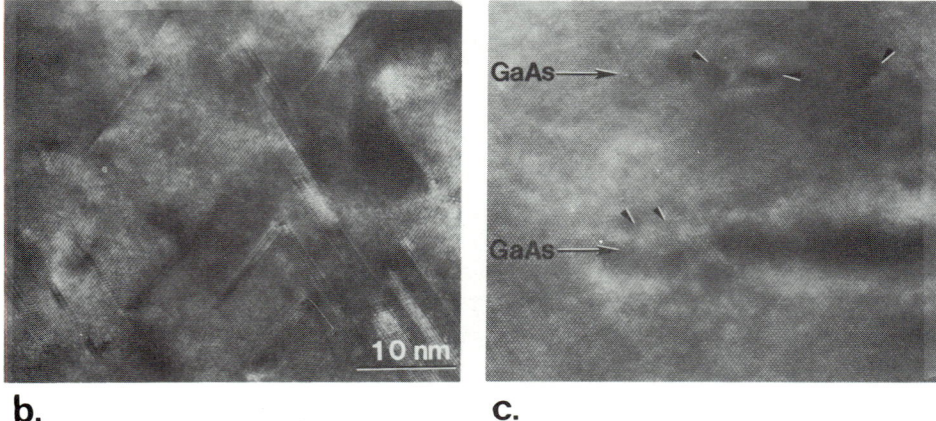

Fig. 2. Cross-sectional images of an unannealed Si-implanted superlattice
showing (a) preferential formation of defects in the GaAs layers.
(b) Lattice image of dynamically recrystallized region and (c)
region with end-on dislocations(arrows) in the GaAs layers.

Table 1.  Width of defect regions in annealed ion-implanted superlattices

| Implant | Region 1 Dynamically Re-crystallized intermixed layer | Region 2 Intermixed layer free of defects | Region 3 Defect Region |
|---------|------------------------|---------------------|---------------|
| Be | 0 | 0 | 400 nm |
| Mg | 0 | 60 nm* | 305 nm |
| Si | 0 | 64 nm* | 306 nm |
| Se | 55 nm | 70 nm** | 150 nm |

(*) Partially mixed; layers are still visible.
(**) Totally mixed layers.

cross-sectional analysis.  One striking fact which has been revealed by cross-sectional TEM is the inhomogeneity in the distribution of defects present in the superlattice of alternating layers of 10 nm thick GaAs and 10 nm thick $Al_xGa_{1-x}As$.  High-resolution microscopy shows that the dislocation loops formed in region 3 were 10-20 nm in diameter and were located on {111}-type planes.  High resolution images support the interpretation that these loops are Frank interstitial loops.  Figure 3 shows the distribution of defect loops in an annealed Be-implanted superlattice.  A clear inhomogeneity in the distribution of dislocation loops was also observed in Si-implanted superlattices [12].  The mechanism causing this preferred loop distribution may be associated with several factors.  It may be influenced by the tetragonal distortion [16] of the superlattice layers which occurs due to the small mismatch between the $Al_xGa_{1-x}As$ and the GaAs, or ion-beam damage may occur preferentially in the GaAs thus favoring the formation of point defect clusters, and hence the formation of dislocation loops in the GaAs layers.  It should be noted that the GaAs layers are in tension due to the mismatch and interstitial loops would therefore be expected to form preferentially in this layer.

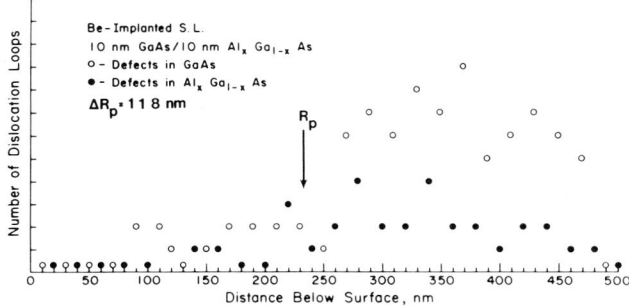

Fig. 3.  Distribution of dislocation loops as a function of depth in an annealed Be-implanted superlattice.

4.  Conclusions

In conclusion, a decrease in the compositional modulation of the topmost
layers is observed for all of the ion-implantations except those using Be
ions.  The extent of ion-beam mixing depends on the ion mass and gives an
alloyed interface between the layers (in the case of Mg and Si implants
where the mixing is by diffusion in the crystal) or a bulk alloy (in the
case of Se implant where the mixing happens in the amorphous region).  In
the case of high dose Se-implantation the top layers are amorphized and
recrystallize during the annealing.  Finally a clear inhomogeneous distribution of small dislocation loops has been observed which suggests that the
GaAs layers are preferred sites for the nucleation of the loops.

Acknowledgments

The authors wish to thank Prof. J. W. Mayer for discussions and Mr. Ray
Coles for his maintenance of the microscope.  The microscope is part of a
Materials Science Center Facility at Cornell which is supported in part by
NSF.  This research was supported by the U.S. Office of Army Research under
contract No. DAAG-29-82-K0148 and by the Joint Services Electronics Program
under contract No. F43620-84-60082.

References

[1]  S.T. Picraux, G.W. Arnold, D.R. Myers, L.R. Dawson, R.M. Biefeld,
     I.J. Fritz, T.E. Zipperian.  Proceedings of Conference on Ion-Beam
     Mixing of Materials (IBMM), Cornell University, Ithaca, NY, July
     16-20, 1984, to be published.
[2]  W.D. Laidig, N. Molonyak Jr., M.D. Camras, K. Hess, J.J. Coleman,
     P.D. Dapkus, J. Bardeen.  Appl. Phys. Lett. $\underline{38}$ (1981), p. 776.
[3]  W.D. Laidig, J.W. Lee, P.K. Chiang, L.W. Simpson, S.M. Bedair.  J.
     Appl. Phys. $\underline{54}$ (11) (1983), p. 6352.
[4]  A.H. Hamdi, $\overline{M}$.-A. Nicolet, J.L. Tandon.  Materials Lett., Vol. 2, N5B
     (1984), p. 437.
[5]  K. Meehan, N. Holonyak, J.M. Brown, M.A. Nixon, P. Gavrilovic, R.D.
     Bwinham.  Appl. Phys. Lett. $\underline{45}$ (5) (1984), p. 549.
[6]  J.J. Coleman, P.D. Dapkus, $\overline{C.G.}$ Kirkpatrick, M.D. Camras, N. Molonyak
     Jr.  Appl. Phys. Lett., Vol. 40, No. 10 (1982), p. 904.
[7]  D.R. Myers, R.M. Biefeld, I.J. Fritz, S.T. Picraux, T.E. Zipperian.
     Appl. Phys. Lett. $\underline{44}$ (11) (1984), p. 1052.
[8]  J.L. Merr, A.S. Barker, Jr., A.C. Gossard.  Appl. Phys. Lett. $\underline{31}$ (2)
     (1977), p. 117.
[9]  M. Ilezems, G.L. Pearson.  Phys. Rev. B, $\underline{1}$ (4) (1970), p. 1576.
[10] P.M. Petroff.  J. Vac. Sci. Technol. $\underline{14}$ ($\overline{1977}$), p. 973.
[11] K.-H. Kuesters, B.C. De Cooman, C.B. Carter.  Journal of Crystal
     Growth, in press (1985).
[12] J.P. Biersack, L.G. Haggmark.  Nucl. Instr. Meth. $\underline{174}$ (1980), p. 257.
[13] J. Ralston, G.D. Wicks, L.F. Eastman, B.C. De Cooman, C.B. Carter
     (1985).  Submitted to Appl. Phys. Lett.
[14] L.L. Chang, A. Koma.  Appl. Phys. Lett. $\underline{29}$ (3) (1976), p. 138.
[15] D.K. Sadana, T. Sands, J. Washburn.  Appl. Phys. Lett. $\underline{44}$ (6) (1984),
     p. 623.
[16] J.M. Brown, N. Holonyak Jr., R.W. Kaliski, M.J. Ludowise, W.T.
     Dietze, C.R. Lewis.  Appl. Phys. Lett. $\underline{44}$ (12) (1984), p. 1158.

*Inst. Phys. Conf. Ser. No. 76: Section 7*
*Paper presented at Microsc. Semicond. Mater. Conf., Oxford, 25–27 March 1985*

307

# The detection of local strains in strained layer superlattices

H L Fraser*, D M Maher**, C J Humphreys***, C J D Hetherington***, R V Knoell** and J C Bean**.

\*   Department of Metallurgy and Mining Engineering, University of Illinois, Urbana, Illinois 61801, USA.

\*\*  AT & T Bell Laboratories, Murray Hill, New Jersey 07974, USA.

\*\*\* Department of Metallurgy and Science of Materials, University of Oxford, Parks Road, Oxford OX1 3PH, UK.

Abstract   The detection and measurement of local strains through changes in high order Laue zone lines within the central disc of convergent beam electron diffraction patterns are discussed. Results are presented for commensurate long-wavelength strained layer superlattices of $Ge_x Si_{1-x}$ and these results show that marked and systematic differences between the Si buffer layer and two dilute alloy layers are observed. These differences are attributed to tetragonal distortions in the alloy layers and a preliminary estimate of the distortions, neglecting surface relaxations, is derived from computer simulated patterns.

## 1. Introduction

Until recently, all semiconductor superlattices have been grown from materials that are closely lattice matched (to within about 0.1%) e.g. GaAs/GaAlAs. In the last two years it has become possible to grow high-quality strained layer superlattices (SLS's) from a variety of lattice mismatched semiconductors. Structures of this type allow the exploitation of band-gap engineering. A knowledge of the local strain in SLS's is essential in order to understand and quantify properties such as strain induced band gap variations and mobilities. In this presentation we report the results of convergent beam electron diffraction (CBED) experiments which were designed to assess the sensitivity and the potential quantification of the CBED technique in the context of a model system, namely a long wavelength, dilute $Si/Ge_x Si_{1-x}$ strained layer superlattice. The lattice parameter of Ge is 4.1% larger than Si and short-wavelength superlattice structures based on these two elements are obviously important in future silicon based technologies.

## 2. Crystal Growth and Macroscopic Characterisation

For this work a commensurate long-wavelength SLS was grown by molecular beam epitaxy (MBE) on a (100) Si substrate by codeposition from two electron-beam evaporation sources (Bean et al., 1984). The growth sequence was as follows: a 500Å Si buffer was grown at 750°C; following cool down to 550°C an additional 500Å of buffer was grown; and then at 550°C three 1μm thick layers were grown, namely Si-5% Ge, Si buffer and Si-10% Ge (see fig.1). Since the nominal composition of the alloy may be uncertain by as much as 25%, Rutherford backscattering relative yield measurements were performed on the 10% Ge layer. Analysis of the results showed that the nominally 10% Ge alloy was in fact 9 ± 0.3% Ge. We assume that the nominally 5% Ge alloy is 4.5 ± 0.15% by interpolation. The weak periodic banding apparent in the images of the alloy layers (figs. 1 and 3) is due to a small periodic compositional variation arising from the rotation of the specimen during growth (Alavi et al., 1983).

## 3. TEM and CBED Experimental Details

Cross-sectional (011) electron microscopy specimens were prepared from cleaved crystals by mechanical lapping, chemical lapping and low-energy, ion-beam thinning. The CBED patterns shown here were taken under the following microscope conditions:- 120kV accelerating voltage; ~ 12 mrad beam convergence and ~ 40Å probe diameter. The specimen was maintained at a temperature of about 85°K, a (013) zone axis orientation was used and a specimen thickness, t, of ~ 2000Å which corresponds to a spatial resolution of ~ 120Å, in accordance with an elastic scattering model of beam broadening.

The high order Laue zone lines within the central disc which are recorded with probe diameters of approximately 40, 100 and 200Å, and at a specimen thickness of 1000 and 3000Å, exhibit the same characteristic features as those shown here.

## 4. Detection of Strains

It should be clear from an examination of the experimental whole central disc patterns which are shown in Fig.2 that marked and systematic changes are observed as one compares the Si buffer pattern to the two alloy patterns. These changes are especially large for first and weak second-order Laue lines whose reciprocal lattice vectors make relatively small angles with respect to the [100] growth direction. On the other hand, the intersections of Holz lines whose reciprocal lattice vectors lie close to projected ± [0$\bar{1}$1] directions (i.e. parallel to the interfaces in fig.1) are approximately invariant. Substrate and buffer patterns were identical, and patterns recorded within ± 0.4μm relative to the centre of each 1μm layer were identical to patterns from the centre of that layer. These observations, using the CBED technique, of differences between buffer and alloy layers we have interpreted as being due to a tetragonal distortion in the [100] direction. Furthermore, the absence of strain in the ± [0$\bar{1}$1] directions has been readily confirmed by dynamical two-beam contrast images for g = ± [0$\bar{2}$2].

The detection of lattice distortions in CBED patterns obtained from small volumes is the central theme of this work, nevertheless it is worth pointing out that CB images contain spatially resolved HOLZ lines which show the same characteristic features as the patterns in fig.2, as well as lattice distortions at interfaces. An example of a CB image is shown in fig.3.

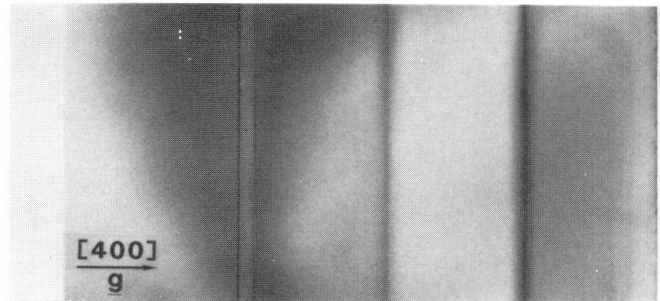

Fig.1 Bright-field cross-sectional image of a
model SLS showing from left to right: Si substrate;
0.1 μm Si buffer; 1 μm 4.5% Ge-Si alloy; 1 μm Si buffer;
and 1 μm 9% Ge-Si alloy.

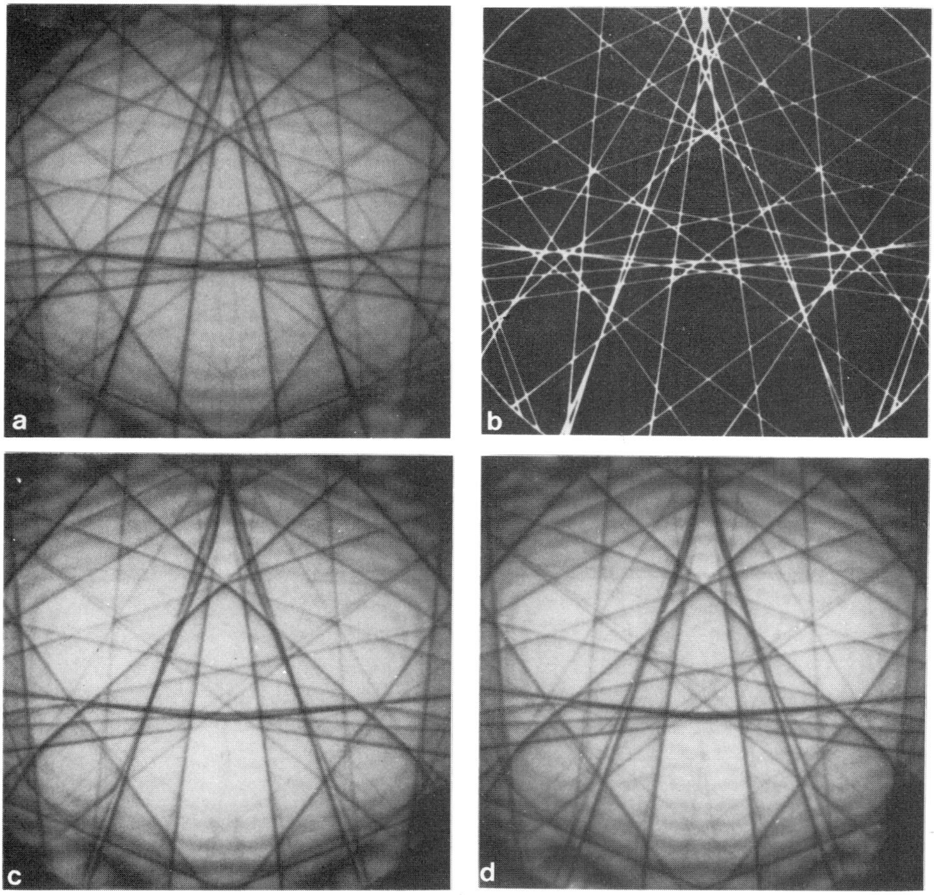

Fig.2  Whole central disc CBED patterns for the 013 zone axis:
(a) Si buffer;  (b) computer simulation of buffer with $a_0$ = 5.4294 Å,
$\lambda$ = 0.03358 Å and 2$\alpha$ = 0.011 rad;   (c) 4.5% Ge-Si; and (d) 9% Ge-Si.

## 5. Measurement of Strains

Our approach to the measurement of a tetragonal strain, $\varepsilon_T$, from CBED patterns has been through a computer simulation matching procedure. In this procedure 34 first-order Laue lines were included in the simulation and these matched to the cubic reference state (i.e. Si buffer). The silicon lattice parameter $a_0$ was taken from the literature (Dismukes et al., 1964) and corrected for temperature changes using measured values of $\Delta a/a_0$ versus temperature (Douloukian et al., 1975). The electron wave length was then varied in the simulation until the best match was obtained. By this procedure only two line pairs differed by more than the full line width of either the simulation or experiment (see whole pattern simulation of the Si buffer, fig.2b). Once the reference state was established, an estimate of the lattice parameter change in the growth direction, $b_\perp - a_0$, where $b_\perp = a_{100}$ of the alloy and $a_0$ is the lattice parameter of the buffer, was obtained by iterative simulations assuming no relaxation (i.e. $a_{010} = a_{001} = a_0$). The results for the two alloys are shown in fig.4, where systematic changes in the $[14\ \overline{4}\ 2]$ and $[\overline{13}\ 5\ \overline{1}]$ FOLZ lines have been arrowed. Whole pattern agreement was again taken as a necessary condition for the match. Values of $\varepsilon_T$ (that is $\{b_\perp - a_0\}/a_0$) which were obtained from these CBED analyses are tabulated below and compared to tetragonal strains which were derived from an X-ray measurement (Fiory et al., 1984) and calculated from elasticity theory (Gibson et al., 1985).

Tabulation of $\varepsilon_T$ for $Ge_x\ Si_{1-x}$

| Origin | x in % | $\varepsilon_T$ |
|---|---|---|
| CBED analyses | 9 ± 0.3 | 0.0026 ± 0.0003 |
| Elasticity theory | | |
|     Bulk crystal | 9 ± 0.3 | 0.005 ± 0.0005 |
|     Relaxed crystal | 9 ± 0.3 | 0.0039 ± 0.0004 |
| X-ray measurement | 10 ± ? | 0.0055 |
| CBED analyses | 4.5 ± 0.15 | 0.0013 ± 0.0001 |
| Elasticity theory | | |
|     Bulk crystal | 4.5 ± 0.15 | 0.0028 ± 0.0001 |
|     Relaxed crystal | 4.5 ± 0.15 | 0.0022 ± 0.0001 |

It is evident that the values of $\varepsilon_T$, obtained from CBED analyses and neglecting relaxation effects, are significantly lower than those obtained from X-ray measurements or from elasticity theory. Clearly lattice relaxation must be modelled and incorporated into the analyses as the next step is interpreting the CBED results.

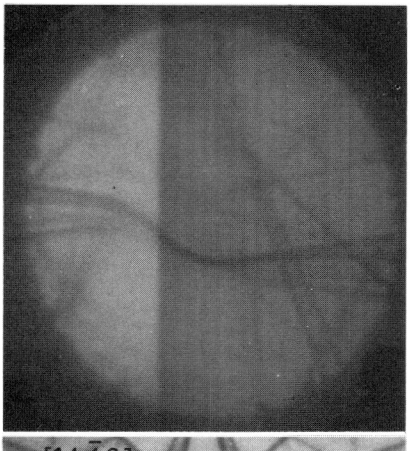

Fig.3 Convergent beam image showing spatially resolved 013 HOLZ line patterns at the Si buffer/9% Ge-Si alloy interface.

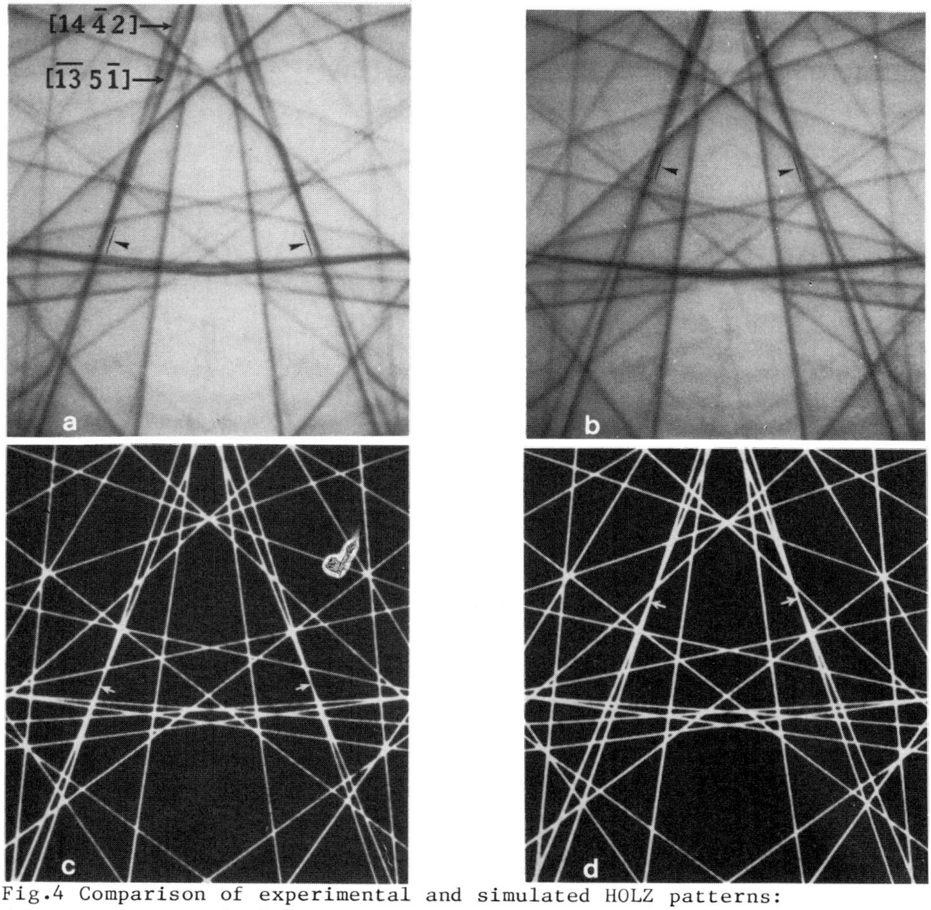

[14 4̄ 2]→

[1̄3 5 1̄]→

Fig.4 Comparison of experimental and simulated HOLZ patterns: (a) experimental 4.5% Ge-Si; (b) experimental 9% Ge-Si; (c) simulated 4.5% Ge-Si; (d) simulated 9% Ge-Si. Note large changes in the intersections of the $\left[14\ \bar{4}\ 2\right]$ and $\left[\bar{1}3\ 5\ \bar{1}\right]$ line pairs.

6. Conclusions

The present work has shown that marked and systematic differences are detected within the (013) central convergent beam discs from an MBE Si buffer layer and long-wavelength commensurate MBE alloys of 4.5% Ge - Si and 9% Ge - Si.  These differences are attributed to tetragonal strains in the growth direction and these strains are estimated to be 0.0013 ± 0.0001 for the 4.5% alloy and 0.0026 ± 0.003 for the 9% Ge alloy.  These strain estimates must only be taken as an indication of the potential to quantify observed changes in higher-order Laue zone lines since the present quantification is based on a kinematic model which neglects surface relaxation and assumes a homogeneous alloy.

Acknowledgements

The authors would like to acknowledge J.M.Gibson for very useful discussions and D.C.Jacobson for his assistance in carrying out the RBS experiment.

References

Alavi K, Petroff P M, Wagner W R and Cho A Y 1983 J. Vac. Sci. Technol. B1(2) 146
Bean J C, Feldman L C, Fiory A T, Nakahara S and Robinson I J 1984 J. Vac. Sci. Technol. A2(2) 436
Dismukes J P, Ekstrom L and Paff R J 1964 J. Phys. Chem. 68 3021
Douloukian Y S, Kirby R K, Taylor R E and Desia P D 1975 Thermal Physical Properties of Matter Vol 13, Non-metallic Solids (New York:Plenum) p154
Fiory A T, Bean J C, Feldman L C and Robinson I J 1984 J. Appl. Phys. 56 1227
Gibson J M, Hull R and Bean J C 1985 Appl. Phys. Lett. 46 649

*Inst. Phys. Conf. Ser. No. 76: Section 7*
*Paper presented at Microsc. Semicond. Mater. Conf., Oxford, 25–27 March 1985*

# Dislocations and strains in PbTe–Pb$_{1-x}$Sn$_x$Te superlattices

P Pongratz, H Clemens[*], E J Fantner[*] and G Bauer[*]

Inst. of Applied and Technical Physics, TU-Vienna , Karlsplatz 13, A-1040 Vienna, Austria
[*]) Inst. of Physics , Montanuniversität Leoben A-8700 Leoben, Austria

Abstract Epitaxial PbTe-Pb$_{1-x}$ Sn$_x$Te (x $\leqslant$ 0.2) multilayers grown on cleaved BaF$_2$ (111) substrates by a modified hot wall evaporation technique were investigated by TEM and by X-ray diffractometry to determine their lattice defect structure and elastic strains. Threading dislocations originating at the BaF$_2$ substrate were found with a density of $10^9$ cm$^{-2}$ and strain measurements indicate that the lattice mismatch of 2.5-4*10$^{-3}$ (for 0.1 $\leqslant$ x $\leqslant$ 0.2) can be almost totally accommodated by elastic strains. Only a small contribution is relieved by misfit dislocations which can be found along the three $\langle 1\ \bar{1}\ 0\rangle$ directions in the (111) growth plane.

## 1. Introduction

Artificial semiconductor superlattices composed of a periodic sequence of crystalline layers of IV-VI compounds with alternating composition (e.g. PbTe-PbSnTe ) exhibit quite different properties as compared to III-V/III-V or elemental IV/IV systems (Esaki 1970, Esaki and Tsu 1970). These differences arise from their different (NaCl) crystal structures, a small energy gap and high static dielectric constant in lead compounds. High quality single crystalline IV-VI semiconductor films were grown either by molecular beam epitaxy (MBE) (Holloway 1980) or by the hot wall epitaxy (HWE) method (Lopez-Otero 1978). Kinoshita et al (1982) and Clemens et al (1983) prepared IV-VI super-lattice systems of PbTe-PbSnTe by HWE. Carrier concentrations n or p of $10^{17}..10^{18}$ cm$^{-3}$ and mobilities up to 2*10$^5$ cm$^2$Vs$^{-1}$ were reported. To achieve a considerable band edge modulation in these superlattices a Sn content of x=0.1-0.2 has been used corresponding to an energy gap of 130-80 meV, as compared to 190 meV for PbTe (T=4.2 K). These Sn concentrations change the lattice constant d and a lattice mismatch $\Delta$d/d equal 2-4*10$^{-3}$ results. To accommodate this lattice mismatch an elastic misfit strain which is parallel to the interface and com-pressive in the PbTe layers and tensile strain in the PbSnTe layers is generated but misfit dislocations could be energetically more favourable for layers thicker than a critical thickness h$_c$ .In single layer films of PbSnTe grown on BaF$_2$ substrates a measurement of strain typically yields 3*10$^{-4}$ compressive strain at T=300K and dislocations are also introduced at the BaF$_2$ interface due to the mismatch of BaF$_2$ (a$_o$ =620pm) and PbTe (a$_o$ =646pm). These dislocations are nucleated as soon as the original islands of PbSnTe grow together and holes between them are closed and the film thickness becomes uniform (Pongratz and Sitter 1984). TEM and strain measurements on heterostructures were performed to understand the magnitude of the elastic and plastic contributions to the misfit relief.

## 2. Experimental

PbTe and $Pb_{1-x}Sn_xTe$ films were deposited by a modified hot wall technique from PbTe and PbSnTe sources on cleaved (111) BaF substrates (Clemens et al 1983). Various samples with individual layer thicknesses of 15, 50-300 nm and up to 50 layers were grown. A buffer layer of PbTe or PbSnTe was initially deposited on $BaF_2$. Samples which were prepared for TEM analysis were grown up to a film thickness of 270nm with a buffer layer of 120nm and 10 (or 3) alternating layers of 15nm (or 50nm) thick PbTe and PbSnTe respectively.

To determine the strain in the PbTe and in the $Pb_{1-x}Sn_xTe$ layers as well as in the substrate, the lattice constants of various lattice planes parallel and inclined to the interface were measured. Using an $\Omega$-diffractometer the $CuK_\alpha$ and $MoK_\alpha$- reflections were measured at the highest order possible. The net accuracy of the technique was better than $2*10^{-4}$ in all cases using the lattice parameter of unstrained single crystalline $BaF_2$ ($a_o$ = 620pm) as internal standard. Measured differences of the lattice constants for various lattice planes of the $\langle 211 \rangle$ zone and for the superlattice constituents (PbTe, $Pb_{1-x}Sn_xTe$) and unstrained $BaF_2$ are shown in Fig.1 for a sample with a superlattice period of 200nm. From the slopes of the lattice constants vs $\sin^2(\psi)$ (Ortner,1983) and correcting anisotropic elastic strains according to Hornstra and Bartels (1978) with elastic moduli of Miller et al (1981), we determined these strains. The 300K data show that PbTe layers are subjected to compressive and the PbSnTe layers to tensile strains in the film plane. Their magnitude is almost equal but of opposite sign and is $(1.3 \pm 0.1)*10^{-3}$. $\psi$ is the angle of inclination between the lattice plane investigated and the interface plane. The same value can be found for three other samples of individual layer thicknesses varying from 30-300nm .For the special symmetry of our samples a linear dependence of $\Delta\psi$ on $\sin(2\psi)$ is valid $\Delta\psi$ being the change if inclination of a lattice plane due to strain relative to the interface orientation. This means that the angle $\psi$ is increased for compressive and decreased for a tensile strain. This can be seen on figure 2.

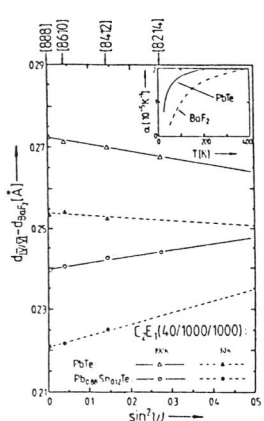

Fig.1   Differences of SL-lattice constants for planes in the $\langle 211 \rangle$ -zone at 300K and 30K.40 layers of 100 nm thick PbTe and PbSnTe were grown. (Insert:thermal expansion coefficients of PbTe,$BaF_2$ vs temperature)

In order to investigate the strain in superlattices by X-ray techniques and by electron microscopy on the same samples, specimens with a total thickness of 270nm were used.Their

Identification and the X-ray diffraction results are given in table I
.An experimental lattice mismatch
of about $3.5*10^{-3}$ was found in both
samples.The strain is shared between
the PbTe and PbSnTe layers proportional to their relative thicknesses and yields the same lattice
constant in the interface planes
of the constituent layers within
the experimental error.

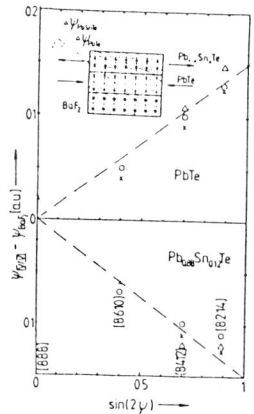

Fig.2 Inclination angles of planes
of the $\langle 211 \rangle$ zone for 3 samples
of different SL-periodicity: $\mathbf{o}$.60nm,
x...200nm, $\Delta$.600nm.

Table I

| Sample No. | P35M2 | P35N2 |
|---|---|---|
| buffer thickness | 120nm | 120nm |
| buffer type | $Pb_{1-x}Sn_xTe$ | PbTe |
| d(PbTe) | 50nm | 50nm |
| d(PbSnTe) | 50nm | 50nm |
| total thickness | 270nm | 270nm |
| Sn content | x=0.175 | x=0.175 |
| lattice constant difference | | |
| a(PbTe)-a(BaF$_2$) | 2.735pm | 2.692pm |
| a(PbSnTe)-a(BaF$_2$) | 2.289pm | 2.226pm |
| a(PbTe)-a(PbTe) unstrained | 1.35pm | 0.92pm |
| a(PbTe)-a(PbSnTe) | 4.46pm | 4.66pm |
| $\varepsilon^{exp}$ (mismatch) | $3.45*10^{-3}$ | $3.56*10^{-3}$ |

(internal standard BaF$_2$   $a_o = 620.0$pm)

Samples for TEM analysis were chemically polished from the BaF$_2$ side
with a solution of $AlCl_3:NH_4Cl:HCl:H_2O=1:1:1.5:8$ (Aronova et al 1982).
The PbSnTe films were covered by a mesh grid and protected with wax
which could be removed in tricloroethylene after polishing.Finally
the samples were carefully rinsed in methanol and water . A Jeol 200CX
microscope was used for TEM work. All PbSnTe films were well oriented
single crystalline with $\langle 111 \rangle$ orientation both in the case of single
layers and as heterostructures. No islands of (100) orientation were
found and films were of uniform thickness if thicker than 100nm. Figure
 3 shows the dislocations which are present in a single layer of
$Pb_{.8}Sn_{.2}Te$. They are crossing the film and originate mainly from the
holes between the islands of PbSnTe which are not covered in the early
state of island growth.These dislocations have a Burgers vector of
type $1/2[110]$ and glide on three equivalent $\{100\}$ glide planes which
are inclined to the $\langle 111 \rangle$ growth plane at 54.7 degrees.In some cases
even two dislocations are found to be generated at the same place
in the interface (marked in Fig.3).The dislocations are usually very
mobile (even by focussing the electron beam on them) but they do not
form any regular network of misfit dislocations. To compare the defect
structure of films which are removed from the BaF$_2$ substrate with one
film where the BaF$_2$ substrate and the interface is imaged together

Fig.3

Dislocations in a homogeneous
epitaxial film of PbSnTe which
was removed from the substrate.
Arrows indicate  points where
two dislocations originated at
the interface. BM₌[111] ,G=(02$\bar{2}$)
foil thickness=270nm, dislocation
density 2*10$^9$ cm$^{-2}$.

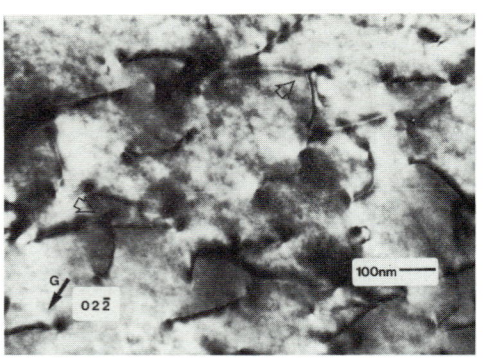

with PbSnTe figures 4 and 5 are instructive. Due to the Moirè fringes
with  the  (02$\bar{2}$)  reflections  of  PbSnTe  (x=0.20)  and  BaF₂, which  are
very sensitive to any local strain field, the dislocations perpendicular
and  inclined  to  the  interface  are  easily  recognized.The  Moirè  period
is 6.0nm which is just the theoretical value for this epitaxial system.
The  Moirè  is  of  the  translation  type  and  gives  evidence  that  PbSnTe
is  not  strained  as  much  as  to  match  the  BaF₂  substrate  lattice.If  one
compares the number of dislocations in PbSnTe on areas with and without
a BaF₂  coverage  not  too  much  difference  is  found.  Therefore  it  is  concluded  that  the  images  which  were  taken  from  semiconductor  film  area
free  of  BaF₂  substrate  are  representative  as  the  number  and  configuration of dislocations is concerned.

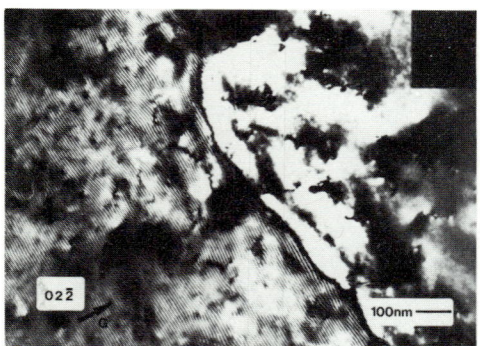

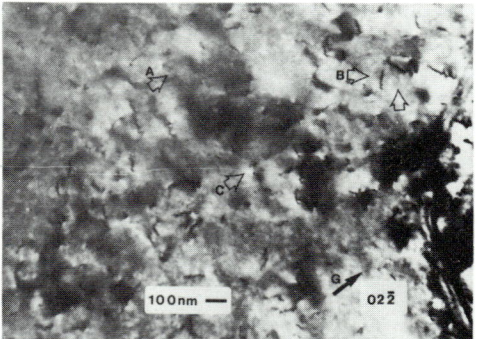

Fig.4.     Area  of  partial  overlap
of  PbSnTe  and  BaF₂.Moirè  fringes
are  due  to  the  (022)  reflections
in    PbTe    and    BaF₂.Dislocations
originating  at  the  interface  can
be  seen.Local  strains  are  visible
by Moirè fringe bending.

Fig.5.     Moirè  fringes  due  to
overlapping    epitaxial    PbSnTe
and BaF₂ crystals. Extra fringes
due  to  dislocations  can  be  seen
at A,B,C (arrows).No misfit dislocation  network  can  be  seen,only
dislocations    inclined    to    the
foil are present.

After having demonstrated where the threading dislocations in homogenous
films originate, the dislocations in heterostructures are investigated.
In  Fig.6  a  superlattice  of  individual  layer  thickness  of  15  nm  is
imaged.  No  difference  of  the  dislocation  structure  in  comparison  to
a single layer of comparable thickness is observed. On the other hand,

Fig.7  is an image of a heterostructure with 50 nm layers (sample P35M2)
   The Reaction of two dislocations from two adjacent glide planes ,according to the following equation can be observed:

$$1/2\,[101]/(010) \; + \; 1/2\,[\bar{1}\bar{1}0]/(001) \; = \; 1/2\,[0\bar{1}1]$$

The reaction lowers the energy of the system and creates a rather
sessile  dislocation along $[100]$ in a glide plane of type $\langle 110 \rangle$ which
is not a primary glide plane in PbTe. Figure 8 shows the glide elements
in PbTe (Gilman, 1959).

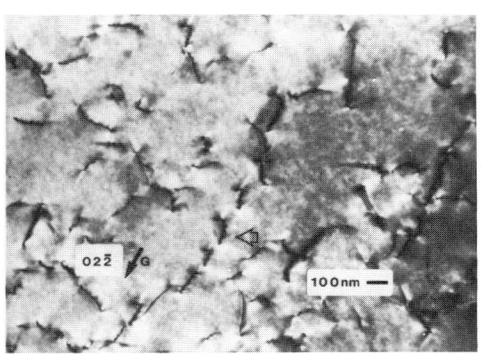

Fig.6.
Dislocation  structure  of a superlattice with  individal layer thickness  of  15  nm.Slight  bending  of
the dislocations due to local strain
can be seen.

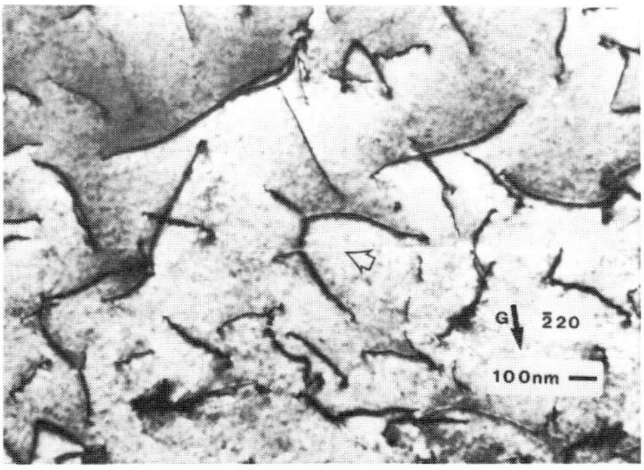

Fig.7.
Dislocation reaction in a heterostructure with 50 nm layer thickness (Sample P35M2).

Fig.8.
Geometry  of  the  glide  elements
in   PbTe:   (100),(010),(001)glide
planes  intersect  (111)  growth plane
along $[1\bar{1}0]$ , $[\bar{1}01]$,and $[01\bar{1}]$ .

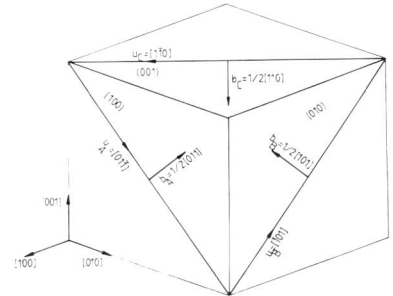

Reactions of dislocations are also seen in a similar heterostructure
(sample P35N2) of 50 nm individual layers,but in comparison with the
former (P35M2) the sequence of layers starts with a PbSnTe buffer
layer on BaF$_2$ substrate followed by PbTe,PbSnTe,PbTe (see Fig.9a,b).
Straight misfit dislocations along $[\bar{1},0,1]$ , $[1,\bar{1},0]$ and $[0\;1\;\bar{1}]$ directions ( $u_A,u_B,u_C$ ) in Fig.8) parallel to the interface (111) are
found in this case. These dislocations show complete extinction for
operating reflections which are parallel to their line direction.From
the fact that  both $\vec{g}\cdot\vec{b} = 0$ ; $\vec{g}\cdot\vec{b}\times\vec{u} = 0$  is valid in case of extinction,
it is evident that these misfit dislocations are pure edge dislocations
with their Burgers vector along  $b_A$ , $b_B$ , $b_C$ as indicated in figure
8.As these Burgersvectors are not in the (111) plane only their projection onto (111) i.e   $b\cdot\cos(54.7) = b/\sqrt{3}$  is effective to relieve
a misfit .

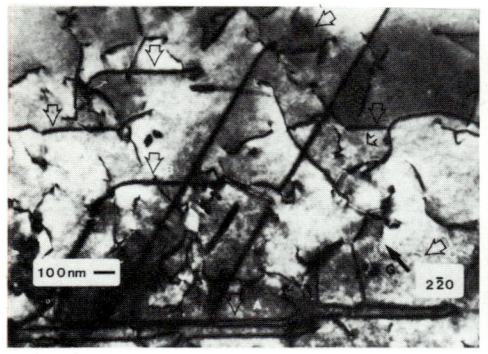 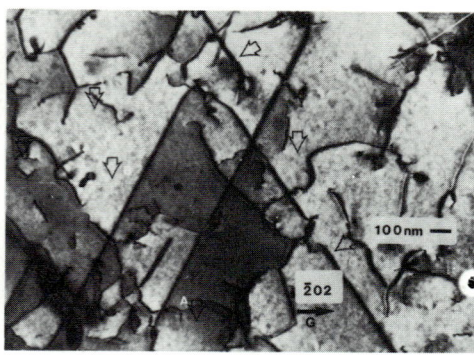

Fig. 9a.                                      Fig. 9b.

Misfit dislocations parallel to the interface along $[\bar{1}01]$ , $[1\bar{1}0]$ , $[01\bar{1}]$ of pure edge type are found in sample P35N2.The operating reflections near the $[111]$ zone give extinction for the misfit dislocations with the same line direction. Compare Figures 9a,9b where arrows and letters A , B indicate identical positions .

## 3.Discussion

According to the X-ray measurements the misfit between PbTe and PbSnTe layers can be accommodated totally by elastic strain, despite the fact that the dislocation density in these layers is of the order of $10^9$ cm$^{-2}$.However, these are not misfit dislocations but threading dislocations originating from the substrate (BaF$_2$ ). If the constituent layer thickness in superlattices exceeds 50nm the strains bend some of the threading dislocations parallel to the interface along $[1\bar{1}0]$ , $[\bar{1}01]$ , $[01\bar{1}]$ directions. These are edge dislocations and are able to relieve ( $\lesssim$ 10%) of the misfit strain present in the superlattice. If one could avoid threading dislocations e.g. by a better or different choice of substrate the PbTe/PbSnTe system could present an example for a dislocation free strained layer superlattice.

## References

Aronova A M, Berezhkova G V, Perstnev P P 1982
    Cryst. Res. &Technol. 17 1331-1334
Clemens H, Fantner E J , Bauer G 1983 Rev.Sci.Instr. 54 685
Esaki L 1984 Proc.Int.NATO School Erice 1983 Amsterdam North Holland
Esaki L,Tsu R 1970 IBM J.Res.Dev. 14 61
Gilman J J 1959 Acta Met. 7 608-613
Holloway H 1980 in Physics of Thin Films   ed. G Haas and M Francombe
    Vol 11 (New York Academic) p106
Hornstra J , Bartels W J 1978 J.Cryst.Growth 44,513-517
Kinoshita H and Fujiyasu H 1980 J.Appl.Phys. 51 5845
Lopez-Otero A 1978 Thin Solid Films 49 3
Miller A J, Saunders G A and Yogutcu Y K 1981
    J.Phys.C 14 1569-1584
Ortner B 1983  Eigenspannungen ed.Macherauch E and Hauk V
    Deutsche Ges. f. Metallkunde Vol.2,p49
Pongratz P, Sitter H 1984 Proc. VIII  Europ. Congr. on Electron
    Microscopy ed. Csanady et al Budapest p1221

*Inst. Phys. Conf. Ser. No. 76: Section 8*
*Paper presented at Microsc. Semicond. Mater. Conf., Oxford, 25–27 March 1985*

# Characterization of defects in semiconductors by combined application of SEM(EBIC) and SDLTS

J Heydenreich and O Breitenstein

Akademie der Wissenschaften der DDR, Institut für Festkörperphysik und Elektronenmikroskopie, DDR-4020 Halle/Saale

Abstract   Scanning Deep Level Transient Spectroscopy (SDLTS) allows the spatially resolved investigation of deep level centres in semiconductor space charge regions. It is therefore a useful method to study the distribution and behaviour of point defects, especially their interaction with extended defects. The basic principle of SDLTS and recent developments are reviewed and its possibilities and limitations are discussed. Various procedures are introduced to obtain reliable quantitative SDLTS results. Several experimental examples demonstrate that SDLTS can be performed most effectively in combination with EBIC imaging techniques.

## 1. Introduction

The strong dependence of the properties of semiconducting crystalline materials on existing crystal defects, on the one hand, requires a reliable characterization of the defects with respect to their geometrical structure and, on the other hand, the estimation of their electrical activity. While for the first-mentioned task transmission electron microscopy is widely used, the electrical activity of individual crystal defects is investigated chiefly by applying the scanning electron microscope (SEM) in the electron beam-induced current (EBIC) mode or also in the cathodoluminescence (CL) mode. Thus, information can be gained on the recombination efficiency(EBIC) or on the radiative/non-radiative behaviour (CL) of individual crystal defects.

The above methods are usually applied to investigate extended crystal defects, like dislocations, planar defects and precipitates. For the determination of the electronic behaviour of these defects it is important to take into account their interaction with existing point defects, especially with impurities, which causes a particular state of "decoration" of the extended defects with foreign atoms.SEM(EBIC) investigations e.g. have shown (see e.g. Ourmazd et al 1981, Heydenreich et al 1982, Kittler and Bugiel 1982) that dislocations decorated with impurity atoms have a higher recombination efficiency than "clean" ones. For this reason, the knowledge of the concentration, the local distribution and

the energy levels of point defects and impurities, resp., is
highly desirable. Scanning deep level transient spectroscopy
(SDLTS) enables relevant investigations to be carried out,
at least for defects lying in the space charge region of a
pn junction or a Schottky barrier in materials under investigation. The combined application of SEM(EBIC) and SDLTS
yields complementary information on the recombination efficiency of a defect region for electron beam-induced minority
charge carriers and on the distribution, concentration and
energy levels of point defects and impurities, resp., involved.

## 2. General and Instrumentation

Scanning deep level transient spectroscopy proposed by Petroff and Lang (1977) is based on the standard technique of
deep level transient spectroscopy (DLTS), which since its
introduction by Lang (1974) has been used as a routine technique for the integral measurement of the concentration and
energy levels of deep level centres in space charge regions
of semiconductor crystals, of silicon as well as of compound
semiconductors. The principle of the method is based on the
measurement of capacitance changes (or of induced currents)
in the space charge region (pn junction or Schottky barrier)
after an excitation pulse as a function of the temperature.
The electrical excitation pulse causes a non-equilibrium
condition, and the following thermal emission process restores the system to thermal equilibrium of the occupation
of the levels. Measuring the magnitude and the time constant
of the capacitance transient as a function of the temperature
and the excitation pulse repetition frequency (rate window)
allows one to estimate the concentration and the energy
level of a deep level centre.

In scanning deep level transient spectroscopy the levels are
excited by the pulsed electron probe of the SEM. Choosing a
special energy level by working at a well-determined fixed
sample temperature (and rate window) and scanning the specimen under these conditions by a pulsed electron beam may
lead to a mapping of the local distribution of deep level
centres of a chosen energy. Accordingly, the SDLTS signal is
proportional to the total of recharged states. In order to
estimate the local concentration from this one should have
a detailed knowledge of the excited sample area. Similar to
the SEM(EBIC) technique the lateral resolution of the SDLTS
method is in the micron-range. For SDLTS, the decisive practical problem is the fact that, contrary to standard DLTS,
the greater part of the sample remains unaffected by the
electron beam, and the area which deep levels are filled in
and where a signal appears is very small. Comparing DLTS and
SDLTS reveals a difference in the extension of these regions
of at least two orders of magnitude (DLTS: some 0.1 mm,
SDLTS: some /um) leading to an SDLTS signal being typically
about 4 orders of magnitude smaller than the corresponding
DLTS signal of the same sample. In order to overcome the
difficulties related to sensitivity, in the first-mentioned
publication on SDLTS (Petroff and Lang 1977) instead of the

more usual capacitance DLTS current DLTS was applied, which
for fast rate windows ($10^5$-$10^6$ sec$^{-1}$) enables a fairly high
sensitivity to be achieved ($10^{15}$ deep levels cm$^{-3}$ $\cong$ 10 000
deep level centres per scanning point of 2 /um $\emptyset$). The disadvantage of the current DLTS lies in its limited rate window range and in the fact that a distinction between electron traps and hole traps is no longer possible. Altogether,
it should be pointed out that the restricted sensitivity of
the SDLTS technique at that time has prevented a wider application; only a few publications have appeared in this
field (see e.g. Petroff et al 1978, Breitenstein 1982, Breitenstein and Wosinski 1983, Breitenstein and Heydenreich
1983).

The following will briefly describe a new SDLTS system,which
has been designed particularly to solve the sensitivity
problems above. In the mean time several authors (see e.g.
Borsuk and Swanson (1980) or Misrachi et al (1980)) have
shown that a very high sensitivity can be attained also in
capacitance-based systems so that the advantage of dis-
tinguishing between electron and hole traps further exists
and that the sensitivity is almost independent of the emission rate. The constructed novel capacitance meter (Breitenstein 1982) is a resonance-tuned LC bridge working at 28 Mc
and having an absolute sensitivity limit of about $10^{-6}$ pF,
which is about 2 orders of magnitude higher than in conventional systems. Note that unlike the standard DLTS, where
the sensitivity is typically given in relative units assuming a 100 pF sample, for SDLTS the absolute sensitivity
is the decisive quantity. The new system enables the detection of already $10^{13}$ cm$^{-3}$ deep level centres for a spatial
resolution of a few microns, which is equivalent to the detection limit of about 100 atoms (traps) per scanning point.

For the combined application of EBIC and SDLTS a microcom-
puter-aided arrangement is used as is schematically shown
in Fig. 1. The system is based on a commercial scanning
electron microscope (TESLA, BS-300) working with accelerating voltages between 10 and 30 kV. The DLTS spectrometer
unit, on the one hand, includes the capacitance meter and,
on the other hand, the necessary EBIC amplifier. The actual
DLTS signal generation is carried out conventionally by
using an analogue correlator and a linear analogue signal
integration. The microcomputer yields the analogue exponential correlation function having a time constant between 20 /us and 1 s and manages the control of the measurement process and the data handling. This means a hybrid
concept is applied in which the actual DLTS signal is generated still by analogue techniques using a special DLTS
hardware. The microcomputer yields – in addition to the
analogue exponential correlation function – an excitation
pulse trigger signal with a pulse width between 100 ns and
10 ms (using this DLTS hardware); furthermore, it is to
digitize and store the measured values, to control the electron beam scan over the sample, to control the sample temperature between 80 K and 400 K (using a temperature control
unit) and to control the display of the results in a Y-mo-

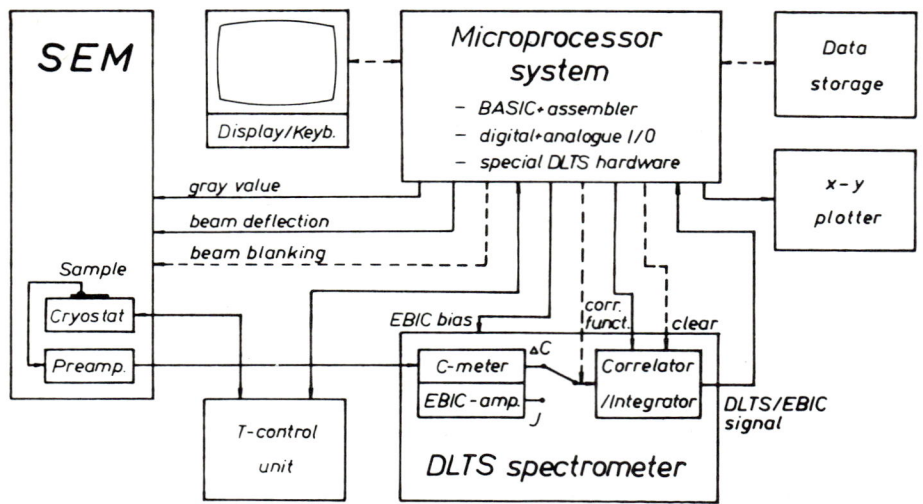

Fig. 1    Functional block diagram of the computer-controlled
          SDLTS system

dulation representation or a quasi 3-dimensional one on the
x-y plotter or as a gray-patch image on the cathode ray
screen of the SEM. The chief advantage of the hybrid SDLTS
system used is the fact that, unlike a purely digitally
working system, it uses almost the whole measure time for
capturing the transient signal, thus guaranteeing the best
possible signal-to-noise ratio.

## 2. Practical Problems

Using the state of the art, qualitative SDLTS information
(e.g. a check if a level is homogeneously distributed or
not, or some correlation to extended defects) can be gained
rather easily. Such qualitative results are often decisive
for the physical problem under investigation. If, however, a
detailed quantitative analysis is necessary (e.g. to measure
the absolute deep level concentration) several parameters
have to be checked, before a direct interpretation of an
SDLTS image in terms of a concentration distribution is possible. There are two principal points to be taken into
account that might prevent such a straightforward interpretation. The first question is whether the SDLTS signal is
indeed only due to the interesting deep level and not influenced by other processes. The second is related to the
problem of knowing the accurate size of the excited sample
area.

The probably disturbing signal sources are mainly extended
crystal defects or adjacent point defects. They can be identified by checking the temperature dependence or the rate
window dependence of the SDLTS signal. It is well-known that
isolated point defects show a strong exponential dependence

of their thermal emission rate on the sample temperature
thus exhibiting a well-defined peak in the dependence of the
DLTS signal on the temperature or the rate window. Extended
defects, however, typically show a non-exponential emission
behaviour leading to very broad DLTS bands. Thus, the simplest way of checking the origin of an SDLTS signal is to
perform a temperature scan in one sample position under
electron beam excitation conditions. The result of such a
procedure on a GaAs Schottky barrier is shown in Fig. 2a.
There is one dominant level peaking at the temperature of
250 K. At this temperature the signal can be assumed to be
mainly due to this level.The practical problem in some cases
rejecting a temperature scan is that it is difficult to con-
struct a cryostat that does not at all shift the sample position under the electron beam during a temperature scan.
Thus, if, for example, the interesting DLTS signal is expected to appear only within a very small region, it is better to perform a rate window scan at a constant temperature
as is demonstrated in Fig. 2b, or to compare SDLTS images
at different sample temperatures.

The second problem in quantitative SDLTS is the knowledge of
the area where the deep level filling takes place under the
electron beam excitation conditions chosen. Note that this
area is typically governed by the extension of the generating
volume and by the spreading of the electron beam-induced
carriers. The problem is complicated by the non-linear dependence of the SDLTS sig-
nal on the exciting elec-
tron beam-induced current.
This non-linearity is in-
herent in the SDLTS tech-
nique itself; once the levels are filled a further
action of the exciting current causes no further increase of the SDLTS signal.
Thus, in each position the
SDLTS signal has an exponential saturation type dependence on the exciting cur-
rent action. This current,
however, depends on the
distance from the centre of
the incident electron beam;
for a focused electron beam
it has a glow curve shape.
Accordingly, if for an
SDLTS measurement the filling pulse length is step-
wise increased, this saturation behaviour occurs
depending on the distance
from the electron beam centre. Fig. 3a shows the expected dependence of the
electron-beam excited DLTS

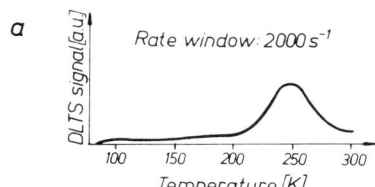

a

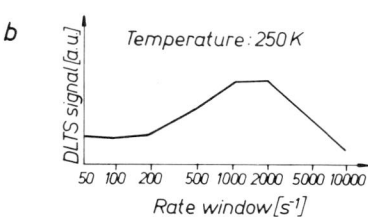

b

Fig.2　SDLTS signal identification by means of a temperature scan (a) or by an
emission rate scan (b) for a
400 meV hole trap level in a
GaAs Schottky barrier

signal on the filling pulse length. At short filling pulse
lengths no saturation occurs; in this regime the SDLTS signal
is linearly dependent on the filling pulse length, and the
local degree of trap filling corresponds to the radial distribution of the electron-beam induced current. In this linear regime (region A) the spatial resolution is best and
equal to that of the EBIC technique. The signal height is,
however, small because of the low degree of trap filling,and
all factors influencing the electron-beam induced current
also influence the SDLTS signal height (excitation contrast!).
If the pulse length is further increased, in the centre of
the excited area trap saturation occurs that is easily detectable by a beginning sublinear dependence of the SDLTS
signal on the filling pulse length. This regime (region B)
can be regarded to be best suitable for SDLTS investigations
because the spatial resolution is only slightly degraded
from that of the linear regime; for this resolution the signal has its maximum height, and because of its sublinear dependence on the beam current the signal is no longer so much
influenced by excitation contrast mechanisms. If the filling
pulse length is further increased (region C) trap filling
will occur already in positions more distant from the incident beam centre, thereby increasing the effectively excited area and decreasing the spatial resolution. The accurate magnitude of the excited area also under these experimental conditions is, however, not yet known. Thus, if
the net doping proves to be homogeneous, the signal can be
expected to be really proportional to the concentration distribution, but, on the other hand, it is not yet possible
to estimate the absolute value of the concentration.

The way out is to employ a definite electron beam defocus,
whereby the excited area becomes well-defined and measurable,
and the signal height also increases, but the spatial resolution decreases, of course. Accordingly, this technique can
be applied only if there are relatively large regions on the
sample with an almost homogeneous deep level distribution,
otherwise it is better to scale
the result in units of detected
atoms rather than in concentration units. If, however, the
defocus dominates the natural
resolution factors, the exciting radial current profile can
be regarded to be rectangular
with a well-defined exponential
trap filling behaviour inside
this area and practically no
trap filling outside. Fig. 3b
shows the corresponding dependence of the SDLTS signal on the
filling pulse length.

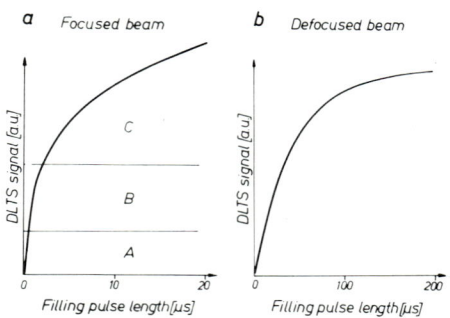

Fig. 3 Typical filling
pulse length dependence
of the SDLTS signal for
a focused (a) and a de-
focused (b) electron beam

## 3. Examples of the Combined Application of EBIC and SDLTS

In comparing EBIC and SDLTS images one should take into account that the physical processes underlying the different imaging techniques are different. In the EBIC technique the sample volume under investigation lies outside the actual space charge layer that only has the task to collect the excess minority carriers.The physical quantity appreciably influencing the EBIC signal is the local minority carrier lifetime in the neutral volume that is governed by the action of local recombination centres. Extended crystal defects are the major objects of EBIC investigations, since they typically exhibit a high local recombination efficiency due to their typically high local density of electron states. There are, however, only a few examples that an EBIC contrast could be uniquely identified to be due to deep electron states of isolated point defects.

Unlike EBIC, SDLTS is especially suitable for the detection and identification of point defect levels having their well-defined energy positions in the gap. Here, the sample volume under investigation is the actual space charge layer. Therefore, it is advantageous to apply a higher reverse bias to SDLTS investigations than to EBIC investigations in order to ensure that the volume of interest is roughly the same in both cases. The physical process leading to the SDLTS signal is not the recombination efficiency of a centre but its thermal emission behaviour, hence the opposite process. These two processes are governed by different defect parameters, viz. the carrier capture cross sections (EBIC) and the thermal activation energy (SDLTS), which generally are physically independent. Particularly for extended crystal defects the thermal activation energy is often not well-defined and the thermal emission process is influenced by carrier interaction processes.Therefore, though on principle they are detectable by SDLTS,they have only little chance to be characterized in more detail by SDLTS. Thus, EBIC and SDLTS contrasts of an object usually cannot be expected to be strictly due to one and the same physical origin. Hence, in many cases SDLTS does not directly image the dominant recombination centres. But, on the other hand, these different points of view favour the combined application of these techniques. Whenever interaction between point defects and extended defects is of interest (e.g. such physically and technologically important processes like decoration, gettering or precipitation) SDLTS can yield valuable information in addition to EBIC investigations as will be demonstrated in the following

The investigation of an n-GaAs:Au Schottky diode (net doping concentration: $(N_D-N_A) = 2 \cdot 10^{15}$ cm$^{-3}$) is used as the first example of the combined application of EBIC and SDLTS (see Fig. 4). The material, grown by vapour phase epitaxy (VPE) under Ga-rich conditions allows one to expect point defects, which should be correlated to As vacancies, Ga interstitials, or Ga$_{As}$ antisite defects. By using electron beam excitation in the SDLTS technique a hole trap level at an energy of $E_V+400$ meV can be detected, which probably is identical with

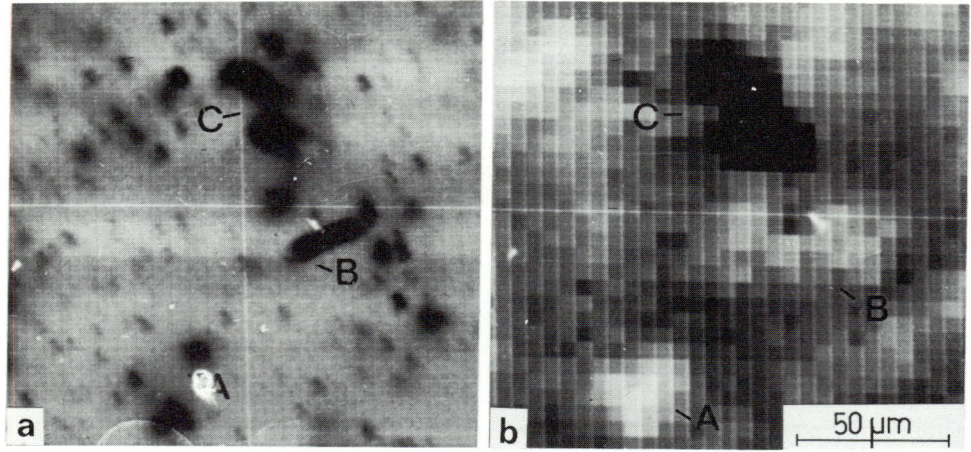

Fig. 4    EBIC image (a) and SDLTS image (b) of a region
of a GaAs:Au Schottky barrier. The SDLTS parameters are
chosen to display the "A" centre

the "A"-level known from the literature. This level should
occur preferentially under Ga-rich growth conditions, which
are given e.g. also for liquid phase epitaxy. According to
investigations by Ledebo and Wang (1983) this centre should
be the mentioned $Ga_{As}$ antisite defect, which is characterized
by a Ga atom on an As lattice site. With the aid of the quantitative SDLTS technique (see Section 2) using a defocused
electron beam for the excitation, an average concentration of
this level of about $1.5 \cdot 10^{14}$ $cm^{-3}$ can be detected. The EBIC
micrograph (see Fig. 4a) shows a large number of irregularly
distributed black dots with relatively weak EBIC contrast,
which probably are related to emergence points of dislocations. In addition to these there are further defects with a
more pronounced EBIC contrast and a larger extension, the
nature of which is a matter of discussion. The SDLTS gray
patch image (Fig. 4b) taken at the temperature and the rate
window of the A-level (250 K, 2000 $s^{-1}$) shows an inhomogeneous incorporation of this level. A bright image contrast corresponds to a higher local concentration of this level. One
recognizes a stronger SDLTS signal (bright contrast) at those
sites (e.g. region A), where an accumulation of dislocations
is found. The changes in the concentration amount to only a
few per cent so that here this contrast might be possibly
determined also by small fluctuations in the net doping concentration. At the site of the strongest EBIC contrast (B)
the corresponding SDLTS signal changes only little. Obviously
there is only a weak interaction of the defect existing in
this region with the A-level. On the other hand, at the sites
of two pronounced and extended dark spots in the EBIC micrograph (region C) the SDLTS image points to a much lower local concentration of the A-level in this region. This defect
may be possibly regarded as a local precipitate of Ga or at
least as a defect at which preferentially Ga atoms are de-

posited. This defect should be the reason why at these sites
the usual Ga excess is not present or is later on reduced
so that the conditions for the $Ga_{As}$ antisite defect to occur
are no longer complied with in this region.

The second example refers to a GaAlAs/GaAs heterojunction
diode. Fig. 5 (top left) shows standard DLTS investigations, revealing major hole trap peaks at temperatures of
$-43^{\circ}C$ and $85^{\circ}C$ (rate window: $1000\ s^{-1}$) corresponding to deep
level centres with energies of 400 and 580 meV, resp.. The
first-mentioned centre represents the A-level discussed above.
If in a fixed position the sample is excited by a pulsed
electron beam depending on the temperature (dashed line),
one more level appears at low temperatures ($-140^{\circ}C$) that obviously cannot be excited electrically. In Fig. 5 the EBIC
micrograph (a) of the sample characterized by DLTS, struc-
tured by mesa etching and mounted onto a transistor holder
is compared with SDLTS gray patch representations (b-d),
taken at those three temperatures relevant to the peaks in
the given DLTS spectrum (top left). The circular area around
the bonding contact visible in the EBIC micrograph is due to
the p contact metallization. The SDLTS representations,
shown in Fig. 5b-d, exhibit well-pronounced inhomogeneities
of the corresponding levels that clearly differ from each
other. Particularly for Fig. 5b, the SDLTS contrast seems
to appear preferentially under the p contact metallization
layer pointing to an influence of the p metallization technology on the incorporation of the A-level. In Fig. 5c and d
the circular diode edge shows a special contrast. This contrast can, however, not be regarded as a strong accumulation
of the corresponding levels but rather as an artefact arising
from the fact that in this region the pn junction reaches
the sample surface. Accordingly, here the electron beam directly reaches the pn junction and the excitation intensity

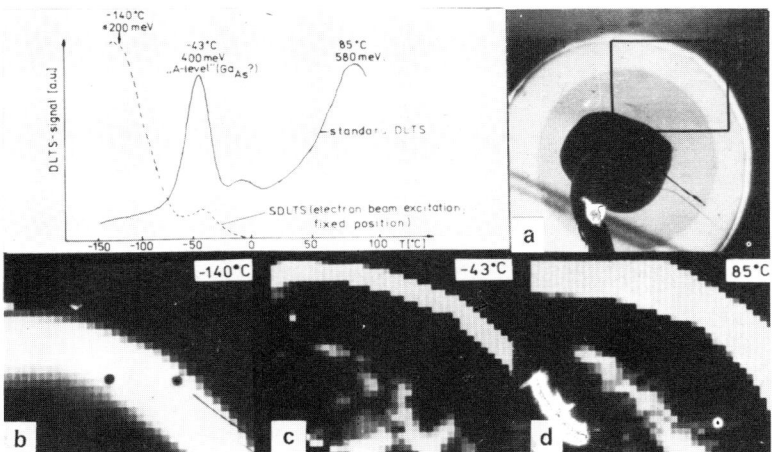

Fig. 5   DLTS spectra, EBIC image and SDLTS images of a
GaAlAs/GaAs heterojunction diode ($\emptyset$ 300 /um). Investigated
SDLTS image region framed in the EBIC image (top right)

is naturally much higher so that a stronger SDLTS signal can be expected, even if the levels are incorporated homogeneously also at the diode edge.

## 4. Conclusion

The SDLTS technique has been established as a useful electron microscope imaging technique for semiconductors. Its major field of application is the investigation of the inhomogeneous distribution of point defects associated with extended crystal defects. It is therefore advantageous to combine SDLTS with EBIC measurements or with other electron microscope techniques. Though the SDLTS method is still in the developing stage, meanwhile some theoretical foundations have been established and some special techniques of investigation have been developed to achieve an interpretation as unique as possible of the SDLTS image in terms of a concentration distribution of a well-defined deep level. Further experience in the application of this technique and progress in instrumentation will help to make this imaging technique more popular than hitherto. The experimental results given were thought to point out the potential possibilities of the combined application of SDLTS and other microscope imaging techniques, especially the EBIC method, to semiconductor physics.

## Acknowledgement

The authors are indebted to Dr. R. Pickenhain and Prof. Dr. G. Oelgart (Leipzig) and to Dr. J. Nowak (Bratislava) for submitting samples for these investigations. The assistance of M. Taege, Th. Nerstheimer, A. Pippel and J.M. Langner (all Halle) in developing the computerized SDLTS system is greatfully acknowledged.

## References

Borsuk J A, Swanson R M 1980 IEEE Trans. Electron Devices 27 2217
Breitenstein O 1982 phys. stat. sol. (a) 71 159
Breitenstein O, Heydenreich J 1983 J. de Physique C4 207
Breitenstein O, Wosinski T 1983 phys. stat. sol.(a) 77 K107
Heydenreich J, Blumtritt H, Gleichmann R, Johansen H 1981
    Scanning Electron Microscopy (Ed Johari O) Chicago,
    SEM Inc 1982, Vol I, p 351
Kittler M, Bugiel E 1982 Cryst. Res. Technol. 17 79
Lang D V 1974 Appl. Phys. Lett. 31 60
Ledebo L A, Wang Zh G 1983 Appl. Phys. Lett. 42 680
Misrachi S, Peaker A R, Hamilton B 1980 J. Phys. E 13 1055
Ourmazd A, Weber E, Gottschalk H, Booker G R, Alexander H
    1981 Inst. Phys. Conf. Ser. No 60 63
Petroff P M, Lang D V 1977 Appl. Phys. Lett. 31 60
Petroff P M, Lang D V, Strudel J L, Savage A 1978
    Proc. 9th Int. Congr. Electr. Microsc. Toronto, Vol I,
    p 130

*Inst. Phys. Conf. Ser. No. 76: Section 8*
*Paper presented at Microsc. Semicond. Mater. Conf., Oxford, 25–27 March 1985*

# New results and an interpretation for SEM EBIC contrast arising from individual dislocations in silicon

P R Wilshaw and G R Booker

Department of Metallurgy & Science of Materials,
University of Oxford, Parks Road, Oxford OX1 3PH

Abstract    The SEM EBIC contrast for individual screw and 60° dislocations formed in high-purity, n-type $10^{15}$ $cm^{-3}$ silicon by deformation
has been measured and found to vary with both specimen temperature and
electron beam current.  A new theory of recombination at dislocations
has been developed and applied to the EBIC method.   The new theory
explains the experimental results and enables parameters associated
with the recombination process at the individual dislocations, e.g.
energy level, density of states, etc., to be deduced.

## 1. Introduction

Numerous investigations have been made of the electrical recombination
behaviour of dislocations in semiconductors.  Experiments were performed
to determine such dislocation parameters as the energy level, density of
states, minority carrier capture cross-section, locally reduced minority
carrier lifetime, etc.  Many techniques were used including Hall measurements, electron spin resonance, photoconductivity, photoluminescence and
deep-level transient spectroscopy.  Most of the dislocations investigated
were introduced into the specimens by deformation.  For a particular dislocation parameter and a particular material, e.g. the dislocation energy
level in Si, a range of sometimes contradictory values was often obtained.
One possible reason for this is that all of the experimental techniques
used were bulk methods, and different types and concentrations of
dislocations were present in each of the specimens investigated.
Furthermore, the electrical behaviour of the dislocations may in some
cases have been modified by impurity decoration.

Clearly, there is a need for a technique to investigate individual dislocation segments that have been previously well characterised so that the
electrical information obtained can be unambiguously related to a single
well-defined defect.  The electron beam-induced current (EBIC) mode of an
SEM provides such a technique in that it enables the recombination of
carriers in semiconductors to be investigated with a spatial resolution of
typically ~ 1μm.  The technique has been used mainly qualitatively to
obtain EBIC micrographs which show the specimen regions corresponding to
enhanced electrical recombination, and to a lesser extent quantitatively
to obtain EBIC line-traces which enable the contrast associated with individual defects to be determined, which is a measure of the efficiency of
the defects as recombination centres.

Quantitative EBIC work has been performed at Oxford during the last few
years with the aim of studying the fundamental properties of individual

dislocations in Si.  In the initial work (Ourmazd and Booker 1979), edge dislocations present in the emitter of a Si bipolar transistor were investigated, while in subsequent work (Ourmazd, Wilshaw and Booker 1983a and b), screw and 60° dislocations in deformed high-purity Si were studied.  For the latter work the dislocation contrast C was measured as a function of specimen temperature T with the aim of moving the Fermi level with respect to the dislocation energy level, thereby changing the charge on the dislocation, and hence the recombination efficiency and the dislocation contrast.  Although pronounced and consistent changes of C with T were obtained, the results could not be explained by the then existing theories.

These theories were in general of two types.  First, there were EBIC contrast theories such as those of Donolato (1978) and Pasemann (1981), which calculated C as a function of the specimen geometry, but did not consider the actual recombination process taking place at the dislocation.  The effect of the dislocation was dealt with by attributing to it a recombination strength $\gamma$, related to a locally reduced minority carrier lifetime $\tau'$ (see below).  Although these theories predicted, for example, how C varied with dislocation depth (geometry effect), they could not predict how C varied with T (mainly recombination effect).  In order to do the latter it would be necessary to calculate how $\gamma$ varied with T, and this was not possible on these theories.  Second, there were recombination theories such as those of Figielski (1978) and Labusch (1979), which considered the process taking place at the dislocation.  In the Figielski theory, which was developed to explain bulk photoconductivity measurements, the minority carrier capture cross-section for the dislocation was taken to be independent of T.  Consequently, if this theory were applied to the EBIC contrast method, then it would not be able to explain the experimental C v T results.  In the Labusch theory, the carrier injection level was taken to be extremely small, and so this theory could not in practice be applied to the EBIC method.  An EBIC contrast theory was proposed by Ourmazd (1981) that took into account both the specimen geometry and the recombination process at the dislocation.  However, a number of simplifying assumptions were made, and this theory was also not able to explain the experimental C v T results.

In the present work, the previous C v T results for individual dislocations in Si (Ourmazd, Wilshaw and Booker 1983a and b) are extended to higher temperatures and more comprehensive measurements are made.  The new results obtained show that the dislocation contrast is a function of not only temperature T, but also electron beam current $I_b$.  A new theory which considers recombination at the dislocation is described and this is presented in a form which is appropriate for the EBIC contrast method.  The dislocation strength $\gamma$ is calculated as a function of both T and $I_b$, and this is then related to the dislocation contrast C by making use of the result obtained by Donolato (1978) that $C \propto \gamma$.  The new theory explains the new dislocation contrast results, and subsequent analysis then enables dislocation parameters such as energy level, density of states, etc., to be deduced.  Due to the limited space available in the present paper, only some of the experimental results that have been obtained are described, and only part of the complete theory is presented.  A fuller account will later be published elsewhere.

## 2. Experimental

The specimens investigated were high-purity, float-zone, n-type, $10^{15}$

$cm^{-3}$, Si that had been deformed under clean conditions by two-stage compression at 850 and 420°C (kindly provided by Professor H Alexander and colleagues, Cologne University, FRG). The specimens contained hexagonal dislocation loops up to ~ 50μm across comprising straight well-characterised screw and 60° dislocation segments lying along <110> directions. In previous studies TEM showed that virtually all of these dislocations were split into pairs of partial dislocations. HREM showed no detectable precipitation or impurity decoration at the dislocation cores. EPR showed only the Si K7 centre, strongly suggesting that the dislocations were free from point defect atmospheres. Thus, it is believed that the dislocations studied are as close as is presently possible to 'ideal' dissociated screw and 60° dislocations in Si.

Slices cut parallel to the (111) active slip plane were polished and surface Schottky barriers were formed by the deposition of Au/Pd layers. Ohmic contacts were made on the back surface of the slices. The resulting specimens were placed in the heating/cooling stage of an SEM and examined using an accelerating voltage of 15kV, a beam diameter of < 0.1μm and a beam current in the range 6 x $10^{-12}$ to 2 x $10^{-9}$ A. The EBIC signal was collected with the Schottky barrier unbiased. Measurements were made for individual dislocations forming parts of loops lying parallel to the slice surface and located in the electrically neutral material beneath the depletion region. The EBIC system used, and the method of measuring the EBIC contrast from individual dislocations, were similar to those described previously (Wilshaw, Ourmazd and Booker 1983). However, the EBIC system was now based on a Philips 505 SEM equipped with a $LaB_6$ gun. Accurate dislocation contrast measurements were consistently obtained.

## 3. Results

The EBIC contrast C is proportional to the dislocation strength $\gamma$ (Donolato 1978). C is given by $(i_b - i_d)/i_b$, where $i_b$ and $i_d$ are the EBIC currents corresponding to the background and dislocation respectively, and $\gamma$ is defined below. However, C also depends on a number of additional parameters such as the specimen geometry, the minority carrier diffusion length and possibly the ohmic contact resistance. Consequently, some initial measurements were made to determine whether, when either the temperature T or the beam current $I_b$ was varied, any of these additional parameters changed sufficiently to modify the contrast C. It was found that any such changes had an insignificant effect on the contrast C over the full range of conditions used in the present work. Hence, the contrast C could justifiably be taken as being proportional to the dislocation strength $\gamma$ for all of the results of the present work. However, the proportionality constant would in general be different for each particular dislocation because the different dislocations were at different depths, and so corresponded to different specimen geometries.

Comprehensive sets of EBIC contrast measurements were made for straight segments of a particular screw dislocation and a particular 60° dislocation. Fig.1 shows how C varied with T for 120K < T < 370K and $I_b$ = 1.1 x $10^{-10}$ A. The two dislocations did not form part of the same loop and so were probably not at the same depth. Consequently, it was not possible to deduce which type of dislocation had the greater strength $\gamma$ from the relative values of C shown in Fig.1. For the two types of dislocation, the following behaviour was observed: (a) as T increased, C increased, (b) at the lower temperatures, the variation was linear, straight lines occurring with the same slope, and (c) at the higher temperatures, the variation was

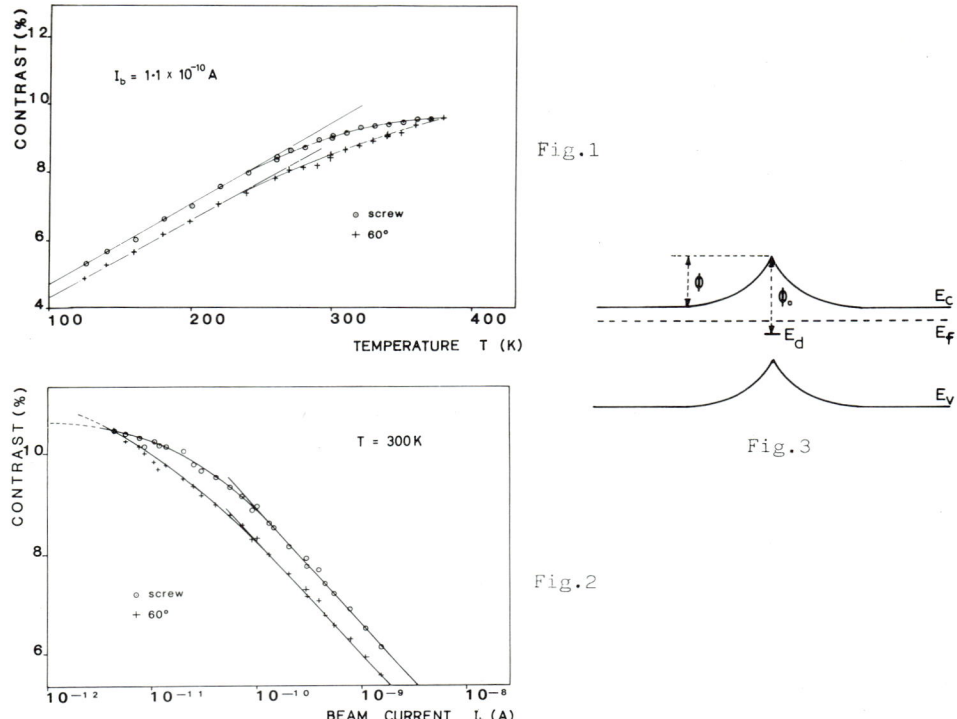

Fig.1

Fig.3

Fig.2

sub-linear, with C increasing less rapidly for the screw dislocation than the 60° dislocation. In the linear region, to a first approximation C ∝ T. Analogous C v T curves were obtained for a number of different values of $I_b$ (not shown).

Fig.2 shows, for the same two dislocation segments, how C varied with I for $6 \times 10^{-12}$ A $< I_b < 2 \times 10^{-9}$ A and T = 300K. For the two types of dislocation, the following behaviour was observed: (a) as $I_b$ decreased, C increased, (b) at the higher beam currents, the variation was linear, straight lines occurring with the same slope, and (c) at the lower beam currents, the variation was sub-linear, with C increasing less rapidly for the screw dislocation than the 60° dislocation. In the linear region, to a first approximation C ∝ $(C_o - \ln I_b)$. Analogous C v $I_b$ curves were obtained for a number of different values of T (not shown).

The fact that the dislocation contrast C depends on the beam current $I_b$ has not been reported previously by EBIC investigators, although the beam currents commonly used for such studies correspond to those used in the present work. The result is unexpected because it has always been assumed that C would be independent of $I_b$ as long as the conditions which were used corresponded to the low-level injection regime, i.e. the maximum excess minority carrier concentration was less than the majority carrier concentration. The maximum excess minority carrier concentration is determined by the beam current, beam voltage and manner in which the carriers diffuse away from the carrier generation volume (Donolato 1978).

All of the present work was performed under conditions corresponding to the low-level injection regime. Clearly, the above assumption is not correct. Furthermore any theory of EBIC contrast as applied to dislocations in Si must be able to take into account the effect of the beam current $I_b$ on the recombination efficiency in addition to any geometrical effects.

4. Theory

It might be expected that the EBIC contrast from individual dislocations in n-type Si would decrease with increasing temperature. This is because the Fermi level would then be lowered, pass through the dislocation level in the band gap, decrease the charge on the dislocation and thereby decrease the recombination efficiency. However, this is contrary to the observed behaviour, and so clearly other factors are important. Because the dislocations in semiconductors are charged, there is a local bending of the band structure at the dislocation, and this will affect the recombination behaviour. Such band bending is a basic concept in recent theoretical treatments of recombination at dislocations (Figielski 1978, Labusch 1979, Ourmazd 1981). The theory now presented initially follows the approach of the Figielski theory, but thereafter differs significantly in the assumptions made, the form in which it is developed and the parameters that are calculated.

Fig.3 shows the band structure of a dislocation in n-type Si where the acceptor states introduced correspond to an energy level below the Fermi level, and are thus occupied giving a negative line charge of Q per unit length. This negative charge produces a positive space charge region of radius $r_d$ around the dislocation where free holes can be captured by the attractive potential into the bound hole states at the top of the bent valence band. This capture of holes results in a reduced minority carrier lifetime $\tau'$ within the space charge region. The hole capture rate per unit dislocation length $J_h$ is related to $\tau'$ by:

$$J_h = \pi r_d^2 \Delta p / \tau',\qquad(1)$$

where $\Delta p$ is the excess minority carrier concentration. It is convenient to describe the recombination occurring at a dislocation by the dislocation strength $\gamma$ (Donolato 1978) given by:

$$\gamma = J_h / \Delta p D_h = \pi r_d^2 / D_h \tau'\qquad(2)$$

An expression for $r_d$ may be obtained by using the condition for charge neutrality at the dislocation, for which it is assumed that the screening is due to the expulsion of majority carriers from the space charge region. This gives:

$$r_d = (Q/\pi n_o q)^{\frac{1}{2}},\qquad(3)$$

where $n_o$ is the majority carrier concentration, and q is the electronic charge.

The electron capture rate per unit dislocation length $J_e$ is determined by detailed balance and is given by:

$$J_e = C_e N_d [ (1-f) n_o \exp(-\frac{q\phi}{kT}) - f N_c \exp(-\frac{q\phi_o}{kT}) ]\qquad(4)$$

where $C_e$ is the probability of an electron transition between the dislocation energy level and the conduction band, $N_d$ is the number of states per unit dislocation length, f is the occupancy factor of the dislocation states, $N_c$ is the density of states in the conduction band, $\phi$ is the potential barrier height and $\phi_o$ is the depth of the dislocation energy level below the bottom of the conduction band (Fig.3).

The expression for $J_e$ consists of two terms. The first represents capture of electrons from the conduction band to the dislocation level by thermal excitation over the potential barrier. The second is due to thermal re-excitation of electrons from the disocation level back into the conduction band. Consideration of equation (4) shows that except for high temperatures or low recombination rates, the first term is much larger than the second term and so the second term will for the present be ignored. It will also be assumed that $f \ll 1$. Equation (4) may then be rewritten to give:

$$\phi = - \frac{kT}{q} \, \ell n \left( \frac{J_e}{C_e N_d n_o} \right) \qquad (5)$$

At steady state conditions, the hole and electron capture rates must be equal to one another and to the recombination rate J:

$$J_h = J_e = J \qquad (6)$$

The barrier height $\phi$ is approximately proportional to the dislocation charge per unit length Q, and so for simplicity it will be assumed that:

$$\phi = A_1 Q \qquad (7)$$

($A_1$, $A_2$ and $A_3$ in equations (7), (8), (9) and (11) are constants). When carriers are being injected, $\Delta p$ is proportional to the electron beam current $I_b$:

$$\Delta p = A_2 I_b \qquad (8)$$

The dislocation strength $\gamma$ can then be obtained by solving equations (2), (3), (5), (6), (7) and (8) as follows:

$$\phi = - \frac{kT}{q} \, \ell n \left( \frac{\Delta p \gamma D_h}{C_e N_d n_o} \right)$$

$$\gamma = \frac{Q}{n_o q D_h \tau'} = \frac{\phi}{A_1 n_o q D_h \tau'}$$

$$= - \frac{1}{A_1 n_o q D_h \tau'} \frac{kT}{q} \, \ell n \left( \frac{\Delta p \gamma D_h}{C_e N_d n_o} \right)$$

$$= \frac{k}{A_1 n_o q^2 D_h \tau'} \, T \left[ \ell n (C_e N_d n_o) - \ell n (A_2 I_b \gamma D_h) \right] \qquad (9)$$

It can be seen that the term $\gamma$ also appears on the right hand side of the expression in a $\ell n$ term. However, in most cases the change in $\gamma$ that is produced experimentally is not large, and so the effect that it has in the $\ell n$ term is relatively small and will for the present be ignored.

In order to predict how the dislocation strength $\gamma$ will depend on temperature T from equation (9), it is initially necessary to consider how the individual parameters in equation (9) vary with temperature. Experiment has shown that $D_h$ varies as $T^{-1.4}$. Consequently, the effect of the variation of $D_h$ in the $\ell n$ term is also small. The parameters $N_d$ and $n_o$ in the present work are constant, and it will for the present be assumed that $C_e$ does not vary with temperature. The parameter $\tau'$ depends on the cascade capture of holes into the potential well surrounding the dislocation, and this has been treated theoretically by Sokolova (1970) who found that $\tau'$ varies as $T^{1.5}$. Consequently, the product $\tau' D_h$ is closely independent of temperature.

On the basis of this simplified theory, which applies to the EBIC method except for high temperatures and low recombination rates, the following predictions can be made regarding the dislocation contrast C:

for constant $I_b$, $C \propto \gamma \propto T$                                          (10)

for constant T, $C \propto \gamma \propto (A_3 - \ell n I_b)$                        (11)

These two theoretical relationships correspond to the two experimental relationships determined for the linear regions of Figs.1 and 2 respectively, and which were found to occur except at high temperatures and low beam currents (low recombination rates).

The experimental values of T and $I_b$ that are commonly used by the majority of investigators performing EBIC contrast experiments correspond to the linear region. From equation (9) it can be seen that the dislocation strength $\gamma$, and hence the dislocation contrast C, depends on several important dislocation recombination parameters, namely, $\tau'$, $C_e$ and $N_d$. However, C does not depend on the dislocation energy level. The reason for this is that when working in the linear region, insignificant numbers of electrons are re-excited from the dislocation energy level back into the conduction band, a process that would have an activation energy corresponding to $\phi_o$, the depth of the dislocation energy level below the bottom of the conduction band. Consequently, when working in the linear region, analysis can provide information about some of the dislocation recombination parameters, but $\phi_o$ cannot be directly deduced.

In the treatment so far, the second term in equation (4), corresponding to electron re-excitation from the dislocation, was ignored because it was considered to be small compared with the first term, corresponding to electron excitation over the potential barrier. Consideration of equation (4) indicates that the second term will become relatively more important on going to either higher temperatures or lower beam currents. This effect is able to explain the sub-linear regions of the experimental curves of Figs.1 and 2. An analysis of these sub-linear regions can provide information about the dislocation recombination parameters, including $\phi_o$. For example, the observation that the deviation from the linear relationship is more pronounced for the screw dislocation than the 60° dislocation indicates that $\phi_o$ is smaller for the screw dislocation. The analysis of this sub-linear region will be described in the full account to be published later.

In the remaining space available in the present paper, an analysis is presented which shows the type of information that can be obtained from the linear regions of the C v T and C v $\ell n I_b$ curves. Equations (2), (5), (6) and (8) give:

$$I_b C \propto I_b \gamma \propto C_e \exp\left(-\frac{q\phi}{kT}\right)$$        (12)

Thus, if it is assumed that $C_e$ is constant, then for any particular value of C, a plot of $\ln I_b$ v $\frac{1}{T}$ should give a straight line and hence enable $\phi$ to be determined for that particular value of C. Data for the screw dislocation taken from the linear regions of the series of C v T and C v $\ln I_b$ curves already obtained showed this predicted behaviour for a particular value of C and so enabled $\phi$ to be determined. This was then repeated for other values of C, and so a complete plot of C v $\phi$ was obtained for the screw dislocation. The procedure was repeated for the 60° dislocation and an analogous plot was obtained. These experimentally determined C v $\phi$ relationships were in excellent agreement with the theoretically determined C v $\phi$ relationship that was obtained using a more detailed form of the above theory.

An analysis of these results gave quantitative data such as the following. For n-type, $10^{15}$ cm$^{-3}$ Si at 300K and a beam current of $\sim 10^{-11}$ A, the potential barrier height $\phi$ for the screw and 60° dislocations was 0.18 and 0.21eV respectively. Because the dislocations are charged, the dislocation energy level $\phi_0$ must be deeper than the Fermi level (0.31eV), and so $\phi_0$ for the screw and 60° dislocations was $\geqslant 0.49$ and $\geqslant 0.52$eV respectively. The line charge Q for the screw and 60° dislocations was 3.4 x $10^{-13}$ and 3.8 x $10^{-13}$ C cm$^{-1}$, and the density of states for the screw and 60° dislocations was $\geqslant 2.2$ x $10^6$ and $\geqslant 2.4$ x $10^6$ cm$^{-1}$ respectively.

In the above analysis it was assumed that $C_e$ was constant. If $C_e$ were to show an exponential dependence on temperature, then the activation energy for $C_e$ would be included in the measured activation energy for $\phi$, and it would not be possible to separate $C_e$ and $\phi$. This difficulty also occurs with the DLTS method for determining energy levels.

## Acknowledgements

The authors wish to thank Dr A Ourmazd and Professor Sir Peter Hirsch for their continued interest in the project and useful discussions, and Professor H Alexander and colleagues at Cologne University, FRG, for kindly supplying the deformed Si specimens and information, and the Science & Engineering Research Council for support.

## References

Donolato C 1978 Optik 52 19
Figielski T 1978 Solid State Electron. 21 1403
Labusch R 1979 J Physique 40 C6-81
Ourmazd A and Booker G R 1979 Phys. Stat. Sol. (a) 55 771
Ourmazd A 1981 Cryst. Res. Techn. 16 137
Ourmazd A, Wilshaw P R and Booker G R 1983a Physica 116B 600
Ourmazd A, Wilshaw P R and Booker G R 1983b J Physique 44 C4-289
Pasemann L 1981 Ultramicroscopy 6 649
Sokolova E B 1970 Sov. Phys. Semiconductors 3 1266
Wilshaw P R, Ourmazd A and Booker G R 1983 J Physique 44 C4-445

*Inst. Phys. Conf. Ser. No. 76: Section 8*
*Paper presented at Microsc. Semicond. Mater. Conf., Oxford, 25–27 March 1985*

# EBIC studies of dislocation contrast and resolution in silicon avalanche photodiodes

M Lesniak* and D B Holt

Department of Metallurgy and Materials Science, Imperial College of
Science and Technology, London    SW7 2BP, U.K.

*  Present address:  B.P. Research Centre, Chertsey Road, Sunbury-on-
                 Thames, Middlesex    TW16 7LN.

Abstract   Systematic measurements were made on the magnitude and
line width of the dark contrast seen at diffusion-induced dislocations
in silicon avalanche photodiodes.  The influence of defect depth,
beam parameters and specimen temperature were investigated.  It is
shown that EBIC measurements can be used to determine the dislocation
depth and the approximate distribution of the energy levels involved.

## 1. Introduction

There is widespread fundamental interest in the core structure and
properties of dislocations in semiconductors at present (see e.g., the
conference proceedings published as supplement to J. de Physique:
Colloque C-6 (1979) and Colloque C-4 (1983).  This paper presents some
results of a study of defects in an important type of Si device.

## 2. Experiment

The specimens were $p^+\,\pi\,pn^+$ reach through avalanche photodiodes (RTAPDs).
The $n^+p$ planar junction depth was 8.2 μm as described previously
(Lesniak and Holt (1983)  which presented EBIC and TEM observations).
Fig. 1 shows a cross section through the active part of the device and
the defects involved.  It was shown (Lesniak and Holt (1983)) that the
noise and soft reverse characteristics of the RTAPDs were not due to the
diffusion-induced misfit dislocations, but to related microplasma effects.
These arose at ragged junctions due to dislocation-retarded diffusion
(Duffy et al (1968), Ashburn and Bull (1979)) which occurred due to
excessive surface densities of the dopant.

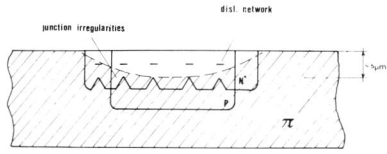

Fig. 1  Schematic illustration of
the misfit dislocation network
and the junction irregularities.

The specimens were examined in a JEOL-35 SEM using the EBIC method. The detection system was described by Lesniak et al (1984).

The RTAPDs were connected so an increased screen brightness was obtained when the $pn^+$ junction charge collection current ($I_{cc}$) increased. The electrically active dislocations appeared as features of reduced $I_{cc}$.

The sample temperature in the SEM was controlled by specially designed "hot" and "cold" stages allowing accurate temperature settings within the 500-300 K and 300-77 K ranges respectively.

3. Results

Varying the SEM beam energy and hence the electron penetration range, the depth of the diffusion-induced misfit dislocations (Fig. 2) was found to be 2.4 μm. This result was confirmed by combining series of SEM EBIC observations with ion beam thinning sequences. By thinning more than 2.4 μm from the top $n^+$ layer, the dislocation dark line contrast was removed leaving only the bright lines representing the junction non-uniformities due to dislocation retarded diffusions.

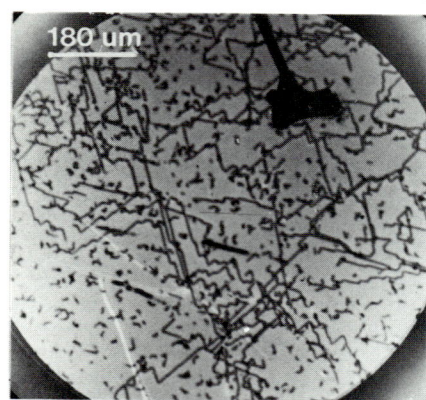

180 um

Fig. 2   SEM EBIC micrograph of the $pn^+$ junction of a RTAPD at OV bias. The dark lines represent single dislocations.

The minority carrier diffusion length ($L_h$) within the $n^+$ region was measured. A cross-section of the device was used for this purpose. $L_h$ was about 1 μm within the phosphorus-rich surface zone.

The influence of the beam energy on the EBIC contrast and line width of dislocations was investigated. A series of EBIC scan lines was taken at accelerating voltages in the range 18 to 33 keV. Line scans representing 20, 28 and 33 keV are shown in Fig. 3. These were collected from a specimen containing a few well spaced dislocations. A low magnification EBIC micrograph of the RTAPD examined is shown in Fig. 2. The half widths of line profiles like those shown in Fig. 3 were measured. Contrast values were calculated as $C = \Delta I_D/I_B$ where $\Delta I_D$ is the reduction in $I_{cc}$ at a dislocation and $I_B$ is the value obtained with the electron beam far from the defect. The calculated values of contrast decrease significantly with increasing electron beam energy as can be seen from Fig. 4. The observed monotonic decrease in the EBIC contrast of dislocations can be understood since recombination at the defect will affect a small fraction of the expanding electron beam energy dissipation volume which will reach nearer to the $pn^+$ junction

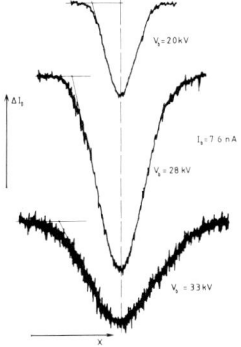

Fig. 3  EBIC profiles of a dislocation taken with beam voltages of 20 keV; 28 keV and 33 keV.

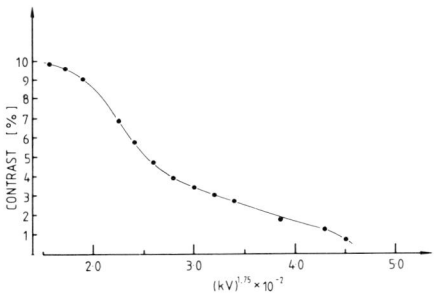

Fig. 4  Contrast of a dislocation vs incident beam voltage.

for higher $E_b$ values and so increase $I_B$.

The dislocations in RTAPDs are thought to be decorated with impurity atoms.  This is based on the facts that:

(a)  the phosphorus doping concentration is high ( $\sim 10^{21} cm^{-3}$ ),

(b)  dislocation-retarded diffusion of the phosphorus was observed to occur,

(c)  TEM observations had showed some of the dislocations to be associated with impurity particles,

(d)  the EBIC contrast observed was relatively large ( $\sim 10\%$ at 18 keV).

The line width of the dislocation images was observed to increase with increase in $E_b$. This accurately followed a $E_b^{1.75}$ dependence as can be seen from Fig. 5.

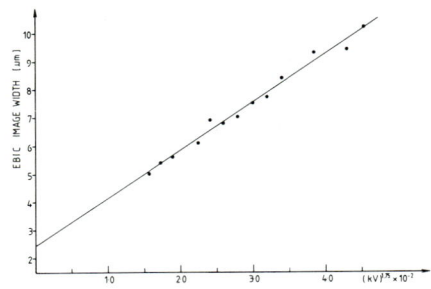

Fig. 5  Dependence of the dislocation image width on the beam voltage.

A similar scaling of EBIC dislocation image widths with $E_b$ was reported for the case of Schottky barriers by Leamy et al (1976).

The experimental values of dislocation line width were used to calculate the parameters of the straight line in Fig. 5 and gave the expression:

$$(1) \quad W = 0.5 \ (0.017 \ E_b^{1.75} + 2.44) =$$

$$0.5 \ (R_e + Z_o) \qquad ( \ \mu m)$$

where $0.017 \times E_b^{1.75}$ is the term dependent on the extent of the energy dissipation volume ie. on the electron range - $R_e$. The constant 2.44 μm defines the image width which could be approached at the smallest possible $E_b$ values and coincides with the value of the depth of the dislocations, $Z_o$, in the RTAPDs.

Fig. 6(a) shows the variation in background current $I_B$ and EBIC contrast C of a dislocation with temperature.

The data were collected from line scans across the defect with a 25 keV,7 nA electron beam. The dislocation contrast decreased with increasing temperature. The broad temperature range of the transition in electrical activity shown in Fig. 6(a) is consistent with a dislocation associated with an impurity atmosphere resulting in a broadened band of energy states (Kimerling et al (1977)). The possibility of a second, shallow state is suggested by the inflection in the C vs T relation occurring at about 140 K. The contrast is constant for temperatures above 260 K. A remarkably different C vs T relation was observed at reduced electron beam powers. This is reflected in Fig. 6(b) representing data taken with a 18 keV, 4 nA beam. Such a difference can be explained since the Fermi-level within the forbidden energy gap is dependent on both temperature and the density of the electron beam injected minority carriers. As the Fermi-level moves down the energy band gap of the n-type material with increasing temperature, more of the dislocation states are emptied. The defect is then less able to act as a recombination centre for the minority carriers (holes). Similarly, as the local density of beam-induced minority carriers changes from low to high, the quasi-Fermi level

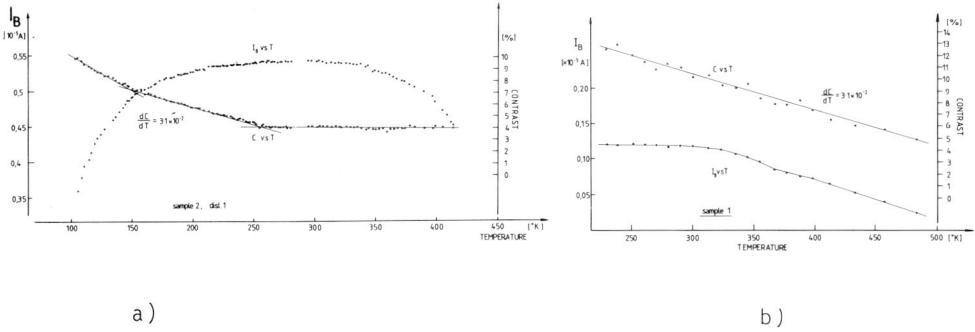

a)                                                    b)

Fig. 6   Temperature dependence of contrast at a dislocation (a) $E_b$ = 25 keV, $I_b$ = 7 nA,   (b) $E_b$ = 18 keV, $I_b$ = 4 nA.

moves further down the energy band gap leaving more of the dislocation states emptied, reducing its electrical activity.  The situation depicted in Fig. 7 could account for the difference between the C vs T relation observed for different beam energies at temperatures above 260 K.

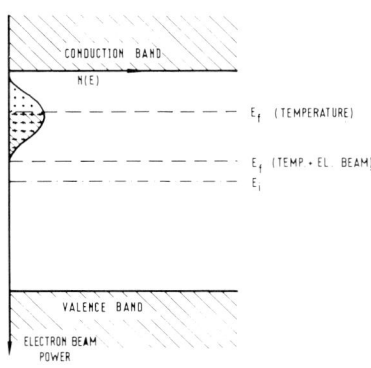

Fig. 7   Energy level diagram of dislocation in the band gap of a RTAPD subjected to heating.  The appropriate position of the quasi-Fermi level for low and high beam excitation levels is indicated.

## 4. Conclusion

A systematic SEM EBIC study of the individual misfit dislocations in RTAPDs provided evidence that the EBIC spatial resolution (dislocation line width) is dependent on instrumental and geometrical factors i.e. the electron range $R_e$ and the depth $Z_o$ of the defects.  No evidence of any dependence of resolution on the minority carrier diffusion length was seen.  The variation of dislocation EBIC contrast with temperature and beam voltage, i.e. the density of the injected carriers, provides information on the location of the states due to dislocations within the energy band gap.

References

Ashburn P and Bull C J (1979) Sol. State Electronics 22, 105-110.
Duffy M C, Barson F, Fairchild J M and Schwuttke G H (1968) J. Electrochem.
   Soc. 115, 84-88.
Kimerling L C, Leamy M J, Patel J R  (1977).  "The Electrical Properties
   of Stacking Faults and Precipitates in Heat Treated Dislocation Free
   Czochralski Silicon".  Appl.  Phys.  Lett.  30, pp. 217-219.
Leamy H J, Kimerling L C, and Ferris S D (1976) Scanning Electron
   Microscopy/1976 (Part IV, Vol, 1) (IIT Research Institute, Chicago,
   ILL), pp. 529-538.
Lesniak M, Holt D B (1983) Microscopy of Semiconducting Materials 1983,
   Conf.  Series No. 67, (A G Cullis,  S M Davidson and G R Booker eds)
   Inst. Phys. Bristol and London pp. 439-444.
Lesniak M, Unvala B A, Holt D B (1984)  J. Microscopy 135, pp. 255-274.

*Inst. Phys. Conf. Ser. No. 76: Section 8*
*Paper presented at Microsc. Semicond. Mater. Conf., Oxford, 25–27 March 1985*

343

# Semiconductor characterization by simultaneous evaluation of electron beam induced current and scanning electron acoustic microscopy in an automated scanning electron microscope

L J Balk, G Richard*, and N Kultscher

Universität Duisburg, Fachgebiet Werkstoffe der Elektrotechnik,
Leiter:Prof.Dr.-Ing.E Kubalek, Kommandantenstraße 60, 4100 Duisburg 1,F.R.G.
*present address:Hewlett Packard, Herrenberger Str. 130, 7030 Böblingen

Abstract   Electron beam induced current measurements and scanning electron acoustic microscopy are carried out simultaneously in an automated
scanning electron microscope. This enables reliable correlation of both
techniques.Due to low signal levels long recording times are usual.
The automation concept employs a new scan method allowing a quick judge-
ment on the chosen experimental parameters and avoiding artefacts due to
system instabilities. Image processing enables reconstruction of micrographs of relevant material parameters.The usefulness of this set-up is
demonstrated by applications to gallium phosphide,indium phosphide, and
poycrystalline silicon.

## 1. Introduction

Both the electron beam induced current technique (EBIC) and the scanning e-
lectron acoustic microscopy (SEAM) have proven to be sensitive tools for the
determination of important semiconductor parameters,such as the localization
of pn-junction and similar barriers,  the evaluation of minority carrier
diffusion lengths or imaging of defects.Unfortunately both these techniques
could not be carried out simultaneously,which would allow a precise correlation of the different parameters gained, especially if very long recording
times become necessary.Consequently,system instabilities like cathode drift
or stage movement are important obstacles to a reliable correlation.The set-
up introduced in this work eliminates the mentioned shortcomings.

## 2. Signal Detection and Amplification

With a new specimen stage
(fig.1) both techniques can
be applied within one single
experiment. By precise positioning of two prober needles
any junction of interest can
be examined in complex doping structures by EBIC. The
SEAM signal is detected by a
piezoelectric transducer attached to the bottom surface
of the specimen (Balk and

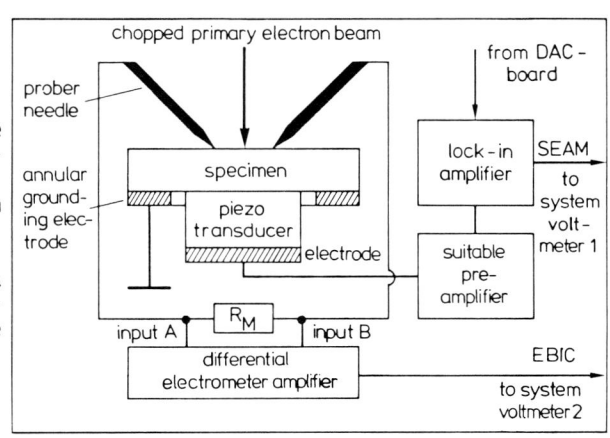

Fig. 1. Schematic of specimen stage and amplification

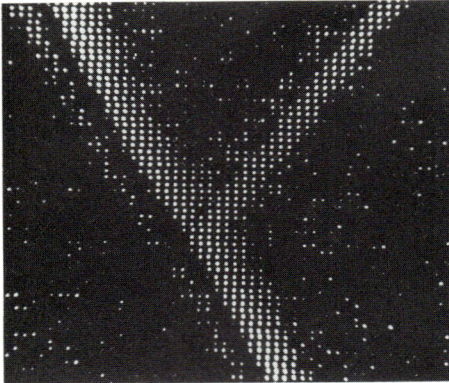

4 records　　　recording time:　26s

16 records　　　recording time: 105s

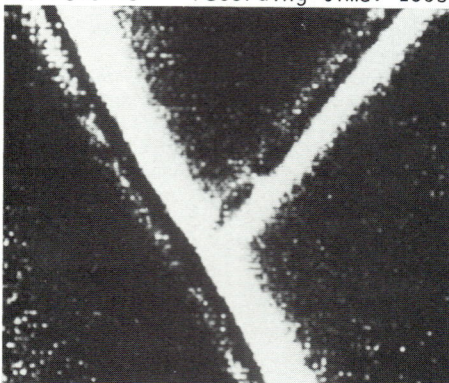

32 records　　　recording time: 209s

Fig. 2. EBIC micrographs of polycrystalline silicon in fast view
scan mode for different record
numbers.　　　　　　　10 /um
1 record = 1024 pixels;
total recording time for
512x512 pixels: 26 minutes

Kultscher 1983a).The two different signals are amplified separately. Lock-in
amplification is used for SEAM,whereas
an electrometer amplifier is mostly
sufficient for EBIC. Both signals are
then digitized by two system voltmeters.

3. Concept of Automation

The whole set-up is automated by means
of a process computer system. The concept for automation has to take into
account both the fact that typically
very long recording times are necessary
(depending on the signal levels even
up to hours) and that the signal processing should deliver micrographs of
relevant material parameters.
The first topic is a main part of the
automation as realized in this work.
Besides the standard scanning mode it
utilizes a special scan method which
is denoted in the following by 'fast
view scan (FVS)' and which optimizes
the requirements for long recording
times. The FVS scans the sampled specimen area in a set of records. Each of
these records contains 1024pixels,which
are uniformly spread over the whole image area. From record to record these
submatrices are slightly shifted to
each other in such a manner that any
time a homogeneous coverage of the area
is achieved.This is repeated,until the
whole area is covered by -at the most-
1024x1024 pixels. The calculation of
the pixel addresses and the control of
this scan mode is done by a Z80-micro-
processor board allowing a much higher
scan speed than could be done by the
process computer itself. As the whole
image area is already covered for the
first record,a fast view on the micrograph can be gained. This enables a
quick judgement(after several seconds)
on the suitability of the chosen experimental parameters, although the total
image recording time may range up to
the order of hours.This is demonstrated
in fig. 2 for an EBIC micrograph of
grain boundaries in polycrystalline
silicon. Furthermore, any long time
instabilities, as for instance due to
cathode drift or even a total failure,
become negligible, as long as at least
25 percent of the image could be taken

at all with reasonable signal-to-noise levels.This is due to the fact that such variations affect all image partitions and thus only introduce some additional noise, but do not harm any special image part ( as in form of black lines or unwanted grey level steps ). As the FVS is quite complicated, the control of the scan coils, done by a special DAC board, has to be timed by the system voltmeters which digitize the signals. If any systems errors have occured, the system can be interrupted and restarted by means of a simple push button without the need for new programming. This is especially useful in case that a computer error happened to stop the scan. For this case a timer module is installed to switch off the primary electron beam automatically to avoid any specimen damage, a feature which is important for series of micrographs taken over night. Finally, in order to speed up the system's capability the images are displayed on a separate storage display. Fig.3 is an overall sketch of the set-up.

## 4. Image Processing

The image processing is optimized for the needs of SEAM and EBIC. In SEAM experiments, especially at high modulation frequencies, quite often only lock-in amplifiers are available which only deliver the outputs $A\sin\emptyset$ and $A\cos\emptyset$,where A is the amplitude of the detected acoustic signal and $\emptyset$ its relative phase position with respect to the modulation waveform of the primary electron beam. For quantitative interpretation, however, A and $\emptyset$ are more suitable (Balk and Kultscher 1983b).Thus the main task for image processing of SEAM micrographs is the calculation of amplitude and phase images. In EBIC, especially for diodes, more processing problems arise. Due to the dependence $EBIC \sim \exp\text{-}L/x$ (with L as diffusion length of minority carriers in the probed region and x the distance to the pn-junction), the shape of the unprocessed EBIC image is quite vague, as is shown in fig.4a for a Zn-doped

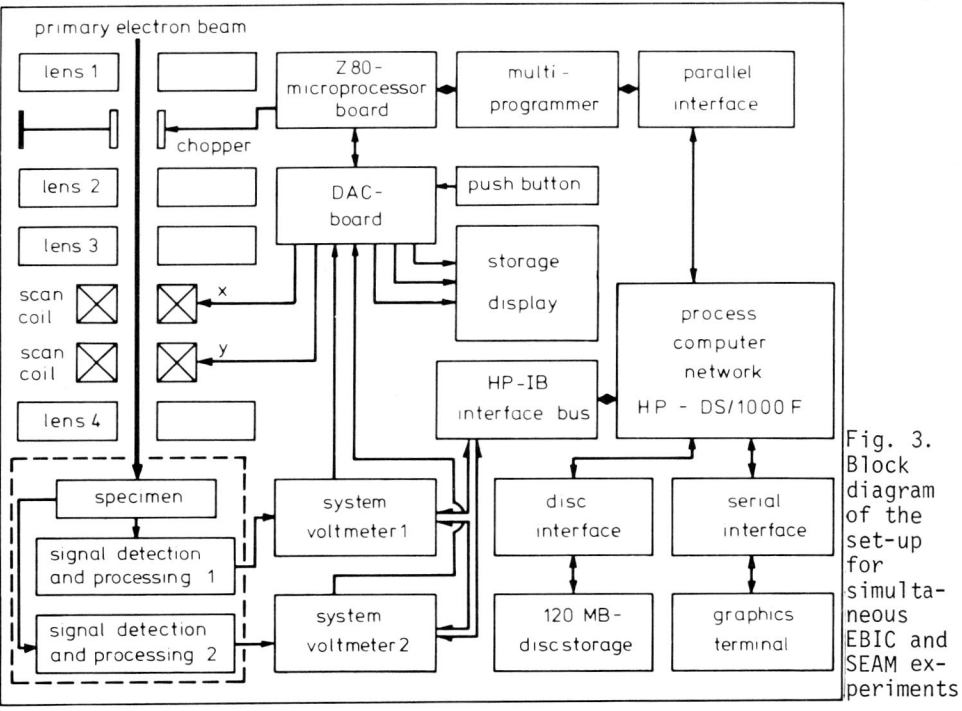

Fig. 3. Block diagram of the set-up for simultaneous EBIC and SEAM experiments

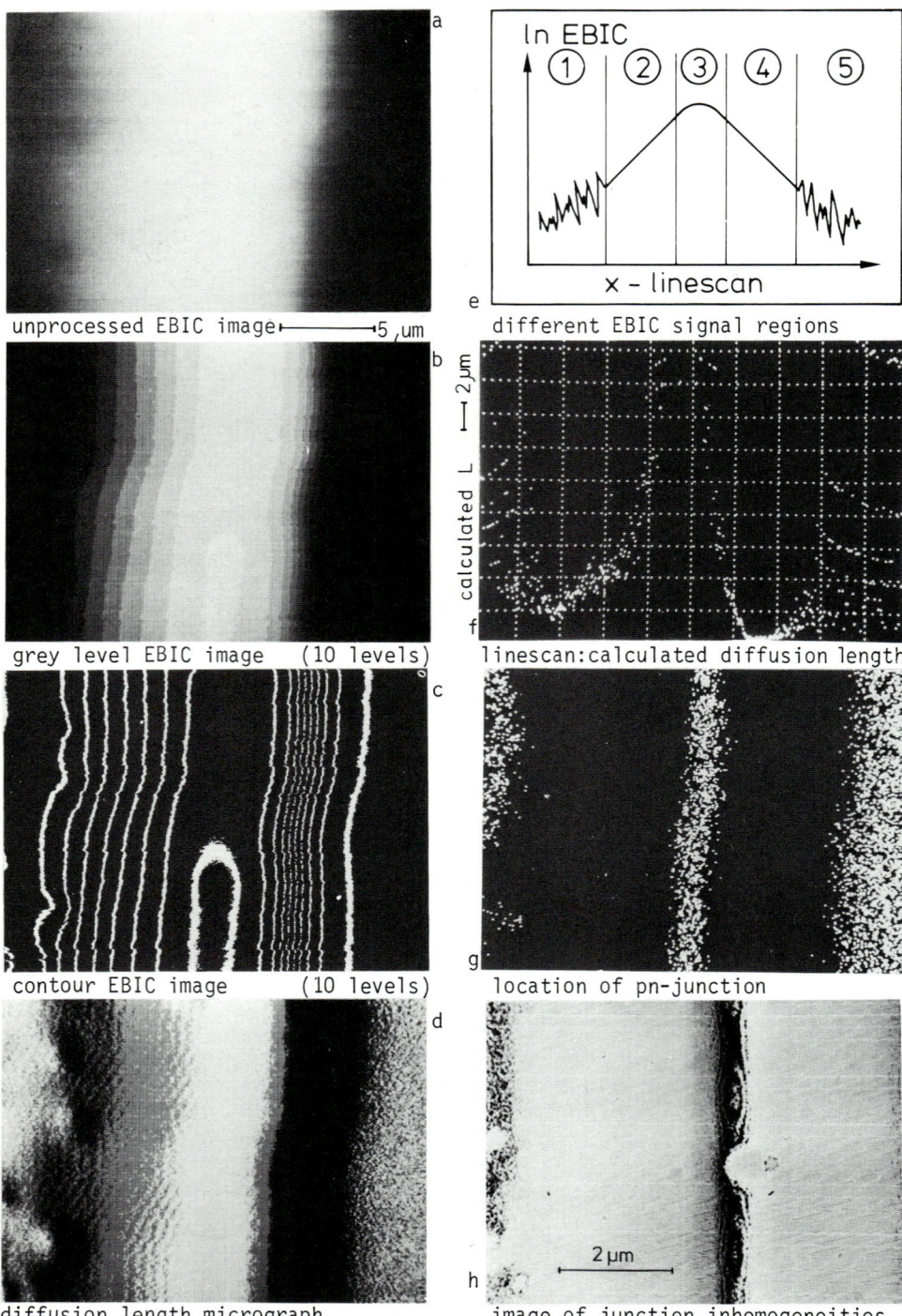

a

unprocessed EBIC image ⊢————→5 μm

e

different EBIC signal regions

b

grey level EBIC image     (10 levels)

f

linescan:calculated diffusion length

c

contour EBIC image     (10 levels)

g

location of pn-junction

d

diffusion length micrograph

h

image of junction inhomogeneities

Fig. 4. Computer processed EBIC for a GaP:Zn diode (primary energy:30keV)

GaP diode.A detailed grey level (fig.4b) or contour (fig.4c) processing
helps to visualize quickly the approximate position of the junction and the
different slopes on each side of it. The local diffusion length can be ob-
tained by calculating $L = - \left| \dfrac{\Delta x}{\Delta \ln(EBIC)} \right|$ .As this operation is very sensitive
for small signals and fluctuations of these, quite often a smoothing of the
obtained data has to be done. This can be achieved by taking the average of
the actually calculated pixel with its direct (eight) neighbouring pixels,
which already gives good smoothing without reducing the spatial resolution
significantly (reduction of linear pixel resolution less than a factor of 2).
Fig.4d is a diffusion length micrograph obtained in this manner.As can be
seen by the sketch of fig.4e, this micrograph can be divided into five different regions (compare Balk et al 1980). In regions 1 and 5 a strong noise
influence due to a very low EBIC signal gives wrong L-values. Region 3 is
characterized by the overlap of the primary electron energy dissipation over
the space charge region, which causes too small EBIC variations within it.
Therefore in region 3 the calculated L-values are too high (due to $\Delta\ln EBIC$
being close to zero). Only in regions 2 and 4 the EBIC signal shows its theoretical dependence and thus only in these regions the diffusion length can
be evaluated. This can be seen clearly from the diffusion length linescan
taken across the pn-junction, the diffusion length values being 1.1/um for
holes and 3.4/um for electrons. By using   contour imaging for the diffusion
length image and setting signal levels to values of very high 'calculated'
diffusion length the actual position of the pn-junction can be located precisely. Furthermore the region width of usable EBIC signal becomes visible
(fig.4g).Finally an additional enhancement of inhomogeneities close to the
junction region can be gained by optimizing the grey level processing for
this region and by additional avoiding of taking the absolute value of L in
the above given formula. As in this region $\Delta\ln(EBIC)$ is very close to zero,
inhomogeneities can cause a change from negative to positive signals and
vice versa, which enables to clearly demonstrate contours of inhomogeneities
(shown in fig.4h for the same sample but for a slightly different specimen
area and a different magnification).

## 5. Comparison of EBIC and SEAM

The usefulness of the described set-up shall be shown for two applications.
In fig.5a a comparison is given for the analysis of grain boundaries in polycrystalline silicon. For this sample relatively long diffusion lengths

EBIC image                    ├──────┤10/um  SEAM image
Fig. 5. Comparison of EBIC and SEAM for grain boundaries in polycrystalline
silicon (primary electron energy: 30keV; total recording time: 29 minutes)

should occur (10 μm and more). Furthermore, the voltage barrier at the grain boundary should be triangularly shaped, and the maximum voltage step should be only of the order of 0.1V. Thus the charge separation efficiency of this barrier is very low. Therefore an open circuit measuring condition has to be used in this case (by using a high $R_M$, compare fig.1; this condition can be attributed to EBI Voltage).In fig.5a black and white areas along the grain boundary are indicating the different slopes of the barrier, the total width of both being the region of sufficient charge separation.In this region the production of SEAM signals necessarily has to be lower than in the grain volume, as for semiconductors SEAM and EBIC are partially competing processes (Kultscher and Balk 1985). Therefore the dark SEAM areas are of identical width to the EBIC structure. Finally, these dark regions are surrounded by bright zones which can be attributed to regions of lower oxygen concentration, which causes a higher acoustic signal due to a lower material stiffness (Yonenaga and Sumino 1984).

EBIC                                                              SEAM                          2 μm ⊢⊣
Fig. 6. Comparison of EBIC and SEAM for selectively doped InP
(primary electron energy: 5keV; total recording time: 29 minutes)

Fig.6 compares EBIC and SEAM results obtained for an FET structure in InP. The prober needles contacted gate and substrate, therefore the pn-junctions defining the gate width show up bright in the EBIC image. The simultaneously recorded SEAM amplitude image again shows a dark contrast at locations of high EBIC due to the same reason as for the silicon sample. Additionally the gate region gives a bright contrast due to a high doping concentration compared with the substrate (the small white lines are only    features caused by metallization layers).

## Acknowledgements

The authors would like to thank Prof. Kubalek for helpful discussions,Mr.J. Elsbrock for assistance in computing. One of the samples was supplied by Prof.Heime of Fachgebiet Halbleitertechnik der Universität Duisburg.

## References

Balk L J and Kultscher N 1983a Inst. Phys. Conf. Ser. 67: Sect.8, 387
Balk L J and Kultscher N 1983b Journal de Physique 45 suppl.no.2, C2-869
Balk L J, Menzel E and Kubalek E 1980 Proc. 8th Int.Conf. ICXOM, Pendell
   Publishing Company, Midland, pp 613 - 624
Kultscher N and Balk L J 1985 J.Scanning Electr.Microsc.,SEM Inc(submitted)
Yonenaga I and Sumino K 1984 J. Appl. Phys. 56 No.8, 2346

*Inst. Phys. Conf. Ser. No. 76: Section 8*
*Paper presented at Microsc. Semicond. Mater. Conf., Oxford, 25–27 March 1985*

349

# Charge-collection memory effect in electron-irradiated hydrogenated amorphous silicon

B G Yacobi

Solar Energy Research Institute, Golden, Colorado  80401

Abstract   The EBIC mode of the SEM is utilized  to characterize the charge collection (CC) inhomogeneities in a-Si:H devices.  Great caution should be exercised in both the raster and spot modes at high electron beam power densities, at which the electron irradiation generates defects that reduce the carrier lifetime in a-Si:H.  In the CC mode of the SEM, irradiated areas appear darker than the unbombarded regions of the device.  This leads to a contrast formation for pattern recognition. Annealing a device at about 200°C restores the CC signal.

## 1.  Introduction

Since the discovery of the effect of hydrogen passivation of defects in amorphous silicon, hydrogenated amorphous silicon (a-Si:H) has been attracting increasing interest in a wide variety of applications.  These are, for example, photovoltaics, electrophotography, vidicons, field effect transistors, high-speed detectors, threshold switching and memory devices (see, for example, a series edited by Pankove 1984).

Although hydrogen passivation is very effective in removing the states from the gap, an a-Si:H random network may still contain a wide variety of bonds with a range of bond lengths and bond angles, internal surfaces  and voids. It is now well established that metastable defects may be generated in this material by irradiation with photons, electrons, ions and x-rays (see, for example, the review by Schade in the series edited by Pankove 1984).

Analyses of a-Si:H devices using a variety of electron probe instruments are usually considered nondestructive, since the physical integrity of the material remains intact.  However, a possible electron-irradiation-induced effect on electronic properties of the material also has to be taken into account.

The major objectives of this study are (i) the characterization of fabrication defects in a-Si:H devices using the electron-beam-induced current (EBIC) mode of the SEM and (ii) the investigation of possible electron-irradiation-induced damage during these observations.

## 2.  Experimental

Hydrogenated amorphous silicon devices used in this study were deposited by three methods:  glow discharge (GD) decomposition of silane, reactive sputtering (RS)  and chemical vapor deposition (CVD).  The GD samples were supplied by SERI's Amorphous Group and the rest were from subcontractors. P-I-N devices were used in most cases, and for several GD samples, Pd was used for Schottky barriers. The types of charge-collection signal that can be used to observe electrical inhomogeneities in semiconductor devices are shown in Fig. 1. In all our experiments, a signal due to the barrier or bulk electron voltaic effects was used (Fig. 1a).

**Types of charge collection signal**

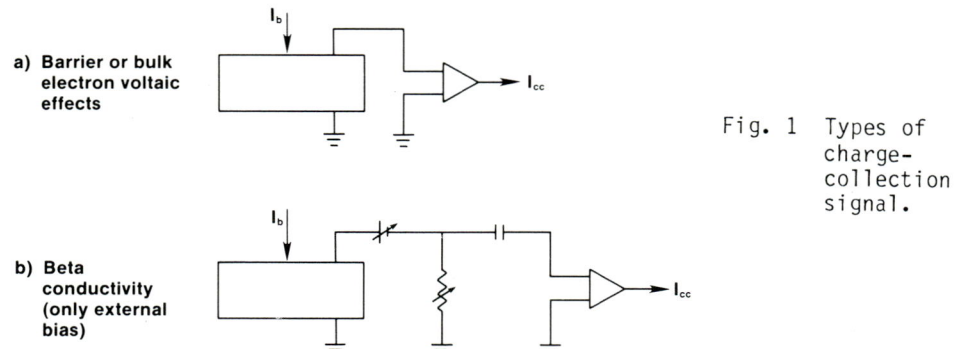

a) **Barrier or bulk electron voltaic effects**

b) **Beta conductivity (only external bias)**

Fig. 1   Types of charge-collection signal.

## 3.   Results and Discussion

Several types of fabrication defects were identified in various samples. These are, for example, pinholes, blistering, and lift-off. In addition, surface roughness of the stainless steel substrate can also lead to inhomogeneities in the charge-collection signal of the device. A detailed description of these fabrication defects was published elsewhere (Yacobi et al 1984). In one particular instance, for example, the secondary electron image (SEI) in Fig. 2 (the left half of a bipartite micrograph) shows no visible fabrication defects. However, the charge collection (or EBIC) image in the right half of Fig. 2 shows dark dots, which are probably caused by the incomplete junction formation due to pinholes propagating only part of the way through the film.

As one can see in Fig. 2, during these observations, darkening (decrease) in the charge collection (EBIC) signal can easily be observed after a few minutes of electron irradiation by, for example, a 30-keV beam of 1 nA in a raster mode of the areas of ~0.1 mm$^2$. An even faster decrease in the charge collection signal was observed in a raster mode of smaller areas (higher magnification) with the same beam current, or at lower magnifications with a higher beam current. Only seconds were sufficient for the

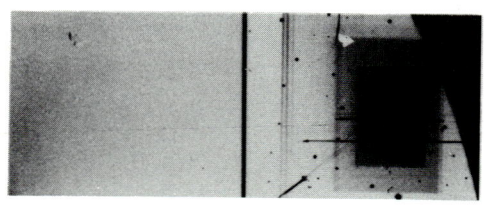

Fig. 2   Effect of the electron irradiation on a p-i-n device: left half of bipartite micrograph is an SEI image, and right half is the charge-collection image showing darkening in both the raster and the line-scan modes (the dark dots correspond to pinholes).

almost complete disappearance of the charge-collection signal in a spot
mode.   Details of the EBIC decay measurements in a spot mode have been
described by Yacobi and Herrington (1985).  It should be emphasized that no
visible damage to the bombarded surface was observed in the SEI mode of the
SEM.     Therefore,  the  electron-beam-induced  decrease  in  the  charge-
collection signal is not caused by the possible surface contamination that
may accompany investigations using electron probe instruments.

The mechanism of the formation of electron-beam-induced microscopic defects
in an a-Si:H matrix may explain the observed phenomena (see, for example,
Schade and Pankove 1981).   According to that mechanism, electron bombardment in a-Si:H introduces dangling bonds by breaking weak Si-Si bonds and,
possibly indirectly, Si-H bonds as well.  Increasing the number of dangling
bonds causes a higher density of gap states, which act as recombination
centers that reduce both the lifetime of carriers and the depletion width
of the device, and thus lead to the decrease in the charge collection signal.  These defects can be removed by annealing a sample in the temperature
range of about $150^{0}$-$200^{0}$C for periods of the order of 10-30 minutes (Schade
and Pankove 1981).     Schade et al (1980, 1983) have shown, using the
scanning Auger microprobe and sputter etching, that the electron damage is
bulk-related and that both the depth and the shape of the damaged volume
can be determined as a function of the electron beam energy.

The charge-collection decay observations were similar in RS and CVD devices
as well.   Some differences in the decay time were found between samples
produced by these three different methods.  However, similar differences
were also observed in several GD samples.  In general, only a few seconds
were required for the charge-collection signal to decay to a saturation
level in the spot mode of the electron beam irradiation.

Using a Tracor Northern digital beam control system attached to the microscope, it is possible to write directly on devices using the observed electron-beam-induced effects in the charge-collection mode.   Such writing was
performed on both Schottky and p-i-n devices for GD samples, and p-i-n
structures prepared by RS and CVD methods (see Figs. 3, 4 and 5).   Thus,
the information storage mechanism is based on the generation of electron-
irradiation-induced microscopic defects (e.g., dangling bonds) that reduce
both the carrier lifetime and the depletion width.   In the charge collection mode of a scanning probe instrument (e.g., SEM), electron-irradiated
areas, or spots, appear darker than the unbombarded regions of the device.
This leads to a contrast formation for pattern recognition.   The charge-
collection contrast depends on the charge-collection efficiency of unbombarded portions of the device and on the extent of the electron-beam-
induced damage in irradiated areas.  The charge collection signal in a spot
mode had always decayed to close to zero in only a few seconds.   However,
the original charge-collection efficiency of the unbombarded device varies
from sample to sample.  Therefore, quantification of the information storage contrast is difficult.

From continuing observations, it appears that the room temperature lifetime
of the information storage is at least one year.   Future observations
should show the long-term behavior of this mechanism.

The spatial resolution of the electron beam writing depends on the probe
size, the generation volume, and the minority carrier diffusion length.
The size of the generation volume of a-Si:H for a beam voltage of 30 kV,
for example, is about 10 μm, which is much larger than the probe size

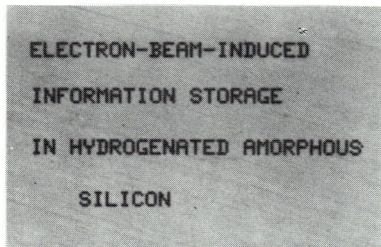

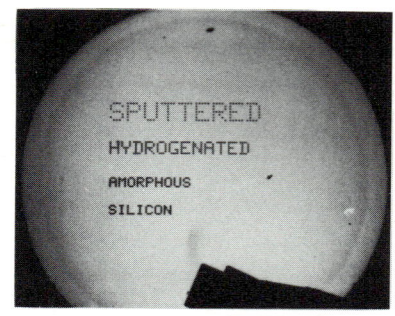

100 μm

1000 μm

Fig. 3    Charge-collection
image of electron beam
writing on a GD p-i-n
device; $E_b$ = 30 keV
and $I_b$ = 1 nA.

Fig. 4    Charge-collection
image of electron beam
writing on a RS
device; $E_b$ = 30 keV
and $I_b$ = 1 nA.

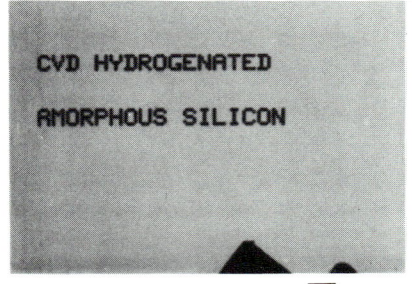

Fig. 5    Charge-collection
image of electron beam
writing on a CVD
device; $E_b$ = 30 keV
and $I_b$ = 1 nA.

100 μm

(~0.1 μm) and the diffusion length (~0.5 μm).  Therefore, the size of the
generation  volume  would  determine  the  spatial  resolution  of  the  beam
writing.  However, since the thickness of a-Si:H devices used in this study
are on the order of only 0.5 μm, a roughly estimated spatial resolution of
the order of a micron can be expected (Yacobi 1984).

As  mentioned  earlier,  annealing  a  device  at  about  200°C  leads  to  the
removal  of  the  irradiation-induced  defects.  In  the  present  study,  annealing of a GD p-i-n device at about 200°C for 15 minutes leads to a complete
recovery of the charge-collection signal, and then the device can be used
again for beam writing (Fig. 6).

Advantages  of  the  electron-irradiation-induced  charge-collection  memory
effect are (i) reversible process (by annealing from about 150° to 200°C);
(ii)  high  spatial  resolution  (on  a  micron,  or  possibly,  submicron  scale  in
thinner  devices);  (iii)  large  physical  size  of  the  information  storage
medium  and  a  relatively  easy  and  cheap  means  of  depositing  a-Si:H;  and
(iv)  ease  of  electron  beam  writing  (ease  of  deflection  and  small  electron
probe size).

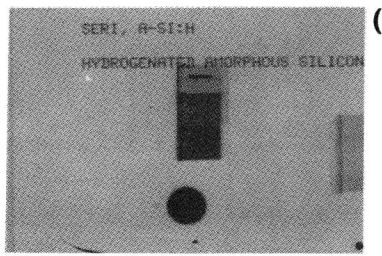

**(a)**

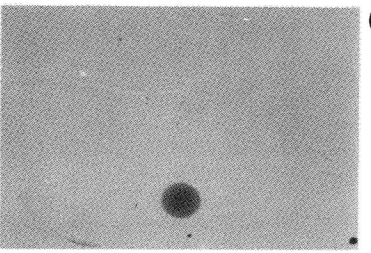

**(b)**

Fig. 6  Effect of annealing on the charge-collection signal in a GD p-i-n device. (a) Initial electron beam writing and pattern formation; (b) after annealing at ~200°C for 15 min (signal recovery); (c) subsequent writing on the same part of the device. Dark dots may be pinholes that helped to identify the area of interest; $E_b$ = 30 keV and $I_b$ = 1 nA.

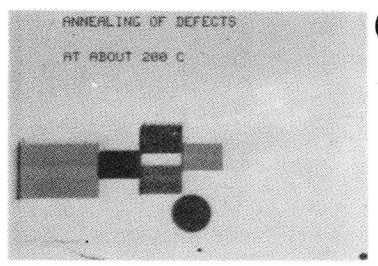

**(c)**

100 $\mu$m

Disadvantages include (i) limitations in electron beam reading in a charge-collection mode at high magnifications (large electron beam power densities) (at low magnifications, note that no electron-beam-induced damage was detected in a raster mode in areas of the order of 5 mm² for a 30-keV electron beam of about 1 nA, for example); and (ii) the need for a scanning probe instrument (electron or optical) for information reading.

Some possible improvements can be expected: (i) the stored information can be read with little or no damage using the LBIC (or OBIC) mode of a laser scanner (instruments with a spatial resolution on a micron scale are commercially available; note also the much slower rate of photon-irradiation-induced damage in this material); (ii) the stored information can also be read with low-energy electrons (below the threshold energy for electron-beam-induced damage of ~0.9 keV in a-Si:H) (the disadvantage in this case is a low signal level, and the advantage is very high spatial resolution in thin devices); and (iii) information can also be read using the beta conductivity technique (Fig. 1b), for which no device structure is required.

## 4. Conclusions

Charge-collection scanning electron microscopy can be effectively utilized to characterize fabrication defects in a-Si:H devices. However, precautions should be taken during these observations. In practice, for example, for typical SEM conditions of a 30-keV electron beam of about 1 nA, no electron-beam-induced damage was detected during observations in a raster mode of areas on the order of 5 mm$^2$ (M $\approx$ 50).

## Acknowledgment

This work was supported by the U.S. Department of Energy under contract No. DE-AC02-83CH10093.

## References

Pankove J I (ed) 1984 Hydrogenated amorphous silicon vol. 21 (Parts A, B, C and D) of Semiconductors and Semimetals eds Willardson and Beer (New York : Academic)
Schade H 1980 Le Vide, Les Couches Minces 201 999
Schade H and Pankove J I 1981 J. Phys. (Paris) Colloq. C4 (Suppl. 10) 42 327
Schade H and Hockings E F 1983 J. Vac. Sci. Technol. A1 592
Yacobi B G 1984 Appl. Phys. Lett. 44 695
Yacobi B G, McMahon T J and Madan A 1984 Solar Cells 12 329
Yacobi B G and Herrington C R 1985 J. Electron Microscopy Technique 2 in press

*Inst. Phys. Conf. Ser. No. 76: Section 8*
*Paper presented at Microsc. Semicond. Mater. Conf., Oxford, 25–27 March 1985*

# Quantitative EBIC imaging by Monte Carlo simulation

David C Joy and C A Pimentel*

AT&T Bell Laboratories, Murray Hill, New Jersey 07974, USA

Abstract   The use of Monte Carlo methods to represent the electron beam specimen interaction permits the accurate modelling of the EBIC imaging process through the generalization of standard point source solutions. Comparison of experimental and computed data allows the measurement of material parameters.

## 1. Introduction

Charge collection microscopy has become a widely used technique for the characterization of semiconductor materials and the investigation of finished devices.  While there is substantial literature on the use of these techniques to measure parameters such as the minority carrier diffusion length, the majority of charge collected images are interpreted in a purely qualitative manner, despite the fact that significant quantitative data should be available.  This situation is not always due to any lack of theoretical understanding of the sources of image contrast, since analytical expressions relating measurable parameters of the image to fundamental constants of the semiconductor are well documented.  Rather, the problem is that in order to provide tractable analytical expressions it is invariably necessary to make major approximations in the representation of the beam interaction with the solid, usually treating it as a point source or a sphere of uniform generation.  This paper discusses a Monte Carlo method (Akamatsu et al 1981) for the quantitation of charge collection data which avoids these limitations and illustrates it by applications to different experimental situations.

## 2. The Simulation

The interaction between an electron beam and a solid can be represented by a Monte Carlo simulation in the following way.  The electron range for a given incident energy E is computed by numerical integration of the Bethe stopping power expression, and this range is then divided into 50 steps of equal length.  At the start of the k-th step the electron is scattered through an angle $\phi$ where

$$\tan(\phi/2) = RND(1)*B/E(k) \qquad (1)$$

where $E(k)$ is the energy at the start of the step, B is a constant which depends on the material and is chosen to match the observed backscattering coefficient, and $RND(1)$ is an equidistributed random number between zero

*Permanent address:  University of Sao Paulo, Brazil

and one.  Scattering can occur in any azimuthal direction $\theta$, where

$$\theta = 2.\pi.\ \text{RND}(2) \tag{2}$$

and RND(2) is an independent random number.  Given these two angles, the step length and the starting coordinates, the end coordinates of the step can be found.  This process is then repeated until the electron reaches the end of its range, or is backscattered out of the sample.

In order to compute charge collection images, the generation and subsequent collection of electron-hole pairs must be incorporated into the simulation. Along the k-th step the energy deposited in the sample is $E(k)-E(k+1)$. If the energy required to form one electron-hole pair is eh, eg 3.6eV in silicon, then the number n of charge carrier pairs formed along this step is

$$n = [E(k)-E(k+1)]/eh \tag{3}$$

Figure 1 shows the computed contours of carrier generation in silicon, plotted in units of the range Re. The detailed form of this generation volume is in good agreement with published experimental data indicating that the simplications used in the simulation are valid.

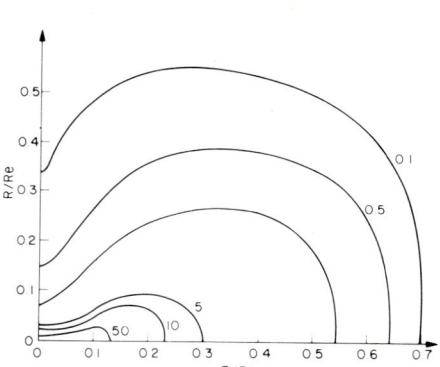

For energies in the range 1 to 20keV the electron range in silicon lies between 0.01 and 5 microns. Thus the maximum step length in the Monte Carlo model is 100nm or less, and to a good approximation the increment of charge carriers produced along the step represents a true point source.  Standard analytical point source solutions for the problem of

Fig. 1.   Contours of generation in Si vs radius and depth Z in units electron range Re.

interest can therefore be used, but by combining this with the Monte Carlo simulation the point source is correctly generalized to a realistic generation volume.  In this way all aspects of the beam interaction including backscattering or transmission can be accounted for, while incorporating specific boundary conditions as required.

### 3. The Schottky Barrier Geometry

The Schottky barrier geometry is widely used for EBIC observations, and a calculation of the gain of this provides an illustration of the application of the Monte Carlo approach.  Assume that the material is depleted to a depth $Z_D$, and has a minority carrier diffusion length L.  A point source of carriers at a depth $Z < Z_D$ will be collected with 100% efficiency, but carriers produced at a depth $Z > Z_D$ will have to diffuse back to the edge of the depleted region to be collected.  The collection efficiency n is then

$$n = \exp(-(Z-Z_D)/L) \tag{4}$$

Using these expressions as appropriate to calculate the fraction of the charge carriers generated along each step of a trajectory, the variation of the gain of a Schottky barrier with the beam energy and material parameters can be computed. Figure 2 shows the result of such a calculation for silicon at 30keV as a function of the depletion depth and diffusion length. Note that the simulation provides a direct numerical evaluation of the gain for specified conditions. Thus a measurement of the gain at, for example, two different beam energies will permit both the diffusion length and depletion depth to be determined from plots such as that in Fig. 2. Alternatively, as proposed by Bell and Hanoka (1981), the variation of the gain in an inhomogeneous material could be used as a high spatial resolution measure of the local diffusion length. For a given sample resistivity, and hence depletion depth, the gain can be instantly converted to a diffusion length once the appropriate simulation has been run. Using the Monte Carlo approach it is even straightforward to make allowance for the effect of energy losses in the metal of the Schottky barrier, an occurrence which significantly affects the gain at low beam energies (Holt 1974). Although an analytical computation of the gain can be made (Donalato 1979) using a one-dimensional approximation to the carrier generation function, such an approach cannot allow for the finite width of the depleted layer.

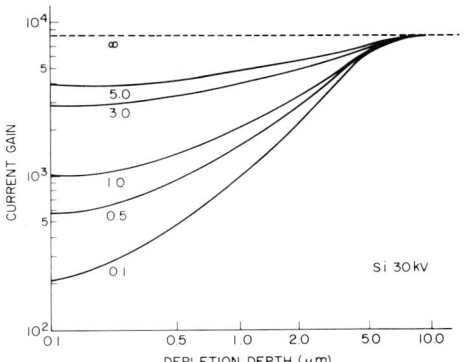

Fig. 2. Gain of Schottky barrier on Si at 30keV vs depletion depth and diffusion length.

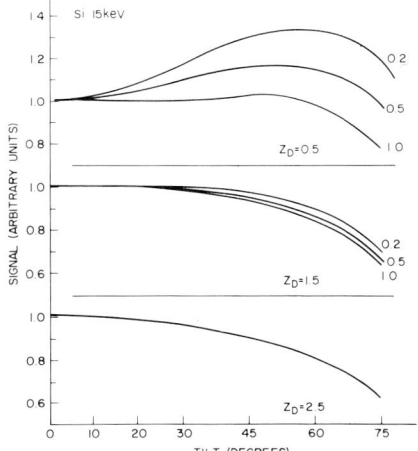

Fig. 3. Relative variation of gain for Schottky barrier vs tilt, depletion depth $Z_D$ and diffusion length.

The flexibility of the Monte Carlo model allows it to be applied to situations where analytical methods are of restricted use. For example the variation in gain of a Schottky barrier for tilted beam incidence is also readily determined as illustrated in Fig. 3. Such an experiment is of interest because it provides a way of varying the penetration of the beam beneath the surface while maintaining other factors constant. The gain at some angle Θ relative to the gain at normal incidence varies in a characteristic fashion with the depletion depth and diffusion length, and this clearly provides an alternative method for determining $Z_D$ and L. In this case the measurement can be speeded by using the electron-optics of the SEM to generate a 'rocking' beam which changes its angle of incidence while remaining fixed on a point (Jakubowicz 1982) while observing the EBIC profile on the display screen.

The accuracy of measurements of $Z_D$ and L derived from comparisons with simulations depends on many factors.  The computations themselves achieve a statistical accuracy of ten percent after only a few hundred trajectories, a figure which can be achieved on most small computers in five to ten minutes computing, and the numerical value of the gain of the Schottky can be determined to within a few percent with simple precautions. The practical limit is set by the sensitivity with which the experimental data can be matched to the computed data.  For values of L which are small in comparison with the electron range the variation of gain with beam energy and tilt is rapid and good precision can easily be achieved.  But for large values of L these variations are small, as seen from Fig. 2, and accurate determinations become more difficult.  These methods therefore have limited application to perfect materials where L is typically tens of microns, but are ideally suited for studies of device grade, or inhomogeneous, materials where L is usually very much shorter.

## 4. Defect Images

The principles of the formation of contrast at electrically active defects was first discussed in detail by Donalato (1979).  In his treatment, as adapted for use with the Monte Carlo method, a point defect is treated as a sink for the recombination of carriers.  This sink has a radius RO (of the order of 0.1 microns) and charge carriers are lost to it at a rate proportional to

$$A.([exp(-R1/L)]/R1 - [exp(-R2/L)]/R2) \qquad (5)$$

where A is a constant containing both the intensity of the point source of carriers and the recombination strength of the defect, L is the minority carrier diffusion length, R1 is the distance from the point source to the outside of the defect and R2 is the corresponding distance from the image of the point source in the surface to the defect.  This expression is then evaluated point by point during the Monte Carlo simulation and a defect profile constructed by varying the linear distance between the beam and the defect.  However, because the process is performed numerically it is possible to make the transition to a line defect by treating this as an assemblage of point defects.  At each step equation (5) is applied and integrated along the length of the dislocation.  Since the recombination rate predicted by equation (5) falls very quickly as the vectors R1, R2 increase relatively few terms are required before the intensity saturates.

Figure 4 shows computed profiles across a line defect lying parallel to the surface but at depths varying from 0.5 to 2.5 microns.  The beam energy is assumed to be 15keV giving an electron range, in silicon, of 3 microns. The variation in depth is accompanied by a change in the width of the profile, which starts narrow, reaches its maximum width at about half of the electron range, and then narrows again.  The contrast reaches its maximum value at a depth of between one quarter and one third the range and then falls steadily towards zero as the depth approaches the range. The exact form of the profile also depends on the diffusion length.  The computations confirm the calculations of Donalato (1979) that an increase in diffusion length does lead to an increase in the contrast and width of the profile, however this increase saturates for large values of L.  Even at large diffusion lengths the profile width is found to be essentially determined by the width of the electron interaction volume confirming recent experimental observations that substantially sub-micron resolution is possible in EBIC provided that the beam energy is kept low enough (Joy and Freeland 1984).

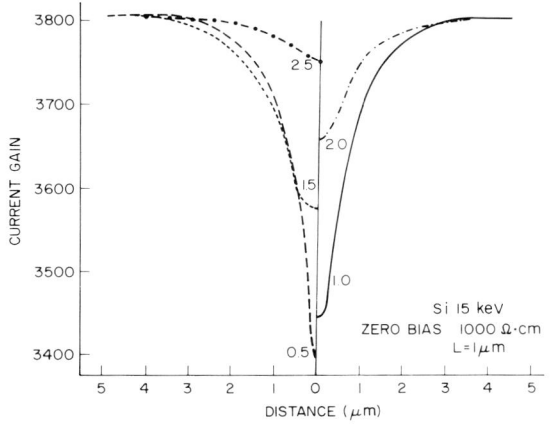

Fig. 4. EBIC profiles across line defects, at depths of 0.5, 1.0, 1.5, 2.0 and 2.5 microns in silicon at 15keV, diffusion length 1 micron. Material is fully depleted.

The magnitude of the contrast depends on the choice of the critical radius RO, which in neutral material was here chosen to be equal to the radius of the single defect electrical barrier (SDEB) as calculated by Mil'shtein and Senderichin (1982). If the defect lies within the depleted region then the depletion field will reduce the effective value of RO by an amount depending on the position of the defect relative to $Z_D$, and the magnitude of any applied bias. This analysis makes it possible to compute the variation in contrast from a defect to be expected as a function of applied bias. Figure 5 shows a sequence of such calculations for a defect at 1.5 microns depth, in materials depleted at zero bias to a depth of about 0.5 microns. At zero bias the defect lies in neutral material and its presence is only revealed by carriers which diffuse back to the depleted zone. The contrast is therefore low. As the bias is increased the contrast rises, reaching a maximum as the edge of the depleted layer sweeps through the defect. Further increases in bias lead to a slow fall in the contrast because the component from the defect has reached its greatest value but

Fig. 5. Variation of contrast from line defect, at depth of 1.5 microns, as function of total bias for diffusion lengths of 0.5, 1.0, 5.0 and 50 microns. Si sample at 15keV.

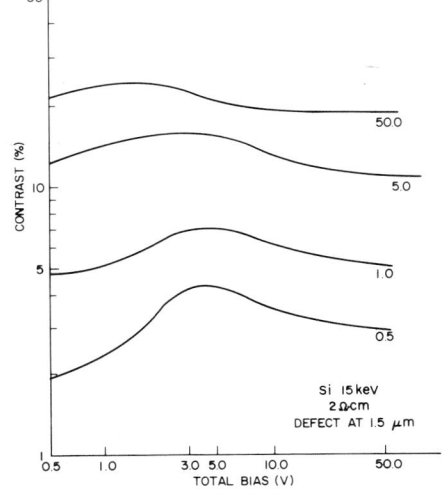

the total collected signal continues to rise until the depleted depth equals the beam range.  At the same time the recombination strength of the defect is reduced due to the contraction of the sink radius.  A measurement of contrast versus bias can therefore unambiguously determine defect depth as experimentally demonstrated by Mil'shtein et al (1984).

## 5. Conclusions

The application of Monte Carlo methods to the computation of EBIC contrast allows point source solutions to be generalized so that the electron beam interaction is fully taken into account.  In addition other experimental parameters such as a finite depletion depth can be included.  Detailed modelling of experimental situations is therefore possible, even in materials where there is inhomogeneity on a scale comparable with the beam inter-action volume.  A comparison of experimental results and the corresponding simulations permits quantitative data to be extracted from EBIC images.

## Acknowledgements

One of us (CAP) thanks CNPq, Brazil, for a Visiting Fellowship.

## References

Akamatsu B, Henoc J and Henoc P 1981 J. Appl. Phys. 52 7245
Bell R O and Hanoka J I 1982 J. Appl. Phys. 53 1741
Donolato C 1979 Optik 52 19
Holt D 1974 Quantitative Scanning Electron Microscopy ed D Holt (London: Academic Press) 235
Jakubowicz J 1982 Solid State Electronics 25 651
Joy D C and Freeland P E 1984 Proc. Int. Conf. Defects in Semiconductors ed J P Parsey 186
Mil'shtein S and Senderichin A 1982 Phys. Stat. Sol.(b) 109 429
Mil'shtein S, Joy D C, Ferris S D and Kimerling L C 1984 Phys. Stat. Sol.(a) 84 363

*Inst. Phys. Conf. Ser. No. 76: Section 8*
*Paper presented at Microsc. Semicond. Mater. Conf., Oxford, 25–27 March 1985*

361

# Investigation of plasma effect on EBIC dislocation contrast in Si

A L Tóth

Res. Inst. for Technical Physics of the Hungarian Academy of Sciences, Budapest, POB 76, H-1325 Hungary

Abstract   Sample bias dependent EBIC dislocation contrast measurements revealed 3 characteristic parts of the contrast/bias curve with different origins (drift, diffusion and plasma).   The existence of electron-hole plasma, and its influence on dislocation contrast was proved by simultaneous plasma loss/bias measurements.

## 1. Introduction

In the conductive mode of scanning electron microscopy the quantitative evaluation of EBIC signals is spreading.  The aim of a typical measurement is to establish the relationship between the EBIC contrast and geometrical or physical properties of well characterized crystal defects.   However to clarify the details of single defect characterization, it is important to examine the influence of various experimental parameters on the EBIC dislocation contrast.

After measuring the effect of beam energy on dislocation contrast and resolution (Toth 1981) supporting the calculations of Donolato (1979), the next parameter to be investigated was the sample bias.

This paper is intended to study the change of EBIC dislocation contrast as a function of sample bias and beam current, and to distinguish its components of different origin.

## 2. Experimental

### 2.1. Samples and Excitation

The samples were Schottky barrier diodes with Al metallization 0.3mm in diameter and 100nm in thickness on As doped Si substrate $N = 5 \times 10^{14}/cm^3$ with a dislocation density of $5000/cm^2$.

The measurements were carried out in a SEM type JSM35 with 25keV beam energy and 1-20nA beam current (measured with a Faraday cage).  Accepting the Gruen range of Everhart and Hoff (1971) as a diameter of a uniformly generated sphere and substituting this value into Berz and Kuiken's expression (1976) for maximum generated minority carrier concentration, we obtain

$$\Delta_{Pmax}(25keV,10nA) = 8.6 \times 10^{15}/cm^3$$

which shows high injection when compared to the doping level.

The Debye length, which is the lower limit of plasma dimension, is 0.2μm. This is less than the typical depletion layer width W(4V) = 3μm and electron range $R_G$(25keV) = 5μm.  As the distribution of electric field in barrier depletion layers always contains a weak field region the occurrence of plasma effect can be expected (Leamy et al 1978).

## 2.2. Apparatus

To measure the dislocation contrast as a function of bias a phase sensitive computer controlled data acquisition and reduction system was used (Labar et al 1983) (Fig. 1).  The detector of the system was a calibrated fast current amplifier (Keithley 427) used as a preamplifier of a phase sensitive (lock-in) amplifier (Keithley Autoloc).  The lock-in amplifier (PSD) separated the DC diode current due to the bias and the AC EBIC signal, modulated by the chopped beam (10kc/s frequency, 50% duty circle).

The EBIC signal was digitized by a voltage-to-frequency converter (VFC) and fed into the multichannel analyser (MCA) memory of an energy dispersive X-ray analyser system (Ortec EEDS-II), where it was treated as the pulses of the original X-ray detector.

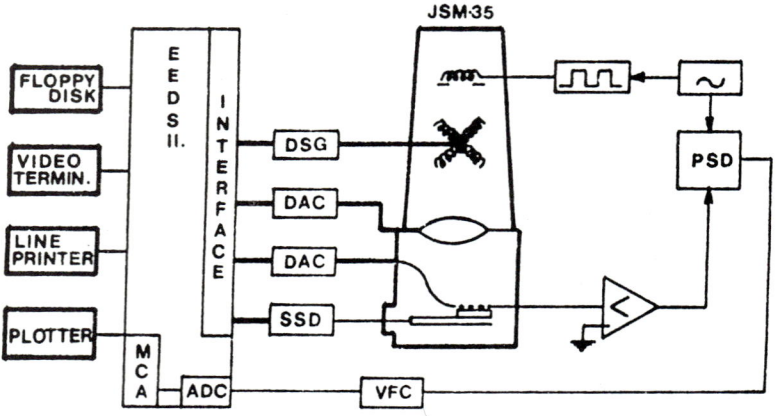

Fig. 1  Block diagram of the system.

The EDS system has been modified both in hardware and software to provide the possibility of stepwise digital line scan measurement with SEM control commands between two steps.  EBIC intensity measurements became possible as a function of the following computer controllable experimental parameters:

beam  X-Y position by digital scan generator (DSG)

sample movement (SSD)

sample bias

objective lens focus/defocus

using digital-to-analog converters (DAC).

## 3. Results

### 3.1. Dislocation Contrast Measurements

The dislocations to be measured were perpendicular to the surface, localized and checked on the conventional (unbiased DC) EBIC image. The contrast
was then measured as the relative difference of the EBIC intensity on the
dislocation and on a reference area free of dislocations

$$C = (I_{ref} - I_{disl})/I_{ref}$$

while the bias was changed from 9V reverse to 1V forward.

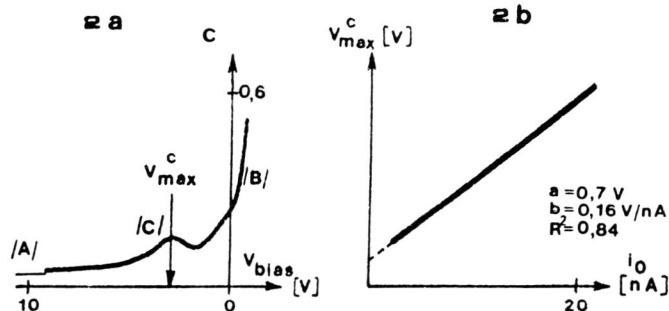

Fig. 2  Typical contrast/bias curve (2a) and the beam
current dependence of $V_{max}^C$ (2b).

The contrast/bias curve has 3 components as shown in Fig. 2:

A:  a constant level observable at high reverse bias

B:  an increasing part as approaching the forward bias

C:  a maximum at $V_{max}^C$

The $V_{max}^C$ bias depends on the beam current (i.e. on the generated excess
charge carrier concentration) linearly (Fig. 2b),

$$V_{max}^C [V] = 0.7 + 0.16 * i_o [nA]$$

### 3.2. Plasma Loss Measurements

A possible hypothesis for the contrast maximum is the enhanced recombination
inside an electron-hole plasma droplet free of collecting electric field
due to its polarization at the edges.

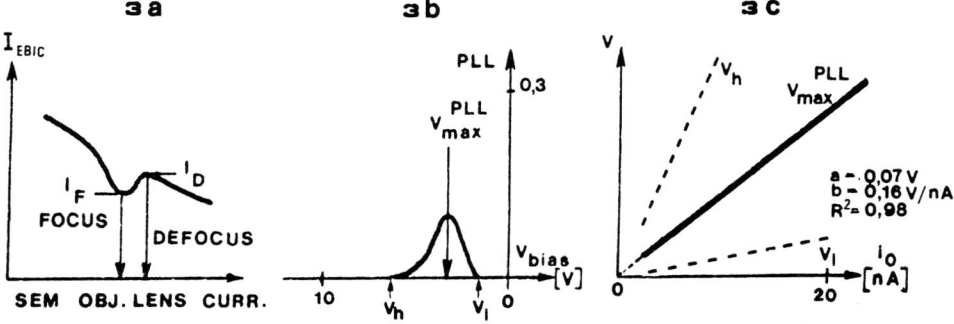

Fig. 3  The definition of PLL (3a), the PLL/bias curve (3b) and the
beam current dependence of $V_{max}^{PLL}$ (3c).

To prove the plasma nature, the plasma loss (PLL) i.e. the relative differ-ence of EBIC intensities generated by focused and defocused beams

$$PLL = (I_{DF} - I_F)/I_{DF}$$

was measured simultaneously with the contrast measurement by computer controlled defocusing of the beam above the dislocation free area (Fig. 3a).

As the plasma exists only under high current density the enhanced bulk recombination measured with a focused beam shows its presence (Fig. 3b). The linear relationship (Fig. 3c) of the $V_{max}^{PLL}$ position of PLL maximum and the beam current is

$$V_{max}^{PLL} \ [V] = 0.07 + 0.16*i_o \ [nA]$$

which shows a good agreement with the C component of the contrast/bias curve.

## 4. Conclusion

As the plasma loss and the maximum in dislocation contrast occur nearly at the same bias, and show the same beam current dependence, the components of the contrast/bias curve can be explained by recombination at the defect

A:  within the depleted layer

B:  under the depleted layer

C:  inside the neutral plasma droplet

during drift and diffusion.

The $V_i$ low bias limit of PLL (Fig. 3c) can be explained by the approach of the plasma dimension to the Debye length, while the $V_h$ high bias limit shows qualitative agreement with the predictions (i.e. linear dependence on beam current) of the Tove and Seibt (1967) model for plasma erosion.

## References

Berz F and Kuiken H K 1976 Solid State Electron. 19 437
Donolato C 1979 Appl. Phys. Lett. 34 81
Everhart T E and Hoff P H 1971 J. Appl. Phys. 42 5837
Labar J L, Vladar A E and Toth A L 1983 Inst. Phys. Conf. Ser. No. 68 181
Leamy H J, Kimerling L C and Ferris S D 1978 Scanning Electron Microscopy
    ed. O Johari (SEM Inc. AMF O'Hare) Vol. 1 pp 717-26
Toth A L 1981 Inst. Phys. Conf. Ser. No. 60 221
Tove P A and Seibt W 1967 Nucl. Instr. and Meth. 51 261

*Inst. Phys. Conf. Ser. No. 76: Section 8*
*Paper presented at Microsc. Semicond. Mater. Conf., Oxford, 25–27 March 1985*

# Spatial variation of dopant concentration in $^{29}Si^+$ implanted Czochralski and metal-organic vapour phase epitaxial GaAs

C A Warwick, S S Gill, P J Wright and A G Cullis

Royal Signals and Radar Establishment,
St Andrews Road, Malvern, Worcs, WR14 3PS, UK

Abstract    Undoped, semi-insulating (001) Czochralski GaAs wafers and
undoped high-resistance metal-organic vapour phase epitaxial layers
were implanted with $^{29}Si^+$ and annealed.  A substantial spatial variation
of the activated dopant concentration near to and away from grown-in
cellular dislocation arrays in the Czochralski material was determined
by cathodoluminescence measurements.  No such variation was detected
in the implanted and annealed epitaxial layers, in which the cellular
dislocation pattern of the substrate is replicated in the layer, but
which have a different point defect distribution.

## 1. Introduction

Nanishi et al (1982) have shown that the threshold voltage of field effect
transistors varies with position on individual GaAs wafers.  In particular
this parameter has been correlated to the distance between the transistor
gate and the nearest cellular dislocation array.  This variation, which is
thought to be due to spatial non-uniformity of the implant activation, is
deleterious to the production of GaAs large scale integrated circuits.

For the present study, we investigated whether the non-uniformity of the
activation is due to the dislocations themselves (eg strain fields etc) or
to the point defect distribution around them and whether greater uniformity
could be obtained by fabrication in an undoped high resistance metal-
organic vapour phase epitaxial (MOVPE) buffer layer.  Stewart et al (1984),
investigating VPE GaAs layers backdoped with Cr, have shown that buffer
layer fabrication can also give improved mobility and lower parasitic
resistance in devices.

## 2. Experimental

### 2.1 Czochralski Growth

The undoped GaAs 50 mm diameter ingot was grown by Cambridge Instruments
by the liquid encapsulation Czochralski (LEC) method, employing high
pressure in-situ near stoichiometric synthesis in a BN crucible.  Dry $B_2O_3$
was used as the encapsulant.  The ingot received post-growth annealing at
950°C for 5 hours.  The ingot was wafered into circular (001) slices,
normal to the growth direction and polished with a colloidal silica/NaOCl
mixture.  The resistivity of this material was ~ 1 x $10^8$ Ωcm (ie semi-
insulating: SI) and the dominant point defects detected were the native
deep donor, EL2, at a concentration of ~(1-2) x $10^{16}$ $cm^{-3}$ as measured by

infra-red absorption (Martin 1981) and $C_{As}$, at a concentration of
~ 5 x $10^{15}$cm$^{-3}$ as measured by local vibrational mode absorption (Woodhead
et al 1983). The material also contained a dislocation density of
$10^4$ - $10^5$ cm$^{-2}$, with a W-shaped density variation across the wafer. The
dislocations were polygonized into cellular arrays, the "cells" being
~ 250 μm in diameter and the dislocation arrays, which formed the "cell
walls", were ~ 20 μm across.

## 2.2  MOVPE Growth

The undoped MOVPE GaAs layers were grown on quadrants of 50 mm diameter
(001), undoped, SI LEC GaAs wafers. The substrate was heated, by radio
frequency induction via a graphite susceptor, to 700°C during growth, which
proceeded by the deposition and reaction of, Ga(CH$_3$)$_3$ and AsH$_3$ from a Pd-
diffused H$_2$ gas flow. The Ga(CH$_3$)$_3$ to AsH$_3$ mole fraction ratio determined
the concentration of shallow acceptor impurities, $N_{AS}$, and shallow donors,
$N_{DS}$, incorporated into the layer. Growing with a relatively high AsH$_3$
concentration gave n-type layers and with a low one gave p-type layers.
For the layers in this present study an intermediate mole fraction ratio
of Ga(CH$_3$)$_3$ to AsH$_3$ equal to 1:8 was used which gave layers in the high
resistivity condition i.e [EL2] >> $N_{AS}$ - $N_{DS}$ >> 0, where [EL2] is the
concentration of the EL2 defect, this being(1-3) x $10^{14}$ cm$^{-3}$ in this type
of layer (Samuelson et al, 1981). The EL2 compensated the excess of $N_{AS}$
over $N_{DS}$ to pin the Fermi level at mid gap and gave a high-resistance
layer. The electrical properties of high resistance, thin layers (~ 5 μm
in the present case) were difficult to measure because of surface depletion
effects. However, capacitance-voltage profiling of these layers at room
temperature showed complete depletion of the layer, giving an upper limit
of $|N_{AS} - N_{DS}|$ << $10^{13}$ cm$^{-3}$ and indicating that the high resistance
criterion was satisfied.

## 2.3  Ion Implantation, Capping, Annealing

Some of the samples of the LEC and MOVPE material were implanted with a
dose of 8 x $10^{12}$ cm$^{-2}$ $^{29}$Si$^+$ ions at an energy of 180 keV. This corresponded to a depletion field effect transistor channel implant, with a depth
of 0.24 μm (projected ion range plus one standard deviation). The peak
carrier concentration was ~ 4 x $10^{17}$ cm$^{-3}$, as measured by Hall effect
profiling (Gill et al 1985). The implanted layers were capped with plasma
chemical vapour deposited SiN$_x$, and then thermal pulse annealed in an AG
Associates Heat pulse 210 system. Various samples were annealed at peak
temperatures between 900°C - 950°C and for anneal durations between 1 - 8
sec. Typical time-temperature curves for this system have been given by
Hodge et al (1984). After annealing the SiN$_x$ cap was removed with HF
solution.

## 2.4  Assessment

The assessment of spatial uniformity of the dopant concentration was performed with a Cambridge Instrument S150 scanning electron microscope (SEM)
with an LaB$_6$ electron gun and a Link Systems wavelength dispersive (WD)
cathodoluminescence (CL) computer, controlling a Bentham Instruments
M300 HR monochrometer. An in-house design of collection optics was used.
The SEM was operated at 10 kV accelerating voltage and a probe current of
45 nA and a probe diameter of 0.2 μm. At this voltage both the diameter
of the generation volume and the electron range had a value of ~ 0.4 μm
in GaAs (Everhart and Hoff 1971, Coslett and Thomas 1964). The samples

under examination were cooled using an Oxford Instruments continuous flow
liquid He cryostat operating at 6K and the sample surface temperature under
excitation was determined to be ~10K. A spectral resolution of 0.6 nm was
used and CL was detected with an S20ER photomultiplier tube. This tube was
sensitive to photon energies > ~1.4 eV i.e it detected the near band edge
CL peaks in GaAs, but not the deep level peaks. The beam was near normal
to the sample surface and figure 1 shows the geometry of the generation
volume relative to the ion implanted layer. Most of the excited region
is in the implanted region, and, because this region has a much higher
luminescence efficiency than the underlying undoped material, virtually
all the CL comes from the ion implanted layer. Figure 2 illustrates that
the lateral resolution is adequate to image the dopant profile; it shows
a CL micrograph of a gateless field effect transistor, the bright region
corresponding to the selected area channel implant. Black regions correspond to contact metallization. As can be seen in figure 2 mask imperfections and misregistrations can be detected by this method, as well as
imaging the local dislocation and dopant arrangements; dislocations show
as dark spots. However, for the study presented here we have studied
material with blanket implants rather than device structures.

## 3. Results and Discussion

Figure 3 shows SEM CL "total-light" micrographs from a sample of the LEC
material that was neither ion implanted nor annealed. The cell walls
show as bright bands ~20 µm wide surrounding the dark cell interiors. The
ratio of the total light CL intensity on the cell wall to the cell centre
was ~5:1. At higher magnification (figure 3b) the dark spots, eg D, corresponding to the intersection of a dislocation with the specimen surface,
can be more clearly seen. Warwick and Brown (1985) have shown a correlation between x-ray topographs and CL images. In the CL micrographs of
figure 4 the dislocations in the MOVPE, again neither implanted nor
annealed, show as dark spots on a uniform background; no bright haloes
surround the arrays of dislocations. It is important to note that in the
case of the LEC material, the cellular arrays are formed by polygonization
of the dislocations at high temperature (>> 700°C) whereas in the MOVPE
material the arrays are merely replicas of the cellular pattern of dislocations on the LEC substrate. The dislocation segments in the layers

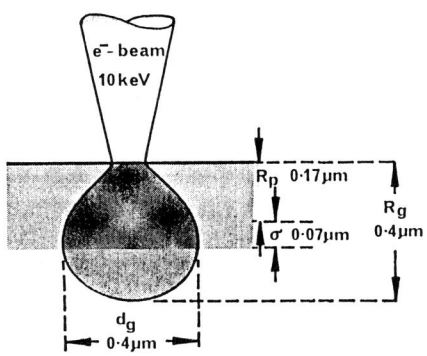

Figure 1. Geometry of the generation volume in relation to the
implanted layer depth

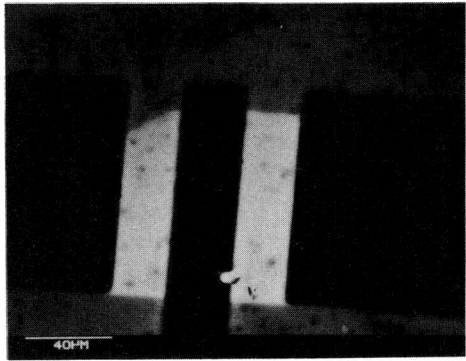

Figure 2. SEM CL micrograph of
gateless transistor in LEC GaAs

have not been exposed to a temperature above 700°C. The difference in CL
image in the two types of material is due to different point defect distributions caused by the different thermal history and growth conditions of
the two processes.

Figure 5 shows SEM CL total light images of a sample of the LEC material
that was implanted and annealed for 1s at 900°C. The average CL intensity
increased by up to a factor of 60 compared to the unimplanted material.
From this we infer that these CL images represent only the implanted
layer, with virtually no contribution from the underlying material. In
figure 5 the bright haloes (eg W in figure 5b), ~60μm wide, and the dark
spots, eg D, are clearly visible. The ratio of CL intensity on the cell
wall (eg W) to the cell centre (eg C) was ~1.6:1. Figure 6 shows analogous images for the MOVPE material that was implanted and annealed for
1s at 900°C. Unlike figure 5, no haloes are detected, but the dislocations are still present giving rise to arrays of dark spots, eg D in
figure 6b.

Near band edge CL spectra from points close to the dislocation arrays
(eg the points marked W in figures 5b and 6b) and close to the cell

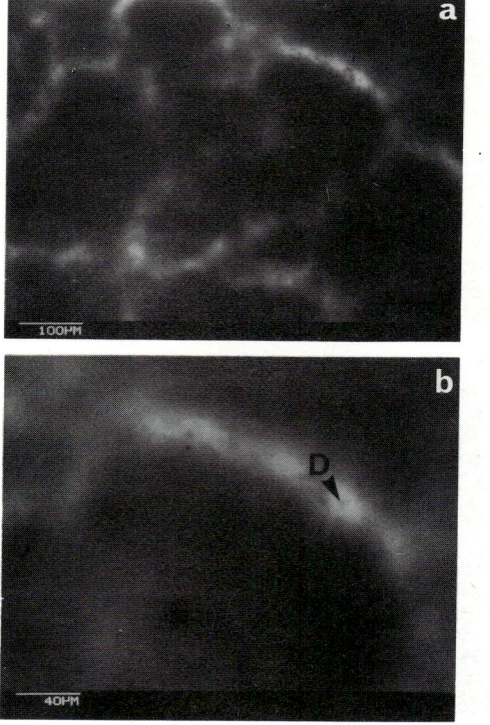

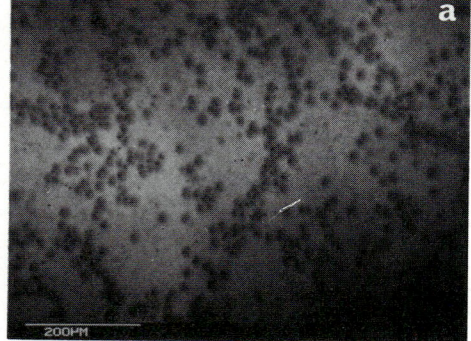

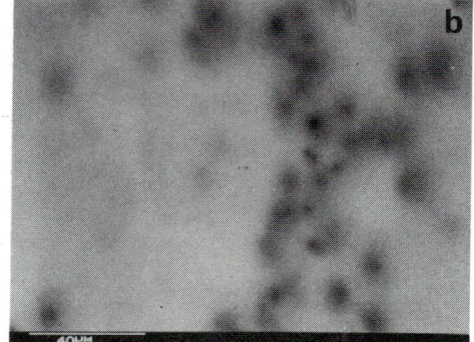

Figure 3.  SEM CL micrograph of
unimplanted LEC GaAs
a) low magnification
b) higher magnification

Figure 4.  SEM CL micrograph of
unimplanted MOVPE GaAs layer
a) low magnification
b) higher magnification

centres (eg the points marked C in figures 5b and 6b) were recorded and
stored digitally, using the WDCL acquisition system. Figure 7 shows part
of the spectra for the implanted and annealed (1s, 900°C) LEC material and
figure 8 those from the corresponding implanted MOVPE layer. All the
specimens show two main peaks, one at 1.512 eV due to free holes recombining with electrons bound the $Si_{Ga}$ shallow donors and one at 1.493 eV
due to a mixture of free electrons and donor-bound electrons recombining
with holes bound to $C_{As}$ shallow acceptors. No evidence of the $Si_{As}$ shallow
acceptor related peak at 1.479 eV is seen in these specimens and this is
consistent with the work of Woodcock et al (1975) who only saw this peak
in high dose (>$10^{14}$ cm$^{-2}$) Si implants. The 1.479 eV portion of the
spectrum and a weak peak at 1.457 eV, due to the longitudinal optical
phonon replica of the 1.493 eV peak, are not shown in figures 7 and 8.

The full width at half maximum (FWHM) of the 1.512 eV peak is dependent on
the Si shallow donor concentration, by the Mott broadening mechanism, and
so can be used to determine the spatial uniformity of activation, by
comparing FWHM of CL peaks taken from selected areas of the specimen. This
method is preferable to comparing peak heights or ratios of peak heights,
since both these latter parameters are affected by competing recombination

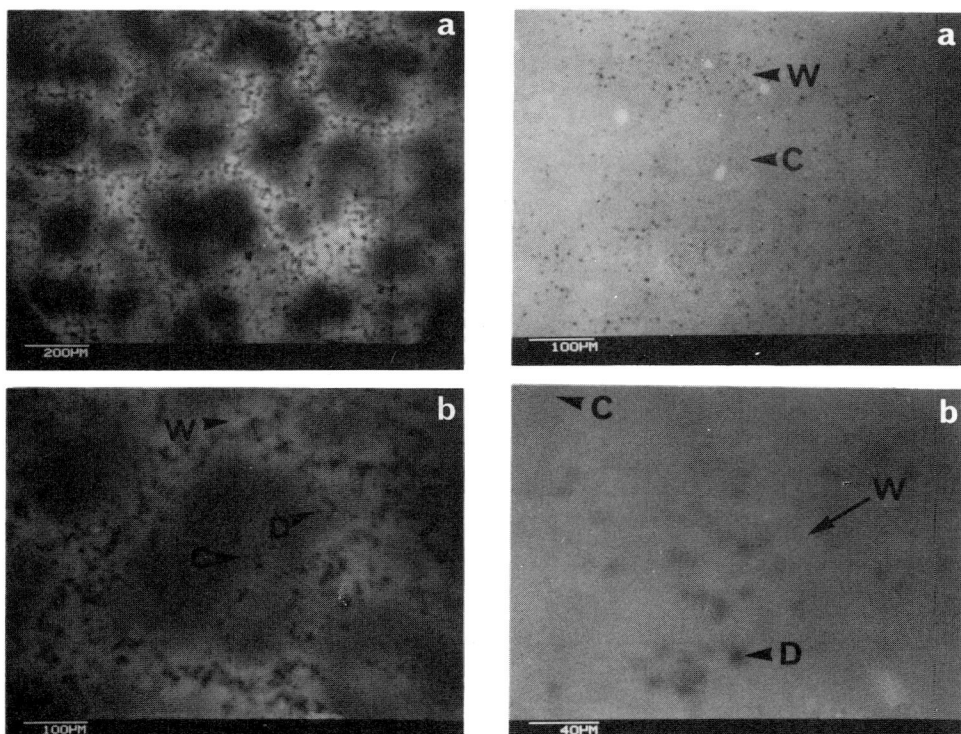

Figure 5.  SEM CL micrograph of
implanted and annealed (900°C,1s)
LEC GaAs

Figure 6.  SEM CL micrograph of
implanted and annealed (900°C,1s)
MOVPE GaAs

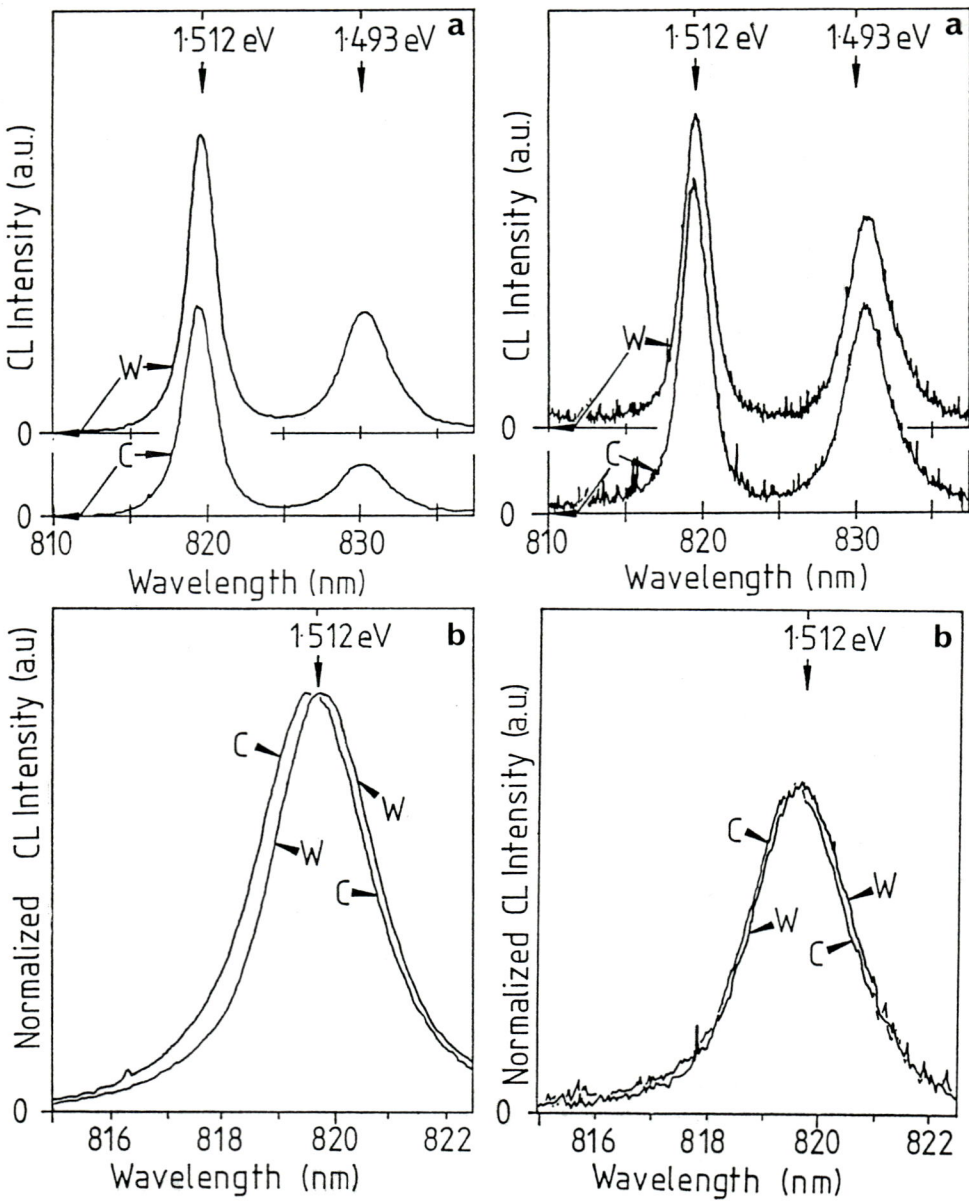

Figure 7. Local WDCL spectra from typical cell wall (trace W) and cell centre (trace C) in implanted and annealed (900°C,1s) LEC GaAs
a) unnormalized showing 1.512 and 1.493 eV peaks
b) normalized to 1.512 eV peak height

Figure 8. Same as for figure 7 except that the material is implanted and annealed (900°C,1s) MOVPE GaAs

mechanisms, which may also be spatially non-uniform. Very accurate FWHM measurements are required for the Mott broadening method and, in order to check the accuracy, an acquisition for a specific point on a sample was repeated many times. The standard deviation on a FWHM of 2.22 nm was ± 0.03 nm. This means a ±2% change in the peak width is just detectable. Table 1 shows the percentage FWHM change of the 1.512 eV peak at the cell centre compared the cell wall (defined as $\Delta\% = (W_c - W_w) \times 100\% \div W_w$ where $W_c$ is the FWHM for the cell centre and $W_w$ that for the cell wall) for various specimens and anneals.

Figures 7b and 8b show enlarged portions of the 1.512 eV peak in figures 7a and 8a respectively, but with peak height of the W and C traces normalized for easier visual peak width comparison. Small peak energy changes can also be seen. No significant FWHM change (1% ± 2%) is seen between trace C and W in figure 8b (implanted MOVPE, 1 sec at 900°C anneal) but an 8% ± 2% change is measured for the corresponding LEC specimen. It is important to note that because the increase in peak width at the cell centre of the LEC material corresponds to an increase in active Si concentration there, then we would expect an increase in the 1.512 eV luminescence at the cell centre (ie bright cells and dark walls). The reverse contrast is observed because of the predominant effect of the heterogenous distribution of non-radiative centres which have a controlling influence on the lifetime of the injected carriers. The results clearly show that the [29]Si activation is much more homogeneous in the case of MOVPE material than for LEC samples, despite the fact that both contain cellular arrays of dislocations. We can therefore eliminate both Si diffusion and interaction of the [29]Si with the dislocation strain field as possible causes for this behaviour. We conclude that different behaviour is due to the different initial point defect distributions in the two types of material, due in turn to the different growth methods. However, the exact nature of the initial point defect distribution and of interaction of the Si with the point defects is not yet clear.

| Anneal Time (S) | Peak width change (%) $\Delta\%$ | | | |
| | LEC | | MOVPE | |
| | 900°C | 950°C | 900°C | 950°C |
|---|---|---|---|---|
| 1 | 8 ± 2 | 21 ± 2 | 1 ± 2 | n/a |
| 5 | 5 ± 2 | n/a | n/a | n/a |
| 8 | 24 ± 2 | n/a | n/a | n/a |

n/a = not assessed

Table 1. Peak width change ($\Delta\%$) for various anneal times and temperatures of implants into GaAs LEC and MOVPE material

## 4. Conclusions

For implantation into SI LEC GaAs and high resistance MOVPE layers, the spatial variation of the CL peak width is interpreted as a measure of the inhomogeneity of the implant activation. The results show that inhomogeneity depends on the point defect distribution in the material to be implanted. In particular LEC material, which has a heterogeneous point defect distribution associated with the grown-in cellular dislocation arrays, shows a heterogeneous activation. In contrast, MOVPE material, which has a homogeneous point defect distribution (despite the presence of a replica dislocation pattern), shows homogeneous activation.

## Acknowledgements

We gratefully acknowledge the help of Dr M L Young and Dr D Lee of RSRE for electrical measurements and Dr H Badawi of STL for $SiN_x$ deposition.

## References

Coslett V E and Thomas R N 1964 Brit. J. Appl. Phys. 15 1283.
Everhart T E and Holt P H 1971 J. Appl. Phys. 42 5837.
Gill S S, Foreman B J, Dawsey J R and Cullis A G 1985 to be published in J. de Physique: Proc. MRS Europe Conf. Strasbourg. May 1985.
Hodge A M, Cullis A G and Chew N G 1984 in Polymicro crystalline and amorphous semiconductors ed P Pinard and S Kalbitzer (Les Ulis: Les editions de Physique) pp 683-688.
Martin G M 1981 Appl. Phys. Lett. 39 747.
Nanishi Y, Ishida S, Honda T, Yamazaki H and Mizazama S 1982 Jpn. J. Appl. Phys. 21 L335
Samuelson L, Omling P, Titze H and Grimmeis H G 1981, J. Crystal Growth 55 164
Stewart C P, Medland J D and Wickenden D K 1984 ed D C Look and J S Blakemore (Orpington:Shiva) pp 410-413.
Warwick C A and Brown G T 1985 Appl. Phys. Lett. 46 574.
Woodcock J M, Shannon J M and Clark D J 1975 Sol. State Electron 18 267.
Woodhead J, Newman R C, Grant I, Rumsby D and Ware R M 1983 J. Phys. C: Solid St. Phys. 16 5523.

Inst. Phys. Conf. Ser. No. 76: Section 8
Paper presented at Microsc. Semicond. Mater. Conf., Oxford, 25–27 March 1985

# Characterisation of semi-insulating LEC GaAs by scanning electron microscopy

B Wakefield and S T Davey

British Telecom Research Laboratories, Martlesham Heath, Ipswich, IP5 7RE

Abstract  Microscopic variations in the resistivity across semi-insulating LEC GaAs have been imaged using a scanning electron microscope. The variations have been correlated with the cellular network of dislocations typical of this type of material.

## 1. Introduction

The liquid encapsulated Czochralski (LEC) technique for growing GaAs is receiving considerable attention because it is capable of producing, at reasonable cost, large diameter semi-insulating crystals without the need to introduce dopants such as Cr. Semi-insulating GaAs has a use in the production of GaAs integrated circuits, and for this application it must have uniform properties over the whole area of a wafer cut from a grown crystal.   Studies   of   this   material   by   X-ray   topography   (Kamejima et al 1982) and defect etching (Clark and Stirland 1981) have shown that it contains a network of dislocations, arranged in a cellular structure, with  the  size  of  the  cells  being  of  the  order  of  a  few  hundred micrometers. A similar cellular structure has also been observed when  the material   has   been   imaged   using   cathodoluminescence (Wakefield et al 1984, Chin et al 1982, Kamejima et al 1982) and transmission infra-red microscopy (Brozel et al 1983). Both the cathodoluminescence intensity and the infra-red absorption appear to be enhanced in the vicinity of the dislocation network.

Semi-insulating GaAs is obtained when the stoichiometry-dependent concentration of the deep donor known as EL2 exceeds that of the uncompensated residual acceptors. (These residual acceptors are mainly carbon.) Consequently the uniformity in the resistivity of the material may depend on the uniformity in the concentrations of the EL2 centres and of the acceptors. From the infra-red transmission studies it has been shown that the distribution of EL2 is not uniform across an LEC wafer, and that   it   appears   to   be   enhanced   at   the   dislocation   network (Holmes  et  al  1983, Brozel et al 1983, Skolnick et al 1984). This was confirmed by Warwick and Brown (1985) using cathodoluminescence imaging and spectroscopy, in which they demonstrated that there is a build up of EL2 at the cell walls. Also in our earlier work on LEC GaAs using cathodoluminescence   at  a  temperature  of  4K  (Wakefield et al 1984) we found that the intensity of the luminescence at 1.494eV due to the residual carbon acceptors in the material relative to the 1.514eV peak (due to bound excitons) was stronger where the dislocation density was high (at the cell boundaries) as compared with the centre of the cells. This suggested that the carbon acceptor concentration was enhanced near the dislocations. From these two observations it seemed likely that the resistivity of the material might also be non-uniform. Attempts have previously been made    to    look for variations in electrical properties across a wafer (Matsumoto et al 1982, Kitahara et al 1983),    but these have used mechanical probes    and so have had rather poor spatial resolution. In this work we wish to report the observation of microscopic resistivity variations, measured usng a scanning electron microscope, with good spatial resolution.

## 2. Experimental Method

Samples of LEC grown GaAs, up to 5mm square, were cut from 5cm diameter commercial wafers. As in our earlier work (Wakefield et al 1984), cathodoluminescence   micrographs were recorded at room temperature and at temperatures near to 4K in a Cambridge Instruments S180 SEM. An example of a cathodoluminescence micrograph is presented in figure 1a, and shows the enhanced cathodoluminescence in the vicinity of the dislocation network. Line scans of the intensities of the 1.514eV (bound exciton) peak and the 1.494eV (carbon acceptor) peak were also measured at temperatures near to 4K, and as before showed that the intensity of the 1.494eV peak relative to the 1.514eV peak was enhanced at the cell boundaries (Wakefield et al. 1984).

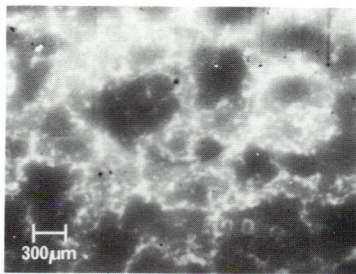

Fig. 1a Cathodoluminescence image of part of an LEC GaAs wafer.

Fig.1b. Secondary electron image of the same sample as fig. 1a, recorded under conditions of electrostatic charging.

Because   the   GaAs   under investigation is semi-insulating, there   will be a tendency for the surface to become electrically charged when being scanned by the electron beam from the SEM. Specimen charging will affect the collection of   secondary   electrons   emitted   from the specimen, and any local variations in the amount of charging will give rise to contrast in the   secondary electron   image. Variations in local electrical conductivity will affect the local surface charging, and so a secondary electron image recorded under conditions such that electrostatic charging of the specimen occurs might   therefore be expected to reveal variations in electrical conductivity.

Samples   of   semi-insulating   LEC GaAs from a number of different suppliers were examined in the SEM, whilst the primary electron probe current was varied. At very low beam currents, such as would be used to obtain high resolution secondary electron images,   the secondary electron images obtained from the surfaces of the specimens were featureless. The same was true when the current was increased up to about 10nA. Above this value however, charging of the specimens became noticeable and contrast began to appear in the images (Fig 1b). Finally when the beam current had reached about luA the specimens charged so strongly that only highly distorted images were obtained.

## 3. Results

Figures 1b and 2b show examples of the secondary electron images recorded under electrostatic charging conditions. Clearly visible is a cellular pattern similar to that observed in cathodoluminescence, X-ray topography and infra-red transmission microscopy examination of this type of material. The cathodoluminescence images recorded from the same areas as Figures 1b and 2b are shown in Figures 1a and 2a, and it can be seen that there is a very good correlation between them. The dislocation network is obviously producing contrast in the secondary electron images when electrostatic charging occurs, and, as discussed above, this suggests that the electrical conductivity is ,different in the vicinity of the dislocation network than in the centres of the cells. As the beam current was increased it was noticed that the centre of the cells charged up before the regions near to the cell walls, suggesting that the resistivity was lowest at the cell walls.

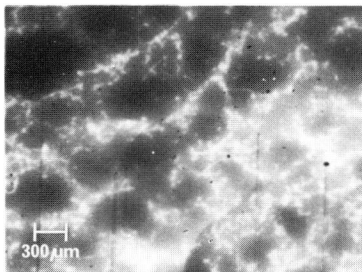 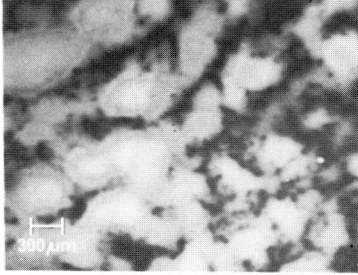

Fig.2a  Cathodoluminescence image

Fig.2b Electrostatically charging secondary electron image.

To check that the contrast was genuinely due to electrostatic charging, the secondary electron collector in the SEM was replaced by a backscattered electron detector. Backscattered electrons are not affected by surface charge to the same extent as secondary electrons. The backscattered electron images so obtained were featureless, just as the low beam current secondary electron images were. Secondly, a sample was given a very thin coating of gold to prevent surface charging. When this was done no cellular structure contrast could be seen in the secondary electron images. Both of these observations support the conclusion that the cellular structure contrast seen in the secondary electron images is caused by local variations in surface charging, and not by topograhic effects.

## 4. Discussion

We have observed what we believe to be local variations in the electrostatic charging of the surface of the semi-insulating LEC GaAs used in this work. These are in the form of a cellular pattern identical to that revealed by cathodoluminescence. From the appearance and behaviour of the images as the primary electron beam current is varied about the value that gives the clearest images, we believe that the charging that occurs at the cell walls is less than that which occurs in the centre of the cells. This implies that the electrical resistivity of the material is lower at the cell boundaries than inside the cells. This is perhaps not too surprising as the dislocations forming the cellular network are believed to getter impurities from the surrounding material. An increase in the impurity concentration around the dislocation could lead to a local increase in the carrier concentration and hence to a decrease in resistivity. Because the diameter of the cells is of the order of a few hundred micrometers such non-uniformities in the resistivity of the material will have serious implications in the use of the material for the fabrication of integrated circuits. This therefore re-emphasises the need to eliminate the dislocation network from the material. All the specimens examined displayed cellular structure contrast in the secondary electron images, so the effect does seem to be general to LEC GaAs.

## 5. Conclusion

By allowing the surface of samples of semi-insulating GaAs to charge up whilst being examined in the SEM, we have observed voltage contrast in the secondary electron images formed by the SEM. This contrast,which has a cellular appearance, has been interpreted as being caused by local variations in the resistivity of the material. The regions of reduced resistivity, so found, correlate with the dislocation network present in this type of GaAs.

## Acknowledgements

The authors thank Dr P.A.Leigh for supplying the specimens. Acknowledgement is also given to the Director of Research of British Telecom for permission to publish this paper.

## References

Brozel M R, Grant I, Ware R M and Stirland D J 1983 Appl. Phys. Lett. 42 610
Chin A K, Von Neida A R and Caruso R 1982 J.Electrochem. Soc. 129 2387
Clark S and Stirland D J 1981 Inst. Phys. Conf. Series 60 339
Holmes D E, Chen R T, Elliott K R and Kirkpatrick C G 1983 Appl. Phys. Lett. 43 305
Kamejima T, Shimura F, Matsumoto Y, Watanabe H and Mitsui J 1982 Japan J. Appl. Phys.21 L721
Kitahara K, Oseki M and Shibatomi A 1983 Fujitsu Sci. Tech. J. 19 279
Matsumoto Y and Watanabe H 1982 Japan J. Appl. Phys. 21 L515
Skolnick M S, Brozel M R, Reed L J, Grant I, Stirland D J and Ware R M 1984 J Electron. Mater. 13 107
Wakefield B, Leigh P A, Lyons M H and Elliott C R 1984 Appl. Phys. Lett.45 66
Warwick C A and Brown G T 1985 Appl.Phys. Lett. 46 574

*Inst. Phys. Conf. Ser. No. 76: Section 8*
*Paper presented at Microsc. Semicond. Mater. Conf., Oxford, 25–27 March 1985*

377

# Application of TEM cathodoluminescence to defect studies of III−V compound semiconductors

S Myhajlenko, H J Hutchinson and J W Steeds

H H Wills Physics Laboratory, Royal Fort, Bristol BS8 1TL.

Abstract     The interpretation of cathodoluminescence (CL) from thin semiconductor specimens suitable for TEM can be complex compared with bulk (SEM) specimens. Factors such as optical interference, surface recombination, specimen thickness variation and their effect on luminescence play a major role in affecting the analysis of CL data. We discuss these factors in relation to a study of luminescence near dislocations in InP, GaP and GaAlAs/GaAs. We also report a shift of 7.5 meV in the exciton related emission from a complex dislocation cluster in InP and some luminescence results from dislocations in GaP.

## 1.  Introduction

The application of CL techniques to the assessment of opto-electronic semiconductors and devices is an established procedure in SEM. In partic- ular, SEM CL has proved useful for the observation of crystal defects, doping inhomogeneities, etc. However, specific information about the defects calls for detailed analysis by TEM. Simultaneous TEM and CL should make this exercise more convenient. Consequently a number of TEM CL systems have been developed such as the one used in this work (Roberts 1982, Myhajlenko et al 1984). Although TEM CL requires more elaborate specimen preparation than SEM CL, it offers better spatial resolution at the expense of luminescence intensity. We describe the application of TEM CL to the study of defects (mainly dislocations) in III-V opto-electronic materials and devices; liquid encapsulated Czochralski (LEC) InP, vapour phase epitaxy (VPE) GaP and double heterostructure (DH) GaAlAs/GaAs lasers. During the course of this work a better appreciation has been gained of factors which are important in transmission studies and we describe some of them. Unless otherwise stated the TEM specimens have been prepared by jet polishing with a solution of chlorine in methanol.

## 2.  Optical Interference

Work performed on DH GaAlAs/GaAs lasers grown by OMCVD (Whiteaway and Thrush 1981) highlights two forms of optical interference; spatial and spectral. Fig.1 shows CL images acquired from a DH laser, with the top p-GaAs capping layer and substrate removed by selective etching, at (a) 1.88 eV (GaAlAs emission) and (b) 1.46 eV (GaAs emission) at 40 K. An example of spectral interference is shown in fig.2. CL spectra acquired from (a) a bulk DH laser and (b) the same stucture as used in fig.1 are shown. The separation of the oscillations in the spectrum of fig.2(b) are consistent with the

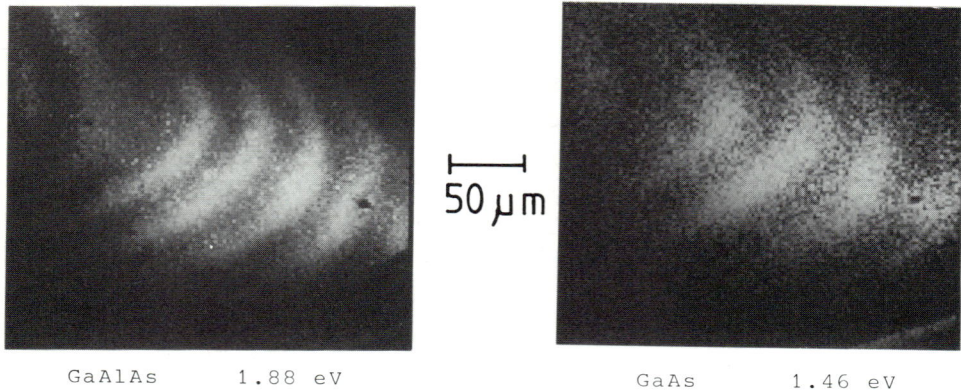

GaAlAs     1.88 eV                    GaAs        1.46 eV

Fig.1  Example of optical interference (fringes) observed
in chemically thinned GaAlAs/GaAs laser structure.

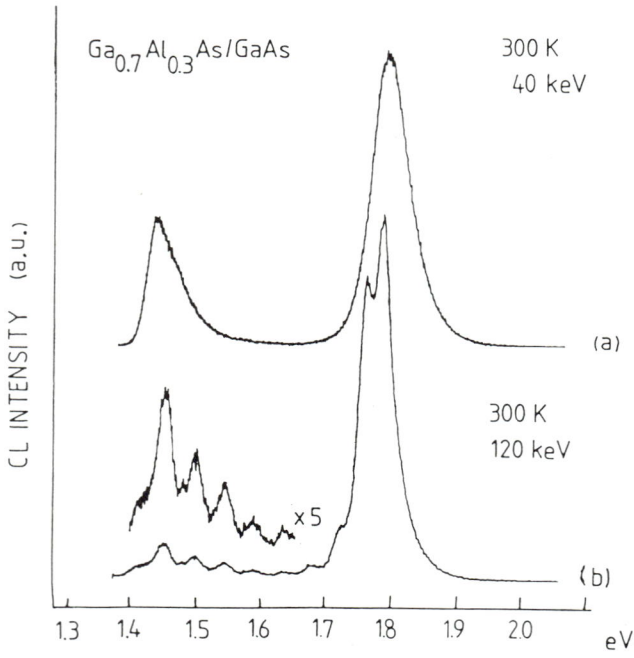

Fig.2  Example of optical interference (oscillations) in the
  emission spectrum from thin semiconductor layers; (a) CL
spectrum from bulk double heterostructure GaAlAs/GaAs laser
and (b) CL spectrum from the same device structure with
p-capping layer and substrate removed by selective etching.

sample thickness (2.25 $\mu$m) and the refractive index (Trommer 1981). Similar interference effects have been observed in TEM specimens of MBE GaAlAs/GaAs (Steeds and Leacy 1985).

## 3. Dislocations in LEC InP

The low surface recombination velocity of InP in comparison with GaAs (Casey and Buehler 1977) makes InP an attractive material for TEM CL studies; for comparable radiative lifetimes there is more luminescence from a given specimen thickness. In the much studied case of silicon the electrical behaviour of dislocations is still not properly established (Ourmazd 1984). In III-V materials the situation is less well understood and the non-centro-symmetric nature of these semiconductors further complicates matters. We have performed CL imaging experiments on bulk n-type LEC InP which has shown dark line and spot contrast, generally ascribed to crystal defects. The spatial extent of some of the dark spots in our specimens was 10 $\mu$m. As this distance is greater than the generation volume or carrier diffusion length it may be the result of impurity clouds or charges associated with the defects. In our initial TEM CL work we were unable to detect luminescence from electron transparent LEC InP at room temperature. CL has been detected subsequently from thinned material only at low temperatures < 100 K. A further advantage of studies at low temperatures is the suppresion of dislocation motion under the influence of a focussed electron probe, though oscillations of dislocation lines between pinning points are still sometimes observed. Preliminary TEM CL imaging and spectral investigations of individual dislocations in LEC InP showed that some locally quenched the exciton related emission, whereas others did not. No systematic correlation between dislocation type and luminescence behaviour was observed (Hutchinson and Myhajlenko 1984). Specimens suitable for investigation by TEM CL in our system are too thick for weak beam analysis of defects even at 300 keV. Therefore, the question of the state of dissociation of these dislocations remains unanswered at present. Subsequent work has shown that it is important to take account of the effect of specimen thickness on the effiency of the relevant luminescence processes (Myhajlenko et al 1984). In particular, the excitonic processes are more easily perturbed than the free to bound transitions by both thickness and electric fields. As both dislocations and surfaces have associated electric fields the interplay between surfaces and defects in affecting luminescence processes is quite complicated. For dislocations these fields may be a result of the core structure, piezoelectricity and the effects of point defects and impurities. The behaviour of dislocations in as-grown (both stoichiometric and In-rich melts) LEC InP and in annealed samples will be summarised elsewhere (Hutchinson et al 1985).

Fig.3 shows some new CL results obtained from as-grown LEC InP (stoichio-metric melt) using 3 meV system band pass. Fig.3(a) shows the CL spectrum at 35 K from bulk material with excitonic (mixture of free and bound) emission (X) and free to bound (Zn acceptor) emission (A), fig.3(b) shows a CL spectrum from thin dislocation free material while in fig.3(c) the CL spectrum was obtained from the dislocation cluster. In addition to the quenching of X at the cluster a shift of 7.5 meV in the X emission peak was observed. There was no observed shift in the A peak. Fig.3(d) shows the CL spectrum from a group of dislocations, all with the same Burgers vector in a thinner region. In this case no shift in X was observed. A similar shift in the exciton emission has been reported at dislocation clusters in GaAs by Petroff et al (1981). These workers also saw this shift at individual dislocations, an effect we have not been able to resolve

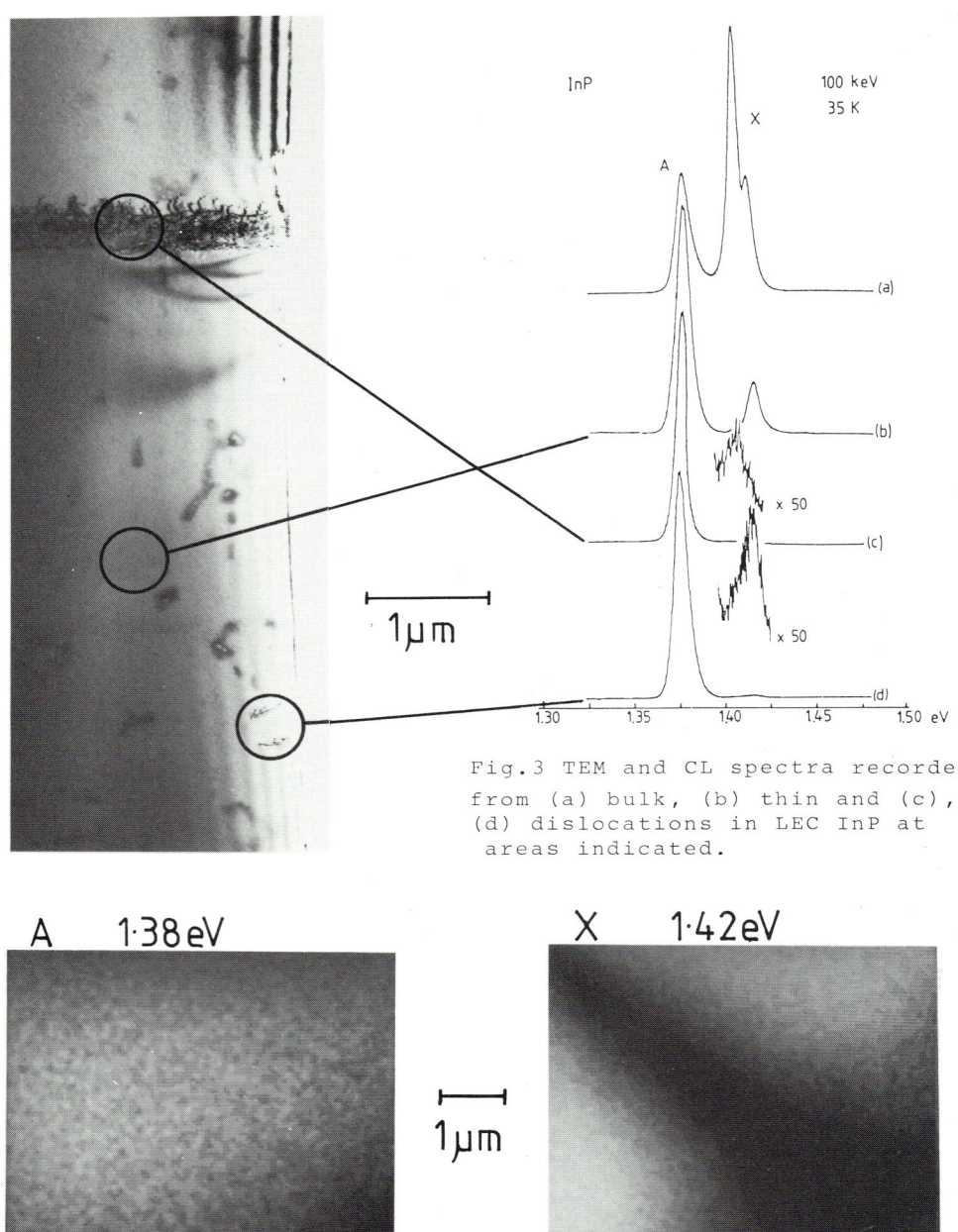

Fig.3 TEM and CL spectra recorded from (a) bulk, (b) thin and (c), (d) dislocations in LEC InP at areas indicated.

Fig.4 Monochromatic CL images of dislocation cluster shown in fig.3 at 1.38 eV (A transition) and 1.42 eV (X transition). Note the local quenching of X emission along the direction of dislocated material.

spectrally. To summarise, TEM CL has shown that the low temperature luminescence behaviour of InP at dislocations is complex. Exciton related transitions are sometimes quenched and/or shifted in energy relative to defect free material, whereas free to bound transitions appear to be unaffected. No deep level luminescence associated with dislocations has been detected in InP, unlike the case of ZnSe where a dislocation correlated emission band has been observed (Batstone and Steeds 1985). Note that these low temperature results may have a different origin from the room temperature quenching of luminescence at defects. The relevance of our findings to the operation of opto-electronic devices operating at room temperature and above has to be investigated in the future.

## 4. Dislocations in VPE GaP

TEM CL studies have been extended to GaP and GaAs and we report some preliminary observations on green VPE GaP (S-doped $5 \times 10^{17}$ cm-3, N-doped $2 \times 10^{18}$ cm-3). These highlight the changes that occur in a complex CL spectrum when the sample thickness is reduced and emphasize how the thickness dependence dominates the observations that can be made at dislocations. Fig.5 shows typical CL spectra at 35 K from bulk and thin (< 0.2 $\mu$m) VPE GaP. The near band edge luminescence in bulk specimens is dominated by exciton recombination associated with sulphur donors ($S_0$) with some background contribution from nitrogen and free hole recombination at neutral donors ($D_0$) (Wight 1977). The specimen also shows a strong deep level emission ($V_{Ga}D_{As}$) thought to be associated with donor-vacancy complexes (Metz and Fritz). In thin material excess carrier recombination via excitonic processes is not favoured thereby enhancing the radiative recombination through other centres, i.e. $D_0$ and $V_{Ga}D_{As}$. Fig.6 shows the near band edge CL in more detail. Fig.7 shows CL spectra acquired at and away from individual dislocations for (a) a screw dislocation lying in material of thickness 0.14 $\mu$m and (b) a 60° dislocation lying in material of thickness 0.27 $\mu$m. The dislocations have no apparent effect on the near band edge luminescence, the difference in $S_0$ intensity being due to the specimen thickness.

## References

Batstone J L and Steeds J W 1985 these proceedings.
Casey H C and Buehler E 1977 APL 30 247.
Hutchinson H J and Myhajlenko S 1984 Phil.Mag. B50 L49.
Hutchinson H J, Myhajlenko S and Steeds (to be published).
Metz S and Fritz W 1977 Inst.Phys.Conf.Ser.No.33a 66.
Myhajlenko S, Batstone J L, Hutchinson H J and Steeds J W 1984 J.Phys.C 17 6477.
Ourmazd A 1984 Contemp.Phys. 25 251.
Petroff P M, Weisbuch C, Dingle R, Gossard A C and Weigmann W 1981 APL 38 965.
Roberts S 1982 Inst.Phys.Conf.Ser.No.61 51.
Steeds J W 1985 Microscopia Electronica Y Biologia Celular, in the press
Trommer R 1981 Inst.Phys.conf.Ser.No.56 353.
Wight D 1977 J.Phys.D 10 431.
Whiteaway J E A and Thrush E J 1981 JAP 52 1528.

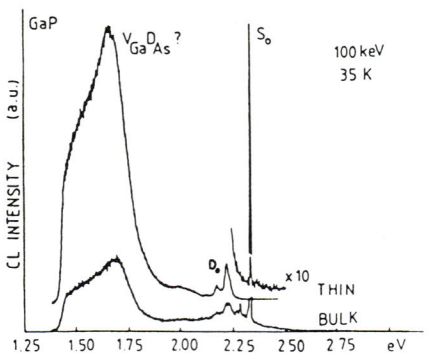

Fig.5 CL spectra from bulk and
thin VPE GaP.

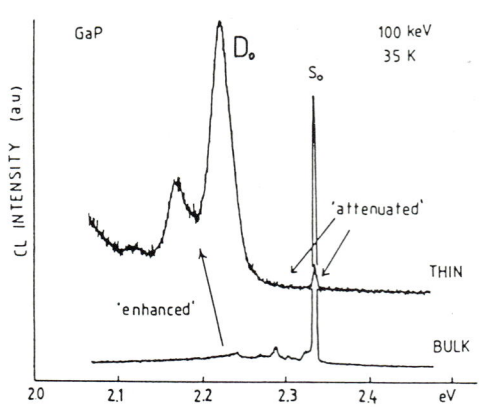

Fig.6 The near band edge CL from
VPE GaP in greater detail.

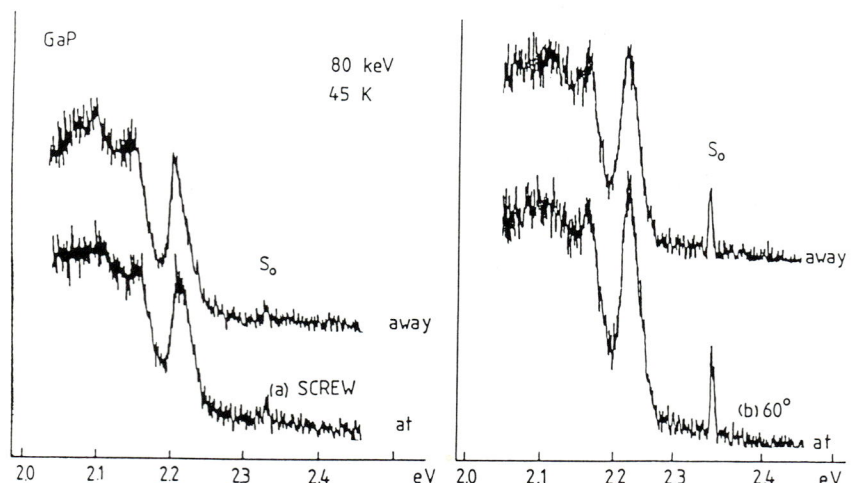

Fig.7 Near band edge CL spectra acquired at and away from (a) screw and
(b) 60° dislocations.

Acknowledgements

The authors would like to thank C A Warwick and G T Brown of RSRE Malvern
for suppling the samples of InP and for useful comments, J E A Whiteaway of
STL Harlow for suppling the GAALAS/GaAs heterostructures and useful
discussions.   This work was supported by Procurement Executive, Components
Valves and Devices.

*Inst. Phys. Conf. Ser. No. 76: Section 8*
*Paper presented at Microsc. Semicond. Mater. Conf., Oxford, 25–27 March 1985*

383

# TEM and CL characterization of dislocations in OMCVD ZnSe

J L Batstone and J W Steeds

H.H. Wills Physics Laboratory, University of Bristol, Tyndall Avenue,
Bristol BS8 1TL

Abstract    Low temperature (30K) CL and TEM studies have enabled the
identification of luminescence bands with specific crystal defects.
Dislocations in OMCVD ZnSe/(100)GaAs are associated with an emission
band at 2.60 eV (Y band). Comparisons between undoped, Al-doped and In-
doped ZnSe show variations in defect type and density which are related
to factors such as the substrate growth temperature and the doping concentration. Dense tangles of dislocations in Al-doped and In-doped ZnSe
show very strong Y emission. Misfit dislocations are regularly observed
at the Al-doped ZnSe/GaAs interface.

## 1. Introduction

Interest in ZnSe as an optoelectronic device material is increasing as a
result of current growth techniques. Organometallic chemical vapour deposition (OMCVD) and molecular beam epitaxy offer low temperature growth of
single crystal thin films with high carrier mobilities and blue room temperature emission. Local variations in the emission have been attributed to
crystal defects which degrade the optical quality of the material. An understanding of the nature and origin of these defects may be gained using
transmission electron microscopy (TEM) with in-situ cathodoluminescence
(CL). Low temperature (30K) CL studies have enabled the correlation of one
luminescence band (Y at 2.60 eV) with dislocations (Myhajlenko et al 1984).

Three types of OMCVD ZnSe/(100)GaAs (Wright and Cockayne 1982) have been
studied; (a) nominally undoped ZnSe, (b) Al-doped ZnSe (Triethylaluminium)
and (c) In-doped ZnSe (Triethylindium). The defect type and density has
been studied as a function of doping concentration for Al-doped ZnSe, and
as a function of growth temperature for undoped ZnSe. Preliminary results
on In-doped ZnSe have also been obtained.

## 2. Experimental

Simultaneous TEM and CL experiments are performed on an extensively modified Philips EM400, equipped with a CL system. The light collection optics
and signal processing facilities have been described elsewhere (Myhajlenko
et al 1984) and will be mentioned only briefly here. CL emitted from the
semiconductor specimens is collected by an ellipsoidal mirror and transferred using a system of mirrors and quartz lenses to a Bentham M300 monochromator. An RCA 31034A photomultiplier operated in single photon counting mode is used for spectrum acquisition and monochromatic imaging. The
microscope scan coils, monochromator stepping motor and data acquisition

are controlled by a Link Systems 860 Series II dedicated minicomputer.

Samples of OMCVD ZnSe/GaAs (ZnSe film thickness 1-3 μm) have been prepared
for TEM-CL by jet-polishing in a solution of 1% $Br_2$ in methanol to remove
the substrate.  CL spectra and images have been obtained from regions of
crystal which are sufficiently electron transparent at 120kV to allow dislocation identification with good CL intensity (0.1-0.5 μm).  TEM analysis
in a Philips EM430 at 300 kV allows dislocation characterization in samples
up to 2μm, and has enabled observations of the ZnSe/GaAs interface without
chemically thinning the ZnSe layer.

## 3. Results

TEM analysis at 120 kV and 300 kV revealed differences between the three
types of ZnSe.  Nominally undoped ZnSe (residual donors are Ga and Cl) contained a high density of stacking faults on $\{111\}$ planes, bounded by $a/_6$<211>
partial dislocations as well as dislocations of type $a/_2$<110>.  No evidence
of precipitation was observed.  The stacking fault density was found to
vary with substrate growth temperature ($T_G$).  The results are summarized in
Table 1.

Table 1.  Nominally undoped ZnSe

| Sample | $T_G$ in °C | stacking fault density/$cm^3$ |
|---|---|---|
| 1224 | 300 | $6 \times 10^{13}$ |
| 1226 | 350 | $3 \times 10^{12}$ |
| 1227 | 400 | $1 \times 10^{12}$ |

As $T_G$ increased, the stacking fault density decreased.  A common measure of
the optical quality of the material is the ratio of band edge to deep level
luminescence, R.  R was found to decrease as $T_G$ increased.  Reducing $T_G$
below 300°C resulted in poor epitaxy and a tendency towards polycrystalline
growth.  Addition of a dopant such as Al or In improved this epitaxy and
allowed good single crystal growth at $T_G$ as low as 275°C.  Lower $T_G$ should
result in reduced in-diffusion from the GaAs substrate.  Examination of Al-
doped ZnSe revealed variations in the stacking fault density with doping
concentration, and the results are summarized in Table 2.

Table 2.  Al-doped ZnSe, $T_G$ = 275°C

| Sample | doping concentration/$cm^3$ | stacking fault density/$cm^3$ |
|---|---|---|
| 986 | $2 \times 10^{16}$ | $2 \times 10^{13}$ |
| 987 | $3 \times 10^{16}$ | $1 \times 10^{13}$ |
| 989 | $6 \times 10^{16}$ | $3 \times 10^{11}$ |

The stacking fault density decreased as the doping concentration increased.
PL data from these samples (Dean 1984) showed that R increased with increasing Al concentration up to $6 \times 10^{16} cm^{-3}$ and then decreased as Al deep
levels were introduced.

TEM at 300 kV in the region of the GaAs interface revealed a network of
misfit dislocations, which are thought to relieve the lattice strain due
to the 0.3% lattice mismatch between ZnSe and GaAs.  Fig. 1 shows the two
sets of interfacial dislocations, (a) $\underline{b} = a/_2$[110] and (b) $\underline{b} = a/_2$[1$\bar{1}$0].  The
incident beam direction was close to [001] and the thickness of the foil
was estimated by the projected width of the stacking fault in Fig. 1(a) to
be ~1.8 μm.

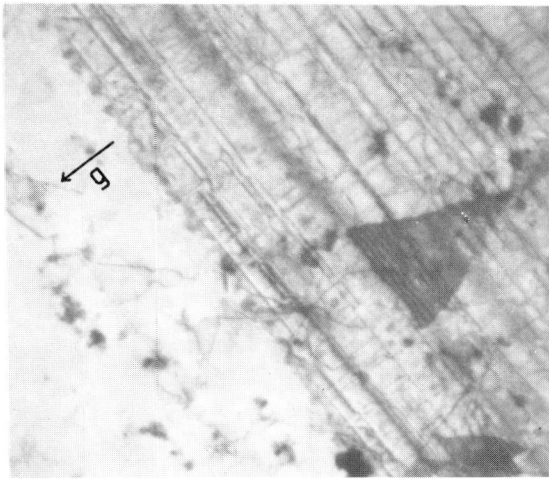

(a) g = 220

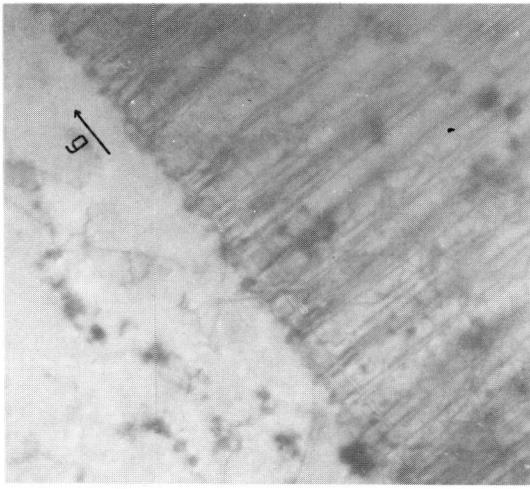

(b) g = 2̄20     ————
                1μm

Fig. 1. Interfacial dislocations in
Al-doped ZnSe

CL studies on these samples
($T_G$ = 275°C) showed very good
optical quality material with
D°X (donor bound exciton) at
2.79 eV and a weak deep level at
2.1 eV. No Y emission at 2.60
eV was observed. Exciton emission was reduced in the vicinity
of dislocations. Partial dislocations also reduced the exciton signal, but the stacking
fault ribbon has no effect. CL
spectra and maps from a sample
with a higher growth temperature
(300°C) revealed Y emission localized both at complex tangles
of dislocations and individual
dislocations. Fig. 2 shows a
typical dislocation tangle
stretching over several microns.
An explanation of the tangle is
still being sought but SEM examination of these areas revealed
slight indentations which may
have formed during growth as a
result of impurities falling off
the walls of the growth chamber.
Fig. 3 shows spectra acquired
from the region shown in Fig. 2.
Spectra were acquired with a
small focussed probe (~0.2 μm)
placed well away from the tangle
in Fig. 3a. D°X at 2.79 eV, FE
(1s-2s) at 2.78 eV and a deep
level at 2.1 eV were observed
away from the tangle. Fig. 3b
shows strong Y emission at 2.60
eV with the probe placed on the
tangle. Fig. 3c shows an additional emission band at 2.68 eV.
This transition has been tentatively assigned as of donor
acceptor pair (DAP) origin, involving Na acceptors and Al
donors (Po series). Fig. 4
shows the monochromatic CL
images obtained from the dislocation tangle. The exciton emission was reduced in the region of the tangle, Y emission was localized along the dislocation tangle, and the DAP emission was found to be extremely localized.
Careful examination of Fig. 2 revealed isolated precipitates and associated
punched-out prismatic dislocation loops. A precise correlation between
precipitates and the 2.68 eV emission has not yet been established due to
difficulties in comparisons between the digital CL maps and the TEM image.
Similar tangles of dislocations were also observed in the In-doped ZnSe
and were associated with Y emission. No evidence for tangles has been
found in undoped ZnSe. CL studies on the undoped ZnSe revealed strong Y
luminescence from individual dislocations. Not all dislocations emitted Y

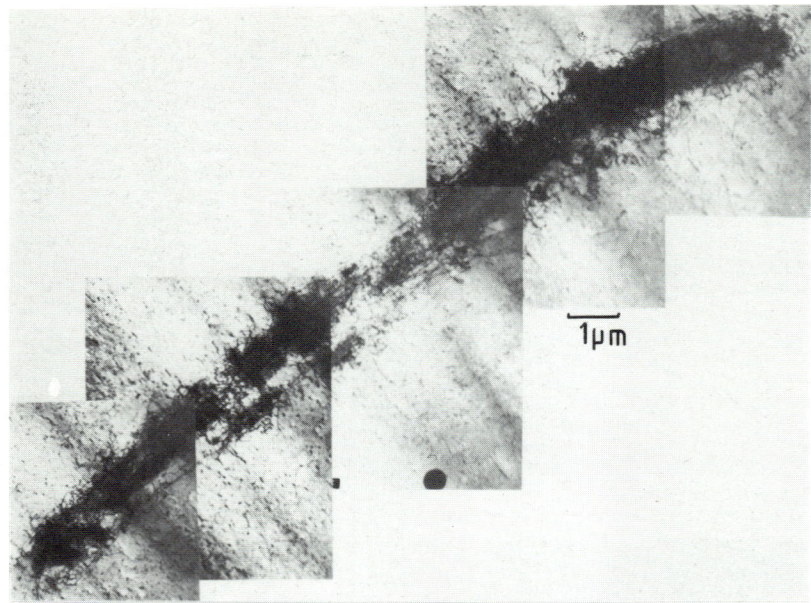

1µm

Fig. 2. A typical dislocation tangle in Al-doped ZnSe.  Punched out
dislocation loops reveal precipitates hidden in the tangle

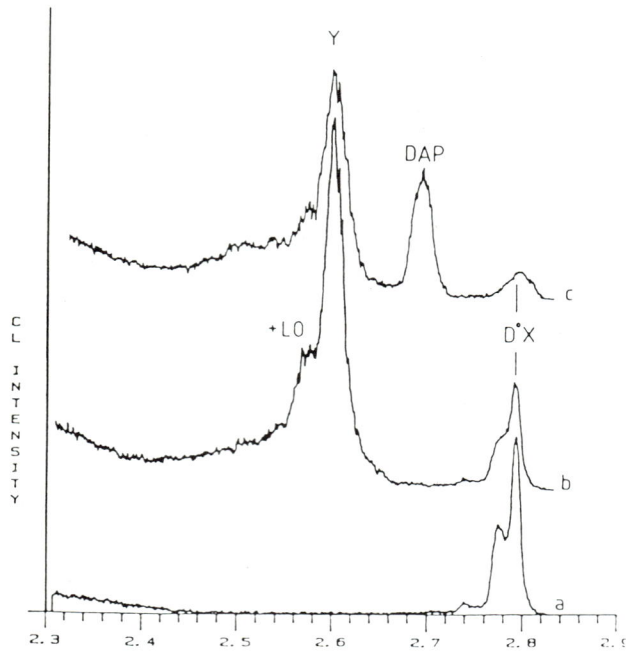

Fig. 3. Spectra (30K)
acquired from the dis-
location tangle in
Fig. 2;  (a) probe away
from tangle, (b) and (c)
probe on tangle.

radiation, but the correlation of dislocation character and core structure with luminescence efficiency has not been established yet. Samples with $T_G$ = 300°C revealed a sharp transition at 2.712 eV, which has been tentatively assigned as DAP (Ro series) involving Li, although the no-phonon line is rather high in energy for this transition. As $T_G$ increased to 400°C, the DAP emission shifted to 2.69 eV which was assigned as the Qo series. Monochromatic images of the DAP transitions showed very localized bright spots which have not been conclusively correlated with specific defects.

In all the samples studied, undoped, Al-doped and In-doped ZnSe, the deep level transitions were unaffected by the presence of stacking faults, dislocations and dislocation tangles. No quenching was observed.

## 4. Discussion

CL results have shown conclusively that Y luminescence is associated with both complex tangles and individual dislocations. However, Y emission was not observed from all dislocations. Comparisons between Al-doped material grown at 275°C and at 300°C revealed an absence of Y emission at the lower growth temperature. Exciton reduction in the vicinity of dislocations was observed in both cases. Factors likely to affect exciton emission at a dislocation are (i) strain fields, (ii) impurities or clouds of point defects surrounding or decorating the dislocation core, (iii) imperfections at the dislocation core such as kinks, and (iv) the dislocation core itself. The extent of exciton quenching in InP (Hutchinson & Myhajlenko 1984) was found to vary with dislocation type and a proposed mechanism for the quenching involved strain-generated electric fields. Impurities in ZnSe may be responsible for both exciton quenching and emission at

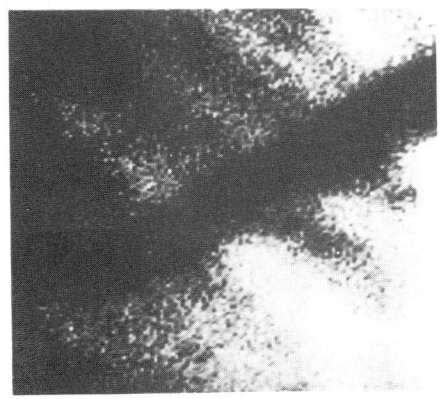

(a) D°X, 2.79 eV          2 µm

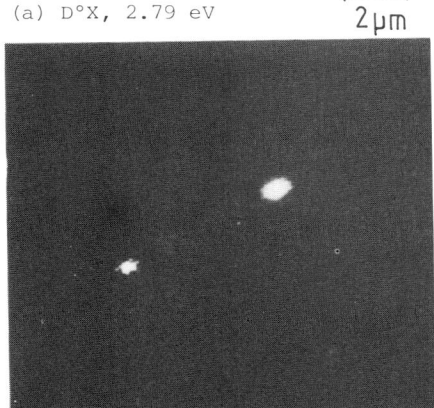

(b) DAP, 2.68 eV

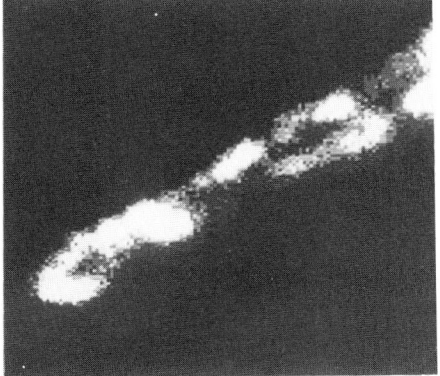

(c) Y, 2.60 eV

Fig. 4. CL images from the dislocation tangle in Fig. 2

2.60 eV. Two major sources of impurities in ZnSe are (a) pipe diffusion along threading dislocations from the substrate, and (b) trace impurities present in the gas flows and reaction chamber during growth. The latter source is the same for all samples, but variations in $T_G$ are likely to affect impurity diffusion rates. If the origin of Y luminescence is impurity related, then a reduction in $T_G$ should result in reduced diffusion and reduced Y emission, as observed.

The 2.68 eV transition observed in Al-doped ZnSe has been assigned as of DAP origin. The energy of the transition corresponds well with the Po DAP series in ZnSe, where the donor is $Al_{Zn}$ and the acceptor is $Na_{Zn}$ (Bhargava et al 1979). A similar DAP transition was observed in undoped ZnSe, with a peak energy of 2.712 eV. A band pass of 3 meV results in an uncertainty in the last decimal place. This has been assigned as the Ro series involving transitions between $Li_{int}$ and $Li_{Zn}$. The Ro series is usually found at 2.708 eV, but the higher value observed here may be consistent with an upshift due to electric fields set up in thin films (Dean 1984). Confirmation of this effect will be established in the future by comparisons of spectra obtained in both thin and bulk material. The shift to lower energies as $T_G$ increased suggested that the transitions observed were the Qo series involving $Al_{Zn} - Li_{Zn}$. Na and Li are common trace impurities, but additional evidence for the presence of Al in these undoped layers has not been observed, and the assignment of Qo remains uncertain.

Aluminium appears to play an important role in the crystal growth process. The quality of the ZnSe improved with the addition of Al which is likely to getter $O_2$ and moisture from the substrate and reaction chamber. The stacking fault density observed varied from $10^{10} - 10^{13}$ cm$^{-3}$ which is lower than has been previously reported (Williams et al 1984) but is greater than could be explained by the density of dislocations in the substrate. A clear network of misfit dislocations seen in the Al-doped case has not yet been observed in undoped ZnSe. A possible explanation for this lies in the increased mobility in undoped layers which may enable dislocations to run out of the thin foil as the substrate is removed. The density of stacking faults decreased as the Al concentration increased. This is consistent with the observed reduction in dislocation mobility on doping which could lead to a smaller degree of dissociation of the partials.

In conclusion, dislocations in Al-doped, In-doped and undoped ZnSe quenched exciton luminescence complex tangles and some individual dislocations were associated with an emission band, Y, at 2.60 eV. Results suggest the role of an impurity in the recombination mechanism. Stacking faults and dislocations were observed to relieve interfacial strain.

Acknowledgements
Thanks are due to P.J. Wright of RSRE Malvern for growth of the ZnSe and for many helpful discussions. Financial support from the SERC is gratefully acknowledged.

References
Bhargava RN, Seymour RJ, Fitzpatrick BJ and Herko SP 1979 Phys.Rev.B20 2407
Dean PJ 1984 Phys.Status Solidi (a) 81 625.
Hutchinson HJ and Myhajlenko S 1984 Phil.Mag.B50 L49.
Myhajlenko S, Batstone JL, Hutchinson HJ and Steeds JW 1984 J.Phys.C:Solid State Physics 17 6477.
Williams JO, Crawford ES, Jenkins JLL, NG TL, Patterson AM, Scott MD, Cockayne B and Wright PJ 1984 J.Mater.Sci.Lett.3 189.
Wright PJ and Cockayne B 1982 J.Cryst.Growth 59 148.

Inst. Phys. Conf. Ser. No. 76: Section 8
Paper presented at Microsc. Semicond. Mater. Conf., Oxford, 25–27 March 1985

389

# A Fourier transform spectrometer for the analysis of near infrared cathodo luminescence in an SEM

F M  Saba and D B  Holt

Department of Metallurgy and Materials Science, Imperial College of Science and Technology, London SW7 2BP, U.K.

Abstract    A simple Fourier transform spectrometer system for the cathodoluminescence mode in a scanning electron microscope was built. It is designed for analyses in the near infrared beyond the spectral response range of S-1 type photomultipliers. The principles of its design and its performance are reported. It is applicable to the study of silicon and the III-V alloys employed in optical fibre communications devices. It has been tested on the spectra of LEDs and InP/InGaAsP layers. The results demonstrate the value of the system and the factors still requiring optimization.

## 1. Introduction

The cathodoluminescent mode of the SEM has proved to be useful in the microcharacterization of optoelectronic materials and devices, because of its sensitivity to variations in composition and its high spatial resolution, and several laboratories have developed cathodoluminescence (CL) detection systems based on grating monochromators and photo-multiplier tubes (PMTs). However, the latest generation of optical communication devices has been developed to use the 1.55 μm minimum in the spectral attenuation curve for silica fibres. This, unfortunately, precludes the use of PMTs because their spectral response range is limited to about 0.2 to 1.2 μm. Moreover silicon has its intrinsic emission at about 1.1 μm and extrinsic Si emission (due to impurities) occurs at greater wavelengths.

At present, solid state detectors must be used to cover infrared wavelengths longer than 1.2 μm. The detectivity of these devices is, at best, only 1% of that of a PMT, and while their quantum efficiencies can be high, they only have gains of about 1 to 100, as opposed to about $10^6$ for PMTs. This necessitates long dwell times to obtain acceptable signal/noise (S/N) ratios.

Holt and Datta (1980) suggested that dispersive instruments should be abandoned in favour of the Fourier transform spectrometer (FTS) for infrared CL analyses because the latter makes more efficient use of the input radiation by detecting all the spectral elements simultaneously rather than detecting each element sequentially. When the noise is independent of the signal level, as in the case of solid state detectors, this leads to a gain of the order of $\sqrt{N}$ where N is the number of spectral elements. This gain is called the multiplex or Fellgett's advantage. The FTS also does not need entry and exit slits so that it has a throughput advantage.

Fellgett's advantage disappears when the noise is proportional to the square root of the signal level, but the throughput or Jacquinot's advantage is not affected so the FTS may show a better performance than dispersive instruments even in the ultraviolet and visible spectral regions (Luc and Gerstenkorn (1978)). For the near infrared, Davidson et al (1981) reported an improvement of about 20 times for the CL analysis of a Cr-doped GaP sample.

Although the FTS technique is widely used for spectrochemical measurements, especially in the far infrared, it has yet to be so extensively applied to CL analyses, and to our knowledge it has only been attempted at two or three laboratories. This is perhaps related to the unavailability of suitable, low-cost, commercial instruments.

In this paper we present details of a compact instrument with a modest resolution ($\sim$15 cm$^{-1}$) which was built to explore the problems of the technique and to attempt the detection of near infrared CL beyond the response range of PMTs.

## 2.  Theory

Horlick (1968), Schopper and Thompson (1974) and Sakai (1977) give good introductions to FT spectrometry.  A more detailed account is given by Connes (1961).

The FTS is, in general, based on the Michelson interferometer or one of its variants.  Figure 1 shows a schematic diagram of a simple instrument. The radiation under analysis is collimated so that a parallel beam falls on the beam-splitter which ideally transmit 50% of the beam to the fixed plane mirror and reflects 50% to the moving mirror. The two beams are then reflected back to the splitter where they recombine.  However, because the moving mirror introduces an optical path difference between the two arms of the interferometer, there is a phase difference between the beams which varies sinusoidally with the position of the moving mirror.  This results in a signal being detected that is given by:

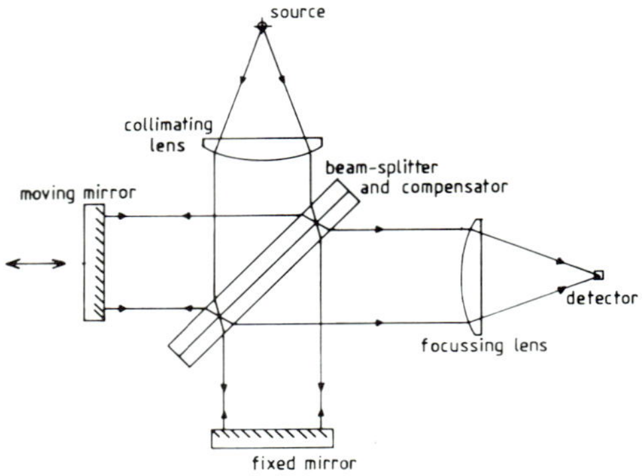

Figure 1.  Schematic diagram of a basic Michelson interferomter.

$$I(x) = \int_{\nu=0}^{\infty} B(\nu) \ (1 + \cos 2\pi\nu x) \ d\nu \qquad (1)$$

where x is the optical path difference, $\nu$ is the reciprocal of the wavelength, the wavenumber, and $B(\nu)$ d$\nu$ is the intensity in the interval $\nu$ to $\nu + d\nu$.

The period of the sinusoidal variation depends on the wavelength of the radiation so that the interferometer effectively codes each wavelength uniquely. Recovery of the spectrum can be achieved by obtaining the Fourier transform of the modulated part of $I(x)$. The unmodulated part gives rise to a D.C. level which can be eliminated quite easily.

A compensator made from the same material and with the same thickness as the beam-splitter substrate is used to keep the optical path lengths through the substrate material the same for both arms of the interferometer. This is necessary to eliminate non-linear phase errors which occur in an asymmetric instrument due to dispersion.

$I(x)$ cannot be determined for all values of x because this would involve sampling an infinite number of points. However, the interferogram needs only to be sampled at intervals of d given by:

$$d = \frac{\lambda m}{2} \qquad (2)$$

where $\nu_M$ is the maximum wavenumber in the detected signal, before different orders overlap. The total path difference also cannot be extended to infinity thus imposing some maximum value, $x_M$, on x. This truncates the interferogram which results in the spectrum being convoluted by sinc $(2 \pi \nu x_M)$, so that the resolution of the FTS is given by:

$$d\nu = 1/2x_M \qquad (3)$$

The number of independent data points N is then given by:

$$N = x_M/d = 2x_M\nu_M \qquad (4)$$

and the resolving power is:

$$R = \nu_M/d\nu = N/2x_Md\nu = N \qquad (5)$$

so that the resolving power depends on the number of independent data points.

## 3. Instrumentation

Our spectrometer is based on a design developed by Horlick and Yuen (1978) for spectrochemical measurements from the ultraviolet to the mid-infrared. It can be considered as three units as shown in figure 2.

The instrument is basically a simple Michelson interferometer with the white light and laser reference systems sharing the same principal optical components as the radiation under analysis (figure 3).

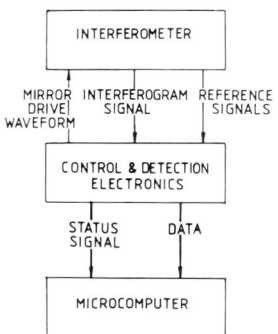

Figure 2.  Block diagram of the FTS system.

This simplifies the alignment procedure, but necessitates the use of baffles to ensure that the reference beams are not scattered into the main signal detector.

The beamsplitter substrate, compensator and collimating lens are made from BK7 glass which restricts the detectable range to about 0.3 to 2.0 µm, but this could be changed by using components made from appropriate materials. The substrate and compensator are 76.2 mm diameter optical windows with λ/10 (λ = 589 nm) wavefront distortion performance. The collimator is a plano-convex lens with a 50.8 mm diameter and 76.2 mm focal length.

Plane mirrors flat to λ/20 (λ = 546.1 nm) are used in preference to retroreflectors on grounds of cost and design simplicity. The latter are less sensitive to misalignments due to tilting, but for a total movement of about 4 mm with an air-bearing to support the moving mirror, it was felt that a plane mirror system would be adequate. The mirrors have 50.8 mm diameters and are made from a glass with a low thermal expansion coefficient to minimise distortions.

The moving mirror is mounted on a pair of steel plates which are separated by three pairs of push-pull screws to give coarse adjustment of the mirror orientation (James and Sternberg (1969)). The entire assembly is attached

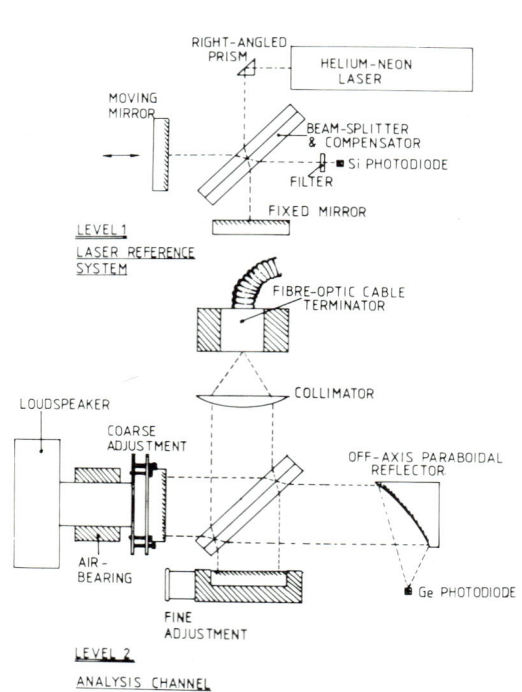

Figure 3a    Schematic diagram of the interferometer bottom and middle levels

to a hollow, cylindrical piston which is floated in an air-bearing supplied with nitrogen gas at 10 p.s.i. The other end of the piston is joined to the coil of a 10 W loudspeaker which has had its paper cone removed. The fabric "spider" was retained to keep the coil centred. Fine adjustment is provided on the fixed mirror by means of a precision mirror mount from Oriel Scientific.

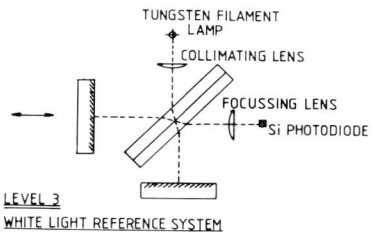

LEVEL 3
WHITE LIGHT REFERENCE SYSTEM

Figure 3.  Schematic diagram of
the interferometer (b) top level.

As in the Horlick and Yuen (1978)
design, the output beam is focussed on
to the detector with an off-axis paraboloidal mirror. This component does
not introduce chromatic aberrations
and ideally the collimator should also
be a paraboloidal reflector, but cost
dictated the choice of a lens.

At present the Ge photodiode with
thermoelectric cooling from Judson
Infrared is used to detect the 0.8 to
1.8 μm range of wavelengths, but with
a suitable adaptor and aperture PMTs
could be used for detecting visible wavelengths.  Si photodiodes are used
to detect the reference signals.

HeNe laser fringes are used to reference the position of the moving
mirror, and a small, tungsten-filament lamp is used to indicate the
position of zero path difference between the two arms of the interferometer by providing white light fringes.

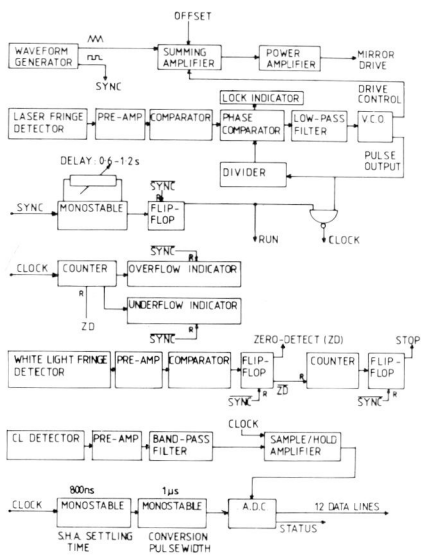

Figure 4.  Block diagram of the control and
detection electronics.

Figure 4 shows a schematic
diagram of the electronics
unit. It provides the 0.2 Hz,
triangular, mirror-drive waveform which is controlled by a
signal fed back from the
phase-locked loop (PLL) monitoring the laser fringe frequency to keep the mirror
velocity constant. The PLL
also doubles the fringe frequency to provide the digitisation signal for the 12-bit
analogue-digital convertor
(ADC) so that the laser wavelength (632.8 nm) becomes the
short wavelength limit instead
of 1.2656 μm. This should
allow observations of GaAs
(∿860 nm) and InP (∿910 nm)
without aliassing, i.e. without overlapping orders.

Second order Bessel filters
are used to limit the bandwidth to 1.8 to 18.0 KHz and
a sample/hold amplifier is
used to provide a stable input to the ADC. Status
signals indicate the con-
version of an interferogram
and whether data are valid.  These signals are sent with the data via a
15-line interface to the Research Machines 380z microcomputer.

The 380z does not control the FTS, but monitors the status signals to determine when to store the incoming data in its memory buffer. With appropriate changes to the interface circuitry almost any microcomputer could be used. The main limitation is the amount of memory available on the machine. In our case the 380Z limits the length of the interferogram to 2048 points with 1024 points on either side of the point of zero path difference, but the FTS electronics can handle up to a total of 8192 points.

Data-processing is controlled by a BASIC program with the Fast Fourier Transform (FFT) written in Z80 assembly language according to the Cooley-Tukey algorithm (Cooley and Tukey (1965)). Floating point operands were used, but transforms were provided 9 times faster than an equivalent program written in BASIC. Using integer operands would probably improve the performance further. At present 2048 points can be converted in about 100s. The resulting spectrum can be displayed on the VDU screen, plotted on a printer, stored on a disc or can be improved by further signal averaging.

## 4.  Results

Initial test results were obtained by detecting the electroluminescence of a red LED (figure 5) and an infrared LED (figure 6).

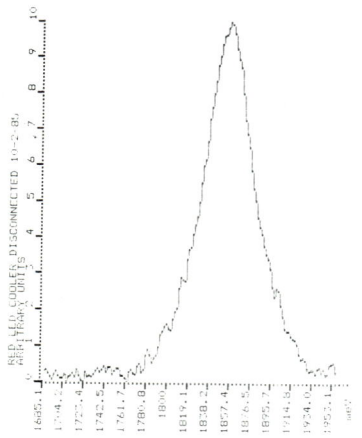
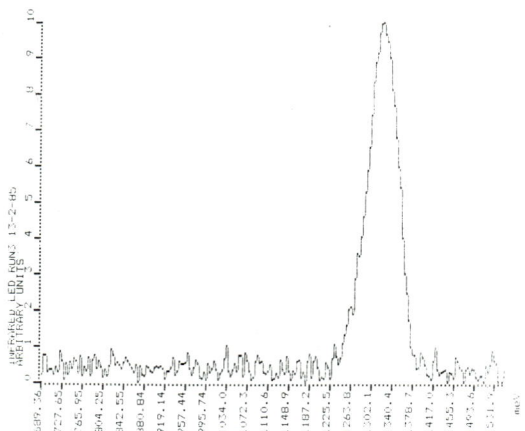

Figure 5.  Spectrum of a red LED recorded by the FTS and Ge photodiode.

Figure 6.  Spectrum of an infrared LED recorded by the FTS and Ge photodiode.

For comparison figure 7 shows the spectrum of the infrared LED obtained with our grating monochromator system and an S-1 PMT. The position of the peak after converting to the appropriate units differs from the spectrum obtained with the FTS because corrections for the spectral responses of the two systems have not been applied.

Figure 7 was obtained for a total observation time of 40 seconds whereas figure 6 was obtained in under 1 second. However, a higher signal level was required by the FTS to produce this result.

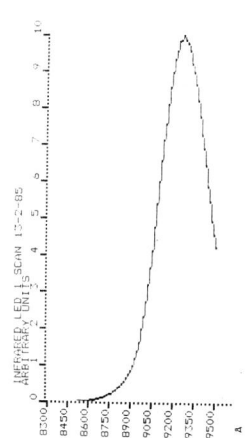

Figure 7. Spectrum of an infrared LED recorded by the monochromator and S-1 photomultiplier tube.

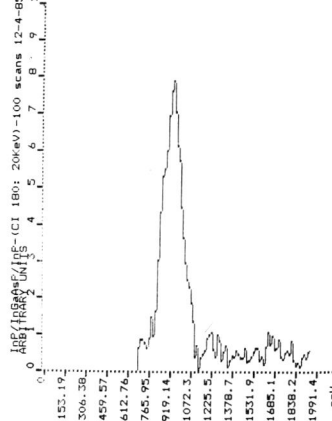

Figure 8. CL spectrum of an InP/InGaAsP/InP layered sample recorded by the FTS and Ge photodiode.

From these tests the performance was found to be less than expected and this together with the low beam currents ($\sim$ 100 nA) available in our SEM, made it impossible to obtain CL spectra in our laboratory. Work is in progress for improving our CL collection stage as well as for optimising the optical alignment and the detection electronics of the FTS. A simple immediate solution, however, was to increase the input signal level.

A Cambridge Instruments 180 SEM at British Telecom Research Laboratories was operated at 20 KeV with the final aperture removed to obtain the CL spectrum shown in figure 8 from an InP/InGaAsP/InP layered sample. Beam currents can be of the order of 100µA in this machine and indeed samples can be destroyed at higher acceleration voltages.

The peak in the spectrum is at about 0.935eV which corresponds to about 1.33 µm and therefore arises from the InGaAsP layer. Again corrections have not been made for the spectral response of the FTS. 100 inter-freograms were signal-averaged to obtain this result and although this represents about 20 seconds of observation time, the travel and fly-back of the moving mirror have not been optimised so 12 minutes were taken to produce the spectrum.

Further work is needed to improve the sensi-tivity of the instrument, but it has proved to be capable of analysing the required spectral range and to be easily transported.

## Acknowledgements

We would like to acknowledge the work of Mr R Belben in machining the mechanical components of the interferometer and to thank Dr BA Unvala and Dr SM Davidson for many useful discussions concerning the design of the system. We are also very grateful to Drs B Wakefield and S Davy for all their help and the use of their SEM at British Telecom Research Laboratories, Martlesham Heath.

References

Connes J 1961 Rev. Opt. 40 45 116 171 231
Cooley J W and Tukey J W 1965 Math. of Comput. 19 297
Davidson S M, Cumberbatch T J, Huang E and Myhajlenko S 1981 Microscopy of
    Semiconducting Materials 1981 eds A G Cullis and D C Joy (Bristol: Inst.
    Phys.) pp 191-196
Holt D B and Datta S 1980 Scanning Electron Microsc. 1980 I 259
Horlick G 1968 Appl. Spectrosc. 22 617
Horlick G and Yuen W K 1978 Appl. Spectrosc. 32 38
James J F and Sternberg R S 1969 The Design of Optical Spectrometers
    (London: Chapman and Hall) pp 188-189
Luc P and Gerstenkorn S 1978 Appl. Opt. 17 1327
Sakai H 1977 Spectrometric Techniques Vol. 1 ed G A Vanasse (New York:
    Academic Press) pp 1-70
Schopper H W and Thompson R J 1974 Methods of Experimental Physics Vol. 12
    Astrophysics:   Part A:   Optical and Infrared ed N Carlton (New York:
    Academic Press) pp 491-529

*Inst. Phys. Conf. Ser. No. 76: Section 9*
*Paper presented at Microsc. Semicond. Mater. Conf., Oxford, 25–27 March 1985*

# STEM applied to semiconductor materials and devices

L M  Brown

Cavendish Laboratory, Madingley Road, Cambridge.  CB3 0HE

Abstract   Characterisation of semiconductors and devices in a high-resolution scanning transmission electron microscope (STEM) offers very complete chemical and structural analysis of small precipitates, non-stoichiometric regions, defects and surfaces.  The technique is less effective at detecting trace elements in solid solution, and cannot follow the dopant levels through a p-n junction.

The STEM also provides images formed from electron-beam-induced-current (EBIC) and cathodoluminescence (CL) which can reveal electrical activity associated with defects.  The resolution in such images is controlled by carrier diffusion lengths which in the thin film are normally of order of the film thickness.  It seems that the very bright, finely-focussed probe produced by a field-emission gun is usually wasted in images formed with these techniques.  However in certain cases high-resolution images can be obtained, and useful information is available.  Information on the distribution of localised states can also be obtained at high spatial resolution from the electron energy-loss spectrum.

## 1.  Routine use of STEM

The art of applying STEM to characterise semiconductor materials and devices is very much a question of using both imaging techniques and the analytical outputs to provide complementary and overlapping information. The mixture of outputs chosen will depend somewhat on the type of microscope available.  A dedicated STEM, with its UHV environment and field-emission gun capable of producing useful probes of sub-nanometre size, is well-equipped to produce X-ray analysis (EDX) of heavy elements and analysis by EELS of light elements for particles in the nanometre size range.  The literature now contains many examples of chemical analysis of particles containing only a few thousand atoms: classic examples include copper particles in amorphous silicon (Craven et al. 1980) clusters of metallic precipitates in epitaxial and CZ-silicon (Fathy and Pennycook (1981), Dlamini, this conference).  Perhaps the most striking example of the success of this approach is the location of nitrogen in sub-monolayer quantities in platelets in diamond (Berger and Pennycook 1982).

Another interesting example is the detection of non-stoichiometry in regions of 'honeycomb' dislocation networks at the core of 'grappe' defects in InP (Augustus et al. 1983).  The interest here lies in the study of compositional changes which occur gradually over a distance of perhaps 100 nm; for such purposes the fine probe of the dedicated STEM is unnecessary and instruments with high total beam current (to improve the counting statistics) but low spatial resolution are superior.

In addition to the outputs providing chemical information, EDX and EELS, the microdiffraction patterns obtainable are very valuable. In microdiffraction, the probe is focussed to its smallest possible size, and a convergent electron beam pattern is recorded. In a dedicated STEM, the field-emission source permits the formation of microdiffraction patterns of sub-nanometre areas. However the microdiffraction facilities of such instruments have been rather rudimentary - although extremely useful - and microanalytical attachments to conventional microscopes, coupled with excellent microdiffration capabilities, have proved very powerful in the hands of the Bristol group (1984). The problem with UHV is that it is not compatible with direct photographic recording, so that it has been difficult to achieve high-quality microdiffraction patterns. Of course, the problem with electron sources of lower brightness than the field-emission source is the limited size of the particle from which useful information can be obtained; the particles must be larger than 10 or 20 nanometres; another problem with a microscope which does not have UHV is that light element analysis from very small probes becomes difficult because of carbon contamination. And so we see an interesting divergence in the type of analytical microscopy performed: those with conventional instruments concentrate on microdiffraction and EDX while those with dedicated STEM instruments concentrate on very small particles, mainly characterised by EELS and EDX.

Although STEM techniques are now several years old (our first instrument was delivered eleven years ago) technical development still continues. The problem of interpreting electron energy-loss spectra has been greatly simplified by the atlas of such spectra published by Ahn and Krivanek (1983) and the collection of convergent beam patterns produced by the Bristol group should prove similarly valuable. The design of electron spectrometers has improved greatly, partly as a result of the correction of aberrations and partly by the use of lenses to aid collection of the electrons. Finally, Rodenburg and McMullan (1984) have succeeded in producing directly-photographed high-resolution microdiffraction patterns in the dedicated STEM. This enables very wide angle (± 12°) diffraction patterns to be recorded in very short exposure times (from milliseconds to fractions of a second) from nanometre probes. One can look forward to much more use being made of microdiffraction in dedicated STEM.

It is important to recognise that although high-resolution STEM is capable of yielding chemical and structural information from very small particles, it is not very effective at detecting small amounts of one element homogeneously distributed in another. Optical emission spectrometry can detect about 5 ppm of carbon in iron, and similar amounts of other elements. Experiments using EELS (Liu and Brown 1981) suggest that with about $10^6$ counts per channel one might be able to detect about $10^4$ ppm or one atomic per cent. So far, however, even this has not been reliably achieved. The EDX output can achieve better results because of the low relative background; an example appropriate to this conference is phosphorous in silicon: according to Fathy and Brown (1981) the detection limit is about 2 ppm, not really high enough to follow the phosphorous profile <u>through</u> a p-n junction, but just up to the junction. At the detection limit the electron probe contains about 80 phosphorous atoms, so the fluctuation in the sample size is about $\sqrt{80}$, amounting to just under 10%. This shows clearly the nature of the problem: small-probe instruments with limited penetration can sample only small volumes, so there is a real physical limit to the fraction of minority element one can detect above the background. In principle, if one can afford an infinite acquisition time, one can always detect a single atom above the background of whatever other atoms there are in the probe. The detection limit on this basis will always be inversely proportional to the square-root of the number of counts per channel in the spectrum (Isaacson and Johnson 1975). Hence in these problems it will always pay to use the largest possible objective aperture consistent with the desired spatial resolution. In practice, there are many complications and although the magnitudes of the detection limits quoted above are likely to apply to many situations, one can expect considerable variation from one problem to another.

One may summarise by saying that the three outputs of EDX, EELS and microdiffraction provide overlapping and complementary information. Normally all three will be required to solve a given problem, but instruments with conventional guns are being used for larger particles and are less effective for EELS but more effective for microdiffraction, whereas the dedicated instruments with field-emission guns are capable of analysing nanometre particles, but generally have less efficient microdiffraction facilities. The dedicated instruments can certainly detect less than 100 atoms of many elements, but because of the very small probe they cannot detect small atomic fractions of trace elements: for this, classical methods based on bulk samples are superior.

## 2. Exploring Electronic Structure by STEM

In addition to routine chemical and structural characterisation, recent work has attempted to explore what can be learnt about electronic structure. The starting point here is the electron energy-loss spectrum which can provide the real and imaginary parts of the dielectric response function $\varepsilon(\omega)$ with an energy resolution limited ultimately by the energy spread in the gun, let us say about 0.4 eV for a field emission gun. The best modern spectrometers mounted in STEM instruments are on the verge of achieving this. Of course, this spectral resolution is not nearly so good as can be achieved by optical spectrometry; the only advantage of the STEM is the parallel characterisation of the material by electron microscopy and by the other STEM outputs. In fact, for many problems, this simultaneous characterisation is not a trivial advantage. Studies of fine structure in energy loss spectra require careful structural and chemical control if the spectra are to be interpreted: see, for example, the article by Colliex et al. (1984), in which the detailed appearance of the edges in the spectra depends upon the polytype of the substance, which in many cases (such as evaporated rare-earth films) is variable over a distance of 50 nm.

The spectral information can be obtained at far higher resolution if the cathodoluminescence (CL) produced by the fast electrons is studied. Experiments on the SEM have a long history; the review by Holt and Datta (1980) contains a useful introduction and a listing of literature. In an attempt to improve the spatial resolution, various groups have developed CL equipment for the STEM. The development of transmission CL and EBIC seems to have been pioneered by Yoffe and his group (1973). Petroff applied the techniques systematically to III-V compounds and devices, with spectacular results (1978, 1981).

At present, the technique is spreading, and work based on light collection using an ellipsoidal mirror is in progress at Bristol, Arizona and Bell

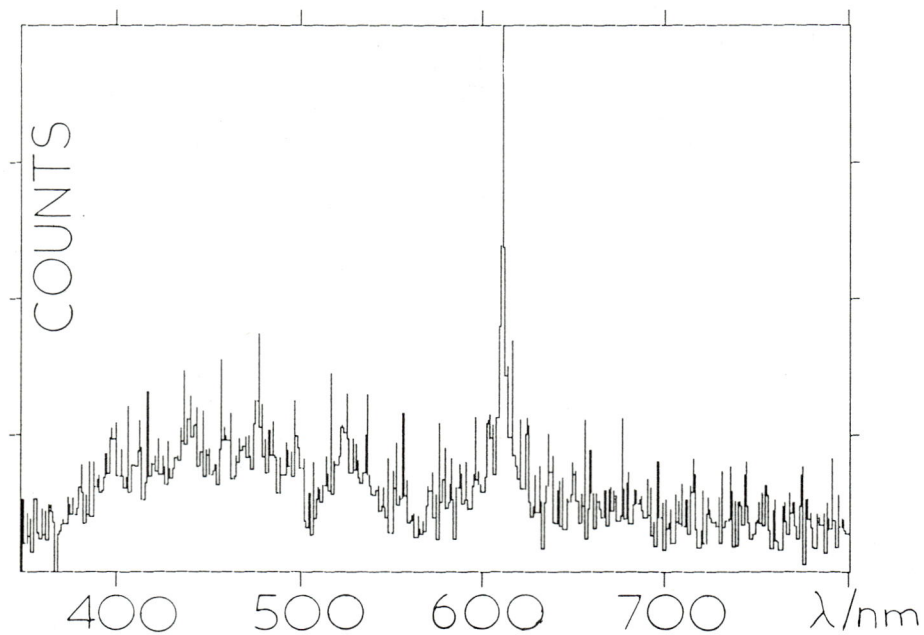

Fig. 1    CL spectrum from an electron-transparent thin film of Eu-doped $Y_2O_3$.    Vertical full scale = 1000.    Acquisition time 8 s.    Spectrum uncorrected for system response.    Courtesy of J. Yuan.

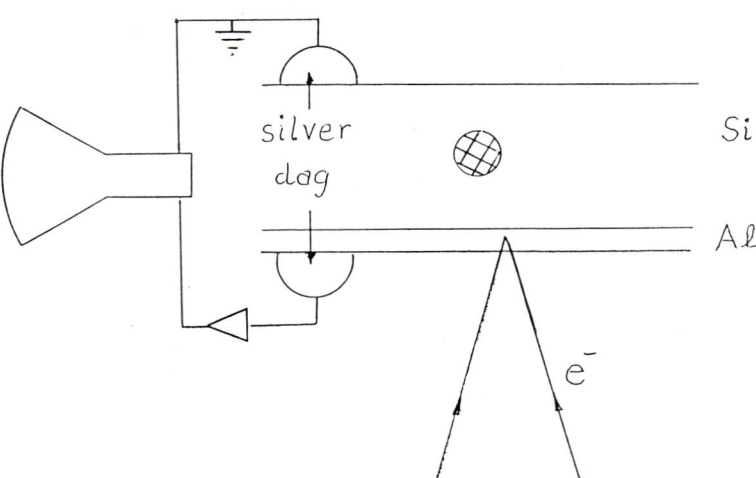

Fig. 2    A schematic diagram of Dlamini's experiment.    The amplifier is a Keithley 427 and the earth is through the microscope column.    The precipitate is about 4 nm in diameter in a silicon foil about 100 nm thick.

Laboratories.  In our group, the light is collected by a lens system. Fig. 1 shows a recent example from a thin film phosphor of europium-doped $Y_2O_3$ currently under study in our group.  The characteristic Eu red line is resolved with a half-width of about 1nm in 700 nm.  This type of study enables one to correlate the europium content with other structural features, in particular with the concentration of other elements determined by EDX spectra.  However, the spatial resolution obtainable with CL is not very good.  The reason for this is that the carriers produced by the fast electrons diffuse long distances before recombining radiatively.  Following Pennycook (1982) we may estimate the effective probe size by adding all the different diffusion lengths in quadrature: we will have the impact parameter associated with the major inelastic scattering mechanism, namely plasmon excitation, which is typically 10 nm; the plasmon decay distance, typically 5 nm; but most important in nearly all materials studied to date, the electron-hole diffusion distance which might be $10^3$ nm or more.  In thin films, the diffusion distance is dominated by the distance to the nearest surface, so thin-film CL images have a resolution comparable to the foil thickness.  This has been neatly demonstrated by Pennycook (1981), using images of dislocations in diamond where one can compare directly the images resulting from diffraction contrast with that resulting from the integrated CL signal.  Berger and Brown (1985) estimate the image width to be about $t/\sqrt{6}$, where t is the foil thickness.

Thus one arrives at the conclusion that CL images offer the possiblity of high spectral resolution but spatial resolution limited by the foil thickness, which in most cases means that the very small probe formed from a high-brightness gun is wasted.

A similar conclusion ought to hold for EBIC images.  If one forms an EBIC image from a thin film, by depositing a Schottky barrier on one surface, one should be able to assess the extent to which defects act as recombination centres, but the resolution of such images will be poor because of the long diffusion lengths of the carriers.  This can be seen in the images of Si produced by Fathy et al (1980): one can see EBIC images of dislocations nearly 1 μm in width in foils of thickness somewhat in excess of 1 μm.  Such foils are fully depleted of carriers, but the dislocation images are associated with a reduction of EBIC signal: the extra recombination at the dislocations reduces the collected current.  The images nevertheless provide unique information because they permit the electrical activity of the dislocations to be correlated directly with the high-resolution transmission images.

In the context of this discussion, it is extremely interesting that the precipitate images produced by Dlamini in his paper at this conference show very high resolution (about 1 nm in a foil at least 100 times thicker) and show that the precipitate generates extra EBIC signals, that is, acts as the source of extra carriers.  This fact has been carefully checked by comparing the signal of the precipitate with the signal in the neighbourhood of the contacts to the specimen.  The experimental set-up is shown in fig. 2., the experiment was performed in a VG HB5 operating at 80 kV.  Clearly there are circumstances in which carrier generation can produce high-resolution images.  The precipitates in this case are tin, of much higher atomic number than silicon, and it seems that the atomic number may play a crucial role.

An explanation for this behaviour may be given following the lines of Pennycook and Howie (1980).  These authors suggest that energetic core electrons resulting from the fast electron impact have short mean free paths and dump their energy into the valence band, therefore creating electron-hole pairs. Such processes will occur with greater probability, the higher the atomic number of the atom.

It is possible to give a simplified account of the calculation performed by Pennycook and Howie. The starting point is the fact that most inelastic core transitions result in Auger electrons. If we use the curves presented by MacDonald (1971), we see that the Auger yield exceeds 90% for all elements with $Z < 20$; that it exceeds 90% for all shells except the K shell for all elements with $20 < Z < 50$; and that it exceeds 90% for all shells except the K and L shell for all elements with $50 < Z < 90$. It follows that it is a very good approximation to regard the inelastic collision as producing Auger electrons with an energy not very different from the energy transfer; such electrons have very short mean-free-paths and will be stopped in a range of a few tens of nanometres. If one assumes that all the Auger electrons are stopped in the thin foil, one can estimate the production of carriers as follows: to a sufficient accuracy, the cross-section for ionisation of a shell containing $Z_i$ electrons bound with energy $E_i$ is proportional to $Z_i \ln(E_0/E_i) /E_0E_i$. The stopping power will be proportional to the same expression, multiplied by $E_i$. This leaves a residual logarithmic dependence on $E_i$, which is rather weak for a fast electron with an energy $E_0 \approx 100$ kV interacting with atomic shells ranging in energy from a few tens of volts to a few kilovolts. It follows that the stopping power for all the shells in the atom depends mainly upon Z, the atomic number. There will of course be energy dissipated in collective excitations, and this does not depend in a systematic way on the atomic number, but rather on the number of electrons in the uppermost band, filled in the case of a semiconductor. But this term again depends linearly on the number of such electrons, apart from a logarithmic factor.

Now what produces contrast in the EBIC signal is the difference between carrier production in the precipitate and carrier production in the matrix. If we say that carrier production depends to a good approximation linearly on Z, the contrast associated with a precipitate of one element with atomic number Z, and atomic volume $\Omega$, embedded in a second element will be equal to

$$C = \frac{Z_1/\Omega_1 - Z_2/\Omega_2}{Z_1/\Omega_1 + Z_2/\Omega_2} \qquad (1)$$

The rather elaborate argument given above can be stated more simply by saying that because of the Auger effect, by far the greatest part of the stopping power goes into electron-hole pair production, and it is well-known that the stopping power for fast electrons is to a very good approximation proportional to Z.

In the case of tin in silicon observed by Dlamini the contrast will be very high, about 0.6, but in the case of Ni and Cu precipitates, observed by Fathy and Pennycook (1981), the contrast will be somewhat lower, about 0.3. This is qualitatively in accord with observations. Dislocations and stacking faults will of course produce no contrast of this sort.

The spatial resolution associated with this contrast mechanism will be the average impact parameter, which in the spirit of the preceding discussion one might estimate to be

$$\bar{b} = \frac{hv}{Z} \left( \sum \frac{Z_i{}^2}{E_i{}^2} \right)^{\frac{1}{2}} \tag{2}$$

(v is the fast electron velocity).  In a typical case, the resolution is thus a few nanometres, as observed.

Thus it seems likely that the mechanism of valence excitations proposed by Pennycook and Howie (1980) can account for the EBIC contrast at precipitates in thin foils.  The question arises: to what extent can this mechanism account for the bright contrast occasionally seen in the SEM, and described as 'microplasma excitation'?  It is most interesting to examine the observations of Fathy et al (1980).  The microplasma sites in this example contain a high concentration of phosphorous, which according to equation (1) should produce a small positive contrast.  The sites show a dark region at the centre, which is most easily explained by recombination of carriers generated at the centre of the site: one expects more carriers to be collected when the Auger electrons are produced near the outside of the defect.  It seems difficult to be sure to what extent the proposed mechanism contributes to this type of contrast.

Acknowledgements

The author is grateful to all members of the Microstructural Physics Group for discussions, particularly Dr. A. Howie FRS, Dr. S.D. Berger and Mr. J. Yuan.

References

Ahn C C and Krivanek O L 1983 EELS Atlas published by Gatan Inc., Warrendale PA 15086
Augustus P D, Stirland D J and Yates M 1983 J. Crystal Growth 64 pp 121-128
Berger S D and Brown L M 1985 to be published
Berger S D and Pennycook S J 1982 Nature pp 635-637
The Bristol Group, (J W Steeds, Director) 1984 Convergent Beam Electron Diffraction of Alloy Phases, Adam Hilger Ltd., Bristol 1984
Colliex C, Manoubi T, Gasgnier M and Brown L M 1984 SEM Conference Philadelphia (SEM Inc.: AMF O'Hare)
Craven A J, Lynch P J, Brown L M and Mistry A B 1980 Inst. Phys. Conf. Series No. 52 (EMAG 79) pp 343-344
Fathy D and Brown L M 1981 Inst. Phys. Conf. Ser. 61 Ch. 11 pp 509-513
Fathy D and Pennycook S J 1981 Inst. Phys. Conf. Series No. 60 (Micros. Semicond. Mater. Conf. 1981) pp 243-248
Fathy D, Sparrow T G and Valdrè U 1980 J. Microscopy 118 264
Holt D B and Datta S 1980 SEM 1980 (SEM Inc.: AMF O'Hare) pp 259-278
Isaacson M, Johnson D 1975 Ultramicroscopy 1 p 333
Liu D R and Brown L M 1981 Inst. Phys. Conf. Ser. 61 Ch 4 pp 201-204
MacDonald N C 1971 SEM 1971 IITRI Chicago
Pennycook S J 1981 Ultramicroscopy 7 pp 99- 104
Pennycook S J and Howie A 1980 Phil. Mag. A 41 pp 809-827
Petroff P M 1981 Inst. Phys. Conf. Ser. 61 (EMAG 81) pp 501-508
Petroff P M, Lang D V, Strudel J L and Logan R A 1978 SEM 1978 (SEM Inc.: AMF O'Hare) pp 325-332
Yoffe A D, Howlett K J and Williams P M 1973 SEM 1973 IITRI Chicago

*Inst. Phys. Conf. Ser. No. 76: Section 9*
*Paper presented at Microsc. Semicond. Mater. Conf., Oxford, 25–27 March 1985*

405

# Crystalline effects in the analysis of semiconductor materials using Auger electrons or X-rays

J F Bullock, C J Humphreys, A J W Mace[2], H E Bishop[1] and J M Titchmarsh[1]

Department of Metallurgy and Science of Materials,
Oxford University, Parks Road, Oxford OX1 3PH.

[1]AERE Harwell, Didcot, Oxon OX11 ORA;

[2]Now at GEC Hirst Research Centre, East Lane,
Wembley, Middlesex HA9 7PP.

Abstract    Calculations of the effect of crystal orientation on X-ray and Auger electron intensities have been made using the many beam dynamical theory of electron diffraction.  In particular the effects of incident beam energy and specimen thickness are considered.  A model of the back scattered electron contribution to the X-ray or Auger intensity is used and the calculations give reasonably good agreement with published experimental data.

## 1. Introduction

Auger electron spectroscopy (AES) and X-ray spectroscopy (XRS) are widely used techniques in the quantitative microanalysis of semiconductors and other materials.  In the interpretation of these spectra the influence of the orientation of the crystal, relative to the incident electron beam, on the production of Auger electrons and characteristic X-rays is usually ignored, although it has been known for some time that a three-fold variation in Auger and X-ray intensities is possible as the crystal is tilted through a zone axis.  The purpose of this paper is to give a more quantitative theory than has been given previously of the effect of electron channelling upon Auger electron and X-ray production, and to compare the results with experiment.

The effect of diffraction of the incident electron beam upon X-ray production was first pointed out by Hirsch et al. (1962), and a more recent treatment has been given by Cherns et al. (1973).  For the case of Auger electron production, the signal received by the detector depends not only upon the diffraction of the incident beam but also on the diffraction of the Auger electrons themselves.  In practice, however, if a detector with a large enough solid angle of detection is used, this averages out the effects of diffraction of the Auger electrons.

The theories of the orientation dependence of Auger electron and X-ray emission are very similar since both result from the orientation dependence of the original ionisation of atoms by the incident electron beam.  We must also take into account the ionisation of atoms by electrons which have been inelastically scattered, including backscattered electrons, and the absorption of the Auger electrons and

X-rays within the crystal.    Auger electrons are, of course, much more strongly absorbed than characteristic X-rays, so that only those Auger electrons produced close to crystal surfaces can escape.

## 2. Calculation of Auger and X-ray Production

If the wavefunction of an incident fast electron in a crystal is $\psi(\underline{r})$, then the probability of Auger electron production by the incident beam in a volume $d\tau$ is proportional to $|\psi(\underline{r})|^2 d\tau$.    The wavefunction can be considered as a linear combination of Bloch waves j, so

$$\psi(\underline{r}) = \sum_{jg} \alpha^{(j)} C_g^{(j)} \exp(2\pi i(\underline{k}^{(j)}+\underline{g}).\underline{r}) \ \exp(-2\pi q^{(j)} z) \tag{1}$$

Here $\underline{g}$ is a reciprocal lattice vector, $\underline{k}^{(j)}$ is the wavevector and $q^{(j)}$ represents the attenuation of the jth Bloch wave.

Ionisation from core states occurs close to the atom sites, so $A(\underline{r})$, the interaction potential for Auger production is, to a good approximation, a set of delta functions, localised at the atomic sites, and broadened by thermal vibration through the Debye-Waller factor B.    It is convenient to write $A(\underline{r})$ as a Fourier series based on the reciprocal lattice (after Cherns et al. (1973)), so

$$A(\underline{r}) = \sum_h A_h \ \exp(2\pi i \ \underline{h}.\underline{r}) = A_0 \sum_h \exp(-Bh^2) \ \exp(2\pi i \underline{h}.\underline{r}) \tag{2}$$

The probability of Auger electron production is proportional to $A(\underline{r})$. If $\lambda$ is the mean escape depth for the Auger electron considered (calculated following Seah and Dench (1979)), then the intensity of Auger electrons produced at a depth z in the crystal will be attenuated by $\exp(-z/\lambda)$ before reaching the surface.    The intensity of Auger electrons leaving the surface is then

$$I^{(A)} = \int |\psi(\underline{r})|^2 \ \exp(-z/\lambda) \ A(\underline{r}) \ d\tau \tag{3}$$

In unit area of crystal the x and y components of this integral yield a constant.    If the crystal thickness is t, and we consider reciprocal lattice vectors only in the (x,y) plane, after application of boundary conditions we obtain

$$I^{(A)} = \sum_{jl} C_0^{(j)} C_0^{(1)*} \sum_{gh} C_g^{(j)} A_h C_{g-h}^{(1)*} \int_0^t \exp(2\pi i(k_z^{(j)}-k_z^{(1)})z) \\ \exp(-(2\pi q^{(j)}+2\pi q^{(1)}+\lambda^{-1})z)dz \tag{4}$$

To simplify, $\lambda$ is usually much smaller than t, so we can take the integral to $\infty$.    Also, $q^{(j)}$ and $q^{(1)}$ are small compared with $\lambda^{-1}$. For a centrosymmetric crystal the Fourier coefficients $C_g^{(j)}$ are real, and $k^{(j)}-k^{(1)} = \gamma^{(j)}-\gamma^{(1)}$, where the $\gamma^{(j)}$ are eigenvalues of the matrix of Fourier coefficients.    This yields

$$I^{(A)} = \sum_{jl} \frac{C_0^{(j)} C_0^{(1)} \lambda A_0 \sum_{gh} C_g^{(j)} C_{g-h}^{(1)} \exp(-Bh^2)}{1+4\pi^2\lambda^2(\gamma^{(j)}-\gamma^{(1)})^2} \tag{5}$$

Two beam theory calculations show that the contrast c, obtained by rocking the crystal through the Bragg reflection g is given by

$$c = 2\pi(\lambda/\xi_g)/\sqrt{(1+4\pi^2(\lambda/\xi_g)^2)} \tag{6}$$

where c is defined as

$$c = (I_{max}-I_{min})/\tfrac{1}{2}(I_{max}+I_{min}) \tag{7}$$

and $\xi_g$ is the two beam extinction distance of the reflection g. This two beam expression agrees with that of Bishop et al. (1984) who used a slightly different formulation of $\psi(z)$. Equation 5 gives the contribution of the incoming primary beam to Auger production, however inelastically scattered, in particular backscattered, electrons will also contribute (Anderson and Howie (1975)).

In their calculations on characteristic X-ray production Cherns et al. (1973) estimated the backscattered contribution as

$$I^{(X)}_{BS} = P_0 \int_0^t 1-\sum_j |C_0^{(j)}|^2 \exp(-4\pi q^{(j)}z)dz \tag{8}$$

where $P_0$ is the interaction potential for X-ray production. Applying equation 8 to the Auger case, and taking into account that electrons produced at a depth z are attenuated, gives

$$I^{(A)}_{BS} = A_0 \int_0^t (1-\sum_j |C_0^{(j)}|^2 \exp(-4\pi q^{(j)}z)) \exp(-z/\lambda)dz \tag{9}$$

Spencer et al. (1972) and Spencer and Humphreys (1979) used a somewhat more detailed inelastic scattering theory and derived the following expression for the backscattered electron intensity leaving the crystal surface;

$$I^{(IN)} = \frac{1}{1+a_1p^{(0)}t} [a_1p^{(0)}t + a_2\sum_j |C_0^{(j)}|^2 \frac{p^{(j)}-p^{(0)}}{\mu^{(j)}}(1-\exp(-\mu^{(j)}t))] \tag{10}$$

where $p^{(j)}$ is the backscattering coefficient of Bloch wave j, and $p^{(0)}$ is the average backscattering coefficient. $a_1$ and $a_2$ are empirical parameters (typically 2·0 and 6·0 respectively) and t for bulk specimens is taken as $0·4R_b$, where $R_b$ is the Bethe range, or typically 1000nm.

$I^{(IN)}$ can be considered to be approximately constant in the thin surface layer where any Auger electron produced will escape to the detector, so

$$I^{(A)}_{BS} = A_0 \int_0^t I^{(IN)} \exp(-z/\lambda)dz \tag{11}$$

The total calculated Auger current is

$$I^{(A)}_T = I^{(A)} + I^{(A)}_{BS} \tag{12}$$

assuming that backscattered electrons are as efficient as the primary beam in producing Auger electrons. In fact the lower energy backscattered electrons will be considerably more efficient.

The use of characteristic X-ray production differs somewhat from that of Auger production. The crystal does not appreciably attenuate the signal as the X-ray absorption distance is so large. Following Cherns et al. (1973) the X-ray production by the incoming beam is given by

$$I^{(X)} = \sum_{j1} C_0^{(j)} C_0^{(1)} P^{j1} \int_0^t \exp(2\pi i(k_z^{(j)} - k_z^{(1)})z) \exp(-2\pi(q^{(j)} + q^{(1)})z) dz \quad (13)$$

where $P^{j1} = \sum_{gh} C_g^{(j)} P_h C_{g-h}^{(1)}$

In addition, X-rays will be produced by inelastically scattered electrons. The intensity, using Spencer et al. (1972) is

$$I^{(X)}_{IN} = P_0 \sum_j \int_0^t I_F^{(j)}(z) + I_B^{(j)}(z) \ dz \quad (14)$$

$I_F^{(j)}(z)$ and $I_B^{(j)}(z)$ are the intensities of inelastically scattered electrons at a depth z, in a forward and backward direction respectively. The total X-ray intensity is then given by

$$I^{(X)}_T = I^{(X)} + I^{(X)}_{IN} \quad (15)$$

## 3. Calculation of the Crystallographic Orientation Effect

A computer program was written to calculate effects of crystal orientation on X-ray or Auger electron intensity. With the beam incident along a zone axis, values of the Block wave excitation amplitudes $C_g^{(j)}$ and, from the expressions given above, the intensity values were calculated. The many beam calculation was then repeated for a series of orientations, passing through the first order Bragg reflection position at $k_g/g$ = 0.5, the second order position at $k_g/g$ = 1.0, etc. The values of the backscattering contribution were calculated and added, total intensity was plotted as a function of orientation, and a contrast figure given by equation 7 was determined.

The effect on the calculated Auger electron intensity of the number of beams used in the calculation is shown in figure 1. This is for the case of KLL Auger electrons in bulk Al ($\lambda$ = 1.97nm) with an incident beam energy of 10keV. $k_g/g$ = 0 corresponds to the beam incident along the (110) zone axis. The beam is then tilted through the (1$\bar{1}$1) Bragg position at $k_g/g$ = 0.5). Figure 2 uses the formulation of the backscattered contribution derived from Cherns et al. (1973), figure 2 uses the form of Spencer and Humphreys (1979) under otherwise identical conditions. Two features are readily apparent. Increasing the number of beams used increases the maximum intensity and the contrast, since the Bloch waves are localised more on the atomic planes. Secondly, figure 2 shows a larger backscattering contribution, by a factor of about two, which reduces the overall contrast.

Figure 3 shows the small effect on the Auger signal of changing the specimen thickness. This can be contrasted with the large effect on the X-ray signal of a thickness change. Primary contrast increases with decreasing beam energy (figure 4), although the backscattered component largely nullifies this effect.

Different Auger lines will show different contrast values (figure 5) as the value of $\lambda$ is energy dependent.     Here we should only compare contrast values, not absolute intensities as $A_0$ for the LMM line is different from that of the KLL line.

Similar calculations for X-ray production in Si for a number of different thicknesses show the expected large variation in intensity (figure 6).

## 4. Quantitative Explanation of the Orientation Effect

Measured values of Auger electron contrast from various workers are compared with our theoretical calculations in Table 1. Some of the experimental results in this table were obtained by scanning polycrystalline samples of random orientation.    These results will underestimate the maximum contrast since the beam is unlikely to be incident exactly down a zone axis.

It is clear from Table 1  that this theoretical many beam treatment tends to overestimate the contrast in the Auger signal;   this could be due to several reasons.    These calculations take no account of the shape and divergence of the incident beam, they assume the beam to be parallel.   In practice, beam divergence will decrease the calculated contrast, and calculations are in progress to determine this effect.   The contribution of inelastically scattered electrons has been added to that of the incident beam in the form $I_T = I + aI_{BS}$ with a = 1.   However Andersen and Howie (1975) suggest that a could be as much as 3.   This weighting would decrease the first value calculated in Table 1 from 1.14 to 0.94.

The calculation of the channelling effect is important in quantitative microanalysis by EDX and AES.    In these techniques the ratio of X-ray or Auger signals from different elements is important.    It has been shown that different elements will show different contrast effects, thus the ratio of signals, too, will have a crystalline orientation effect.   X-ray microanalysis with very high spatial resolution requires the use of very thin films, and here the orientation effect is expected to be very high, and must be taken into account.    The channelling effect has been exploited in the ALCHEMI technique of Spence and Taftø (1983) to determine atomic site location, and we propose to extend our calculations to a quantitative evaluation of ALCHEMI, using both X-rays and Auger electrons.

## Table 1

| Material | Line | kV | $\lambda$(nm) | c measured | c calculated |
|---|---|---|---|---|---|
| Al single crystal[1] | KLL | 10 | 1.97 | 1.0 | 1.14 |
| | LMM | 10 | 0.46 | 0.1 | 0.39 |
| Cu polycrystalline[1] | LMM | 10 | 1.35 | 0.34 | 0.61 |
| | LMM | 4 | 1.35 | 0.4 | 0.62 |
| | MVV | 10 | 0.38 | 0.1 | 0.29 |
| | MVV | 4 | 0.38 | 0.1 | 0.39 |
| Cu single crystal[2] | LMM | 1.5 | 1.35 | 1.0 | 0.65 |
| | MVV | 1.5 | 0.38 | 0.3 | 0.49 |
| Cu single crystal[3] | LMM | 1.5 | 1.35 | 0.52 | 0.65 |
| | MVV | 1.5 | 0.38 | 0.22 | 0.49 |

[1]Bishop et al. (1984); [2]Armitage et al. (1980); [3]Carr (1979).

6. References

Andersen S K and Howie A 1975 Surf. Sci. 50 197
Armitage A F, Woodruff D P and Johnson P D 1980 Surf. Sci. 100 L483
Bishop H E, Chornik B, Le Gressus C and Le Moel A 1984 Surf. Interface
    Anal. 6 116
Carr P 1979 D.Phil Thesis, University of Oxford
Cherns D, Howie A and Jacobs M H 1973 Z.Naturforsch. 28a 565
Hirsch P B, Howie A and Whelan M J 1962 Phil. Mag. 7 2095
Seah M P and Dench W A 1979 Surf. Interface Anal. 1 2
Spence J C H and Taftø J 1983 J.Microsc. 130 147
Spencer J P, Humphreys C J and Hirsch P B 1972 Phil. Mag. 26 193
Spencer J P and Humphreys C J 1979 Phil. Mag. A, 42 433

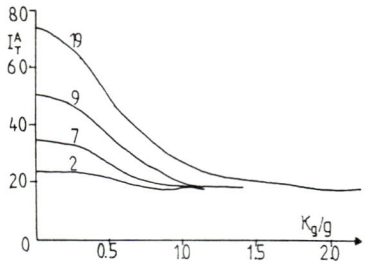

Fig.1 Calculated Auger intensity variations using 2,7,9 and 19 beams.

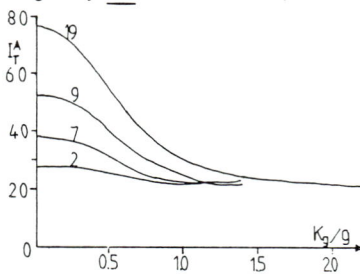

Fig.2 As Fig.1, but using alternative backscattering term.

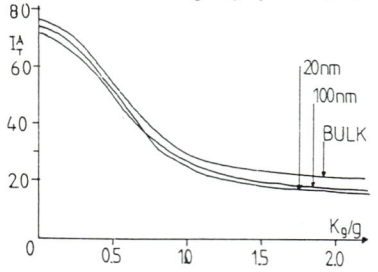

Fig.3 Calculated variation of Al KLL with t. Calculated as Fig.2 (19 beams).

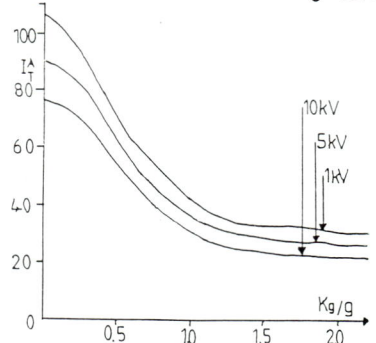

Fig.4 As Fig.3, but for three values of kV (bulk thickness).

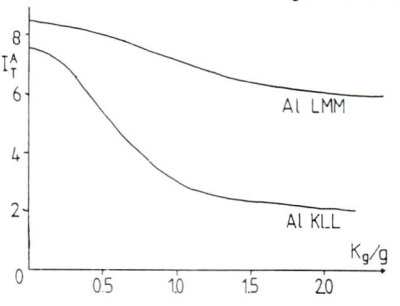

Fig.5 Calculated contrast variation in two Al Auger lines.

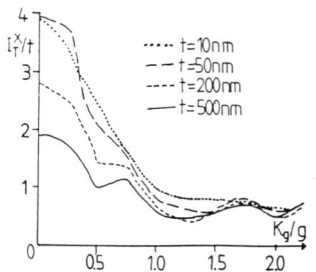

Fig.6 Calculation of Si X-ray intensity (g = 220, 100kV).

*Inst. Phys. Conf. Ser. No. 76: Section 9*
*Paper presented at Microsc. Semicond. Mater. Conf., Oxford, 25–27 March 1985*

# Z-contrast imaging and electron channeling analysis of dopants in semiconductors

S J Pennycook, R J Culbertson and E Fogarassy[+]

Solid State Div., Oak Ridge National Laboratory, Oak Ridge, TN 37831 USA
[+]Centre Nationale de la Recherche Scientifique, BP-20, Strasbourg, France

Abstract    The determination of dopant distribution and lattice location
are key elements in the microanalysis of semiconductor materials.  Here
we describe two techniques for this purpose.  The first allows the
imaging and elemental mapping of heavy dopants in light semiconductors,
and the second is a means for determining the substitutional fraction
of dopants or impurities in any crystal structure.

## 1.  Introduction

Transmission electron microscopy is a powerful technique for characterizing
the structure and defect distribution in semiconductor materials, but
dopants only contribute to the image when they form a separate phase, by
which time they are electrically inactive.  The important parameters of
dopant concentration and lattice location have usually been provided by ion
channeling analysis, which although it has depth resolution can give no
lateral resolution.  Here we describe two microanalysis techniques which
provide this information on the microscopic scale.  We show how the distribution of heavy dopants in light semiconductors can be quantitatively
imaged using scanning transmission electron microscopy (STEM) by detecting
Rutherford-scattered transmitted electrons, and how planar and axial
electron channeling can determine substitutional fractions of dopants.

## 2.  Z-Contrast Imaging of Dopant Distributions

With STEM it is possible to form images with Rutherford-scattered transmitted electrons using a high-angle annular detector.  Such images show
strong Z-contrast combined with minimum contrast due to structural features, and have been used to image heavy catalyst particles on light
amorphous or polycrystalline supports (Treacy et al. 1980, Pennycook et al.
1983).  The technique was generally very successful, although with polycrystalline supports, contrast due to channeling remained in crystallites
oriented close to a Bragg condition.  With a single crystal sample it can
be oriented in a "random" direction, far from Bragg reflections, so that
structural features are almost invisible, and the remaining contrast then
reflects the dopant concentration (Pennycook and Narayan, 1984).  No ratio
is used (Crewe et al. 1975), so that the image gives the maximum possible
contrast, approaching $Z^2$ for a sufficiently high inner collection angle,
and the samples are not restricted to a thickness suitable for energy loss
spectroscopy.  Thickness variations will then contribute to the image, but
are generally small with a single crystal sample.

*Research sponsored by the Division of Materials Sciences, USDOE under
contract DE-AC05-840R21400 with Martin Marietta Energy Systems, Inc.

Figure 1 shows cross-section images of $^{121}Sb^+$ (80 keV, 1.35 x $10^{16}$ cm$^{-2}$) implanted Si {100} following solid phase epitaxial (SPE) growth at 550°C/ 40 mins. Good regrowth occurred to within 30 nm of the surface, the remaining material being polycrystalline with some twins. The structural detail is visible in the TEM image, Fig. 1(a), and in the STEM image using diffracted electrons, Fig. 1(b). It is not seen in the STEM image using Rutherford-scattered electrons, Fig. 1(c), where the contrast is due to the Gaussian distribution of implanted Sb. A small segregation effect is seen at the point where epitaxial growth broke down. The image was taken using black level to enhance the contrast. A line scan across the image is shown in Fig. 2, from which the ratio of the Sb signal to Si signal is measured 0.39. The expected image contrast for collection angles of 65—150 mrad using cross-sections for elastic scattering given by Treacy (1982) is 0.34, in good agreement with experiment. For dopant concentrations which do not significantly increase the multiple scattering out of the annular detector, the dopant image contrast should be linearly proportional to concentration. The minimum detectable concentration is estimated to be 0.5 atomic % of Sb. Lower concentrations of heavier elements can be imaged (0.2 atomic % Bi) but higher concentrations of lighter elements are required. The limit depends on dopant distribution and the thickness uniformity of the sample. The dopant image provides a high level signal which can be used quantitatively for elemental mapping of low concentrations of impurities in thin specimens, where the x-ray signal is orders of magnitude too weak. A few x-ray spectra from representative points are required however to calibrate the image.

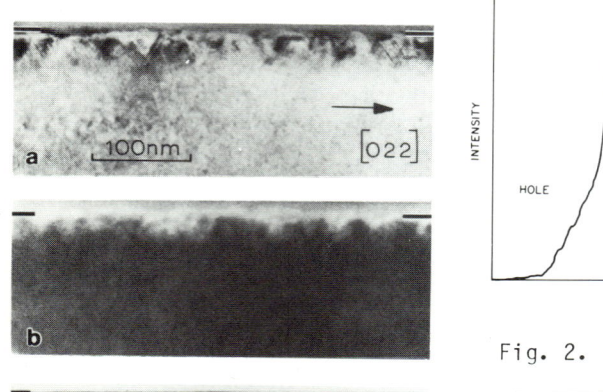

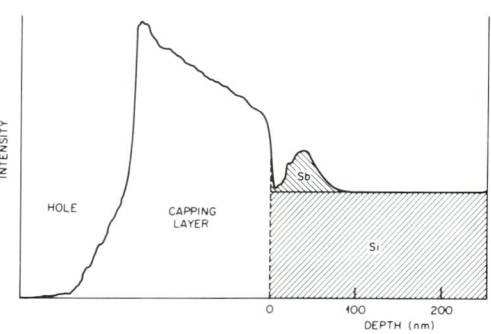

Fig. 2.   Line trace across Fig. 1(c).

Fig. 1.   Cross-section micrographs of Sb$^+$ implanted Si after SPE growth: (a) TEM, (b) STEM diffraction contrast, (c) STEM dopant image using Rutherford-scattered electrons.

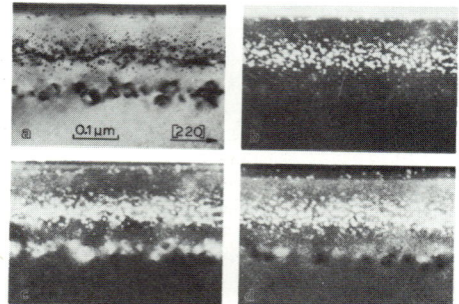

Fig. 3.   Dopant imaging under channeling conditions.

If images from Rutherford scattered electrons are taken under channeling conditions, then contrast returns at structural features. Figure 3 shows an example of Si {100} implanted with $^{209}Bi^+$ (250 keV, 5 x 10$^{15}$ cm$^{-2}$) recrystallized at 575°C/40 mins. and annealed at 900°C/1 hr. The Bi has precipitated out and a band of dislocation loops marks the original amorphous/crystalline interface. These loops are practically invisible in the dopant image taken in a random orientation Fig. 4(b), but show contrast when the incident beam is outside (c) or inside (d) the {220} Kikuchi line. This is not diffraction contrast, which would be similar in these two conditions but is channeling contrast. If the defects are thought of as disordered regions of material, then they will scatter at the random level regardless of whether the scattering from the crystal is reduced (c) or enhanced (d) by channeling. They will therefore appear respectively bright or dark. In defect-free regions of crystal, it should be possible to use dopant images taken under channeling conditions for lattice location information, forming in effect a lattice location map. Preliminary investigations indicate, however, that the change in multiple scattering between two different incident beam directions can affect the analysis.

## Dopant Atom Location by Electron Channeling Analysis

The orientation dependence of characteristic x-ray emission close to a Bragg reflection has long been regarded as a hindrance to accurate microanalysis and it has only recently been realized that it can form the basis of a powerful lattice location technique. The first studies of this kind located trace elements in layer structure minerals, where it was known that the only possible impurity sites were within the layer planes A and/or B (Tafto and Spence 1982, Spence and Tafto 1983). Here we describe a generalization to any crystal structure based on the analysis and definitions used for ion channeling analysis, a lattice location technique which has been widely applied for many years.

The basis for electron channeling analysis remains the standing wave electron intensity profiles set up near Bragg reflecting conditions. The maxima can be located predominantly on or between the reflecting planes by a beam incident respectively inside or just outside the exact Bragg condition. This results in a change in the yield from any close encounter process such as thermal diffuse scattering or inner shell excitation. If, when switching between channeling conditions, the characteristic x-ray emission of an impurity or dopant follows that of the matrix, then it necessarily lies within the matrix planes. If there is no variation in the dopant emission then it is randomly located in the lattice (Pennycook et al. 1984a). Channeling along a number of different planes is required to prove substitutionality in general. Figure 4(a) shows the planar electron channeling analysis of Si {100} implanted with $^{75}As^+$ (100 keV, 2 x 10$^{16}$ cm$^{-2}$) and recrystallized by SPE growth. The analysis was performed using a 50 nm diameter beam located on the peak of the As profile in a cross-section sample, switching the electrical beam tilts between channeling conditions. It is clear that the As lines follow the Si lines, at least qualitatively, as expected since ion channeling analysis of the bulk material showed the As to be highly substitutional.

Greatly enhanced yield variations are observed near a zone axis, since standing waves can be set up in several directions simultaneously, channeling the electron current into columns which can be located on or between the atomic rows in the crystal. Figure 4(b) shows spectra obtained for <100> axial electron channeling of the same sample region. The increased

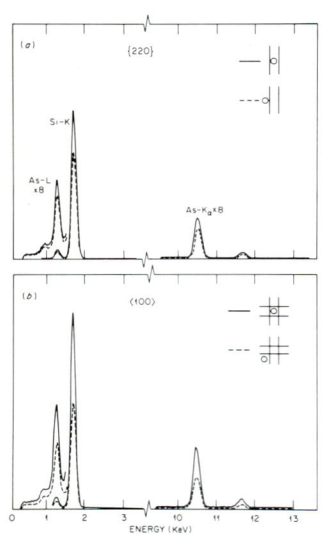

Table 1

Values of $b^{RMS}$ and C in Si.

| | $b^{RMS}$ (nm) | {220} | <100> | <111> |
|---|---|---|---|---|
| Sb-L$_\alpha$ | 0.012 | 1.10±0.04 | 1.04±0.03 | 1.05±0.02 |
| As-K$_\alpha$ | 0.005 | 1.12±0.05 | 1.10±0.04 | |
| As-L | 0.034 | 0.64±0.04 | 0.71±0.04 | |
| Si-K | 0.025 | | | |

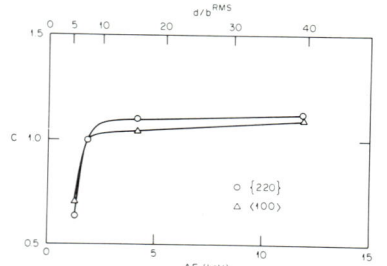

Fig. 4. X-ray spectra for planar and axial electron channeling analysis of As in Si. Inserts show beam tilt conditions.

Fig. 5. Delocalization correction factors as a function of inner shell binding energy ΔE.

yield variation is a great advantage in practice, allowing smaller quantities of dopant to be analyzed since the accuracy of the analysis depends on the measurement of the difference between two spectra. An even larger yield variation is observed with <111> axial channeling (Pennycook and Narayan 1985).

It should be possible to quantify the analyses and define a substitutional fraction $F_s$ from the x-ray yield ratios ΔX, in a manner analagous to the analysis used for ion channeling, as

$$F_s = \frac{(\Delta X_{As}-1)}{(\Delta X_{Si}-1)} \qquad (1)$$

However, there are some important provisos. Firstly, since the variation of the matrix peak is used as a measure of the channeling effect, and the electron intensity profiles change with depth, it is important that the dopant is distributed uniformly through the sample thickness. This is easily achieved with cross-section samples, but can introduce errors of the order of ±10% in plan-view samples of Gaussian implants (Pennycook et al., 1984b). Secondly, for low energy lines, and/or small interplanar spacings, the effects of delocalization may be significant. Inner shell electrons can be excited by fast electrons passing within a distance known as the impact parameter b. An average impact parameter for x-ray emission is given by Pennycook (1982) as

$$b^{RMS} = \frac{\hbar v}{\Delta E}\left[\ln\left(\frac{4E}{\Delta E}\right)\right]^{-1/2} \qquad (2)$$

where v is the fast electron velocity, E is the fast electron energy, and ΔE is the inner shell binding energy. The impact parameter increases for lower energy excitations, giving smaller variations in x-ray yield with channeling conditions. In analyses such as Fig. 4., the variation of the As-$K_\alpha$ line systematically exceeded that of the Si-K line, which in turn was always greater than the As-L variation. Provided a reasonable channeling effect is still observed, this effect can be corrected for by defining

$$F_s = \frac{1}{C} \frac{(\Delta X_{As}-1)}{(\Delta X_{Si}-1)} \qquad (3)$$

where C is a delocalization correction factor. Table 1 shows values of $b^{RMS}$ and C determined for various channeling conditions in Si, and Fig. 5 shows these as a function of ΔE. The values are reduced for axial channeling but become very large for lines approaching 1 keV, where $b^{RMS}$ exceeds d/5. A reduced scale of $d/b^{RMS}$ is shown which should allow the correction factors for other elements and interplanar spacings to be obtained by taking the ratio of the C values with respect to Si for the impurity and the matrix.

As with ion channeling analysis, the quantitative interpretation of fractional values of $F_s$ should be undertaken with care. Firstly, the yield ratios ΔX should be for a channeled beam to a random beam direction, and secondly, it must be shown that the nonsubstitutional dopant is randomly distributed with respect to the matrix, such as in the form of randomly oriented precipitates. This is easily checked by standard TEM techniques, and a quantitative analysis of coherent Sb precipitates has shown the relative location of Sb planes with respect to the Si matrix planes (Pennycook et al., 1984a).

## Application to Studies of Pulsed Laser Annealing

The dopant imaging technique has been applied to Si-Sb alloys produced by a variety of excimer laser annealing treatments (Fogarassy et al., 1985). TEM images show cell formation after single shot laser annealing of implanted (Fig. 6a) or deposited (Fig. 7a) Sb, deeper cells occurring in the implanted material. The dopant images (b) reveal that the distribution

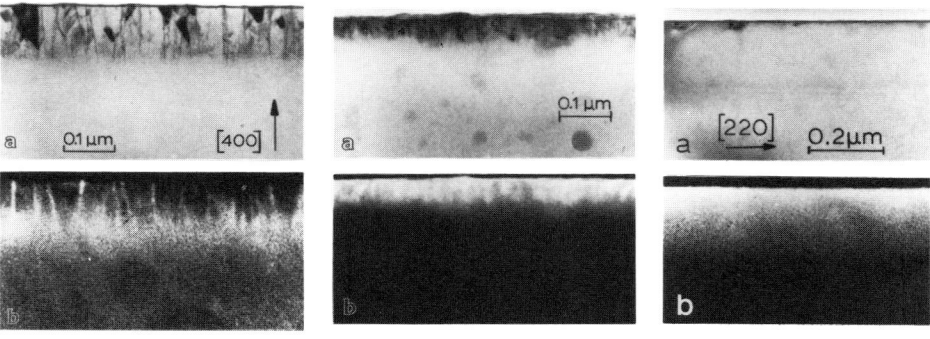

Fig. 6. Sb (150 keV, $4\times10^{16}$ cm$^{-2}$) implanted Si {100}, one shot 1.0 J cm$^{-2}$.

Fig. 7. 8-nm Sb deposit on Si {100}, one shot 1.0 J cm$^{-2}$.

Fig. 8. 8-nm Sb deposit on Si {100}, 20 shots 1.0 J cm$^{-2}$.

of dopant around and between the cells is very different, in particular that the Sb concentration near the surface between the cells is low in the implanted sample, but high in the deposited sample. After 20 shots, good regrowth occurred (Fig. 8a) and the deposited Sb layer was distributed more uniformly through the melted layer (Fig. 8b).

With Si-Sb alloys, the average dopant profile and solubility limit under the above conditions could be measured by ion channeling analysis. Laser annealing of Al implanted Si with a single shot also results in a cell structure, althought precipitates form at the base of the cells (Fig. 9). The profile could not be obtained by ion channeling analysis, and as measured by STEM x-ray analysis using line scans parallel to the sample surface to average over several cells (Fig. 10). Much of the Al has been swept to the surface during recrystallization, but a peak remains corresponding to the precipitates. <100> electron channeling analysis with 50 nm depth resolution was used to determine the substitutional profile. The solubility limit was found to be $2.25 \pm 0.75 \times 10^{20}$ cm$^{-3}$.

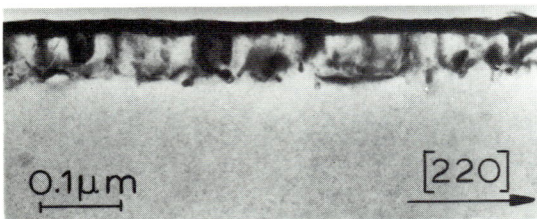

Fig. 9. Cell Structure in laser annealed Si-Al alloy.

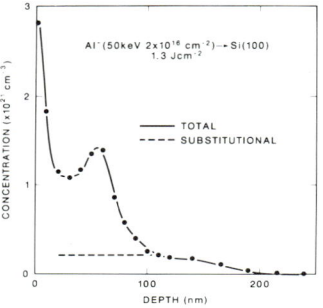

Fig. 10. Al profile and electron channeling analysis.

# References

Crewe A V, Langmore J P and Isaacson M S 1975 Physical Aspects of Electron Microscopy and Microbeam Analysis, eds Siegel B M and Beaman D R (New York: Wiley) p 45

Fogarassy E 1985 Materials Research Society Proceedings Symp. A (to be published)

Pennycook S J 1982 Contemp. Phys. 23 371

Pennycook S J, Howie A, Shannon M D and Whyman R 1983 J. Mol. Catalysis 20 345

Pennycook S J, Narayan J and Holland O W 1984a Appl. Phys. Lett. 44 547

Pennycook S J, Narayan J and Holland O W 1984b Electron Microscopy of Materials (New York: North Holland) pp 97-104

Pennycook S J and Narayan J 1984 Appl. Phys. Lett. 45 385

Pennycook S J and Narayan J 1985 Phys. Rev. Lett. 54 1543

Spence J C H and Tafto J 1983 J. Microscopy 130 147

Tafto J and Spence J C H 1982 Science 218 49

Treacy M M J, Howie A and Pennycook S J 1980 Electron Microscopy and Analysis 1979 (London: Inst. of Physics) pp 261-264

Treacy M M J 1982 J. Microsc. Spectrosc. Electron 7 511

*Inst. Phys. Conf. Ser. No. 76: Section 9*
*Paper presented at Microsc. Semicond. Mater. Conf., Oxford, 25–27 March 1985*

# High resolution microanalysis of bulk specimens in the scanning electron microscope (SEM)

M D Hill*

Department of Metallurgy & Science of Materials,
University of Oxford, Parks Road, Oxford OX1 3PH

Abstract    Sub $0.1\mu m$ resolution backscattered electron images have been obtained from bulk $Cd_xHg_{1-x}Te$ in the SEM. By suitable processing of the data chemical microanalyses have been performed to an accuracy of one atom percent.

## 1. Introduction

Routine microanalysis of bulk specimens is performed using X-ray electron-probe microanalysis (X-EPMA) and scanning Auger electron microscopy (SAM). X-EPMA data, suitably corrected for matrix effects, can provide excellent quantitative results. However, the spatial resolution of the technique is limited not only by the physics of the X-ray production but also by the fact that quantitative linescan profiles and X-ray maps are obtained on commercial instruments, such as the Cameca Camebax, by stepping the stage through, typically, one micron intervals. Recently digital beam-scanning has been used to obtain semi-quantitative analysis with a higher sampling frequency, but the resolution attained remains scarcely better than one micron.  SAM possesses a much better intrinsic lateral resolution (typically of the order of $0.1\mu m$ on commercial instruments), but the surface sensitivity, avantageous in certain applications, and low signal levels seriously limit its chemical accuracy.  Both techniques suffer from the further limitation that a large dwell-time per pixel is necessary, which not only makes data acquisition a time consuming process but also raises the doubt that the accuracy of the analyses may be adversely affected by electron beam damage and beam and stage instabilities over a period of time, which may run to several hours.

Thus, particularly in microelectronic and semiconductor applications, there is a need for an alternative technique for the microanalysis of sub-micron structures in the bulk which has high spatial resolution without drastic loss of chemical discrimination and which can be performed in a conventional SEM within a relatively short period of time.

The purpose of this paper is to describe recent work which has been carried out in an attempt to meet this objective, by interpreting the compositional information carried by the back scattered electron signal in a quantitative fashion.

---

*Now at British Telecom Research Laboratories, Martlesham Heath, Ipswich

## 2. Apparatus

The experiments were carried out on a JEOL JSM-35X SEM subject to several modifications to improve the resolution of backscattered electron imaging. Most importantly the standard twin channel solid state backscattered electron detector was replaced by a new system consisting of a 75mm diameter disc of NE102A plastic scintillator material (Nuclear Enterprises) secured directly to the objective lens polepiece using a threaded dural insert (Fig.1). The design of the detector, similar to that used by Robinson (1974), was constrained by the objective of maximising detection efficiency without serious disruption of the standard operation of the microscope with the specimen at the 15mm working distance required for high-resolution work. In practice this has been achieved by adopting a planar, rather than hemispherical, annular geometry, resulting in a $1.7\pi sr$ cone of acceptance. The excluded volume about the beam is only $0.16\pi sr$ thus further improving efficiency, but at the expense of partial occlusion of low magnification (< 50X) images. Charging effects are overcome by coating the front surface of the detector with a thin ($\sim 300\text{Å}$) layer of evaporated aluminium and light losses minimised by the application of a high reflectivity titania emulsion to the remaining exposed surfaces. With the detector in place there is no degradation of the secondary electron resolution nor significant increase in astigmatism of the microscope

The efficiency of this system, relative to the solid state device it replaces, is clearly demonstrated by a marked improvement in backscattered electron imaging performance. At a beam energy of 35keV high quality images may now be obtained using probe currents as low as $2 \times 10^{-12}$ A and at all scanning rates, including TV rate (whereas, previously, at least $5 \times 10^{-9}$ A was required and only slow scan rates were possible due to bandwidth limitations), with signal and noise levels comparable to those of the corresponding secondary electron image. At 10keV the resolution of the backscattered electron image is not significantly different from that of the secondary electron image and is worse by only a factor of 1.5-2 at 35keV using probe currents in the range $10^{-10}$-$10^{-9}$ A compatible with microanalysis.

The twin objectives of digitisation and noise-reduction of the video signal have been achieved by the use of a digital framestore (Boyes et al). The 0-5V analogue output from the microscope supplementary detector unit is digitised to 8-bit precision and fed into an Intellect 200 system via a Slowscan controller (both from Microconsultants) which drives the framestore raster from the blanking pulses generated by the microscope (Fig.1). Images of 512x256 pixels are integrated in hardware via the recursive video processor loop and stored in the framestore with 16-bit resolution. An eight bit window of the accumulated image is fed via a set of three programable output processors to both a monochrome monitor and a high resolution colour monitor. The system is controlled by a DEC LSI 11/23 microcomputer with 128 kbyte RAM which also enables data to be dumped onto hard-disk storage media or to undergo subsequent manipulation. Software has been written, utilising the Microconsultants FORTRAN-callable Slowscan and Framestore handlers, to acquire backscattered electron images from the microscope and to extract and display, in appropriate format, quantitative chemical information.

## 3. Application

The use of the system is illustrated by consideration of the analysis of the interface formed by the liquid-phase epitaxial growth of $Cd_{1-x}Hg_xTe$ on CdTe. The intensity of the backscattered electron signal is not characteristic of the elements present but is, rather, a measure of the mean atomic number of the region of the sample probed. Moreover, given the problems associated with the accurate measurement of absolute signal intensity, quantitative back scattered electron probe microanalysis (BS-EPMA) can best be achieved by comparison of data from samples, relative to those from standard materials, obtained under exactly identical conditions. In the case of $Cd_{1-x}Hg_xTe$ suitable standards are CdTe and HgTe. The former is present as the substrate material and consequently samples were prepared by mounting a thin slab of HgTe together with a series of 90° cross-sections, accurately cut from the wafers to be tested, in a copper impregnated resin. Since the total backscattered electron signal also contains topographic information extreme care was taken during mechanical polishing of the sample surfaces. After cleaning, the samples were given a thin carbon coat to allow X-EPMA to be carried out.

For BS-EPMA the following experimental procedure was adopted:- high quality images were first obtained on the microscope CRT at 10, 15 and 35keV beam energies and probe currents ranging from $2x10^{-12}-1x10^{-9}$ A. The SDU black-level and gain controls were then adjusted interactively with the computer, such that the video levels and associated noise from the CdTe and HgTe standards just filled the 0-5V output range, thus allowing the maximum discrimination in the digitised signal. Images from each standard, in turn, were accumulated over, typically, ten frames at a rate of 10 secs/frame and an average intensity per pixel per frame evaluated from the centre 512-pixel block. The coefficients of the relationship betwean mean atomic number and recorded intensity were calculated from the standards' data and stored to enable subsequent calibration of the images from the samples.

The theoretical upper limit on the number (n) of frames of eight bit data which may be integrated into sixteen bits is 256. In practice, both because of the $n^2$ dependence of the improvement in signal-to-noice ratio and to overcome any problems of stability to the necessary degree of accuracy, images were accumulated over, typically, 64-150 frames (Fig.2).

For speed of computation and ease of display one or a series of 512-point linetraces were extracted from the images (Fig.3a). At magnifications in excess of ~20,000X the sampling frequency greatly exceeds the possible reoslution of the image and thus, since the data is in digital format, a 3- or even 5-point linear or binomial filter may be passed over the data to further reduce residual noise without loss of genuine information. Using the stored calibration coefficients the traces were redisplayed with an ordinate scale in units of mean atomic number. In the case of the stoichiometric ternary $Cd_{1-x}Hg_xTe$ the mean atomic number may readily be converted into an atomic percentage of mercury and the trace displayed in this final form (Fig.3b).

## 4. Discussion

In the case of the analysis of the complete range of $Cd_{1-x}Hg_xTe$ compositions the 8-bit resolution of the signal digitisation limits the chemical discrimination to approximately one atom percent mercury (Fig.3b). An even better resoltuion might be obtained by examining a reduced range of compositions and/or replacing the Slowscan input to the framestore with a twelve-bit ADC.   In terms of chemical precision, the compositions in Fig.3b differ from those determined by X-EPMA of the same sample (Fig.4) by between 0 and 5 atom percent, which is an acceptable level of agreement given the uncertainties introduced during the processing of the data.   The true value of BS-EPMA is clearly illustrated by comparison of Figs.3 and 4 and shows that the 512 point backscattered trace corresponds to only 2 or 3 points in the X-ray data.   Whilst being of high chemical precision, X-ray data obtained in this way can reveal little about the nature of the interface.   By using digital beam scanning the sampling frequency may be increased (Fig.5) but at the expense of quantitation, since it is impracticable to perform an accurate background subtraction and full ZAF correction for all 256 data points.   However, Fig.5 does show that compositional grading occurs across the interface and to a greater extent on the epilayer than the substrate side, in agreement with the higher interdiffusion coefficient of $Cd^{2+}$ than $Hg^{2+}$.   The BS-EPMA data offer the dual advantages of allowing these changes to be monitored with an order of magnitude higher intrinsic spatial resoltuon and with a high sampling frequency whilst retaining chemical precision.   Traces similar to Fig.3b have been obtained which demonstrate the usefulness of BS-EPMA in the study of the effect of varied growth conditions on interface quality. Ultimately BS-EPMA may prove an attractive alternative to radio-tracer techniques in the measurement of relative interdiffusion coefficients.

## 5. Conclusions

Work has been carried out to characterise the performance of a high efficiency scintillator detector to obtain relative compositional data on a sub-micron scale using digital analysis of the bakcscattered electron signal.   Using a beam energy of 10 or 15keV and a probe current of $2x10^{-10}$ A, back scattered electron images with a resolution better than $0.1 \mu m$ have been obtained routinely from semiconductor materials.   This is an order of magnitude better than that achieved in X-EPMA and comparable with that in SAM.   BS-EPMA offers the additional advantages that it may be peformed in a conventional SEM and that compositional information can be obtained in a relatively short period of time in two spatial dimensions and with a higher sampling frequency than is practical for either X-EPMA or SAM.

In terms of chemical discrimination the fact that the signal detected is not characteristic of the elements present is a severe limitation to the technique.   Uncertainties may also arise from inadequate sample preparation (residual topography) and the algorithms used in processing.   In the $Cd_{1-x}Hg_xTe$ work mean atomic numbers have been calculated using the formula of Saldick and Allen (1954).   Recent work (Herrmann and Reimer 1984) has shown that this formula produces consistently high values for the backscattering coefficient.   Hoever, since BS-EPMA is a relative rather than an absolute technique this does not represent a serious problem.   Similarly, in the $Cd_{1-x}Hg_xTe$ work a linear relationship between Z and intensity has been assured, whereas an approximately $Z^{\frac{1}{2}}$ dependence of the back scattered yield from bulk specimens is known to exist (Bishop, 1966). However, this curvature will be most significant when dealing with low mean atomic number materials (e.g. AlGaAs) and when considering wide compositon ranges.   In the case of $Cd_{1-x}Hg_xTe$ the material has relatively high Z and the complete compositon range spans only twenty daltons. Calculation shows that the error due to the linear approximation, in this context, is comparable with the experimental error.

As will be discussed at length elsewhere, significant improvements can undoubtedly be made to the system described here. Nevertheless, these preliminary measurements indicate that BS-EPMA should prove a most valuable addition to the range of quantitative microanalytical techniques available for the study of bulk specimens.

## Acknowledgements

The author would like to acknowledge the contribution to this work of AERE, Harwell in funding an EMR Fellowship and to thank Drs G R Booker, E D Boyes and M J Goringe and Messrs G Dixon-Brown, J Stead and M Lyster for their assistance, advice and many fruitful discussions. The $Cd_{1-x}Hg_xTe$ samples were supplied by Dr M Astles and Mrs V Steward, RSRE, Malvern.

## References
Boyes E D, Goringe M J, Gill J J, Muggridge B, Northover J P and Salter C J Inst. Phys. Conf. Ser. No.68 (1984) Ch.7 p211
Bishop H E, Ph D Dissertation (1966) Cambridge
Herrmann R and Reimer L Scanning 6 (1984) 20
Robinson V N E J. Phys. E. 7 (1974) 650
Saldick J and Allen A O J. Chem. Phys. 22 (1954) 438

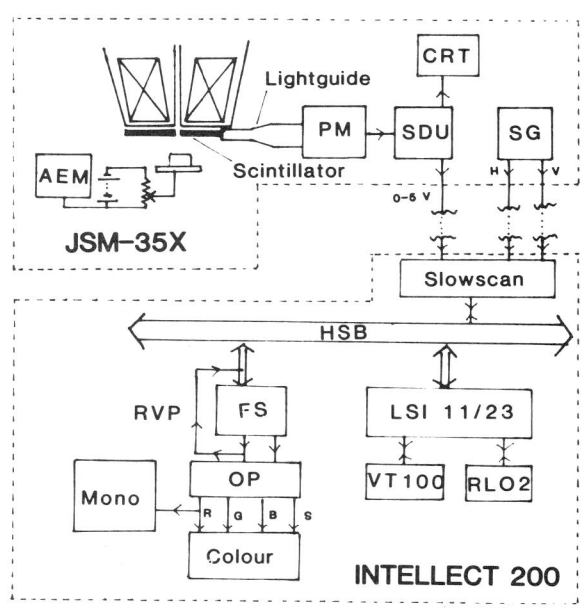

Fig. 1. Schematic representation of the BS-EPMA system. (PM-photomultiplier, SDU-supplementary detector unit, SG-scan generator, FS-framestore, RVP-recursive video loop, OP-output processor).

Fig. 2. Typical BS electron image from a 2.5x1.7 um area of the $Cd_x Hg_{1-x}Te/CdTe$ interface.

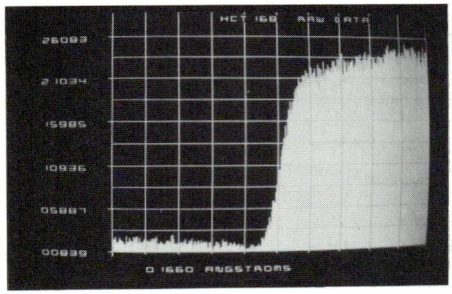

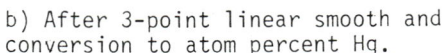

Fig. 3. Linetrace data from Fig.2.

a) Raw intensity data.

b) After 3-point linear smooth and conversion to atom percent Hg.

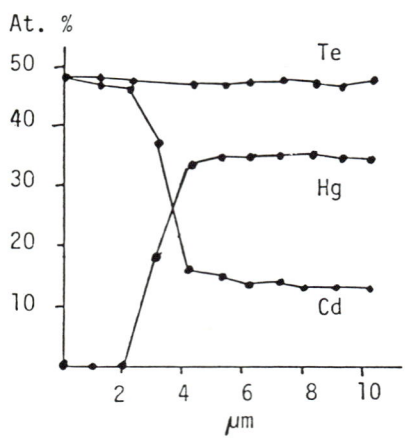

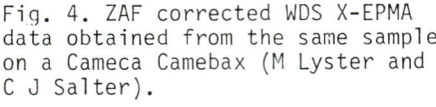

Fig. 4. ZAF corrected WDS X-EPMA data obtained from the same sample on a Cameca Camebax (M Lyster and C J Salter).

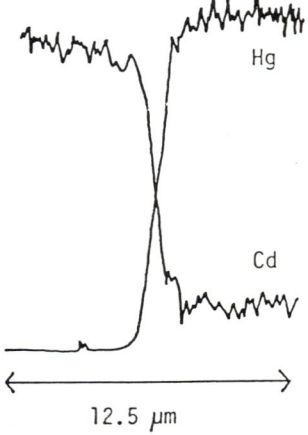

Fig. 5. Digital EDS X-EPMA linetrace data from the sample obtained using a Tracor TN 2000 system fitted to a Cameca Camebax.

Inst. Phys. Conf. Ser. No. 76: Section 9
Paper presented at Microsc. Semicond. Mater. Conf., Oxford, 25–27 March 1985

# Recent developments in the study of semiconductors by atom probe microanalysis

C R M Grovenor A Cerezo and G D W Smith

Department of Metallurgy and Science of Materials,Parks Road,Oxford,UK.

Abstract  Pulsed Laser Atom Probe (PLAP) analysis has recently been applied to a range of semiconductor samples in Oxford.  It has been shown for the first time that the stoichiometry of III–V semiconductors and of thin amorphous silicon layers, can be accurately analysed by the use of the PLAP.  The composition of the Si/thermal oxide interface, and of native oxide layers, has also been investigated, showing that the native oxide is stoichiometric SiO and that an intermediate SiO layer exists at the Si/thermal oxide interface

## 1.   Introduction

Pulsed Laser Atom Probe microanalysis (Kellogg and Tsong 1980) is a recently developed technique for the study of the chemistry of materials with very high spatial resolution and absolute chemical specificity. The use of a nanosecond laser pulse to evaporate surface atoms from a field ion tip held at high dc voltage has allowed the analysis of the composition of semiconducting and insulating materials.

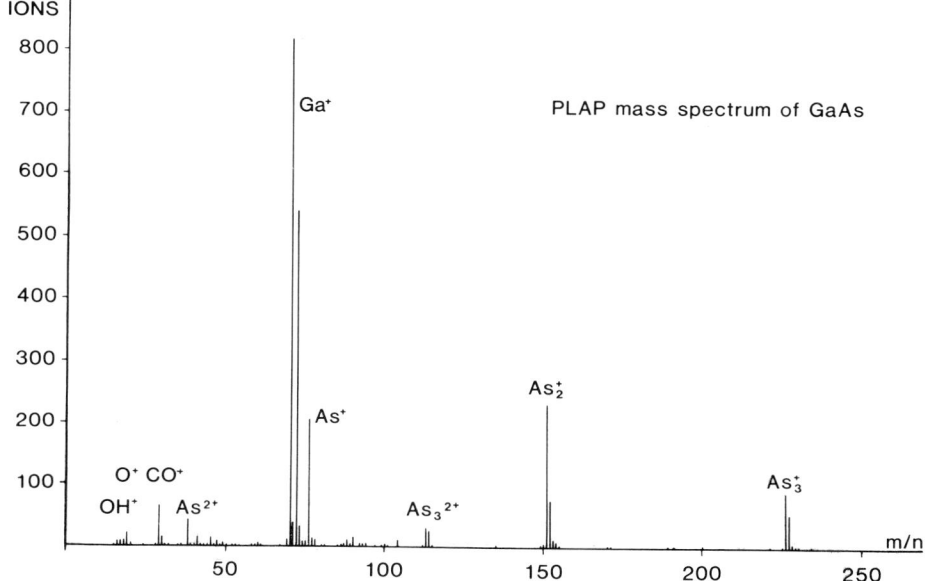

Fig. 1. A 5000 ion PLAP mass spectrum from a GaAs whisker. The stoichiometry of the whisker calculated from this data is Ga/As=0.99±0.03

The application of PLAP analysis to both intrinsic and doped semiconductors has been demonstrated in a number of recent publications (e.g. Grovenor and Smith 1982, Grovenor et al 1983), and the details of the technique are fully reported in these references. This paper will present some new results on the analysis of III-V materials and silicon/oxide interfaces. These results illustrate that the technique of PLAP analysis can give unique information on the stoichiometry of important semiconducting materials.

## 2. Stoichiometry of III-V Semiconductors

One of the applications of Atom Probe analysis that seems most attractive is the study of the stoichiometry of compound semiconductors, and of metal/semiconductor interfaces. A number of workers have investigated the potential of the Atom Probe technique for the analysis of III-V semiconductors; GaAs (Ohno et al 1978, Tsong et al 1978 and Sakurai et al 1984) and GaP (Ohno et al 1978 and Yamamoto et al 1982). Although these papers have demonstrated that conventional voltage pulsed Atom Probe analysis is sucessful in the sense that ions of both species are collected the stoichiometry of the semiconductors is very inaccurately measured. For instance, Ga to As ratios between 0.5 and 2 have been measured in experiments where pulse fraction, gas pressure and crystallographic area of analysis have been varied. Even when the experimental parameters were varied well outside the ranges normally used for routine analyis the correct stoichiometry of these compound semiconductors was not obtained. However, our PLAP analysis of GaAs and InAs whisker specimens (polished to sharp FIM tips by dipping in $H_2SO_4$/$H_2O_2$ solution at $60°C$ has given the correct stoichiometry on every occasion, under normal analysis conditions. Figure 1 shows a typical PLAP mass spectrum from a GaAs whisker. It can be seen that the Ga isotopes are fully resolved, and that the arsenic often desorbs as molecular species such as $As_2^+$ and $As_3^+$. (A similar high concentration of molecular species has been seen in PLAP analysis of silicon specimens by Kellogg (1982) using high laser powers). The stoichiometry of the GaAs whisker measured from this spectrum is Ga/As = $0.99 \pm 0.03$. A similar experiment on InAs whiskers has measured a stoichiometry ratio of In/As = $0.97 \pm 0.05$ (Cerezo et al 1985).

Fig.2 A TEM dark field image from a 200nm thick amorphous silicon layer deposited onto a tungsten field ion tip.

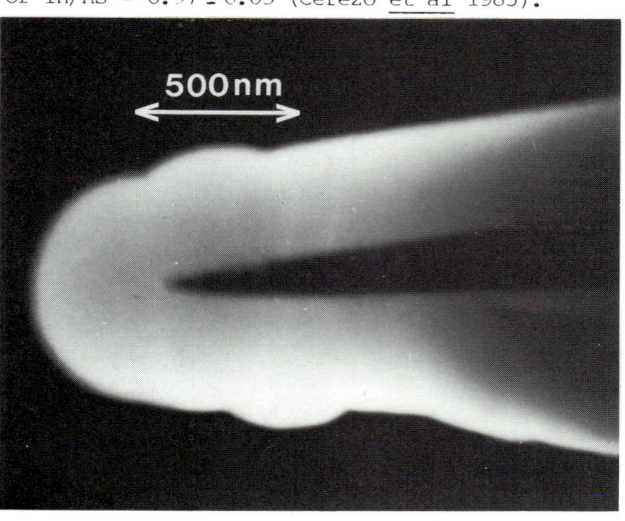

500nm

For both these materials we have also shown that conventional voltage pulsed Atom Probe analysis gave apparant bulk stoichiometries close to Ga/As or In/As = 0.7. We were not able to obtain the correct stoichiometry in this analysis mode by changing the pulse fraction (Cerezo et al 1985). These results clearly demonstrate that the PLAP allows the accurate analysis of compound semiconductor compositions. It is particularily important to have established this before the analysis of the composition of metal/semiconductor interfaces can be meaningfully attempted.

3. Amorphous Silicon Composition Analysis

Although amorphous silicon is currently being used in a wide range of device applications, the basic chemistry of the material is still rather poorly characterised. The variations in Si/O and Si/H ratios with changes in the deposition conditions may have important effects on the electrical properties of the material. PLAP analysis of the composition of amorphous silicon layers deposited under a variety of conditions is considered to be a useful practical application of the technique. Experiments by Krishnaswamy et al (1980) attempted to analyse the composition of RF sputtered silicon layers on metal field ion tips by conventional Atom Probe techniques. These met with only limited success however, and no clear picture of the composition of the layers was obtained. We have recently shown that amorphous silicon layers can easily be grown on tungsten field ion tips in conventional CVD equipment, and that these specimens can be used for PLAP analysis without further preparation.

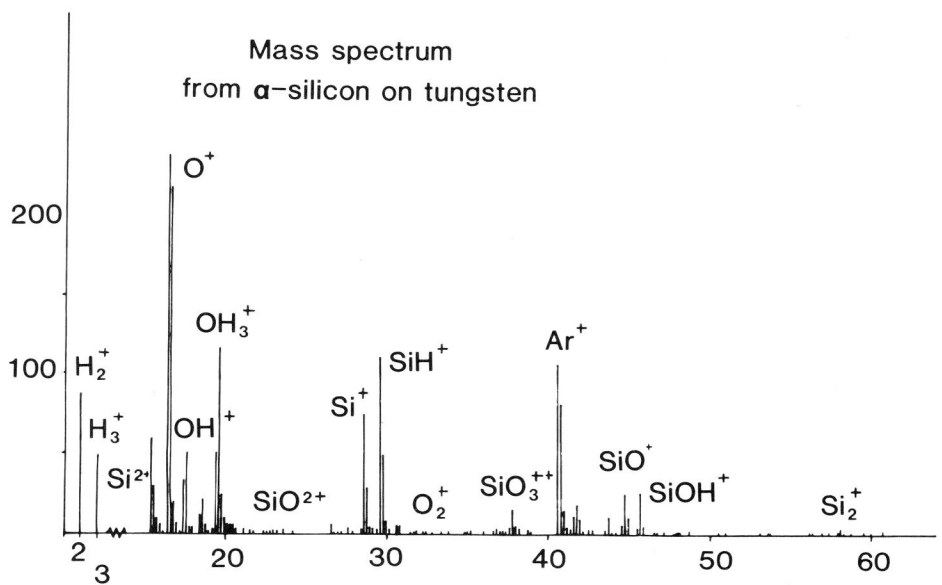

Fig. 3 A mass spectrum from a 5nm amorphous silicon layer deposited onto a tungsten substrate tip. A very high concentration of oxygen and hydrogen is clearly present in the silicon.

Figure 2 shows a TEM micrograph of a tungsten field ion specimen that has 200nm of amorphous silicon deposited on the surface. The smooth surface of the layer and its faithful reproduction of the hemispherical form of the underlying tip render such specimens very amenable to Atom Probe analysis. Figure 3 shows a 2000 ion mass spectrum from a 5nm amorphous silicon layer on a similar tungsten tip. It can be clearly seen that the hydrogen and oxygen contents in this nominally pure silicon layer are very high.   From this spectrum the composition of the layer can be calculated to be roughly 33% silicon, 40% oxygen and 27% hydrogen.   This highly contaminated layer is probably created by reaction of the amorphous silicon with atmosphere between deposition and analysis.   The spectrum also shows the strong absorption of Ar (the imaging gas in these experiments) and contaminant hydroxyl species by the amorphous silicon layer.   Experiments are in progress to analyse the thicker silicon layers to obtain a 'bulk' composition, and to measure the thickness of this highly contaminated layer.   It is further intended to investigate the effects of deposition conditions, and subsequent heat treatments, on the layer composition.

## 4. Native and Thermal Oxide Layers on Silicon

Recent experiments on the stoichiometry of oxide layers on silicon have demonstrated the feasibilty of studying insulating materials with PLAP techniques (Grovenor et al 1985). The native oxide layer on silicon has often been referred to as $SiO_x$, since its composition was unknown.
These layers are often important in device manufacture since they retard the reaction of metals with the silicon surface, and so information on the

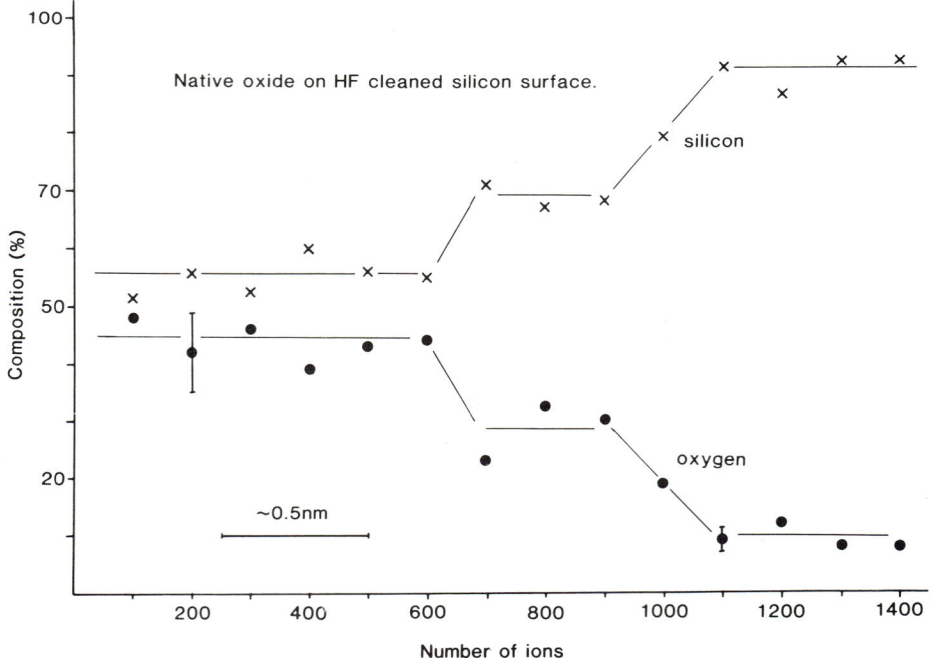

Fig 4. A composition profile through a native oxide layer grown at room temperature on an HF cleaned silicon field ion tip surface.

composition is potentially significant. Native oxides have been grown on silicon field ion tips (chemically polished from bars cut from Wacker single crystal boules) after removal of any anodic oxide that remains with dilute HF solutions. A typical PLAP composition profile through a native oxide layer is shown in Figure 4.  The stoichiometry of the bulk of the native oxide is very close to SiO, with indications of a thin region of composition roughly $Si_2O$ before the silicon substrate is reached.  The thickness of the oxide layer is not easy to calculate precisely from Atom Probe profiles, but an approximate depth scale is shown on the figure indicating that the native oxide is ~1.5nm thick after 3 hours air exposure following the HF dip.  This result shows that most of the native oxide layer grown in this manner has composition SiO, not $SiO_2$ as is often assumed.  The thickness of the oxide layer is in good agreement with direct TEM observations on oxide specimens prepared in a similar way (Mazur et al 1983).  Experiments on the stoichiometry of native oxides grown on atomically clean silicon surfaces (after field evaporation show the same bulk composition, SiO (Grovenor et al 1985).

Thermal oxides have also been grown on silicon field ion tips at $800^{o}C$ in

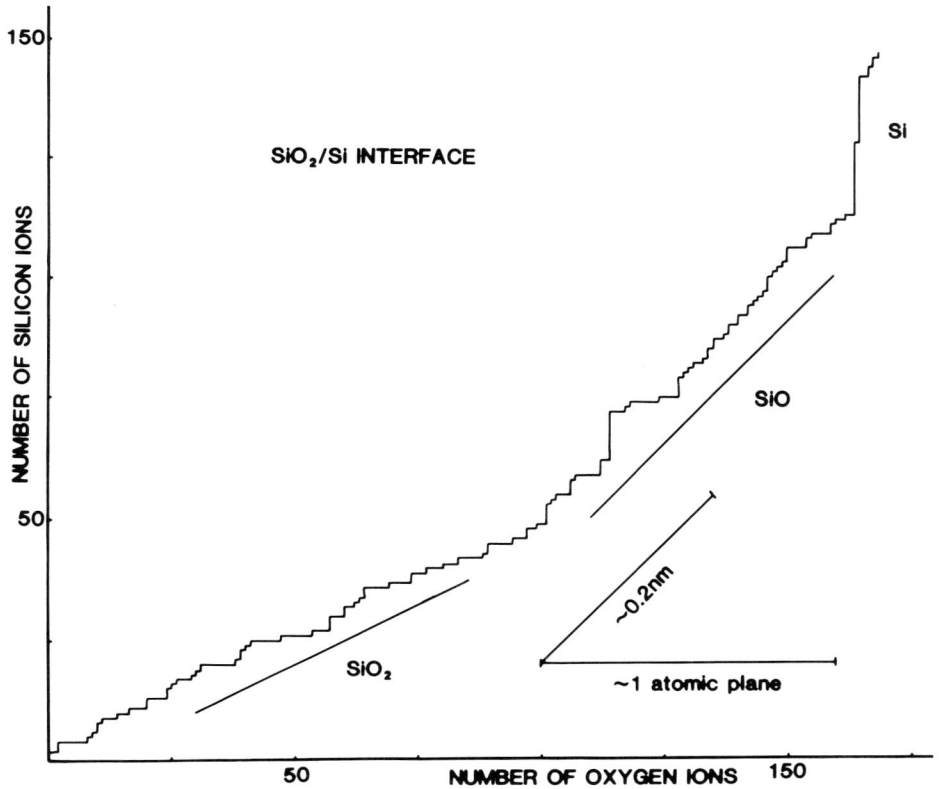

Fig. 5. A ladder diagram through the silicon oxide/silicon interface showing the discrete layer of composition SiO.

wet oxygen. As expected the bulk composition of the thermal oxide as measured by PLAP analysis is $SiO_2$, but as the oxide/silicon interface is approached a thin region of different stoichiometry is apparent. It is tempting to interpret this data as showing a region of composition SiO betwen the thermal oxide and the silicon substrate. Figure 5 shows a 'ladder diagram' of this data where the collection of individual silicon or oxygen atoms are recorded in sequence. This figure shows that between the thermal oxide layer (where the slope of the ladder diagram, which gives directly the local stoichiometry, is 1/2) and the silicon, a sharply defined region of unitary slope (SiO) is seen. The thickness of this layer is estimated to be 0.3nm, or 2-3 atomic planes. This is the first time that the composition and thickness of this interfacial layer have been measured in such a direct manner, although numerous experiments have suggested that a layer of different stoichiometry exists at the interface (Grovenor et al 1985).

## 5. Conclusions

The PLAP has been applied to a number of semiconducting materials, and the results presented here demonstrate that the technique can accurately analyse the local stoichiometry of bulk semiconducting materials and important semiconductor interfaces. It is expected that in the future this kind of chemical analysis, with its very high spatial resolution and unique sensitivity to all elements concurrently, will be used more often to investigate the composition of compound semiconductors and interfaces.

## Acknowledgements

The authors are especially grateful to Dr.B Jones of GEC for her help in the deposition of the amorphous silicon layers and Mr P Augustus for providing the GaAs and InAs whiskers. Professor Sir Peter Hirsch is gratefully acknowledged for the provision of laboratory facilities. CRMG is supported by the Warren Research fund of the Royal Society, AC acknowledges VG Scientific for the support of a CASE studentship. The development of the PLAP was supported by the Paul Instrument Fund of the Royal Society.

## References

Cerezo A Grovenor C R M and Smith G D W 1985  Appl.Phys.Lett.  46
Grovenor C R M and Smith G D W 1982  Surf.Sci. 123  1686
Grovenor C R M Cerezo A and Smith G D W 1983 in Microscopy of
    Semiconducting Materials 1983 (eds. Cullis A G Davidson S M and Booker
    G R) Inst. Phys. London  p.109
Grovenor C R M Cerezo A and Smith G D W 1985 Proc. MRS meeting on
    Interfaces and Epitaxy  Boston Nov. 1984 To be published.
Kellogg G L and Tsong T T 1980  J.Appl.Phys. 51 1184
Kellogg G L 1982  Appl.Surf.Sci. 11/12 186
Krishnaswamy S V Messier R Wu C S McLane S B and Tsong T T 1981
    J.Vac.Sci.Tech. 18 309
Mazur J H Gronsky R and Washburn J 1983 in Microscopy of Semiconducting
    Materials 1983 (eds. Cullis A G Davidson S M and Booker G R) Inst.Phys.
    London p.77
Ohno Y Kuroda T and Nakamura S 1978 Surf.Sci. 75 689
Sakurai T Hashizumi T Jimbo A and Sakata T 1984 J.Phys (Fr) 44 C9.453
Tsong T T Ng Y S and Melmed A J 1978 Surf.Sci. 77 L187
Yamamoto M Seidman D N and Nakamura S 1982 Surf.Sci. 118 555

*Inst. Phys. Conf. Ser. No. 76: Section 10*
*Paper presented at Microsc. Semicond. Mater. Conf., Oxford, 25–27 March 1985*

# X-ray double crystal topographic studies of III–V compounds

B K Tanner, S J Barnett and M J Hill

Department of Physics, University of Durham, South Rd., Durham DH1 3LE

Abstract   The X-ray optics of dispersive geometry double crystal X-ray topography are discussed with particular reference to the use of synchrotron radiation. Double crystal topography studies of semi-insulating GaAs are reviewed and the use of quasi-forbidden reflections for the determination of stoichiometry variations is critically examined. Following a brief review of topography studies of epitaxial layers, the use of rocking curve analysis to determine compositional variations with depth in epitaxial layers is described. The use of double crystal diffraction at grazing incidence to study very thin epitaxial layers is also discused.

## 1. Introduction

Despite enormous investment in instrumentation for imaging and analysis of crystal structure , the semiconductor industry has never made extensive use of X-ray diffraction and topography. This neglect is important, as X-ray techniques are complementary to electron diffraction methods of imaging lattice strain. Transmission electron microscopy (TEM) can be used to image defects with high spatial resolution because the strong scattering of electrons gives rise to a low sensitivity to lattice strain. On the other hand, the weak scattering of X-rays gives an extremely high sensitivity to lattice strain, resulting in very wide images of lattice defects. Due to the strong absorption of electrons, TEM specimens must be very thin whereas the weaker absorption of X-rays permits integrated circuit wafers to be studied non-destructively. Thinning techniques limit the area of sample which can be studied by TEM, while X-ray topography can be used to study whole wafers. Use of the two techniques can yield significantly more data than can be obtained from each individually.

It is unfortunate that practitioners of the topographic art have concentrated on exploiting the last feature rather than the first. Most attention has been paid to single crystal techniques, where the integrated nature of the X-ray wavefront gives poor sensitivity to long range strains and too little to double and multiple crystal techniques where the strain sensitivity can be exploited. To a large extent this arose from the lack of a commercially available high precision double axis diffractometer capable of handling large wafers, a defect now happily remedied. In this paper we describe the use of such double crystal techniques in the study of III-V compounds preceeded as is appropriate at a conference on microscopy by a brief discussion of X-ray optics. Due to the relative imperfection of III-V compounds compared with silicon, we restrict our discussion of X-ray optics to the dispersive geometry where Bragg plane spacings of reference and specimen crystals are mismatched by a few

percent. Readers interested in the application of extremely high strain
sensitivity non-dispersive settings are referred both to the topographic
studies of Hart (1968a),Kohra & Matsushita (1977), Bonse and Hartmann
(1981) and to lattice parameter comparators developed by Hart (1968b) and
Buschet et al (1983). As much of our work has been performed with
synchrotron radiation, consideration is also given to the peculiarities
of double crystal topography with synchrotron radiation.

## 2. X-ray Optics

Fig. 1a shows a schematic diagram of a typical double crystal topography
setting, with an asymmetrically cut reference crystal. This has the
advantages of increasing the beam size and reducing the reflecting range
of X-rays diffracted from the first crystal. An obvious penalty to be paid
for beam expansion is a reduction in flux and hence increase in exposure
time for a topograph. As seen in the DuMond diagram (Fig 1b.) simultaneous
Bragg reflection occurs only over a band of wavelengths. In the (+m-m)
geometry the slopes of the specimen and reference curves are identical and
all wavelengths diffract within the bandwidth determined by the incident
beam slit geometry and the rocking curve is just the convolution of the
two perfect crystal reflection curves. In the dispersive geometry, as the
second crystal is rotated the band of wavelengths diffracted (i.e. the
intersection of the two curves cross-hatched in Fig. 1b) moves. Provided
that the mismatch in lattice parameters is not too great, typically 5-
10%, quite high strain sensitivity can be obtained because the cross-
hatched region will move outside the range of wavelengths set by the
collimating slit for a relatively small angular rotation.

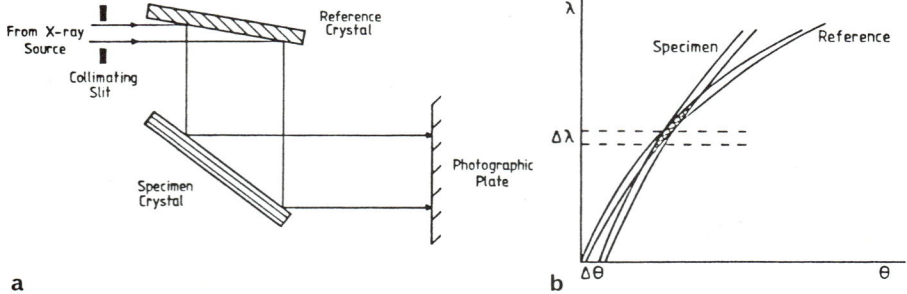

a                                          b

Fig.1 (a) Schematic diagram of a typical energy dispersive double
crystal arrangement. (b) DuMond diagram of the +m-n setting.

Strain sensitivity is determined, at least in part, by incident beam
geometry. The use of asymmetric reflections enables the beam divergence
to be kept sufficiently low that topographs from only one characteristic
line are taken (Jones et al 1981), removing the troublesome double images
from the $K\alpha_1$ and $K\alpha_2$ lines encountered in the non-dispersive (+m-m)
setting.

A major advantage of the dispersive setting for work with III-V compounds
is that a dislocation free silicon or germanium crystal of highly uniform
lattice parameter can be used as a reference. Lattice strains in both
crystals contribute to the topographic image although in the laboratory
images from the first crystal appear geometrically blurred due to the
large reference crystal to plate distance compared to the source to
reference crystal distance (Tanner 1976). With synchrotron radiation, the

source to diffractometer distance is large (65m on the SRS at Daresbury Laboratory) and the geometric resolution at the photographic plate of defects in either crystal is similar. Fig. 2a shows an example of a topograph of GaAs taken in the laboratory with a highly perfect reference crystal compared with one taken at the SRS (Fig. 2b) using a 'black box' double reflection beam conditioner (Bowen and Davies 1983). The regions of lattice strain in the reference crystals (Fig. 2b) dominate over the contrast of any defects. This particular beam conditioner can not be used for topography. These defects were not identified on an initial assessment and concern still exists that strains, possibly due to surface hydrocarbon formation, are increasing under prolonged exposure to the synchrotron radiation beam.

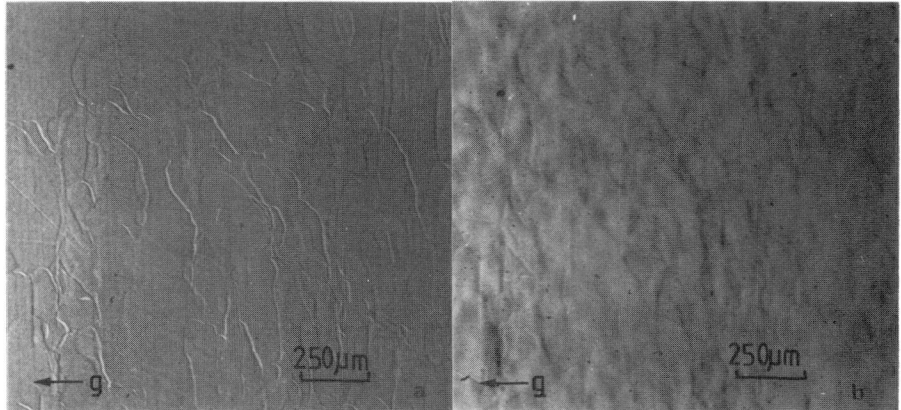

Fig.2 (a) 422 topograph of a {111} cut slice of GaAs taken from the neck of a [100] boule. The topograph was taken using Cu $K\alpha_1$ radiation and a {111} cut Si first crystal from which the 422 reflection was used. (b) Topograph of the same sample taken on the SRS at 1.5Å using a 'black box' double reflection Si {111} beam conditioner.

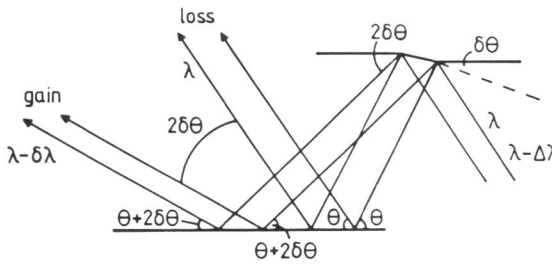

Fig.3 Schematic diagram showing the concept of displacement orientation contrast.

We note, that with SR an additional orientation contrast effect becomes significant in addition to that discussed by Bonse (1962). This latter model, in which the relative change in intensity on the flank of the rocking curve is simply proportional to the effective misorientation works extremely well. With SR, however, the different directions in space of beams diffracted from perfect and misorientated regions are significant in that intensity is displaced. Fig. 3 shows a schematic diagram of how the displacement orientation contrast arises. Note that the misorientation $\delta\theta$ is less than the perfect crystal reflecting range so that a significant fraction of the intensity is diffracted on the second reflection. In the figure different wavelengths from the parallel beam are diffracted, although in practice there will, be a much more complicated situation due to the small divergance of the SR beam. In addition, there will usually be a continuous change in lattice distortion which will give a continuous displacement of the diffracted beam. Nevertheless, we have demonstrated that this effect occurs. In Fig. 4 we show two topographs of the beam conditioner pair (a +m-m reflection) taken at two different specimen to plate distances. Significant differences in the two images can be detected.

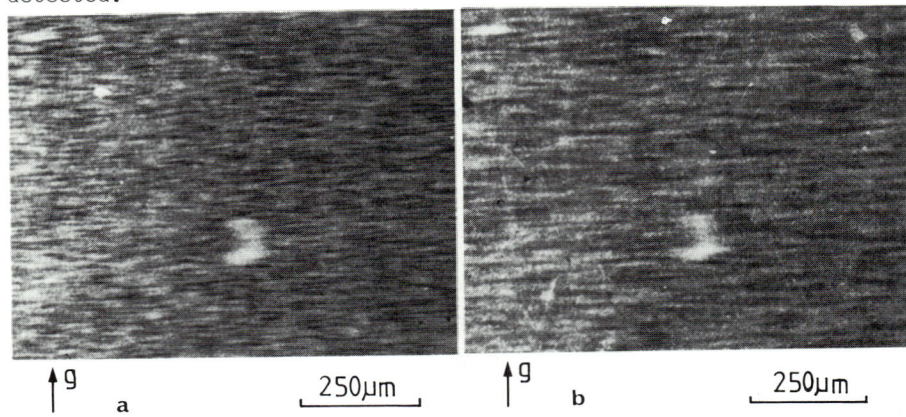

Fig.4   Topographs of the {111} double reflection beam conditioner (a) sample to plate distance ≈5cm. (b) sample to plate distance ≈15cm.

Further, due to the large source to specimen distance, the beam divergence with SR is very much less than in the laboratory. For example, for a 50mm wafer, the divergence is $7 \times 10^{-4}$ radians at station 7.5 at Daresbury and broadening of the rocking curve due to beam divergence is given approximately by $7 \times 10^{-5}$ radians even for a relative misorientation of 5° (Schwarzschild 1928). This arises because, for tilted Bragg planes, beams not exactly in the plane of incidence make different angles with the reference and specimen crystals. A band of double diffracted beams is thus formed which moves up and down the crystals as the specimen is rocked about the $\omega$ axis.

For laboratory work it is essential to have accurate, ideally automated, goniometers to optimise the relative tilts of specimen and reference crystals about the $\phi$ axis. When parallel, the height of the rocking curve is maximised and the half-width minimised, although the integrated intensity will remain constant. For the case of sychrotron radiation the effects of beam divergence are negligible, greatly simplifying experimental procedure.

## 3. Studies of Semi-Insulating GaAs

There is presently intense interest in the production of high perfection semi-insulating (SI) gallium arsenide for fast electronic devices. Of great importance are the deep level defects, known as EL2, found in SI GaAs about which a number of studies have been reported. Spectroscopic evidence suggests that the EL2 is an anti-site defect associated with the interchange of Ga and As atoms. Etch pit studies (Brozel et al 1983) and our own double crystal X-ray topography work (Brown, et al 1984 and Barnett, Tanner and Brown 1985) have shown that EL2 concentration can be correlated with the dislocation structure. In particular EL2 maps produced by infra-red absorption show a maximum EL2 concentration at tilt boundaries. The ability to map dilation and tilt over the whole area of the sample is a major feature of double axis topography. If the lattice strains are larger than the intrinsic rocking curve width, not all of the specimen will be imaged at once. The regions of diffracted intensity define contours of effective misorientation. A series of topographs taken with different ω settings of specimen with respect to the reference crystal yields a set of equal misorientation contours. Rotation of the specimen by 180° about the diffraction vector enables tilts and dilations to be distinguished (Lefaucheux and Robert 1984, Barnett et al 1985) An example of tilt and dilation maps for a cross-section of a 2 inch <100> SI LEC GaAs crystal is shown in Fig. 5.

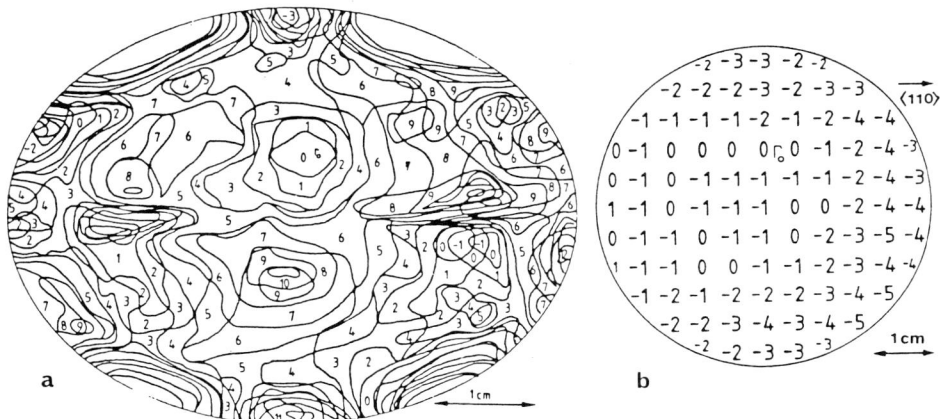

Fig.5 (a) Relative tilt map of a typical 2 inch slice of SI LEC GaAs in units of seconds. (b) Relative lattice parameter map of the same sample in units of $\Delta d/d = 3.9 \times 10^{-6}$ .

The association of EL2 concentration with tilt boundaries together with our earlier correlation of EL2 with slip bands (Brown et al 1984) suggest that EL2 defects may be formed during dislocation climb. If this is so, plastic deformation experiments on suitably orientated samples should provide an important insight. Studies by X-ray topography of plastically deformed samples are now underway to attempt to correlate dislocation type, conservative and non-conservative movement and overall defect configuration with EL2 maps.

Studies of stoichiometry variations have been performed using the 200 'forbidden' reflection. For silicon, such reflections are totally forbidden, but due to the different scattering factors of Ga and As, significant, though weak, intensity is observed. Clearly, the reflected intensity will be proportional to the stochiometry of the samples and Fujimoto (1984) recently described single crystal experiments in which the integrated intensity was measured as a function of position across the crystal. For small deviations from stoichiometry the intensity of the 200 reflection varies linearly with the fractional concentration difference of Ga and As, $\Delta c$. Fujimoto (1984) showed that the intensity did indeed vary radially in a <100> undoped LEC GaAs crystal, the variation in intensity being about 1% when using CuKα radiation, corresponding to Δc≈5x10⁻⁴. Unfortunately, however, the intensity variation correlated very well with the variation with etch pit density, that is to say, the dislocation density. In his paper Fujimoto devotes considerable space to his claim that the 200 reflection is insensitive to lattice strains, a requirement that the intensity changes are proportional to Δc. Our own experiments confirm that dislocations visible in 400 Lang topographs are invisible in 200 reflections but only when care is taken to ensure the elimination of harmonic contamination from the 400 at half the wavelength of the fundamental. With the X-ray generator running at 30kV, dislocations were faintly visible in the 200/400 reflections at 1.54Å and 0.77Å. Reducing the voltage to 15kV, (below the excitation voltage for 0.77Å radiation) led to disappearance of the images. We note that Fujimoto ran his generator at 60 kV and harmonic contamination will have been significant, probably accounting for the correlation with etch pit density. Further, although nowhere is the specimen thickness quoted, it is stated that experiments were performed on "wafers". As the intensity in the kinematical limit is proportional to thickness, for crystals less than typically a quarter of an extinction distance thick, it is essential to have surfaces of thin wafers parallel to much better than 1% if the intensity variations of this order are to be significant. In order to enhance the sensitivity of the 200 integrated intensity to fractional atomic concentration differences Δc we have performed double crystal topography at a wavelength of 1.16Å, between the Ga and As Kα absorption edges, close to the position where the structure factor changes sign due to the anomalous dispersion close to F=0, the extinction distance becomes extremely large. Although the strain sensitivity in a double crystal topograph goes up, the integrated intensity becomes much less sensitive to strain as the scattering becomes kinematical. As the structure factor changes sign due to competing terms associated with Ga and As, the scattered intensity close to F=0 becomes very sensitive to Δc. Further, variations in anti-site defect concentration should also lead to a change in diffracted intensity. Use of synchrotron radiation enables the wavelength tuning to be performed. A double crystal experiment was set up initially at the Ga absorption edge (detected by rotating the reference crystal until a change was detected in the fluorescence yield from the specimen). The diffractometer (Bowen and Davies 1983) was then adjusted sequentially until a minimum in diffracted intensity was obtained. Rocking curves were taken at different points across the

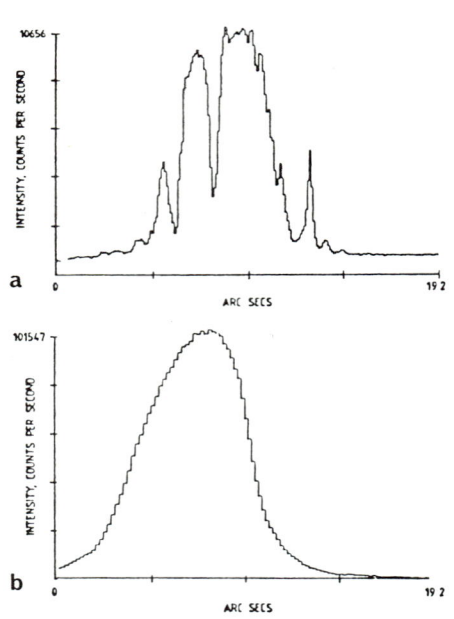

Fig. 6. (a) 200 rocking curve at 1.16Å. (b) 200 rocking curve at 1.2Å, above the Ga edge.

crystal using a translation stage similar to that described by Halliwell, Lyons, Tanner and Ilczyszyn (1983). We have used a dislocation free, 2mm thick crystal where, paradoxically, specimen thickness becomes unimportant. The distance over which the intensity is reduced to 1/e by normal absorption is only 28 µm at 0.7Å. Thus, for very thick crystals and very long extinction distances, there is effectively no contribution from the back surface of the specimen. Fig. 6 shows an example of a rocking curve taken in the region close to F=0 compared with a similar rocking curve at 1.195Å. The overall curve widths are similar, but complex structure (which as can be seen from the background, is not the result of statisticsl fluctuations) appears in the rocking curve when the extinction distance is long. We interpret this as due to fine mosaic structure revealed by the narrow perfect crystal reflecting range. The integrated intensity should be insensitive to such structure but measurements show that the changes across the crystal are not radially symmetric. Further work is needed before definitive conclusions can be drawn from our data.

## 4. Heteroepitaxial Structures

X-ray topography and diffractometry have been used extensively for work on heteroepitaxial structures of III-V compounds. Topographic studies identified misfit dislocations as being responsible for poor lifetimes of laser devices and a number of studies have been conducted into the mechanism of formation of misfit dislocations (Petroff and Sauvage 1978, Hagen 1978). Sauvage (1981) has reviewed such studies in a paper earlier in this series. An important feature of double crystal topography is that it is easy to seperate substrate and layer reflections and hence identify defects only present in the epitaxial layer.In their simplest form rocking curve studies can be used to measure lattice mismatch between substrate and epilayer directly from the angular separation of the two Bragg peaks. Indeed, this is now becoming a routine assessment procedure in many electronic companies, several of whom take double crystal rocking curves of all layers grown with a throughput of up to 15 samples per day per diffractometer. Recently much interest has been shown in rocking curve profile analysis for determination of composition variation with depth through the layer. Since epitaxial layers of III-V compounds are often thick in comparison with the extinction distance, use of dynamical theory is necessary. We have collaborated with

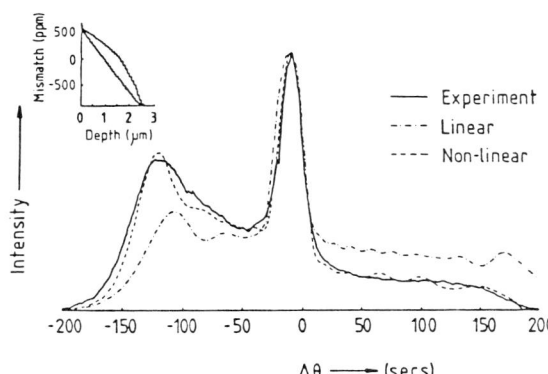

Fig. 7 Experimental and theoretical 004 rocking curves from an inhomogeneous layer of GaInAs. Theoretical curves are for linear and non-linear composition profiles. Wavelength CuKα₁ .

Halliwell's group at British Telecom Research Laboratories in the development of a model based on solution of the Takagi-Taupan equations. Studies on linearly graded layers have shown an excellent correlation between In concentration deduced from rocking curve analysis and from X-

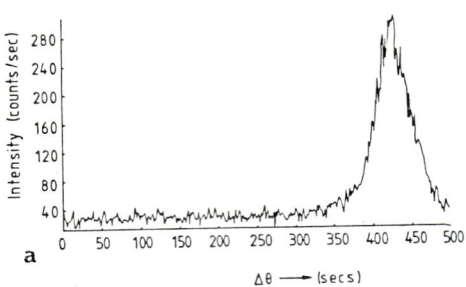

a

$\Delta\theta \longrightarrow (secs)$

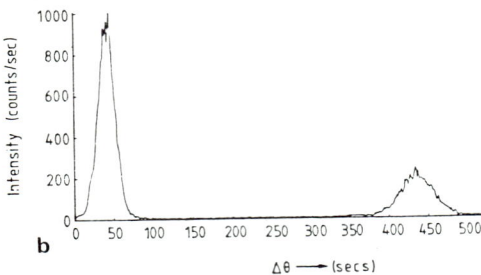

b

$\Delta\theta \longrightarrow (secs)$

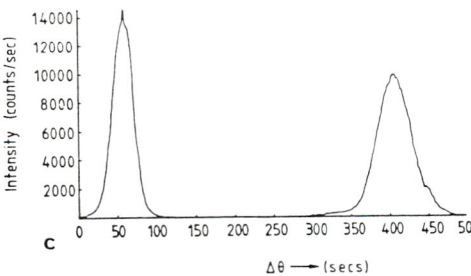

c

$\Delta\theta \longrightarrow (secs)$

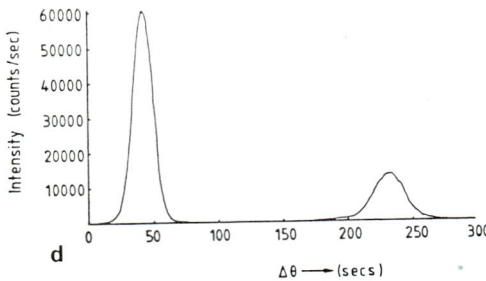

d

$\Delta\theta \longrightarrow (secs)$

Fig. 8 404 rocking curves from a
0.5μm layer of GaInAs at (a)
1.40Å, (b) 1.80Å and (c) 1.52Å.
(d) 004 rocking curve from the
same sample at 1.50Å.

ray fluorescence analysis on
bevelled sections (Halliwell, Juler
and Norman 1983). Good fits to the
rocking curve structure are
generally only found when a non-
linear lattice parameter variation
is assumed (Fig.7). Using
synchrotron radiation we have
studied the change in rocking curve
profile as a function of
wavelength. A single lattice
parameter profile gives the best
fit to rocking curves taken at all
wavelengths (Hill,Tanner,Halliwell
and Lyons 1985). This gives
confidence in the use of a single
X-ray measurement at one wavelength
as an analytical tool for
determining composition profiles.
The model has also been used to
calculate rocking curves from
multiple layer structures and
recent studies have concentrated on
multiquantum well (MQW) structures
consisting of up to 100 alternating
layers typically 50Å thick. Even
with thick encapsulating layers
present, very good agreement is
found between theoretical and
experimental rocking curves of both
MBE and MOCVD grown layers (Hill,
Tanner and Halliwell 1985).
Extension of the model to the case
of asymmetric reflections is
straightforward and yields
additional data on the lattice
distortions (Lyons and Halliwell
1985). We have recently
investigated the use of Bragg
diffraction at very low angles of
incidence in order to study sub-
micron thickness epitaxial layers.
For (001) orientated samples the
404 reflections are ideal for such
studies as a Bragg angle of 45°
gives grazing incidence and a
diffracted beam normal to the
surface. We have exploited the
tunability of synchrotron radiation
to examine the optimum conditions
for studies of very thin layers. As
seen in Fig. 8a at very low angles
of incidence, total external
reflection occurs and three
dimensional Bragg scattering is not
observed. For high angles of
incidence the X-ray beam penetrates

the thin layer and shows a strong diffracted peak for the substrate
(Fig. 8b). In between, (Fig. 8c), conditions exist where there is
strong diffraction from the substrate and the layer.
This optimum condition is for $\lambda=1.52\mathring{A}$, which we note to be fortunately
close to the $CuK\alpha_1$ wavelength. Some broadening of the rocking curve
inevitably occurs due to sample curvature as even with an 80μm square
beam, a 1.5mm length of sample is irradiated at a 3° incidence angle.
Nevertheless this is not too serious and we note the anisotropy in the
grazing incidence rocking curve not seen in the high angle surface
symmetric rocking curves (Fig. 8d). This asymmetry is predicted in
Hartwig's (1978) theory when the amplitudes of reflected waves at the
surface are included in the calculations, and not neglected as is the case
for the classical dynamical theory.

## 5. Conclusions

X-ray topography and diffractometry in the double crystal mode are
powerful tools for the study of long range strains at high sensitivity in
semiconductor crystals. Their use has increased dramatically over the past
few years as interest in the III-V semiconductors has grown. The
development of interest in III-V epitaxial structures and continued demand
for ever higher perfection silicon will certainly continue to see growth
in both the development and application of double crystal techniques.

## Acknowledgements

The work on III-V compounds at Durham is supported by SERC, British
Telecommunications plc and RSRE Malvern. Particular thanks are expressed
to M.A.G.Halliwell, M.H.Lyons and G.T.Brown for their stimulating
collaboration. We would also like to thank D.A.Hope and D.A.Andrews for
the supply of experimental samples.

## References

Barnett S.J., Tanner B.K. and Brown G.T. 1985 Materials Research Soc.
  Symposium Proceedings 41 pp83-88.
Bonse U. 1962 Direct Observations of Imperfections in Crystals ed
  J.B.Newkirk and J.H.Wernick (New York: Wiley) pp431-460
Bonse U. and Hartman I. 1981 Z.Kristallographie 156 265
Bowen D.K. and Davies S.T. 1983 Nucl.Instr.Meth. 208 725
Brown G.T., Skolnick M.S., Jones G.R., Tanner B.K. and Barnett S.J. 1984
  Semi Insulating III-V Materials Kahneeta ed D.C.Cook and J.S.Blakemore
  (Shiva) pp76-82
Brozel M.R., Grant I., Ware R.M. and Stirland D.J. 1983 Appl. Phys. Lett.
  42 610
Buschet R., Meyer A.J., Kauffman D.S. and Gotwals J.K. 1983 J. Appl.
  Cryst.16 599
Fujimoto I. 1984 Jap. J. Appl. Phys. 23 L287
Hagen W. and Queisser H.J. 1978 Appl. Phys. Lett. 32 269
Halliwell M.A.G., Juler J. and Norman A.G. 1983 Inst.Phys. Conf. Ser.
  67 : Microsc. of Semicond. Mater. Conf. Oxford 1983, 365
Halliwell M.A.G., Lyons M.H., Tanner B.K. and Ilczysyn 1983 J. Crystal
  Growth 65 672
Hart M. 1968a Science Progress (Oxford) 56 429
Hart M. 1968b Proc. Roy. Soc. A309 281
Hart M. 1975 J. Appl. Cryst. 8 436

Hart  M.  1980  Characterisation  of Crystal Defects by X-ray  Methods  ed
    B.K.Tanner and D.K.Bowen (New York:Plenum) p241
Hart M. 1981 J. Crystal Growth 55 409
Hartwig J. 1978 Exp. Techn. Physik 26 447
Hill  M.J.,  Tanner B.K.  and Halliwell M.A.G.  1985  Proc.  MRS  November
    Meeting, Boston 1984 (in press)
Hill M.J.,  Tanner B.K., Halliwell M.A.G. and Lyons M.H. 1985 Submitted to
    J. Appl. Cryst.
Jones G.R.,  Young I.M., Cockayne B. and Brown G.T. 1981 Inst. Phys. Conf.
    Ser. 60 : Microsc. of Semicond. Mater. Conf. Oxford 1981, 265
Kohra  K.  and  Matsushita M.  1977 Semiconductor Silicon ed H.R.Huff  and
    E.Sirtl (Princeton: Electrochemical Society) pp441-455
Lyons M.H. and Halliwell M.A.G. 1985 this volume
Petroff J.F. and Sauvage M. 1978 J. Cryst. Growth 43 628
Robert M. and Lefaucheux F. 1983 J. Crystal Growth 65 637
Sauvage M.  1981 Inst.  of Phys. Conf. Ser. 60 : Microsc. Semicond. Mater.
    Oxford 1981 249
Schwarzschild M.M. 1928 Phys. Rev. 32 162
Tanner B.K. 1976 X-ray Diffraction Topography (Oxford:Pergamon)

Inst. Phys. Conf. Ser. No. 76: Section 10
Paper presented at Microsc. Semicond. Mater. Conf., Oxford, 25–27 March 1985

439

# Dislocation mobilities in indium phosphide investigated by synchrotron X-ray topography

A George, A Jacques and R Coquillé(*)

Laboratoire de Physique du Solide, ENSMIM, Parc de Saurupt, 54042 Nancy Cedex, France
LURE, Université de Paris Sud, Bat. 209C, 91405 Orsay Cedex, France
(*) CNET, Centre de Lannion B, Route de Trégastel, 22301 Lannion, France

Abstract   The velocity of screw and $\alpha$ dislocations in Sulfur doped InP was measured using synchrotron topography in the range 548 K $\leqslant$ T $\leqslant$ 648 K, $\tau \cong$ 9 MPa. The velocity of $\alpha$ dislocations seems to be strongly affected by their orientation relatively to the crystal surface. The temperature dependence is characterized by an activation energy close to 1.7 eV.

## 1. Introduction

Indium Phosphide is of great technological importance in optoelectronics as the substrate material for lasers emitting at 1.3 or 1.55 µm. Since dislocations are known to be detrimental to device performance and since presently grown InP crystals contain a non negligible density of these defects, it seems convenient to study their mobilities as functions of stress and temperature.

Dislocation velocity measurements were performed in many semiconducting materials but data for InP are very scarce (Maeda and Takeuchi 1983). Measurements used either double etching or diffraction techniques to localize successive positions of dislocations. X-ray topography, as a transmission technique, is best suited to get rid of possible surface effects and to measure velocities of dislocations with different characters. In III-V compounds it is expected that the six segments of a hexagonal loop move with different mobilities as they differ from each others by their character (screw or 60°), the core structure ($\alpha$ or $\beta$ types) or the sequence of partial dislocations (leading or trailing). X-ray topography could, in principle, allow measurement of these six mobilities in the same sample, with the reservation that it will be difficult anyway to measure them simultaneously if they are very different.

Thanks to synchrotron radiation, the strong absorption of X-rays by In atoms is no longer prohibiting and in situ observations of moving dislocations are possible, which may be useful to prevent impurity segregation at resting dislocations, leading to their subsequent pinning (Imai and Sumino 1983). This paper reports on the application of synchrotron topography to velocity measurements in InP and the first results recently obtained at LURE-DCI.

This work is supported by the Centre National d'Etudes des Télécommunications under contract n°836 B 054 PAB.

## 2. Experimental

Tensile samples of 15 x 4 x $\sim$ 0.5 mm$^3$ gauge length dimensions (Fig.1) were machined from LEC grown InP crystals doped with Sulfur ($\sim 10^{19}$ at.cm$^{-3}$). The tensile axis is [230] and large faces (001). The gauge length is crossed by two twins, a very narrow one AA' and a second, $\sim$ 4 mm wide, BB'CC'. In the twins, the stress axis is [4 1 10]. In Fig.1 and hereafter the sign of indices follows the convention that the (111) surface is made of In atoms. In and P surfaces were distinguished in the used ingot by their etching behaviours. Samples were diamond lapped (1/4 μm) and surface damage was removed in 1.5 % Br$_2$ methanol.

In situ observations were performed with a heating tensile stage (George and Michot 1982) mounted on the two-crystal spectrometer of the DCI topography beam port (Sauvage 1978). The matrix and twins were imaged simultaneously using $\vec{g}$ = 220. The beam was first monochromatized by reflection on a (110) germanium crystal. The selected wavelength $\lambda$ was 0.8 Å. (With the spectrum delivered by DCI, $\lambda \sim$ 0.44 Å just above the K absorption edge of In did not lead to shorter times of exposure and better contrast). Exposures were typically $\sim$ 1 mn long for recording on Ilford L4 nuclear plates with DCI run at 1.85 GeV, $\sim$ 200 mA. High absorption conditions are realized ($\mu$t = 6, t specimen thickness). Though the Borrmann effect is reduced at high temperature, good topographs could be obtained up to 650 K.

Burgers vectors were determined after cooling by Lang topographs with $\vec{g}$ of 220, 400 and 422 types. A pure screw segment is always present in half-loops and is clearly out of contrast when $\vec{g}.\vec{b}$ = 0. For velocity measurements, real positions of dislocations in the sample were determined assuming that the image results from two successive projections, the one on the X-ray exit surface along the reflecting planes and the other on the photographic plate along the diffracted beam.

## 3. Results

Hereafter $\alpha$ dislocations are those whose extra-half plane is terminated in the core by a row of P atoms, assuming the dislocation to be of the glide set. Conversely the extra half plane of $\beta$ dislocations will be terminated by a row of In atoms if they belong to the glide set (Hirth and Lothe 1968).

An example of dislocation development and motion is given by the synchro-

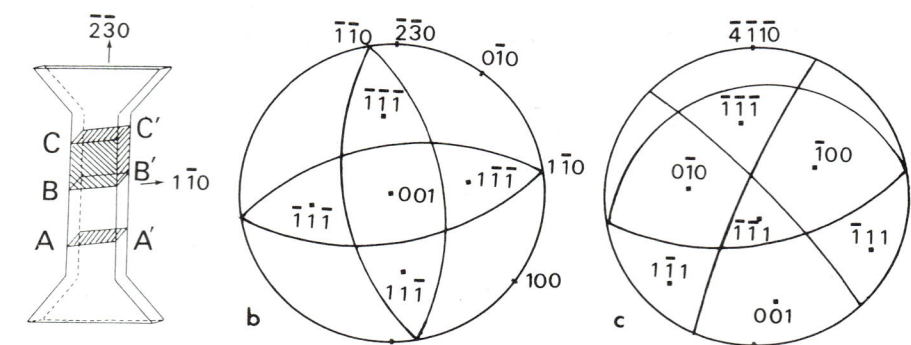

Fig.1 : Sample geometry (a) and crystallographic orientation of the parent crystal (b), of the twins (c).

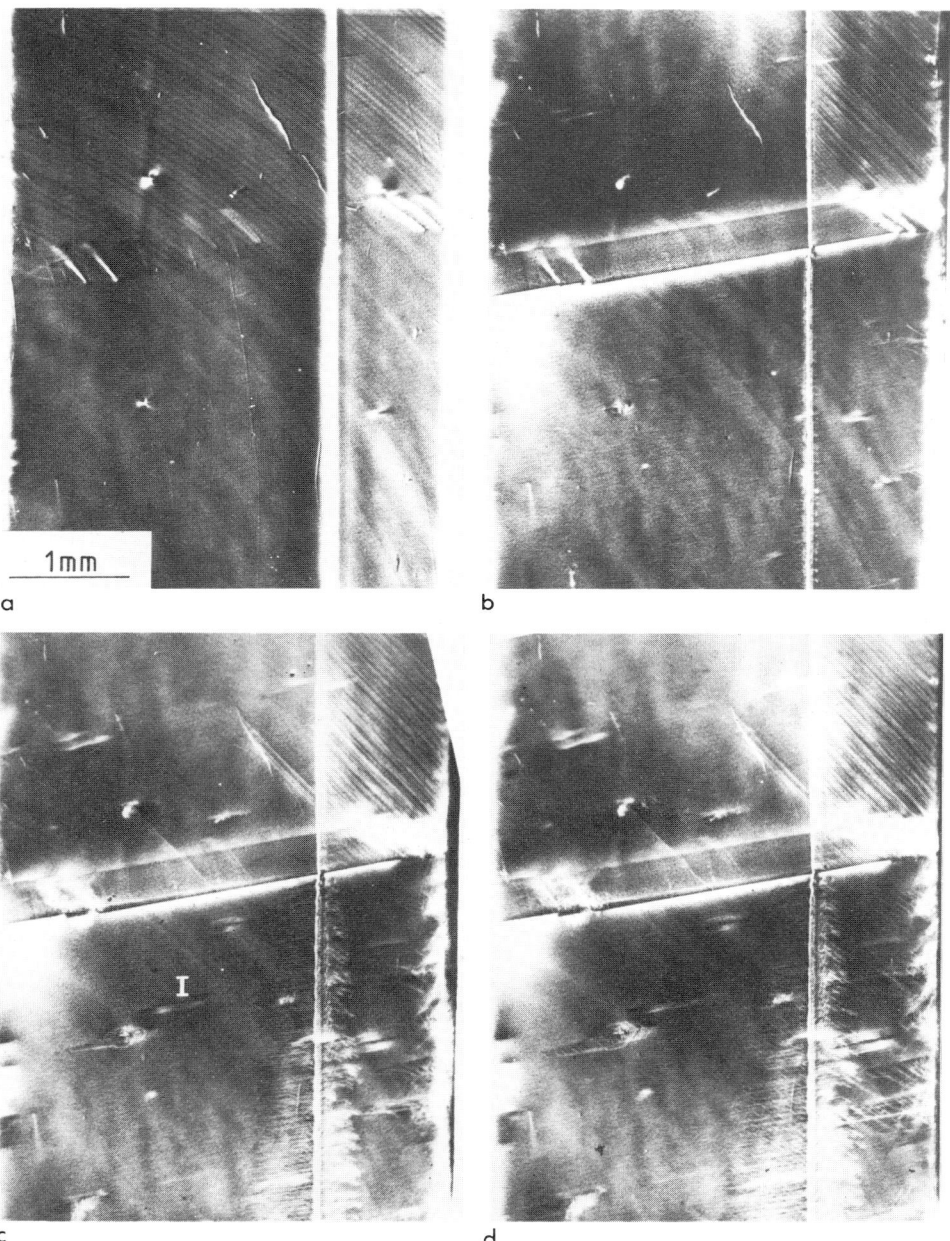

Fig.2 : Motion of dislocation half-loops. Sequence of synchrotron $\bar{2}20$
        topographs. (a) initial state ; (b) after loading at T = 623 K,
        under a nominal stress σ = 15 MPa during Δt = 20 mn ; (c) after
        T = 648 K, σ = 15 MPa, Δt = 15 mn ; (d) after T = 623 K,
        σ = 15 MPa, Δt = 15 mn.

tron topographs of Fig.2. The main features are as follows :

(i) At T $\leq$ 650 K, dislocations were developed from scratches made at room temperature with a diamond needle under a load of 5 g or from Vickers indentations (5 to 25 g) only at $\tau \geq$ 7 MPa, where $\tau$ is the resolved shear stress.

(ii) Dislocations created in the matrix belonged to the two slip systems of highest Schmid factors (111)[01$\bar{1}$] and (11$\bar{1}$)[011] (s : 0.471). When created from the (00$\bar{1}$) surface, half-loops consisted of one screw segment and two 60° dislocations of the $\alpha$ type. When initiated at the (001) surface, they consisted of one screw segment and two 60° segments of the $\beta$ type. The sequence of partial dislocations, assuming an intrinsic stacking fault (Gottschalk et al. 1978), is depicted in Fig.3. In the twin, dislocations either belonged to the (1$\bar{1}$1)[011] system (s : 0.5) or to the (111)[01$\bar{1}$] (s : 0.471). In the former case, half-loops consisted of a screw segment nearly parallel to the free surface and two outcropping 60° dislocations of different types (Fig.3).

(iii) A striking feature is the very different behaviour of $\alpha$ dislocations of different slip systems : at several instances, $\alpha$ dislocations of the (1$\bar{1}$1)[011] system in the twin were observed to move more than one order of magnitude faster than $\beta$ and screw dislocations of the same systems and than all dislocations, including $\alpha$, of other systems. (Accurate measurements of fast $\alpha$ dislocation velocities were not possible since $\alpha$ segments run from their source to crystal edges in less than the interval of time between two consecutive topographs).

(iv) The measured velocities of screw and "slow" $\alpha$ dislocations are presented in Fig.4. The data are rather scattered and must be completed by further work. According to these first results, in the narrow range of temperature and stress that has been investigated :

$$v_{30\alpha/90\alpha} > v_{30\alpha/30\beta} \gtrsim v_{90\alpha/30\alpha} \text{ (where the leading partial is given first)}$$

$$v \sim \exp(-Q/kT) \qquad \text{with} \quad Q \cong 1.7 \text{ eV}$$
$$v \sim (\tau/\tau_0)^m \qquad \text{with} \quad 1 < m \leq 2$$

There are indications that m and Q slightly depend on the dislocation character.

(v) The velocity of $\beta$ dislocations could not be accurately measured since dislocations close to the X-ray entrance surface have a very poor contrast on topographs (see the group labelled I in Fig. 2, 3). It is clear, however, that $\beta$ dislocations move with velocities very similar to that of screw or "slow" $\alpha$ dislocations.

## 4. Discussion

The measured activation energy for dislocation glide is in accord with the results of Maeda and Takeuchi (1983) who also mentioned an anomalous behaviour of $\alpha$ dislocations. As a general trend in III-V compounds $\alpha$ dislocations move much faster than $\beta$ and screws, except for some particular doping concentrations (Ninomiya 1979, Steinhardt and Haasen 1978) but the coexistence in one and the same test of "fast" and "slow" $\alpha$ dislocations does not seem to have been reported previously. We have no definite explanation for

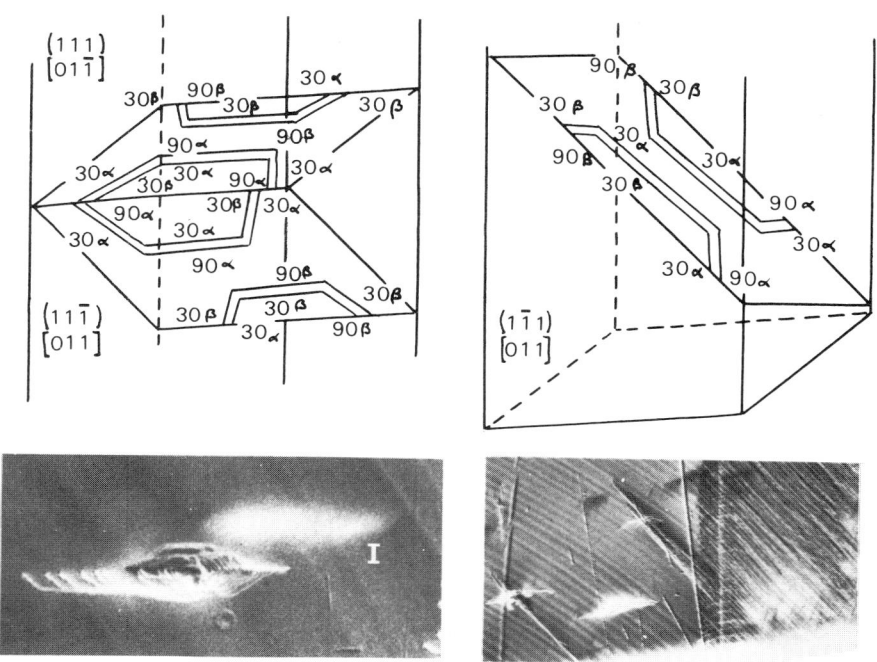

Fig.3 : Geometry of dislocation half-loops of the primary slip systems in
the matrix (a) and in the twin (b).

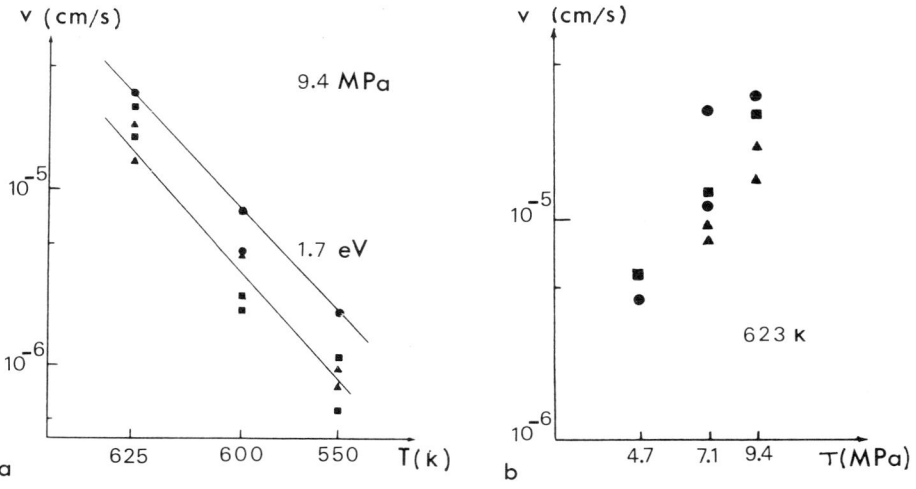

Fig.4 : Dislocation velocity in S doped InP (a) Temperature dependence (b)
stress dependence. ▲ : 30α/30β dislocations, ■ : 90α/30α dislocations, ● : 30α/90α dislocations.

it. The effect cannot be explained by the small differences of Schmid factors and artefacts due to parasitic bending or twisting can be ruled out. Also it is not clear how impurity pinning could explain our observations. Present results rather point to a surface effect : a detailed observation of dislocations of the $(1\bar{1}1)[011]$ slip system shows that fast motion on the $\alpha$ side is maintained only very close ($\leq$ 20 µm) to the surface. If the surface is taken into account, several differences can be found between loops exhibiting "fast" and "slow" motion of $\alpha$ segments :

(i) fast $\alpha$ dislocations were all found in the twin. On all topographs (Fig. 2) it could be noticed that growth striations were best visible in the twin than in the matrix. This could indicate a difference in the state of the surface in these two parts of the sample, due to the orientation dependence of the efficiency of chemical polishing.

(ii) If partial dislocations rotate towards screw orientation at the surface (Hazzledine et al. 1975, George and Champier 1980) at the emerging point the stacking fault ribbon is widened for "slow" $\alpha$ dislocations and narrowed for fast ones.

(iii) If on the other hand a bend is formed beneath the surface, as often observed in Si for example, such a bend should be of the $\alpha$ type for fast dislocations and of the screw type for slow ones.

Further work is in progress to clarify this matter.

Acknowledgements

We would like to thank Dr. M. Sauvage, the staff of LURE and the group "anneaux" of the Accélérateur Linéaire d'Orsay.

References

George A and Champier G 1979 Phys. Stat. Sol. (a) 53 529
George A and Michot G 1982 J. Appl. Cryst. 15 412
Gottschalk H, Patzer G and Alexander H 1978 Phys. Stat. Sol. (a) 45 207
Hazzledine PM, Karnthaler HP and Wintner E 1975 Philos. Mag. 32 81
Hirth JP and Lothe J 1968 Theory of Dislocations (New York : Mc Graw Hill)
    pp 353-62
Imai M and Sumino K 1983 Philos. Mag. 47A 599
Maeda K and Takeuchi S 1983 Appl. Phys. Lett. 42 664
Ninomiya T 1979 J. de Physique 40 C6-143
Sauvage M 1978 Nucl. Instr. and Meth. 152 313
Steinhardt H and Haasen P 1978 Phys. Stat. Sol. (a) 49 93

*Inst. Phys. Conf. Ser. No. 76: Section 10*
*Paper presented at Microsc. Semicond. Mater. Conf., Oxford, 25−27 March 1985*

# Double-crystal diffractometry of III−V semiconductor device structures

M H Lyons and M A G Halliwell

British Telecom Research Laboratories, Martlesham Heath, Ipswich IP5 7RE, UK

Abstract X-ray double-crystal diffractometry is widely used for the non-destructive assessment of layers. Computer simulations based on dynamical X-ray theory have greatly assisted the interpretation of results. In this paper the simulation method is extended to more complex structures and asymmetric reflections. Conventional methods of recording rocking-curves from GaInAs layers on InP require layers thicker than 0.2 μm. A method involving skew beam paths is proposed which should reduce this minimum by at least an order of magnitude.

1.Introduction

The increasing use of hetero-epitaxial layers in devices for optical communications systems places new demands on analytical techniques. The performance of devices is strongly affected by the crystallographic quality of the layers which in turn is determined in large part by the lattice parameter difference between substrate and layer. A key method for measuring the lattice parameter difference is double-crystal diffractometry. In this method, the X-ray beam is reflected by a reference crystal and then by the sample. This sample is rotated through the Bragg condition and the intensity recorded. The plot of intensity as a function of angle is a 'rocking curve' and contains much information about the layer. As well as giving a direct measure of the lattice parameter difference, analysis of the peak shapes can reveal the thickness of the layers and variations in the lattice parameter through the layer arising from compositional grading. Extraction of this information requires a suitable method for modelling the rocking curves.

A previous paper (Halliwell et al., 1983) described the application of the Takagi-Taupin equations (Takagi 1962, Taupin 1964, Takagi 1969) to single layers of GaInAs on InP. Equations were derived for the 004 symmetric reflection and used to simulate the rocking curves for compositionally graded layers of GaInAs on (001) InP substrates. This paper describes the extension of the model to other reflections and its application to more complex device structures. Particular areas of interest are multi-quantum well (MQW) structures and the extension of the theory to reflections where the beam path is not symmetric with respect to the sample surface. Such reflections give information about strains both normal and parallel to the interface, whereas the 004 reflection used in our previous work only gave information about the strains normal to the interface. In addition the use of skew asymmetric reflections should give an order of magnitude decrease in the thinnest single layer that can be measured.

## 2. Theory

### 2.1 Asymmetric Reflections

The 004 reflection previously modelled represents a special case because the beam path is symmetrical with respect to the surface, and the structure factors for the 004 and 00$\bar{4}$ reflections are identical. For a general reflection the beam path may be asymmetrical and the hkl and $\overline{hkl}$ structure factors are not usually the same. This increases the complexity of the equations required. Reflections for which the beam path is asymmetric with respect to the sample surface will be refered to as asymmetric reflections. For an asymmetric reflection the basic equation describing the variation of the amplitude ratio of reflected and incident beams (X) with depth (Z) is

$$dX/dZ = (i\pi/\lambda\gamma_o)(\psi_{\bar{g}}X^2 + (p\psi_o + q)X + s\psi_g) \qquad (1)$$

The subscripts o,g and $\bar{g}$ refer to the 000, hkl and $\overline{hkl}$ reflections respectively. $\psi$ is the complex polarisability which is related to the structure factor F by the formula $\psi = -(\lambda^2/\pi V)r_eF$ and $\gamma_o$ is a direction cosine. $p = (1-\eta)$, $q = a\eta$ and $s = -\eta$ where $a = -2\lambda(\theta-\theta_o)\cos\theta_o/d$ is a measure of the departure of the incident beam from the Bragg angle $\theta_o$ and $\eta$ is the ratio of the direction cosines of the incident and reflected beams and takes the value $-1$ for a symmetric Bragg reflection. Note that the polarisabilities for the hkl and $\overline{hkl}$ reflections are involved and so in the general case equation (1) must be used even for symmetric reflections.

Equation (1) which can be rewritten in the form:

$$dX/dZ = iD(AX^2 + 2BX + C) \qquad (1a)$$

can be solved analytically to give a function for the ratio of the amplitudes at the top of a layer given the boundary condition X=x at depth Z=z:

$$X = xS + i(Bx + C)\tan(DS(Z-z)/(S - i(Ax + B)\tan(DS(Z-z))) \qquad (2)$$

where $S^2 = (B^2 - AC)$. When z is large so that x=0 equation (2) simplifies to

$$X = -B + S(\text{Sign}(\text{Im}(S)))/A \qquad (3)$$

The calculation procedure is then straightforward. The amplitude ratio of the substrate is calculated using equation (3). This is then the boundary condition for the first layer. The amplitude ratio at the top of this layer is calculated from equation (2) and this ratio is used as the boundary condition for the calculation of the amplitude ratio at the top of the second layer. The procedure is repeated for all the layers in the structure. The reflectivity is given by $\text{Mod}(X)^2$. Rocking curves from layers with compositional grading may be simulated by dividing the graded layer into a number of laminae of constant composition (Halliwell et al. 1983).

### 2.2 Multi-Quantum Well Structures

Simulation of rocking curves for MQW structures is a trivial problem as the calculation procedure involves dividing the layer into a number of laminae. Each layer of the MQW is identified with a lamina in the calculation procedure. The simulated rocking curve shows the relative intensities of the peaks from the mean lattice, the substrate and the satellites. An advantage of the computational procedure is that effects of confining layers or irregularities in the super-lattice can readily be included in the simulation. Experimentally it has been found more satisfactory to use the 002 reflection to assess MQWs. Although the main peaks are weaker than those from the 004 reflection, the satellites are very much stronger. Simulation of symmetric 002 reflections is straight-forward and simply involves changing the reflectivities and Bragg angle in the equations published previously (Halliwell et al., 1983; Halliwell et al., 1984).

## 3. Applications

### 3.1 Symmetric Reflections: Multi-Quantum Well Structures

Rocking curves from MQW structures have been successfully simulated. Satellite positions and intensities agree well with experimental curves (Hill et al 1984). An example of an 002 reflection can be seen in fig 1.

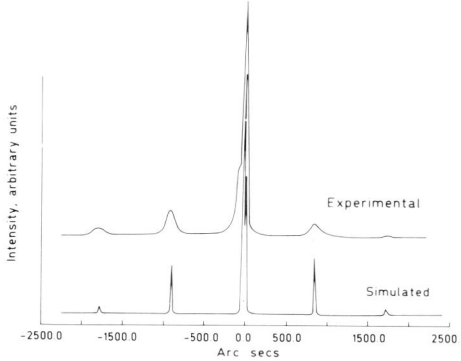

This shows experimental and simulated rocking curves for a superlattice consisting of 100 alternating layers of GaInAs and InP grown by MOVPE (Moss and Spurdens, 1984). The superlattice period was 180Å and both first and second satellites can be seen. The presence of second satellites was a clear indication that the GaInAs and InP layers were of unequal thickness and TEM studies indicated that the GaInAs layers were 50% thicker than the InP layers. The simulated curve in fig 1 was obtained with 110Å GaInAs and 70Å InP. The experimental peaks are broader than the simulation, reflecting irregularities in the superlattice

Fig 1 Experimental and simulated 002 rocking curves of MQW structure.

spacing. This particular superlattice was a trial run in which the growth times of the individual layers were known to be irregular and TEM micrographs confirmed variations in the thickness of the layers forming the superlattice.

### 3.2 Asymmetric Reflections: Validation of Extended Model

In this section experimental rocking curves are compared with simulated curves derived using equations (2) and (3). The model assumes that the only strain present is a tetragonal distortion of the layer unit cell. Plastic accomodation of strain was not expected in the samples used because the lattice parameters of the layers were all within 0.1% of the substrate value. All available experimental evidence supported the assumption that all the strain due to lattice parameter differences was accommodated elastically. Few real samples are free from additional sources of strain, such as substrate dislocations, where localised core strains may be of a similar magnitude to the epitaxial strains. The samples studied had average substrate dislocation densities of $10^3$

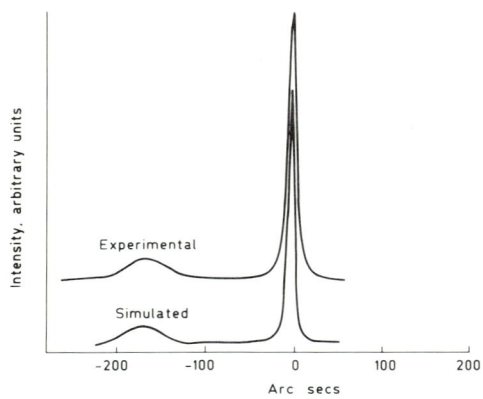

Fig 2 Experimental and simulated 115 rocking curves for a 0.5μm GaInAsP layer on InP.

dislocations/sq cm, equivalent to 20 dislocations in the area sampled by the X-ray beam. Hence some minor discrepancies in rocking curve shapes may be expected. Fig 2 shows simulated and experimental rocking curves using the 115 reflection for a single layer of GaInAsP on InP grown by LPE. Such layers have been shown by TEM to have very low dislocation densities and should therefore have almost ideal rocking curves. The layer thickness, as determined by an electrochemical profiling method, was $0.45 \pm .05$ μm. The layer had a mismatch of 350ppm. Layer thickness may be estimated from the relative intensities of the layer and substrate peaks (Halliwell et al. 1984). The best fit with experimental results was obtained by assuming a 0.5μm layer which is consistent with the independent thickness measurement. To demonstrate that the 115 simulations were consistent with the symmetric 004 simulations, an $InP/Q_{1.3}/Q_{1.5}/Q_{1.3}$ structure was examined. Both 004 and 115 rocking curves were recorded. A good fit with the 004 curve was obtained using the layer details shown in table 1. The 115 rocking curve was then simulated using the same inputs and compared with the experimental curve. The results are shown in fig 3. Agreement between the simulated and experimental 115 curves is reasonable within the limits of the experiment and clearly shows the self consistency of the method.

Table 1

| Layer | | Thickness(μm) | Mismatch(ppm) |
|---|---|---|---|
| 1 | $Q_{1.3}$ | 0.3 | -730 |
| 2 | $Q_{1.5}$ | 0.4 | 690 |
| 3 | $Q_{1.3}$ | 0.8 | -730 |

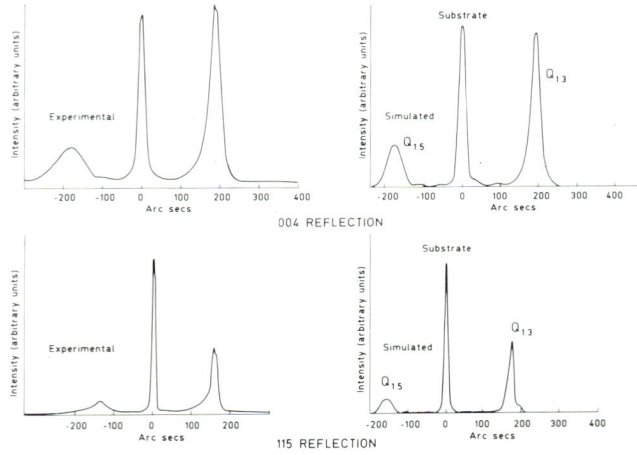

004 REFLECTION

115 REFLECTION

Fig 3 Experimental and simulated rocking curves for $InP/Q_{1.3}/Q_{1.5}/Q_{1.3}$ structure for 004 (upper curves) and 115 reflections.

### 3.3 Skew-Asymmetric Reflections: Sub-micron Layers.

Many devices involve very thin layers and an assessment of these layers is often required. However, as layers become thinner the layer peak not only becomes weaker but is greatly broadened. For instance, layers thinner than 0.2μm cannot be measured using the 004 reflection. Often the only method of assessing the quality of thin device layers is to grow much thicker layers under the same growth conditions specifically for assessment purposes. However, this can be misleading as the bulk of the layer may not be a good guide to the quality of the first 500-1000Å grown.

A way of overcoming this problem is to use asymmetric reflections. This has two advantages. Firstly, the penetration of X-rays can be greatly reduced thus increasing the signal from the layer relative to the substrate. Secondly, the peak widths can be reduced. This is because peak width is proportional to the square root of the ratio of the direction cosines of the emerging and incident beams. If the emerging beam makes a glancing angle with the surface very narrow peaks are obtained which offsets the broadening due to the thinness of the layer. Fig 4 shows the Bragg cone construction for an asymmetric reflection. The usual experimental arrangement (assumed in section 3.2) is for the incident and emergent beams to lie in a plane normal to the sample surface. This is shown in fig 4 to correspond to the angles of incidence $(\theta+\phi)$ and $(\theta-\phi)$ where $\theta$ is the Bragg angle and $\phi$ is the angle the reflecting plane makes with the surface. The number of reflections possible for a given radiation and substrate orientation is very restricted if only beam paths

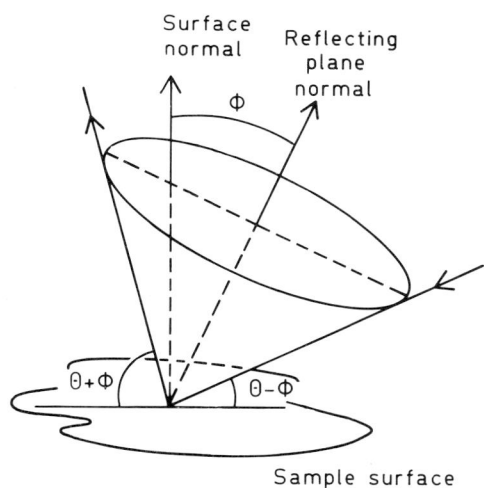

of this type are considered. However the cone shown in fig 4 represents many other beam paths which also satisfy the Bragg diffraction condition. By allowing the plane of incidence to form an arbitrary (skew) angle with the surface, a whole range of incidence angles is possible. In order to obtain the conditions required to assess very thin layers, we consider a reflection where $\phi>\theta$. The Bragg cone will now cut the surface and some beam paths will not be accessible. However, there will still be a restricted range of accessible diffraction conditions. These will include angles of incidence or emergence down to $0°$ thus covering the region of total external reflection (critical angle about $0.4°$ for InP).

Fig 4 Bragg cone construction for asymmetric reflection.

To demonstrate the potential of this method, which we call skew asymmetrical rocking curve analysis (SARCA), rocking curves have been simulated for a 0.1 μm layer of GaInAs on InP. The layer mismatch was assumed to be -600 ppm. Fig 5 shows the simulated curves for the symmetric 004 reflection and for a skew 333 reflection where the beam path has been chosen to give an emergence angle of $1°$. In both cases an (001) surface is assumed. It can be seen that in the 004 reflection the layer

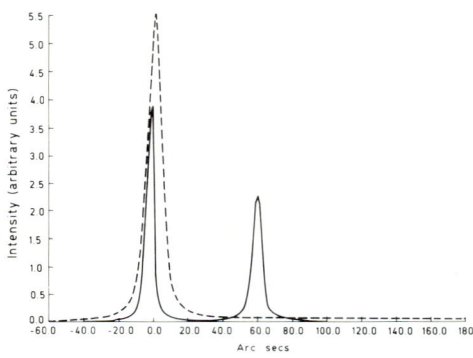

Fig 5 Simulated rocking curves for 0.1 μm GaInAs on InP for 004 reflection (dashed line) and skew asymmetric 333 reflection.

peak is so broad and weak that it can hardly be distinguished from the background. The contrast in the case of the skew 333 reflection is striking. In this case the 0.1μm layer gives a distinct peak well separated from the substrate. Preliminary calculations suggest that this particular reflection could be used satisfactorily for layers down to 0.02μm (200Å). Other reflections may prove to be even more sensitive and furthermore, the skew asymmetrical beam path opens up the possibility of finding a suitable beam path for very thin layers on any orientation substrate.

## 4. Conclusion

In this paper a model for simulating rocking curves has been extended to MQW structures and to all reflections. To demonstrate the theory rocking curves taken using the 115 reflection were simulated. There was good agreement with the experimental curves. The model was applied to skew asymmetric reflections in order to demonstrate the potential of X-ray methods for assessing sub-micron layers.

### Acknowledgements

Acknowledgement is made to the Director, British Telecom Research Laboratories for permission to publish this paper. We would like to thank P C Spurdens, A K Chatterjee and A S M Ali for providing samples and M R Taylor and M Hockly for the TEM results. We have had the benefit of many useful discussions with B K Tanner and M J Hill of Durham University.

### References

Halliwell M A G 1981 Inst Phys Conf Ser No 60 271
Halliwell M A G, Juler J and Norman A 1983 Inst Phys Conf Ser No 67 365
Halliwell M A G, Lyons M H and Hill M J 1984 J Crystal Growth 68 523
Hill M J, Tanner B K and Halliwell M A G 1984 Mater Res Soc Fall Meeting, in press.
Moss R H and Spurdens P C 1984 Electronics Letts 20 978
Takagi S 1962 Acta Cryst 15 1311
Takagi S 1969 J Phys Soc Japan 26 1239
Taupin D 1964 Bull Soc Franc Mineral Crist 87 469

*Inst. Phys. Conf. Ser. No. 76: Section 11*
*Paper presented at Microsc. Semicond. Mater. Conf., Oxford, 25–27 March 1985*

# Recent advances in HVEM and HREM analyses of process-induced defects in silicon

C Claeys and H Bender*

IMEC, Kapeldreef 75, B-3030 Leuven, Belgium
*IMEC, c/o Universiteit Antwerpen, RUCA, Groenenborgerlaan 171,
 B-2020 Antwerpen, Belgium

Abstract   For almost three decades process-induced defects in silicon
have been extensively studied.   Only a limited number of systematic
approaches are reported.   This paper gives a general overview of the
defect spectra observed in single and multistep Czochralski silicon, and
highlights some recent HVEM and HREM analyses leading to a better understanding of the defect modelling.

## 1. Introduction

In the late 1950's the work on process-induced defects was mainly concentrating on diffusion-induced dislocations, precipitates due to metallic
contamination and thermally-induced dislocations. By improving the processing conditions these problems could mostly be overcome.  Later on it became
clear that the quality of the starting material plays a dominant role in
the generation of defects such as thermal donors, new donors, oxidation-
induced stacking faults and oxygen precipitates.  Therefore for more than
twenty years the role of grown-in defects (e.g. swirls), interstitial oxygen
and substitutional carbon has been extensively studied. However, by using
a larger variety of more sophisticated analytical techniques, the number
and the complexity of the observable defects also strongly increased. A good
understanding of the nucleation and the growth kinetics of these lattice
defects has become very crucial as the defects can have both a beneficial
or a harmful influence on the electrical performance of the devices.  Only
in recent years very systematic analyses taking into account the quality
of the starting material (oxygen, carbon, thermal history ...) and the
processing complexity (number and sequence of heat treatments, temperature,
ambient ...) have been reported (Maher et al 1976, Yang et al 1978,
Tempelhoff et al 1979,1981, Claeys et al 1983a, Bender 1984a,b).

This paper gives a general overview of the defect spectra observed in
single and multistep annealed Czochralski silicon, revealed by high voltage
and high resolution electron microscopy.   Special attention is given to
some recent advances in this field, leading to a better insight in the
defect modelling.

## 2. Experimental Procedure

The experiments are performed on [001] and [111] wafers coming from the
same [001] Czochralski-grown silicon ingot.  Although most of the wafers
had a high interstitial oxygen content of $11.5-11.7 \times 10^{17}$ atoms/cm$^3$ and a
low carbon content of $7.0 \times 10^{15}$ atoms/cm$^3$, some results for wafers with

a high carbon content of $3.3-4.7 \times 10^{17}$ atoms/cm$^3$ and a moderate oxygen content of $8.1-8.6 \times 10^{17}$ atoms/cm$^3$ are also reported.

After a standard RCA cleaning step different annealing steps are performed in respectively a nitrogen atmosphere in the temperature range 550 to 1000°C for 3 to 100h and in wet oxygen between 1000°C and 1150°C. The high temperature wet oxidation is preceded by a low temperature nitrogen anneal.

The wafers are investigated by means of high voltage electron microscopy (HVEM) in a JEOL 1250kV microscope and by means of high resolution electron microscopy (HREM) in a JEOL 200CX instrument. A special cross-section sample preparation technique is applied in order to obtain specimens in any orientation required (Vanhellemont et al 1983).

## 3. Results and Discussion

A typical defect spectrum as observed after respectively a low temperature nitrogen anneal (650-1000°C) and a two-step heat treatment (nitrogen anneal followed by a wet oxidation at 1150°C) has recently been discussed in detail by Bender (1984a,b) and is shown in Fig. 1. It can be noticed that in the low temperature region, rod-like coesite precipitates, amorphous platelike SiO$_x$ precipitates, 60° and 90° dislocation dipoles, and prismatic dislocation loops are found. By the subsequent wet oxidation step the defect distribution changes drastically and other types of defects such as e.g. Frank stacking faults, a/6 <114> stacking faults, octahedral SiO$_x$ precipitates and prismatic punching systems are observed. These defects nucleate heterogeneously on the defects generated by the low temperature anneal. As a systematic investigation of the different types of defects has been discussed

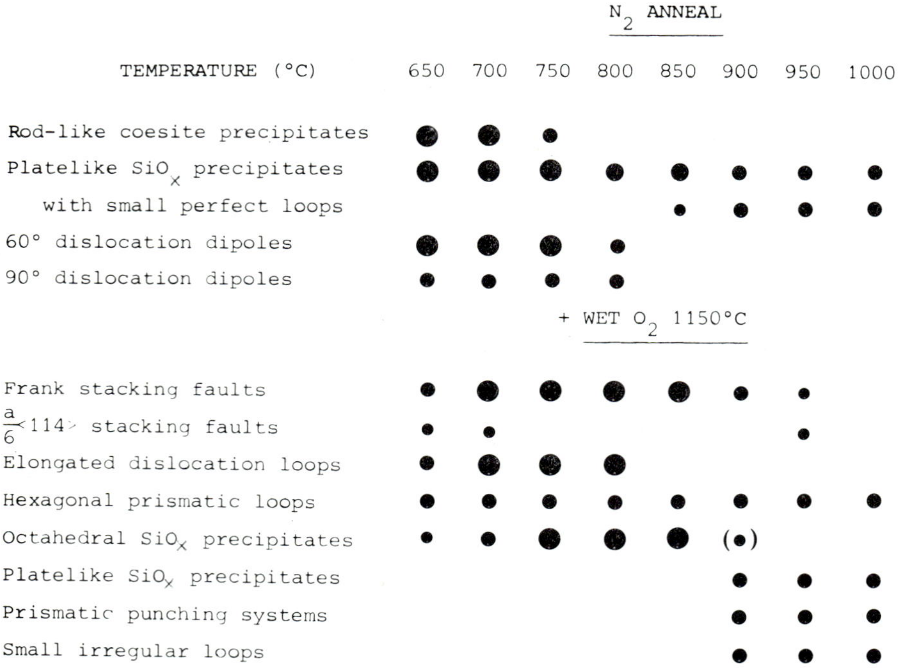

N$_2$ ANNEAL

| TEMPERATURE (°C) | 650 | 700 | 750 | 800 | 850 | 900 | 950 | 1000 |
|---|---|---|---|---|---|---|---|---|
| Rod-like coesite precipitates | ● | ● | • | | | | | |
| Platelike SiO$_x$ precipitates | ● | ● | ● | • | • | • | • | • |
| with small perfect loops | | | | | | • | • | • |
| 60° dislocation dipoles | ● | ● | ● | • | | | | |
| 90° dislocation dipoles | • | • | • | • | | | | |

+ WET O$_2$ 1150°C

| | 650 | 700 | 750 | 800 | 850 | 900 | 950 | 1000 |
|---|---|---|---|---|---|---|---|---|
| Frank stacking faults | • | ● | ● | ● | ● | • | • | |
| $\frac{a}{6}$<114> stacking faults | • | • | | | | | • | |
| Elongated dislocation loops | • | ● | ● | ● | | | | |
| Hexagonal prismatic loops | • | • | • | • | • | | • | • |
| Octahedral SiO$_x$ precipitates | • | • | ● | ● | ● | (•) | | |
| Platelike SiO$_x$ precipitates | | | | | | • | • | • |
| Prismatic punching systems | | | | | | • | • | • |
| Small irregular loops | | | | | | • | • | • |

Fig. 1 Defect distribution after a single and two-step anneal of silicon wafers with a high interstitial oxygen content. The size of the dots is for each type of defects a rough measure for the density distribution.

previously (Bender 1984a,b), only a few defect structures are further discussed by taking into account some recent advances in understanding the
defect behaviour.

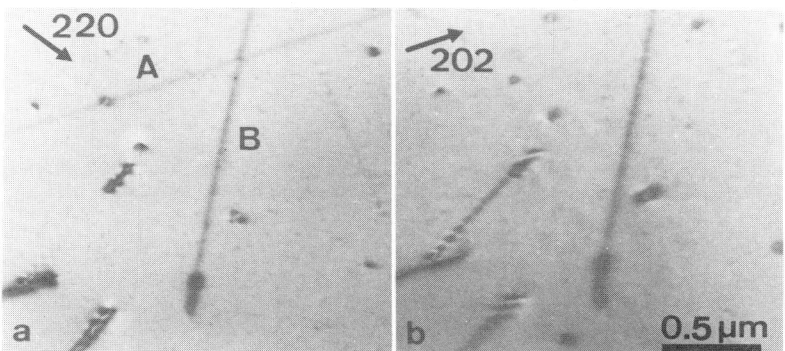

Fig. 2   Two rod-like defects (A and B) elongated in
<011> directions.   In b the defect A is in extinction.

## 3.1 Precipitates

### 3.1.1 Rod-like Coesite Defects

This type of defect is illustrated in Fig. 2, showing two rod-like defects
parallel with the specimen surface.  The defects are elongated in the <011>
directions and show a weak diffraction contrast independent of the imaging
condition.  A high resolution image of such a rod-like defect oriented along
the [011] axis is given in Fig. 3.  It can be noticed that the defect consists of a second phase in the silicon lattice.  Detailed analyses pointed
out that the second phase ribbons have the features of the coesite phase
(Bourret 1983,1984, Bender 1984a,b).  For the reported experiments, these
defects are present after annealings in the 550 to 750°C temperature range.
Recently, Bergholz et al (1984) noticed that the initial stage of the coesite
formation already occurs during annealing at 450°C and that such a preanneal

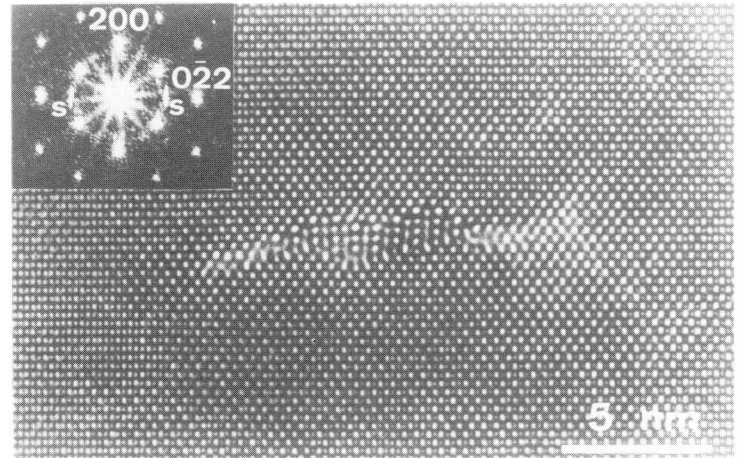

Fig. 3   A HREM image and optical diffraction pattern
of a coesite precipitate along the [011] zone axis.

strongly influences the rod-like defect density after a 650°C anneal. Due to the strong similarity with the behaviour of thermal donors in silicon, it seems likely that the thermal donors are related to the nucleation of the coesite ribbons.

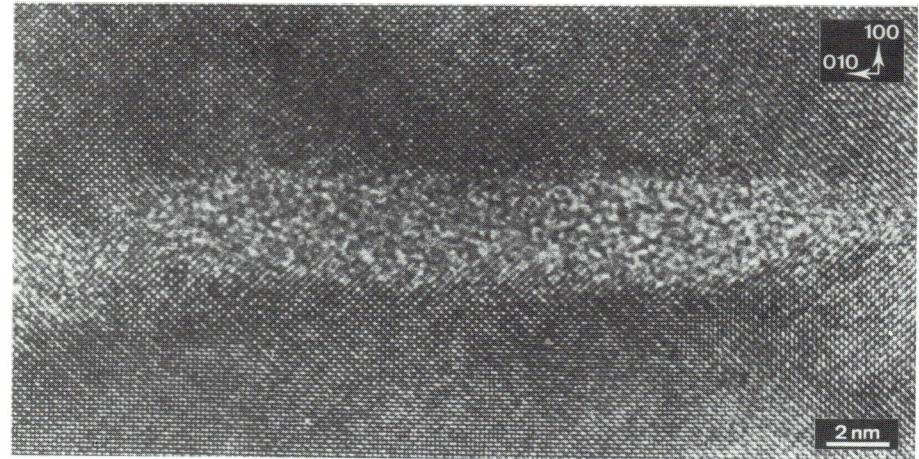

Fig. 4   A HREM image of a platelike precipitate in a [001] oriented specimen as present after an anneal at 850°C for 100h.

### 3.1.2 Amorphous Precipitates

After single step heat treatments square-shaped platelike precipitates are observed over the whole temperature region, whereas for two-step treatments the formation of these precipitates strongly depends on the preanneal temperature. Preanneal temperatures below 900°C result in the formation of truncated octahedral oxygen precipitates during the subsequent high temperature step.

At low preanneal temperatures the square-shaped precipitates are seen as small black dots in the two beam imaging mode. More detailed information on the exact shape and structure can only be obtained by high resolution imaging. At the lowest anneal temperatures (550°C-650°C) the precipitates are seen as black dots in the silicon matrix, which are caused by the associated stress field (Bender et al 1983a). More information can be obtained at higher temperatures. An example of such a case is given in Fig.4 for a sample annealed at 850°C for 100h. Inside the precipitate the amorphous phase is clearly visible. For large enough precipitates the two beam imaging technique indicates that these platelike precipitates are parallel to the (100) planes with <011> edges. For increasing anneal temperatures their saturation density decreases, while their edge size increases. The precipitate thickness saturates around 3.5-4nm. It can be remarked that for anneals at 485°C Bergholz et al (1985) also observed some dark regions on their TEM micrographs, which are however unstable under electron irradiation in the microscope. Most likely these dark blobs are the initial stage of very small $SiO_x$ precipitates. Due to their small size little lattice distortion is expected.

For two-step anneals either platelike or octahedral precipitates are found depending on the preanneal conditions. In the case of platelike precipitates they have the same nature as the ones found after single heat treatments, although their thickness increases to about 10nm. Mostly the precipitates are accompanied by prismatic punching systems in order to release the strain

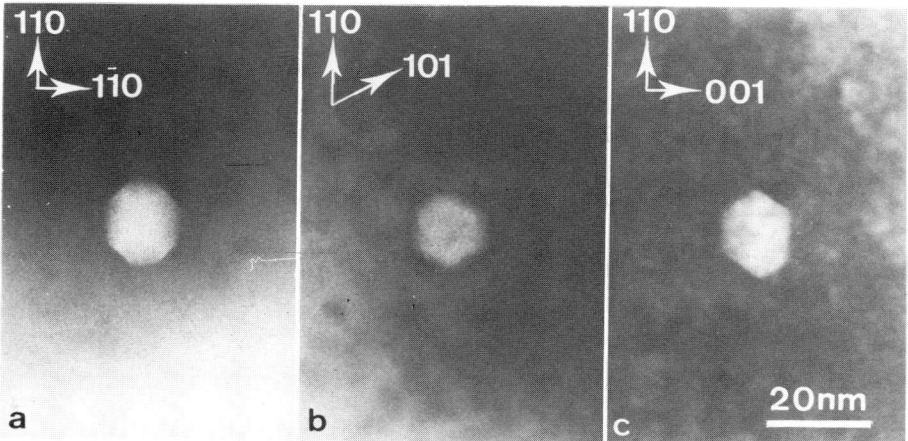

Fig. 5 Truncated octahedral precipitates observed for respectively [00$\bar{1}$](a), [$\bar{1}$11](b) and [$\bar{1}$10](c) pole orientations.

field. Octahedral shaped precipitates are found for preanneal temperatures below 900°C. Typical images for different pole orientations of such a precipitate are shown in Fig. 5. These shapes can be correlated with the projections of a truncated octahedron. The high resolution image in Fig. 6 clearly visualizes the amorphous character of the precipitates. Octahedral precip-

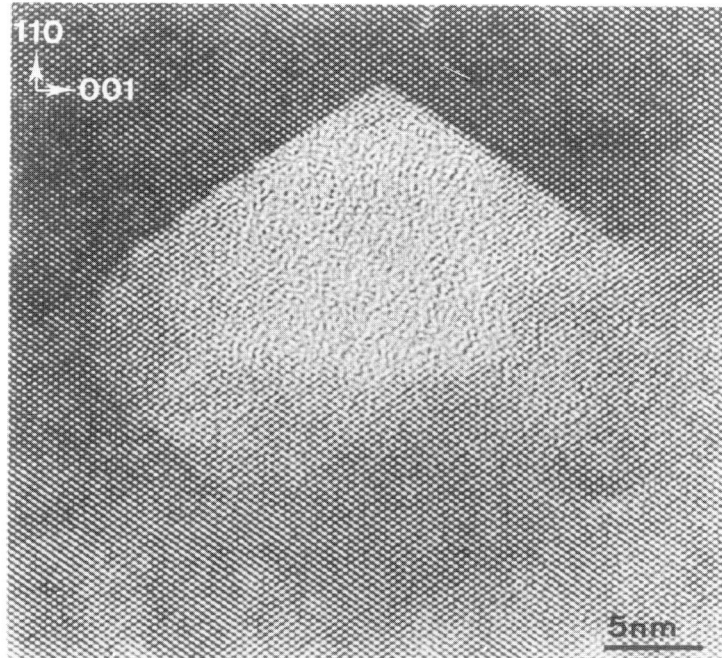

Fig. 6 In very thin specimen regions the truncated octahedral precipitates are seen to consist of amorphous material.

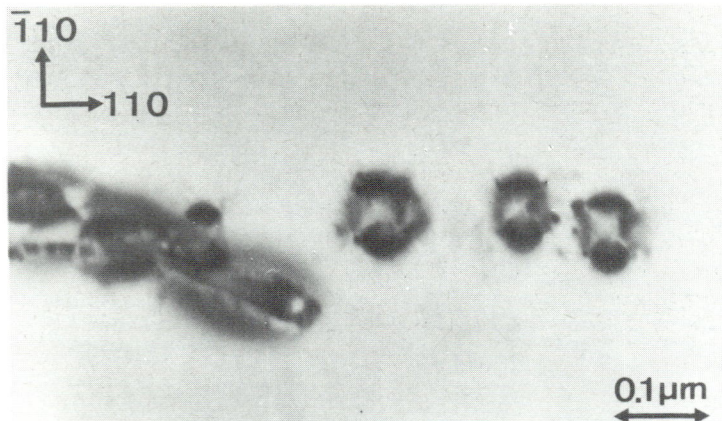

Fig. 7    Row of octahedral precipitates resulting from the break-up of a coesite ribbon during the high temperature treatment.

itates have also been observed by respectively Shimura (1981) and Matsushita (1982) and truncated ones by Ponce et al (1983).    The oxidation of wafers with a preanneal in the temperature range in which ribbon coesite precipitates form, results in the presence of rows of octahedral precipitates as shown in Fig. 7.

Recently Bender (1984a,b) proposed a model to explain the formation of the different types of oxygen precipitates.    During the low temperature anneal

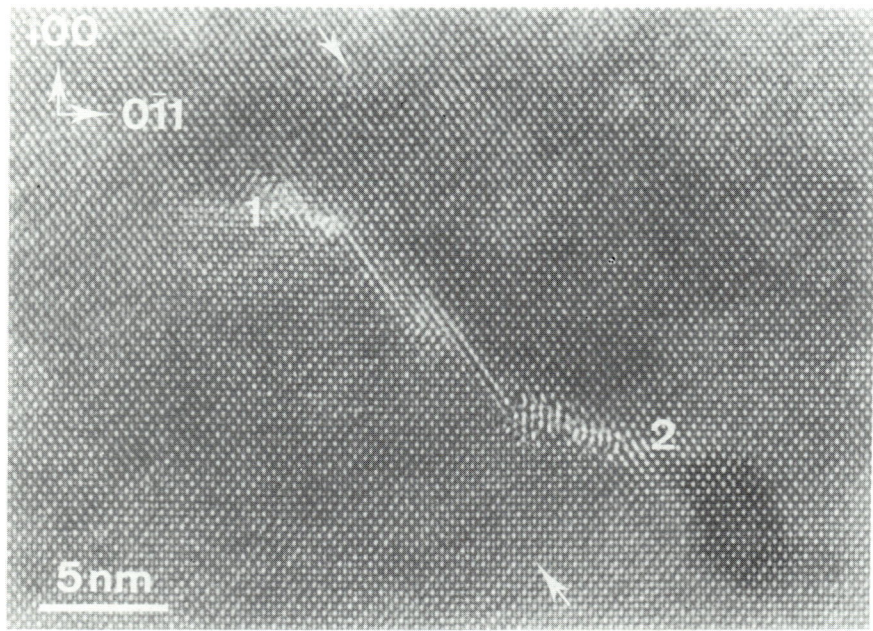

Fig. 8    A mixed 60° dipole-coesite precipitate defect complex. The arrows indicate the position of the extra half planes.

the homogeneous precipitation of oxygen results in the formation of amorphous platelike precipitates. Only for the lowest anneal temperatures rod-like coesite precipitates are also generated. During the subsequent high temperature anneal the precipitates will either break-up or change their shape. Annihilation of ribbon defects can generate rows of octahedral precipitates. Platelike precipitates accompanied by a dislocation system will remain platelike, while isolated platelikes are reshaped into truncated octahedral precipitates. The silicon interstitials associated with the oxygen precipitation processes are the basis for the generation of the other types of defects mentioned in Fig. 1.

## 3.2 Dislocation systems

For single heat treatments dislocation dipoles or prismatic punching systems are frequently observed. In the temperature region 550-800°C a large density

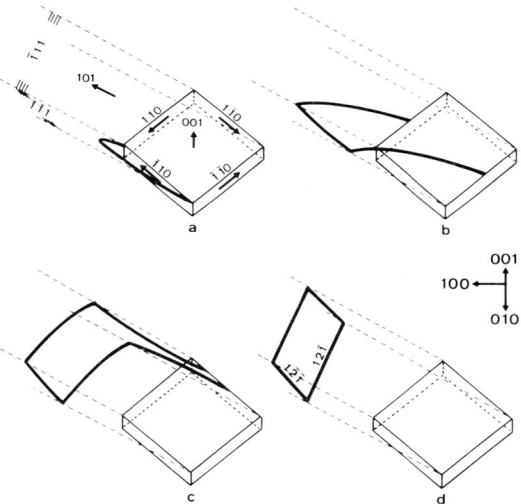

Fig. 9   Schematic representation of the generation mechanism of prismatic punched loops on platelike precipitates.

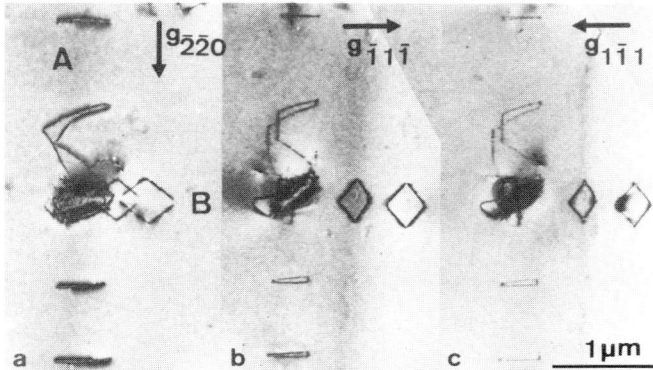

Fig. 10 Prismatic punching complex, showing the generation mechanism for different imaging conditions.

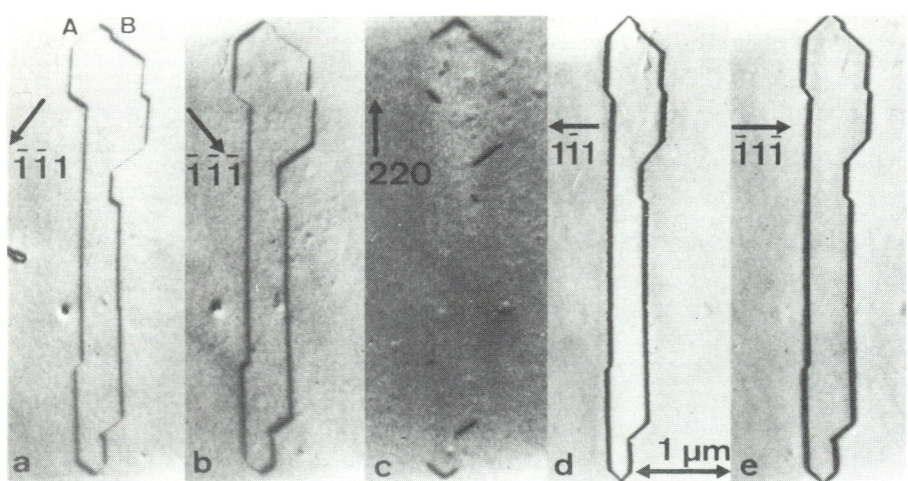

Fig.11 Elongated prismatic dislocation loop with interstitial character.

of 60° dipoles and a low density of 90° dipoles are found. These dipoles are undissociated excepted for the 60° dipoles observed in samples annealed above 750°C. The 60° dipoles are generally undecorated. They occur in alternating sequences with coesite precipitates along the <110> directions. Some mixed dipole-precipitate structures are also observed (Fig. 8). The undissociated 60° dipoles have their habit plane close to the (113) plane with a spacing between the dislocations of about 25nm. At higher temperatures the habit plane becomes the equilibrium (001) plane. The nucleation of the dislocation dipoles can be explained by the formation of so-called intermediate defect complexes as was suggested by Tan (1981).

At higher anneal temperatures the amorphous platelike precipitates are generating prismatic dislocation loops in order to release the stresses. A schematic representation of the different stages of the generation of the perfect dislocation loops is illustrated in Fig. 9, while Fig. 10 shows a prismatic punching system under different two-beam imaging conditions. The loops are punched out in the [110] and [1$\bar{1}$0] directions to which the loops are orthogonal. The size of the loops for a same defect system does not change as a function of the distance from the central precipitate, which is in contrast with the observations reported by Olivier (1975). Prismatic punching in silicon has been analysed in detail by Tan and Tice (1976). For the two-step experiments prismatic punching is only observed when the preanneal was done at temperatures above 900°C. At lower preanneal temperatures elongated 60° and 90° (Fig. 11) dislocation loops are found. Most likely these loops are formed on the 90° dislocation dipoles, although the smaller loops can also nucleate directly on the oxygen precipitates. The two-step treatments result in the formation of Frank stacking faults, which nucleate according to the Bardeen-Herring mechanism. The density and size of the stacking faults strongly depend on the preanneal treatment and are inversely proportional to each other (Claeys et al 1983a). As will be pointed out later, the carbon content has a strong influence on the distribution curves. For low temperature preanneals a/6 <114> stacking faults can be observed. These stacking faults are formed by the transformation of double Frank stacking faults (Bender et al 1983b, Claeys et al 1983b).

3.3 High carbon concentrations

For wafers with a high carbon and a moderate oxygen content the defect spectrum is different from the one given in Fig. 1 (Bender et al 1984c). After the low temperature anneal no defects could be detected. An overview of the major defects observed in wafers preannealed for 15h and subsequently heat treated for 20h in nitrogen at 1050°C is given in Fig. 12. Stacking faults (SF), prismatic punching systems (PP), large platelike precipitates (LP1) and tiny precipitates (prec) are found. The latter type of defect is

Defect distribution after the treatments
$N_2$ $T_1$ 15h + $N_2$ 1050°C 20h

| $T_1$(°C) | center of the wafer | | | edge of the wafers | | |
|---|---|---|---|---|---|---|
| 550 | - | PP | | - | PP | |
| 600 | - | PP | | - | PP | LP1 |
| 650 | SF | PP | | SF | PP | LP1 |
| 700 | SF | - | prec | SF | - | prec |
| 750 | - | - | prec | SF | - | prec |
| 800 | - | - | - | SF | - | prec |
| 850 | - | - | - | SF | - | prec |
| 875 | SF | - | prec | SF | - | prec |
| 900 | SF | PP | - | SF | PP | |
| 950 | SF | PP | LP1 | SF | PP | LP1 |
| 975 | - | PP | LP1 | - | PP | LP1 |
| 1000 | - | PP | LP1 | - | PP | LP1 |

Fig. 12 Overview of the different defect types as a function of the preanneal temperature in wafers with a high carbon concentration. Near the edge of the wafer the carbon content drops to a low value.

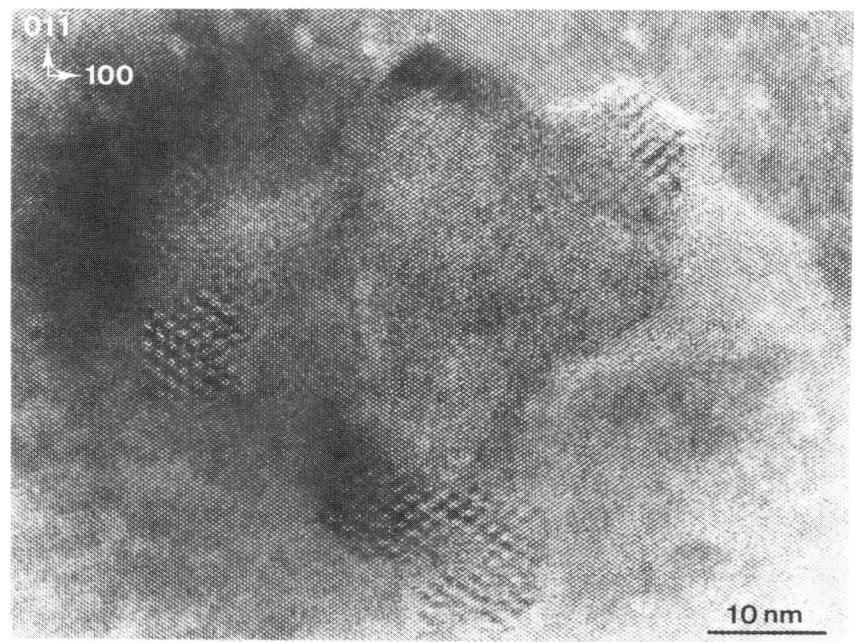

Fig. 13 HREM image of a precipitate in a silicon wafer with a high carbon content ($N_2$ 875°C 15h + $N_2$ 1050°C 20h).

illustrated in Fig. 13. The stacking fault density distribution versus the preanneal temperature goes through a minimum around 850°C. It is assumed that during the high temperature step the defects nucleate heterogeneously on the non-detectable defects generated during the low temperature preanneal. Most likely these nucleation centers are correlated to "C-O" complexes. This hypothesis is in agreement with the observations of Leroueille (1981) concerning the influence of the carbon content on the formation of new donors. These new donors are generated in the temperature range 600-900°C with a maximum generation rate at 800°C, which increases with increasing carbon content.

## Acknowledgement

This work has been supported by the Nationaal Fonds voor Wetenschappelijk Onderzoek (I.I.K.W.)

## References

Bender H, Claeys C, Van Landuyt J, Declerck G, Amelinckx S and Van Overstraeten R 1983a J. de Physique 44 C4-261
Bender H, Claeys C, Van Landuyt J, Declerck G, Amelinckx S and Van Overstraeten R 1983b Defect Complexes in Semiconductor Structures eds Giber J, Beleznay F, Szep I C and Laszko J (Berlin: Springer Verlag) pp 134-139
Bender H 1984a Ph.D-thesis Universitaire Instelling Antwerpen
Bender H 1984b Phys. Stat. Sol.(a) 86 245
Bender H, Claeys C, Van Landuyt J, Declerck G, Amelinckx S and Van Overstraeten R 1984c Proc. 13th Int. Conf. on Defects in Semiconductors eds Kimerling L C, Parsey J M (Pennsylvania: The Metallurgical Soc. of AIME) pp 587-593
Bergholz W, Pirouz P and Hutchison J L 1984 Proc. 13th Int. Conf. on Defects in Semiconductors eds Kimerling L C, Parsey J M (Pennsylvania: The Metallurgical Soc. of AIME) pp 717-723
Bergholz W, Hutchison J L and Pirouz P 1985 J. Appl. Phys. to be published
Bourret A 1983 J. de Physique 44 C4-227
Bourret A, Thibault-Desseaux J and Seidman D N 1984 J. Appl. Phys. 55 825
Claeys C, Bender H, Declerck G, Van Landuyt J, Van Overstraeten R and Amelinckx S 1983a Physica 116B 148
Claeys C, Bender H, Declerck G, Van Landuyt J, Van Overstraeten R and Amelinckx S 1983b Aggregation Phenomena of Point Defects in Silicon eds Sirtl E and Goorissen (Pennington: The Electrochem. Soc. Softbound Ser.) pp 74-88
Leroueille J 1981 Phys. Stat. Sol.(a) 67 177
Maher D M, Staudinger A and Patel J R 1976 J. Appl. Phys. 47 3813
Matsushita Y 1982 J. Crystal Growth 56 516
Olivier M 1975 Ph.D-thesis Grenoble
Ponce F A, Hahn S, Yamashita T, Scott M and Carruthers J R 1983 Inst. Phys. Conf. Ser. 67 65
Shimura F 1981 J. Crystal Growth 54 588
Tan T Y and Tice W K 1976 Phil. Mag. 34 615
Tan T Y 1981 Defects in Semiconductors eds Narayan J and Tan T Y (New York, Oxford: North Holland) pp 163-172
Tempelhoff K, Spiegelberg F, Gleichmann R and Wruck D 1979 Phys. Stat. Sol. (a) 56 213
Tempelhoff K, Hahn B and Gleichmann R 1981 Semiconductor Silicon eds Huff H R, Kriegler R J and Takeishi Y (Pennington: The Electrochem. Soc. Softbound Ser.) pp 244-253
Vanhellemont J, Bender H, Claeys C, Van Landuyt J, Declerck G, Amelinckx S and Van Overstraeten R 1983 Ultramicroscopy 11 303
Yang K H, Kappert H F and Schwuttke G H 1978 Phys. Stat. Sol.(a) 50 221

*Inst. Phys. Conf. Ser. No. 76: Section 11*
*Paper presented at Microsc. Semicond. Mater. Conf., Oxford, 25–27 March 1985*

# TEM for support of VLSI technology

H Oppolzer

Siemens AG, Research Laboratories, Munich, Fed. Rep. Germany

Abstract TEM methods applied to thin cross sections represent a powerful tool for studying technological problems during development of
VLSI circuits. On the one hand, these problems involve details in the
geometrical configuration of IC structures like sites of reduced oxide
thickness or the lateral extent of shallow pn junctions. On the other
hand, material problems like microstructure and interface texture of
the thin films used for interconnections can be investigated. As examples for such problems, contact reactions of $TaSi_2$ on Si and arsenic
segregation at the poly-Si/Si interface will be discussed.

## 1. Introduction

Device miniaturization recently resulted in the fabrication of integrated
circuits (ICs) with $10^5$ or more components per chip (very large scale
integration: VLSI). The minimum device dimensions are now approaching the
1 μm level. Advances in IC technology came about not only from the development of better processes for patterning by photolithography and etching
but also from improvement of material properties, like the resistivity of
interconnections. The latter, e. g., involved the use of new materials as
in the case of silicides. Examination of the various technological processes, besides electrical characterization, requires sophisticated analytical techniques.

The morphology of ICs is usually studied in the SEM by imaging the surface
or a cross section. The small device dimensions which can be as small as
10 nm in the case of thin gate oxides, however, call for the higher resolution of the transmission electron microscope (TEM). The possibility to
prepare thin cross sections through IC structures has been opening a wide
field of new TEM applications. This stems from the fact that TEM cross
sections allow to delineate the interfaces of the various thin films an IC
is composed of, with a resolution down to an atomic scale. Furthermore non
planar configurations as well as laterally confined phenomena can be studied. Since TEM cross sections offer both good depth perception as well as
good lateral resolution, they provide more information and  so  complement
depth profiling techniques like AES or RBS.

To apply TEM methods including analytical electron microscopy to problems
in IC technology in a routine type manner, elaborate specimen preparation
techniques are necessary. Therefore, a method to prepare thin cross sections through IC structures in actual devices is described shortly. The
technological problems of interest can be devided into two types (Oppolzer
and Huber 1983). The first type concerns details in the geometrical configuration of IC structures. Such details of electrical significance, e.

g., are sites of reduced gate oxide thickness. The second type involves material problems like microstructure and interface reactions of the thin films used for gate electrodes and interconnections. Examples of both types of problems will be presented.

## 2. Experimental

### 2.1 Cross Sections Through IC Structures

The preparation of cross sectional specimens of large area samples has become a routine method in recent years. As the first step, usually two pieces of a wafer or a stack of several thinned pieces are glued together. Cutting of about 150 µm thick slices from this stack by means of a diamond dicing saw as used for separating wafers into chips, has proven to be a valuable method as will be shown below. Mechanical thinning of the slices to a thickness of about 20 µm saves ion milling time and provides large transparent specimen areas.

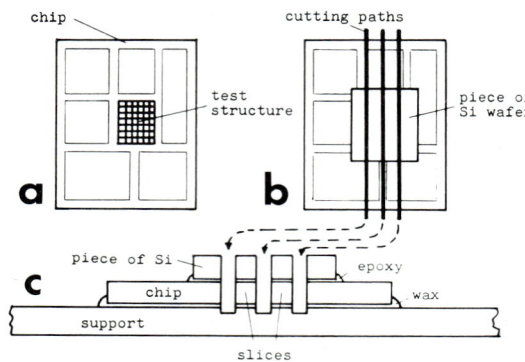

Fig. 1 Preparation of cross sectional specimens from IC structures:
(a) chip with test structure of interest,
(b) a piece of a Si wafer is glued upon the test structure and slices are cut with a diamond dicing saw,
(c) cross section through sample of Fig. (b) showing slices.

### 2.2 TEM Methods

For TEM imaging a JEOL 200 CX microscope was used. On bright field (BF) images of sufficiently thin specimens, the interfaces of thin films or IC structures can be delineated with a resolution of about 0.5 nm. Although this is sufficient in most cases, HREM images provide more information (e.g. Fig. 9d).

Energy dispersive x-ray microanalysis (EDX) was performed in a dedicated STEM equipped with a field emission gun, of type ELMISKOP ST 100F (Oppolzer and Knauer 1979). An electron probe with a diameter of about 1.5 nm and a current of 1 nA, together with a high collection angle allows analysis on thin specimen areas with good counting statistics. In this case beam broading is small and a spatial resolution for x-ray analysis of 5 to 10 nm is obtained.

## 3. Thin Dielectrics

Details in the geometrical configuration of IC structures which cannot be resolved in the scanning electron microscope (SEM), mainly involve thin dielectrics. Two examples of gate oxide thinning will be given below. Due to its higher dielectric constant compared to $SiO_2$, also thin $Si_3N_4$ layers will be used as gate dielectric in VLSI circuits.

For the preparation of thin cross sections through IC structures, the easiest and therefore best method is to use specially designed test patterns consisting of periodic long stripes (Marcus and Sheng 1983). Such TEM test patterns, however, are frequently not available when actual devices have to be investigated. IC structures, on the other hand, usually have periodic patterns as, e.g., in the cell matrix of memories. Test chips for device development also contain periodic structures as multi-transistors or arrays of contact windows. Their size, however, may be as small as, e.g., 100 x 100 $\mu m^2$. In this case a small piece of a Si wafer is glued upon the test structure on the chip (Fig. 1a,b). Then, the surrounding part of the chip can be used to adjust the position of the cutting paths for sawing (Fig. 1b,c). In this way it is assured that the test structure of interest is located within the slice cut from the stack. During final ion milling the section plane has to be controlled within the period of the structure such, that the cross section cuts through a specific element of the structure. Knowing the design of the test structure, this can be achieved by inspecting both surfaces of the specimen in a good optical microscope. When the desired section plane is reached on one surface of the cross sectional specimen, ion milling is continued only on the other side until transparency is obtained. By this procedure the section plane can be controlled with an accuracy down to about 1 $\mu m$.

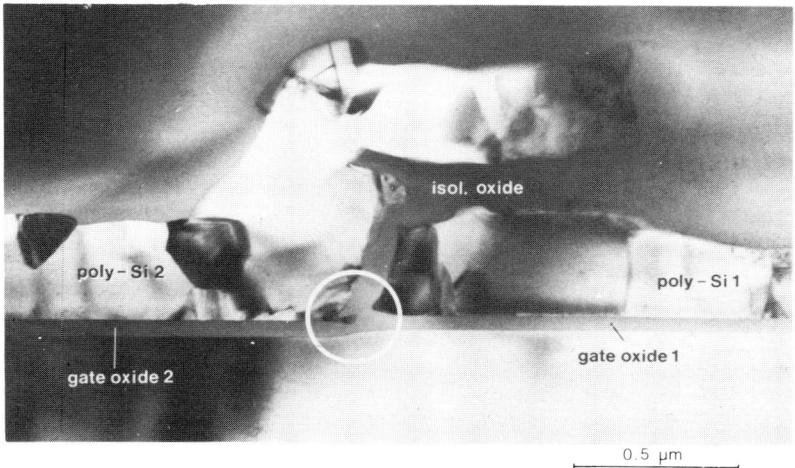

Fig. 2 TEM cross section through memory cell of RAM chip showing site of reduced gate oxide 2 thickness at polysilicon 1 edge (circle).

Fig. 2 shows a cross section through a memory cell of a random access memory (RAM) chip. Breakdown measurements of the gate oxide 2 yielded lower values than corresponding to the nominal oxide thickness. The site of reduced oxide thickness could be located to lie at the edge of the lower level poly silicon (poly-Si 1, circle). The undercutting of the poly-Si 1 edge which is formed by wet-chemical etching of the gate oxide 1, is not closed during gate oxide 2 growth, but filled in during poly-Si 2 deposition. Thus, a 50 nm long poly-Si "nose" is formed wich is analogous to the "cat's paw" described by Sinha et al (1979). The reduction of the gate oxide 2 thickness from 50 nm to 30 nm at the poly-Si 2 nose corresponded well to the reduction in breakdown voltage. In a new cell design this configuration resulting in reduced gateoxide 2 thickness was avoided.

The bird's beak configuration which is formed at the edge of the field oxide during selective oxidation reduces the available active surface area. Fig. 3a shows a bird's beak after an additional oxidation (and oxide stripping), and gate oxidation. The length of the bird's beak depends on various process parameters (e.g. Lemme and Oppolzer 1981). Compared to the bird's beak of Fig. 3b (detail of beak like in Fig. 3a), the beak length in Figs. 3c,d is smaller due to a greater thickness of the $Si_3N_4$ mask and also due to overetching of the field oxide. The latter was especially strong for the beak in Fig. 3d. Besides the problem of beak shape which can be studied with sufficient accuracy in the SEM, there is a reduction of the gate oxide thickness at the tip of the bird's beaks. This gate oxide thinning is a serious problem, especially when the gate oxides are very thin (around 10 nm). For the long beak of Fig. 3a with a small beak angle of only 20°, the gate oxide thickness of 11 nm is reduced by 22 % (Fig. 3b). The larger beak angles of about 60° in Figs. 3c,d result in an increased gate oxide thinning of 35%. This is analogous to the increase of oxide thinning with the slope of etched field oxide walls (Sheng and Marcus 1978). The amount of oxide thinning could not be explained fully by modelling the two dimensional oxidation process (Wilson 1982). Probably, mechanical stress at the oxide edge reduces the oxidation rate.

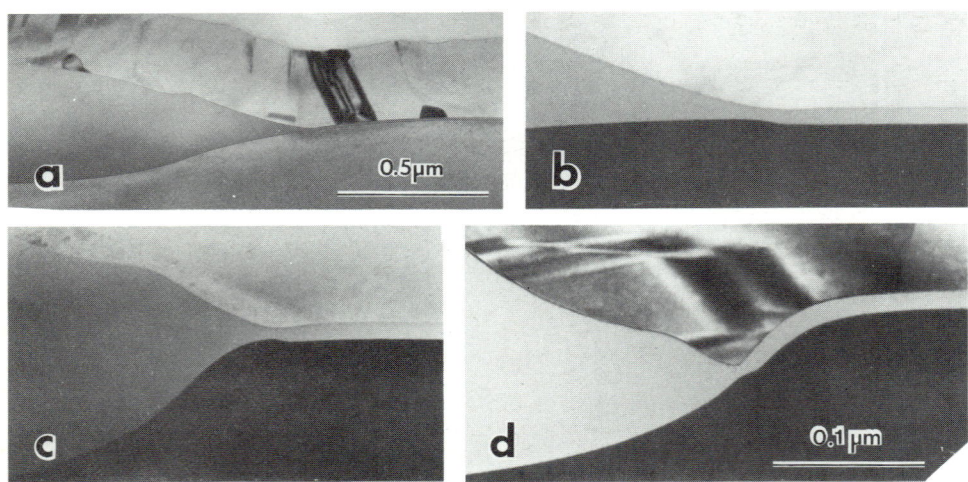

Fig. 3 (a) Bird's beak configuration at field oxide edge; (b) detail of beak like in Fig. (a); (c) and (d) larger beak angles result in increased gate oxide thinning.

Gate oxides of only 10 nm thickness are liable to yield and reliability problems. Therefore, the use of $Si_3N_4$ layers as gate dielectric is promising since their higher dielectric constant allows to achieve the same capacitance with larger film thicknesses than with $SiO_2$. By plasma-enhanced thermal nitridation $Si_3N_4$ films with a thickness of several nm can be grown (Ito et al 1981). To improve their dielectric strength they were annealed in dry oxygen. The 4 nm thick $Si_3N_4$ film of Fig. 4, however, contained pores at which thermal oxidation took place during oxygen annealing. Similar to the selective oxidation process with a $Si_3N_4$ mask, small oxide islands grew (Fig. 4b).

To maintain the well proven Si-SiO$_2$ interfaces, it is favourable to use a triple layer dielectric consisting of oxide - nitride - oxide (Sunami et al 1984). In this case the Si$_3$N$_4$ layer is deposited by chemical vapour deposition (CVD). Fig. 4c shows such a triple layer between the Si substrate and the polysilicon. The darker nitride is clearly to be distinguished from the oxides. The edge configuration is determined by lateral oxidation resulting in a bird's beak below the nitride edge, and another oxide beak at the lower edge of the polysilicon.

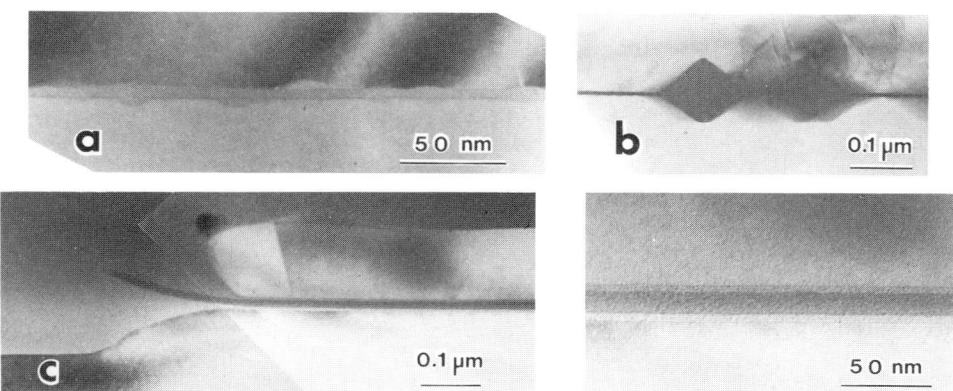

Fig. 4  Si$_3$N$_4$ layers as gate dielectric;  (a)  and  (b)  plasma-enhanced thermally grown and oxygen annealed Si$_3$N$_4$; (c) triple layer dielectric: SiO$_2$ - CVD Si$_3$N$_4$ - SiO$_2$.

## 4.  Geometrical Transistor Parameters

The effective channel length of MOS transistors is determined by the lateral extent of the source and drain regions beneath the gate electrode. The knowledge of this overlap is especially relevant for short-channel devices. Fig. 5a shows a schematic cross section of a transistor employing a double layer of polysilicon and TaSi$_2$ as gate electrode. The geometrical parameters of interest are indicated. The parameters (except the junction depth $x_j$) can be determined by capacitance measurements on a multitransistor test structure (Vitanov et al 1984). To check these measurements with a more direct method, cross sections through individual transistors of test structures on several chips which have been characterized electrically, were imaged in the TEM (Oppolzer et al 1985). The junctions of the arsenic implanted source and drain regions were delineated by preferential etching in a mixture of 0.5 % HF in HNO$_3$ (Sheng and Marcus 1981). Fig. 5b shows an example of numerous TEM images. Within the source/drain regions, residual damage from the arsenic implantation is visible. In the table below, the mean values from TEM cross sections of several transistors in one test structure are compared to the results from the electrical measurements. The parameters listed are the effective channel length $L_{eff}$, the overlap d, and the junction depth $x_j$:

| Method | $L_{eff}$ | d | $x_j$ |
|---|---|---|---|
| TEM cross sections | 2.04 µm | 0.28 µm | 0.28 µm |
| capacitance meas. | 1.91 µm | 0.30 µm | - |

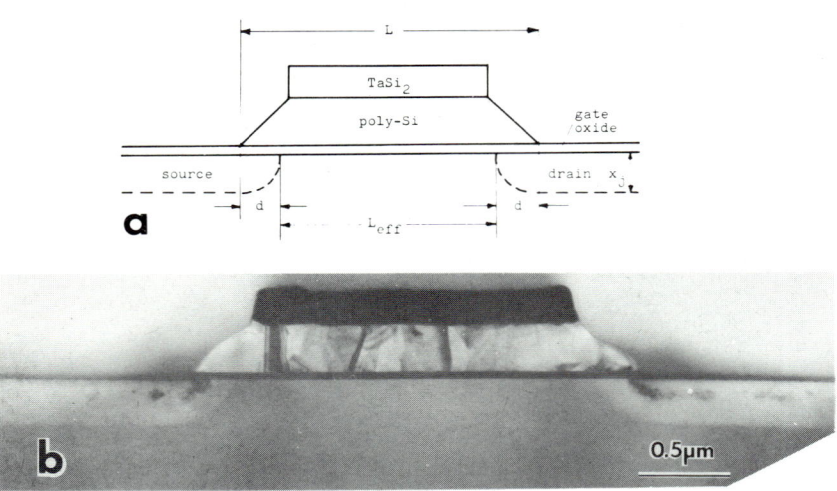

Fig. 5 Cross section through MOS transistor: (a) diagram showing geometrical transistor parameters (L: gate length, $L_{eff}$: effective channel
length, d: overlap, $x_j$: junction depth); (b) TEM cross section with preferentially etched source/drain regions.

The ratio of overlap to junction depth $d/x_j = 1$. Because of the small
slope (~45°) of the polysilicon edge, implanted As ions could penetrate
the thin parts of the edge as can be seen from the residual damage. Thus,
the overlap is increased. For steeper mask edges smaller values of $d/x_j$
around 0.8 were found by Sheng and Marcus (1981) and also by Roberts et al
(1983). Comparison with a SIMS depth profile showed that the delineation
depth corresponds to an arsenic concentration of $10^{19}$ cm$^{-3}$ and is therefore smaller than the junction depth since the channel doping concentration is much lower. Because of the large gradient of arsenic profiles this
difference, however, is small (about 0.05 μm for a channel doping concentration of $10^{16}$ cm$^{-3}$) but, nevertheless, shows up in the table: the TEM
value for the overlap d is slightly smaller than the value from the electrical measurements, and vice versa for $L_{eff}$. The very good agreement
between both methods implies that the capacitance measurements can be
effectively used for process control.

## 5. Interface Reactions of TaSi$_2$ in Contact Windows to Si and of TaSi$_2$ on SiO$_2$

Because of their smaller sheet resistance, double layers of TaSi$_2$ on polysilicon are recently used at the gate level of VLSI circuits (Fig. 5). As
first level interconnections also single layers of TaSi$_2$ appear promising,
especially for CMOS devices. After annealing at 900°C, the contacts of
such TaSi$_2$ interconnections to the Si substrate, however, showed poor
electrical behaviour. This was supposed to be related to interface reactions in the contact windows caused by slight deviations from stoichiometry during deposition. Such slight deviations cannot be avoided during
cosputtering of Ta and Si. To study this effect, Ta-Si films with ±2 at%
deviation from stoichiometry were deposited, and TEM cross sections
through contact windows with a size of 4 x 4 μm$^2$ were investigated (Oppolzer et al 1984).

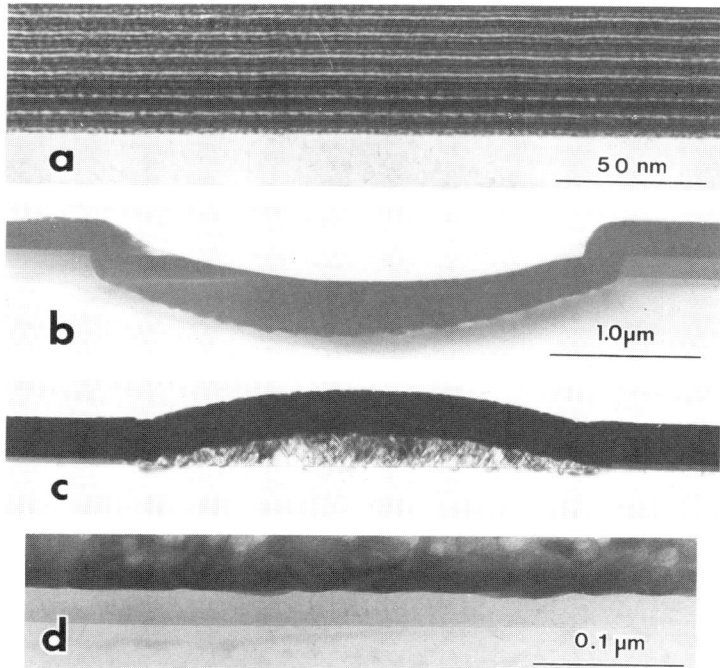

Fig. 6 (a) Multilayer structure of Ta-Si film after cosputtering; (b) and
(c) interface reactions in contact windows resulting in pits for Ta rich
films (b) and Si hillocks for Si rich films (c); (d) interface reaction of
$TaSi_2$ film with gate oxide.

Fig. 6a shows the multilayer structure of a Ta-Si film which is amorphous
after cosputtering. The thickness of a Ta-Si double layer is 1.5 nm. During annealing of laterally unconfined Ta-Si films on Si, stoichiometry is
obtained by reaction with the Si substrate involving - as in the case of
$TaSi_2$ on polysilicon - diffusion only in vertical direction. $TaSi_2$ interconnections, however, mainly lie on $SiO_2$, and the area of the contact
window is small compared to the total area. Since the reactions are restricted to the contact windows they will be enhanced. For Ta rich films,
Si from the substrate is consumed and deep pits are formed in the contact
windows (Fig. 6b) causing junction leakage. For Si rich films, the excess
Si precipitates epitaxially in the contacts (Fig. 6c) resulting in non
linear I-V characteristics. The Si hillocks contain numerous microtwins
and the interfacial oxide is broken up into small oxide particles similar
to the one shown in Fig. 7b. The strong contact reactions imply a huge
lateral mass transport which was found to extend over several 10 µm and
which is supposed to take place during the initial stage of recrystallization of the Ta-Si films. Barrier layers like TaN could not prevent the
detrimental contact reactions at 900°C. Possibly, this can be achieved by
depositing inherently stoichiometric and already crystalline films, as it
is feasible by CVD.

When $TaSi_2$ is used as gate material, during annealing slight reactions
with the gate oxide can occur depending on the deposition parameters (Fig.
6d). In case of not optimized deposition conditions, individual $TaSi_2$

grains at the interface protrude into the gate oxide. Obviously, the resulting interface roughness degrades the breakdown strength of thin gate
oxides.

## 6. Polysilicon on Silicon

In modern bipolar processes poly-Si is deposited directly onto the Si
substrate. Arsenic doped poly-Si then serves as diffusion source and con-
tacting layer for shallow emitter bipolar transistors, and boron doped
poly-Si is used to diffuse the base connections. Depending on the surface
treatment prior to poly-Si deposition, the interfacial oxides remain continuous during annealing or can break up resulting in epitaxial regrowth
of the poly-Si (Wilson et al 1982). The interfacial oxides not only control the dopant diffusion into the substrate, but can also have a favourable influence on bipolar transistor parameters (Graul et al 1976). Fig.
7a shows a B-implanted poly-Si film after 60 min annealing at 950 °C.
After a threshold time of several minutes the residual interfacial oxide
formed after etching in buffered hydrofluoric acid (HF dip) breaks up into
small particles (Fig. 7b), and epitaxial regrowth proceeds with a velocity of about 2 nm/min. For high dose As implantation which causes grain
growth in poly-Si, regrowth is much faster leaving only a few defects
(Fig. 7c).

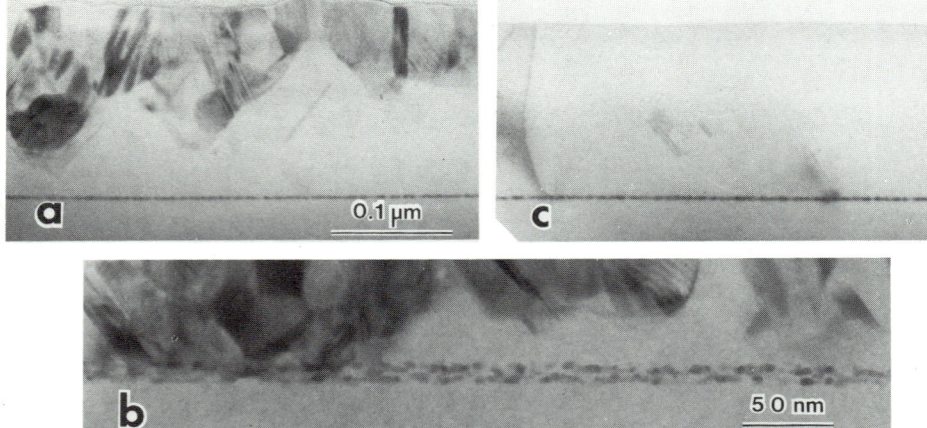

Fig. 7 Epitaxial regrowth of polysilicon films on Si (with HF dip) after
60 min annealing at 950°C; (a) boron doped poly-Si; (b) breaking up of
interfacial oxide (interface tilted); (c) fully regrown arsenic doped
poly-Si.

By depth profiling techniques As was shown to segregate at the interface
poly-Si/Si substrate (e.g. Wilson et al 1982). To compare the As segregation for various interfacial oxides with the segregation at the poly-Si/
$SiO_2$ interface and at grain boundaries within the poly-Si, As profiles
were measured in a STEM allowing EDX with high spatial resolution (Oppolzer et al 1985). As segregation to grain boundaries in poly-Si was recently studied by Grovenor et al (1984). Fig. 8b shows a typical As profile across a grain boundary as indicated at A in Fig. 8a. At the grain
boundary the As concentration is higher by a factor of about 3. For measuring the amount of impurity segregation, the increase in As concentra-

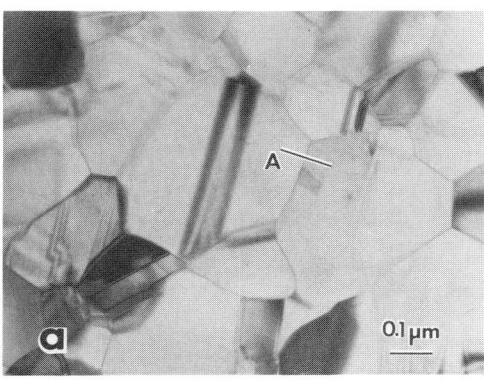

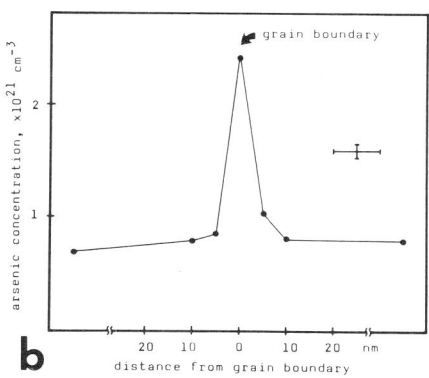

Fig. 8 (a) TEM image of highly arsenic doped polysilicon film; (b) Arsenic profile across grain boundary (like at A in Fig.(a)) measured by STEM-EDX.

tion when analyzing a well defined specimen volume containing a grain boundary, was compared to the As concentration in the grain volume (without grain boundary). Values of 0.7 to $1.5 \cdot 10^{15}$ cm$^{-2}$ As atoms at the grain boundaries were found, the mean value of which corresponds to about one monolayer. From the grain size of 0.3 μm follows that 30 % of the total As fraction of $8 \cdot 10^{20}$ cm$^{-3}$ is segregated to the grain boundaries.

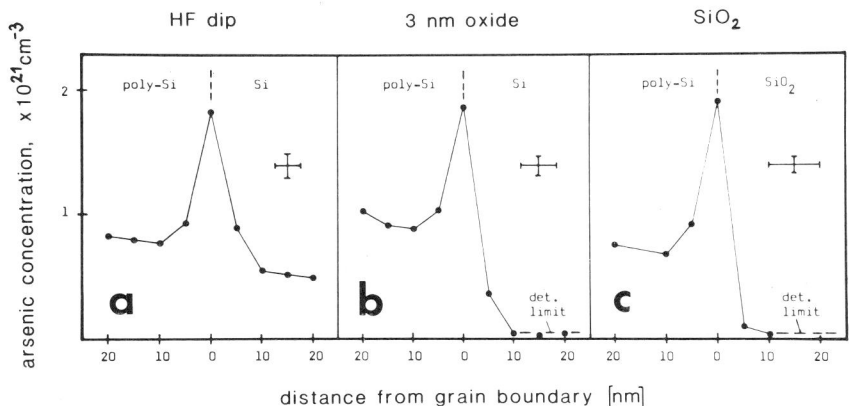

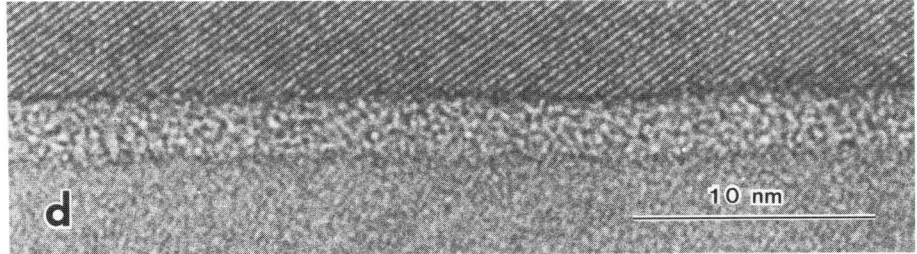

Fig. 9 (a) and (b) Arsenic profiles across polysilicon/silicon interfaces with interfacial oxide broken up (a), resp. 3 nm thermal oxide (b); (c) Arsenic profile across polysilicon/$SiO_2$ interface; (d) HREM image of 3 nm thermal oxide between poly-Si (top) and Si substrate (bottom).

In Figs. 9a,b,c the As segregation at the various interfaces is compared. In the sample of Fig. 9a (with HF dip) the interfacial oxide was broken up. The 3 nm thick thermal oxide in the sample of Fig. 9b is shown in the HREM image of Fig. 9d. The profile of Fig. 9c stems from a poly-Si/SiO$_2$ interface. Surprisingly, the maximum concentrations at the various interfaces are about equal. Since they reach the same level as the maxima at the grain boundaries, also about one monolayer is segregated to the various interfaces. The width of the region with increased As concentration is $<5$ nm in all cases.

## 7. Conclusions

During experiments for IC process development, electrical characterization of the samples is the first step. In many cases, however, more direct analytical methods are required, providing a better insight into the chemical and physical processes involved. This insight allows advances in technology development. Because cross sectional TEM allows to study small device geometries and the interfaces of thin films, this technique represents a powerful tool for support of VLSI technology.

## Acknowledgements

The author is indebted to W. Eckers, V. Huber, and S. Schild for assistance in the TEM work, to L. Reidt for excellent photographic work, and to his colleagues from the technology group for very good cooperation and discussions.

## References

Graul J, Glasl A and Murrmann H 1976 IEEE J. Solid State Circuits SC-11 491
Grovenor C R M, Batson P E, Smith D A and Wong C 1985 Phil. Mag. A 50 409
Ito T et al 1981 Appl. Phys. Lett. 38 370
Lemme R and Oppolzer H 1981 Semiconductor Silicon 1981 (The Electrochem. Soc.) 81-5 811
Marcus R B and Sheng T T 1983 TEM of silicon VLSI circuits and structures (New York: Wiley) pp 26-8
Oppolzer H and Knauer U 1979 Scanning Electron Microsc. 79/I 111
Oppolzer H and Huber V 1983 Inst. Phys. Conf. Ser. No.67: Sect.10 461
Oppolzer H, Neppl F, Hieber K and Huber V 1984 J.Vac.Sci.Technol. B 2 630
Oppolzer H, Eckers W and Schaber H 1985a J. de Physique (Colloque C4) in press
Oppolzer H, Hiergeist P and Schild S 1985b to be published
Roberts M C, Booker G R, Davidson S M and Yallup K J 1983 Inst. Phys. Conf. Ser. No.67: Sect.10 467
Sheng T T and Marcus R B 1979 J. Electrochem. Soc. 125 432
Sheng T T and Marcus R B 1981 J. Electrochem. Soc. 128 881
Sinha A K et al 1979 Proc. 17th Ann. Reliability Phys. Symp. (San Francisco) p 35
Sunami et al 1984 IEEE Trans. Electron Devices ED-31 746
Vitanov P, Schwabe U and Eisele I 1984 IEEE Trans. Electron Devices ED-31 96
Wilson L O 1982 J. Electrochem. Soc. 129 831
Wilson M C et al 1982 J. de Physique 43 Suppl. No.10 C1-253

*Inst. Phys. Conf. Ser. No. 76: Section 11*
*Paper presented at Microsc. Semicond. Mater. Conf., Oxford, 25–27 March 1985*

471

# TEM investigation of the effect of anneal temperature and arsenic concentration on the polysilicon/thin oxide/single-crystal-silicon emitter of a new high-performance bipolar transistor

N Jorgensen[1], J C Barry[1], G R Booker[1], P Ashburn[2], G R Wolstenholme[2], M C Wilson[3] and P C Hunt[3]

[1]Department of Metallurgy & Science of Materials,
 University of Oxford, Parks Road, Oxford OX1 3PH
[2]Department of Electronics and Information Engineering,
 The University, Southampton SO9 5NH
[3]Plessey Research (Caswell) Ltd, The Allen Clark Research Centre,
 Caswell, Towcester, Northants NN12 8EQ

Abstract   A new type of Si bipolar transistor is being fabricated by a conventional ion-implantation process but with the emitter consisting of a polysilicon/thin-oxide-film/single-crystal Si combination.   TEM lattice-image and weak-beam examinations were performed to investigate the effect of different annealing temperatures and different dopant concentrations.   The observed structures are correlated with the measured transistor gains.

## 1. Introduction

Bipolar transistors are being fabricated so that higher gains and/or operating frequencies can be obtained without having to reduce device dimensions.   The fabrication is based on standard ion implantation processes except that the emitter consists of a combination of polysilicon, thin oxide film and single-crystal silicon (Fig.1).   The presence of the oxide film modifies the flow of electrons and holes through the device so that the base current is reduced.   For the device to be successful, the oxide film ideally needs to be continuous, uniform and of the correct thickness, both when initially formed and at the end of the fabrication process. Initial oxide film thicknesses in the range ~0.5 to ~2.5nm are being investigated.

The standard fabrication procedure used for such devices is as follows. (100) n-type Si slices are $B^+$ ion implanted to form the base.   The slices are given a chemical solution 'cleaning' treatment to produce the thin oxide film.   A polysilicon layer is deposited, and the layer is $As^+$ ion implanted and given an oxidation drive-in.   The drive-in causes the As to diffuse rapidly down the grain boundaries of the polysilicon, through the oxide film and into the underlying single-crystal Si to form the e/b junction.

Previous TEM examinations of such structures (Wilson et al 1982, Albu-Yaron et al 1984) using an $As^+$ ion dose of $5 \times 10^{15}$ $cm^{-2}$ and a drive-in at 1000°C showed that the oxide became increasingly non-uniform and eventually 'balled-up'.   As this occurred the underlying single-crystal silicon progressively 'penetrated' the oxide film and grew into the polysilicon layer, the latter eventually becoming entirely single-crystal.   An oxidation drive-in temperature of 1000°C is clearly too high for such device fabrication, and so for the continuation of the project the drive-in temperature was reduced to 900°C.

In the present work two separate experiments were performed, each based on a slightly different fabrication procedure (Table). Firstly, the Si slices were 'cleaned' using an RCA solution (HCl, $H_2O$, $H_2O_2$). After the polysilicon layer deposition, and before the $As^+$ ion implantation, the slices were pre-annealed at either 900, 950 or 1000°C. The aim of this experiment was to 'break-up' the oxide film by the pre-anneal, and then to see what effect this had on the subsequent device fabrication. Control specimens were prepared using an HF 'clean'. Secondly, the Si slices were 'cleaned' using an HF/SP solution (HF followed by $H_2SO_4$, $H_2O_2$). The slices were implanted with $As^+$ ions to doses of either $5x10^{15}$, $1x10^{16}$ or $2x10^{16}$ cm$^{-2}$. The aim of this experiment was to see what effect different As concentrations had on the device fabrication procedure. Control specimens were prepared using an HF/SP/HF solution (HF followed by SP followed by HF).

TEM examinations were made of the various slices corresponding to the stage after the oxidation drive-in. Plan-view and cross-sectional specimens were prepared by mechanical polishing and $Ar^+$ ion-beam milling. Two-beam, weak-beam and (110) lattice-imaging examinations were performed using Philips EM300 and JEOL 200CX instruments. The observed structures were correlated with the measured transistor gains (Table).

## 2. Experiment 1

For the RCA/900°C specimen, the TEM lattice-image cross-section micrographs (Fig.2a) showed that the oxide film was slightly non-planar, but was nevertheless continuous and of uniform thickness 1.4±0.2nm. For the RCA/950°C specimen, the oxide film was more irregular and in some regions the underlying single-crystal Si was just beginning to penetrate into the polysilicon (Fig.2b). For the RCA/1000°C specimen, many regions were now penetrated, the regrown areas extending up to ~100nm into the layer (Fig.2c).

For the HF/no-pre-anneal specimen, an oxide film of thickness 0.8±0.2nm was present, but this was penetrated in many places, the regrown areas extending up to ~2nm into the layer(Fig.3a). For the HF/1000°C specimen, the oxide film was extensively 'balled-up', large areas of the polysilicon layer being regrown (Fig.3b). The oxide 'balls' were up to ~5nm across.

## 3. Experiment 2

For the HF/SP/no-implant specimen, the oxide film was planar and of uniform thickness 1.0±0.2nm (Fig.4a). For the HF/SP/$5x10^{15}$ cm$^{-2}$ specimen, the oxide film was slightly non-planar but was nevertheless still continuous (Fig.4b). For the HF/SP/$1x10^{16}$ cm$^{-2}$ specimen, the oxide film was penetrated in many places (Fig.4c). For the HF/SP/$2x10^{16}$ cm$^{-2}$ specimen, the penetration was greater with the regrown areas extending up to ~35nm into the layer (Fig.4d).

For the HF/SP/HF/no-implant specimen, the oxide film was very thin and not continuous (Fig.5a). For the HF/SP/HF/$5x10^{15}$ cm$^{-2}$ specimen, the oxide film was balled-up and the regrown areas extended ~50nm into the layer (Fig.5b). For the HF/SP/HF/$1x10^{16}$ cm$^{-2}$ specimen, and also the

HF/SP/HF/$2\times10^{16}$ cm$^{-2}$ specimen, the behaviour was similar except that all of the polysilicon layer had regrown (Fig.5c). The oxide 'balls' were up to ~4nm across.

## 4. Discussion

The RCA, HF/SP and HF 'cleans' produced initial oxide films of thickness ~1.4, ~1.0 and ~0.8nm respectively, while the HF/SP/HF 'clean' produced an initial oxide film which was still thinner and possibly discontinuous in places.

For all of the specimens investigated, a consistent trend occurred. When the oxide film was continuous, the transistor gain was high (for the particular type of transistor). When the oxide film was initially penetrated, the gain decreased. When the oxide film 'balled-up', the gain further decreased. The 'break-up' of the oxide film was more rapid when the temperature was higher, and when the As concentration was higher.

The reason for the 'break-up' of the oxide film is considered to be due mainly to a reduction in energy due to a decrease in the total area of the oxide/silicon interfaces. Our previous analysis (Albu-Yaron et al 1984) indicated that this 'break-up' occurred without significant change taking place in the total volume of the oxide, i.e. the behaviour was associated with local diffusion. A possible reason why the increased As concentration increased the oxide film 'break-up' is that As segregated at the oxide film interface, and this As either increased the local diffusion rate or decreased the oxide/silicon interfacial energy.

## Acknowledgments

We wish to acknowledge support by the Science and Engineering Research Council and the Alvey Directorate.

## References

Wilson M C, Ashburn P, Soerwirdjo B, Booker G R and Ward P 1982 J Physique 43 C1-253
Albu-Yaron A, Barry J C and Booker G R 1984 Proc. 8th European Congress on Electron Microscopy ed A Csanady (Budapest) p521

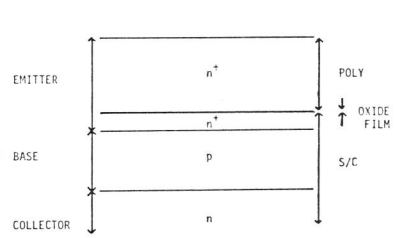

Fig.1 Schematic of a device structure.

| Clean | Poly (nm) | Anneal (°C) | As⁺ dose (cm⁻²) | Drive-in (°C) | Gain |
|---|---|---|---|---|---|
| RCA | 400 | 900 | $1\times10^{16}$ | 900 | 1300 |
| RCA | 400 | 950 | $1\times10^{16}$ | 900 | 360 |
| RCA | 400 | 1000 | $1\times10^{16}$ | 900 | 90 |
| HF | 400 | - | $1\times10^{16}$ | 900 | 250 |
| HF | 400 | 1000 | $1\times10^{16}$ | 900 | 50 |
| HF/SP | 200 | - | 0 | - | |
| HF/SP | 200 | - | $5\times10^{15}$ | 900 | 22 |
| HF/SP | 200 | - | $1\times10^{16}$ | 900 | 12 |
| HF/SP | 200 | - | $2\times10^{16}$ | 900 | 5 |
| HF/SP/HF | 200 | - | 0 | - | |
| HF/SP/HF | 200 | - | $5\times10^{15}$ | 900 | 2 |
| HF/SP/HF | 200 | - | $1\times10^{16}$ | 900 | 4 |
| HF/SP/HF | 200 | - | $2\times10^{16}$ | 900 | 5 |

Preparation conditions and device gains.

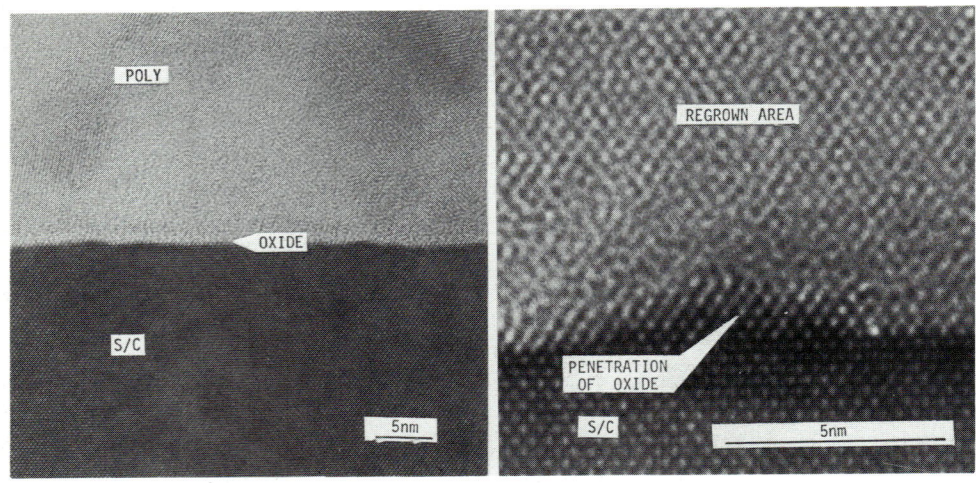

Fig.2a RCA cleaned slice pre-
annealed at 900°C for 10 min.

Fig.2b RCA cleaned slice pre-annealed
at 950°C for 10 min.  Note the
epitaxial re-alignment of the polysilicon layer.

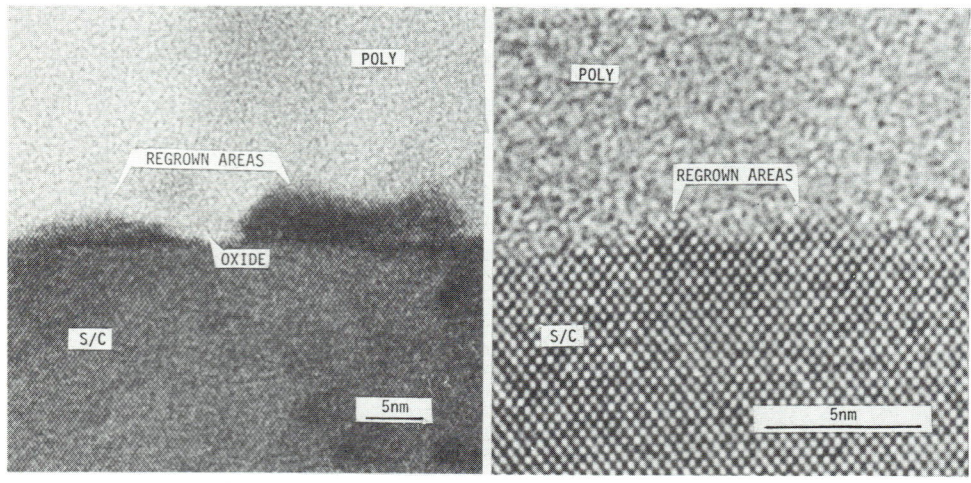

Fig.2c RCA cleaned slice pre-
annealed at 1000°C for 10 min.
Note the areas of epitaxial
re-alignment and the oxide
break up.

Fig.3a HF cleaned slice not pre-
annealed.  Note the small areas of
epitaxial re-alignment.

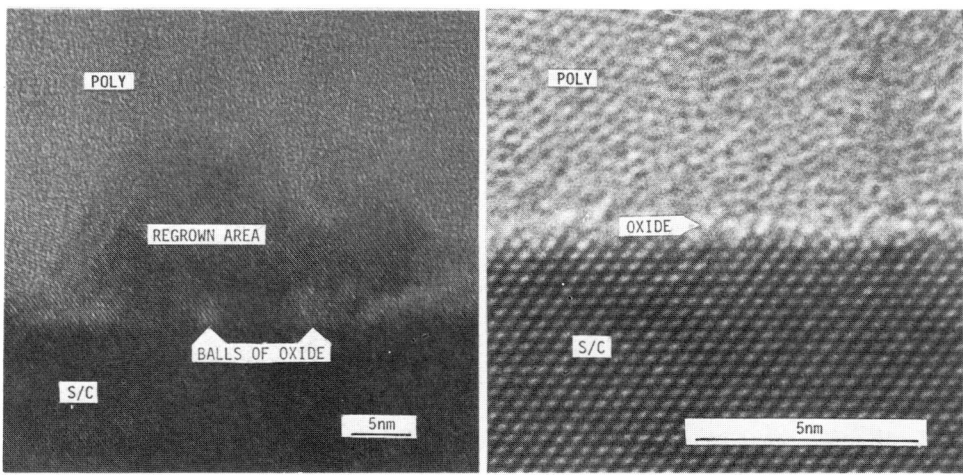

Fig.3b HF cleaned slice pre-
annealed at 1000°C for 10 min.  Note
the large areas of epitaxial re-
alignment, and the balled-up oxide.

Fig.4a HF/SP cleaned slice which had
not been implanted.  Note the continuous oxide film.

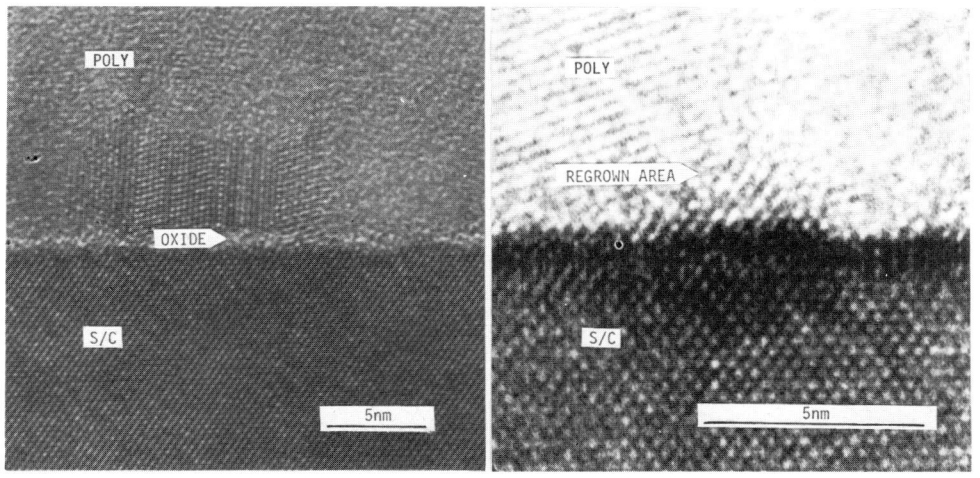

Fig.4b HF/SP cleaned slice after
$5 \times 10^{15}$ cm$^{-2}$ As$^+$ and 900°C for
15 min.  Note the uneven, but continuous oxide film.

Fig.4c HF/SP cleaned slice after
$1 \times 10^{16}$ cm$^{-2}$ As$^+$ and 900°C for
15 min.  Note the small area of
epitaxial re-alignment.

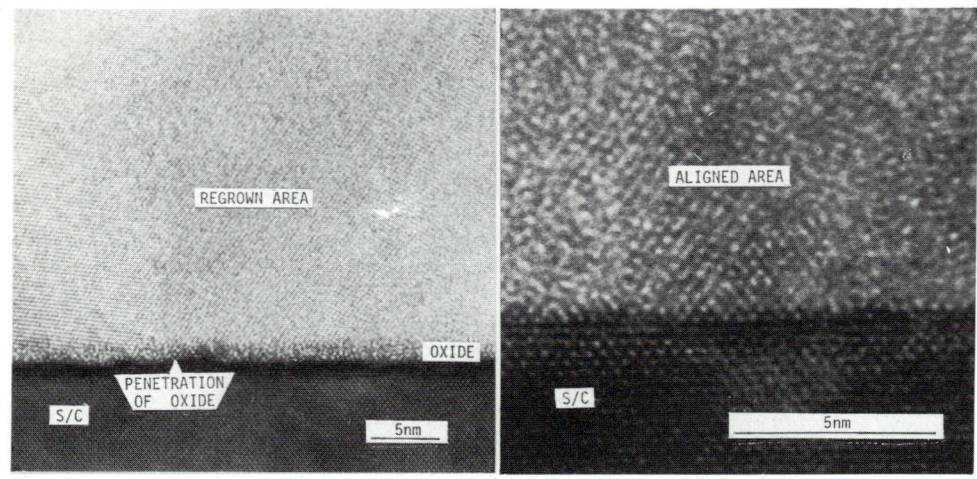

Fig.4d HF/SP cleaned slice after
2x10$^{16}$ cm$^{-2}$ As$^+$ and 900°C for
15 min.  Note the large area of
epitaxial re-alignment, and the
point of penetration of the
oxide film.

Fig.5a HF/SP/HF cleaned slice which
had not been implanted.  Note the
aligned region in the polysilicon.

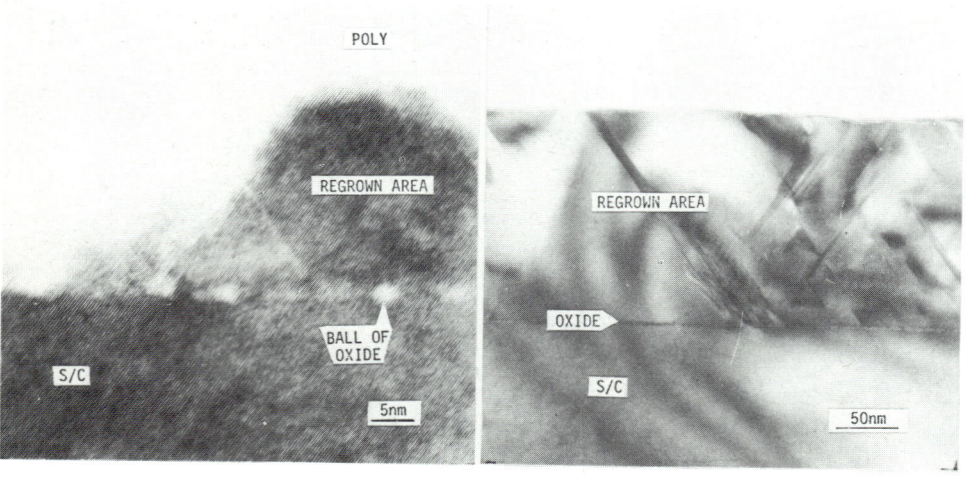

Fig.5b HF/SP/HF cleaned slice
after 5x10$^{15}$ cm$^{-2}$ As$^+$ and 900°C
for 15 min.  Note the large
epitaxially re-aligned area and
the balling-up of the oxide film.

Fig.5c HF/SP/HF cleaned slice after
1x10$^{16}$ cm$^{-2}$ As$^+$ and 900°C for 15 min.
The polysilicon layer has regrown
completely.

*Inst. Phys. Conf. Ser. No. 76: Section 11*
*Paper presented at Microsc. Semicond. Mater. Conf., Oxford, 25–27 March 1985*

477

# Determination of arsenic distribution in silicon by a thermal oxidation replica technique

C Hill, P D Augustus and A Ward*

Plessey Research (Caswell) Ltd., Caswell, Towcester, Northants., NN12 8EQ

Abstract   Characterisation of dopant distributions in one-micron geometry silicon integrated circuit structures urgently needs a rapid, two and three dimensional, analytical technique of medium spatial resolution (100 Å). A new technique for mapping arsenic distributions at this resolution is described, based on converting dopant concentration variations into thickness variations in an oxide replica of the silicon surface. Fabrication of replicas by thermal oxidation at 700°C in dry and in wet oxygen is examined, and the technique is applied to determination of arsenic distributions at the edge of implanted windows.

## 1. Introduction

The technology development for one- and sub-micron geometry very large scale integrated circuits has thrown up many challenges to analytical techniques, particularly in its requirement for medium resolution (10-1000 Å) maps of the two (and sometimes three) dimensional dopant distributions within the device components (Hill and Butler 1983). A typical structure requiring analysis of the n-type dopant (arsenic) is shown in Fig.1. Existing techniques are either insufficiently sensitive (e.g. Auger analysis) or of insufficiently high resolution (e.g. SIMS analysis). Newer techniques are being developed to address this problem (Grovenor et al 1984, Grovenor et al 1985, Roberts 1985) but although high spatial resolution and reasonable sensitivity are achieved, the techniques are very time consuming and allow examination of only a very small (and thus potentially unrepresentative) volume of the silicon. A complementary rapid technique of medium resolution (~100 Å) for arsenic distribution determination has been developed, enabling large numbers of similar structures to be analysed in one examination: the technique is based on producing a thin-film replica of the silicon surface by thermal oxidation.

## 2. Low Temperature Thermal Oxidation

The oxidation rate of silicon at 700°C is increased by the presence of arsenic doping, as shown in Fig.2. The enhancement is greater in wet oxygen than in dry oxygen ambient, but is detectable in both down to arsenic doping levels of about $5 \times 10^{19}$/cc. Data at the high doping levels is approximate, since uniformly doped samples are not available, and oxidation rates were measured over shallow implanted regions. The tempera-

* now at Somerville College, Oxford

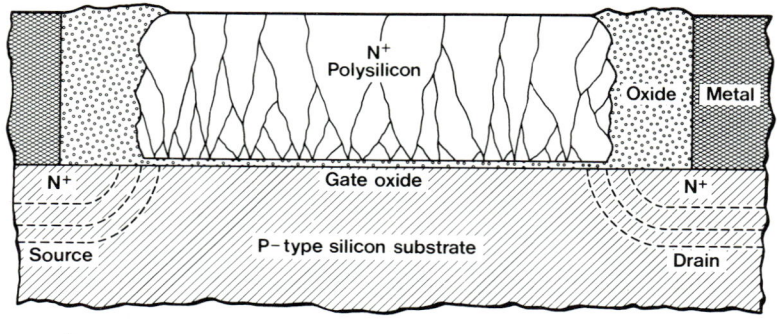

1000 Å

Fig.1 Vertical section through a Schematic 1 μm N-MOS transistor, showing typical dimensions & configuration of polycrystalline silicon microstructure, gate oxide & dopant distribution contours in source & drain regions.

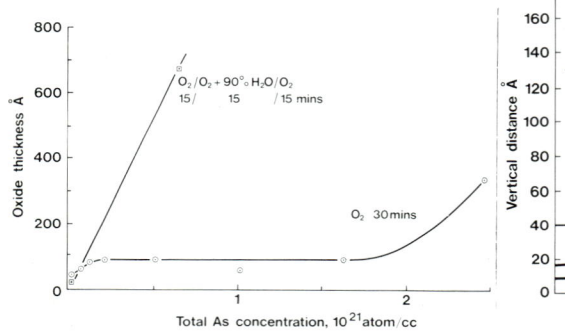

Fig.2 Oxide thickness as a function of Arsenic concentration, for oxidation of an As-doped (100) silicon surface @ 700°C in dry and wet $O_2$

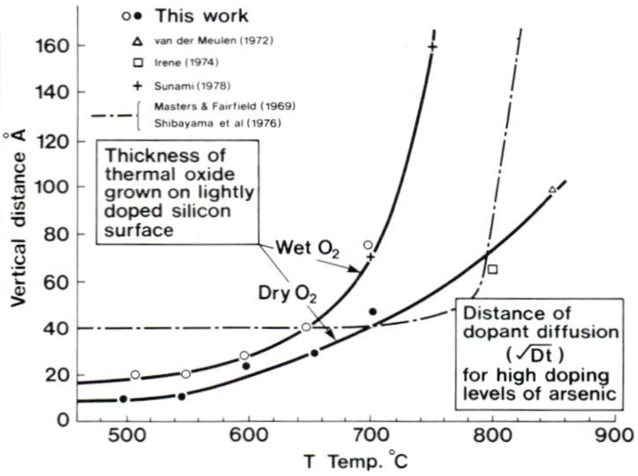

Fig.3 Distances of dopant diffusion & thermal oxidation near a (100) silicon surface after 60mins as a function of temperature & oxidising ambient.

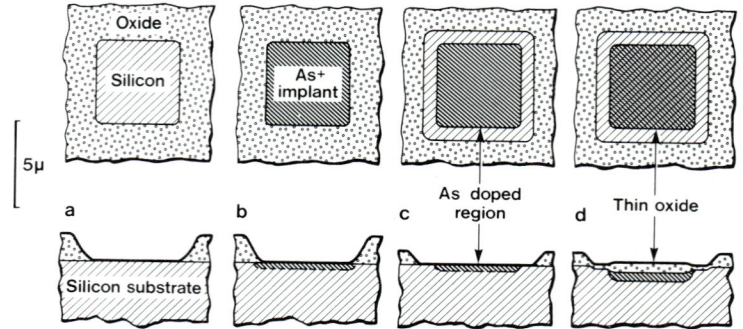

Fig.4 Process sequence to fabricate a thermal oxide replica of the arsenic distribution across a silicon surface implanted through an oxide window: (a)Etch window in oxide (b)implant arsenic (c)enlarge oxide window and thin oxide (d)oxidise the silicon surface at 700°C.

ture of 700°C was chosen to give a sufficiently thick oxide film to survive subsequent handling (~50Å) while minimising the solid state diffusion of arsenic to less than the oxide thickness. This optimum temperature window can be seen clearly in Fig.3. Both for sensitivity to doping and for producing a robust film on undoped material, wet oxygen ambient is superior to dry (Figs.2,3) and this ambient was used in subsequent replica fabrications.

## 3. Replica Fabrication

Oxide replicas of the arsenic distribution at the edge of an arsenic implanted and annealed window were made as shown schematically in Fig.4. A thick oxide (8950Å) was grown by thermal oxidation on 2 ohm-cm n-type (100) CZ silicon at 1100°C, and photoengraved with an array of 5 micron windows with a wet etch (Fig.4a). Arsenic ions were then implanted at 40keV energy and $10^{16}$ ions/sq.cm. dose, and the sample annealed for 30 mins at 620°C (4b). The thick oxide was then thinned to about 1000Å in 7:1 $H_2O$/HF: the oxide also etched laterally, leaving the doped region isolated from the oxide edge by an annular region of 2 ohm-cm silicon about two microns wide (4c). A thin oxide was then grown over the whole of the silicon exposed in the windows by thermal oxidation at 700°C in dry oxygen (10 mins), wet oxygen (15 mins) and dry oxygen (5 mins) (4d). The silicon was completely removed from over an area of about 0.3sq.mm. by jet etching with HF/$HNO_3$ from the back of the sample, leaving a free-standing oxide across the hole.

## 4. TEM Examination

A low magnification view of a typical oxide film is shown in Fig.5. A large number of windows are visible and the varying oxide thickness across each can be seen to be similar but not identical. The thin oxide rim around the edge of each window (resulting from oxidation of 2 ohm cm n-type silicon) is only about 40Å thick, and occasionally a membrane drops out, as at A. In Fig.6, a higher magnification view of one of these displaced membranes overlapping an intact membrane is shown. In both membranes, the 5 micron diameter circle of thicker oxide and the 9 micron diameter concentric rim of thin oxide can clearly be seen. These correspond respectively to the implanted and unimplanted regions of the original underlying silicon surface, and were expected from the doping dependence of oxidation rate of Fig.2. An unexpected feature of the replicas was the dark narrow transition zone between the inner and outer regions (Z, Fig.6). The strong attenuation of the beam by this zone shows clearly in a STEM line-scan across the oxide, shown in Fig.7; the inclusion shown of regions with 0 and 100% transmission allows this type of scan to be used for quantitative measurements of the oxide membrane thickness. At higher magnification, shown in Fig.8, the transition zone can be seen to be of constant width even though it does not lie on a perfectly smooth curve. The deviations of about ±200Å are attributed to the roughness of the original photoengraving mask edge and material variations in resist and oxide that defined the original implant edge. At much greater resolution, (Fig.10), a superimposed linescan of transmitted electron current and normal TEM image both reveal a thick zone of about 600Å width: the line-scan shows that either side of this the oxide tapers down to constant thickness over about 400Å. More accurate transmission measurements can be made at lower beam voltages as can be seen by the line-scan of Fig.9, made on the opposite side of the same doped window.

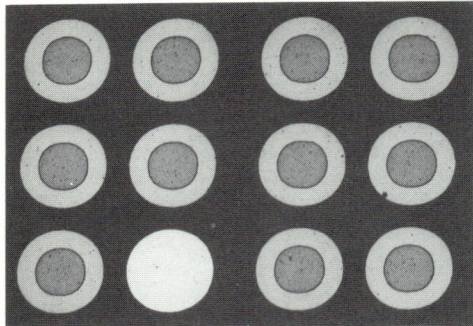

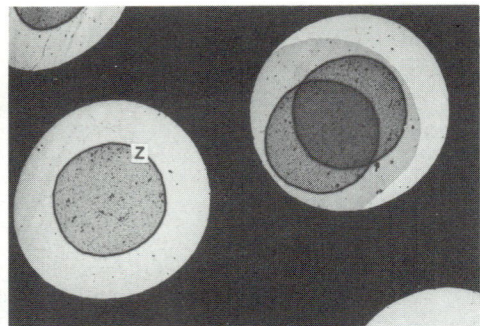

Fig.5 Plan view TEM (40keV) of a thick oxide layer (black) with an array of 9μm windows, each with a thin oxide membrane, replicating in its thickness the As doping level of the original underlying silicon.

Fig.6 Detail of overlapping oxide membranes from Fig.5, showing the reduced transmission of the double thickness of both the thin oxide outer rim, (diameter 9μm) and the thicker inner circle (dia. 5μm).

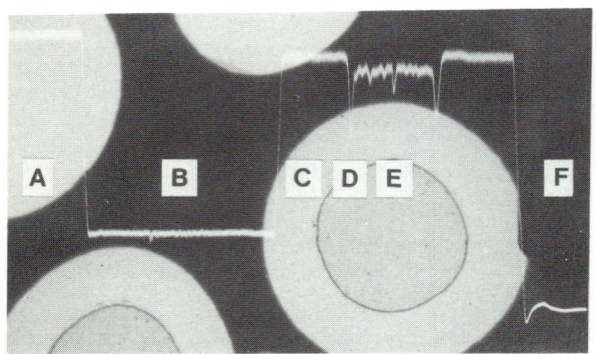

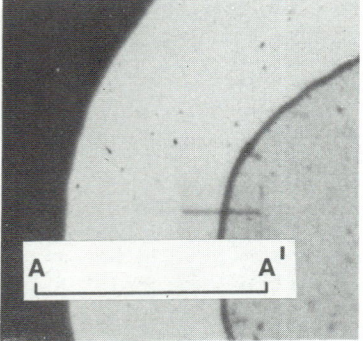

Fig.7 Plan view STEM (40keV) of oxide membranes across thick oxide windows with line scan of transmitted electron current in STEM mode super-imposed. Transmissions are A 100%, B 28%, C 91%, D 64%, E 87%, F 0%.

Fig.8 Plan view STEM(40keV) of oxide membrane showing position of line scan used for Fig.9.

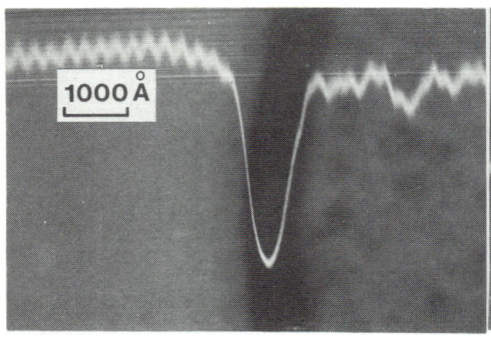

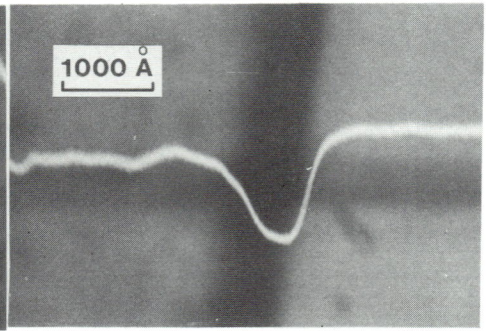

Fig.9 Plan view STEM(20keV) of transition zone between the very thin (left) and thicker (right) oxide membrane with superimposed STEM line scan of transmitted electron current.

Fig.10 Plan view STEM(40keV) of transition zone on opposite side of window in Fig.8.

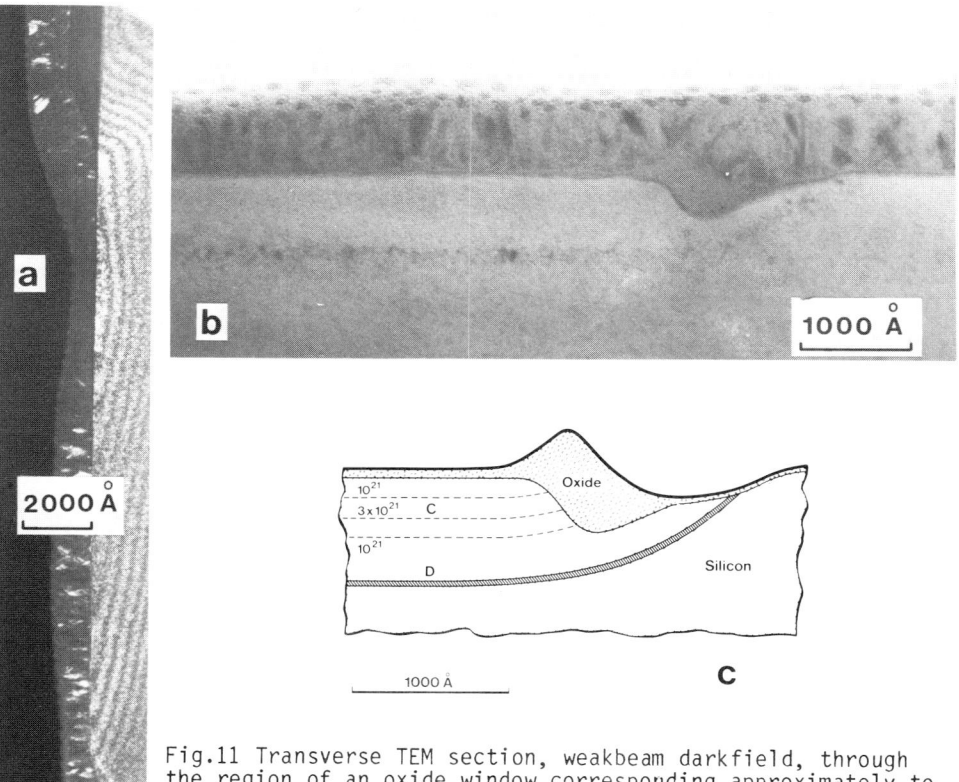

Fig.11 Transverse TEM section, weakbeam darkfield, through the region of an oxide window corresponding approximately to that indicated by the line A-A' in Fig.8. (a)Low magnification of the whole section (b)High magnification micrograph of the transition zone (c)Sectional diagram showing the important features of (b): the oxide topography, the damage layer delineating the original amorphous-crystalline interface (D) and the estimated arsenic concentration contours (C).

In order to determine the nature of the transition zone, a transverse TEM section was prepared from the same sample, but with the silicon substrate still intact and a capping layer of polysilicon over the oxide. A low magnification view of the section, corresponding to the section line AA in Fig.8, is shown in Fig.11a. This shows clearly the 1000 Å thick oxide and its tapered edge down to the thin oxide (about 40 Å), which extends for 2 microns across the silicon surface to a localised thick region, at the same spatial location as the transition zone in Fig.8. An enlarged view of this region is shown in Fig.11b. The spatial location of the original implant is conveniently delineated by the defects left at the position of the original amorphous-crystalline interface after arsenic implantation. The defect boundary lies parallel to the surface except in the vicinity of the thick oxide zone, where it turns up towards the surface at an angle of 25-30°, corresponding approximately to the angle of slope of the original oxide mask edge. The thickened oxide lies just within the implanted zone, and close to the point where the peak arsenic concentration contour, esti- mated from one-dimensional arsenic concentration profiles measured normal to the silicon surface (Hill and Butler 1983), intersects the silicon surface. The relative positions of these features are more clearly seen in the enlarged scale diagram, Fig.11c. The thick transition oxide thus corresponds to a very local region of high arsenic concentration (about $3 \times 10^{21}$ atom/cc) and delineates this contour faithfully. The tapered inner edge shows that measurable changes in oxide thickness occur down to about $10^{20}$ atoms/cc, the approximate surface concentration of arsenic. Furthermore, the enhanced oxidation rate is very spatially localised: where the buried peak arsenic contour is within 300Å of the silicon surface, it has no effect on surface oxidation rate. This supports the theory of Ohkawa and Nakajima (1978) that enhancement involves a surface reaction rate increase in which the arsenic atoms are directly involved, and is not simply due to increased carrier concentration. Thus the spatial selectivity of the oxidation process is not a limit to the resolution of the technique. An unexpected feature of Fig.11c is the "bird's beak" profile of the oxide, corresponding to an original silicon surface with a depression in it as shown. The location of this depression suggests that it can only have resulted from etching of the heavily doped silicon in the long HF etch required to thin the oxide. The extension of the depression beyond the doped region is not understood. The transversal TEM of Fig.11 and the line scan of Fig.9 agree well as to the spatial extent of the transition oxide (1350 and 1450Å respectively), and qualitatively as to the variation of thickness with distance.

## 5. Conclusions

This preliminary exploration of a novel analytical technique has shown that it is a useful tool for rapid two-dimensional surveys of the surface location of arsenic at a resolution of 100Å and a sensitivity of about $10^{20}$ atoms/cc. Considerable developments of the technique are possible including quantitative calibration of doping level with oxide thickness, oxidation of sectioned silicon to reveal 3 dimensional arsenic distributions, and higher sensitivities by using emission (e.g. EDX) rather than absorption methods to measure oxide thickness. The technique will aid characterisation of advanced VLSI structures, especially for arsenic distributions in polysilicon for which no simple technique at present exists.

## Acknowledgements

The authors gratefully acknowledge the experimental assistance of Dennis Boys and David Bellamy, and also financial support and permission to publish from Plessey Research (Caswell) Ltd.

## References

Grovenor C R M, Batson P E, Smith D A, and Wong C (1984)
   Phil Mag A 50 No.3 409
Grovenor C R M, Cerezo A and Smith G D W (1985) Paper G6. This Conference.
Hill C and Butler A L (1983) Solid State Devices 1983 ed Rhoderick E H
   (London. Inst. of Physics) pp 161-180
Irene E A (1974) J. Electrochem. Soc. 119 No.4 530
Masters B J and Fairfield J M (1969) J. Appl. Phys. 40 No.6 2390
Ohkawa S and Nakajima Y (1978) J. Electrochem. Soc. 125 No.12 1997
Roberts M C, Yallup K J and Booker G R (1985) Paper L5. This Conference.
Shibayama H, Masaki H, Ishikawa H & Hashimoto H (1976) J.E.C.S.123 742
Sunami H (1978) J. Electrochem. Soc. 125 No.6 892-897
Van der Meulen Y J (1972) J. Electrochem. Soc. 119 No.4 530

*Inst. Phys. Conf. Ser. No. 76: Section 11*
*Paper presented at Microsc. Semicond. Mater. Conf., Oxford, 25–27 March 1985*

483

# A new TEM technique for evaluating 2-D dopant distributions in silicon with high spatial resolution

M C Roberts[1][*], K J Yallup[2] and G R Booker[1]

[1] Dept. of Metallurgy and Science of Materials,
University of Oxford, Parks Road, Oxford, OX1 3PH

[2] GEC Research Laboratories, Hirst Research Centre,
East Lane, Wembley, Middx. HA9 7PP

[*] Now at Plessey Research (Caswell) Ltd., Allen Clark Research Centre,
Caswell, Towcester, Northants. NN12 8EQ

Abstract  A new technique which determines the 2-D n- and p-type dopant distribution in ion implanted and annealed Si wafers is described.  The dopant distribution is measured by monitoring the thickness variation in a chemically etched (0.3% HF, 99.7% $HNO_3$ at 5°C) TEM cross-sectional specimen.  The relationship between the etch rate and dopant concentration is independently measured enabling the thickness variation in the TEM specimen to be directly related to the dopant concentration.  By using this technique the position of the iso-concentration contours can be determined to a depth accuracy of ~10nm.

## 1. Introduction

As the dimensions of silicon microelectronic devices are reduced to micron and sub-micron geometries, there is an increasing need to be able to determine the 2-D distrbution of dopants.  The lateral diffusion of dopants can have a significant effect on the performance of small geometry devices.  For example, in an MOS transistor the lateral spreading of the source and drain reduces the electrical channel length and the lateral spreading of the boron and phosphorus field implants reduces the width and also shifts the threshold voltage of the MOS transistor.  There are a number of high resolution techniques that can evaluate the 1-D dopant distribution, e.g. SIMS (secondary ion mass spectroscopy) and SR (spreading resistance).  However, at present there are no high resolution 2-D techniques.  In this paper a new 2-D technique is described.

## 2. Experimental technique

The experimental technique is based on the work of Sheng and Marcus (1981) who showed that it is possible to delineate heavily doped $n^+$ regions in silicon by chemically etching TEM cross-sectional specimens in an aqueous solution of 0.5% hydrofluoric acid (40% volume) and 99.5% nitric acid (69.5% volume) at room temperature.  In this present work the etching solution of Sheng and Marcus is used as well as a solution containing 0.3% HF (40% vol.) and 99.7% $HNO_3$ (69.5% vol.) used at a temperature of 5°C. The concentration of HF and the temperature of the etch solution have been

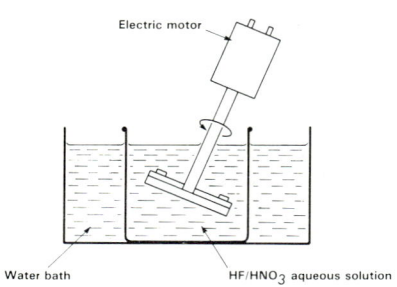

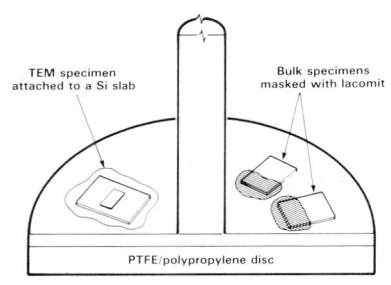

Figure 1
Experimental apparatus used to etch Si specimens at a constant stirring frequency and temperature

Figure 2
Diagram showing how the Si specimens are mounted on the etching disc (diameter ~5cm)

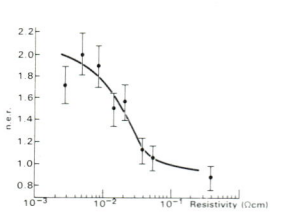

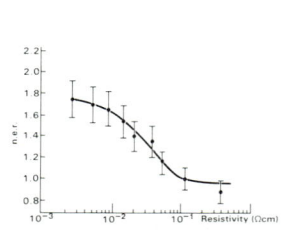

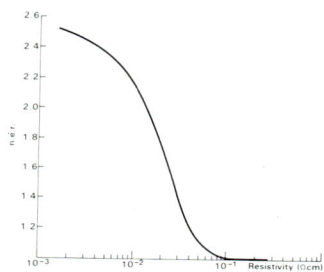

Figure 3
n.e.r. vs. resisivity (n and p-type Si) for an etch solution of 0.3%HF, 99.7%: $HNO_3$ at 5°C. Stirring frequency = 1.8hz (etch rate = n.e.r.x0.80nm/sec)

Figure 4
n.e.r. vs. resistivity (n and p-type Si) for an etch solution of 0.5%HF, 99.5% $HNO_3$ at 21°C. Stirring frequency = 1.8hz (etch rate = n.e.r.x2.35nm/sec)

Figure 5
Relationship between n.e.r. and resistivity when etching the TEM cross-sectional specimen (B4351/4) in an etch solution of 0.3%HF, 99.7%$HNO_3$ at 5°C. Stirring frequency = 1.8hz (etch rate = n.e.r.x0.97nm/sec)

reduced from the values used by Sheng and Marcus in order to lower the etch rate, enabling the TEM specimens to be etched with greater control. It is also found that the new etch has a greater sensitivity to doping variations. In order to be able to determine the doping variations in the silicon specimens, the relation between dopant concentration and etch rate must be determined. This relation is evaluated by etching both n- and p-type uniformly doped bulk Si specimens of different resistivity using the apparatus shown in Figs.1 & 2. The apparatus enables the temperature and stirring frequency to be kept constant. The etch rate was determined by measuring with a surface profilometer the height of the step formed on the Si surface, the step arising because one half of the specimen is masked from the chemical etch. The results of the experiments are shown in Figs.3 & 4. The etch rates have been normalized to the etch rate of Si specimens which have a resistivity of ≥1Ωcm. It is found that the etch rate is independent of the dopant concentration if the resistivity is greater than 1Ωcm. A mechanism to explain the observed dependence of the etch rate on resistivity will be published later (Roberts 1985).

Having established the relationship between the etch rate and dopant concentration, one can then proceed to determine the dopant distribution in
an etched cross-sectional TEM specimen. In order to investigate the
possibility of using this new technique a p-type (5Ωcm) Si wafer (B4351/4)
was implanted with As$^+$ (100keV, $10^{16}$cm$^{-2}$) which was then driven in at
1100°C in an inert ambient for 480 minutes to produce an n-type doping
profile. A cross-sectional TEM specimen was prepared from wafer B4351/4
using the technique described by Fletcher (1979). Prior to the specimen
being mounted on a TEM grid, the TEM specimen is attached to the etching
disc together with bulk specimens of different resistivity; $10^{-2}$Ωcm and
2Ωcm (see Fig.2). The bulk specimens enable the behaviour of the etch to
be monitored since it is found that there is some slight degree of fluctuation in the etch rates from one experiment to another. This is indicated by the error bars in the graphs in Figs.3 and 4. The TEM specimens and
bulk specimens were then etched for 90 seconds in the etch solution
(0.3%HF, 99.7% HNO$_3$ at 5°C). The etch rates of the bulk specimens were
measured, enabling the relation between resistivity and etch rate to be
determined for this particular experiment (see Fig.5). The curve in Fig.5
(ner'(ρ)) was determined by transforming the curve in Fig.3 (ner(ρ)) using
the following equation:

$$ner'(\rho) = \frac{ner'(10^{-2})-1}{ner(10^{-2})-1} \cdot ner(\rho)$$

where ρ is the resistivity and ner'($10^{-2}$) is the normalized etch rate of
the etched bulk specimens (ρ = $10^{-2}$Ωcm). An optical micrograph of the TEM
specimen taken under sodium light illumination before and after etching
enables the amount of material etched from the Si substrate, i.e. the
p-type region of the wafer, to be determined. After the TEM specimen has
been etched it is then examined in the electron microscope at an accelerating voltage of 100keV. The specimen is tilted to the [110] pole so that
the Si wafer surface is 'edge on' to the beam and also so that the many-
beam thickness fringes can be observed. Calculation using the numerical
method of Skarnulis (1979) shows that the separation between two fringes
corresponds to a Si thickness difference of 23nm. The TEM micrograph of
specimen B4351/4 is shown in Fig.6 together with the position of the many
beam thickness fringes. For such a specimen the position of the thickness
fringes can be determined to an accuracy of ±10nm. Using the information
in Fig.5, one can calculate the doping concentration as a function of the
distance from the Si wafer surface. The calculated values are shown in
Table 1. It is assumed in the calculation that the etch is anisotropic,
i.e. the etch only attacks the Si in a direction which is normal to the
surface. The etch is, in fact, isotropic; however, one can show that the
above assumption does not lead to any large discrepancies when calculating
the doping profile (Roberts 1985). The resistivity values in Table 1 have
been converted to n-type doping concentration values using the results
provided by the American Society for Testing of Materials. The values in
Table 1 are plotted in Fig.7 together with values obtained from another
specimen which was prepared from another region of wafer B4351/4. As well
as the values obtained from the TEM specimen, the SR profile and SIMS
profile of wafer B4351/4 are also plotted. The SR profile was measured by
M Pawlik and R D Groves (GEC Hirst Research Centre) and the SIMS profile
was measured by M Dowsett (City of London Polytechnic). As one can see
from Fig. 7, there is excellent agreement between the three techniques.

TABLE 1

| Many-beam fringe | Depth below Si surface ($\mu$m) | Si removed (nm) | n.e.r. | $\rho$($\Omega$cm) | N(cm$^{-3}$) |
|---|---|---|---|---|---|
| 1 | 0.93 | 380 | 2.53 | $1.5 \times 10^{-3}$ | $5 \times 10^{19}$ |
| 2 | 1.23 | 357 | 2.38 | $4.8 \times 10^{-3}$ | $1.2 \times 10^{19}$ |
| 3 | 1.36 | 334 | 2.23 | $8 \times 10^{-3}$ | $6 \times 10^{18}$ |
| 4 | 1.45 | 311 | 2.07 | $1.2 \times 10^{-2}$ | $4 \times 10^{18}$ |
| 5 | 1.5 | 288 | 1.92 | $1.5 \times 10^{-2}$ | $2 \times 10^{18}$ |
| 6 | 1.54 | 265 | 1.76 | $2 \times 10^{-2}$ | $1.3 \times 10^{18}$ |
| 7 | 1.56 | 242 | 1.61 | $2.4 \times 10^{-2}$ | $8 \times 10^{17}$ |
| 8 | 1.58 | 218 | 1.46 | $2.8 \times 10^{-2}$ | $7 \times 10^{17}$ |
| 9 | 1.62 | 196 | 1.31 | $3.2 \times 10^{-2}$ | $5 \times 10^{17}$ |
| 10 | 1.65 | 173 | 1.15 | $4.5 \times 10^{-2}$ | $2.6 \times 10^{17}$ |
| Si substrate | $\geqslant$1.65 | 150 | 1.0 | | |

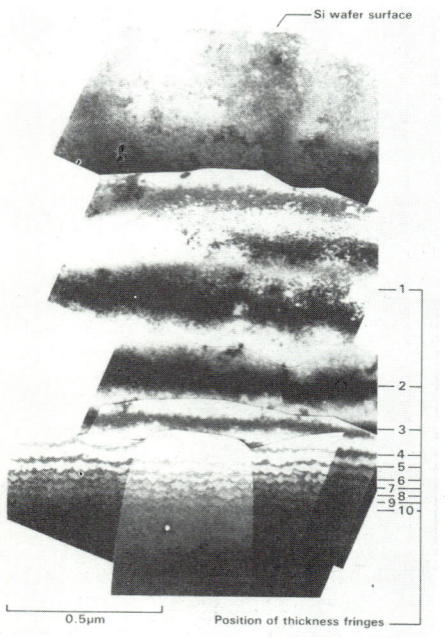

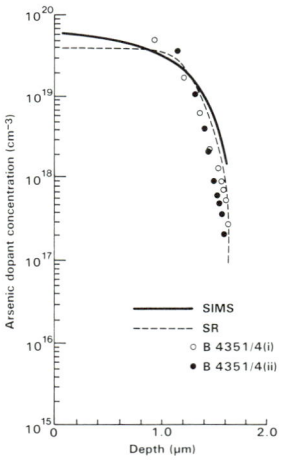

Figure 6
TEM cross-sectional micrograph of specimen B4351/4 showing the position of the many-beam thickness fringes

Figure 7
Diagram showing the doping profiles calculated from the TEM specimens (B4351/4) and also the dopant profiles obtained from SIMS and SR

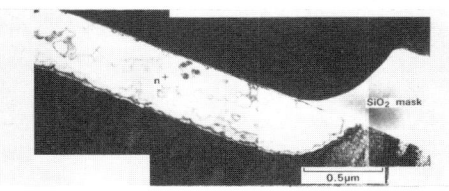

Figure 8
TEM cross-sectional micrograph of
the As$^+$ implanted and diffused
specimen (40keV, $10^{16}$cm$^{-2}$, inert
drive-in, 950°C, 240min) showing
the 2-D iso-concentration contour
($2\times10^{17}$cm$^{-3}$). The specimen was
etched in 0.5%HF, 99.5%HNO$_3$ at 21°C
for 20sec.

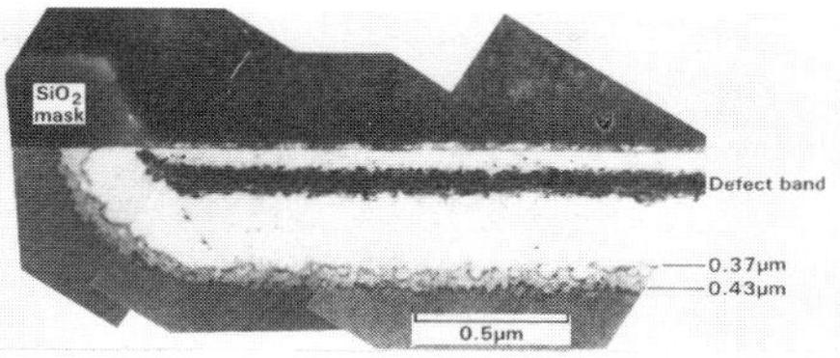

Figure 9
TEM cross-sectional micrograph of
the B+ implanted and diffused
specimen (25keV, $5\times10^{15}$cm$^{-2}$, inert
drive-in 950°C, 30min) showing the
2-D iso-concentration contour
($6\times10^{17}$cm$^{-2}$). The specimen was
etched in 0.5%HF, 99.5%HNO$_3$ at 21°C
for 25sec.

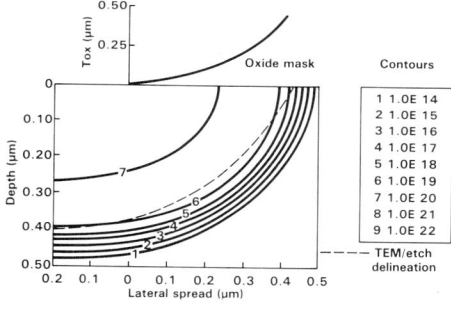

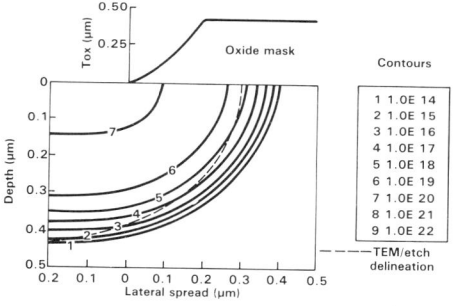

Figure 10
Diagram showing the result of the
2-D model for the As$^+$ implanted and
diffused specimen near a mask edge
(40keV, $10^{16}$cm$^{-2}$, inert drive-in,
950°C, 240min). The TEM/etch
result is shown by the broken
line.

Figure 11
Diagram showing the result of the
2-D model of the B implanted and
diffused specimen near a mask edge
(25keV, $5\times10^{15}$cm$^{-2}$, inert drive-in,
950°C, 30min). The TEM/etch result
is shown by the broken line.

The SR and SIMS techniques, however, can only give 1-D dopant distributions. However, the TEM/etch technique can be extended to 2-D. TEM
cross-sections of As$^+$ and B$^+$ implanted Si wafers were prepared and etched.
The As was implanted and diffused into p-type Si (5 Ωcm) which had an oxide
mask patterned into 14 μm wide tracks separated by 8 μm gaps. Similarly the
B was implanted and diffused into n-type Si (2 Ωcm) which had an identical
oxide mask patterned on the wafer. The etched TEM cross-sectional specimens are shown in Fig.8 and Fig.9. These two specimens were not etched
for a sufficient length of time to reveal a sufficient number of thickness
fringes to enable a full 2-D doping profile to be obtained. However, the
results in Fig.4 enable the dopant concentration at the first iso-
concentration contour to be accurately determined. The delineation is
observed in the TEM specimen when the thickness of the specimen has
changed by ~12nm, i.e. half the thickness variation between two many-beam
fringes. For the specimens shown in Fig.8 and Fig.9, ~100nm has been
etched from the Si substrate. Therefore the TEM delineation is observed

when the n.e.r. has increased to ~1.1. From Fig.4 one can see that this corresponds to a resistivity of ~6x10$^{-2}$Ωcm. For the As$^+$ implanted specimen this corresponds to a doping concentration of 2x10$^{17}$cm$^{-3}$ and for the B$^+$ implanted specimen to a doping concentration of 6x10$^{17}$cm$^{-3}$ at the TEM delineation. These values for the doping concentration at the delineation are much lower but more accurate than the value quoted by Sheng and Marcus (1981) for n$^+$-type Si (~10$^{19}$cm$^{-3}$). Their value was estimated by comparing the depth of the TEM/delineation to a SUPREM model of the ion implantation and diffusion of As. Due to the steep slope of the As profile in their experiment, it is very difficult to accurately determine the doping concentration at the delineation by this technique. By using bulk specimens to measure the etch rate as a function of resistivity, a far more accurate determination of the doping concentration at the delineation is obtained.

In Figs.10 and 11 the results from the TEM/etched specimens are compared to results calculated from a 2-D model (K J Yallup 1985) of the B and As distribution near a mask edge. The precise shape of the mask edge as determined by TEM cross-sectional examination has been inputted into the model. As one can see, there is a reasonable agreement between the 2-D model and the TEM/etch for the As$^+$ implanted and diffused specimen. However, for the B$^+$ implanted and diffused specimen, the agreement is not as good. This is due to the difficulty involved in accurately modelling B diffusion rather than to any inadequacy in the experimental technique since as one can see from Fig.7, good agreement is obtained between the TEM/etch technique and the other experimental techniques.

## 3. Summary and Conclusions

It has been shown that it is possible to obtain the dopant distribution in an etched TEM cross-sectional specimen provided that the relationship between the etch rate and Si resistivity is known. The chemical etch used is sensitive to the acceptor and donor impurities provided that the doping concentration is greater than 2x10$^{17}$cm$^{-3}$ in n-type Si and 6x10$^{17}$cm$^{-3}$ in p-type Si. The position of the iso-concentration contours in the TEM etched specimen can be determined to a depth accuracy of 5-10nm. By etching the TEM cross-section for a sufficient length of time, it is possible to obtain the position of more than one iso-concentration contour enabling a full doping profile to be obtained. Good agreement is obtained between the TEM/etch technique and the SIMS and SR doping profiles and also with a 2-D model. However, further work is required to evaluate the reproducibility, accuracy and general applicability of the TEM/etch technique for determining 2-D dopant distributions in Si.

## Acknowledgements

We would like to thank SERC for the financial support of the work at Oxford.

## References

Fletcher J 1979 D.Phil Thesis
Roberts M C 1985 to be published
Sheng T T and Marcus R B 1981 J. Electrochem. Soc. 128 881
Skarnulis A J 1979 J. Appl. Cryst. 12 636
Yallup K J, Roberts M C, Edwards S P, Booker G R 1985 to be published in Physical Review B.

*Inst. Phys. Conf. Ser. No. 76: Section 11*
*Paper presented at Microsc. Semicond. Mater. Conf., Oxford, 25–27 March 1985*

# TEM, RBS and SIMS investigations of buried nitride layer structures in silicon formed by high-dose N[+] ion implantation

C D Meekison[1], G R Booker[1], K Reeson[2], P L F Hemment[2], R J Chater[3], J A Kilner[3] and R P Arrowsmith[4]

[1]Department of Metallurgy & Science of Materials,
 University of Oxford, Parks Road, Oxford OX1 3PH
[2]Department of Electronic & Electrical Engineering,
 University of Surrey, Guildford, Surrey GU2 5XH
[3]Department of Metallurgy & Materials Science,
 Imperial College of Science & Technology, London SW7 2BP
[4]British Telecom Research Laboratories, Martlesham Heath, Ipswich IP5 7RE

Abstract    Buried nitride layers have been formed by high-dose implantation of nitrogen ions into silicon. Observations by cross-sectional
TEM, RBS and SIMS before and after high temperature annealing are
reported. A good quality upper silicon layer was present after annealing. Plan-view TEM of the annealed specimens showed that the nitride
was mainly crystalline, with a spherulitic structure. A double buried
nitride layer was produced by implantation at two energies.

## 1. Introduction

The implantation of high doses of oxygen or nitrogen ions of sufficiently
high energy into silicon leads to the formation of buried dielectric
layers, which may after high temperature annealing produce silicon-on-
insulator (SOI) substrates suitable for VLSI device circuits. The
majority of work in this field has used oxygen implantation (see Lam and
Pinizzotto 1983 for a review). However, device operation has been demonstrated on substrates using buried nitride (Zimmer and Vogt 1983).
Investigations of the material have been carried out by a number of groups
(Belz and te Kaat 1984, Bourguet et al 1980, Kreissig et al 1983, Skorupa
et al 1984, Maeyama and Kajiyama 1982). The present paper reports on an
investigation by TEM, RBS and SIMS aimed at determining the optimum conditions for forming SOI by nitrogen implantation.

## 2. Experimental

Device grade (100) 75mm diameter silicon wafers were implanted with $^{14}N^+$
ions using the 400keV implanter at Surrey University. The implanted area
was defined by a 25mmx25mm silicon aperture and the beam was electrostatically scanned. The wafers were thermally isolated and beam heating was
used to maintain a sample temperature of $500\pm10°C$. The implantation doses
used were 0.25 and $0.75\times10^{18}$ ions $cm^{-2}$ at 200keV, and also one sample was
implanted with $0.75\times10^{18}$ ions $cm^{-2}$ at 350keV followed by an equal dose at
200keV. Portions of the samples were given a silica capping layer and
annealed at 1200°C for 2 hours in dry flowing nitrogen.

Cross-sectional and plan-view TEM specimens were prepared using ion

thinning.    TEM examination was carried out using a Philips EM300 microscope.    Other pieces from the same samples were analysed by RBS using
1.5MeV He$^+$ ions, and by SIMS.    The SIMS was performed in an Atomika
instrument, the ions monitored being Si$^+$ and Si$_2$N$^+$ under 10keV Ar$^+$ ion
bombardment.

## 3. Results

### 3.1    $0.75 \times 10^{18}$ N$^+$ cm$^{-2}$, 200keV, before annealing

Figure 1 shows a cross-sectional micrograph of the sample implanted at
200keV to $0.75 \times 10^{18}$ N$^+$ cm$^{-2}$ without subsequent processing.    A layer of
single crystal silicon is seen from the surface to a depth of $0.29\,\mu$m, most
of which is heavily damaged;    however the lower part ($\approx 0.07\,\mu$m) is relatively defect-free.    Below this is an apparently structureless region,
$0.21\,\mu$m thick, which selected area diffraction shows to be amorphous.
Beneath this is single crystal silicon again, with a heavily damaged
layer extending approximately $0.11\,\mu$m below the amorphous region.    No polycrystalline material or precipitates were found.    The dark/light band at
the upper crystalline/amorphous interface is strain contrast.

Random and channelled RBS spectra are shown in Figure 2.    The high
channelled yield confirms the high defect density.    The dip is consistent
with a nitrogen-containing amorphous zone, but it does not reach the level
expected for stoichiometric Si$_3$N$_4$.    The SIMS analysis (Figure 3) also
shows a broad peak in the nitrogen distribution.

### 3.2    $0.75 \times 10^{18}$ N$^+$ cm$^{-2}$, 200keV, annealed

Figure 4 shows a cross-section of the sample implanted at 200keV to
$0.75 \times 10^{18}$ N$^+$ cm$^{-2}$ and annealed at 1200°C for 2 hours.    A buried layer of
silicon nitride is seen in single crystal silicon.    The upper layer of
silicon is $0.32\,\mu$m thick.    The layer below is $0.18\,\mu$m thick and in the
Figure is divided laterally into two regions.    One of these is polycrystalline, with a grain size of about 1000Å, and was identified as $\alpha$-
Si$_3$N$_4$ by electron diffraction.    The other appears to consist mainly of
amorphous material, but crystallites ~100Å in diameter are embedded in it.
Other TEM specimens from this material indicated that the polycrystalline
structure is the more common.

No dislocations or other defects were observed in the silicon layers,
indicating a dislocation density $\leqslant 10^7$ cm$^{-2}$.    However, the nitride layer
was found by diffraction and dark field imaging to contain randomly oriented silicon crystals ~300Å in diameter, these representing a small
fraction of the total volume of the layer.    The interfaces between the
buried layer and the single crystal silicon are abrupt but not perfectly
planar.

The RBS results (Figure 5) are in accordance with the cross-sectional TEM,
indicating a sharply defined buried layer of composition close to Si$_3$N$_4$,
and a low defect density in the upper silicon layer ($\chi_{min}$=0.043).    The
sharpness of the interfaces is shown by the SIMS data (Figure 6), which
also indicates the depletion of nitrogen from the upper and lower silicon
layers (compare Figure 3).

The same material is shown in plan-view micrographs in Figures 7 and 8.
The silicon nitride layer (here entirely polycrystalline) consists of

spherulitic regions ~5-10 μm in diameter. At higher magnification (Figure 8) the grain structure seen in Figure 4 is apparent. Within each spherulite the crystallographic orientations of the grains are the same to within a few degress.

## 3.3  $0.25 \times 10^{18}$ $N^+$ $cm^{-2}$, 200keV, annealed

Figure 9 shows a cross-section of the specimen implanted at 200keV to $0.25 \times 10^{18}$ $N^+$ $cm^{-2}$ and annealed at 1200°C for 2 hours. As for the specimen implanted to $0.75 \times 10^{18}$ $N^+$ $cm^{-2}$ and annealed, a buried layer of $\alpha$-Si$_3$N$_4$ in single crystal silicon is found. However, the nitride layer does not form a continuous barrier between the upper and lower silicon regions. The geometry of the layer is intermediate between that of a continuous layer and one of isolated particles. Plan-view micrographs (not shown) indicate a spherulitic structure similar to that seen in the material implanted to $0.75 \times 10^{18}$ $N^+$ $cm^{-2}$ and annealed.

## 3.4  $0.75 \times 10^{18}$ $N^+$ $cm^{-2}$, 350keV, followed by $0.75 \times 10^{18}$ $N^+$ $cm^{-2}$, 200keV, annealed

Figure 10 is a cross-sectional micrograph of the sample implanted first at 350keV to $0.75 \times 10^{18}$ $N^+$ $cm^{-2}$, then at 200keV to $0.75 \times 10^{18}$ $N^+$ $cm^{-2}$, and finally annealed at 1200°C for 2 hours. Two buried nitride layers of unequal thickness are seen, each similar to the polycrystalline part of the layer in Figure 4. The material above, between and below these two buried layers is single-crystal silicon, but unlike the previous annealed specimens has a significant dislocation density in the upper layer, ~3x10$^8$ $cm^{-2}$. The lower surface of the upper nitride layer is very irregular, with outgrowths extending in some cases to the lower nitride layer. The RBS spectra (not shown) confirm the presence of two buried layers of silicon nitride and the relatively high defect density in the upper silicon layer.

## 4. Discussion

Results in the literature (Belz and te Kaat 1984, Bourguet et al 1980, Josquin and Tamminga 1982) show that when silicon is implanted with nitrogen to a dose less than that required to form stoichiometric Si$_3$N$_4$, redistribution occurs during high temperature annealing to form Si$_3$N$_4$ and nitrogen-depleted silicon. The present results are in accordance with this. In contrast to the results of oxygen implantation (e.g. Taylor et al 1983) precipitates are not formed in the silicon layers. The dislocation densities in the annealed materials are considerably lower than those found in oxygen-implanted materials, in the case of the two single-energy implants here, and comparable in the case of the double-energy implant. The spherulitic structure of the nitride layers suggests that during annealing the layer crystallises from nuclei spaced by about 5-10 μm. The detailed mechanism by which this occurs, leading to grains ~1000Å in diameter, is however not yet clear.

## 5. Conclusions

It has been demonstrated that a dose of $0.75 \times 10^{18}$ $N^+$ $cm^{-2}$ under the implantation and annealing conditions used here is sufficient to produce a continuous buried layer of silicon nitride while maintaining good crystal quality in the silicon overlayer. However, the presence of silicon crystallites in the buried layer may present a leakage current problem, so that it may be necessary to use higher doses.  We have also shown that it is possible to produce two distinct buried layers by implantation at two energies.

## Acknowledgments

We should like to acknowledge support from the Science & Engineering Research Council.

## References

Belz J and te Kaat E 1984 8th European Congress on Electron Microscopy, Budapest, p967
Bourguet P, Dupart J M, Le Tiran E, Auvray P, Guivarch A, Salvi M, Peious G and Henoc P 1980 J. Appl. Phys. 51 6169
Josquin W J M J and Tamminga Y 1982 J. Electrochemical Soc. 129 8
Kreissig U, Skorupa W and Hensel E 1983 Thin Solid Films 100 L25
Lam H W and Pinizzotto R F 1983 J. Crystal Growth 63 554
Maeyama S and Kajiyama K 1982 Japan. J. Appl. Phys. 21 744
Skorupa W, Kreissig U, Hensel E and Bartsch H 1984 Electronics Letters 20 427
Taylor M R, Tuppen C G, Arrowsmith R P, Dobson R M, Glaccum A E, Wilson M C, Booker G R and Hemment P L F 1983 Inst. Phys. Conf. Ser. No.67 p485
Zimmer G and Vogt H 1983 IEEE Transactions on Electron Devices ED-30 1515

0.5µm

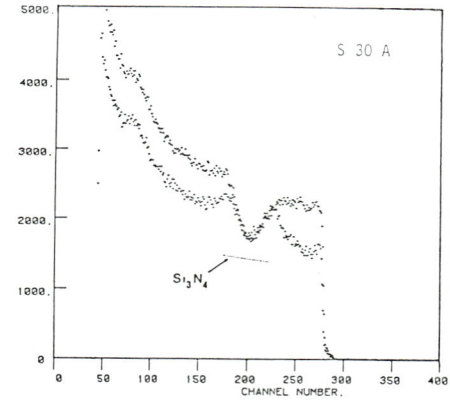

Fig.1  Silicon implanted with 0.75 x 10¹⁸ N⁺ cm⁻², unannealed.  TEM 90° cross section, bright field.  The white marker indicates the slice surface.

Fig.2  Random and [100] channelled RBS spectra from sample shown in Fig.1, using 1.5MeV He⁺ ions.

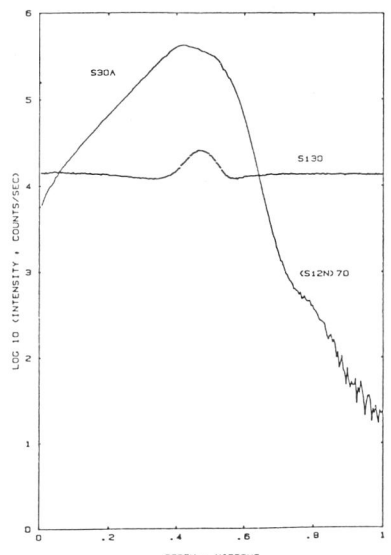

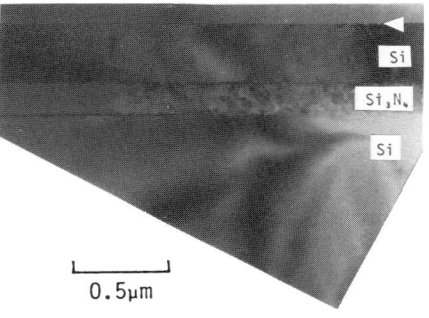

Fig.3  SIMS profile of sample shown in Fig.1.  The silicon and nitrogen concentrations are approximately proportional to the $Si^+$ and $Si_2N^+$ intensities respectively.

Fig.4  As Fig.1, but annealed at 1200°C for 2 hours.

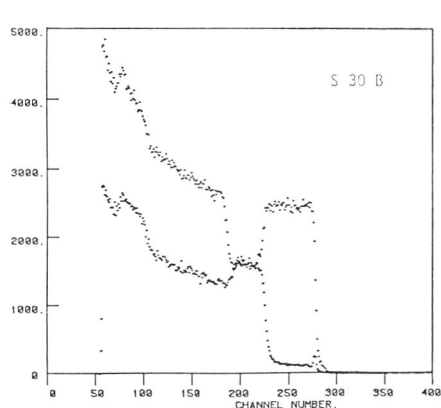

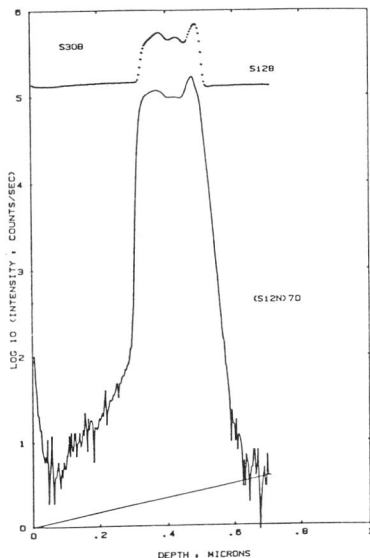

Fig.5  Random and [100] channelled RBS spectra from sample shown in Fig.4.

Fig.6  SIMS profile of sample shown in Fig.4.

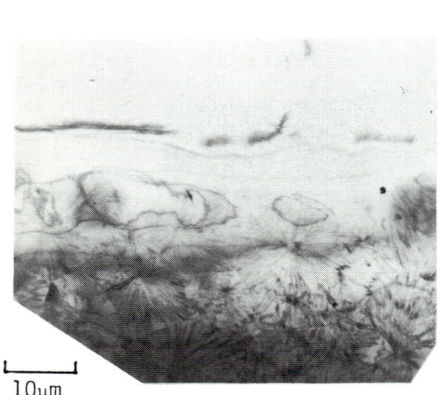

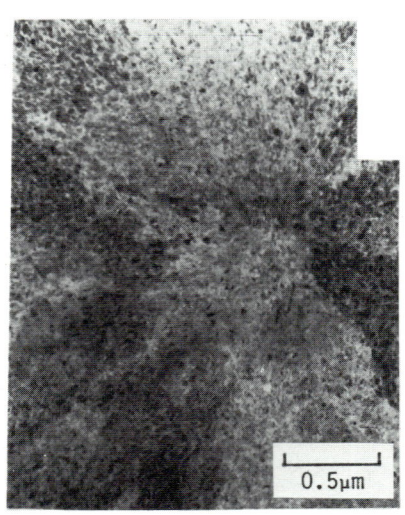

Fig.7  Sample shown in Fig.4, TEM plan view, bright field.  In the upper part, only the top silicon layer is present after the thinning.

Fig.8  As Fig.7, region containing silicon nitride at higher magnification.

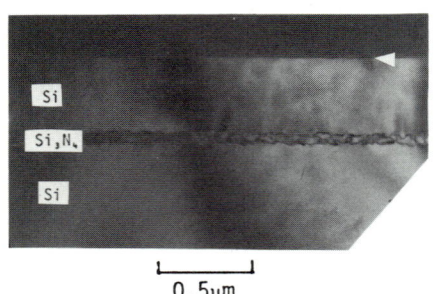

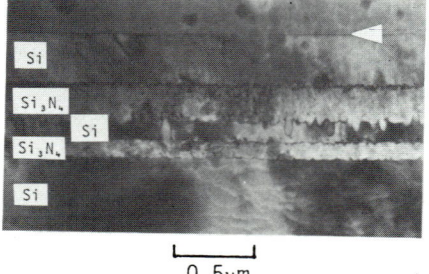

Fig.9  Silicon implanted with 0.25 x $10^{18}$ $N^+$ $cm^{-2}$, annealed at 1200°C for 2 hours.  TEM 90° cross section, bright field.

Fig.10  Silicon implanted with 0.75 x $10^{18}$ $N^+$ $cm^{-2}$ at 350keV, followed by 0.75 x $10^{18}$ $N^+$ $cm^{-2}$ at 200keV, and then annealed at 1200°C for 2 hours.  TEM 90° cross section, bright field.

*Inst. Phys. Conf. Ser. No. 76: Section 11*
*Paper presented at Microsc. Semicond. Mater. Conf., Oxford, 25–27 March 1985*

# The effect of fluorine on oxidation induced stacking faults in silicon

F G Kuper, J Th M De Hosson and J F Verwey

Department of Applied Physics, Materials Science Centre,
University of Groningen, Nijenborgh 18, NL-9747 AG Groningen,
The Netherlands.

Abstract    Shrunken stacking faults were studied by transmission
electron microscopy. A difference between the effect of fluorine and
that of chlorine on the stacking fault was observed. A possible
explanation is given involving gettering of copper contamination.

## 1. Introduction

Once stacking faults in silicon have been nucleated, they grow if the
silicon self interstitial concentration exceeds the equilibrium concentration around the bounding Frank partial (R. B. Fair 1981, S. M. Hu
1981). Stacking faults grow in (100) wafers during dry oxidation up to
1240 °C and shrink during oxidation at higher temperatures (S. M. Hu
1975). The production of interstitials at the $Si/SiO_2$ interface can be
lowered effectively by adding chlorine to the oxidizing ambient. This
results in an increase of the retrogrowth region, as was observed by H.
Shiraki (HCL 1976), T. Hattori (TCE 1977), C. L. Claeys et al. (C 33
1977) and S. Kawado (Cl implantation 1979).

In the case of HCl oxidations it has been shown that the chlorine piles
up at the interface (B. C. Beard et al. 1983, B. E. Deal et al. 1978).

As interstitials are involved in diffusion of substitutional dopants one
can expect that chlorine oxidation restrains oxidation enhanced diffusion. This is observed by e.g. Y. Hokari (1979) applying HCl oxidation
and by S. Solmi and P. Negrini (1984), with implanted chlorine.

It was shown by S. Isomae et al. (1980) that another halide - fluorine -
also has an effect on the growth of stacking faults. Length measurements
after etching of stacking faults and a SIMS depth profile led to the
conclusion that chlorine and fluorine react in a similar way. Comparing
the results of S. Kawado (1979) and S. Isomae et al. (1980) one must
conclude that chlorine is a little more efficient.

Fluorine is smaller than silicon, so we might expect it to diffuse quite
easily in silicon. No diffusion coefficients are available, but there
have been some reports on the decoration of dislocations (damage) in
silicon by fluorine (M. Y. Tsai et al. 1979, S. Prussin et al. 1984),
though not supported by electron micrographs. Our research is aimed at
investigating the influence of fluorine in the silicon substrate,

especially near the bounding dislocation of a stacking fault.


## 2. Experiments

Float zone (100) boron doped 1-30 $\Omega$ cm Si wafers, supplied by Wacker were
mechanically polished with 6 $\mu$m diamond polish paste to create stacking
fault nucleation sites. The wafers were given a standard cleaning and
were oxidized at 1200 °C for 30 minutes in a dry oxygen flow of 1l/min.
The oxide thickness obtained is about 150 nm, the stacking fault length
is up to 15 $\mu$m; the density of the SF's differs among the various samples
but is usually large enough for TEM examination.

At this stage there are two main defects: extrinsic stacking faults and
dislocations with 60° and edge character. Often the edge part is etched
out in the TEM specimen, leaving two (dissociated) 60° dislocations to be
seen. See figures 1 and 2.

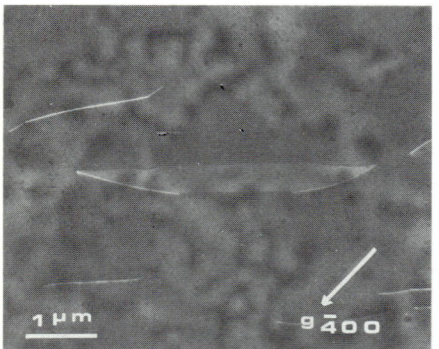

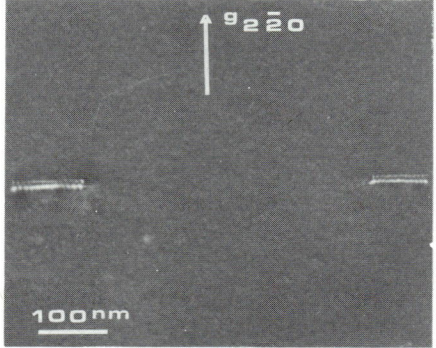

Fig. 1  Stacking fault and dislocations in oxidized silicon.

Fig. 2    Dissociated 60° dislocations.


Subsequently fluorine (or chlorine) is implanted in the silicon-dioxide
layer. The accelerating voltage for fluorine is 40 keV (projected range
67 nm) and for chlorine it is 78 keV (67 nm as well).
The wafers were cleaved into several samples for subsequent annealing in
dry $O_2$ at temperatures up to 1200 °C. From these samples TEM specimens
were made by backside chemical etching of squares with a ribbon of 2 mm.

We used a JEOL 200 CX transmission electron microscope operating at 120
kV, just below the Frenkel pair formation energy. Parts of the samples
were etched in Wright Jenkins etch in order to measure stacking fault
length with an optical microscope.


## 3. Results

The effects of the halides are very strong at 1200 °C. The no growth dose
for fluorine is only about $3.10^{15}$ ions per cm² and for chlorine (S.
Kawado 1979) about $2.10^{14}$. In this paper we concentrate on the effects at
this particular temperature, to begin with fluorine. Some stacking fault
growth curves are shown in fig. 3, where the largest stacking fault
length is plotted versus reoxidation time.

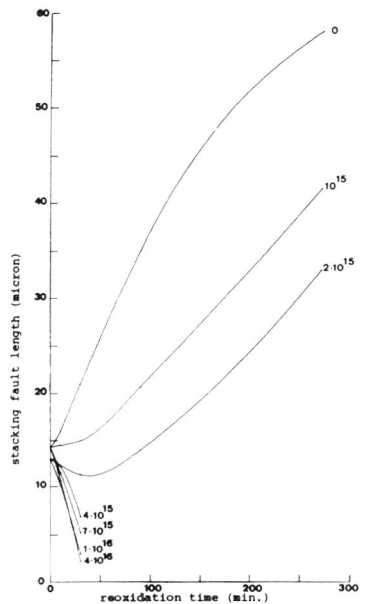

Fig. 3 Stacking fault growth curves at 1200 °C in dry $O_2$ with the fluorine dose as a parameter.

Details of the decoration are given in figures 5, 6 and 7. Stereo electron micrographs revealed loops on the {111} stacking fault plane and on {011} planes.

The curve of dose $2.10^{15}$ $F^+/cm^2$ suggests that fluorine diffuses out of the sample during reoxidation. This is confirmed in a study using nuclear reaction analysis, which will be published elsewhere. Some of the shrunken stacking faults were examined by transmission electron microscopy and it appeared that all faults being examined were heavily decorated along the bounding Frank partial. This decoration stretches out along the total edge as one can conclude from the etch "bow" between the ends of the stacking fault. (See figure 4).

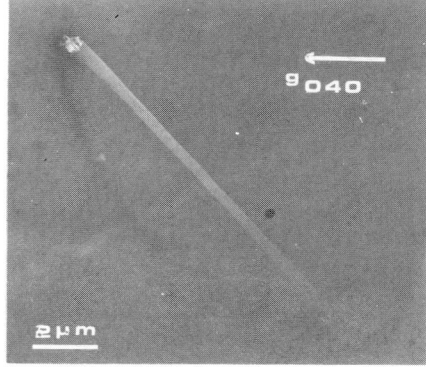

Fig. 4 Stacking fault and preferential etch pattern. Dose: $4.10^{16}$ $F/cm^2$, reoxidation time: 10 minutes at 1200 °C.

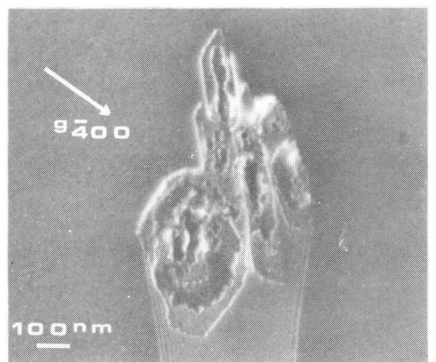

Fig. 5 Dark field/weak beam electron micrograph of the decoration of a stacking fault $4.10^{16}$ $F/cm^2$, 30 minutes at 1150 °C.

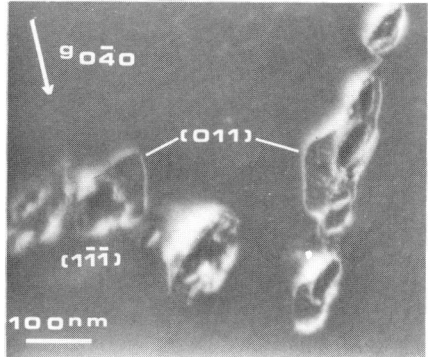

Fig. 6 {111} and {011} planes, decorating Frank partials $4.10^{15}$ $F/cm^2$, 30 minutes at 1200 °C.

The {111} loops tend to accept a hexagonal shape with edges along <022> and do not exhibit stacking fault fringes. This is quite clearly seen in fig. 5 which shows a sample reoxidized at 1150 °C for 30 minutes.

In figures 7 and 8 one sees regions exhibiting parallel Moiré fringes. These regions are only found in {111} loops and could only be seen with <220> and <400> reflections. The Moiré spacings are about 39 Å with <220> reflections and 27 Å with <400> reflections ($\pm$ 10%). More information about the lattice parameter of these precipitates was not obtainable as microdiffraction did not show extra spots. The {011} planes have a rectangular shape (see figure 6) and contain a few dislocation loops, but no Moiré fringes.

With standard invisibility criteria it was not possible to determine the Burgers vectors of the decoration edges. Contrast experiments revealed that the loops are in a compressive state.

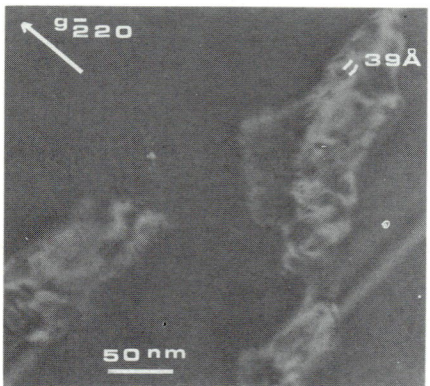

Fig. 7   Moiré fringes in the decorated area. Same specimen as depicted in figure 6.

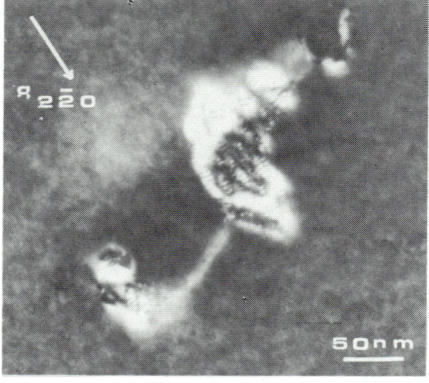

Fig. 8   Moiré fringes along an invisible stacking fault, as $\mathbf{g}.\mathbf{R}=0$. $4.10^{16}$ F/cm$^2$, 10 minutes 1200 °C.

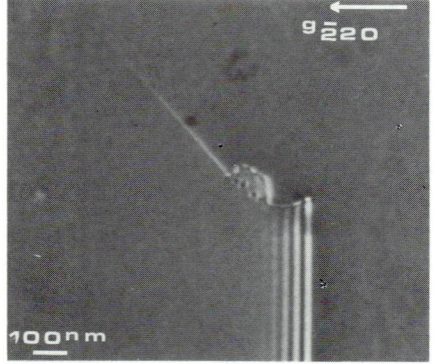

Fig. 9   A stacking fault shrunken by influence of chlorine. $4.10^{15}$ Cl/cm$^2$ 30 minutes 1200 °C.

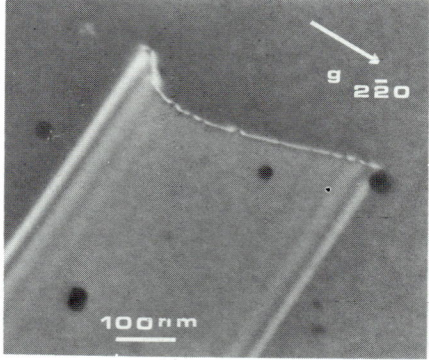

Fig. 10   Same specimen as depicted in figure 9.

The effect of chlorine on the Frank partial is illustrated in figures 9 and 10. The shrunken faults are hardly decorated!

## 4. Discussion

The Moiré fringes indicate that there are ordered structures in the silicon with a constant lattice parameter and a fixed orientation with respect to the matrix. Knowing the Moiré spacings and that the loops are in a compressive state, the lattice parameter of these precipitates can be calculated: $5.72 \pm 0.03$ Å. This is the same as the lattice parameter of CuSi precipitates with a $B_3$ structure (G. Das 1973, E. Nes and J. Washburn 1971).

An element analysis of these precipitates was tried, but proved to be very difficult to perform, as fluorine is too light to be detected by our EDS analyzer equipped with a beryllium window. Moreover, detection of copper is difficult because of the large copper background signal in our microscope.

Recently, A. Ohsawa et al. (1984) presented evidence of the gettering of copper at the $Si/SiO_2$ interface during HCl oxidation. The same may have happened here: Copper contamination is gettered at the interface after chlorine implantation but not after fluorine implantation. The other possibility, decoration with fluorine, can however not be excluded.

## 5. Conclusion

In conclusion one can say that the influence of fluorine and chlorine is not the same. Apparently fluorine stimulates rather than prevents condensation of impurities at the bounding Frank partial of the stacking fault. This might be due to a difference in diffusion between fluorine and chlorine or to a difference in gettering capabilities.

## Acknowledgements

This work is part of the research program of the Foundation for Fundamental Research on Matter (F.O.M. - Utrecht) and has been made possible by financial support from the Netherlands Organization for the Advancement of Pure Research (Z.W.O. - The Hague).

## References

Beard B C, Titcomb S, Butler S R 1983 J. Electrochem. Soc. 130 (9) 1959
Claeys C L, Laes E E, Declerck G J, Van Overstraeten R J 1977 in: "Semiconductor Silicon", Huff H R and Sirtl E eds. The Electrochem. Soc. Princeton p. 773
Das G 1973, J. Appl. Phys. 44 (10) 4459
Deal B E, Hurrle A, Schulz M J 1978 J. Electrochem. Soc. 125 (12) 2024
Fair R B 1981 J. Electrochem. Soc. 128 (6) 1360
Hattori T 1977 Appl. Phys. Lett. 30 (7) 312
Hokari Y 1979 Japan J. Appl. Phys. 18 (5) 873
Hu S M 1975 Appl. Phys. Lett. 27 (4) 165

Hu S M 1981 in: "Defects in Semiconductors" Narayan J and Tan T Y eds.
    North Holland 333
Isomae S, Tamura H, Tsuyama H 1980 Appl. Phys. Lett. 36 (4) 293
Kawado S 1979 Japan J. Appl. Phys. 18 (2) 225
Nes E, Washburn J 1971 J. Appl. Phys. 42 (9) 3562
Ohsawa A, Honda K, Toyokura N 1984 J. Electrochem. Soc. 131 (12) 2964
Prussin S, Mangolese D I, Tauber R N, Hewitt W B 1984 J. Appl. Phys.
    56(4) 915
Shiraki H 1976 Japan J. Appl. Phys. 15 (1) 83
Solmi S, Negrini P 1984 Appl. Phys. Lett. 45 (2) 157
Tsai M Y, Day D S, Streetman B G, Williams P, Evans C A jr 1979 J. Appl.
    Phys. 50 (1) 188

*Inst. Phys. Conf. Ser. No. 76: Section 12*
*Paper presented at Microsc. Semicond. Mater. Conf., Oxford, 25–27 March 1985*

501

# Advances in electron beam testing equipment

S M Davidson

Lintech Instruments, Cambridge Science Park, Cambridge

Abstract    Recent advances in electron beam testing
equipment are described, with particular reference to
electron guns, specimen stages, image processing, electron
collectors and beam blanking systems. The current state
of the art is discussed, with indications where development will be needed to meet the test requirements of the
next generations of integrated circuits.

## 1.  Introduction

Over the last five years, electron beam testing has been
transformed from a  research laboratory technique which was
being investigated by a few electronics companies (notably
Siemens and IBM) into a method of routinely assessing and
analysing semiconductor devices which is in widespread use
around the world.  With reference to the position described
by Wolfgang (1983) in these proceedings two years ago, few
dramatic developments have occurred.  However the means of
applying the technique to real devices in the semiconductor
industry are now much more readily available.

A range of equipment has appeared in the marketplace
specifically  designed for electron beam testing.  Lintech
Instruments have been pioneers in this field, having supplied
electron beam testing equipment since 1979.  However, more
recently other manufacturers – Cambridge Instruments, JEOL,
ICT and Applied Beam Technology (ABT) – have also been
offering E-beam testing systems.  The main aim in this paper
is to describe the main advances in equipment currently
available for electron beam testing, and to indicate what
further developments will be needed for the successful
testing of the next generations of integrated circuits.

The  basis  of electron beam testing, ie the  measurement  of
voltages and waveforms on integrated circuits by combining
high speed electron beam pulsing with a voltage measuring
electron collector has been well described elsewhere, eg
Wolfgang et al (1979), Gopinath et al (1978).  Here we will
concentrate on the major items of equipment used to perform
these measurements, to indicate where improvements may be
necessary.

## 2.  Electron Optics

Integrated circuits, in particular MOS devices, are easily
damaged by electron beam irradiation.  For electron beam
testing, it usually essential to operate at low beam voltages
- typically 0.5 to 1.5kV.  The performance of the SEM at low
beam voltages is thus particularly crucial.  High spatial
resolution is not required, 100nm - 500nm sufficing for most
measurements. However it is important for the beam current
into these probe sizes to be as high as possible.

High gun brightness at low kV is very important, and there
have been a number of developments over the last few years
(Figure 1).  The options are (a) conventional tungsten
filament (b) tungsten filament with a low KV gun assembly
(Wehnelt and anode with smaller spacings) (c) tetrode gun,
where an auxiliary extractor electrode between the Wehnelt
and anode  increases the field at the filament (d) any
combination of (a) - (c) with a LaB6 filament and (e) a field
emission gun.  Low KV guns with LaB6 filaments now have
reported brightnesses of 10-4 to 2x10-4, almost as high as
the 20kV electron guns of a few years ago.  One factor
possibly limiting future development is the Boersch effect,
ie the mutual repulsion of electrons in high density electron
probes.  Some recent work has shown that the improvement in
brightness with LaB6 at low kV appears to be limited by this
effect.  The use of field emission guns may be valuable when
spatial resolution in the sub-100nm region is needed, but
these will not give higher total probe currents than LaB6
guns.

Most other aspects of current SEM technology are adequate for
electron beam test applications.  However, one facility needed
if the SEM is to be used as a building block in a larger test
system is remote control.  Most SEMs now have microprocessors
controlling the basic electronics functions, but full remote
control is not yet available.  Instrumental developments such
as automatic focussing and automatic astigmatism correction
are bringing this day significantly nearer.

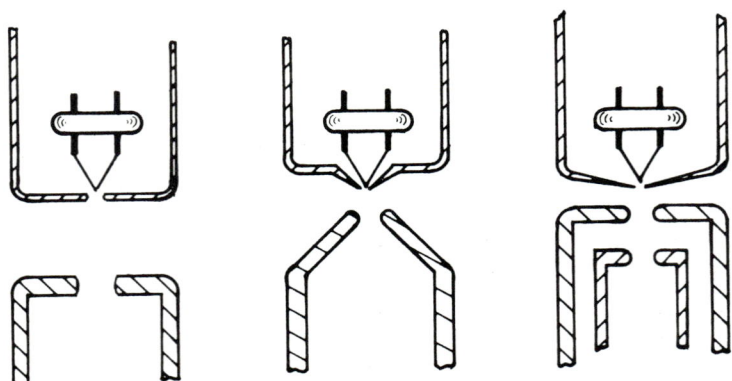

Fig 1 (a) conv. electron gun   (b) low KV gun   (c) tetrode gun

## 3. Image Processing

One area where significant developments have appeared is
image processing. It is now accepted that non-expert
operators of scanning electron beam instruments prefer to
observe an image at TV rates rather than a slow scan image.
Even with the improvements in gun brightness at low KV
outlined above, there are still occasions where the real time
(TV rate) image is too noisy for analysis. This is
particularly the case with stroboscopic imaging, where the
beam duty cycle (ratio of beam-on time to beam-off time) is
often less than 1%. Image processors, which perform noise
reduction by recursively averaging the image over a number of
frames, are now being considered as an essential part of an
electron beam testing system. The market leaders in this
field are Micro Consultants, but other manufacturers with
similar products include Cambridge Instruments and Lintech.

The virtues of image processing become even more apparent
when one considers future developments. Stroboscopic voltage
contrast images effectively give the user a "frozen phase"
view of an operating IC, ie the state of all the nodes at one
instant during its operating cycle. Since this information
(for a correctly operating device) is available to the device
designer, it should be possible to simulate the expected
stroboscopic image. It is then a short step to perform an
automatic comparison with the recorded image, rapidly
pinpointing the components on the chip responsible for poor
performance. This operation is made much easier if the image
is already stored in a suitable digital form.

## 4. Device Handling

Another component of an electron beam testing system which
has to be optimised for IC examination is the specimen stage.
The integrated circuits being examined using electron beam
testing are usually either part of the wafer or mounted in
conventional packages. Occasionally it is also necesary to
examine bare chips. The type of testing which can be
performed also depends on the form of the mounting. If the
device is packaged, then it should be possible to test the
device at full speed.

The device mounting in the E-beam test system should not
limit its performance. With a conventional specimen stage,
the electrical connections must be sufficiently long both to
allow freedom of movement and to reach the vacuum feedthrough
fitted to the wall or door of the chamber. A lead length of
200-300 mm would not be uncommon. This may significantly
degrade the device performance by altering the shape of the
driving clocks and causing unwanted coupling between signals.
One solution to this problem, adopted first by Siemens and
subsequently by Lintech (Fig 2) is to vacuum seal the device
socket into the floor of the chamber. Connections from the
test electronics to the device can then be a short as 10 mm.

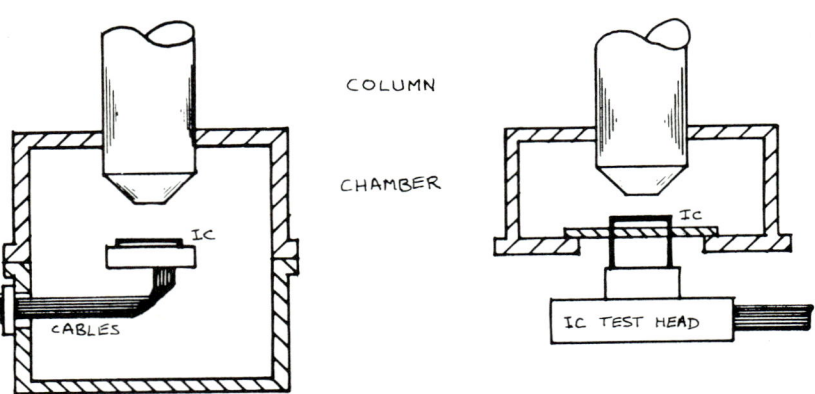

COLUMN

CHAMBER

Fig 2 (a) conventional specimen chamber (b) IC testing stage

Other considerations apply if the device has to be tested on the wafer. High speed is not so important, since the probe card imposes its own limitations; however not only have the wafer and probecard to be accommodated in the SEM chamber, but also the means for aligning any chip on the wafer with the probe needles. Most SEM chambers are too small to allow the fitting of a wafer prober, so special chambers or chamber/stage combinations are needed. Alignment of the probecard with the wafer is surprisingly easy, the most important SEM requirement being a large, unobstructed field of view. This can however be compromised by the type of voltage measuring collector in use (see later).

## 5. Voltage Measuring Collectors

Electron beam testing relies on the variation of the secondary electron energy spectrum with specimen voltage. It is usual to operate on the integral of the spectrum, known as the S-curve. In a well behaved system, the position of the S-curve along the voltage axis will accurately track the specimen voltage. Analysing or filtering the secondary electron distribution is the function of the voltage measuring collector, sometimes called the spectrometer (Fig 3). In fact, all electron "spectrometers" used for voltage measurement are high pass energy filters. Two types of voltage measuring collector are currently available, the grid type and the aperture type. In both cases, the secondary electrons are first accelerated away from the device with an extraction electrode before being filtered and collected.

An example of the grid type is the Feuerbaum (1979) collector used by Siemens, ICT and Cambridge Instruments. This gives a uniform, high extraction field across the width of the grids (8 mm on the Cambridge Instruments version); however the grid bars themselves obscure the image so much that only the central area of the device (typically 300um) can be seen

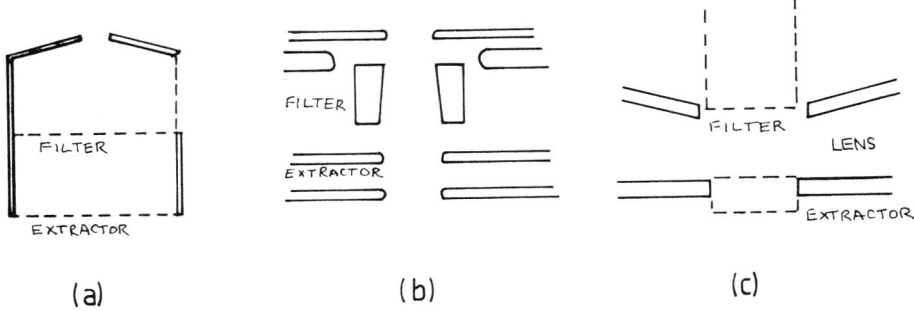

Fig 3   Detectors (a) Feuerbaum   (b) Lintech   (c) in-lens (ABT)

clearly.  Waveform measurements are OK, but large area
stroboscopic imaging is difficult.  The second type of
spectrometer is typified by the Lintech voltage measuring
electron collector (VMEC).  Here all the electrodes have an
annular, cylindrically symmetric geometry.  This give a clear
unobstructed view of the device over a large area, albeit
with some distortion at the edges of the field.  The maximum
extraction field is, however, not as high as for the grid
analyser.

Spectrometers are normally positioned below the SEM final
lens.  However good low kV operation requires the working
distance to be as small as possible. A recent development,
pioneered by ABT, is to incorporate the spectrometer inside
the SEM final lens.  In this case the working distance can be
as low as a few mm.  This approach requires a large bore
final lens, such as fitted to some ISI and Hitachi SEMs.  The
lens field contains the secondary electrons, and there is
some evidence that it can reduce fringing field errors when
measurements are made on closely spaced tracks. The drawbacks
are (a) the need for a special lens/detector combination and
(b) the small field of view when operating at a short working
distance.  Other manufacturers, eg Lintech, are also working
on this approach.

The electronics used to control the energy filter has also
undergone a transformation over the past two years.  In
earlier systems, the electrode voltages and amplifiers were
set up manually, requiring a skilled operator if the best
results were to be achieved.  Different settings were often
required for the various "modes" in which the system was
operated, eg real time image, stroboscopic image, open loop
waveform, closed loop waveform, S-curve, etc.  Most current
systems are computer controlled, making it a simple matter to
switch from one mode to another, store and recall waveforms
and perform automatic measurement sequences.

## 6. Beam Blanking

The final component of any electron beam tester is the beam
pulsing system. In stroboscopic operation the time
resolution is determined by the shortest beam pulse which can
be generated. This is turn is partially determined by the
blanking sensitivity, ie the voltage required to blank the
beam. All commercially available E-beam test sytems employ
electrostatic plate deflection blanking, using either single
sided or push-pull deflection. This gives high sensitivity at
low beam voltages but can cause resolution degredation,
principally dependant on the position of the plates in the
column. Blanking sensitivity at a given KV depends simply on
plate spacing and length. The maximum allowable plate length
is determined by the transit time of the electrons. At 1kV
the electron velocity is 10+7 m/sec; if pulses 0.2ns long are
required the plates should not be longer than 2mm. The plate
separation is limited by practical considerations,
principally the need to avoid contact between the beam and
the plates. Minimum separation is 0.5 - 1mm. Under these
circumstances beam blanking at low kV can be achieved with a
few volts.

Resolution degredation occurs during blanking if the virtual
electron source moves laterally before blanking occurs. This
movement causes a smearing of the image in the direction of
the source movement. The extent of the smearing depends on
the blanking sensitivity, the drive voltage available, the
position of the plates in the column and the currents through
the SEM condenser lenses. Smearing can be avoided if the
blanking plates are located in a conjugate plane. Cambridge
Instruments position their plates (in fact, wires) below the
second lens, while in the Lintech EBT they appear at the
image plane of the first lens.

In most accessory systems the plates come between the gun and
first lens. An adjustable aperture or knife edge can limit
any resolution degradation, but not eliminate it. An
additional complication with this arrangement occurs if the
condenser lens is a symmetrical double lens with a common
winding. If the blanking plates are positioned less than two
focal lengths above the lens, the blanking sensitivity
rapidly decreases and can reverse. Concomitant with this is
a marked degredation in spatial resolution. Work is in
progress to quantify this behaviour.

## References

Feuerbaum H P 1979 Proc SEM 1979 pp 285-296 (SEM Inc., AMF
O'Hare)
Gopinath A, Gopinathan K G Thomas P R 1978 Proc SEM 1978
pp 375 - 380 (SEM Inc., AMF O'Hare)
Wolfgang E Lindner R, Fazekas P, Feuerbaum H P 1979 IEEE J
Solid State Circ. Sc-14 471
Wolfgang E 1983 Proc Mic. Semicond. Mat. Conf. IOP Conf Ser.
No 67 pp 407-414

*Inst. Phys. Conf. Ser. No. 76: Section 12*
*Paper presented at Microsc. Semicond. Mater. Conf., Oxford, 25–27 March 1985*

507

# On the primary electron energy dependence of radiation damage in passivated NMOS transistors

W Reiners, S Görlich and E Kubalek

Universität Duisburg Fachgebiet Werkstoffe der Elektrotechnik
Leiter: Prof. Dr.-Ing. Erich Kubalek
Kommandantenstraße 60, 4100 Duisburg 1, F.R.G.

Abstract  Threshold voltage shift in passivated transistors caused by
electron irradiation is measured for primary energies from 0.7...35keV
and for irradiation doses from $10^{-9}..10^{+2}C/cm^2$. Using existing theories
a model is described explaining damage by primary electron energy absorption within the gate oxide. However, radiation damage is caused also for
energies too low to enable such a penetration depth. Modification of
theory shows that for low primary energies X-ray generation and subsequent absorption in the gate oxide become dominant. Thus irradiation
damage can be understood for the whole energy range mentioned above.

## 1. Introduction

The conditions enabling nondestructive electron beam testing of passivated
NMOS devices are still investigated until now. Two different techniques
are used at present : The first one, using high electron energies is based
on the electron beam induced conductivity in insulators shorting the passivation and enabling a voltage measurement at the covered conductor tracks
(Taylor 1971). The application of this method leads to severe radiation
damage, which can only be reduced to an acceptable amount by using the
window scan mode (Görlich and Kubalek 1985). The second one, using low
electron energies, is based on the capacitive coupling voltage contrast
(Görlich et al. 1984), which to a first approximation enables a nondestructive testing, though irradiation damage is not excluded totally.

A theoretical model for the radiation damage, based on a change of the
space charge (Mitchell 1967) due to primary electron energy absorption
(Everhart and Hoff 1971), predicts no radiation damage for primary electron energies below a critical energy. However, experiments show the contrary (Miyoshi et al. 1982), i.e. the initial model has to be modified
for low energies. Thus fundamental investigations are necessary to understand the irradiation damage at low primary electron energies. Therefore,
the threshold voltage shift, as a measure for radiation damage, is determined for an energy range from 35keV down to 0.7keV, i.e. for energies
above and below the critical energy.

## 2. Experimental

The electron irradiation experiments are performed at a test structure
consisting of NMOS-transistors with different dimensions of gate width w
and gate length l. They are classified in type A : transistor 0 (w=50 μm/
l=10 μm) 1 (50/5) and 2 (10/5) and in type B : transistor 3(10/50), 4(5/50)

and 5(50/50). Fig. 1 shows the schematic cross sections of the two types of transistors. The gate oxide between the drain and source diffusion zone is thermally grown on a p-type silicon substrate and has a thickness of 90nm. On top of the gate oxide there is a poly-silicon gate covered by a reflowglass. Two aluminium layers represent the drain and source contacts. In contrast to transistor type A, type B is provided with an additional aluminium layer above the reflow-glass, which covers part of the gate. All types are protected by a passivation layer of 1.1μm silicon dioxide. The total distance between the surface and the gate oxide thus amounts to 2.3μm for the transistors of type A and 3.3μm for the transistors of type B.

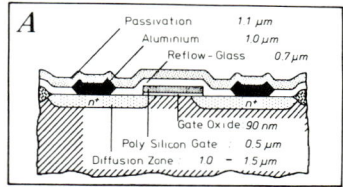

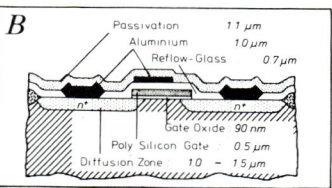

Fig. 1  Schematic cross sections of the NMOS transistors type A and type B

The shift of the threshold voltage $\Delta U_{TH}$ is measured for different electron doses D and primary electron energies $E_{PE}$. Source, drain, gate and substrate of the transistors are held at ground potential during the irradiation. The electron beam is scanned over a part of the test structure including all of the investigated transistors. Starting with the threshold voltage of the nonirradiated transistor the threshold voltage $U_{TH}$ is determined after each irradiation of a certain dose and then the difference $\Delta U_{TH}$ is calculated. This procedure is repeated several times for a certain primary electron energy and a certain test structure, but simultaneously for all transistors, thus avoiding unintentional changes of irradiation parameters.

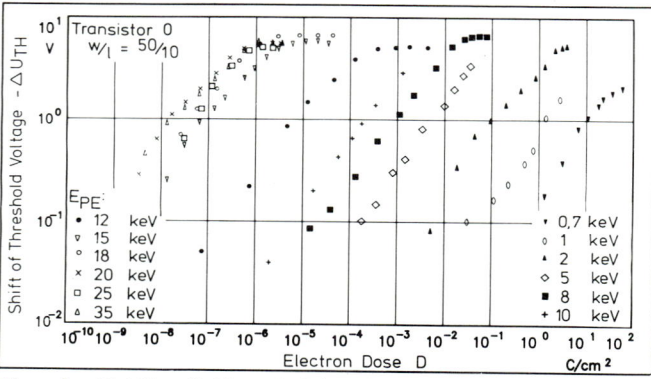

Fig. 2  Shift of threshold voltage of transistor 0 vs. electron dose with primary electron energy as parameter

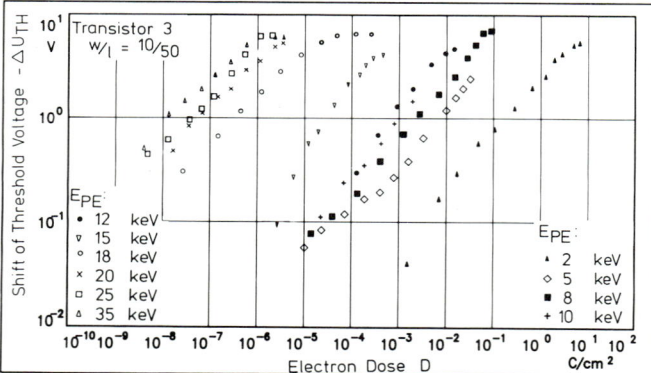

Fig. 3  Shift of threshold voltage of transistor 3 vs. electron dose with primary electron energy as parameter

The results are shown for transistor 0 and transistor 3 in fig. 2 and 3, representing the results for the two types. For all transistors and all primary energies the shift of the threshold voltage is negative. Furthermore always the same principal dependence of the threshold voltage shift versus the irradiation dose is found, consisting of a linear part for lower electron doses and a saturation at higher doses. A decrease of the primary electron energy from 35keV to 0.7keV results in a shift of the $\Delta U_{TH}(D)$-curve along the dose axis of nine orders of magnitude, i.e. $10^9$ times higher doses can be applied before causing the same radiation damage. However, there is an essential difference between the transistors of type A and B in the primary electron energy range from 20keV to 10keV, indicating that for this energy range the transistors of type A without the additional aluminium layer are more sensitive to radiation than those of type B with the additional aluminium layer.

## 3. Theory

A mathematical description of the experimental results is introduced consisting of two parts. In the first part the correlation between the space charge generated within the gate oxide during the irradiation time and the subsequent threshold voltage shift is discussed. The second part describes the calculation of the absorbed energy within the gate oxide that generates the space charge.

### 3.1 Correlation between shift of threshold voltage and irradiation time

Based on Mitchell's (1967) model for radiation damage and Maxwell's equations the dependence of the normalized threshold voltage shift $\Delta U_{TH}/\Delta U_{THsat}$ on the space charge generated within the gate oxide is derived (Reiners 1984). The space charge itself is determined by the generation rate g for electron beam induced electron-hole pairs within the irradiation time intervall t.

$$\frac{\Delta U_{TH}}{\Delta U_{THsat}} = 1 + \frac{A_2}{g \cdot t}\left(\exp\frac{g \cdot t}{A_1} - \exp\left(\frac{g \cdot t}{A_2} - \frac{g \cdot t}{A_1}\right)\right) \qquad (1)$$

where

$\Delta U_{THsat}$ = saturation value of $\Delta U_{TH}$

$A_1, A_2$ = parameters.

The ratio $A_1/A_2$ is determined by the thickness $d_{ox}$ of the gate oxide and the depth $d_{sc}$ of radiation induced space charge.

$$\frac{A_1}{A_2} = 1 - \frac{d_{sc}}{d_{ox}} \qquad (2)$$

In fig. 4 the graphs of the normalized shift of threshold voltage versus the irradiation time after eqt.(1) are shown.

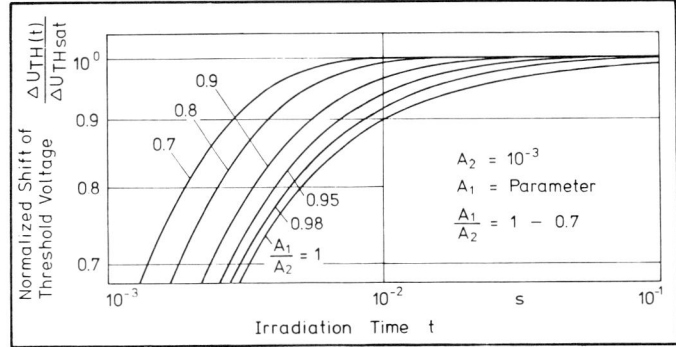

Fig. 4  Normalized shift of threshold voltage vs. irradiation time

The theory predicts a linear dependence for short irradiation times and a saturation for longer irradiation times. With decreasing ratio $A_1/A_2$ the slope of the function next to the saturation increases. Furthermore, a multiplication of $A_1$ and $A_2$ by the same constant leads to a shift of the graph along the time axis without any other changes.

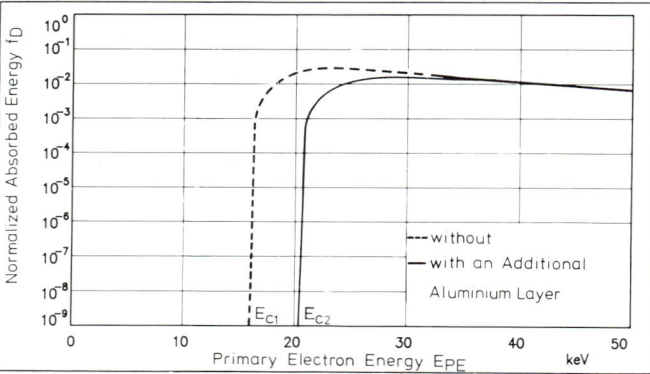

Fig. 5 Normalized absorbed energy vs. primary electron energy for transistor type A and type B

In this manner an adjustment of the theoretical curve to the experimental results for different energies is possible.

3.2 Energy absorption in an MOS transistor by electron irradiation

When an MOS transistor is irradiated with a dose D of primary electrons of an energy $E_{PE}$, the generated electron-hole pairs are determined by the part of the energy, which is absorbed within the gate oxide:

$$g \cdot t = A_3 \cdot f_D \cdot E_{PE} \cdot D \qquad (3)$$

where $f_D$ is the normalized absorbed energy in the gate oxide and $A_3$ is a constant of proportionality taking into account the geometry of the transistor.

The energy dissipation in MOS structures has already been studied by Everhart and Hoff (1971). Based on their calculations, which are valid for $5 keV < E_{PE} < 25 keV$, the normalized absorbed energy within the gate oxide is determined for a given transistor for various primary electron energies. Fig. 5 shows the result of the calculation for transistor type A without and for type B with an additional aluminium layer. The essential feature of these graphs is the existence of so-called critical energies $E_{C1}$ and $E_{C2}$ below which it is impossible for primary electrons to reach the gate oxide. Furthermore, there is a lower critical energy for the transistor type A than for the type B. In addition, the graphs take a maximum next to $E_{C1}$ or $E_{C2}$ and decrease slightly with increasing primary electron energy.

Therefore, this model predicts no radiation damage for a given transistor, if the primary electron energy is below the critical energy.

4. Comparison of experimental and theoretical results

The theoretical curve of the normalized threshold voltage shift versus electron irradiation dose is compared in fig. 6 with the experimental result of transistor 3 at the primary electron energy $E_{PE}=15 keV$. The parameters $A_1$ and $A_2$ of eqt.(1) are chosen to achieve a maximum agreement between theory and experiment. This curve fitting is possible for all transistors by just one set of constants $A_1$ and $A_2$, because for all transistors the same principal function is found (compare fig. 2,3). After this adjustment a determination of an experimental value of $f_D$ is achieved by

eqt.(3). Thereby the parameter $A_3$ is the same one for
all transistors of one type
A or B, respectively. Re-
peating this procedure for
all transistors and primary
electron energies the values
of $f_D$ are obtained over the
whole energy range.

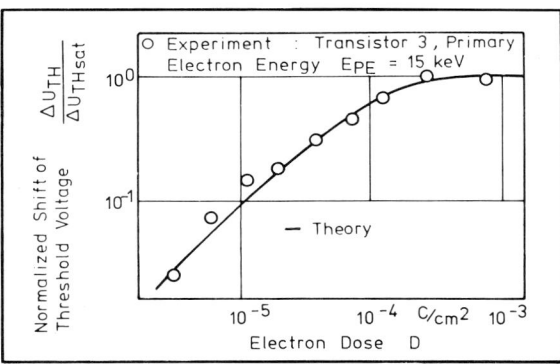

Fig. 7 shows the graph of
$f_D$ for all transistors in
an energy range from 0.7keV
to 35keV. All transistors
of one type behave identical-
ly, but the results are different for the two types.
In addition the known
Everhart and Hoff curves
are shown (compare fig.5).

Fig. 6  Comparison of theoretical and experimental normalized threshold voltage shift
vs. electron dose

The comparison with the experimental values results in a good agreement
for both transistor types. Additionally, the difference in $f_D$ for the two
transistor types in the primary electron energy range from 10keV to 20keV,
is described by the different critical energies $E_{C1}$ and $E_{C2}$, as caused by
the additional aluminium layer. However, though the experimental values
below the critical energies $E_{C1}$ and $E_{C2}$ are very small, they do not vanish
as predicted by the Everhart and Hoff curves.

This discrepancy for low primary electron energies is caused - as proposed by Miyoshi et al. (1982)- by secondary X-rays, especially by the
bremsstahlung. The validity of this assumption will be proven.

The X-rays are generated by primary electrons in the upper layer (for
simplicity    production at the surface, z=0, is assumed) and penetrate
towards the gate oxide. There they can generate electron-hole pairs which
cause a shift of threshold voltage. In order to estimate the effect due
to the secondary X-rays their generation at the surface and their absorption in the gate oxide must be calculated. The total intensity of the generated X-rays is
given by Stephenson
(1957):

$$I(z=0) \sim E_{PE}^{2} \qquad (4)$$

The intensity of the
bremsstrahlung per
unit energy interval
takes a maximum at
$E_{Xm}$. Stephenson(1957)
determined $E_{Xm}$ to:

$$E_{Xm} = \frac{2}{3} E_{PE} \qquad (5)$$

X-rays passing
through the upper
layers will suffer
an absorption, which
is described by

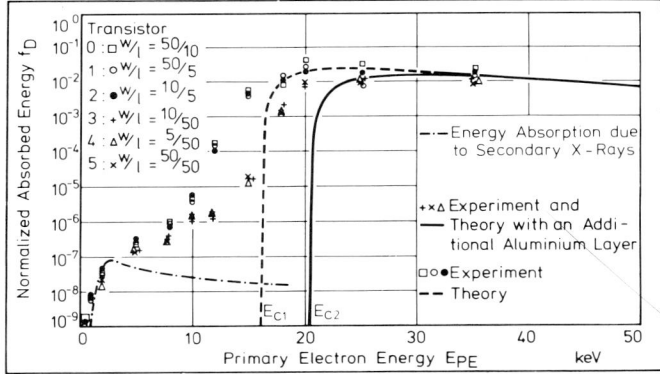

Fig. 7  Comparison of theoretical and experimental
normalized absorbed energy vs. primary electron
energy for all transistors

Theisen and Vollath (1967):

$$dI(z) = - I(z) \cdot \mu \cdot dz \qquad (6)$$

where      $I(z)$ = intensity in the depth $z$,
           $\mu$ = attenuation coefficient

The decrease X-ray intensity within the gate oxide is correlated with an energy absorption in it. In order to obtain this absorbed energy the X-ray intensity loss must be calculated. Considering the attenuation coefficient at the X-ray energy $E_{Xm}$ (compare eqt.(5)) after Theisen and Vollath(1967) the absorbed energy is given by integrating the X-ray intensity $I(z)$ in eqt.(6) over the thickness of the gate oxide. The result is shown in fig. 7 by the chained curve in agreement with the experimental values.

## 5. Conclusion

Electron irradiation causes  damage in passivated MOS transistor which is explained by an energy absorption in the gate oxide. For energies above a critical energy the damage can be interpreted by calculations based on the investigations of Everhart and Hoff (1971), which consider the energy absorption due to the primary electrons. However, for energies below the critical energy the experimental results are in contradiction to this interpretation. In order to explain this discrepancy an interaction with secondary X-rays is proposed. Especially the bremsstrahlung is considered which is generated by primary electrons in the upper layer  and which penetrates into the gate oxide causing the damage. In this manner irradiation damage can be theoretically understood for the whole range of primary electron energy from 0.7keV to 35keV.

## 6. Acknowledgement

The authors wish to thank Dr. L.J. Balk for helpful discussions and for reading the manuscript, and Dr. Sieber, AEG-Telefunken, for supply of test structures. Part of the work was supported by the federal ministry of research and technology (BMFT) of the F.R.G.

## 7. References

Everhart T E, Hoff P H 1971 J. Appl. Phys. 47 5337
Görlich S, Herrmann K D, Kubalek E 1984 Proc.Int. Conf. Microlithography :
   Microcircuit Engineering Berlin(London: Academic Press) in press
Görlich S, Kubalek E 1985 Scanning Electron Microscopy I 87
Mitchell J P 1967 IEEE-ED 14 764
Miyoshi M, Ishikawa M, Okumura K 1982 Scanning Electron Microscopy IV 1507
Reiners W 1984 Diplomarbeit University of Duisburg
Stephenson S T 1957 in: Encyclopedia of Physics XXX Flügge S (Ed.)
   (Berlin: Springer) 337-370
Taylor D M 1978 J. Phys. D 11 2443
Theisen R, Vollath D 1967 Tables of X-ray Mass Attanuation Coefficients
   (Düsseldorf:Stahleisen)

*Inst. Phys. Conf. Ser. No. 76: Section 12*
*Paper presented at Microsc. Semicond. Mater. Conf., Oxford, 25–27 March 1985*

# A fast frequency tracing method using scanning speed modulation

F Fox[+], H-D Brust[*] and J Otto[+]

[+]Forschungslaboratorien der Siemens AG, Otto-Hahn-Ring 6,
 D-8000 München 83, F.R. Germany
[*]Universität des Saarlandes, Physikalisch-elektronische Meßtechnik,
 Im Stadtwald B38, D-6600 Saarbrücken, F.R. Germany

Abstract  The recently developed frequency tracing method allows the
visualization of interconnections in VLSI chips carrying signals of
a selected frequency. With this technique, the scanning speed is
limited by the bandwidth of the frequency tracing signal chain. On
the other hand, a small bandwidth is necessary to obtain high sensitivity. Using scanning speed modulation, high spatial resolution and
high sensitivity were obtained together with reduction of the recording time by a factor of 5. This leads to significantly lower
specimen loading by electron beam irradiation.

## 1. Introduction

The constantly increasing complexity of integrated circuits requires that
measurements are carried out inside the ICs as early as the design phase.
Measurements of this kind are also needed for failure analysis. Since the
ever greater reduction of structures precludes the use of mechanical
probes or makes them difficult to apply, the contactless measuring method
of electron beam testing is gaining increasingly in importance (Menzel
and Kubalek 1983, Feuerbaum 1983).

Apart from voltage measurements at a single measuring point and the mapping of logic states, it is frequently necessary to trace a signal inside
the IC. This becomes particularly easy if we succeed in making visible
only those interconnections which carry a signal of a specific sought-for
frequency.

A possible approach is to extract the sought-for frequency directly from
the secondary electron (SE) signal, possibly with the aid of a lock-in
amplifier (Collin 1983). Unfortunately however, not only is this method
very slow and thus subjects the specimen to undue loading, it is restricted to detecting signals of relatively low frequency, since the scintillator present in the SE detector and the signal chain of the scanning electron microscope (SEM) can only transmit signals of up to about 3 MHz.

These important drawbacks have led to the recent development of the frequency tracing method (FTR) (Brust et al. 1984, Brust et al. 1985), which
overcomes them by using an intermediate frequency procedure. It can thereby be used up to frequencies of 500 MHz. Although FTR allows a significantly shorter image recording time than earlier methods, it can under certain

circumstances lead to charging of the specimen and other disturbing effects due to electron beam irradiation. Up to now,however,reduction of the
measuring time, which is also desirable with a view to routine application
of the methods, could be attained only at the cost of the reduced sensitivity and resolution. The improved FTR presented here for the first time,
which we call fast frequency tracing (FFTR), solves this problem by introducing scanning speed modulation. This allows fast image recording with
unchanged high spatial resolution and sensitivity.

Before presenting the new method in section 3, the basic principle of FTR
will first be explained. The experimental set-up will then be described in
section 4 and a practical application presented in section 5.

## 2. Frequency Tracing

In order to overcome the bandwidth limitation imposed by the scintillator
and the SEM signal chain, the FTR makes use of the heterodyne principle.
In this, the frequency actually sought for in the IC, which may be very
high, is mixed down to a low fixed intermediate frequency (IF) which can
easily be transmitted by the scintillator and the SEM signal chain. Since
the SE current is firstly porportional to the primary electron (PE) current and secondly depends upon the electrical voltage at the measuring
point, this effect can also be used for mixing. This is done by pulsing
the PE current with a frequency $f_B$ which is shifted with respect to the
sought-for frequency $f_S$ by the desired IF $f_{IF}$. A signal component of IF
$f_{IF}$ occurs in the SE current whenever a signal of frequency $f_S$ is present
at the measuring point. Fig. 1 clearly shows the principle of the method.
Two signals with frequencies $f_1 = f_S$ and $f_2 \neq f_S$ are present at the two
interconnections. The pulsed electron probe with a frequency $f_B = f_S + f_{IF}$
scans the surface from B to E line by line. If the electron probe strikes
the upper interconnection where a signal with frequency $f_S$ is present, the
resulting secondary electron signal of IF $f_{IF}$ is extracted and amplified,
thus causing this interconnection to appear bright.

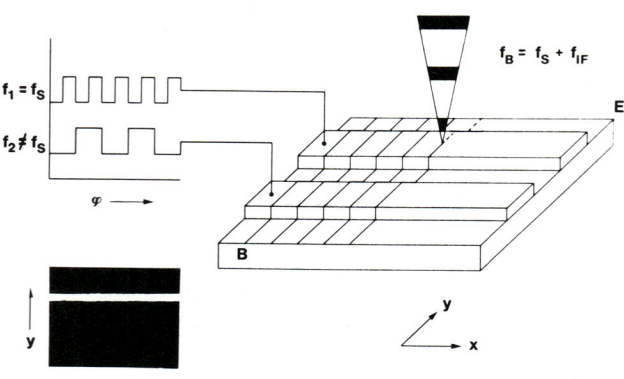

Fig. 1: Operation principle of frequency tracing

FTR is very similar in design and mode of action to a superhet radio recei-
ver which is set to a specific transmit frequency.Fig.2 shows the requisite
experimental set-up for FTR. The PE beam is pulsed with a frequency $f_B$ by
the beam blanking generator with the aid of the beam blanking system. A
signal component of frequency $f_{IF}$ occurs behind the detector and photomultiplier due to the already mentioned multiplicative relationship if the
electron beam strikes an interconnection carrying the frequency $f_S$. The
succeeding band-pass filter (see Fig. 2 (2)) extracts precisely this frequency. If in contrast, the PE beam strikes an interconnection carrying a
frequency which deviates from $f_S$, then corresponding mixed frequencies
occur here too and consequently appear in the SE signal (Fig. 2 (1)).These
mixed frequencies are, however, not transmitted by the band-pass filter.
Since the SEM signal chain needs transmit only the relatively low fixed IF,
signals of very high frequency can be detected in an IC. The amplitude of
the IF signal (Fig. 2 (3)) is then determined with the aid of an envelope
demodulator and compared with a variable threshold voltage in the succeeding comparator. The output signal of the comparator (Fig. 2 (4)) controls
the brightness of a CRT. If the value of the IF signal exceeds the set
threshhold value, the CRT is unblanked. The threshold voltage is used not
only for sensitivity adjustment but also for suppressing interference
signals and in addition to this provides an excellent image contrast.

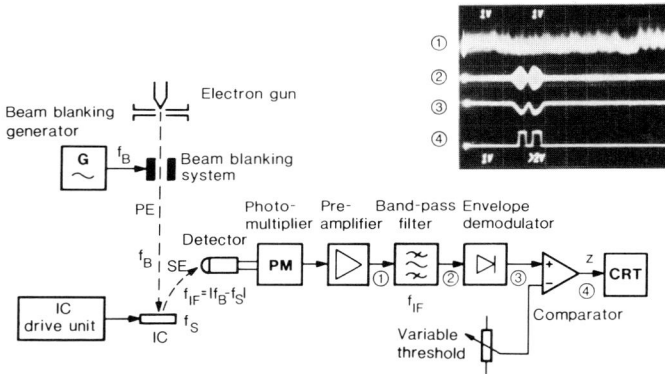

Fig. 2: Experimental setup for frequency tracing with measured signals
        (at the marked points in the signal chain)

In contrast to the previously known methods, this procedure does not require synchronization of the frequency of the pulsed PE beam $f_B$ with the
frequency of the tested IC $f_S$. It can therefore also be used to test asyn-
chronously operating components or free-running circuits.

In the frequency tracing method (FTR), the maximum speed with which the
electron beam can scan the specimen is, for a prespecified spatial resolution, limited by the bandwidth of the IF filter $\Delta f_{IF}$. In order to permit
routine use of this measuring method, the measuring time should be as
short as possible. This is also desirable with a view to ensuring minimum
loading of the specimen by the electron beam (charging, electron beam
damage). For taking measurements therefore, $\Delta f_{IF}$ should be as broad as
possible.

However, the bandwidth of the IF filter determines not only the image recording time but also the sensitivity of the measuring method. A good
signal-to-noise ratio at the filter output and thus a highly sensitive
procedure can be attained only by means of an IF filter of minimum bandwidth. Up to now therefore, a short measuring time could be attained only
when a reduction of spatial resolution or sensitivity ( $\Delta f_{IF}$ greater) was
excepted as a trade-off.

## 3. Fast Frequency Tracing

Only by introducing scanning speed modulation can both requirements be met
at the same time. We call this method fast frequency tracing (FFTR). It is
based on a very simple idea: normally the frequency sought for by the FTR
method is carried by only very few interconnections of the IC. It is therefore sufficient in principle to scan only these interconnections at relatively low speed and thus higher sensitivity and spatial resolution, whereas the scanning speed in the other regions of the image can be significantly higher. This would allow a considerable reduction of image recording
time to be achieved.

The solution of the problem lies therefore in modulating the scanning
speed as a function of the signal at each measuring point. This does, however raise the difficulty of having to determine with some certainty
whether the sought-for frequency is present at the measuring point or not.
This can be done only with the measuring chain used in traditional FTR
(see Fig. 2). In order to do justice to the two opposing requirements for
high sensitivity and small IF bandwidth on the one hand and for high speed
and large IF bandwidth on the other, two separate signal chains of the same
basic design and effecting parallel measurements must be used. The first
is a slow chain of narrow bandwidth and very high sensitivity and the other
a fast chain of wide bandwidth and relatively low sensitivity. A configuration of this kind is shown in Fig. 3. The signal chain of wide bandwidth
is used only to register, at all scanning speeds, whether the sought-for
signal frequency is at all present at each measuring point, and correspondingly to control the scanning speed by means of the scan generator. As long

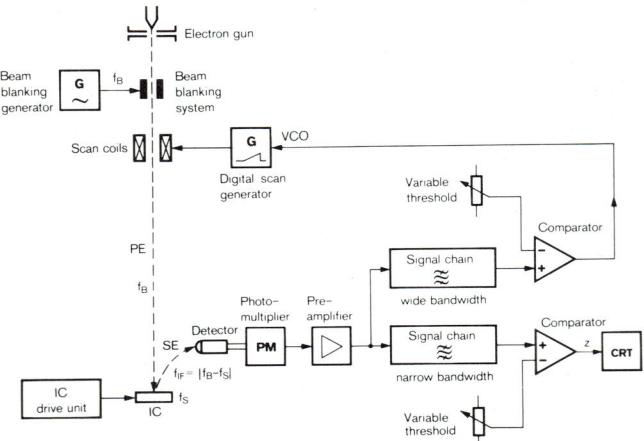

Fig. 3: Experimental setup for fast frequency tracing

as the sought-for frequency is not present, the scanning speed corresponds
to the maximum value prespecified by the wideband IF filter. As soon as
the sought-for frequency is detected however, the speed is reduced to the
value prespecified by the IF filter, and the measurement is made with high
sensitivity and spatial resolution due to the lower bandwidth. In this procedure too,therefore,the FTR image is recorded with the sensitive measurement chain. The function of the wideband signal chain is only to make the
preliminary decision whether a more accurate measurement is worth doing
at all.

As yet, this decision represents no final statement as to whether the
sought-for frequency is actually present, but merely that a certain probability exists that this is the case. The corresponding probability value
is set with the aid of the threshold value of the wideband signal chain.
The time gained depends upon what proportion of the overall image area is
taken up by those circuit regions which carry signals of the sought-for
frequency.

## 4. Experimental Setup

The basic equipment used in our experiments was a modified ETEC AUTOSCAN
scanning electron microscope (Fazekas et al. 1983). The primary electron
energy was 2.5 keV, the probe current 0.1-0.5 nA and the working distance
10-20 mm. A HP 8660C frequency synthesizer was used as beam blanking generator. Our SEM contained a digital scan generator whose clock signal for
scanning speed modulation was supplied by a voltage controlled oscillator
(HP 8116A). The core of the narrow-band signal chain was formed by a mechanical band-pass filter with 50 kHz center frequency, 3 kHz bandwidth
and steep edges. The wideband signal chain contained an adjustable filter
whose bandwidth was, in our experiments, selected between 5 and 15 kHz.
The fast and slow scanning speeds varied accordingly by a factor of between 2 and 5.

## 5. Application

The FFTR method was subjected to a practical test by investigating a NMOS
multiplier. Fig. 4a shows the SE image of the IC. The device was operated

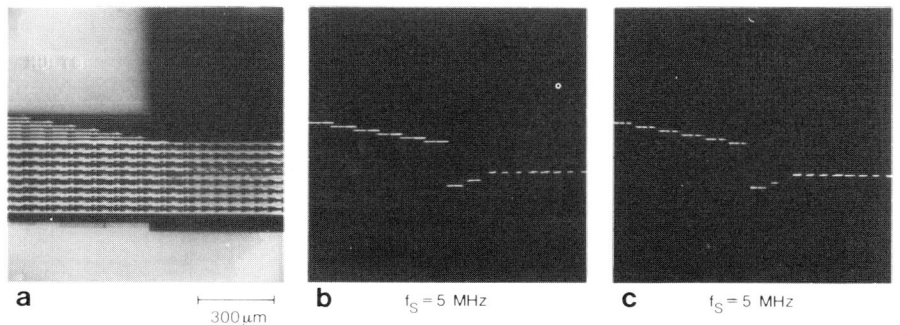

a          300μm          b   $f_S$ = 5 MHz          c   $f_S$ = 5 MHz

Fig. 4: NMOS multiplier circuit (one data input was switched periodically
        with a frequency of 5 MHz)
        a) normal secondary electron micrograph
        b) frequency tracing micrograph (recording time 110 s)
        c) fast frequency tracing micrograph (recording time 23 s)

at a clock frequency of 50 MHz and one of the data inputs was switched periodically with a frequency of 5 MHz. The aim was to trace this 5 MHz signal in the device with the aid of FTR. Fig. 4b shows the traditional FTR image and Fig. 4c the image recorded with the aid of the new FFTR. The path of the 5 MHz signal can be clearly seen in both images. No practical difference is apparent between them, (slight differences are due to little deviations of the two threshold voltages), i.e. spatial resolution and sensitivity are the same in both cases. But whereas it took 110 s to record the image in Fig. 4b, the FFTR image was recorded in a mere 23 s. This corresponds to a measurement speedup by a factor of 5.

## 6. Conclusions

FFTR permits a signal of a specific frequency to be traced in the interior of an IC. In contrast to most previously known electron beam testing methods, no synchronization is required with the signals in the device. This allows the applications area of electron beam testing to be extended to include asynchronous and free-running circuits.

The introduction of scanning speed modulation permits a considerable reduction of the measuring time and its concommitant specimen loading, without - as is the case in traditional FTR - a simultaneous impairment of the sensitivity or the spatial resolution. The image recording time may be reduced to between a third and a fifth of the time required by the traditional FTR method.

## Acknowledgements

The authors wish to thank Mr. G. Luplow for providing the mechanical filter, Mr. H. Mulatz for setting up electrical circuits,Mr. T. Noll for providing the multiplier, Mrs. L. Reidt and Mr. G. Schulz for their photographic work, Mrs. M. Kaczmarek for typing the manuscript, Dr. E. Wolfgang for fruitful discussions and Dr. H.-J. Pfleiderer for his general support.

## References

Brust H-D, Fox F and Wolfgang E 1984 Proc. Microcircuit Engineering Conf. (London: Academic Press)
Brust H-D and Fox F 1985 Microelectronic Engineering, to be published
Collin J P 1983 Proc. of Journée d'Electronique (Lausanne: Presses Polytechniques Romandes)
Fazekas P, Fox F, Papp A, Widulla F and Wolfgang E 1983/IV Scanning Electron Microscopy 1595
Feuerbaum H P 1983 Scanning 5 14
Menzel E and Kubalek E 1983 Scanning 5 103

*Inst. Phys. Conf. Ser. No. 76: Section 12*
*Paper presented at Microsc. Semicond. Mater. Conf., Oxford, 25–27 March 1985*

# A novel measurement method of VLSI pattern linewidth

M Miyoshi, M Kano, H Yamaji and K Okumura

Toshiba Corporation, Integrated Circuit Division
72 Horikawacho, Saiwaiku, Kawasaki-city, 210 Japan.

Abstract   The present paper deals with an automatic linewidth
measurement method of the photoresist pattern linewidths with high
accuracy, which we call "the linear regression method".  The absolute
accuracy of 0.009 μm was obtained by this method.  A practical VLSI
linewidth measuring system is also described.

1. Introduction

High accuracy measurement of pattern linewidth is particularly important
in the VLSI (Very Large Scale Integrated Circuits) manufacturing.  Since
the pattern linewidth becomes smaller and smaller, even a small change in
pattern linewidth gives rise to remarkable change in the device
characteristics.  Therefore, it is extremely important to measure the
pattern size with high accuracy.

The measurement of pattern linewidth has been done by optical method.
However, the optical methods have severe problems.  As the measured values
depend on slope angle at pattern edge, thickness and optical property of
film and also substrate, there usually exists a large difference in size
(0.3 μm) between the defined edge and the true edge.  This problem would
become most critical for photoresist linewidth measurement which is most
important in VLSI manufacturing.

Electron beam can ensure the high spatial resolution image, and the
secondary electron (SE) signal can sensitively reflect surface topography,
which enables precise determination of the pattern edge.

The linewidth measurement should be made automatically.  The commonly
accepted threshold method in which the cross point of line profile and
threshold level is defined as the edge, may produce certain error due to
the difference between the true edge and the cross point.

The present paper deals with an accurate and automatic pattern
linewidth measurement method for the photoresist pattern which we call
"the linear regression method".  A practical VLSI pattern linewidth
measuring system for manufacturing, particularly, in-process is also
described.

2. Principle of the Linear Regression Method and the Appraisal of its
   Accuracy

In VLSI processing, pattern linewidth measurements are mostly for
photoresist patterns and post etching film patterns.  Of particular
importance is the photoresist pattern measurement and linewidth control.
A cross section of photoresist is approximately trapezoidal.  In practice,

the bottom linewidth of the trapezoid is important, because it determines the pattern linewidth subsequently to be etched. Accordingly, the bottom of pattern edge should be precisely detected.

Figure 1 shows the schematic illustration of secondary electron (SE) profile at photoresist pattern edges. When the electron beam scans across the pattern, signal intensity gives a shallow bottom as shown in Fig. 1 which is close to the bottom edge and starts increasing linearly, along the slope of pattern edge, and then decreasing, after passing the top of the pattern edge. This characteristic curve is conserved even when the slope angle of photoresist pattern edge changes. Since the increasing point P almost corresponds to the bottom edge, the bottom edge can be accurately determined by the signal waveform analysis for identifying the increasing point P.

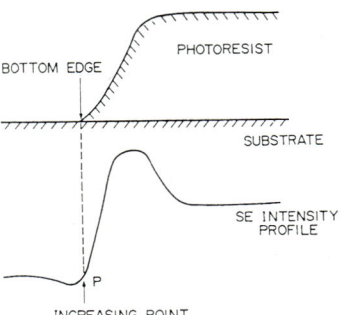

Fig. 1 Schematic illustration of secondary electron profile at photoresist pattern edge.

A typical SE signal and the principle of the linear regression method is shown in Fig. 2. For detection of the increasing point of SE profile, the line profile is approximated by two lines at pattern edge. One is the average line of SE profile for the substrate and another is the slope line at edge slope. The cross point of two lines corresponds to the bottom edge of photoresist pattern.

Figure 3 shows the flow chart of the measurement procedure. A desired pattern to be measured is appointed by two cursors on the SE still video image of scan converter. The cursor address data are transmitted to the scan generator of SEM, and the electron beam is scanned at the appointed address. The electron beam is scanned repeatedly, and the SE signal is averaged. After averaging, 9-point convolute smoothing is carried for further improvement of the signal to noise ratio.

In the linear regression method, the maximum values, MAXL, MAXR and the minimum values, MINL, MINR are detected, and the SE signal around the edges of photoresist is divided into the base line region (L-AMINL), (AMINR-R) and the slope

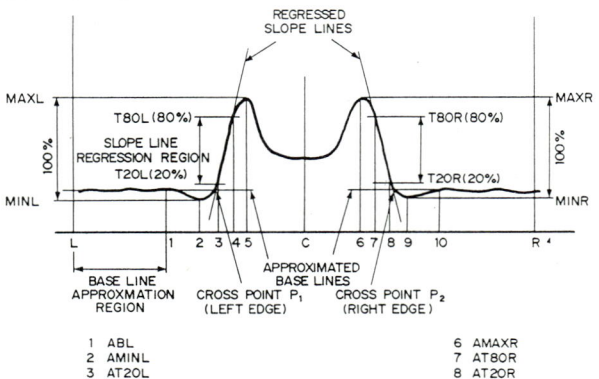

Fig. 2 A typical secondary electron signal on the photoresist pattern and the principle of the linear regression method.

line region (AMINL-AMAXL), (AMAXR-AMINR), as shown in Fig. 2. and Fig. 3. In the slope line region, the slope line regression region is defined by excluding upper 20% and lower 20%, and it is approximated by the least square method. In the base line region, the bottom region (ABL-AMINL), (ABR-AMINR) is excluded and the base line approximation region is approximated by the average line.

Same calculations are carried out for both left and right edges, and the pattern linewidth can be obtained by calculating the distance of two detected edges $(P_1-P_2)$. After the linewidth calculation, the calibration is carried out. These operations are repeated for eight lines, and the measured result is obtained by averaging eight lines.

As an appraisal of the accuracy of the linear regression method, we compared measured linewidths with linewidths by the cross sectional view of SEM. A linewidth at the bottom of photoresist pattern measured from the cross sectional view was defined as the true value.

The photoresist patterns for this examination were fabricated on the silicon substrate. The slope angle of photoresist pattern ranges from 72° to 85°. Both SEMs for the cross sectional view and the linewidth measurement were calibrated by the same NBS standard calibration photomask (Jensen et al 1980) with the accuracy of 0.01 µm. The measurement was carried out at the acceleration voltage of 1 kV.

Figure 4 shows the correlation between the measured values by the linear regression method and the true values. Measured and true values coincide well, as shown in the figure. In this examination, we defined that the absolute accuracy was the mean value of differences between measured values and true values ($\overline{\Delta W}$). As a result, the absolute accuracy of 0.009 µm has been obtained, for different slope angle of photoresist.

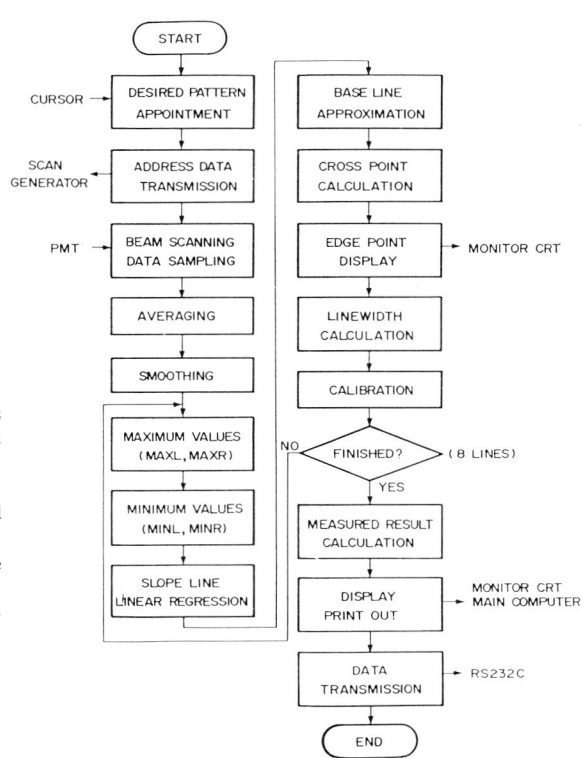

Fig. 3 Flow chart of the automatic linewidth measurement procedure.

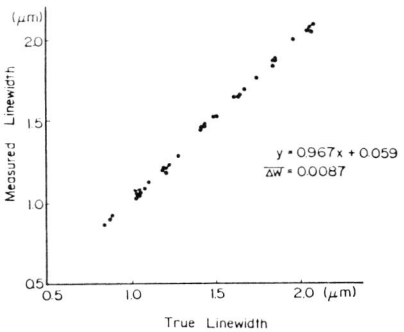

Fig. 4 Correlation between measured values by the linear regression method and true values.

Figure 5 shows the descripancies of measured values from true values with respect to slope angles for the two different methods, i.e, the linear regression method and the threshold method. In the threshold method, the threshold level was set at 50% of SE signal intensity profile. In the figure, measured values by the linear regression method are not influenced by the slope angle variation, while, the values by the threshold method depend upon the slope angle.

Figure 6 shows a photograph of SE image and measured result on the monitor CRT of the scan converter. In the figure, central photoresist pattern is appointed by two cursors and white dots are calculated edges.

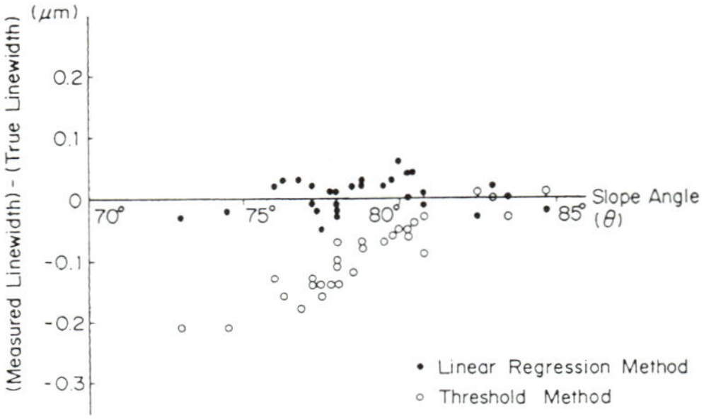

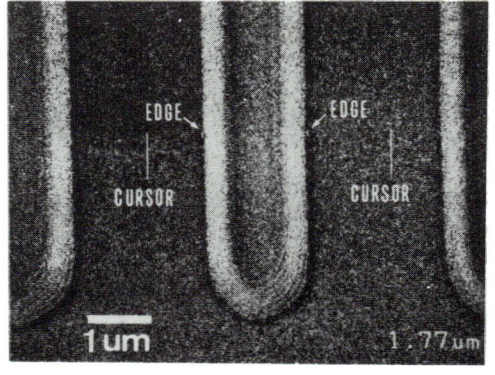

Fig. 5 Relationship between the differences of measured values and true values and the photoresist slope angle.

## 3. Linewidth Measuring Apparatus

Figure 7 shows the block diagram of linewidth measuring apparatus. It consists of the scan converter image memory and the automatic measuring unit. The linewidth measuring CPU(Z80A) controls data sampling, signal processing and linewidth calculation.

SE image can be observed as the still video image on the monitor CRT which is stored in the frame memory (8-bit x 512 x 512 pixels) of the scan converter. Desired pattern is appointed by two cursors on the monitor CRT.

Fig. 6 A photograph of secondary electron image and measured result on the monitor CRT of the scan converter.

In the automatic measurement, the graphic CPU reads the appointed cursor address which is transmitted to the scan generator of SEM, and the electron beam starts scanning the appointed address. The scanning resolution is 1024 dots/line, which means 0.01 μm/line at the magnification of about 10,000X.

The obtained SE signal profiles are stored and accumulated in the data memory. After averaging, CPU and ALU (Arithmetic Logic Unit) which is bus-interfaced with CPU, make 9-point convolute smoothing and linewidth calculation. ALU processes the convolute smoothing and the least square calculations with high speed. The calculated linewidth is transmitted to the graphic CPU and displayed on the monitor CRT or to the main personal computer.

As the time consuming calculation is done by the hardware (ALU), the total measurement time is 2.34 sec. which is quite practical for

application of this
system.  The total
measurement time consists
of the data sampling and
beam scanning, 1.05 sec,
signal processing and
linewidth calculation,
0.79 sec, and data
output, 0.50 sec.
   Construction of
software in the linewidth
measuring apparatus is
shown in Fig. 8.

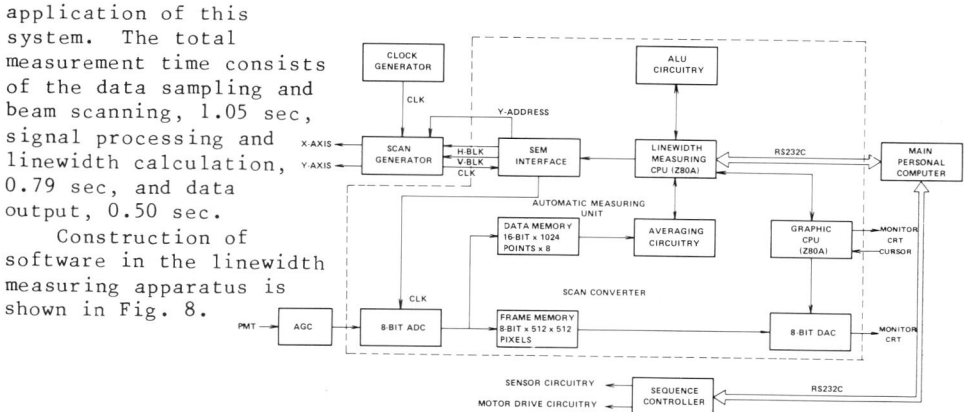

Fig. 7 Block diagram of the linewidth
measuring apparatus.

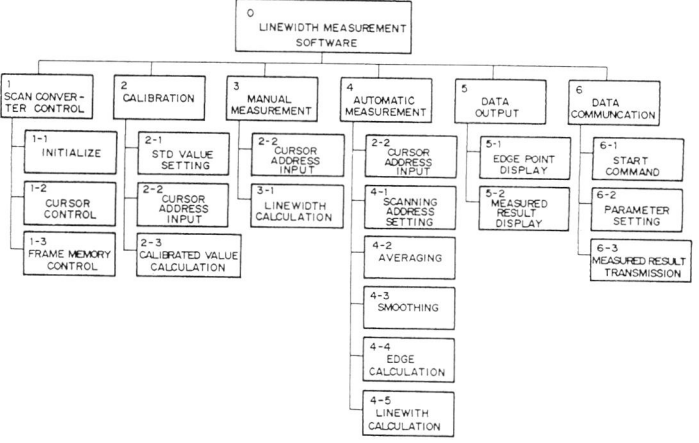

Fig. 8 Construction of the linewidth measuring software.

## 4.  Pattern Linewidth Measuring System

Figure 9 shows a photograph of a newly developed pattern linewidth
measuring system which is particularly utilized for in-process VLSI
manufacturing monitoring.  Operations such as wafer conveying, stage
shifting and location and linewidth measurements are automatically
controlled by the main personal computer (8-bit Z80A based personal
computer of 16-bit HP-9816).
   By employing a cassette system, the equipment is capable of
automatically conveying up to 25 wafers up to six inches in diameter.  The
location of measurement point on a wafer is selected by a computer
controlled positioning system.
   Whole area of 6" wafer can be observed without rotation, with the
maximum shifting speed of 20mm/sec., the minimum step of 2 μm and the

stopping accuracy of 5 μm at the
maximum shifting mode.

In order to use an electron beam in
the semiconductor manufacturing process
for in-process monitoring, it is
necessary to supply a low acceleration
voltage of around 1 kV to avoid the
electro-static charging-up of
insulating materials and the radiation
damage of device.

By employing a newly designed
electro-optical column system with low
acceleration voltage for the electron
gun fabrication and lens system, the
equipment achieves the high resolution
of 10 nm the high quality image, and
the accurate measurement without
electro-static charging-up.

Fig. 9 A photograph of the
linewidth measuring system
MEA-3000.

### 5. Summary

A novel photoresist pattern linewidth measurement method which we call
"the linear regression method" has been described.  In this method, the
absolute accuracy of 0.009 μm has been demonstrated which was not
influenced by the photoresist slope angle variation.

The linewidth measuring apparatus which consists of the scan converter
and CPU combined with ALU can achieve the practically short meausurement
time, and convenience of operation for the complicated algorithm has been
designed to realize this novel method.

High quality secondary electron image with extremely low acceleration
voltage can be obtained and it realizes the nondestructive linewidth
measurement in VLSI manufacturing.

By employing the CPU controlled automatic wafer conveying, stage
shifting and location of desired pattern, a practical VLSI linewidth
measuring system for VLSI manufacturing, capable of in-process monitoring,
has been developed.

### Acknowledgements

The authors would like to thank Dr. Y Nishi for his critical reading of
the manuscript, M Nakamura and J Matsumoto for their encouragement,
K Norimatsu and H Furukawa for their helpful technical discussion.

### Reference

Jensen SW and Swyt DA 1980 Scanning Electron Microscopy Vol 1 (IL, USA;
SEM Inc) pp 393 - 406

# Author Index

# Subject Index†

†Page numbers refer to the first pages of the papers in which the citations occur.